Nanoscale MOS Transistors

Written from an engineering standpoint, this book provides the theoretical background and physical insight needed to understand new and future developments in the modeling and design of n- and p-MOS nanoscale transistors. A wealth of applications, illustrations, and examples connect the methods described to all the latest issues in nanoscale MOSFET design. Key areas covered include:

- Transport in arbitrary crystal orientations and strain conditions, and new channel and gate stack materials;
- All the relevant transport regimes, ranging from low field mobility to quasi-ballistic transport, described using a single modeling framework;
- Predictive capabilities of device models, discussed with systematic comparisons to experimental results.

David Esseni is an Associate Professor of Electronics at the University of Udine, Italy.

Pierpaolo Palestri is an Associate Professor of Electronics at the University of Udine, Italy.

Luca Selmi is a Professor of Electronics at the University of Udine, Italy.

Cover illustration: the images represent the k-space carrier distributions at the end of the channel of nanoscale n- and p-MOSFETs biased in the saturation region of operation.

"In this comprehensive text, physicists and electrical engineers will find a thorough treatment of semiclassical carrier transport in the context of nanoscale MOSFETs. With only a very basic background in mathematics, physics, and electronic devices, the authors lead readers to a state-of-the-art understanding of the advanced transport physics and simulation methods used to describe modern transistors."

Mark Lundstrom, Purdue University

"This is the most pedagogical and comprehensive book in the field of CMOS device physics I have ever seen."

Thomas Skotnicki, STMicroelectronics

"This is a modern and rigorous treatment of transport in advanced CMOS devices. The detailed and complete description of the models and the simulation techniques makes the book fully self sufficient."

Asen Asenov, University of Glasgow

Nanoscale MOS Transistors

Semi-Classical Transport and Applications

DAVID ESSENI, PIERPAOLO PALESTRI, and LUCA SELMI
University of Udine, Italy

CAMBRIDGE
UNIVERSITY PRESS

CAMBRIDGE
UNIVERSITY PRESS

Shaftesbury Road, Cambridge CB2 8EA, United Kingdom

One Liberty Plaza, 20th Floor, New York, NY 10006, USA

477 Williamstown Road, Port Melbourne, VIC 3207, Australia

314–321, 3rd Floor, Plot 3, Splendor Forum, Jasola District Centre, New Delhi – 110025, India

103 Penang Road, #05–06/07, Visioncrest Commercial, Singapore 238467

Cambridge University Press is part of Cambridge University Press & Assessment, a department of the University of Cambridge.

We share the University's mission to contribute to society through the pursuit of education, learning and research at the highest international levels of excellence.

www.cambridge.org
Information on this title: www.cambridge.org/9780521516846

First published 2011

A catalogue record for this publication is available from the British Library

ISBN 978-0-521-51684-6 Hardback

Contents

Preface

The traditional geometrical scaling of the CMOS technologies has recently evolved in a generalized scaling scenario where material innovations for different intrinsic regions of MOS transistors as well as new device architectures are considered as the main routes toward further performance improvements. In this regard, high-κ dielectrics are used to reduce the gate leakage with respect to the SiO_2 for a given drive capacitance, while the on-current of the MOS transistors is improved by using strained silicon and possibly with the introduction of alternative channel materials. Moreover, the ultra-thin body Silicon-On-Insulator (SOI) device architecture shows an excellent scalability even with a very lightly doped silicon film, while non-planar FinFETs are also of particular interest, because they are a viable way to obtain double-gate SOI MOSFETs and to realize in the same fabrication process n-MOS and p-MOS devices with different crystal orientations.

Given the large number of technology options, physically based device simulations will play an important role in indicating the most promising strategies for forthcoming CMOS technologies. In particular, most of the device architecture and material options discussed above are expected to affect the performance of the transistors through the band structure and the scattering rates of the carriers in the device channel. Hence microscopic modeling is necessary in order to gain a physical insight and develop a quantitative description of the carrier transport in advanced CMOS technologies.

In this context, our book illustrates semi-classical transport modeling for both n-MOS and p-MOS transistors, extending from the theoretical foundations to the challenges and opportunities related to the most recent developments in nanometric CMOS technologies. Moreover, we describe relevant implementations of the semi-classical models which rely on the momentum relaxation time approximation and on the Monte Carlo approach for solution of the transport equations. The book aims at giving a description of the models that, without sacrificing the rigor of the treatment, can be accessible to both physicists and electronic engineers working in the electron device community. In this spirit, the selection of topics is driven by the innovations recently introduced in the semiconductor industry and by the trends in CMOS technology forecast by the International Roadmap for Semiconductors. Furthermore, since the CMOS technologies make inherently equal use of n-type and p-type MOSFETs, and because the physically based transport modeling is far more complicated for p-MOS than for n-MOS transistors, we describe the models for the two devices separately and in the same detail, thus avoiding leaving the reader with the misleading impression that modeling of p-MOS devices is a trivial extension of the n-MOS case.

With respect to implementations, we have highlighted the multi-subband Monte Carlo approach because of some distinct features compared to other methods. These are its generality (with a suitable choice of boundary conditions all transport regimes can be explored, including the uniform and the non-uniform, the low field and the high field regimes), accuracy (the Boltzmann transport equation is solved without a-priori assumptions about the carrier distribution functions), modularity (new scattering mechanisms can be added without changing the core of the Monte Carlo solver) and completeness (all the scattering mechanisms claimed to be relevant for nanoscale MOSFETs can be accounted for).

As for the modeling methodologies alternative to the semi-classical approach illustrated in this book, quantum transport and its application to nanoscale MOSFETs has recently made important progress, especially thanks to the non-equilibrium Green's function formalism. However, we believe that semi-classical transport will remain for a long time the reference framework to understand the transport and support the design and innovation of MOS transistors, because it is an adequate approach for both uniform transport in long devices and strongly non-local, quasi-ballistic transport in nanoscale MOSFETs. These characteristics fit well with the path to innovation followed in the CMOS technologies, which typically starts from observation of possible improvements in low field mobility and then tries to translate them into enhancements of the on-current for nanoscale transistors.

At the time of writing, several alternative devices are being investigated as complements to the traditional MOSFETs, such as nanowires, carbon nano-tubes, graphene nano-ribbon transistors, and tunnel-FETs, to name a few. Nevertheless, we believe that devoting a book to nanoscale MOS transistors is a well defendable choice, because on the one hand in the foreseeable future none of the above devices is expected to replace MOSFETs for mainstream applications, and, furthermore, we know from experience that the semi-classical transport methodologies described in this book can be extended quite naturally also to devices with a different carrier gas dimensionality or with different channel materials.

Due to the volume of literature related to semi-classical transport in MOSFETs, the references included in the book could not be exhaustive. Rather, for each topic we have tried to include a selection of the most relevant journal papers, books and also papers presented at the leading conferences, which are frequently the most dynamic vehicles for introduction of the latest developments into the electron device community.

We wrote this book to serve as a reference for graduate student courses devoted to the theoretical foundations of, and recent developments in, carrier transport in nanoscale CMOS technologies, and also as a reference book for researchers and practitioners working in development and optimization of advanced MOS devices.

The prerequisite knowledge of physics for this book is limited to the basic concepts of classical electrostatics and electrodynamics, to the basic notions and methods of quantum mechanics and, in particular, to a familiarity with the Schrödinger equation and with the meaning of the corresponding eigenvalues and wave-functions. A previous basic knowledge of the band structure in crystals would be useful for the reader, however, the second chapter aims at making the book self-contained also in this respect. The mathematical prerequisite knowledge is instead related to matrix algebra and to differential equations and differential eigenvalue problems. The book also assumes that the reader has a basic acquaintance with the working principle of semiconductor devices and, in particular, of MOS transistors.

The book was written to be as much as possible self-contained, so that most of the derivations are included in detail, also by resorting to appendixes in the cases where we thought that they resulted in too long a digression from the main flow of the discussion. The availability of the derivations allows the reader to trace back the origin and understand the validity limits of some results which may be very widely quoted and used in the literature but not as often fully justified and explained.

Essentially all the models described in the book have been implemented by the authors in benchmark codes or in complete simulators, so that it has been possible to include many simulation results in order either to clarify some theoretical aspects or to exemplify the insight provided by the models in practically relevant case studies.

David Esseni
Pierpaolo Palestri
Luca Selmi

Acknowledgements

Many people contributed to this book and to the work which is behind it. Among them, we would like to express our sincere gratitude to M.De Michielis, F.Conzatti, N.Serra, P.Toniutti, L.Lucci, Q.Raphay, and M.Iellina for their contributions to the development of the simulation tools used to obtain many of the results included in the book, for their help in producing some of the figures and also for their careful reading of the manuscript. M.Bresciani, A.Cristofoli, A.Paussa, M.Panozzo, and E.Beaudoin helped us with the bibliographic entries in order to make the style of the references uniform throughout the book and also with editing some of the figures.

We are also in debt to our colleagues F.Driussi, A.Gambi, and P.Gardonio for the critical reading of some sections of the book, that was really invaluable for correcting mistakes and improving the text clarity.

This work has benefited substantially from interactions with colleagues with whom we have had a fruitful and stimulating collaboration over the years; among them, we would like to thank E.Sangiorgi, A.Abramo, C.Fiegna, and R.Clerc.

Our special thanks go also to J.Lancashire and S.Matthews at Cambridge University Press for following the progress of our work in all its phases, and to S.Tahir for support with all the LaTeX related troubles that inevitably occurred during the writing.

The understanding of our families for our devoting to this project much of our supposedly free time during the last two years has been at least as necessary as all the previously mentioned contributions in making possible the completion of the writing. To our families we gratefully dedicate this book.

David Esseni
Pierpaolo Palestri
Luca Selmi

Terminology

Abbreviations and acronyms

BTE	Boltzmann transport equation
DG	Double gate
DIBL	Drain induced barrier lowering
DoS	Density of states
EMA	Effective mass approximation
EOT	Equivalent oxide thickness
EPM	Empirical pseudo-potential method
ITRS	International technology roadmap for semiconductors
MC	Monte Carlo
MOS	Metal-oxide-semiconductor
MOSFET	MOS field effect transistor
CMOS	Complementary metal-oxide-semiconductor
MSMC	Multi-subband Monte Carlo
MRT	Momentum relaxation time
SG	Single gate
SOI	Silicon on insulator
SS	Subthreshold swing
TCAD	Technology computer-aided design
VLSI	Very large scale integration
VS	Virtual source

Notation

x	Scalar
$\mathrm{x}^{\dagger}$	Complex conjugate of the scalar x
$\mathrm{x} + (\mathrm{c.c})$	A scalar plus the complex conjugate, namely $(\mathrm{x} + \mathrm{x}^{\dagger})$
$\mathbf{x}$	Vector, matrix or multi-dimensional tensor
x_{ij}	Element of the matrix $\mathbf{x}$
$\mathbf{x}^{\mathrm{T}}$	Transpose of the vector or matrix $\mathbf{x}$
$\mathbf{x}^{\dagger}$	Transpose conjugate of the vector or matrix $\mathbf{x}$
$\mathbf{x}\cdot\mathbf{y}$	Scalar product between vectors $\mathbf{x}$ and $\mathbf{y}$

$\hat{\mathbf{e}}_x, \hat{\mathbf{x}}, \hat{\mathbf{e}}_y, \hat{\mathbf{y}}, \hat{\mathbf{e}}_z, \hat{\mathbf{z}}$	Unit vectors along the direction x, y and z
$\hat{\mathrm{H}}$	Operator: typically consisting of a differential and an algebraic part
$\mathrm{H}_v(\mathrm{x})$	Heaviside function: 0 for negative x values and 1 otherwise
$F\{f(\mathrm{x})\} = F(q)$	Fourier transform of the function $f(\mathrm{x})$
$(f*g)(x)$	Convolution of the functions $f(\mathrm{x})$ and $g(\mathrm{x})$
∇ or $\nabla_{\mathbf{R}}$	Gradient with respect to real space three-dimensional coordinates $\mathbf{R}$
∇ or $\nabla_{\mathbf{r}}$	Gradient with respect to real space two-dimensional coordinates $\mathbf{r}$
$\nabla_{\mathbf{K}}$ or $\nabla_{\mathbf{k}}$	Gradient with respect to wave-vectors $\mathbf{K}$ or $\mathbf{k}$
$[hkl]$	Miller indices that specify a crystal direction
$\langle hkl \rangle$	Miller indices that specify equivalent crystal directions
(hkl)	Miller indices that specify the crystal plane normal to $[hkl]$
$\{hkl\}$	Miller indices that specify the equivalent crystal planes normal to $\langle hkl \rangle$

Symbols:

a_0	Direct lattice constant of a crystal	m
E_F	Fermi level	J
$g(E)$	Density of the states for a d dimensional carrier gas	$\mathrm{m}^{-d}\,\mathrm{J}^{-1}$
n_{sp}	Spin degeneracy factor: can be either 1 or 2	unitless
$\mathbf{F}$	Electric field	$\mathrm{V\,m}^{-1}$
F_x, F_y, F_z	Electric field components in the x, y and z direction	$\mathrm{V\,m}^{-1}$
F_{eff}	Effective electrical field in an inversion layer	$\mathrm{V\,m}^{-1}$
$\mathbb{F}$	Driving force for carrier motion	Newton
$\mathbf{V}_g, \mathbf{v}_g$	Group velocity for a 3D or a 2D carrier gas	$\mathrm{m\,s}^{-1}$
m_x, m_y, m_z	Effective electron masses in the x, y and z direction	kg
Ω	Normalization volume	m^3
A	Normalization area	m^2
ϕ	Electrostatic potential	V
U	Potential energy	J
T	Temperature	K
V	Voltage at device terminals	V
V_{GS}	Intrinsic terminal voltage difference from gate to source	V
V_{DS}	Intrinsic terminal voltage difference from drain to source	V
L_G	Gate length	m
I_{ON}	Drain current per unit width at $\|V_{GS}\| = \|V_{DS}\| = V_{DD}$	A/m
I_{OFF}	Drain current per unit width at $V_{GS} = 0$, $\|V_{DS}\| = V_{DD}$	A/m
t_{ox}	Physical oxide thickness	m
N_{inv}	Electron inversion layer density	m^{-2}
P_{inv}	Hole inversion layer density	m^{-2}
N^+	Inversion density of carriers moving from source to drain	m^{-2}
v^+	Average velocity of carriers moving from source to drain	m/s
N^-	Inversion density of carriers moving from drain to source	m^{-2}

v^-	Average velocity of carriers moving from drain to source	m/s
v_{sat}	Saturation velocity	m/s
r	Back-scattering coefficient	unitless

Physical constants

h	Planck's constant	6.626075×10^{-34} Js
$\hbar$	Reduced Planck's constant	$h/(2\pi)$
K_B	Boltzmann's constant	1.380662×10^{-23} JK^{-1}
e	Positive electron charge	1.602189×10^{-19} C
m_0	Electron rest mass	9.109390×10^{-31} kg
ε_0	Dielectric constant of vacuum	8.854188×10^{-12} CV^{-1}m^{-1}

1 Introduction

1.1 The historical CMOS scaling scenario

Complementary Metal Oxide Semiconductor (CMOS) technology is nowadays the backbone of the semiconductor industry worldwide and the enabler of the impressive number of electronic applications that continue to revolutionize our daily life. The pace of growth of CMOS technology in the last 40 years is clearly shown in the so-called Moore's plot (see Fig.1.1 [1]), reporting the historical trend in the number of transistors per chip, as well as in the trends of many other circuit performance metrics and economic indicators.

Key to the success of CMOS technology is the extraordinary scalability of the Metal Oxide Semiconductor Field Effect Transistor (MOSFET). The word *scaling* denotes the possibility, illustrated in Fig.1.2 and Table1.1, of fabricating functional devices with equally good or even improved performance metrics but smaller physical dimensions. The design of scaled transistors starting from an existing technology has been driven initially by simple similarity laws aimed to maintain essentially unaltered either the maximum internal electric field (hence, to a first approximation, the device reliability) or the supply voltage (hence the system integration capability) [2].

According to these two scaling strategies, defined in Table1.1, all the lateral (primarily the gate width, W, and length, L_G) and the vertical physical dimensions (the thickness of the gate dielectric, t_{ox}, and the junction depth, x_j) should decrease from one technology generation to the next by a factor α, thus yielding an increase of the number of transistors per unit chip area by a factor of α^2. In order to proportionally reduce the channel depletion depth, the doping concentration in the substrate should increase by no less than a factor α. The intrinsic switching delay $\tau = CV/I$ is consequently reduced by a factor ranging between α^{-1} and α^{-2} in the constant field and constant supply voltage scaling scenario, respectively.

The constant field and constant supply voltage scaling rules are derived from quite simple one-dimensional models of the MOSFET electrostatics. These models and the rules above became inadequate to the design of MOS transistors as the gate length (L_G) approached one micron, thus leading to development of more sophisticated criteria. As an example, Table1.1 reports the mixed scaling rules proposed in [3] to design 0.25 μm MOSFETs, where different reduction factors are introduced for the geometrical dimensions (α) and the voltages (λ).

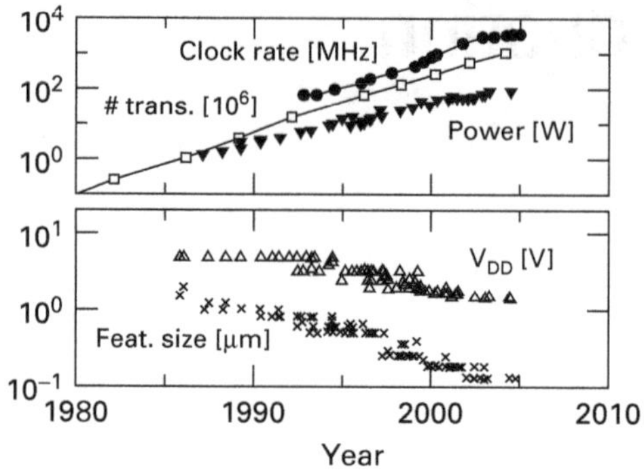

Figure 1.1 Progress in CMOS technology. Number of transistors in memory chips, clock rate, power supply voltage, power consumption, and minimum feature size.

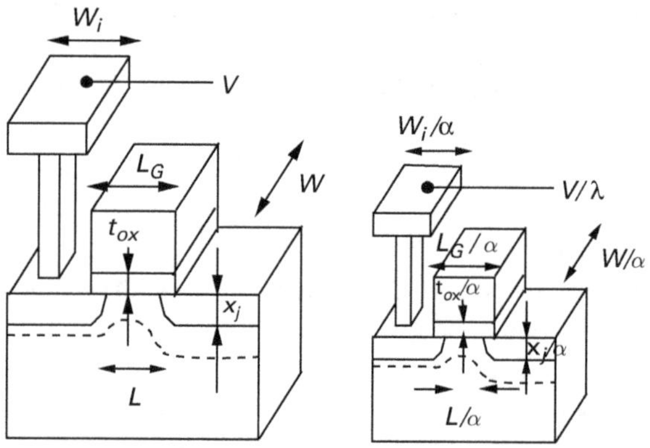

Figure 1.2 Bulk MOSFET scaling principles and corresponding scaling factors for geometrical dimensions (α) and voltages (λ). Note that L_G and L denote the gate length and the effective channel length, respectively.

Table 1.1 Scaling rules for CMOS technology. Note that α and λ denote the geometry and voltage scaling factors, respectively.

Parameter	Const.field scenario	Const.voltage scenario	Mixed scenario
Dimensions	$1/\alpha$	$1/\alpha$	$1/\alpha$
Voltages	$1/\alpha$	1	$1/\lambda$
Fields	1	α	α/λ
Doping	α	α^2	α^2/λ
Current	$1/\alpha$	α	α/λ^2
Capacitance	$1/\alpha$	$1/\alpha$	$1/\alpha$
Interconnect resistance	α	α	α
Switching delay	$1/\alpha$	$1/\alpha^2$	λ/α^2
Interconnect delay	1	1	1
Power delay product	$1/\alpha^3$	$1/\alpha$	$1/\alpha^2\lambda$
Power area-density	1	α^3	α^3/λ^3

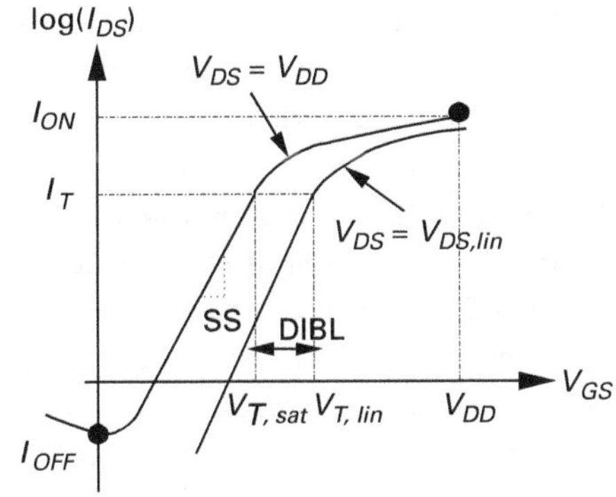

Figure 1.3 Definition of the main static performance metrics of a MOSFET. V_{DD} is the power supply voltage, I_T is a threshold drain current (typically 1 μA/μm). $I_{ON} = I_{DS}$ at $V_{GS} = V_{DS} = V_{DD}$; $I_{OFF} = I_{DS}$ at $V_{GS} = 0$ V and $V_{DS} = V_{DD}$; $V_{T,lin} = V_{GS}$ at $I_{DS} = I_T$ and $V_{DS} = V_{DS,lin}$; $V_{T,sat} = V_{GS}$ at $I_{DS} = I_T$ and $V_{DS} = V_{DD}$; subthreshold swing $\mathrm{SS} = dV_{GS}/d[\log(I_{DS})]$; $\mathrm{DIBL} = (V_{T,lin} - V_{T,sat})/(V_{DD} - V_{DS,lin})$.

In particular, since the thermal voltage $K_B T/e$, the band gap and the junction built-in voltage do not scale, the subthreshold swing (SS) of the transfer characteristic and the flatband voltage of poly-silicon gate MOSFETs remain almost invariant to scaling [4]. As a result, the two-dimensional distribution of the electrostatic potential inside the scaled device is distorted compared to that of the parent technology generation and so-called Short Channel Effects (SCE) become apparent as:

- a decrease of the linear and saturation threshold voltages ($V_{T,lin}$, $V_{T,sat}$, Fig.1.3) at short channel lengths, due to the penetration of the source and drain electric field lines in the channel region;
- a large sensitivity of the threshold voltage to the drain voltage (an effect denoted as DIBL, Drain Induced Barrier Lowering);
- an increase of the subthreshold swing SS.

Narrow channel effects, detrimental to control of the threshold voltage, also appear in the scaled technology.

An optimum choice of channel doping, junction depth and thickness of the gate dielectric is crucial to keep SCE under control. Accurate tailoring of the source and drain extensions below the spacers and reduction of parasitic source/drain resistances contribute as well to achieving good performance and high I_{ON}/I_{OFF} ratios. As a consequence of the increased complexity of this optimization task, during the eighties two- and three-dimensional CAD tools for numerical device simulation (mostly based on the Drift-Diffusion semiconductor device model [5–9]) have found widespread use in the semiconductor industry to assist process engineers in analysis and tuning of the doping profiles to counteract the short channel effects.

Starting from the early nineties, foresight studies on the scaling of CMOS technology have emerged from the joint efforts of associations such as the US Semiconductor

Industry Association (SIA) and later the International Technology Roadmap for Semiconductors (ITRS). The guideline documents on MOSFET scaling prepared by the ITRS [10] aim at the early identification of risk factors in the developments of the microelectronics industry, as well as at steering research toward the so called "red brick walls" which may impede further progress of this strategic technology.

In recent years, diversification of microelectronic applications has led to a differentiation of the ITRS for High Performance (HP), Low Power (LP) and Low STand-by Power (LSTP) applications [11]. Nevertheless, regardless of the specific market area, the semiconductor industry has steadily pursued the scaling of the device footprint, that is the area scaling, in spite of the increased complexity of the fabrication technology and growing fabrication costs.

To a different extent, all the roadmaps for the bulk MOSFET architecture nowadays share a common difficulty in finding the balance in the trade-off involving the containment of SCE (which demands high channel doping and gate dielectrics with small equivalent oxide thickness, EOT), the quest for high on-current (which requires high carrier mobility and low threshold voltage), and the need for low subthreshold leakage (which requires high threshold voltage, low subthreshold swing and relatively thick gate dielectrics). The performance metrics of the bulk MOSFET technology have steadily improved [12] but, as the minimum channel length entered the sub 0.1μm range, it became increasingly difficult to maintain the historical scaling trends by mere optimization of the conventional architecture. Due to complexity and cost, however, the introduction of significant innovations has always been deferred till the time when no real alternative was possible.

A prominent example in this respect is the replacement of SiO_2 (with its nearly ideal interface properties, large band gap, low trap density, etc.). In an effort to prolong the usability of the most popular dielectric in silicon microelectronics, nitrided SiO_2 layers (SiON) were adopted first [13–17], with undebatable advantages in terms of increased dielectric constant and beneficial effects against boron penetration in p-MOSFETs. It is only with the advent of 45 nm technology that the first breakthrough innovation at the heart of the bulk MOSFET architecture, namely the introduction of high-κ dielectrics, has started to become a reality [18–20].

Recently, the number of technology challenges putting at risk the scaling of the conventional bulk MOS transistor has increased. Fundamental studies suggest that the evolution of CMOS technology, as outlined in the ITRS, is leading the MOSFET to nearly achieve the ultimate performance expected for charge transfer switches [21–25]. However, it is also becoming clearer and clearer that significant innovations will be necessary to make the ultimate CMOS a reality.

Consistently, new options (the so called *technology boosters*) and new device concepts have been identified by the ITRS to flank the traditional dimension, doping and voltage scaling. These new options could give significant advantages in terms of intrinsic device performance, thus allowing microelectronics to maintain progress along the so called Moore's law. Recent developments in CMOS technology are thus outlining a *generalized scaling* scenario, which is briefly illustrated in the next section.

1.2 The generalized CMOS scaling scenario

For decades the basic architecture of the MOS transistor has not changed dramatically, although a large number of innovations, including new materials (e.g., new metals, low-κ dielectrics for interconnects, etc.) and new processes (e.g., shallow trench isolation, source/drain silicidation, lightly doped extensions, etc.), have been introduced to enable controlled device scaling to smaller dimensions. In recent years, however, CMOS scaling has become in a sense a definitely more diversified exercise.

To illustrate this point, Fig.1.4 shows a few of the advanced MOSFET architectures envisioned by the ITRS for future MOSFET scaling scenarios toward the ultimate limits.

In order to contain static power dissipation in the off state and guarantee the device reliability, gate leakage currents must be kept under control. The simultaneous need to increase the effective gate capacitance has led to exploration of the use of alternative gate insulators with a relative dielectric constant κ higher than that of SiO_2 and SiON [26, 27], which can provide a given equivalent oxide thickness (EOT) with a larger physical thickness with respect to SiO_2 and thus reduce the gate leakage. The introduction of metal gate electrodes (Fig.1.4.a) eliminates poly-silicon depletion, thus contributing increased capacitance, but generates Fermi level pinning issues [28, 29]. Unfortunately, almost all eligible high-κ materials degrade the channel mobility [26, 27, 30, 31] unless a thin SiO_2 interfacial layer is left above the channel, which conversely limits the increase of the gate capacitance. Completely new reliability problems are raised as well by the introduction of the high-κ insulators [32].

Reduction of the EOT is not enough to maintain a good electrostatic integrity, because the penetration of the drain field in the channel increases the DIBL and the subthreshold

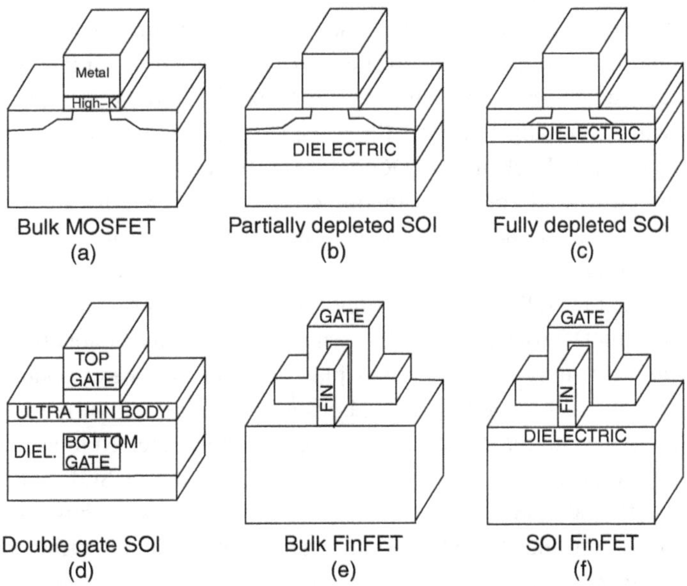

Figure 1.4 MOSFET architectures proposed for present and future CMOS technologies.

swing, unless the substrate doping is increased as well. In this respect, studies in the mid-nineties showed that improved control of the threshold voltage roll-off and low values of the subthreshold swing could be achieved at short channel lengths with ground plane architectures and, even better, with Silicon On Insulator (SOI) technologies [33]. Partially depleted SOI devices (PD-SOI, Fig.1.4.b) demonstrated some advantages over bulk MOSFETs, but the relatively large kink effect and the degradation of static and dynamic performance due to transient charge storage and self-heating effects impeded the blossoming of this technology. Moreover, SOI was and still is a costly technology option; except in a few cases, the portability of bulk designs to a SOI platform is not straightforward [34, 35].

The advent of the SIMOX and Unibond Smart-Cut processes [36] revitalized SOI as a credible technology option [37] and boosted research on high quality aggressively scaled SOI films [38–45]. For small enough silicon thickness the body of the transistor becomes fully depleted (FD-SOI, Fig.1.4.c); consequently, short channel effects, DIBL and subthreshold swing remarkably improve. The impact ionization induced kink effect disappears and good electrostatic integrity is achieved. The source/drain parasitic capacitance is also reduced because of the underlying buried oxide layer.

The SOI technology also facilitates the realization of double gate (DG, Fig.1.4.d) and gate all around (GAA) architectures that can bring CMOS even closer to its ultimate scaling limits by offering nearly optimum control of the gate over the channel [42, 46–48]. In fact, provided the film thickness is at least about 2.5 times the channel length, SCE are suppressed and nearly ideal subthreshold swing is observed (SS $\approx$ 60 mV/dec at room temperature) even in undoped channel transistors. Therefore a reduced fluctuation of V_T due to the discrete doping can be achieved but, at the same time, new means other than channel doping must be devised to tailor the V_T (e.g. workfunction engineering).

Another advantage of DG and GAA architectures is that, in the direction perpendicular to the transport, the average electric field at given inversion charge per channel is reduced compared to bulk devices because good electrostatic control can be achieved with essentially undoped films; hence, the carrier mobility is larger. Moreover, due to the double channel a DG device provides the same total inversion charge at lower effective field compared to single gate SOI; hence it can achieve the same I_{ON} of a single channel device at smaller gate voltages: a clear advantage in view of low voltage operation.

The FinFET technology (Figs.1.4.e and 1.4.f) provides an alternative approach to fabricating DG transistors [49]. In narrow FinFETs the conduction takes place mostly along sidewalls normal to the wafer plane and, in essence, a double gate device is obtained [50]. If the fin is large, instead, a significant fraction of the current flows along the top interface and the device is more appropriately referred to as a triple gate transistor.

The process complexity, variability, and cost of SOI and FinFET technology tend to offset the advantages offered in terms of scaling, thus leaving room for prolonged efforts on bulk MOSFET optimization. In particular, strained silicon technology and optimization of the crystal orientation are very effective means of boosting the mobility and I_{ON} of both n-MOS and p-MOS devices [51–57]. Indeed, the strain in the crystal lattice has a remarkable impact on the band structure, hence on the electrostatics and the transport properties of the device. With an appropriate combination of strain type, magnitude and orientation with respect to the crystal axes and the transport direction, on-current enhancements of up to 20–30% for sub-50 nm channel lengths have been demonstrated [58–60]. The remarkable success of strained silicon technology is keeping bulk MOSFET architecture competitive; as a result, the year of expected introduction of advanced SOI technology options has recently been postponed by the ITRS [61, 62].

To improve the device performance further it has also been proposed to replace the silicon channel with alternative semiconductors characterized by enhanced transport properties. As an example, bulk III-V materials are known to have superior electron mobility with respect to silicon, whereas hole mobility is high in bulk germanium. These considerations have led to a search for new ways to locally grow islands of different semiconductors on silicon substrates [63–65] and to develop compatible high quality gate stacks [66–70]. Studies have flourished aimed at assessing if alternative channel materials can bring real advantages in terms of inversion layer mobility and overall device performance [63, 71–75].

Last but not least, we emphasize that extrinsic parasitic components (source/drain resistances and overlap capacitances) may jeopardize the advantage of having smaller and faster intrinsic transistors. This is especially true for FinFETs and ultra-thin body fully depleted SOI MOSFETs, where the limited SOI film thickness implies a high series resistance. Elevated source/drain technology and non-overlapped devices alleviate these issues [76–83]. To boost the device performance even further, metallic source and drain technology has been proposed. By exploiting doping segregation, a pile-up of the dopants at the metal–semiconductor junction is obtained which relieves the detrimental effects of Schottky barrier formation [84]. Careful selection of the metal can possibly lead to achieving high current drive [85]. Variability due to fluctuations of the tail of dopants in the channel is also expected to decrease thanks to these technology improvements.

1.3 Support of modeling to nano-scale MOSFET design

As illustrated in the previous section, new materials and device architectures are expanding the design space to be explored for future CMOS and nano-electronic technologies. Single gate SOI, double gate SOI, FinFET, MuGFET, gate all around and nanowire device architectures are being investigated as possible successors of the conventional planar bulk MOSFET [86]. Gate metal workfunction, silicon body thickness, stress–strain distribution, gate stack composition, source, drain and channel material are only a few of the additional variables that it is necessary to engineer for the existing and future MOSFET generations.

The design and optimization of nano-transistors exploiting these new options demand general purpose models to describe electrostatic and transport phenomena at the nanoscale in an unprecedented variety of materials, with a reasonably predictable degree of accuracy and with affordable computation time. The established Drift-Diffusion model available in conventional TCAD tools is presently inadequate for the purpose.

In this respect, it is important to consider the substantial quantum mechanical effects in the direction perpendicular to the transport plane which are emphasized by size induced confinement in ultra-thin body architectures with silicon thickness below 10 nm. Carrier quantization decreases the effective gate capacitance (due to the combined effects of finite inversion layer thickness and dead spaces at the SiO_2 interfaces [87–89]) and reduces the inversion charge for a given gate voltage, thus altering the threshold voltage. The appearance of subbands affects the carriers' scattering as well, with remarkable implications for both the low and the high field transport characteristics of the inversion layer.

Quantum confinement is especially strong at the top of the potential energy barrier that governs carrier injection from the source to the channel region (the so called *virtual source*, [90, 91]). Since high levels of charge are desired in the on-state, the carrier gas becomes highly degenerate and the average carrier velocity becomes gate bias dependent.

Another relevant aspect concerning transport is that when the gate length L_G scales below a few tens of nanometers the mean free path in the channel is expected to become comparable to the device length [92, 93]. The fraction of the carrier population that reaches the drain without suffering scattering events tends to increase and the effects related to far from equilibrium transport become important. However, even if rare, scattering events in the channel cannot be neglected, because they affect the carrier density and thus the potential profile along the channel and contribute to shaping the potential energy barrier at the source and to setting the I_{ON} [94]. A sound description of transport in MOSFETs should cover the transition between conventional drift-diffusion and purely ballistic transport, and should obviously include all the relevant scattering mechanisms, especially those related to the introduction of new dielectric or semiconducting thin films.

Tunneling through the source barrier and band-to-band tunneling at the drain end of the channel may also become relevant, especially in the low band gap, small tunneling mass semiconductors being considered for ultimate CMOS [95–100]. Degraded I_{OFF} and subthreshold swing SS are expected if these leakage mechanisms are not kept under control.

The design and optimization of future nano-transistors require us to understand and master all these physical effects and their interrelations in an increasingly large number of materials and device architectures. A broad matrix of combinations must be evaluated and the device simulation can considerably facilitate this process, provided that predictive models are available to reduce the risk and cost of fabrication trials and errors.

Historically the attention of the industry toward the field of modeling and simulation has been mostly driven by the need to steer the selection of process and device parameters for incremental improvements of existing technologies. The broad spectrum of present day scaling scenarios has raised new interests in device modeling and simulation. New theories and new models to describe the links between the band structure of the materials, the device electrostatics, the transport and the performance have become of utmost importance. This new perspective is well expressed by the ITRS roadmap

[10], which devotes a full chapter to modeling and simulation and reiterates the quest for renewed efforts in the modeling of MOSFETs incorporating all the technology boosters of interest.

In this respect it is worth noting that the band gap, the density of states, the carriers' mobility and the other physical properties of the thin, possibly strained semiconductor layers used in fully depleted single or double gate SOI and FinFETs cannot be simply extrapolated from the corresponding properties of the bulk material. The widespread exploitation of stress and strain, and the possible use of alternative channel materials (germanium, silicon–germanium alloys, gallium arsenide) demand models to describe the subband structure and the transport parameters of quantized inversion layers (group velocity, effective mass, scattering rates, mobility, etc.) for both electrons and holes. These models should be general enough to tackle various substrate crystal orientations with respect to the quantization and the transport directions, and accurate enough to predict the stress-strain, film thickness and bias dependencies.

It is clear then, that exploring by simulation the design space of new nano-scale CMOS transistors demands a large innovative effort in physically based and in TCAD oriented modeling, which for decades has been mainly focused on unstrained silicon transistors fabricated almost exclusively on (001) wafers. Physically sound, modular and robust device modeling frameworks are necessary, where new physical effects can be added and related to the device performance, possibly starting from the physical properties of new materials. These frameworks should be general enough to include quantization effects on both electrostatics and carrier transport and to encompass all conduction regimes from drift-diffusion to fully ballistic.

1.4 An overview of subsequent chapters

Stimulated by recent developments in nano-electronics and inspired by the scenario outlined in the previous sections, we wrote this book to describe in detail the semi-classical modeling of carrier transport in modern nanoscale MOSFETs, accounting for the significant quantization effects that enforce the formation of electronic subbands in the transistors inversion layer. In particular, in the framework of this semi-classical model, the Schrödinger equation is used to calculate the quantum energy levels and the wave-functions of the inversion layer while a system of coupled Boltzmann transport equations describes the transport in the subbands. The Poisson equation is solved iteratively with the Schrödinger and the Boltzmann equations until convergence is reached to a fully self-consistent solution of the whole electrostatic and transport problems. Moreover, we illustrate a relevant implementation of the model, which we concisely denote as *multi-subband Monte Carlo* because it relies on use of the Monte Carlo method to solve the Boltzmann equations in the inversion layer subbands.

We have enhanced the book with a broad set of simulation results mostly obtained with the multi-subband Monte Carlo implementation of the model. These were selected to illustrate in detail how the physical elements of the semi-classical transport model in inversion layers affect the operation of modern MOSFETs.

With these objectives in mind, the book begins by recalling in Chapter 2 the elements of the semi-classical treatment of carrier transport in bulk crystals. In particular, we introduce the fundamental results regarding electrons in periodic crystalline lattices and the band structure of bulk crystals. We then describe a few methodologies to compute the conduction and valence band structure in bulk semiconductors and the simplest analytical approximations commonly used to model the dispersion relation in the proximity of the band edges. The last paragraphs of the chapter introduce the foundations of the semi-classical model of carrier transport by using a wave-packet representation of the electrons. We derive the semi-classical equations of motion under the action of slowly varying potentials and introduce the Fermi golden rule for the treatment of carrier scattering due to the action of rapidly fluctuating potentials.

Chapter 3 develops the effective mass approximation and the $\mathbf{k}\cdot\mathbf{p}$ quantization models for, respectively, electron and hole inversion layers. A full band quantization model based on the linear combination of bulk bands method is described as well, since it can serve as a useful reference to check the validity of the simpler quantization models in conditions of strong confinement, such as those present in ultra-thin semiconductor films. From there, the chapter moves to the calculation of carrier densities accounting for the density of states in a two-dimensional carrier gas and finally to self-consistent solutions of the Poisson and Schrödinger equations.

Chapter 4 contains an extensive theoretical treatment of scattering for carriers in inversion layers. Starting from the envelope eigenfunctions and eigenvalues and exploiting the Fermi golden rule, Coulomb, surface roughness, and phonon scattering mechanisms are analyzed in detail for both electrons and holes. The static and dynamic screening of the scattering potential produced by the inversion layer charge is also addressed. We have tried to provide a clear and pedagogical presentation of these topics. Particular attention was devoted to justifying and discussing the approximations behind the mathematical developments.

After Chapters 3 and 4, which provide the quantum mechanical foundations for the treatment of the two-dimensional carrier gas, we continue with Chapter 5 aimed at a description of the set of coupled BTEs for the subbands in the inversion layer. The case of free electrons and holes is treated first, to underline the connections to the semi-classical transport concepts explained in Chapter 2. Several examples clarify the expression of the driving force for carriers' motion in cases of practical relevance.

Chapter 5 describes also the solution of the BTE in inversion layers by means of the widely used Momentum Relaxation Time (MRT) approximation, whose usefulness and validity limits are discussed. The recently proposed ballistic and quasi-ballistic MOSFET models are then derived from the solution of the BTE where the terms related to scattering are neglected. These derivations allow us to clarify the approximations behind these popular models and are instrumental in introducing many concepts useful for interpretation of numerical simulations.

The discussion of solution methods for the BTE continues with Chapter 6, which is devoted to the Monte Carlo method. Here again the free carrier gas is treated first, but the multi-subband case is also specifically addressed at the end of the chapter. Many non-trivial technical details arising in the practical implementation of the method are discussed. Moreover, general methods for stability analysis of the self-consistent coupling between the Monte Carlo and the Poisson equation are introduced.

Having established the theory of inversion layer modeling and having described the methodologies to solve the relevant equations (the Schrödinger and Poisson equations in Chapter 3, the BTE in Chapter 6), we let the reader take a breath to appreciate a large number of simulation results, mostly obtained with the multi-subband Monte Carlo method, that illustrate the ability of the semi-classical model to clarify the physics of transport in inversion layers of bulk and SOI silicon MOSFETs fabricated on (001) wafers. To this purpose, Chapter 7 initially compares simulations and measurements of effective mobility in inversion layers. High field transport in nanoscale MOSFETs is described as well, by illustrating and discussing the behavior of many internal quantities such as charge density, velocity, occupation functions, and their relation to the on-current of the device.

The last three chapters of the book address the most relevant technology boosters presently implemented in production level CMOS technologies; namely, optimized crystal orientation, strained silicon, and high-κ gate dielectrics. Alternative channel materials such as germanium and gallium arsenide are described as well in the last chapter. Differently from the previous chapters, here we show the effect of each booster with appropriate simulation results immediately after the development of the related theory.

In particular, Chapter 8 first addresses solution of the Schrödinger equation for crystal orientations other than (001) and generalizes the results from previous chapters to these more complex cases. The impact of the channel orientation on the effective mobility and on the on-current of MOSFETs is illustrated with a selected set of simulation results.

The notations and methods developed in Chapter 8 are also instrumental in the calculation of band structures for inversion layers in strained materials. In this respect, after a short description of technologically relevant means to induce strain in silicon channels, Chapter 9 sets out the definitions and notations for stress, strain, and their relation. The impact of uniaxial and biaxial strain on the silicon conduction and valence bands is shown by means of theory and simulations. Due to the relevance of this booster for present days CMOS technologies, some results concerning the impact of strain on the mobility and the I_{ON} of MOSFETs are also reported.

Chapter 10 completes the coverage of the technology boosters by addressing a selection of topics related to the use of new materials alternative to those employed in the past 40 years by mainstream CMOS technologies. In this respect, the chapter provides a detailed treatment of remote phonon and remote Coulomb scattering due to high-κ gate dielectrics, since these are believed to be responsible at least in part for the effective mobility degradation observed in real devices. Then, transport in alternative channel materials such as germanium and gallium arsenide is explored.

It is apparent from the contents outlined above that our book is mostly focused on modeling of the mobility and on-current in advanced CMOS transistors and does not attempt to be a textbook addressing all the relevant aspects related to the operation of nanoscale MOSFETs. This choice has led us to exclude a priori many physical effects which are certainly very relevant for the optimization of nano-MOSFETs, such as for instance the gate leakage current. The quantum mechanical treatment of the inversion layer charge distribution inherent to the model described here, however, naturally lends itself to gate current calculations.

The focus on semi-classical transport implicitly excludes as well treatment of source to drain tunneling, which might become relevant in deeply scaled CMOS transistors below 10nm channel length. It is worth noting, however, that recent developments in the effective potential corrections to the semi-classical model indicate that the model can be extended to include phenomenologically the effects of S/D tunneling. Also the selection of scattering mechanisms discussed in the book has given prominence to those that have the largest impact on channel mobility and the I_{ON}.

As a final remark, we observe that variability, noise, and reliability are also extremely important aspects to consider in the optimization of devices in aggressively scaled CMOS technology. Once again, however, we reiterate that it was not our intention to cover all aspects of nanoscale MOSFET operation; a choice that well justifies the absence of these topics from the present book.

References

[1] G. Moore, "Cramming more components onto integrated circuits," *Electronics*, vol. 38, no. 8, 1965.

[2] R.H. Dennard, F.H. Gaensslen, L. Kuhn, *et al.*, "Design of ion-implanted MOSFETs with very small physical dimensions," *IEEE Journal of Solid State Circuits*, vol. 9, pp. 256–268, 1974.

[3] G. Baccarani, M.R. Wordeman, and R.H. Dennard, "Generalized scaling theory and its application to a 1/4 micrometer MOSFET design," *IEEE Trans. on Electron Devices*, vol. ED-31, pp. 452–462, 1984.

[4] Y. Taur and T. Ning, *Fundamentals of Modern VLSI Devices*. Edinburgh: Cambridge University Press, 1998.

[5] S. Selberherr, A. Schutz, and H. Potzl, "MINIMOS – A two dimensional MOS transistor analyzer," *IEEE Journal of Solid State Circuits*, vol. 15, no. 4, pp. 605–615, 1980.

[6] E.M. Buturla, P.E. Cottrell, B.M. Grossman, and K.A. Salsburg, "Finite-element analysis of semiconductor devices: The FIELDAY program," *IBM Journal of Research and Development*, vol. 25, no. 4, pp. 218–231, 1981.

[7] P. Ciampolini, A. Gnudi, R. Guerrieri, M. Rudan, and G. Baccarani, "Three-dimensional simulation of a narrow-width MOSFET," in *Proc. European Solid State Device Res. Conf.*, pp. 413–416, 1987.

[8] M. Thurner and S. Selberherr, "Comparison of long- and short- channel MOSFET's carried out by 3D-MINIMOS," in *Proc. European Solid State Device Res. Conf.*, pp. 409–412, 1987.

[9] W. Hansch and S. Selberherr, "MINIMOS 3: A MOSFET simulator that includes energy balance," *IEEE Trans. on Electron Devices*, vol. 34, no. 5, pp. 1074–1078, 1987.

[10] *International Technology Roadmap for Semiconductors*. Public home page *http://www.itrs.net*.

[11] D.J. Frank, R.H. Dennard, E. Nowak, *et al.*, "Device scaling limits of Si MOSFETs and their application dependencies," *Proc. of the IEEE*, vol. 89, pp. 259–288, 2001.

[12] E.J. Nowak, "Maintaining the benefits of CMOS scaling when scaling bogs down," *IBM Journal of Research and Development*, vol. 46, no. 2/3, pp. 169–180, 2002.

[13] D.A. Buchanan, "Scaling the gate dielectric: materials, integration, and reliability," *IBM Journal of Research and Development*, vol. 43, no. 3, p. 245–264, 1999.

[14] T. Sorsch, W. Timp, F. Baumann, *et al.*, "Ultra-thin, 1.0-3.0 nm, gate oxides for high performance sub-100nm technology," in *IEEE Symposium on VLSI Technology – Technical Digest*, pp. 222–223, 1998.

[15] E.P. Gusev, H.-C. Lu, E.L. Garfunkel, T. Gustaffson, and M.L. Green, "Growth and characterization of ultrathin nitrided silicon oxide films," *IBM Journal of Research and Development*, vol. 43, pp. 265–285, 1999.

[16] M.L. Green, E.P. Gusev, R. Degraeve, and E.L. Garfunkel, "Ultrathin (<4 nm) SiO_2 and Si-O-N gate dielectric for silicon microelectronics: Understanding the processing, structure and physical and electrical limits," *Journal of Applied Physics*, vol. 90, pp. 2057–2122, 2001.

[17] M. Togo, K. Watanabe, T. Yamamoto, *et al.*, "Low-leakage and highly-reliable 1.5 nm SiON gate-dielectric using radical oxynitridation for sub-0.1 μm CMOS," in *IEEE Symposium on VLSI Technology - Technical Digest*, pp. 116–117, 2000.

[18] R. Chau, S. Datta, M. Doczy, J. Kavalieros, and M. Metz, "Gate dielectric scaling for high-performance CMOS: from SiO2 to high-k," in *Extended Abstracts of International Workshop on Gate Insulator (IWGI), Tokyo, Japan*, pp. 124–126, 2003.

[19] K. Henson, R.J.P. Lander, M. Demand, *et al.*, "45 nm nMOSFET with metal gate on thin SiON driving 1150 μA/μm and off-state of 10nA/μm," in *IEEE IEDM Technical Digest*, pp. 851–854, 2004.

[20] R. Chau, J. Brask, S. Datta, *et al.*, "Application of high-κ gate dielectrics and metal gate electrodes to enable silicon and non-silicon logic nanotechnology," *Microelectronic Engineering*, vol. 80, pp. 1–6, 2005.

[21] V.V. Zhirnov and R.K. Cavin III, "Scaling beyond CMOS: Turing-Heisenberg rapprochement," in *Proc. European Solid State Device Res. Conf.*, pp. 16–22, 2009.

[22] V. Zhirnov, R.K. Cavin III, J. Hutchby, and G. Bourianoff, "Limits to binary logic switch scaling – A Gedanken model," *Proc. of the IEEE*, vol. 91, pp. 1934–1939, 2003.

[23] G. Cerofolini, "Realistic limits to computation. I. Physical limits," *Applied Physics A*, vol. 86, pp. 23–29, 2007.

[24] G. Cerofolini, "Search for realistic limits to computation. II. The technological side," *Applied Physics A*, vol. 86, pp. 31–42, 2007.

[25] S. Lloyd, "Ultimate physical limits to computation," *Nature*, vol. 406, pp. 1047–1054, 2000.

[26] M. Houssa, L. Pantisano, L.-A. Ragnarsson, *et al.*, "Electrical properties of high-κ gate dielectrics: Challenges, current issues, and possible solutions," *Materials Science and Engineering, R: Reports*, vol. 51, no. 4-6, pp. 37–85, 2006.

[27] M. Baklanov, M. Green, and K. Maex, *Dielectric Films for Advanced Microelectronics*. Wiley, 2007.

[28] C.C. Hobbs, L. Fonseca, A. Knizhnik, *et al.*, "Fermi-level pinning at the polysilicon/metal oxide interface," *IEEE Trans. on Electron Devices*, vol. 51, no. 6, pp. 971–977, 2004.

[29] K. Shiraishi, H. Takeuchi, Y. Akasaka, *et al.*, "Theory of Fermi level pinning of high-κ dielectrics," in *Proc.SISPAD*, pp. 306–313, 2006.

[30] M. Fischetti, D. Neumayer, and E. Cartier, "Effective electron mobility in Si inversion layers in metal-oxide-semiconductor systems with a high-κ insulator: The role of remote phonon scattering," *Journal of Applied Physics*, vol. 90, no. 9, pp. 4587–4608, 2001.

[31] X. Garros, M. Cassé, G. Reimbold, *et al.*, "Performance and reliability of advanced high-κ/metal gate stacks," *Microelectronic Engineering*, vol. 7–9, pp. 1609–1614, 2009.

[32] A. Oates, "Reliability issues for high-κ gate dielectrics," in *IEEE IEDM Technical Digest*, pp. 923–926, 2003.

[33] C. Fiegna, I. Iwai, T. Wada, *et al.*, "Scaling the MOS transistor below $0.1\mu m$: Methodology, device structures and technology requirements," *IEEE Trans. on Electron Devices*, p. 941–951, June 1994.

[34] G. Shahidi, C. Anderson, B. Chappell, *et al.*, "A room temperature 0.1 μm CMOS on SOI," IEEE Trans. on Electron Devices, vol. 41, no. 12, pp. 2405–2412, 1994.

[35] G. Shahidi, "SOI technology for the GHz era," *IBM Journal of Research and Development*, vol. 46, no. 2/3, pp. 121–131, 2002.

[36] M. Bruel, "Silicon on insulator material technology," *Electronics Letters*, vol. 31, p. 1201, 1995.

[37] J.P. Colinge, *Silicon on Insulator Technology: Materials to VLSI.* New York: Kluwer Academic Publishing, 1990.

[38] K. Uchida, J. Koga, R. Ohba, and S. Takagi, "Experimental evidences of quantum-mechanical effects on low-field mobility, gate-channel capacitance and threshold voltage of ultrathin body SOI MOSFETs," in *IEEE IEDM Technical Digest*, pp. 633–636, 2001.

[39] K. Uchida, H. Watanabe, A. Kinoshita, *et al.*, "Experimental study on carrier transport mechanisms in ultrathin-body SOI n- and p-MOSFETs with SOI thickness less than 5nm," in *IEEE IEDM Technical Digest*, pp. 47–50, 2002.

[40] D. Esseni, M. Mastrapasqua, G. Celler, *et al.*, "Low field electron and hole mobility of SOI transistors fabricated on ultra-thin silicon films for deep submicron technology application," *IEEE Trans. on Electron Devices*, p. 2842–2850, 2001.

[41] D. Esseni, M. Mastrapasqua, G.K. Celler, *et al.*, "An experimental study of mobility enhancement in ultrathin SOI transistors operated in double-gate mode," *IEEE Trans. on Electron Devices*, vol. 50, no. 3, pp. 802–808, 2003.

[42] B. Doris, M. Ieong, T. Kanarsky, *et al.*, "Extreme scaling with ultra-thin Si channel MOSFETs," in *IEEE IEDM Technical Digest*, p. 267–270, 2002.

[43] T. Ernst, S. Cristoloveanu, G. Ghibaudo, *et al.*, "Ultimately thin double gate SOI MOSFETs," *Microelectronic Engineering*, vol. 50, no. 3, p. 830–838, 2003.

[44] K. Uchida, R. Zednikand, C.H. Lu, *et al.*, "Experimental study of biaxial and uniaxial strain effects on carrier mobility in bulk and ultrathin-body SOI MOSFETs," in *IEEE IEDM Technical Digest*, pp. 229–232, 2004.

[45] C. Fenouillet-Beranger, S. Denorme, P. Perreau, *et al.*, "FDSOI devices with thin box and ground plane integration for 32nm node and below," in *Proc. European Solid State Device Res. Conf.*, pp. 206–209, 2008.

[46] F. Balestra, S. Cristoloveanu, M. Benachir, J. Brini, and T. Elewa, "Double-gate silicon-on-insulator transistor with volume inversion: A new device with greatly enhanced performance," *IEEE Electron Device Lett.*, vol. 8, pp. 410–412, 1987.

[47] L. Chang, S. Tang, T.-J. King, J. Bokor, and C. Hu, "Gate length scaling and threshold voltage control of double-gate MOSFETs," in *IEEE IEDM Technical Digest*, pp. 719–722, 2000.

[48] A. Amara and O. Rozeau, *Planar Double Gate Transistor: From Technology to Circuit.* Springer, 2009.

[49] D. Hisamoto, W.C. Lee, J. Kedzierski, *et al.*, "FinFET a self-aligned double-gate MOSFET scalable to 20 nm," *IEEE Trans. on Electron Devices*, vol. 47, no. 12, pp. 1320–1325, 2000.

[50] T. Rudenko, V. Kilchytska, N. Collaert, *et al.*, "Carrier mobility in undoped triple-gate FinFET structures and limitations of its description in terms of top and sidewall channel mobilities," *IEEE Trans. on Electron Devices*, vol. 55, no. 12, pp. 3532–3541, 2008.

[51] J. Welser, J.L. Hoyt, S. Takagi, and F. Gibbons, "Strain dependence of the performance enhancement in strained-Si n-MOSFETs," in *IEEE IEDM Technical Digest*, pp. 373–376, 1994.

[52] K. Rim, J.L. Hoyt, and F. Gibbons, "Fabrication and analysis of deep submicron strained-Si n-MOSFETs," *IEEE Trans. on Electron Devices*, vol. 47, p. 1406, 2000.

[53] T. Mizuno, N. Sugiyama, T. Tezuka, T. Numata, and S. Takagi, "High-performance strained-SOI CMOS devices using thin film SiGe-on-insulator technology," *IEEE Trans. on Electron Devices*, vol. 50, no. 4, pp. 988–994, 2003.

[54] T. Mizuno, N. Sugiyama, T. Tezuka, *et al.*, "[110]-surface strained-SOI CMOS devices," *IEEE Trans. on Electron Devices*, vol. 52, no. 3, pp. 367–374, 2005.

[55] M. Yang, V.W.C. Chan, K.K. Chan, *et al.*, "Hybrid-orientation technology (HOT): Opportunites and challenges," *IEEE Trans. on Electron Devices*, vol. 53, no. 5, pp. 965–978, 2006.

[56] M. Horstmann, M. Wiatr, A. Wei, *et al.*, "Advanced SOI CMOS transistor technology for high performance microprocessors," in *Proc. Int. Conf. on Ultimate Integration on Silicon (ULIS)*, pp. 81–84, 2009.

[57] N. Serra, F. Conzatti, D. Esseni, *et al.*, "Experimental and physics based modeling assessment of strain induced mobility enhancement in FinFETs," in *IEEE IEDM Technical Digest*, pp. 71–74, 2009.

[58] Q. Xiang, J.-S. Goo, J. Pan, *et al.*, "Strained silicon NMOS with nickel-silicide metal gate," in *IEEE Symposium on VLSI Technology - Technical Digest*, pp. 101–102, 2003.

[59] F. Boeuf, F. Arnaud, M.T. Basso, *et al.*, "A conventional 45 nm CMOS node lowcost platform for general purpose and low power applications," in *IEEE IEDM Technical Digest*, pp. 425–428, 2004.

[60] A. Thean, D. Zhang, V. Vartanian, *et al.*, "Strain-enhanced CMOS through novel process-substrate stress hybridization of super-critically thick strained silicon directly on insulator (SC-SSOI)," in *IEEE Symposium on VLSI Technology – Technical Digest*, pp. 130–131, 2006.

[61] "*International Technology Roadmap for Semiconductors: 2008 Update*," 2008.

[62] H. Iwai, "Roadmap for 22 nm and beyond," *Microelectronic Engineering*, vol. 86, no. 7–9, pp. 1520–1528, 2009.

[63] M.K. Hudait, G. Dewey, S. Datta, *et al.*, "Heterogeneous integration of enhancement mode $In_{0.7}Ga_{0.3}As$ quantum well transistor on silicon substrate using thin ($< 2\mu m$) composite buffer architecture for high-speed and low-voltage (0.5 V) logic applications," in *IEEE IEDM Technical Digest*, pp. 625–628, 2007.

[64] T. Krishnamohan and K. Saraswat, "High mobility Ge and III–V materials and novel device structures for high performance nanoscale MOSFETS," in *Proc. European Solid State Device Res. Conf.*, pp. 38–46, 2008.

[65] H.-Y. Yu, M. Kobayashi, W.S. Jung, *et al.*, "High performance n-MOSFETs with novel source/drain on selectively grown Ge on Si for monolithic integration," in *IEEE IEDM Technical Digest*, pp. 685–688, 2009.

[66] Y.I. Nissim, J.M. Moison, C. Licoppe, and G. Post, "High temperature LPCVD of dielectrics on III–V substrates for device appliclations," in *Proc. European Solid State Device Res. Conf.*, pp. 173–176, 1989.

[67] D.A.J. Moran, R.J.W. Hill, X. Li, *et al.*, "Sub-micron, metal gate, high-κ dielectric, implant-free, enhancement-mode III-V MOSFETS," in *Proc. European Solid State Device Res. Conf.*, pp. 466–469, 2007.

[68] H.-S. Kim, I. Ok, M. Zhang, *et al.*, "Germanium passivation for high-k dielectric III-V MOSFETs and temperature dependence of dielectric leakage current," in *Device Research Conference*, pp. 87–88, 2006.

[69] L. Weber, J. Damlencourt, F. Andrieu, *et al.*, "Fabrication and mobility characteristics of SiGe surface channel *p*MOSFETs with a HfO_2/TiN gate stack," *IEEE Trans. on Electron Devices*, vol. 53, no. 3, pp. 449–455, 2006.

[70] D. Kuzum, T. Krishnamohan, A.J. Pethe, *et al.*, "Ge-interface engineering with ozone oxidation for low interface-state density," *IEEE Electron Device Lett.*, vol. 29, no. 4, pp. 328–330, 2008.

[71] T. Low, Y.T. Hou, M.F. Li, *et al.*, "Investigation of performance limits of germanium double-gated MOSFETs," in *IEEE IEDM Technical Digest*, p. 691–694, 2003.

[72] S.E. Laux, "Simulation study of Ge n-channel 7.5nm DGFETs of arbitrary crystallographic alignment," in *IEEE IEDM Technical Digest*, p. 135–138, 2004.

[73] A. Rahman, G. Klimeck, and M. Lundstrom, "Novel channel materials for ballistic nanoscale MOSFETs: bandstructure effects," in *IEEE IEDM Technical Digest*, pp. 615–618, 2005.

[74] A. Pethe, T. Krishnamohan, D. Kim, *et al.*, "Investigation of the performance limits of III–V double-gate n-MOSFETs," in *IEEE IEDM Technical Digest*, pp. 619–622, 2005.

[75] M. De Michielis, D. Esseni, and F. Driussi, "Analytical models for the insight into the use of alternative channel materials in ballistic nano-MOSFETs," *IEEE Trans. on Electron Devices*, vol. 54, no. 1, pp. 115–123, 2006.

[76] M. Orlowski, C. Mazure, and M. Noell, "A novel elevated MOSFET source/drain structure," *IEEE Electron Device Lett.*, vol. 12, no. 11, pp. 593–595, 1991.

[77] S. Kimura, H. Noda, D. Hisamoto, and E. Takeda, "A 0.1 μm-gate elevated source and drain MOSFET fabricated by phase-shifted lithography," in *IEEE IEDM Technical Digest*, pp. 950–952, 2001.

[78] F. Boeuf, T. Skotnicki, S. Monfray, *et al.*, "16 nm planar NMOSFET manufacturable within state-of-the-art CMOS process thanks to specific design and optimisation," in *IEEE IEDM Technical Digest*, pp. 637–640, 2001.

[79] A. Vandooren, A. Barr, L. Mathew, *et al.*, "Fully-depleted SOI devices with TaSiN gate, HfO2 gate dielectric, and elevated source/drain extensions," *IEEE Electron Device Lett.*, vol. 24, no. 5, pp. 342–344, 2003.

[80] W. Jeamsaksiri, M. Jurczak, L. Grau, *et al.*, "Gate-source-drain architecture impact on DC and RF performance of sub-100 nm elevated source/drain NMOS transistors," *IEEE Trans. on Electron Devices*, vol. 50, no. 3, pp. 610–617, 2003.

[81] K.D. Seong, C.-M. Park, and J.C.S. Woo, "Advanced model and analysis for series resistance in sub-100 nm CMOS including poly-depletion and overlap doping gradient effect," in *IEEE IEDM Technical Digest*, pp. 723–726, 2000.

[82] R. Gusmeroli, A. Spinelli, A. Pirovano, *et al.*, "2D QM simulation and optimization of decanano non-overlapped MOS devices," in *IEEE IEDM Technical Digest*, pp. 225–228, 2003.

[83] H. Wakabayashi, T. Tatsumi, N. Ikarashi, *et al.*, "Improved sub-10 nm CMOS devices with elevated source/drain extensions by tunneling Si-selective-epitaxial-growth," in *IEEE IEDM Technical Digest*, pp. 145–148, 2005.

[84] A. Kinoshita, Y. Tsuchiya, A. Yagishita, K. Uchida, and J. Koga, "Solution for high-performance Schottky-source/drain MOSFETs: Schottky barrier height engineering with dopant segregation technique," in *IEEE Symposium on VLSI Technology - Technical Digest*, pp. 168–169, 2004.

[85] J.M. Larson and J.P. Snyder, "Overview and status of metal S/D Schottky-barrier MOSFET technology," *IEEE Trans. on Electron Devices*, vol. 53, no. 5, pp. 1048–1058, 2006.

[86] H.S.Wong, "Beyond the conventional transistor," *IBM Journal of Research and Development*, vol. 46, no. 2/3, pp. 133–168, 2002.

[87] G. Baccarani and M.R. Wordeman, "Transconductance degradation in thin oxide MOSFET's," *IEEE Trans. on Electron Devices*, vol. 30, pp. 1295–1304, 1983.

[88] A. Pacelli, A.S. Spinelli, and L.M. Perron, "Carrier quantization at flat bands in MOS devices," *IEEE Trans. on Electron Devices*, vol. 46, no. 2, pp. 383–387, 1999.

[89] A.S. Spinelli, A. Pacelli, and A.L. Lacaita, "Polysilicon quantization effects on the electrical properties of MOS transistors," *IEEE Trans. on Electron Devices*, vol. 47, no. 12, pp. 2366–2371, 2000.

[90] M.S. Lundstrom, "Elementary scattering theory of the Si MOSFET," *IEEE Electron Device Lett.*, vol. 18, no. 7, pp. 361–363, 1997.

[91] M. Lundstrom and Z. Ren, "Essential physics of carrier transport in nanoscale MOSFETs," *IEEE Trans. on Electron Devices*, vol. 49, no. 1, pp. 133–141, 2002.

[92] G. Timp, J. Bude, K.K. Bourdelle, *et al.*, "The ballistic nano-transistor," in *IEEE IEDM Technical Digest*, p. 55, 1999.

[93] P.M. Solomon and S.E. Laux, "The ballistic FET: Design, capacitance and speed limit," in *IEEE IEDM Technical Digest*, pp. 95–98, 2001.

[94] P. Palestri, D. Esseni, S. Eminente, *et al.*, "Understanding quasi-ballistic transport in nano-MOSFETs. Part I: Scattering in the channel and in the drain," *IEEE Trans. on Electron Devices*, vol. 52, no. 12, pp. 2727–2735, 2005.

[95] H. Kawaura, T. Sakamoto, and T. Baba, "Observation of source-to-drain direct tunneling current in 8 nm gate electrically variable shallow junction metaloxidesemiconductor field-effect transistors," *Applied Physics Letters*, vol. 76, no. 25, pp. 3810–3812, 2000.

[96] J. Wang and M.S. Lundstrom, "Does source to drain tunneling limit the ultimate scaling of MOSFETs ?," in *IEEE IEDM Technical Digest*, pp. 707–710, 2002.

[97] M. Bescond, J.L. Autran, D. Munteanu, N. Cavassilas, and M. Lannoo, "Atomic-scale modeling of source-to-drain tunneling in ultimate Schottky barrier double-gate MOSFETs," in *Proc. European Solid State Device Res. Conf.*, pp. 395–398, 2003.

[98] J. Lolivier, X. Jehl, Q. Rafhay, *et al.*, "Experimental characterization of source-to-drain tunneling in 10 nm SOI devices," in *Proc. IEEE International SOI Conference*, pp. 34–35, 2005.

[99] K.C. Saraswat, C.O. Chui, D. Kim, T. Krishnamohan, and A. Pethe, "High mobility materials and novel device structures for high performance nanoscale MOSFETs," in *IEEE IEDM Technical Digest*, pp. 1–4, 2006.

[100] Q. Rafhay, R. Clerc, G. Ghibaudo, and G. Pananakakis, "Impact of source to drain tunneling on the I_{on}/I_{off} trade-off of alternative channel material MOSFETs," in *International Semiconductor Device Research Symposium*, pp. 1–2, 2007.

2 Bulk semiconductors and the semi-classical model

The channel of modern nano-scale MOSFETs is made of crystalline material, shaped in bulk or thin film layers. Since most of the basic electronic properties of crystals can be understood by considering the quantum mechanical behavior of electrons in an infinite periodic arrangement of atoms (bulk crystal), it is reasonable to begin the technical part of this book with a description of the electronic properties of bulk semiconductors.

To this purpose, we start with a short introduction to the basic notions regarding crystal structures and electrons in a strictly periodic potential. The concept of band structure is thus briefly developed. The reader can refer to excellent textbooks for a more detailed treatment of these basic topics [1–3]. We then describe a few methodologies to compute the band structure of electrons and holes in semiconductors and the simplest analytical approximations commonly used to represent the energy relation in proximity to the band edges.

The last sections of the chapter illustrate the effective mass approximation and the foundations of the semi-classical model of carrier transport; namely, the motion of wave-packets in slowly varying potentials and the basics of scattering by rapidly fluctuating potentials.

The chapter sets the stage for the detailed treatment of situations where additional built-in or external potentials cause non-negligible quantum mechanical confinement of the carriers in at least one physical space direction, as, for instance, in the case of the MOSFET inversion layer.

2.1 Crystalline materials

2.1.1 Bravaix lattice

A crystal structure (infinite crystal) is an ordered state of matter described by a *Bravaix lattice* and a *basis*. The Bravaix lattice denotes an infinite periodic array of points placed in such a way that it has exactly the same appearance (arrangement and orientation) from whichever lattice point it is viewed. The *basis* denotes the physical arrangement of the atoms within one period of the Bravaix lattice.

From a mathematical standpoint we can describe a three dimensional Bravaix lattice as the set of points $\mathbf{R_n}$ satisfying the condition:

$$\mathbf{R_n} = \sum_{i=1}^{3} n_i \mathbf{a}_i, \tag{2.1}$$

where the $\mathbf{a}_i$ are non-coplanar *primitive vectors*, the n_i span all integer values and $\mathbf{n} = (n_1,n_2,n_3)$. The choice of the $\mathbf{a}_i$ is in general not unique and some relevant examples in this regard will be given below. Note that Eq.2.1 can be interpreted as the definition of an infinite set of translations where the sum of two translations is also a translation of the same set.

We define *primitive unit cell* as a volume of space such that, when translated by all the vectors of the form (2.1), it fills all the space without voids or overlaps. A simple way to identify a primitive unit cell is to consider all points with coordinates

$$\mathbf{R} = \sum_{i=1}^{3} x_i \mathbf{a}_i, \tag{2.2}$$

with the x_i ranging between 0 and 1. Since the $\mathbf{a}_i$ vectors are not unique, also the primitive unit cell is not unique. It is always possible to choose a primitive cell with the symmetry of the Bravaix lattice and the most popular such choice is the *Wigner–Seitz cell.* The Wigner–Seitz cell of a lattice is made up of the points that are closer to a given lattice point than to any other point. It can be generated by bisecting all the segments joining one lattice point to its nearest neighbors with a plane perpendicular to the segment and then taking the region of space included in these planes.

Relevant examples of Bravaix lattices are shown in Fig.2.1. The cubic lattice (Fig.2.1.a) is made up of cubic cells of edges $a_0\hat{\mathbf{x}}$, $a_0\hat{\mathbf{y}}$ and $a_0\hat{\mathbf{z}}$, where $\hat{\mathbf{x}}$, $\hat{\mathbf{y}}$, $\hat{\mathbf{z}}$ are three cartesian orthogonal unit vectors and a_0 is the lattice constant. The simplest choice for the primitive vectors is

$$\mathbf{a}_1 = a_0\hat{\mathbf{x}}, \qquad \mathbf{a}_2 = a_0\hat{\mathbf{y}}, \qquad \mathbf{a}_3 = a_0\hat{\mathbf{z}}. \tag{2.3}$$

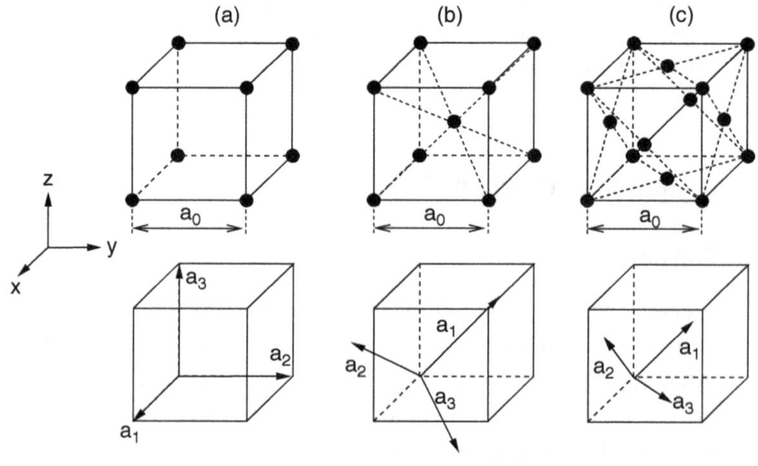

Figure 2.1 Top row: cubic (a), body-centered-cubic (b) and face-centered-cubic (c) Bravaix lattices. Bottom row: corresponding sets of primitive vectors.

The body-centered-cubic (bcc) lattice (Fig.2.1.b) is constructed by placing two lattice points located in the origin (0, 0, 0) and at $\frac{a_0}{2}(\hat{\mathbf{x}}, \hat{\mathbf{y}}, \hat{\mathbf{z}})$ of each vertex of a cubic cell with edge a_0.

The face-centered-cubic (fcc) lattice (Fig.2.1.c), instead, has four lattice points in the cube located at the origin (0, 0, 0) and at $a_0(\hat{\mathbf{x}}+\hat{\mathbf{y}})/2$, $a_0(\hat{\mathbf{y}}+\hat{\mathbf{z}})/2$, $a_0(\hat{\mathbf{z}}+\hat{\mathbf{x}})/2$. It is apparent that the cubic cell is not an elementary cell of the bcc and fcc lattices.

An appropriate choice for the primitive vectors is

$$\mathbf{a}_1 = \frac{a_0}{2}(\hat{\mathbf{y}}+\hat{\mathbf{z}}-\hat{\mathbf{x}}), \qquad \mathbf{a}_2 = \frac{a_0}{2}(\hat{\mathbf{z}}+\hat{\mathbf{x}}-\hat{\mathbf{y}}), \qquad \mathbf{a}_3 = \frac{a_0}{2}(\hat{\mathbf{x}}+\hat{\mathbf{y}}-\hat{\mathbf{z}}), \tag{2.4}$$

for the bcc lattice, and

$$\mathbf{a}_1 = \frac{a_0}{2}(\hat{\mathbf{y}}+\hat{\mathbf{z}}), \qquad \mathbf{a}_2 = \frac{a_0}{2}(\hat{\mathbf{z}}+\hat{\mathbf{x}}), \qquad \mathbf{a}_3 = \frac{a_0}{2}(\hat{\mathbf{x}}+\hat{\mathbf{y}}), \tag{2.5}$$

for the fcc lattice. Note also that the vectors

$$\mathbf{a}_1 = a_0\hat{\mathbf{x}}, \qquad \mathbf{a}_2 = a_0\hat{\mathbf{y}}, \qquad \mathbf{a}_3 = \frac{a_0}{2}(\hat{\mathbf{x}}+\hat{\mathbf{y}}+\hat{\mathbf{z}}), \tag{2.6}$$

describe a bcc lattice, thus demonstrating that the choice of primitive vectors is not unique. The cubic, bcc, and fcc lattices and the corresponding primitive vectors are depicted in Fig.2.1. The volume of the primitive unit cell is

$$\Omega_{cell} = \mathbf{a}_1 \cdot (\mathbf{a}_2 \times \mathbf{a}_3), \tag{2.7}$$

that is, a_0^3, $a_0^3/2$ and $a_0^3/4$, for the simple cubic, bcc, and fcc lattices, respectively.

The most common semiconductors (silicon, germanium, gallium arsenide) crystallize in the so-called diamond (Si, Ge) or zinc-blende (GaAs) structure. The diamond lattice is an fcc lattice with a basis of two identical atoms located at the origin and at one fourth of the cube diagonal, that is at (0, 0, 0) and at $a_0(\hat{\mathbf{x}}+\hat{\mathbf{y}}+\hat{\mathbf{z}})/4$. Alternatively, we can think of the diamond lattice as a structure where identical atoms occupy all the points of two interleaved fcc lattices displaced by one fourth of the cube diagonal. The lattice constant a_0 is 0.54309 nm for silicon and 0.56461 nm for germanium. The interatomic distance is $[a_0\sqrt{3}/4] = 0.2352$ nm and 0.2445 nm for Si and Ge, respectively. The gallium arsenide structure is also given by a fcc Bravaix lattice with a basis of one Ga and one As atom; that is the two interleaved fcc lattices are occupied by Ga and As atoms, respectively. The lattice constant of GaAs is $a_0 = 0.5653$ nm and the interatomic distance is 0.2448 nm.

2.1.2 Reciprocal lattice

The *reciprocal lattice* of the Bravaix lattice of points $\mathbf{R_n}$ is the set of wave-vectors

$$\mathbf{G_m} = \sum_{i=1}^{3} m_i \mathbf{b}_i, \tag{2.8}$$

where $\mathbf{m} = (m_1, m_2, m_3)$ with integer m_i, such that for all possible $\mathbf{R_n}$ and $\mathbf{G_m}$

$$\exp(\mathrm{i}\, \mathbf{G_m} \cdot \mathbf{R_n}) = 1. \tag{2.9}$$

Equation 2.9 is satisfied by vectors $\mathbf{G_m}$ of the form (2.8) with primitive vectors $\mathbf{b}_i$ such that

$$\mathbf{b}_i \cdot \mathbf{a}_j = 2\pi\delta_{ij}, \tag{2.10}$$

where δ_{ij} is the Kronecker delta and the $\mathbf{a}_j$ are the primitive vectors of the direct lattice. In three dimensions:

$$\mathbf{b_1} = \frac{2\pi}{\Omega_{cell}}(\mathbf{a_2} \times \mathbf{a_3}), \qquad \mathbf{b_2} = \frac{2\pi}{\Omega_{cell}}(\mathbf{a_3} \times \mathbf{a_1}), \qquad \mathbf{b_3} = \frac{2\pi}{\Omega_{cell}}(\mathbf{a_1} \times \mathbf{a_2}), \tag{2.11}$$

where Ω_{cell} is the volume of the primitive cell in the direct lattice given by Eq.2.7.

For the simple cubic lattice we immediately find

$$\mathbf{b_1} = \frac{2\pi}{a_0}\hat{\mathbf{x}}, \qquad \mathbf{b_2} = \frac{2\pi}{a_0}\hat{\mathbf{y}}, \qquad \mathbf{b_3} = \frac{2\pi}{a_0}\hat{\mathbf{z}}. \tag{2.12}$$

Hence, the reciprocal lattice unit vectors are aligned to the real space lattice vectors (Fig.2.2.a).

Note that the reciprocal lattice is itself a Bravaix lattice, and that the reciprocal of the reciprocal lattice (often referred to as the *direct lattice*) coincides with the original Bravaix lattice. In particular, the reciprocal of a bcc lattice is a fcc lattice (Fig.2.2.b) with unit vectors

$$\mathbf{b}_1 = \frac{2\pi}{a_0}(\hat{\mathbf{y}} + \hat{\mathbf{z}}), \qquad \mathbf{b}_2 = \frac{2\pi}{a_0}(\hat{\mathbf{z}} + \hat{\mathbf{x}}), \qquad \mathbf{b}_3 = \frac{2\pi}{a_0}(\hat{\mathbf{x}} + \hat{\mathbf{y}}). \tag{2.13}$$

Conversely, the reciprocal lattice of a fcc lattice has primitive vectors

$$\mathbf{b}_1 = \frac{2\pi}{a_0}(\hat{\mathbf{y}} + \hat{\mathbf{z}} - \hat{\mathbf{x}}), \qquad \mathbf{b}_2 = \frac{2\pi}{a_0}(\hat{\mathbf{z}} + \hat{\mathbf{x}} - \hat{\mathbf{y}}), \qquad \mathbf{b}_3 = \frac{2\pi}{a_0}(\hat{\mathbf{x}} + \hat{\mathbf{y}} - \hat{\mathbf{z}}). \tag{2.14}$$

This is easily recognized as a bcc lattice (see Eq.2.4 and Fig.2.2.c) with a cubic cell of edge $4\pi/a_0$.

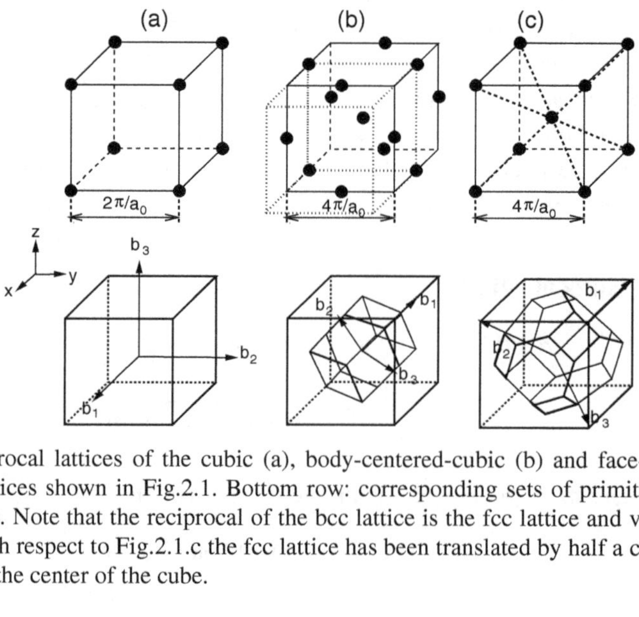

Figure 2.2 Top row: reciprocal lattices of the cubic (a), body-centered-cubic (b) and face-centered-cubic (c) Bravaix lattices shown in Fig.2.1. Bottom row: corresponding sets of primitive vectors and Brillouin zones. Note that the reciprocal of the bcc lattice is the fcc lattice and vice-versa. Note as well that with respect to Fig.2.1.c the fcc lattice has been translated by half a cell to place one lattice point at the center of the cube.

The primitive unit cell of the reciprocal lattice is the volume of the reciprocal space consisting of the points

$$\mathbf{K} = \sum_{i=1}^{3} K_i \mathbf{b}_i, \tag{2.15}$$

with the K_i ranging between 0 and 1.

The Wigner–Seitz cells of the reciprocal lattice are denoted *Brillouin zones*; the *first Brillouin zone* is the one that includes the origin $\mathbf{K} = \mathbf{G_0} = \mathbf{0}$. The first Brillouin zone of the fcc lattice has the same shape as the Wigner–Seitz cell of the bcc lattice, that is a truncated octahedron as illustrated in Figs.2.2.c and 2.3. The first Brillouin zone of the bcc lattice is instead shaped as a rhombic dodecahedron (Fig.2.2.b). High symmetry points and segments within the Brillouin zone are usually labelled with capital roman and greek letters, respectively. The point $\mathbf{K} = \mathbf{0}$ is the Γ point. The standard notation for the lines and directions is also given in Fig. 2.3.

The extension of the zone is defined by the geometrical conditions

$$|K_x| \leq \frac{2\pi}{a_0}, \quad |K_y| \leq \frac{2\pi}{a_0}, \quad |K_z| \leq \frac{2\pi}{a_0}, \tag{2.16a}$$

$$|K_x| + |K_y| + |K_z| \leq \frac{3}{2}\left(\frac{2\pi}{a_0}\right), \tag{2.16b}$$

where K_x, K_y, and K_z are three cartesian axes aligned with the $\hat{\mathbf{x}}$, $\hat{\mathbf{y}}$, and $\hat{\mathbf{z}}$ directions. The volume of the Brillouin zone is easily calculated as

$$\Omega_{BZ} = \frac{(2\pi)^3}{\Omega_{cell}}. \tag{2.17}$$

For the diamond and zinc blende lattices $\Omega_{cell} = a_0^3/4$, hence $\Omega_{BZ} = 4(2\pi/a_0)^3$.

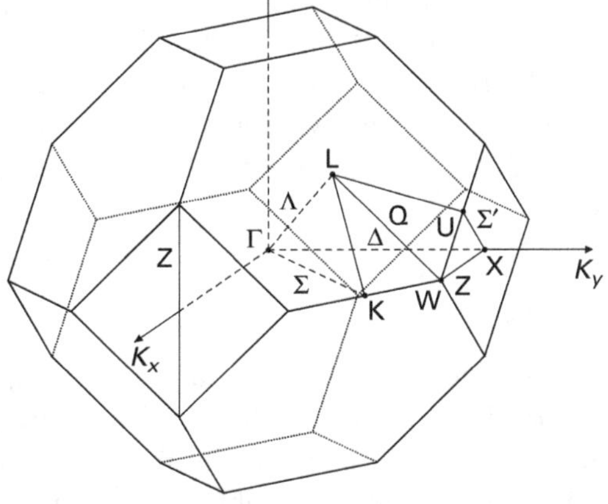

Figure 2.3 First Brillouin zone of silicon with an indication of the most important symmetry points and directions. $L = (\pm 1/2; \pm 1/2; \pm 1/2)$, $X = (\pm 1; 0; 0)$, $(0; \pm 1; 0)$ and $(0; 0; \pm 1)$ in units of $[2\pi/a_0]$.

The usefulness of the reciprocal space can be appreciated by noting that, by definition, plane waves of the form $\exp(i\mathbf{K}\cdot\mathbf{R})$ have the periodicity of the Bravaix lattice if and only if $\mathbf{K}$ belongs to the reciprocal lattice. This is easily seen noting that if $\mathbf{K} = \mathbf{G_m}$ then Eq.2.9 implies

$$\exp(\mathrm{i}\,\mathbf{G_m}\cdot\mathbf{R}) = \exp(\mathrm{i}\,\mathbf{G_m}\cdot(\mathbf{R}+\mathbf{R_n})), \tag{2.18}$$

for every $\mathbf{R}$ and for every $\mathbf{R_n}$. Therefore, a function of the real space $f(\mathbf{R})$ with the periodicity of the Bravaix lattice, $f(\mathbf{R}+\mathbf{R_n}) = f(\mathbf{R})$ for any $\mathbf{R_n}$ value, can be expanded in a Fourier series of plane wave components of the form given in Appendix A.2

$$f(\mathbf{R}) = \sum_{\mathbf{G_m}} C(\mathbf{G_m})\exp(-\mathrm{i}\,\mathbf{G_m}\cdot\mathbf{R}), \tag{2.19}$$

$$C(\mathbf{G_m}) = \frac{1}{\Omega_{cell}}\int_{\Omega_{cell}} f(\mathbf{R})\exp(\mathrm{i}\,\mathbf{G_m}\cdot\mathbf{R})\,d\mathbf{R}. \tag{2.20}$$

In other words, the vectors joining the origin with the reciprocal lattice points can be seen as the wave-vectors of all the plane waves necessary to represent in Fourier series the functions with the periodicity of the direct lattice.

Let us now consider a plane identified by three non-collinear points of the Bravaix lattice and located at a distance d from the origin, which we take at one lattice point. Due to the symmetry and periodicity of the crystal, the plane will contain an infinite number of lattice points arranged in a two-dimensional lattice and, furthermore, there will be an infinite set of such planes. Since the reciprocal lattice vectors can be seen as the wave-vectors of plane waves propagating in the direct lattice, we can find reciprocal lattice vectors $\mathbf{G_m}$ normal to the chosen plane. We can thus identify the orientation of the plane by selecting the shortest $\mathbf{G_m}$ vector perpendicular to that plane. Such a vector takes the form $\mathbf{G_m} = h\mathbf{b}_1 + k\mathbf{b}_2 + l\mathbf{b}_3$ where the integers h, k, l are called the *Miller indices* of the plane. The notation $\overline{h}$ is used for $-|h|$.

Miller indices provide a very effective means to identify planes and directions in a crystal. Following a standard notation, we denote (h, k, l) the plane of Miller indices h, k, l, and $[h, k, l]$ the direction perpendicular to the plane (h, k, l). Symmetry considerations often lead to identifying sets of equivalent planes and directions in a crystal. For example, the [1,0,0] [0,1,0] and [0,0,1] directions are all equivalent in a cubic crystal. Using standard notations we will refer to these sets of equivalent planes and directions as $\{h, k, l\}$ and $\langle h, k, l\rangle$, respectively. Figure 2.4 gives a few examples of the use of Miller indices.

2.1.3 Bloch functions

The study of electrons in a bulk crystal should be approached from the viewpoint of many-body problems. For a complex physical system, such as a crystal, the Hamiltonian operator contains all terms due to the kinetic energy of the individual particles (nuclei and electrons) and all possible interaction terms due to internal forces among the nuclei, among the electrons and between the nuclei and the electrons. Adopting this description is an unmanageable and daunting task, since the number of atoms per cubic

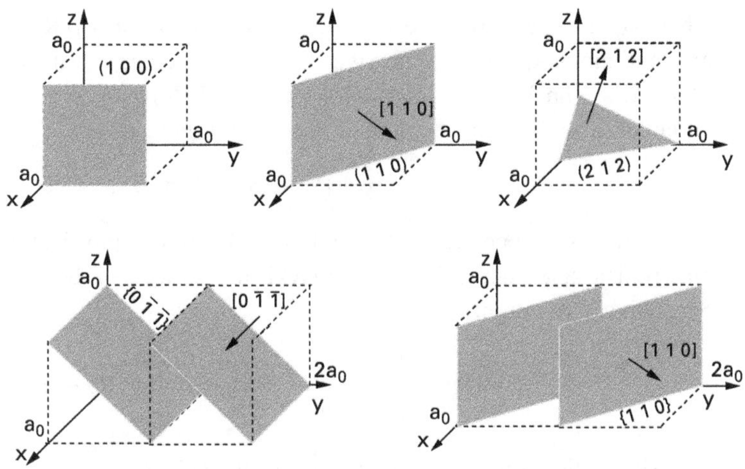

Figure 2.4 Illustration of the use of Miller indices to identify individual planes (parentheses, ()) or families of planes (braces, {}) in a cubic lattice. The notation $h,\bar{k},l$ stands for $h,-k,l$.

centimeter is extremely large ($\approx 10^{23}$) and the number of interaction terms to account for is prohibitive.

Fortunately, in most cases of interest we can embrace a single particle or *independent particle* approximation [4], which leads us to write a single-particle Hamiltonian where the electron's interactions are represented by an effective single-particle potential energy $U(\mathbf{R}, t)$ describing the average forces exerted by the nuclei and by the other electrons on the particle.

The so called *adiabatic* (Born–Oppenheimer) approximation [4] further suggests considering very slowly moving nuclei, with essentially frozen positions, and to treat the interaction of the nuclei with the electron through a static and periodic potential energy term in the expression of the Hamiltonian. This approximation is justified by the fact that the atom's nucleus is much heavier than the surrounding electrons. In the absence of external forces, $U(\mathbf{R}, t)$ thus reduces to

$$U(\mathbf{R}, t) = U_C(\mathbf{R}), \tag{2.21}$$

where $U_C(\mathbf{R})$ is the *crystal potential* featuring the periodicity of the Bravaix lattice due to the perfectly periodic arrangement of the atoms in their rest positions. The choice of the form for the potential energy $U_C(\mathbf{R})$ is definitely not trivial but a detailed discussion of this topic goes well beyond the scope of this book [3]. Fortunately, as we will see in Section 2.2, we do not always need to know explicit expressions for $U_C(\mathbf{R})$ in order to study the basic properties of electrons in crystals.

According to the above approximations, the time-dependent Schrödinger equation governing the behavior of the single electron wave-function reads

$$\mathrm{i}\hbar \frac{\partial \Psi_t(\mathbf{R}, t)}{\partial t} = -\frac{\hbar^2}{2m_0} \nabla^2 \Psi_t(\mathbf{R}, t) + U_C(\mathbf{R})\, \Psi_t(\mathbf{R}, t), \tag{2.22}$$

where m_0 is the electron's rest mass. The wave-function $\Psi_t(\mathbf{R}, t)$ and its first derivative are continuous functions of $\mathbf{R}$. Since the potential energy term is stationary, the solution of Eq.2.22 can be factorized as $\Psi_t(\mathbf{R}, t) = \Psi(\mathbf{R})\psi(t)$ and solved by separation of variables. The time-dependent part of the solution turns out to be simply given by

$$\psi(t) = \exp(-\mathrm{i}E_B t/\hbar), \tag{2.23}$$

where E_B is independent of t and $\mathbf{R}$ and has the physical meaning of electron's energy. The space dependent term $\Psi(\mathbf{R})$ is the solution of the stationary, single-electron Schrödinger equation:

$$\hat{H}_C \Psi(\mathbf{R}) = \left[-\frac{\hbar^2}{2m_0}\nabla^2 + U_C(\mathbf{R})\right]\Psi(\mathbf{R}) = E_B\,\Psi(\mathbf{R}). \tag{2.24}$$

The single-electron crystal Hamiltonian operator $\hat{H}_C$ in Eq.2.24 sets an eigenvalue problem; E_B and $\Psi(\mathbf{R})$ represent the eigenvalues and eigenfunctions of the operator, respectively.

According to Bloch's theorem, the normalized eigenfunctions $\Psi(\mathbf{R})$ of the crystal Hamiltonian $\hat{H}_C$ take the form [1]:

$$\Psi_{n\mathbf{K}}(\mathbf{R}) = \exp(\mathrm{i}\,\mathbf{K}\cdot\mathbf{R})u_{n\mathbf{K}}(\mathbf{R}), \tag{2.25}$$

that is, they are given by the product of a plane wave $\exp(\mathrm{i}\mathbf{K}\cdot\mathbf{R})$ with wave-vector $\mathbf{K}$ and a function $u_{n\mathbf{K}}(\mathbf{R})$ with the periodicity of the Bravaix lattice, i.e. $u_{n\mathbf{K}}(\mathbf{R}) = u_{n\mathbf{K}}(\mathbf{R}+\mathbf{R_n})$. Here we assume that the periodic part $u_{n\mathbf{K}}(\mathbf{R})$ of the Bloch function $\Psi_{n\mathbf{K}}(\mathbf{R})$ is normalized over a reference volume Ω, that is

$$\int_\Omega |u_{n\mathbf{K}}(\mathbf{R})|^2 \mathrm{d}\mathbf{R} = 1, \tag{2.26}$$

so that the Bloch function $\Psi_{n\mathbf{K}}(\mathbf{R})$ is normalized as well.

The continuity of $\Psi_{n\mathbf{K}}(\mathbf{R})$ implies that $u_{n\mathbf{K}}(\mathbf{R})$ and its first derivative are continuous. Note that Eq.2.25 and the periodicity of $u_{n\mathbf{K}}(\mathbf{R})$ imply

$$\Psi_{n\mathbf{K}}(\mathbf{R}+\mathbf{R_n}) = \exp(\mathrm{i}\,\mathbf{K}\cdot\mathbf{R_n})\Psi_{n\mathbf{K}}(\mathbf{R}), \tag{2.27}$$

which is an alternative form of Bloch's theorem.

The function $u_{n\mathbf{K}}(\mathbf{R})$ describes the microscopic distribution of the electron charge on the atomic scale of distance, whereas the plane wave $\exp(\mathrm{i}\,\mathbf{K}\cdot\mathbf{R})$ is a phase factor with unitary square modulus. The electron's probability density, which is proportional to the square modulus of the wave-function, is thus periodic in the direct lattice. Figure 2.5 illustrates in qualitative form the terms composing a Bloch wave and the corresponding square modulus, which determines the spatial distribution of the probability finding the electron in the crystal, hence the corresponding charge density.

Given the general form of the Bloch functions in Eq.2.25, the selection of the allowed $\mathbf{K}$ values derives from the boundary conditions. For an infinite crystal the common choice is to select a reference normalization volume $\Omega = N\Omega_{cell}$ consisting of $N = N_1N_2N_3$ primitive unit cells and to impose the periodicity of the wave-function at the

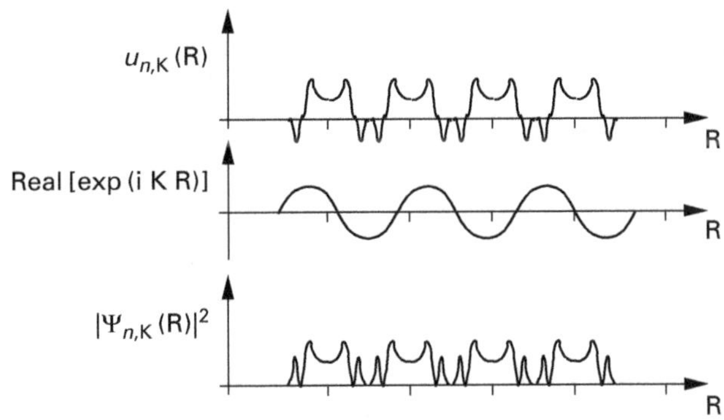

Figure 2.5 Representation of the components of a Bloch wave $\Psi_{n\mathbf{K}}(\mathbf{R})$. The grid strokes mark the atoms' rest positions. Note the periodicity of $u_{n\mathbf{K}}(\mathbf{R})$.

boundaries of such a macroscopic volume. These are the so-called *Born–von Karman boundary conditions* which can be expressed as:

$$\Psi_{n\mathbf{K}}(\mathbf{R} + N_1\mathbf{a}_1 + N_2\mathbf{a}_2 + N_3\mathbf{a}_3) = \Psi_{n\mathbf{K}}(\mathbf{R}), \tag{2.28}$$

where the $\mathbf{a}_j$ are the primitive unit vectors. Combining Eq.2.27 and Eq.2.28 it follows that

$$\exp(\mathrm{i}N_j\mathbf{K}\cdot\mathbf{a}_j) = 1, \qquad j = 1,\ 2,\ 3. \tag{2.29}$$

If we now express $\mathbf{K}$ as a linear combination of the reciprocal lattice primitive vectors ($\mathbf{K} = K_1\mathbf{b}_1 + K_2\mathbf{b}_2 + K_3\mathbf{b}_3$, Eq.2.15), Eq.2.29 gives $\exp(\mathrm{i}2\pi N_j K_j) = 1$ for j=1, 2, 3. Hence the allowed $\mathbf{K}$ values are

$$\mathbf{K} = \sum_{j=1}^{3} \frac{m_j}{N_j}\mathbf{b}_j, \qquad m_j = 0,\ \pm 1,\ \pm 2\ldots \tag{2.30}$$

As will be seen throughout the book, all physical properties of the system are independent of the choice made for the volume Ω. Here we just note that in the limit of an infinitely large crystal ($\Omega \to \infty$ or $N_j \to \infty$) we tend to a continuum distribution of allowed $\mathbf{K}$ values.

Given the general expression for $\Psi_{n\mathbf{K}}(\mathbf{R})$ in Eq.2.25, for each $\mathbf{K}$ Eq.2.24 is an eigenvalue problem whose non-null eigenfunctions exist only for well defined discrete energy values $E_{B,n\mathbf{K}}$. The eigenvalues and the corresponding eigenfunctions will thus be identified by an integer index n and by the $\mathbf{K}$ vector, which justifies the notation $\Psi_{n\mathbf{K}}(\mathbf{R})$. If the crystal volume is very large with respect to the crystal unit cell, then $\mathbf{K}$ can be treated as a continuous variable and each eigenvalue describes a branch of a multi-valued energy relation $E_{B,n}(\mathbf{K})$ as a function of $\mathbf{K}$.

Figure 2.6 sketches the main general properties of the multi-valued function $E_{B,n}(\mathbf{K})$ for a simple cubic lattice. As shown in Section 2.2.1, given n and $\mathbf{K}$ we have

$$E_{B,n}(\mathbf{K} + \mathbf{G_m}) = E_{B,n}(\mathbf{K}), \tag{2.31}$$

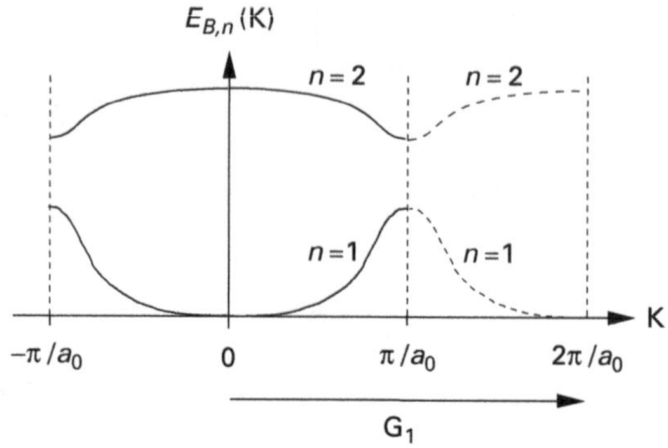

Figure 2.6 A sample band structure $E_{B,n}(\mathbf{K})$ and the first translation vector $\mathbf{G_1}$ for a cubic lattice with lattice constant a_0. Note the periodicity of the energy bands in **K**-space and the choice of numbering of the bands.

and

$$\Psi_{n(\mathbf{K}+\mathbf{G_m})}(\mathbf{R}) = \Psi_{n\mathbf{K}}(\mathbf{R}), \tag{2.32}$$

for all $\mathbf{G_m}$ vectors in the reciprocal lattice. Equation 2.32 implies

$$u_{n(\mathbf{K}+\mathbf{G_m})}(\mathbf{R}) = \exp(-\mathrm{i}\,\mathbf{G_m}\cdot\mathbf{R})u_{n\mathbf{K}}(\mathbf{R}), \tag{2.33}$$

a property which will become useful later on in the book (see Chapter 4 and Appendix D).

As a consequence of the relations above, the **K** vectors necessary to identify the distinct electron states can always be taken within one primitive unit cell of the reciprocal lattice space. Among all primitive cells, the Brillouin zone appears the natural domain to contain these **K** values and to represent $E_{B,n}(\mathbf{K})$. The multi-valued $E_{B,n}(\mathbf{K})$ is thus a family of continuous periodic functions with the periodicity of the reciprocal lattice, also known as the *band structure* of the material. Thus, $E_{B,n}(\mathbf{K})$ represents the dispersion relation of the stationary electron waves in the crystal. Each n value identifies an *energy band* and it is usual to enumerate the energy bands inside each Brillouin zone from the bottom energy up, as shown in Fig.2.6.

Since the crystal Hamiltonian $\hat{H}_C$ is hermitian, if $\Psi_{n\mathbf{K}}$ is an eigenfunction then its complex conjugate $\Psi^{\dagger}_{n\mathbf{K}}$ is also an eigenfunction with the same eigenvalue $E_{B,n}(\mathbf{K})$. For the Bloch states this is equivalent to exchanging **K** with $-\mathbf{K}$; hence, the band structure is always an even function of **K**

$$E_{B,n}(\mathbf{K}) = E_{B,n}(-\mathbf{K}). \tag{2.34}$$

Moreover, the band structure reflects all the symmetries of the Bravaix lattice. Therefore, it is not necessary to calculate it for all the points in the Brillouin zone, but only for a limited set of points that are not equivalent under all symmetry operations in the Bravaix lattice. For the fcc lattice there are 48 symmetry operations, so that all the band structure calculations can be limited within the so called *irreducible wedge* of the Brillouin zone, that is the region of the Brillouin zone delimited by Eq.2.16 and by the additional condition.

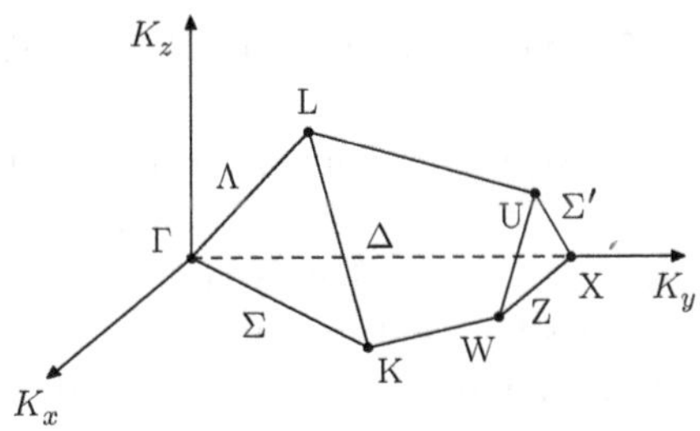

Figure 2.7 Irreducible wedge of the first Brillouin zone of silicon and germanium. The wedge is limited by the planes $K_x = K_y$; $K_x = K_z$; K_z=0; K_y=1; $K_x+K_y+K_z$=3/2 in units of $[2\pi/a_0]$. The symmetry points are: $L = (1/2; 1/2; 1/2)$, $X = (0; 1; 0)$, $W = (1/2; 1; 0)$, $K = (3/4; 3/4; 0)$, $U = (1/4; 1; 1/4)$ in units of $[2\pi/a_0]$.

$$0 \leq K_z \leq K_y \leq K_x \leq \frac{2\pi}{a_0}. \tag{2.35}$$

The irreducible wedge of the silicon and germanium Brillouin zones is shown in Fig.2.7. The computational burden of a numerical calculation of the band structure is reduced considerably by exploiting the symmetry properties of the lattice.

2.1.4 Density of states

We have seen in Section 2.1.3 that the electron states of the crystal are identified by an integer index n and a wave-vector $\mathbf{K}$ which can be taken inside the first Brillouin zone. Equation 2.30 points out that the allowed $\mathbf{K}$ vectors are evenly spaced along the $\mathbf{b}_i$ directions. The volume of reciprocal space corresponding to one $\mathbf{K}$ vector is thus given by

$$(\Delta K)^3 = \frac{\mathbf{b}_1 \cdot (\mathbf{b}_2 \times \mathbf{b}_3)}{N} = \frac{\Omega_{BZ}}{N}. \tag{2.36}$$

It follows that for each band n, the number of allowed $\mathbf{K}$ states in one reciprocal lattice cell (e.g. the Brillouin zone) is equal to $\Omega_{BZ}/(\Delta K)^3 = N$, that is, it is equal to the number of reticular cells in the normalization volume Ω of the crystal. In other words, the number of allowed electron states in the crystal scales linearly with the crystal volume.

It is thus useful to introduce the concept of *Density of States* (DoS) as the number of $\mathbf{K}$ states in a given band per unit volume in real space and per unit volume in reciprocal space. A crystal made of N cells occupies the real space volume Ω and has N distinct electron states in each band which fill the volume Ω_{BZ} of the Brillouin zone. From Eqs.2.17 and 2.36 we compute the DoS as

$$g_{\mathbf{R},\mathbf{K}} = n_{sp} \frac{N}{\Omega\, \Omega_{BZ}} = \frac{n_{sp}}{\Omega_{cell}\, \Omega_{BZ}} = \frac{n_{sp}}{(2\pi)^3}, \tag{2.37}$$

where the factor n_{sp} takes into account spin degeneracy and has a value of 2 for bulk semiconductors.

2.2 Numerical methods for band structure calculations

A knowledge of the band structure of a material is the first necessary step to understanding and predicting its electronic properties. The band structure plays a fundamental role in both the electrostatics and the dynamics of charge carriers in semiconductors. As such, methodologies for band structure calculations have been and still are the subject of extensive research in solid state physics [2, 3]. The first difficulty with all band structure calculation methods based on the independent electron approximation is that the potential energy of the crystal $U_C(\mathbf{R})$ is unknown, since it represents the complex result of interactions between the nuclei and the electrons.

Many techniques have been devised to overcome this difficulty and calculate the dispersion relation $E_{B,n}(\mathbf{K})$ [1, 3]. It is not the purpose of this chapter to illustrate all of them. Instead, we limit our discussion to a practical overview of two methodologies (the pseudo-potential method and the $\mathbf{k}\cdot\mathbf{p}$ method) widely employed to calculate the conduction and the valence band of bulk semiconductors and which can be extended to determine the band structure of the two-dimensional carrier gas in MOSFET inversion layers.

2.2.1 The pseudo-potential method

As discussed in Section 2.1.3, the electronic states in a crystalline semiconductor are the solutions of the single particle Schrödinger equation, Eq.2.22. The stationary part of the wave-function is the solution of Eq.2.24. The pseudo-potential method aims to solve this equation with a drastic simplification by which the core states are omitted from the calculation, thus achieving a remarkable complexity reduction. The basic idea behind the pseudo-potential method is that the electron states can be calculated over a range of energies and with a sufficient degree of approximation by replacing the actual single particle crystal potential energy $U_C(\mathbf{R})$ (which is generated by the nuclei's charge and the distributed charge of all other electrons) with a pseudo-potential where the rapid variation of $U_C(\mathbf{R})$ in the core region is canceled out. A practical introduction to the simplest form of pseudo-potential band structure calculations (the so called *local empirical pseudo-potential* method) is derived below.

We start from the stationary Schrödinger equation, Eq.2.24, with eigenfunctions given by the Bloch waves, Eq.2.25

$$\left[\frac{-\hbar^2}{2m_0}\nabla^2 + U_C(\mathbf{R})\right]\Psi(\mathbf{R}) = E_B\,\Psi(\mathbf{R}), \tag{2.38}$$

and we impose the periodic boundary conditions (Eq.2.29) at the edges of a large normalization volume Ω made up of many instances of the unit cell volume Ω_{cell}. The wave-functions $\Psi(\mathbf{R})$ are then periodic with period Ω and thus we can expand them in a Fourier series of the form

$$\Psi(\mathbf{R}) = \sum_{\mathbf{K}'_t} C(\mathbf{K}'_t)\exp(-\mathrm{i}\mathbf{K}'_t \cdot \mathbf{R}) = \sum_{\mathbf{K}',\mathbf{G}'_\mathbf{m}} C(\mathbf{K}' + \mathbf{G}'_\mathbf{m})\exp[-\mathrm{i}(\mathbf{K}' + \mathbf{G}'_\mathbf{m}) \cdot \mathbf{R}], \tag{2.39}$$

where $\mathbf{G}'_\mathbf{m}$ is a reciprocal lattice vector, $\mathbf{K}'$ is a reciprocal space vector given by Eq.2.30 with a constraint on the m_j values such that $\mathbf{K}'$ belongs to the first Brillouin zone (or any equivalent unit cell) and

$$\mathbf{K}'_t = \mathbf{K}' + \mathbf{G}'_\mathbf{m}. \tag{2.40}$$

The crystal potential is periodic over Ω_{cell}; therefore we can expand it as (Eq.2.19)

$$U_C(\mathbf{R}) = \sum_{\mathbf{G}''_\mathbf{m}} U_C(\mathbf{G}''_\mathbf{m})\exp(-\mathrm{i}\mathbf{G}''_\mathbf{m} \cdot \mathbf{R}), \tag{2.41}$$

where $\mathbf{G}''_\mathbf{m}$ is a reciprocal lattice vector. Substituting Eqs.2.39 and 2.41 in Eq.2.38 we find

$$\begin{aligned}\sum_{\mathbf{K}',\mathbf{G}'_\mathbf{m}} &\left[-\frac{\hbar^2}{2m_0}|\mathbf{K}' + \mathbf{G}'_\mathbf{m}|^2 + \sum_{\mathbf{G}''} U_C(\mathbf{G}''_\mathbf{m})\exp(-\mathrm{i}\mathbf{G}''_\mathbf{m} \cdot \mathbf{R}) \right] \\ &\times C(\mathbf{K}' + \mathbf{G}'_\mathbf{m})\exp[-\mathrm{i}(\mathbf{K}' + \mathbf{G}'_\mathbf{m}) \cdot \mathbf{R}] \\ &\qquad = E_B \sum_{\mathbf{K}',\mathbf{G}'_\mathbf{m}} C(\mathbf{K}' + \mathbf{G}'_\mathbf{m})\exp[-\mathrm{i}(\mathbf{K}' + \mathbf{G}'_\mathbf{m}) \cdot \mathbf{R}]. \end{aligned} \tag{2.42}$$

Following a well established methodology, we select one specific $(\mathbf{K}, \mathbf{G_m})$ pair and we project Eq.2.42 on the basis of plane waves, that is we multiply both sides of Eq.2.42 by $\exp[\mathrm{i}(\mathbf{K} + \mathbf{G_m}) \cdot \mathbf{R}]$ and we integrate over Ω. The term containing the crystal potential can be written as

$$\begin{aligned}&\sum_{\mathbf{K}',\mathbf{G}'_\mathbf{m}} C(\mathbf{K}' + \mathbf{G}'_\mathbf{m}) \sum_{\mathbf{G}''} U_C(\mathbf{G}''_\mathbf{m}) \\ &\quad \times \int_0^{L_x}\int_0^{L_y}\int_0^{L_z} \exp[-\mathrm{i}(\mathbf{K}' + \mathbf{G}'_\mathbf{m} - \mathbf{K} - \mathbf{G_m} + \mathbf{G}''_\mathbf{m}) \cdot \mathbf{R}]\,\mathrm{d}x\,\mathrm{d}y\,\mathrm{d}z. \end{aligned} \tag{2.43}$$

It is easy to recognize that if $\mathbf{K} = \mathbf{K}'$ and $\mathbf{G}''_\mathbf{m} = (\mathbf{G_m} - \mathbf{G}'_\mathbf{m})$, then the exponential term is equal to 1 and the integral sums to Ω. If these conditions are not simultaneously met the integral sums to zero. In fact, since $\mathbf{K}$ and $\mathbf{K}'$ belong to the same Brillouin zone and $\mathbf{G_m}$ and $\mathbf{G}'_\mathbf{m}$ are both reciprocal lattice vectors, the integration domain covers an integer number of periods of sine and cosine functions. Consequently the expression 2.43 reduces to

$$\Omega \sum_{\mathbf{G}'_\mathbf{m}} C(\mathbf{K} + \mathbf{G}'_\mathbf{m})\, U_C(\mathbf{G_m} - \mathbf{G}'_\mathbf{m}). \tag{2.44}$$

Similar arguments can be followed to eliminate the contributions for $\mathbf{K}' \neq \mathbf{K}$ and $\mathbf{G}'_\mathbf{m} \neq \mathbf{G_m}$ from those terms of Eq.2.42 that do not contain the crystal potential. As a result Eq.2.42 reduces to

$$-\frac{\hbar^2}{2m_0}(\mathbf{K}+\mathbf{G_m})^2\, C_{\mathbf{K}}(\mathbf{G_m}) + \sum_{\mathbf{G}'_{\mathbf{m}}} C_{\mathbf{K}}(\mathbf{G}'_{\mathbf{m}})\, U_C(\mathbf{G_m}-\mathbf{G}'_{\mathbf{m}}) \\ = E_B(\mathbf{K})\, C_{\mathbf{K}}(\mathbf{G_m}), \tag{2.45}$$

where the notations $C_{\mathbf{K}}(\mathbf{G_m}) = C(\mathbf{K}+\mathbf{G_m})$ and $E_B(\mathbf{K}) = E_B$ have been introduced.

Equation 2.45 is a restatement in **K**-space of the original Schrödinger equation, Eq.2.24. Remembering that **K** is a fixed reciprocal space vector in the first Brillouin zone that plays the role of a fixed parameter, it becomes clear that Eq.2.45 defines a linear system of homogeneous equations with eigenvalues $E_{B,n}(\mathbf{K})$, where the suffix n has been added to enumerate the eigenvalues. The components of the corresponding eigenvector are the coefficients $C_{n\mathbf{K}}(\mathbf{G_m})$. We can thus rewrite Eq.2.39 as

$$\Psi(\mathbf{R}) = \exp(-\mathrm{i}\mathbf{K}\cdot\mathbf{R}) \sum_{\mathbf{G_m}} C_{n\mathbf{K}}(\mathbf{G_m})\, \exp(-\mathrm{i}\mathbf{G_m}\cdot\mathbf{R}). \tag{2.46}$$

Comparing this last expression with Eq.2.25, it is straightforward to see that the eigenvectors of the pseudo-potential method yield the Fourier components of the periodic part of the Bloch function for the wave-vector $(-\mathbf{K})$. Equation 2.45 can also be interpreted as the projection of the Schrödinger equation on the plane wave basis $\exp(-\mathrm{i}\,\mathbf{G_m}\cdot\mathbf{R})$.

It is interesting to observe that exactly the same system of equations 2.45 is obtained for all **K** vectors which differ by any reciprocal lattice vector $\mathbf{G_m}$. Therefore the **K** vector can always be taken within the first Brillouin zone and the distinct electron states of the crystal are only those whose **K** vector belongs to one (e.g., the first) Brillouin zone. This observation proves the validity of Eqs.2.31 and 2.32.

In principle an infinite number of equations, hence, an infinite number of waves and $\mathbf{G_m}$ vectors, should be considered in Eq.2.45. The largest $\mathbf{G_m}$ vectors are obviously necessary to represent the rapid change of the crystal potential and the fast oscillations of the wave-functions in the proximity of the atom cores. The smallest $\mathbf{G_m}$ vectors, instead, are needed to describe the relatively smooth long range components of the potential energy and the smooth fluctuations of the wave-functions in the inter-atomic regions.

It is reasonable to assume that in the expansion of the crystal potential a limited number of terms $U_C(\mathbf{G_m}-\mathbf{G}'_{\mathbf{m}})$ with small $|\mathbf{G_m}-\mathbf{G}'_{\mathbf{m}}|$ can be sufficient to describe accurately the potential in the interatomic region. Since the Fourier components of the potential appear in Eq.2.45 in the form $U_C(\mathbf{G_m}-\mathbf{G}'_{\mathbf{m}})$, setting a limit on the maximum $|\mathbf{G_m}-\mathbf{G}'_{\mathbf{m}}|$ does not mean that the number and modulus of the $\mathbf{G_m}$ vectors are equally limited. In other words, the number of equations in the system Eq.2.45 is not restricted to the number of non-zero coefficients $U_C(\mathbf{G}''_{\mathbf{m}})$ in the expansion 2.41.

Due to the deliberately limited accuracy in the representation of the crystal potential, the solutions of Eq.2.45 do not reflect the real crystal potential but rather a smoothed version of it, which is called *pseudo-potential*. Provided that a sufficiently large number of $\mathbf{G_m}$ vectors and equations is considered, an adequately accurate description should be achieved for the electron states whose wave-functions extend far from the atom cores and overlap significantly with those of the neighboring atoms.

In order to understand in at least one practical case how the pseudo-potential can be identified, we start by observing that $U_C(\mathbf{R})$ is periodic with the periodicity of the direct lattice, hence, besides using the expansion in Eq.2.41, we can also represent $U_C(\mathbf{R})$ as a superposition of the total single electron potential energy in the unit cell $U_{cell}(\mathbf{R})$ over the direct lattice vectors

$$U_C(\mathbf{R}) = \sum_{\mathbf{R_n}} U_{cell}(\mathbf{R} - \mathbf{R_n}). \tag{2.47}$$

For materials having the diamond and zinc blende structure it is convenient to take the origin ($\mathbf{R} = \mathbf{0}$) at the mid-point between two adjacent atoms, which will then be located at $\mathbf{T} = \pm a_0(1/8, 1/8, 1/8)$. In silicon the two atoms are identical; therefore $U_{cell}(\mathbf{R}) = [U_a(\mathbf{R} - \mathbf{T}) + U_a(\mathbf{R} + \mathbf{T})]$, where $U_a(\mathbf{R})$ is the individual atom potential energy. Remembering the properties of the Fourier series (Appendix A.1) and Eq.2.41 we get

$$U_{cell}(\mathbf{G_m}) = \left[\exp(\mathrm{i}\mathbf{G_m} \cdot \mathbf{T}) + \exp(-\mathrm{i}\mathbf{G_m} \cdot \mathbf{T})\right] U_a(\mathbf{G_m}) = \cos(\mathbf{G_m} \cdot \mathbf{T}) U_a(\mathbf{G_m}), \tag{2.48}$$

where the Fourier components of the pseudo-potential

$$U_a(\mathbf{G_m}) = \frac{1}{\Omega_{cell}} \int_{\Omega_{cell}} U_a(\mathbf{R}) \exp(\,\mathrm{i}\,\mathbf{G_m} \cdot \mathbf{R})\,\mathrm{d}\mathbf{R}, \tag{2.49}$$

are called by atomic form factors and are usually functions of the modulus $|\mathbf{G_m}|$. The term $\cos(\mathbf{G_m} \cdot \mathbf{T})$, instead, is a structural factor which depends only on the crystal geometry.

As already mentioned, the summation over $\mathbf{G_m}$ is truncated to a limited number of the smallest reciprocal lattice vectors. Assuming that the atomic potentials are spherically symmetric, that is $U_a(\mathbf{R}) = U_a(|\mathbf{R}|)$, the form factors only depend on the magnitude of the wave-vector, that is $U_a(\mathbf{G_m}) = U_a(|\mathbf{G_m}|)$. In this respect, we note that for $\mathbf{m} = \mathbf{0}$ the term $U_a(\mathbf{G_m})$ is a constant rigid shift of the energy scale and it is therefore set to zero in the calculation. As for the remaining terms of the summation in Eq.2.45, a careful examination of the reciprocal lattice geometry shown in Fig.2.2.b and consideration of Eqs.2.8 and 2.14 indicates that, in units of $(2\pi/a_0)^2$, the square modulus of the reciprocal lattice vector of the first neighbor unit cells is $|\mathbf{G_m}|^2 = 3, 4, 8, 11, \ldots,$ corresponding, respectively, to $\mathbf{m} = (\pm 1, \pm 1, \pm 1)$, $\mathbf{m} = (\pm 2, 0, 0)$, $\mathbf{m} = (\pm 2, \pm 2, 0)$, $\mathbf{m} = (\pm 3, \pm 1, \pm 1)$, and to all the other $\mathbf{G_m}$ vectors obtained exchanging the K_x, K_y, and K_z directions.

Since $U_C(\mathbf{G})$ typically decays as $|\mathbf{G}|^2$, the terms beyond $|\mathbf{G_m}|^2 = 11$ are usually neglected. Moreover, since $\cos(\mathbf{G_m} \cdot \mathbf{T}) = 0$ for the $\mathbf{G_m}$ vectors with $|\mathbf{G_m}|^2 = 4$, we are eventually left with only three terms, denoted $U_{cell}(\sqrt{3})$, $U_{cell}(\sqrt{8})$, and $U_{cell}(\sqrt{11})$. These coefficients are taken as adjustable parameters and in the so called empirical pseudo-potential method (EPM) they are chosen to fit photon emission and reflectivity spectra [5, 6]. Table 2.1 reports the EPM parameter values for silicon and germanium.

The methodology described above is the so called *local empirical pseudo-potential* method. Along the same lines, a more accurate description of the pseudo-potential and of the band structure is obtained accounting for *non-local* effects by the method in [5].

Table 2.1 Fourier coefficients of the empirical pseudo-potential for band structure calculations in silicon and germanium [5].

Form factor	Silicon [eV]	Germanium [eV]
$U_{cell}(\sqrt{3})$	–3.0475	–3.0067
$U_{cell}(\sqrt{8})$	+0.7483	+0.2585
$U_{cell}(\sqrt{11})$	+0.9796	+0.7619

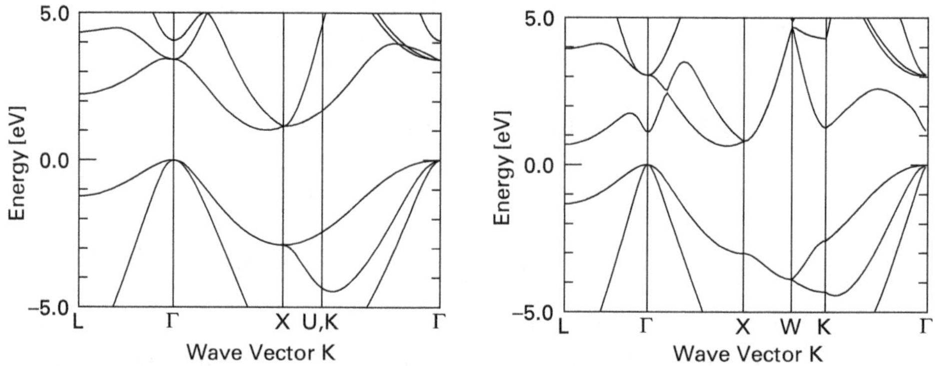

Figure 2.8 Dispersion relation $E_{B,n}(\mathbf{K})$ of the valence and the conduction bands of silicon (left) and germanium (right) calculated with the local empirical pseudo-potential method. Spin-orbit coupling is neglected, resulting in three degenerate branches of the valence band at the Γ point. The band gap values have been adjusted to the nominal values. Pseudo-potential parameters as in Table 2.1.

It is also apparent from the calculations that our derivation neglects spin-orbit coupling. The calculations can be refined to introduce this effect also, as explained in [7].

As an example of empirical pseudo-potential calculations, Fig.2.8 shows the dispersion relation of the valence and conduction bands of silicon and germanium. The energy band gap region is clearly visible. We observe that the maximum of the valence band occurs at the Γ point ($\mathbf{K} = \mathbf{0}$) for both silicon and germanium, whereas the bottom of the conduction band is located along the Δ directions for silicon and at the L point of the Λ directions for germanium. Other features of these band structures are described in Section 2.3.

2.2.2 The k·p method

The **k**·**p** method is a general technique based on expansion of the Schrödinger equation of the crystal in the neighborhood of a **K** point, which is taken to be the Γ point for the description of the valence band. The **k**·**p** method is well suited to calculating the valence band structure of silicon and germanium that, in the absence of spin-orbit coupling, is known to have a three times degenerate local maximum at the Γ point (neglecting spin degeneracy) as shown in Fig.2.8.

The starting point for deriving the $\mathbf{k}\cdot\mathbf{p}$ equation for the band energies is the observation that in a perfect crystal the solutions of the stationary Schrödinger equation, Eq.2.24 take the form of Bloch waves (Section 2.1.3). Substituting Eq.2.25 into Eq.2.24 and using the notation $\hat{\mathbf{P}} = -i\hbar\nabla_{\mathbf{R}}$ for the momentum operator we get

$$\left[\frac{\hat{\mathbf{P}}\cdot\hat{\mathbf{P}}}{2m_0} + \frac{\hbar}{m_0}(\mathbf{K}\cdot\hat{\mathbf{P}}) + U_C(\mathbf{R})\right] u_{n\mathbf{K}}(\mathbf{R}) = \left[E_{B,n}(\mathbf{K}) - \frac{\hbar^2 K^2}{2m_0}\right] u_{n\mathbf{K}}(\mathbf{R}). \tag{2.50}$$

The term $[\hat{\mathbf{P}}\cdot\hat{\mathbf{P}}/2m_0 + U_C(\mathbf{R})] = [-\hbar^2\nabla^2_{\mathbf{R}}/2m_0 + U_C(\mathbf{R})]$ is the crystal Hamiltonian. If the eigenvalues and the wave-functions at $\mathbf{K} = \mathbf{0}$ are known, we can treat the term $[\hbar(\mathbf{K}\cdot\hat{\mathbf{P}})/m_0]$ as a perturbation. In the diamond lattice the first order matrix elements of the $\mathbf{K}\cdot\hat{\mathbf{P}}$ perturbation vanish if the zero-order wave function is taken at $\mathbf{K} = \mathbf{0}$. It is therefore necessary to employ a second-order perturbation theory to calculate the degenerate branches of the band structure at the top of the valence band.

Denoting by $u_{i\mathbf{0}}(\mathbf{R})$ the complete set of orthonormal, cell-periodic wave functions at $\mathbf{K} = \mathbf{0}$ [8] (where i runs over all the degenerate and non-degenerate energy values at $\mathbf{K} = \mathbf{0}$), we can expand the periodic part of the Bloch function for a non-null $\mathbf{K}$ value as

$$u_{n\mathbf{K}}(\mathbf{R}) = \sum_i C'_{i,\mathbf{K}}\, u_{i\mathbf{0}}(\mathbf{R}), \tag{2.51}$$

where the $C'_{i,\mathbf{K}}$ are the expansion coefficients. The energies and the wave-functions for $\mathbf{K}\neq\mathbf{0}$ are found solving the eigenvalue problem

$$\mathbf{H}'_{\mathbf{k}\cdot\mathbf{p}}\, \mathbf{C}'_{\mathbf{K}} = \left[E_{B,n}(\mathbf{K}) - \frac{\hbar^2 K^2}{2m_0}\right] \mathbf{C}'_{\mathbf{K}}, \tag{2.52}$$

where $\mathbf{C}'_{\mathbf{K}}$ is the vector of expansion coefficients $C'_{i,\mathbf{K}}$. The elements of $\mathbf{H}'_{\mathbf{k}\cdot\mathbf{p}}$ coupling the degenerate states i and m at $\mathbf{K} = \mathbf{0}$ with the same energy $E_{B,j}$ are given by [8]

$$H'_{\mathbf{k}\cdot\mathbf{p},(i,m)} = \sum_{l,E_{B,l}\neq E_{B,j}} \frac{\langle u_{i\mathbf{0}}|(\hbar/m_0)\mathbf{K}\cdot\hat{\mathbf{P}}|u_{l\mathbf{0}}\rangle\langle u_{l\mathbf{0}}|(\hbar/m_0)\mathbf{K}\cdot\hat{\mathbf{P}}|u_{m\mathbf{0}}\rangle}{E_{B,j} - E_{B,l}}. \tag{2.53}$$

The expansion in Eq.2.51 is usually truncated at the first three terms corresponding to the three highest degenerate energy values for the valence band at $\mathbf{K} = \mathbf{0}$. In this case, with an appropriate choice of the basis function (that we denote $|1\rangle$, $|2\rangle$, $|3\rangle$) and taking into account the symmetries of the lattice, it is possible to state explicitly the $\mathbf{K}$ dependence of the matrix elements defined in Eq.2.53 as [8–10]

$$\mathbf{H}'_{\mathbf{k}\cdot\mathbf{p}} = \begin{bmatrix} L'K_x^2 + M'(K_y^2 + K_z^2) & N'K_xK_y & N'K_xK_z \\ N'K_xK_y & L'K_y^2 + M'(K_x^2 + K_z^2) & N'K_yK_z \\ N'K_xK_z & N'K_yK_z & L'K_z^2 + M'(K_x^2 + K_y^2) \end{bmatrix}, \tag{2.54}$$

Table 2.2 Parameter values of the **k·p** valence band model and split-off band energy separation for common semiconductors [11, 12].

Constant	Units	Si	Ge	GaAs	InP
L	[eV Å^2]	−25.83	−119.45	−69.69	−59.44
M	[eV Å^2]	−16.92	−22.29	−14.59	−11.89
N	[eV Å^2]	−32.92	−130.08	−74.98	−63.09
Δ	[meV]	44	296	340	130

where the three scalars L', M', and N' are parameters of the model that can be calibrated by means of comparison with cyclotron resonance measurements. Consistently with [8, 10], in order to derive an eigenvalue problem leading to the direct calculation of the band energies, and for the sake of a direct comparison with the results in Sections 3.3.1 and 9.4.1, we define the matrix

$$\mathbf{H}_{\mathbf{k}\cdot\mathbf{p}}^{3\times 3} = \begin{bmatrix} LK_x^2 + M(K_y^2 + K_z^2) & NK_xK_y & NK_xK_z \\ NK_xK_y & LK_y^2 + M(K_x^2 + K_z^2) & NK_yK_z \\ NK_xK_z & NK_yK_z & LK_z^2 + M(K_x^2 + K_y^2) \end{bmatrix}, \tag{2.55}$$

where we have performed the substitution

$$L = L' + \frac{\hbar^2}{2m_0}, \quad M = M' + \frac{\hbar^2}{2m_0}, \quad N = N'. \tag{2.56}$$

Table 2.2 reports the values of L, M, and N for common semiconductors.

Until now we have neglected spin orbit coupling, but this is not negligible for valence band states. It is therefore necessary to assume a basis of six wave-functions $|1\uparrow\rangle$, $|2\uparrow\rangle$, $|3\uparrow\rangle$, $|1\downarrow\rangle$, $|2\downarrow\rangle$, $|3\downarrow\rangle$. The resulting 6×6 **k·p** eigenvalue problem and the corresponding Hamiltonian matrix are then expressed as [8–10]

$$[\mathbf{H}_{\mathbf{k}\cdot\mathbf{p}}(\mathbf{K}) + \mathbf{H}_{so}]\,\mathbf{C}_{\mathbf{K}} = E\,\mathbf{C}_{\mathbf{K}}, \tag{2.57}$$

$$\mathbf{H}_{\mathbf{k}\cdot\mathbf{p}} = \begin{bmatrix} \mathbf{H}_{\mathbf{k}\cdot\mathbf{p}}^{3\times 3} & \mathbf{0} \\ \mathbf{0} & \mathbf{H}_{\mathbf{k}\cdot\mathbf{p}}^{3\times 3} \end{bmatrix}, \tag{2.58}$$

$$\mathbf{H}_{\mathrm{so}} = -\frac{\Delta}{3}\begin{pmatrix} 0 & \mathrm{i} & 0 & 0 & 0 & -1 \\ -\mathrm{i} & 0 & 0 & 0 & 0 & \mathrm{i} \\ 0 & 0 & 0 & 1 & -\mathrm{i} & 0 \\ 0 & 0 & 1 & 0 & -\mathrm{i} & 0 \\ 0 & 0 & \mathrm{i} & \mathrm{i} & 0 & 0 \\ -1 & -\mathrm{i} & 0 & 0 & 0 & 0 \end{pmatrix}, \tag{2.59}$$

where E is the eigenvalue for the valence band and Δ is the split-off energy between the heavy- and light-hole bands (which are degenerate) and the split-off band. Values of Δ for common semiconductors are reported in Table 2.2. Note that Eq.2.57 with $\mathbf{H}_{\mathbf{k}\cdot\mathbf{p}}$ defined in Eq.2.58 and the L, M, and N values in Table 2.2 yield an energy E which is a decreasing function of $K = |\mathbf{K}|$. In the remainder of the book, however, unless otherwise specified, we adopt an electron-like convention for the hole energy. Therefore, the hole energy is shown as an increasing function of K, as discussed in detail in Section 5.1.2.

It is useful to recall that the energy E in Eq.2.57 is not periodic over the first Brillouin zone. If $\mathbf{K}$ is spanned beyond the limits of the first Brillouin zone higher energy branches of the energy relation are found, which in principle should be folded back into the first zone. In practice, however, the accuracy of these high energy branches is questionable whenever the **k·p** method is developed as an expansion around the Γ point, as is the case in this book for the purpose of valence band calculations.

The **k·p** methodology has been extended over the years to an even larger number of bands than shown here [13–15], and it has proved useful in the calculation of unstrained as well as strained semiconductor band structures of a variety of materials featuring direct or indirect bandgap.

2.3 Analytical band structure models

2.3.1 Conduction band

Numerical calculations such as those shown in the left plot of Fig.2.8 make clear that the bottom of the conduction band of unstrained bulk silicon is formed by six equivalent valleys, hereafter indexed with the suffix ν, whose minima are located at 0.85 of the normalized distance between the Γ and the X points of the first Brillouin zone. In the proximity of the minima, the equi-energy surfaces in $\mathbf{K}$ space can be approximated by rotational ellipsoids elongated in three perpendicular directions (K_x, K_y, and K_z) which define the so-called *ellipsoid coordinate system* (ECS). Figure 2.9 locates these ellipsoids in the first Brillouin zone of the reciprocal lattice.

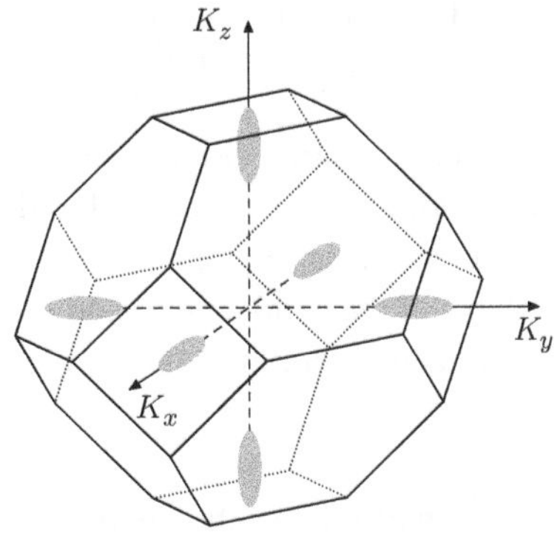

Figure 2.9 Ellipsoidal equi-energy surfaces in the proximity of the conduction band minima for silicon. The ellipsoids lie along the Δ directions.

Table 2.3 Parameters of the six ellipsoids describing the bottom of the bulk silicon conduction band. Note that m_l and m_t represent the longitudinal and transverse effective mass of the ellipsoids. The effective mass values are given in Table 2.4.

Valleys	$\mathbf{K}_\nu$ $[2\pi/a_0]$	m_x	m_y	m_z
Δ_x	$(\pm\,0.85,0,0)$	m_l	m_t	m_t
Δ_y	$(0,\pm\,0.85,0)$	m_t	m_l	m_t
Δ_z	$(0,0,\pm\,0.85)$	m_t	m_t	m_l

Table 2.4 Parameters of the analytical conduction and valence band models for silicon. Data from [16]. m_0 denotes the electron's rest mass.

Conduction band		Valence band	
Parameter	Value	Parameter	Value
m_l $[m_0]$	0.919	A	-4.1 ± 0.2
m_t $[m_0]$	0.190	B	1.6 ± 0.2
α $[\text{eV}^{-1}]$	0.5	C	3.3 ± 0.5

Consistently, in a second order parabolic approximation we can write the dispersion relation close to the ν minima as

$$E_B^{(\nu)}(\mathbf{K}) = E_{\nu 0} + \frac{\hbar^2}{2}\left[\frac{(K_x - K_{\nu,x})^2}{m_x} + \frac{(K_y - K_{\nu,y})^2}{m_y} + \frac{(K_z - K_{\nu,z})^2}{m_z}\right], \tag{2.60}$$

where $E_{\nu 0} = E_B(\mathbf{K}_\nu)$ is the energy at the bottom of the valley ν, while m_x, m_y, and m_z are implicitly defined as

$$\frac{1}{m_i} = \frac{1}{\hbar^2}\left(\frac{\partial^2 E_B^{(\nu)}(\mathbf{K})}{\partial K_i^2}\right)_{\mathbf{K}=\mathbf{K}_\nu} \qquad i \in \{x, y, z\}. \tag{2.61}$$

Table 2.3 gives the combinations of $\mathbf{K}_\nu$ values and of m_x, m_y, and m_z triplets, these latter expressed in terms of the so-called longitudinal m_l and transverse m_t effective masses, while Table 2.4 gives the numerical values of m_l and m_t. The $E_{\nu 0}$ value of all the equivalent valleys of the conduction band is often referred to as E_C. Note that, if we take as the reference zero energy the vacuum level, then

$$E_{\nu 0} = E_C = -\chi, \tag{2.62}$$

where χ is the electron affinity of the material.

For sufficiently high energy values (approximately 100 meV and 1 eV above E_C in silicon if the $\mathbf{K}$ direction is perpendicular or parallel to the ellipsoid elongation axis, respectively), the parabolic expression Eq.2.60 no longer approximates accurately the band structure. Non-parabolicity effects become important and are usually modeled at first order by introducing a non-parabolicity correction. In practice Eq.2.60 is replaced by the expression

$$\left(E_B^{(\nu)}(\mathbf{K}) - E_{\nu 0}\right) \cdot \left[1 + \alpha \left[E_B^{(\nu)}(\mathbf{K}) - E_{\nu 0}\right]\right]$$
$$= \frac{\hbar^2}{2}\left[\frac{(K_x - K_{\nu,x})^2}{m_x} + \frac{(K_y - K_{\nu,y})^2}{m_y} + \frac{(K_z - K_{\nu,z})^2}{m_z}\right], \quad (2.63)$$

where α is a non-parabolicity coefficient. The value of α is given in Table 2.4.

For each **K**, the analytical conduction band models of Eqs.2.60 and 2.63 yield only one energy value that increases monotonically with **K**. The analytical energy band obviously extends even beyond the boundary of the first Brillouin zone. In numerical calculations such as those with the empirical pseudo-potential method, instead, many eigenvalues are obtained for each **K** in the first Brillouin zone. If **K** is allowed to exit this zone, the EPM method yields the same energies that would be obtained folding back to the first Brillouin zone the chosen **K** by subtraction of an appropriate $\mathbf{G_m}$ vector (reduced zone scheme). The **k·p** method instead provides different energies (extended zone scheme).

Therefore, in order to compare the analytical expressions Eqs.2.60 and 2.63 with the results of numerical conduction band calculations, in the analytical model we must take into consideration the energy bands stemming from the minima of all Brillouin zones. In other words, those parts of the analytical energy band that extend beyond the first Brillouin zone must be folded back inside the first zone. The comparison should furthermore take into consideration the specific assumptions behind each method; in particular, while the EPM is an approximation valid over the entire Brillouin zone, the **k·p** method is an expansion in a small interval around a specific point (usually the Γ point). Therefore, a limited accuracy is expected from the **k·p** method for **K** vectors far from the expansion point.

Figure 2.10 compares the parabolic and non-parabolic silicon conduction band models in Eqs.2.60 and 2.63 with the results of full band EPM calculations. For ellipsoids elongated in the [001] direction, as that in the figure, $m_x = m_y = m_t$. Therefore, the analytical models predict isotropic equi-energy contours in the x-y plane, whereas the full band structure clearly is not, as is apparent from the much smaller E values found along the [110] directions. This discrepancy becomes especially relevant for calculation of the density of states and in semiconductor materials such as gallium arsenide (as discussed in Chapter 10).

2.3.2 Valence band

The valence bands of bulk unstrained silicon and germanium have three branches of the dispersion relation with maxima at the Γ point, as shown in Fig.2.8. Two of them (the so called *light-hole* and *heavy-hole* bands) are degenerate at the Γ point, while the *split-off* band features a small energy separation of 44 meV and 296 meV in silicon

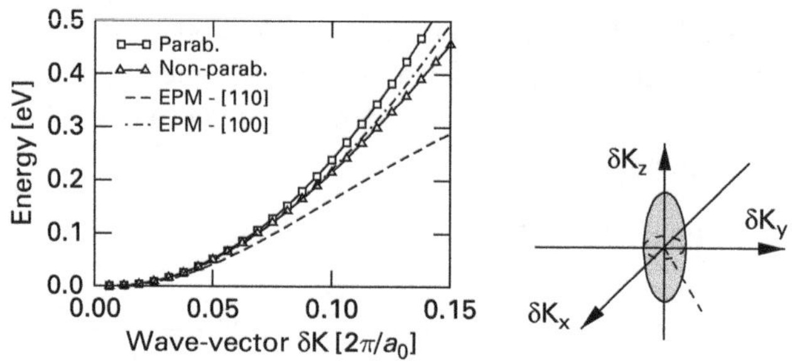

Figure 2.10 Comparison between the full band structure of silicon calculated with the empirical pseudo-potentials method (dashed line) and the parabolic (□) and non-parabolic (△) expressions as a function of the wave-vector δK referred to the minimum of the conduction band valleys along the [001] directions.

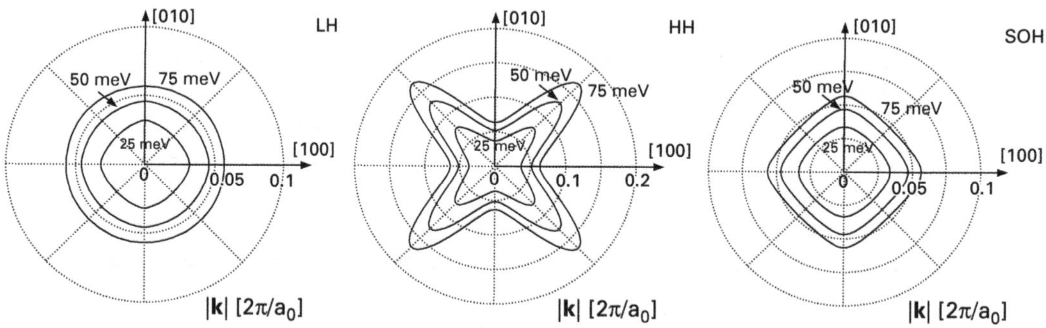

Figure 2.11 Equi-energy contours of the light, heavy and split-off hole bands of silicon in the K_x, K_y plane ($K_z = 0$) calculated with the $\mathbf{k}\cdot\mathbf{p}$ method. In all cases $E = 0$ at the top of the band.

and germanium, respectively. This separation does not appear in Fig.2.8 because the calculations neglect spin-orbit interactions.

The interaction of electron states belonging to the degenerate bands in Γ produces a significant amount of band distortion and warping with respect to the quasi-free-electron model. This is clearly seen in Fig.2.11, which shows the equi-energy contours of the heavy hole, light hole and split-off hole bands in bulk silicon for $K_z = 0$ calculated with the methodologies outlined in Section 2.2.2 and in Appendix C.

A common analytical expression to represent the valence band in the proximity of the Γ point valid for the light and heavy hole bands of unstrained bulk silicon and germanium is [17]

$$E_B(\mathbf{K}) - E_V = \frac{\hbar^2}{2m_0}\left[A|\mathbf{K}|^2 \pm \sqrt{B^2|\mathbf{K}|^4 + C^2\left(K_x^2K_y^2 + K_y^2K_z^2 + K_z^2K_x^2\right)}\right], \tag{2.64}$$

where E_V denotes the energy of the maximum of the valence band and the plus/minus sign refers to the heavy hole and light hole bands, respectively.

Table 2.4 gives the A, B, and C values for the most common semiconductors. These values imply that $[E_B(\mathbf{K}) - E_V] \leq 0$ in Eq.2.64; therefore, the energy decreases for increasing $|\mathbf{K}|$, consistently with the notation adopted in Section 2.2.2 for discussion of the $\mathbf{k}\cdot\mathbf{p}$ method. In the rest of the book, however, we adopt an electron like convention for the hole energy, which is represented as an increasing function of $\mathbf{K}$ (see also Section 5.1.2).

2.4 Equivalent Hamiltonian and Effective Mass Approximation

2.4.1 The equivalent Hamiltonian

Until now our discussion has focused on an infinite and perfect crystal where all atoms occupy a periodic arrangement of rest positions. The periodicity of the crystal potential $U_C(\mathbf{R})$ implies that the steady state solutions of the Schrödinger equation are Bloch waves. In real semiconductor devices, however, lattice vibrations perturb the periodicity of $U_C(\mathbf{R})$ while ionized impurities and defects deform its shape locally on the scale of the atomic distances. Moreover, external voltages applied to the terminals produce electric fields that extend over macroscopic dimensions compared to the lattice constant a_0. These stimuli may give rise to off equilibrium conditions, forcing free carriers to drift over large distances compared to the interatomic distance, but still small enough to be considered infinitesimal on the macroscopic scale. A full quantum mechanical treatment of such a complex case where smooth external and/or rapidly fluctuating potentials add to the crystal Hamiltonian, if at all possible, would be impractical for a description of state-of-the-art semiconductor devices.

It is the purpose of this section to derive an equivalent Hamiltonian by which band structure theory and calculations for perfect unperturbed crystals can be exploited to tackle the problems where internal or external, possibly time dependent, perturbations add to $U_C(\mathbf{R})$. In the general case, these perturbations are represented by a Hamiltonian operator $\delta\hat{H}$ but in most situations of practical interest the operator reduces to a potential energy term $U(\mathbf{R},t)$. For the sake of simplicity, in order to outline the derivation of the equivalent Hamiltonian we restrict ourselves to a stationary potential $U(\mathbf{R})$.

We start by considering the time-dependent wave-functions given by the solution of the time-dependent Schrödinger equation

$$\mathrm{i}\hbar\frac{\partial}{\partial t}\Psi_t(\mathbf{R},t) = \left[\frac{-\hbar^2}{2m_0}\nabla^2_{\mathbf{R}} + U_C(\mathbf{R}) + U(\mathbf{R})\right]\Psi_t(\mathbf{R},t). \quad (2.65)$$

A powerful technique to solve Eq.2.65 can be developed exploiting the completeness of the eigenfunctions of the crystal Hamiltonian, that is, expressing the solution $\Psi_t(\mathbf{R},t)$ of Eq.2.65 as a linear combination of Bloch waves $\Psi_{n\mathbf{K}}(\mathbf{R})\exp\left(-\mathrm{i}E_{B,n}(\mathbf{K})t/\hbar\right)$. Since the $E_{B,n}(\mathbf{K})$ are not eigenvalues of the perturbed Hamiltonian $\hat{H} = \left[\hat{H}_C + \delta\hat{H}\right]$, time-dependent coefficients $C_{n\mathbf{K}}(t)$ must appear in the expansion

$$\Psi_t(\mathbf{R},t) = \sum_{n,\mathbf{K}} C_{n\mathbf{K}}(t)\Psi_{n\mathbf{K}}(\mathbf{R})\exp\left(-\mathrm{i}E_{B,n}(\mathbf{K})t/\hbar\right). \quad (2.66)$$

We now make the additional assumption that the only non-zero coefficients $C_{n\mathbf{K}}(t)$ belong to a single branch of the energy dispersion relation and we simplify the notation by dropping the suffix n and the summation over n and we rewrite Eq.2.66 as

$$\Psi_t(\mathbf{R},t) = \sum_{\mathbf{K}} C_{\mathbf{K}}(t)\Psi_{\mathbf{K}}(\mathbf{R}) \exp\left(-\mathrm{i}E_B(\mathbf{K})t/\hbar\right). \tag{2.67}$$

We consistently indicate with $E_B(\mathbf{K})$ the energy relation in the selected branch. The above assumption is well justified if the branches of the dispersion relation are well separated from each other (as is the case near the minima of the silicon conduction band), and when the change of the perturbation potential, on the distance scale where $|\Psi_t(\mathbf{R},t)|^2$ is non-negligible, is small compared to the energy separation between the bands. In other words, $U(\mathbf{R})$ should be a slowly varying function of space over a crystal unit cell. This is typically the case if the perturbation is due to external forces applied on a macroscopic distance scale, but it may not be the case for microscopic interactions between particles in the semiconductor. Clearly this assumption fails, regardless of the form of the potential energy $U(\mathbf{R})$, in the proximity of the band crossing points. The reader is referred to [4, 18, 19] for a detailed treatment of cases where the conditions above may not hold.

Since $E_B(\mathbf{K})$ is periodic in reciprocal space, we can expand it in a Fourier series of direct lattice vectors

$$E_B(\mathbf{K}) = \sum_{\mathbf{R_n}} E_{\mathbf{R_n}} \exp(\mathrm{i}\mathbf{K}\cdot\mathbf{R_n}). \tag{2.68}$$

We now define the operator $\hat{E}_B\left(-\mathrm{i}\nabla_{\mathbf{R}}\right)$ obtained by replacing $\mathbf{K}$ with $-\mathrm{i}\nabla_{\mathbf{R}}$ in the expression for $E_B(\mathbf{K})$. Expanding in Taylor series the exponentials in the summation we have

$$\begin{aligned}\hat{E}_B\left(-\mathrm{i}\nabla_{\mathbf{R}}\right) &= \sum_{\mathbf{R_n}} E_{\mathbf{R_n}} \exp(\mathbf{R_n}\cdot\nabla_{\mathbf{R}}) \\ &= \sum_{\mathbf{R_n}} E_{\mathbf{R_n}} \left\{\sum_{j=0}^{\infty}\frac{1}{j!}\left(R_{nx}\frac{\partial}{\partial x}+R_{ny}\frac{\partial}{\partial y}+R_{nz}\frac{\partial}{\partial z}\right)^j\right\},\end{aligned} \tag{2.69}$$

and we see that for any function $f(\mathbf{R})$, we can interpret the sum over j in Eq.2.69 as the Taylor series expansion of $f(\mathbf{R}+\Delta\mathbf{R})$ about the point $\mathbf{R}$ evaluated for $\Delta\mathbf{R} = \mathbf{R_n}$, so that

$$\hat{E}_B\left(-\mathrm{i}\nabla_{\mathbf{R}}\right) f(\mathbf{R}) = \sum_{\mathbf{R_n}} E_{\mathbf{R_n}} f(\mathbf{R}+\mathbf{R_n}). \tag{2.70}$$

Hence, if $f(\mathbf{R})$ is a Bloch function $\Psi_{\mathbf{K}}(\mathbf{R})$ for the selected band n, then by recalling Eq.2.27, we find

$$\hat{E}_B\left(-\mathrm{i}\nabla_{\mathbf{R}}\right)\Psi_{\mathbf{K}}(\mathbf{R}) = E_B(\mathbf{K})\Psi_{\mathbf{K}}(\mathbf{R}). \tag{2.71}$$

In view of Eq.2.67 and of the fact that the application of the unperturbed crystal Hamiltonian $\hat{H}_C$ to the wave-function $\Psi_{\mathbf{K}}(\mathbf{R})$ yields $E_B(\mathbf{K})\Psi_{\mathbf{K}}(\mathbf{R})$, we can rewrite the time dependent Schrödinger equation 2.65 as

$$i\hbar\frac{\partial}{\partial t}\Psi_t(\mathbf{R},t) = \left[\hat{E}_B(-i\nabla_{\mathbf{R}}) + U(\mathbf{R})\right]\Psi_t(\mathbf{R},t). \tag{2.72}$$

In this expression $\left[\hat{E}_B(-i\nabla_{\mathbf{R}}) + U(\mathbf{R})\right]$ represents an *equivalent Hamiltonian* operator where the crystal potential no longer appears. All the effects of $U_C(\mathbf{R})$ are accounted for through the operator $\hat{E}_B(-i\nabla_{\mathbf{R}})$, whose form can be derived by solution of the unperturbed problem set by Eq.2.24.

The solution of the Schrödinger equation 2.65 with the perturbation term $U(\mathbf{R})$ is thus decoupled from the solution of the unperturbed problem containing only the periodic potential $U_C(\mathbf{R})$, and from a perturbed problem where $U_C(\mathbf{R})$ does not appear explicitly. It is worth noting that since we consider only one branch of the dispersion relation, the validity of Eq.2.72 is limited to potential energy terms $U(\mathbf{R})$ which are slowly varying in a crystal unit cell or, equivalently, which have Fourier components $U(\mathbf{G_m})$ in $\mathbf{K}$ space that are negligible for all $\mathbf{m} \neq \mathbf{0}$.

2.4.2 The Effective Mass Approximation

The usefulness of Eq.2.72 can be fully appreciated if we focus our attention on the impact of a stationary perturbation $U(\mathbf{R})$ on the electron states whose $\mathbf{K}$ vector lies in the proximity of the minimum of one of the conduction band valleys. In fact, for wave-vectors in close proximity to the νth minimum ($\mathbf{K}_\nu$) of the nth branch of the dispersion relation it is legitimate to neglect the $\mathbf{K}$ dependence of the periodic part of the Bloch function and then look for solutions of Eq.2.72 of the form

$$\Psi_t(\mathbf{R},t) = u_\nu(\mathbf{R})A_{t,\nu}(\mathbf{R},t) = u_\nu(\mathbf{R})\exp(i\mathbf{K}_\nu\cdot\mathbf{R})\Phi_{t,\nu}(\mathbf{R},t), \tag{2.73}$$

where $u_\nu(\mathbf{R}) = u_{\mathbf{K}_\nu}(\mathbf{R})$ is the periodic part of the Bloch wave at the νth minimum of the selected branch of the dispersion relation and, following Eq.2.67, $\Phi_{t,\nu}(\mathbf{R},t)$ is implicitly defined as

$$\Phi_{t,\nu}(\mathbf{R},t) = \sum_{\mathbf{K}} C_{\mathbf{K}}(t)\exp\left[i(\mathbf{K}-\mathbf{K}_\nu)\cdot\mathbf{R}\right]\exp\left(-iE_B(\mathbf{K})t/\hbar\right). \tag{2.74}$$

Equation 2.74 emphasizes that for $\mathbf{K}$ vectors close to $\mathbf{K}_\nu$, $\Phi_{t,\nu}(\mathbf{R},t)$ is a slowly varying function in real space, hence it represents the real space *envelope* of the total wave-function $\Psi_t(\mathbf{R},t)$ defined in Eqs.2.67 and 2.73, where the rapid oscillations of $u_\nu(\mathbf{R})$ on the atomic scale also appear.

Equations 2.70, 2.73, and the periodicity of $u_\nu(\mathbf{R})$ imply

$$\begin{aligned}
&\hat{E}_B^{(\nu)}(-i\nabla_{\mathbf{R}})\{u_\nu(\mathbf{R})\exp(i\mathbf{K}_\nu\cdot\mathbf{R})\Phi_{t,\nu}(\mathbf{R},t)\} \\
&\quad = \sum_{\mathbf{R_n}} E_{\mathbf{R_n}}^{(\nu)} u_\nu(\mathbf{R}+\mathbf{R_n})\exp(i\mathbf{K}_\nu\cdot(\mathbf{R}+\mathbf{R_n}))\Phi_{t,\nu}(\mathbf{R}+\mathbf{R_n},t) \\
&\quad = u_\nu(\mathbf{R})\hat{E}_B^{(\nu)}(-i\nabla_{\mathbf{R}})\{\exp(i\mathbf{K}_\nu\cdot\mathbf{R})\Phi_{t,\nu}(\mathbf{R},t)\},
\end{aligned} \tag{2.75}$$

where the suffix ν reminds us that we are now considering an equivalent Hamiltonian valid in proximity to the νth extreme of the band.

It follows from Eqs.2.72, 2.73, and 2.75 that the auxiliary wave-function

$$A_{t,\nu}(\mathbf{R},t) = \exp(\mathrm{i}\mathbf{K}_\nu \cdot \mathbf{R})\Phi_{t,\nu}(\mathbf{R},t) \tag{2.76}$$

satisfies the so called *single band Effective Mass Equation*

$$\mathrm{i}\hbar\frac{\partial}{\partial t}A_{t,\nu}(\mathbf{R},t) = \left[\hat{E}_B^{(\nu)}(-\mathrm{i}\nabla_{\mathbf{R}}) + U(\mathbf{R})\right]A_{t,\nu}(\mathbf{R},t). \tag{2.77}$$

Given the stationary nature of the perturbation $U(\mathbf{R})$ we can solve Eq.2.77 by separation of variables. The corresponding stationary single band effective mass equation reads

$$\left[\hat{E}_B^{(\nu)}(-\mathrm{i}\nabla_{\mathbf{R}}) + U(\mathbf{R})\right]A_{\nu,m}(\mathbf{R}) = E_{\nu,m}A_{\nu,m}(\mathbf{R}). \tag{2.78}$$

Equation 2.78 sets an eigenvalue problem with eigenfunctions $A_{\nu,m}(\mathbf{R})$ and eigenvalues $E_{\nu,m}$. For each eigenstate with energy $E_{\nu,m}$ we have

$$A_{t,\nu,m}(\mathbf{R},t) = A_{\nu,m}(\mathbf{R})\exp(-\mathrm{i}E_{\nu,m}t/\hbar). \tag{2.79}$$

Moreover, by analogy with Eq.2.76, we can express the stationary part of $A_{t,\nu}(\mathbf{R})$ as

$$A_\nu(\mathbf{R}) = \exp(\mathrm{i}\mathbf{K}_\nu \cdot \mathbf{R})\Phi_\nu(\mathbf{R}). \tag{2.80}$$

The effect of the confining potential energy $U(\mathbf{R})$, then, is to split each branch $E_B(\mathbf{K})$ of the energy relation of the bulk material $E_{B,n}(\mathbf{K})$ into an infinite number of energy eigenvalues $E_{\nu,m}$. The reasons for denoting Eqs.2.77 and 2.78 as effective mass equations will become clear in the following discussion.

In fact, we can now proceed one step further if we recall that in the neighborhood of the band extremes the choice of the x, y, z axis according to the ellipsoid coordinate system defined in Section 2.3.1 allows us to approximate $E_B^{(\nu)}(\mathbf{K})$ as in Eq.2.60. Then, we can look for solutions of the stationary Eq.2.78 based on the approximate expression for the band structure in Eq.2.60. A direct evaluation shows that

$$\begin{aligned}&\hat{E}_B^{(\nu)}(-\mathrm{i}\nabla_{\mathbf{R}})\{\exp(\mathrm{i}\mathbf{K}_\nu\cdot\mathbf{R})\Phi_\nu(\mathbf{R})\}\\ &\quad= \exp(\mathrm{i}\mathbf{K}_\nu\cdot\mathbf{R})\left[E_{\nu0} - \frac{\hbar^2}{2}\left(\frac{1}{m_x}\frac{\partial^2}{\partial x^2} + \frac{1}{m_y}\frac{\partial^2}{\partial y^2} + \frac{1}{m_z}\frac{\partial^2}{\partial z^2}\right)\right]\Phi_\nu(\mathbf{R}).\end{aligned} \tag{2.81}$$

Hence, the allowed energy levels $E_{\nu,m}$ and the corresponding stationary envelope wave-functions $\Phi_{\nu,m}(\mathbf{R})$ are the solutions of the eigenvalue problem

$$\left(-\frac{\hbar^2}{2}\left(\frac{1}{m_x}\frac{\partial^2}{\partial x^2} + \frac{1}{m_y}\frac{\partial^2}{\partial y^2} + \frac{1}{m_z}\frac{\partial^2}{\partial z^2}\right) + U_\nu(\mathbf{R})\right)\Phi_{\nu,m}(\mathbf{R}) = E_{\nu,m}\Phi_{\nu,m}(\mathbf{R}), \tag{2.82}$$

where m is an integer index for the eigenvalues and eigenfunctions and

$$U_\nu(\mathbf{R}) = E_{\nu0} + U(\mathbf{R}). \tag{2.83}$$

Equation 2.82 can be interpreted as the stationary Schrödinger equation of a particle with an anisotropic *effective mass* with direction dependent components m_i,

$i \in \{x, y, z\}$ implicitly defined by Eq.2.61. Equation 2.82 is sometimes called the *parabolic effective mass equation.*

As already mentioned, the *envelope function* $\Phi_{\nu,m}(\mathbf{R})$ does not contain the rapid fluctuations of $u_\nu(\mathbf{R})$ on the short scale of the atomic distances, but rather represents a smoothly varying envelope reflecting only the slow variations of the perturbation potential $U(\mathbf{R})$. The effective masses, instead, embody the effect of the rapidly fluctuating crystal potential on the slowly varying components of the probability density of electrons in the proximity of the band extremes.

In view of Eqs.2.73, 2.76, and 2.79 the complete wave-function for an electron state with well-defined energy $E_{\nu,m}$ in proximity to the band extreme ν reads

$$\Psi_{t,m}(\mathbf{R}, t) \simeq u_\nu(\mathbf{R}) \exp(\mathrm{i}\mathbf{K}_\nu \cdot \mathbf{R})\, \Phi_{\nu,m}(\mathbf{R}) \exp\left(-\mathrm{i}E_{\nu,m}t/\hbar\right). \tag{2.84}$$

We finally note that we can rewrite Eq.2.82 in the form

$$[\hat{E}_{cb}^{(\nu)}(-\mathrm{i}\nabla_{\mathbf{R}}) + U(\mathbf{R})]\, \Phi_{\nu,m}(\mathbf{R}) = E'_{\nu,m}\, \Phi_{\nu,m}(\mathbf{R}), \tag{2.85}$$

where $E'_{\nu,m} = \left(E_{\nu,m} - E_{\nu 0}\right)$ is the energy eigenvalue referred to the conduction band minimum, and the parabolic effective mass approximation (EMA) operator $\hat{E}_{cb}^{(\nu)}(-\mathrm{i}\nabla_{\mathbf{R}})$ is obtained substituting the wave-vector referred to the minimum ($\delta\mathbf{K} = \mathbf{K} - \mathbf{K}_\nu$) with $-\mathrm{i}\nabla_{\mathbf{R}}$ in the expression

$$E_{cb}^{(\nu)}(\delta\mathbf{K}) = \frac{\hbar^2}{2}\left(\frac{(\delta K_x)^2}{m_x} + \frac{(\delta K_y)^2}{m_y} + \frac{(\delta K_z)^2}{m_z}\right). \tag{2.86}$$

In the presence of small deviations of the band structure from a parabolic profile, such as those observed at energies significantly above the minimum, we can assume that Eq.2.85 remains valid and generalize the expression of the operator as suggested by Eq.2.63, that is we can substitute Eq.2.86 with

$$E_{cb}^{(\nu)}(\delta\mathbf{K})\left(1 + \alpha E_{cb}^{(\nu)}(\delta\mathbf{K})\right) = \frac{\hbar^2}{2}\left(\frac{(\delta K_x)^2}{m_x} + \frac{(\delta K_y)^2}{m_y} + \frac{(\delta K_z)^2}{m_z}\right), \tag{2.87}$$

where α is the non-parabolicity coefficient. Equation 2.87 yields

$$E_{cb}^{(\nu)}(\delta\mathbf{K}) = \frac{1}{2\alpha}\left[-1 + \sqrt{1 + 2\alpha\hbar^2\left(\frac{(\delta K_x)^2}{m_x} + \frac{(\delta K_y)^2}{m_y} + \frac{(\delta K_z)^2}{m_z}\right)}\right]. \tag{2.88}$$

2.5 The semi-classical model

2.5.1 Wave-packets and group velocity

Quantum mechanics represents electrons as waves. In particular, in a periodic potential energy field such as that of a perfect crystal, these waves are the Bloch states $\Psi_{n\mathbf{K}}$ (Eq.2.25) with energy $E_{B,n}(\mathbf{K})$ and wave-vector $\mathbf{K}$. The probability density of a single

Bloch state is modulated as $|u_{n\mathbf{K}}(\mathbf{R})|^2$ on the microscopic scale, but it is otherwise uniformly distributed on the macroscopic scale, because of the plane wave term appearing in the expression for $\Psi_{n\mathbf{K}}$ (Eq.2.25).

Classical electrodynamics, instead, represents electrons as point charges with rest mass m_0; once the initial electron position $\mathbf{R}(t=0)$ and velocity $\mathbf{V}(t=0)$ are assigned, the electron dynamics is unequivocally determined by the force $\mathbb{F}$ exerted on the electron and by the equations of motion

$$\mathbf{P} = m_0\mathbf{V} = m_0\frac{\mathrm{d}\mathbf{R}}{\mathrm{d}t}, \tag{2.89a}$$

$$\mathbb{F} = \frac{\mathrm{d}\mathbf{P}}{\mathrm{d}t} = m_0\frac{\mathrm{d}\mathbf{V}}{\mathrm{d}t}. \tag{2.89b}$$

The success of classical electrodynamics in explaining a large number of physical observations on the macroscopic scale poses the issue of reconciling the quantum mechanical and classical pictures. This can be accomplished by examining the behavior of an electron wave-packet, which is addressed below.

The starting point for discussing electron wave-packets in a bulk crystal is the observation that, thanks to the completeness of the Bloch functions, we can represent the general solution of the time-dependent Schrödinger equation, Eq.2.22, as a superposition (commonly denoted *wave-packet*) of Bloch states

$$\Psi_t(\mathbf{R}, t) = \sum_{n,\mathbf{K}} C_{n\mathbf{K}}(t)\Psi_{n\mathbf{K}}(\mathbf{R})\exp\left(-\mathrm{i}E_{B,n}(\mathbf{K})t/\hbar\right). \tag{2.90}$$

As already noted with reference to Eq.2.66, in general the summation runs over $\mathbf{K}$ states belonging to different bands n and the weights $C_{n\mathbf{K}}(t)$ are functions of time. The dynamic behavior of these weights will be discussed with an increasing level of detail in the rest of this chapter. Here we simply observe that in the absence of forces perturbing the perfectly periodic crystal potential energy the $C_{n\mathbf{K}}(t)$ will be independent of time. With an appropriate choice of the $C_{n\mathbf{K}}(t)$ we can obtain a probability density $|\Psi_t(\mathbf{R}, t)|^2$ for the wave-packet of Eq.2.90 that, on the macroscopic scale, appears centered in the neighborhood of a given position $\mathbf{R}_0$ at a given time t, hence being suited to represent a classical point charge located at $\mathbf{R}_0$.

Each Bloch wave in Eq.2.90 has its own evolution in time due to the specific value of $E_{B,n}(\mathbf{K})$ in the exponential term. Therefore, even in the absence of external forces (that is, even if $C_{n\mathbf{K}}$ is independent of time) the wave-packet is a non-stationary solution of the time-dependent Schrödinger equation, Eq.2.22, meaning that the probability density $|\Psi_t(\mathbf{R}, t)|^2$ changes in time and the centroid of the wave-packet moves in real space.

For the sake of simplicity, we hereafter adopt the same assumption made in Section 2.4.1, that is we suppose that the Bloch states necessary to represent the electron belong to the same branch n of the dispersion relation; then we drop n in the notation. Moreover, we choose a set of initial weights $C_{\mathbf{K}}(t=0)$ that vanish rapidly outside a small region of radius ΔK_{wp} centered in $\mathbf{K}_0$, that is for $|\Delta\mathbf{K}| = |\mathbf{K} - \mathbf{K_0}| > \Delta K_{wp}$.

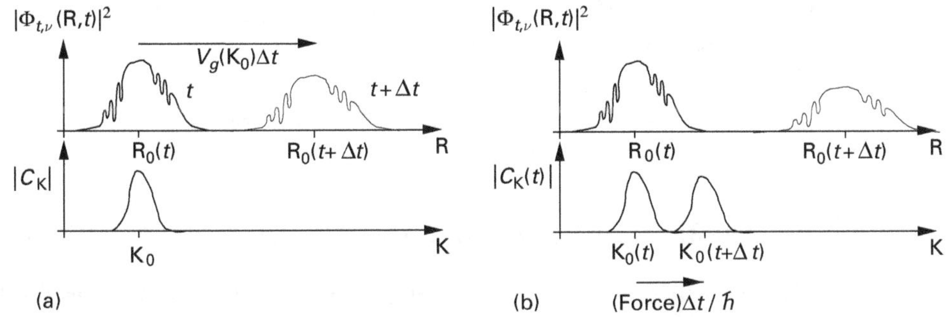

Figure 2.12 Representation of a wave-packet probability density and of the corresponding distribution of the $C_{n\mathbf{K}}(t)$ in wave-vector space at two different times. (a) is the case where only the crystal potential acts on the electron, while (b) is the case where an external force adds to the crystal potential.

Consequently $\mathbf{K_0}$ and ΔK_{wp} provide an estimate of the mean value (centroid) and of the spread of the wave-packet in reciprocal space. This situation is sketched in Fig.2.12.a.

Reciprocity between the real and the $\mathbf{K}$ space implies that the wave-packet probability density $|\Psi_t(\mathbf{R}, t)|^2$ is localized in real space within a region of approximate dimension $|\Delta\mathbf{R}| \approx 1/\Delta K_{wp}$. If we assume ΔK_{wp} small on the scale of the Brillouin zone extension, so that the wave-packet has a rather well-defined wave-vector approximately given by $\mathbf{K}_0$, the wave-packet will extend over many primitive cells in real space.

If we consider an electron device occupying, for instance, one cubic micron of semiconductor, we find that the number of unit cells in real space is in the order of 10^{12}. Recalling Eq.2.36, we see that the number of allowed $\mathbf{K}$ states per band will then be equally large. A wave-packet with small ΔK_{wp} on the scale of the Brillouin zone (e.g. $\Delta K_{wp} = 1/100$ in units of $2\pi/a_0$) will still contain a very large number of $\mathbf{K}$ components ($\approx 10^6$) which give rise to substantial interference effects and wave-packet localization in real space. Such a wave-packet can then be modeled as a point charge with well-defined position, as long as we are not interested in the behavior of the electron wave-function on the scale of a few atomic distances.

Wave-packet dynamics in an unperturbed crystal

To understand the dynamic behavior of the wave-packet, let us consider first the situation in which no perturbation affects the perfectly periodic crystal potential $U_C(\mathbf{R})$. We can linearize $E_B(\mathbf{K})$ around the stationary mean value $\mathbf{K_0}$, that is

$$\begin{aligned}\Psi_t(\mathbf{R}, t) &\simeq \sum_{\mathbf{K}} C_{\mathbf{K}} u_{\mathbf{K}}(\mathbf{R}) \exp(\mathrm{i}\mathbf{K}\cdot\mathbf{R}) \\ &\quad \times \exp\{-\mathrm{i}[E_B(\mathbf{K_0}) + \nabla_{\mathbf{K}} E_B(\mathbf{K}_0)\cdot(\mathbf{K}-\mathbf{K}_0)]\, t/\hbar\} \\ &= \exp[-\mathrm{i}\,\phi(t)] \sum_{\mathbf{K}} C_{\mathbf{K}} u_{\mathbf{K}}(\mathbf{R}) \exp\left[\mathrm{i}\,\mathbf{K}\cdot(\mathbf{R} - \nabla_{\mathbf{K}} E_B(\mathbf{K}_0) t/\hbar)\right], \end{aligned} \tag{2.91}$$

where

$$\phi(t) = [E_B(\mathbf{K}_0) - \nabla_{\mathbf{K}} E_B(\mathbf{K}_0)\cdot\mathbf{K_0}]\, t/\hbar. \tag{2.92}$$

Recalling that $u_{\mathbf{K}}(\mathbf{R})$ is periodic in the Bravaix lattice, we observe that at all the times t_j that satisfy the condition

$$\frac{1}{\hbar}\nabla_{\mathbf{K}} E_B(\mathbf{K}_0) \cdot t_j = \mathbf{R}_{\mathbf{n}}, \tag{2.93}$$

we have

$$|\Psi_t(\mathbf{R}, t_j)|^2 = |\Psi_t\left(\mathbf{R} - \nabla_{\mathbf{K}} E_B(\mathbf{K}_0) t_j/\hbar, 0\right)|^2 = |\Psi_t\left(\mathbf{R} - \mathbf{R}_{\mathbf{n}}, 0\right)|^2. \tag{2.94}$$

Therefore, the wave-packet probability density $|\Psi_t(\mathbf{R}, t)|^2$ periodically recovers its original shape $|\Psi_t(\mathbf{R}, 0)|^2$ at locations that translate in space with an effective velocity

$$\mathbf{V}_g(\mathbf{K}) = \frac{1}{\hbar}\nabla_{\mathbf{K}} E_B(\mathbf{K}), \tag{2.95}$$

calculated at the wave-packet mean wave-vector $\mathbf{K}_0$. Since we know from the derivation of the stationary Schrödinger equation for the crystal potential that $E_B(\mathbf{K})$ is proportional to the radial frequency of a Bloch wave (Eq.2.23), by analogy with wave propagation problems in physics we recognize that $\mathbf{V}_g(\mathbf{K})$ is the so called *group velocity* of the quantum wave-packet. Note that Eqs.2.34 and 2.95 imply that $\mathbf{V}_g(\mathbf{K})$ is odd in $\mathbf{K}$-space:

$$\mathbf{V}_g(-\mathbf{K}) = -\mathbf{V}_g(\mathbf{K}). \tag{2.96}$$

It is clear from the derivation above that the wave-packet periodically recovers its original shape only if $E_B(\mathbf{K})$ is a linear function of $\mathbf{K}$. However, in general this is not the case, neither for electrons in crystals (Bloch waves) nor for free electrons (plane waves). Consequently, even in this relatively simple case of zero external forces and constant coefficients $C_{\mathbf{K}}$, the wave-packet progressively spreads out with increasing time and its position in real space becomes increasingly less determined. The time evolution of a wave-packet in the absence of external forces is sketched in Fig.2.12.a.

Wave-packet dynamics in the presence of external forces

Let us now consider a general case where an additional perturbation expressed by the potential energy $U(\mathbf{R})$ adds to the periodic crystal potential $U_C(\mathbf{R})$, hence to the crystal Hamiltonian. The Schrödinger equation, Eq.2.65, governs the wave-packet dynamics. We assume that the perturbation is smooth on the distance scale where $|\Psi_t(\mathbf{R}, t)|^2$ is non-negligible, so that, following arguments similar to those developed in Section 2.4.1, the expansion 2.90 can be restricted to states in the same band n.

By virtue of the results discussed above, we know that in an unperturbed crystal the wave-packet spectrum in wave-vector space is stationary but the wave-packet moves in real space with the group velocity $\mathbf{V}_g(\mathbf{K}_0)$. In the presence of an external perturbation, instead, the coefficients $C_{\mathbf{K}}(t)$ depend on time and the wave-packet centroid in reciprocal space $\mathbf{K}_0(t)$ also changes with time, as schematically depicted in Fig.2.12.b. It can thus be inferred that if a *weak* perturbation potential is superimposed on the crystal potential, the coefficients $C_{\mathbf{K}}(t)$ remain approximately constant in a time interval long enough for the wave-packet to move over a distance much larger than its extension $|\Delta\mathbf{R}|$.

We can re-write the approximated expression of the wave-packet Eq.2.91 as

$$\Psi_t(\mathbf{R},t) \simeq \exp\left\{\mathrm{i}\left[\mathbf{K}_0\cdot\mathbf{R} - E_B(\mathbf{K}_0)t/\hbar\right]\right\} \times \sum_{\Delta\mathbf{K}} C_{(\mathbf{K}_0+\Delta\mathbf{K})}(t)u_{(\mathbf{K}_0+\Delta\mathbf{K})}(\mathbf{R})\exp\left[\mathrm{i}\Delta\mathbf{K}\cdot(\mathbf{R}-\nabla_{\mathbf{K}}E_B(\mathbf{K}_0)t/\hbar)\right]. \quad (2.97)$$

In this expression $u_{(\mathbf{K}_0+\Delta\mathbf{K})}(\mathbf{R})$ oscillates rapidly on the interatomic distance scale whereas the term $\exp\{\mathrm{i}[\Delta\mathbf{K}\cdot(\mathbf{R}-\nabla_{\mathbf{K}}E_B(\mathbf{K}_0)t/\hbar)]\}$ changes smoothly in real space because $|\Delta\mathbf{K}|$ is small. Therefore, if we are not interested in the oscillations of $\Psi_t(\mathbf{R},t)$ on the atomic scale of distance and we consider a short time interval such that the coefficients $C_{\mathbf{K}}(t)$ and thus $\mathbf{K}_0(t)$ are approximately constant, we can neglect the $\mathbf{K}$ dependence of $u_{\mathbf{K}}(\mathbf{R})$ assuming that $u_{(\mathbf{K}_0+\Delta\mathbf{K})}(\mathbf{R}) \approx u_{\mathbf{K}_0}(\mathbf{R})$, and approximate $\Psi_t(\mathbf{R},t)$ as

$$\Psi_t(\mathbf{R},t) \simeq \psi_t\left(\mathbf{R}-\mathbf{V}_g(\mathbf{K}_0)t,t\right)u_{\mathbf{K}_0}(\mathbf{R})\exp(\mathrm{i}\mathbf{K}_0\cdot\mathbf{R})\exp(-\mathrm{i}E_B(\mathbf{K}_0)t/\hbar), \quad (2.98)$$

where the space and time dependence of the auxiliary function

$$\psi_t = \sum_{\Delta\mathbf{K}} C_{(\mathbf{K}_0+\Delta\mathbf{K})}(t)\exp\left[\mathrm{i}\Delta\mathbf{K}\cdot(\mathbf{R}-\nabla_{\mathbf{K}}E_B(\mathbf{K}_0)t/\hbar)\right] \quad (2.99)$$

have been expressed in the form $\psi_t\left(\mathbf{R}-\mathbf{V}_g(\mathbf{K}_0)t,t\right)$ to emphasize that time appears implicitly as well as explicitly. In fact, time enters Eq.2.99 first in the $\mathbf{R}$ dependence of ψ_t through the term $[\mathbf{R}-\mathbf{V}_g(\mathbf{K}_0)t]$, which is easily interpreted as a rigid translation of ψ_t with an effective velocity $\nabla_{\mathbf{K}}E_B(\mathbf{K}_0)/\hbar$. Then, the function ψ_t has an additional, explicit time dependence as well, which is only due to the change in time of the coefficients $C_{\mathbf{K}}(t)$. We reiterate here that, since we assume that $|\Delta\mathbf{K}|$ is small on the scale of the Brillouin zone, the term $\exp(\mathrm{i}\Delta\mathbf{K}\cdot\mathbf{R})$ in the expression for ψ_t (Eq.2.99) is a smooth function of $\mathbf{R}$ on the scale of the interatomic distance. Consequently ψ_t envelopes the rapid spatial fluctuations of the $u_{\mathbf{K}_0}(\mathbf{R})\exp(\mathrm{i}\mathbf{K}_0\cdot\mathbf{R})$ term in Eq.2.98. The shape of the envelope changes with time only because of the change of the weights $C_{\mathbf{K}}(t)$ and, if the perturbation is sufficiently small, this change will be fairly slow. Thus, the wave-packet is well approximated by the product of the Bloch wave corresponding to its centroid $\mathbf{K}_0$ in $\mathbf{K}$-space times an envelope function $\psi_t\left(\mathbf{R}-\mathbf{V}_g(\mathbf{K}_0)t,t\right)$ that shifts almost rigidly with velocity $\mathbf{V}_g(\mathbf{K}_0)$ and whose shape changes quite slowly in time. The external forces exerted on the electron and represented by $U(\mathbf{R})$ result in a time dependent $\mathbf{K}_0$ because the coefficients $C_{\mathbf{K}}(t)$ change in time.

In close-to-equilibrium conditions the electrons occupy regions of $\mathbf{K}$-space in the proximity of band extremes, therefore $\mathbf{K}_0$ remains close to the wave-vector at the extreme $\mathbf{K}_\nu$. If we are not interested in the rapid fluctuations of the full wave-function related to $u_{\mathbf{K}_0}(\mathbf{R})$, then we can adopt the approximation $u_{\mathbf{K}_0}(\mathbf{R}) \simeq u_\nu(\mathbf{R})$. Equations 2.73 and 2.98 immediately allow us to write

$$\Phi_{t,\nu}(\mathbf{R},t) = \psi_t\left(\mathbf{R}-\mathbf{V}_g(\mathbf{K}_0)t,t\right)\exp\left[\mathrm{i}(\mathbf{K}_0-\mathbf{K}_\nu)\cdot\mathbf{R}\right]\exp\left[-\mathrm{i}E_B(\mathbf{K}_0)t/\hbar\right], \quad (2.100)$$

which is just an alternative form of Eq.2.74. If we now recall Eq.2.76 and the fact that $A_{t,\nu}(\mathbf{R}, t)$ satisfies Eq.2.77, it follows that, in the parabolic band approximation, $\Phi_{t,\nu}(\mathbf{R}, t)$ satisfies the time-dependent Schrödinger equation

$$i\hbar\frac{\partial}{\partial t}\Phi_{t,\nu}(\mathbf{R}) = \left[-\frac{\hbar^2}{2}\left(\frac{1}{m_x}\frac{\partial^2}{\partial x^2} + \frac{1}{m_y}\frac{\partial^2}{\partial y^2} + \frac{1}{m_z}\frac{\partial^2}{\partial z^2}\right) + U_\nu(\mathbf{R})\right]\Phi_{t,\nu}(\mathbf{R}), \tag{2.101}$$

where $U_\nu(\mathbf{R}) = E_{\nu 0} + U(\mathbf{R})$ (Eq.2.83).

Equation 2.100 is an especially meaningful expression of the wave-packet wave-function. In fact, since $|\delta\mathbf{K}| = |\mathbf{K}_0 - \mathbf{K}_\nu|$ is small and $\psi_t\left(\mathbf{R} - \mathbf{V}_g(\mathbf{K}_0)t, t\right)$ is a weak function of $\mathbf{R}$, the envelope of the wave-packet $\Phi_t(\mathbf{R}, t)$ contains only components that vary smoothly in space; the components that change rapidly on the scale of a crystal unit cell are described by $u_{\mathbf{K}_0}(\mathbf{R}) \simeq u_\nu(\mathbf{R})$. The wave-packet envelope takes a plane-wave-like form where the propagation vector is $\delta\mathbf{K} = (\mathbf{K}_0 - \mathbf{K}_\nu)$, the radial frequency of oscillation is $E_B(\mathbf{K}_0)/\hbar$ and the $\mathbf{R}$-dependent amplitude is $\psi_t(\mathbf{R} - \mathbf{V}_g(\mathbf{K}_0)t, t)$. The wave-vector and the energy of the monochromatic plane-wave-like term $\exp\{i[(\mathbf{K}_0 - \mathbf{K}_\nu)\cdot\mathbf{R} - E_B(\mathbf{K}_0)t/\hbar]\}$ vary with time because the forces expressed by the perturbation potential $U(\mathbf{R})$ change $\mathbf{K}_0$ in reciprocal space. The equation that links the dynamics of $\mathbf{K}_0(t)$ to the external forces is derived in the following section.

To conclude, we note that assuming the wave-function Ψ_t is properly normalized, the microscopic probability density of the electron described by the wave-packet of Eq.2.98 is given by

$$|\Psi_t(\mathbf{R}, t)|^2 \simeq |\psi_t(\mathbf{R}, t)|^2\,|u_{\mathbf{K}_0}(\mathbf{R})|^2 \simeq |\Phi_t(\mathbf{R}, t)|^2\,|u_{\mathbf{K}_\nu}(\mathbf{R})|^2. \tag{2.102}$$

Since the $u_\mathbf{K}(\mathbf{R})$ functions are normalized, the contribution of such electrons to the *macroscopic* space charge density is given by

$$\rho(\mathbf{R}, t) \simeq -e|\psi_t(\mathbf{R}, t)|^2 \simeq -e|\Phi_t(\mathbf{R}, t)|^2. \tag{2.103}$$

2.5.2 Carrier motion in a slowly varying potential

We have seen in the previous sections that in crystals we can represent electrons with limited uncertainty in position and wave-vector by means of a superposition of Bloch states. The probability density on the distance scale of several unit cells is represented by the wave-packet envelope $\Phi_{t,\nu}(\mathbf{R}, t)$ in Eq.2.100. We can now ask ourselves if we can interpret the electron dynamics under the action of external forces in terms of the motion of point charges. To this purpose, since point charges are represented by wave-packets of the form expressed by Eqs.2.74 and 2.100, we are led to start again from the time dependence of the envelope function $\Phi_{t,\nu}(\mathbf{R}, t)$ and of the weights $C_\mathbf{K}(t)$.

In order to derive the equations of motion for the wave-packet, we start by recalling that for a particle in the state represented by the wave-function $\Phi(\mathbf{R}, t)$, the expectation value $\langle A\rangle$ of a physical quantity expressed by the real number A associated to the Hermitian operator $\hat{\mathcal{A}}$ is

$$\langle A \rangle = \frac{\langle \Phi | \hat{A} | \Phi \rangle}{\langle \Phi | \Phi \rangle} = \frac{\int \Phi^\dagger(\mathbf{R}, t)\, \hat{A}\, \Phi(\mathbf{R}, t) \mathrm{d}\mathbf{R}}{\int \Phi^\dagger(\mathbf{R}, t) \Phi(\mathbf{R}, t) \mathrm{d}\mathbf{R}}. \tag{2.104}$$

In the parabolic effective mass approximation, to which we adhere in the following, the dynamics of $\Phi_{t,\nu}(\mathbf{R}, t)$ is governed by the time-dependent Schrödinger equation, Eq.2.101. If we assume that $\Phi_{t,\nu}(\mathbf{R}, t)$ is properly normalized, then based on Eq.2.104 the expectation value of the x component of the wave-packet velocity is

$$\frac{\mathrm{d}}{\mathrm{d}t} \langle x \rangle = \int_\Omega \Phi^\dagger_{t,\nu}(\mathbf{R}, t)\, x\, \frac{\partial \Phi_{t,\nu}(\mathbf{R}, t)}{\partial t} \mathrm{d}\mathbf{R} + \int_\Omega \frac{\partial \Phi^\dagger_{t,\nu}(\mathbf{R}, t)}{\partial t}\, x\, \Phi_{t,\nu}(\mathbf{R}, t)\, \mathrm{d}\mathbf{R}, \tag{2.105}$$

where Ω is the normalization volume, $\mathrm{d}\mathbf{R} = [\mathrm{d}x\, \mathrm{d}y\, \mathrm{d}z]$ and x is one of the three integration variables; hence, its time dependence should not be considered in the derivation with respect to time. Expressing the time derivatives according to Eq.2.101 we have

$$\frac{\mathrm{d}}{\mathrm{d}t} \langle x \rangle = \frac{\mathrm{i}\hbar}{2m_x} \int_\Omega \left[\Phi^\dagger_{t,\nu}(\mathbf{R}, t)\, x\, \frac{\partial^2 \Phi_{t,\nu}(\mathbf{R}, t)}{\partial x^2} - \frac{\partial^2 \Phi^\dagger_{t,\nu}(\mathbf{R}, t)}{\partial x^2}\, x\, \Phi_{t,\nu}(\mathbf{R}, t) \right] \mathrm{d}\mathbf{R}. \tag{2.106}$$

A double integration by parts and the observation that the envelope of the wave-packet $\Phi_{t,\nu}$ is spatially localized and rapidly decays to zero at the integration boundaries allow us to transform the second term in the integral at the r.h.s. of Eq.2.106 as

$$\int_\Omega \left(\frac{\partial^2 \Phi^\dagger_{t,\nu}(\mathbf{R}, t)}{\partial x^2} \right) x\, \Phi_{t,\nu}(\mathbf{R}, t) \mathrm{d}\mathbf{R} = \int_\Omega \Phi^\dagger_{t,\nu}(\mathbf{R}, t)\, \frac{\partial^2}{\partial x^2} \left(x\, \Phi_{t,\nu}(\mathbf{R}, t) \right) \mathrm{d}\mathbf{R}, \tag{2.107}$$

so that we eventually get

$$\begin{aligned} \frac{\mathrm{d}\langle x \rangle}{\mathrm{d}t} &= \frac{\mathrm{i}\hbar}{2m_x} \int_\Omega \Phi^\dagger_{t,\nu}(\mathbf{R}, t) \left[x\, \frac{\partial^2}{\partial x^2} \Phi_{t,\nu}(\mathbf{R}, t) - \frac{\partial^2}{\partial x^2} \left(x\, \Phi_{t,\nu}(\mathbf{R}, t) \right) \right] \mathrm{d}\mathbf{R} \\ &= -\frac{\mathrm{i}\hbar}{m_x} \int_\Omega \Phi^\dagger_{t,\nu}(\mathbf{R}, t) \frac{\partial \Phi_{t,\nu}(\mathbf{R}, t)}{\partial x} \mathrm{d}\mathbf{R}. \end{aligned} \tag{2.108}$$

Since the momentum operator is $\hat{\mathbf{P}} = -\mathrm{i}\hbar \nabla_\mathbf{R}$, from Eq.2.104 with $\hat{\mathcal{A}} = \hat{\mathbf{P}}$ we have

$$\langle P_x \rangle = -\mathrm{i}\hbar \int_\Omega \Phi^\dagger_{t,\nu}(\mathbf{R}, t) \frac{\partial \Phi_{t,\nu}(\mathbf{R}, t)}{\partial x} \mathrm{d}\mathbf{R}. \tag{2.109}$$

Comparing Eq.2.108 and 2.109 yields

$$V_x = \frac{\mathrm{d}\langle x \rangle}{\mathrm{d}t} = \frac{\langle P_x \rangle}{m_x}, \tag{2.110}$$

where we have introduced the symbol V_x for the time derivative of $\langle x \rangle$. Similar equations hold for the y and z components of $\mathrm{d}\langle \mathbf{R} \rangle / \mathrm{d}t$, which we denote V_y and V_z, respectively. Therefore, by defining the diagonal *effective mass matrix*

$$\mathbf{M} = \begin{bmatrix} m_x & 0 & 0 \\ 0 & m_y & 0 \\ 0 & 0 & m_z \end{bmatrix}, \tag{2.111}$$

its inverse

$$\mathbf{W} = \mathbf{M}^{-1} = \begin{bmatrix} \frac{1}{m_x} & 0 & 0 \\ 0 & \frac{1}{m_y} & 0 \\ 0 & 0 & \frac{1}{m_z} \end{bmatrix}, \tag{2.112}$$

and the column vectors $\mathbf{V} = (V_x, V_y, V_z)^T$ and $\langle \mathbf{P} \rangle = (\langle P_x \rangle, \langle P_y \rangle, \langle P_z \rangle)^T$ we can then simply write

$$\mathbf{V} = \mathbf{M}^{-1} \cdot \langle \mathbf{P} \rangle, \tag{2.113}$$

or $\langle \mathbf{P} \rangle = \mathbf{M} \cdot \mathbf{V} = \mathbf{M} \cdot \mathrm{d}\langle \mathbf{R} \rangle / \mathrm{d}t$, which is the equivalent of Eq.2.89a for the quantum wave-packet. It is worth recalling here that the simple diagonal form taken by $\mathbf{M}$ and $\mathbf{W}$ in Eqs.2.111 and 2.112, respectively, is a direct consequence of Eq.2.100, that is of the choice of the x, y, z axes aligned with the axes of the ellipsoid coordinate system. The general case where the ellipsoids are not elongated in the x, y, or z directions is discussed in Chapter 8.

With the same assumption on the normalization of $\Phi_{t,\nu}$ set out above, substitution of Eq.2.100 in Eq.2.109 yields

$$\langle P_x \rangle = -\mathrm{i}\hbar \int_{\Omega} \mathrm{i}\,\delta K_x\, \psi_t^{\dagger}(\mathbf{R}, t)\, \psi_t(\mathbf{R}, t)\, \mathrm{d}\mathbf{R} - \mathrm{i}\hbar \int_{\Omega} \psi_t^{\dagger}(\mathbf{R}, t) \frac{\partial \psi_t(\mathbf{R}, t)}{\partial x}\, \mathrm{d}\mathbf{R}, \tag{2.114}$$

where δK_x is the x component of $\delta\mathbf{K} = [\mathbf{K}_0 - \mathbf{K}_\nu]$. Integration by parts and the observation that the wave-packet vanishes at the integration boundaries show that the last integral in Eq.2.114 is zero. Hence, considering all directions, we obtain

$$\langle \mathbf{P} \rangle = \hbar\delta\mathbf{K} = \hbar(\mathbf{K}_0 - \mathbf{K}_\nu). \tag{2.115}$$

Equation 2.115 represents a simple but relevant result. It shows that the expectation value of the momentum of a wave-packet moving in the proximity of a band extreme is given by the so-called *crystal momentum* referred to the minimum of the band, $\hbar\delta\mathbf{K}$. If we now calculate the group velocity as defined in Eq.2.95 for the parabolic expression of the dispersion relation $E_B(\mathbf{K})$ (see Eq.2.60) we readily find

$$\mathbf{V}_g = \mathbf{W} \cdot \hbar\delta\mathbf{K} = \frac{\mathrm{d}\langle \mathbf{R} \rangle}{\mathrm{d}t}, \tag{2.116}$$

thus demonstrating the consistency of the definition given in Eq.2.95 with the usual definition of the velocity as the time derivative of the position (Eq.2.105).

Let us now consider the time derivative of the expectation value of the momentum. Since the envelope $\Phi_{t,\nu}(\mathbf{R}, t)$ satisfies the time-dependent Schrödinger equation, Eq.2.101, the Ehrenfest theorem implies that [20]

$$\frac{\mathrm{d}}{\mathrm{d}t} \langle \mathbf{P} \rangle = \langle \nabla_{\mathbf{R}} U_\nu(\mathbf{R}) \rangle. \tag{2.117}$$

Then, Eqs.2.83, 2.115, and 2.117 lead to

$$\frac{\mathrm{d}}{\mathrm{d}t} (\hbar\delta\mathbf{K}) = -\langle \nabla_{\mathbf{R}} [E_{\nu 0} + U(\mathbf{R})] \rangle. \tag{2.118}$$

Therefore, recalling that $\mathbf{K}_\nu$ is constant in time we see that, if the space partial derivatives of the potential energy term $U_\nu(\mathbf{R}) = [E_{\nu 0} + U(\mathbf{R})]$ are constant in the whole region occupied by the wave-packet, and if the wave-packet envelope is duly normalized, we have

$$\hbar \frac{\mathrm{d}\mathbf{K}}{\mathrm{d}t} = -\nabla_{\mathbf{R}} [E_{\nu 0} + U(\mathbf{R})] . \tag{2.119}$$

Equation 2.119 is the equivalent of Eq.2.89b for the quantum wave-packet and can be interpreted as the equation of motion for a particle subject to the driving force

$$\mathbb{F} = -\nabla_{\mathbf{R}} [E_{\nu 0} + U(\mathbf{R})] . \tag{2.120}$$

We also note that in a uniform bulk crystal, $E_{\nu 0}$ is constant and makes no contribution to the force acting on the electrons. However, in all cases where the energy of the extreme of the band changes with position (e.g. because of changes in the composition or in the strain of the material, see Chapters 9 and 10), we have $E_{\nu 0} = E_{\nu 0}(\mathbf{R})$. Consequently a quasi-field $-\nabla_{\mathbf{R}} E_{\nu 0}(\mathbf{R})$ adds to the electric field $\mathbf{F} = -\nabla_{\mathbf{R}} \phi(\mathbf{R})$ in determining the driving force.

Based on the results discussed above we can generalize the equations of motion for the wave-packet centroid in real and wave-vector space in the simple and symmetric form:

$$\frac{\mathrm{d}\mathbf{R}}{\mathrm{d}t} = +\frac{1}{\hbar} \nabla_{\mathbf{K}} E(\mathbf{R}, \mathbf{K}), \tag{2.121a}$$

$$\frac{\mathrm{d}\mathbf{K}}{\mathrm{d}t} = -\frac{1}{\hbar} \nabla_{\mathbf{R}} E(\mathbf{R}, \mathbf{K}), \tag{2.121b}$$

where we have introduced the total energy

$$E(\mathbf{R}, \mathbf{K}) = E_{B,n}(\mathbf{K}) + U(\mathbf{R}). \tag{2.122}$$

The Eqs.2.121 coincide with the classical equations of motion of a particle with momentum equal to the *crystal momentum* $\hbar\mathbf{K}$. As shown in detail in Section 5.2.3 with reference to carriers in inversion layers, Eqs.2.121 represent conservation of energy for a classical particle whose energy takes the mixed classical and quantum mechanical form $E(\mathbf{R}, \mathbf{K})$. In proximity to the valley minima, Eq.2.122 reduces to

$$E(\mathbf{R}, \mathbf{K}) = E_B^{(\nu)}(\mathbf{K}) + U(\mathbf{R}) = E_{cb}^{(\nu)}(\mathbf{K} - \mathbf{K}_\nu) + E_{\nu 0} + U(\mathbf{R}). \tag{2.123}$$

By virtue of Eqs.2.122 and 2.123 and of the fact that in the unperturbed case the potential energy U is constant, Eq.2.95 appears to be analogous to the classical result that the velocity is the derivative of the total energy with respect to momentum except that the crystal momentum plays the role of classical momentum. Moreover, we can read Eq.2.119 as analogous to the classical result that the time derivative of momentum is opposite to the spatial gradient of the total energy $E(\mathbf{R}, \mathbf{K})$.

Equations 2.121 and 2.122 represent the foundations of the semi-classical description of charged particle dynamics in crystals.

Summarizing what has been shown so far we can state that, if we are interested in a macroscopic view of charge transport phenomena (so that we can neglect the finite extension of the wave-packet in real and momentum space and the slow change of the shape of its envelope with time), then the quantum mechanical behavior of the electron resembles that of a classical particle following the classical Newton's laws of motion under the action of the driving force in Eq.2.120. The crystal momentum $\hbar\mathbf{K}$ plays the same role as momentum for a classical particle, but we emphasize that the crystal momentum *is not* the momentum of Bloch electrons. In fact, assuming the normalization in Eq.2.26, the expectation value of the momentum for a Bloch state is

$$\langle \mathbf{P} \rangle = -i\hbar \int_\Omega \Psi^\dagger_{n\mathbf{K}}(\mathbf{R})\nabla_\mathbf{R}\Psi_{n\mathbf{K}}(\mathbf{R})d\mathbf{R} = \hbar\mathbf{K} - i\hbar \int_\Omega u^\dagger_{n\mathbf{K}}(\mathbf{R})\nabla_\mathbf{R} u_{n\mathbf{K}}(\mathbf{R})d\mathbf{R}, \quad (2.124)$$

and the last integral involving the periodic part of the Bloch function is in general not zero.

The crystal momentum is sensitive only to external forces, whereas the actual electron momentum is changed by both internal (atomic) and external forces. According to the parabolic effective mass approximation, the crystal momentum is related to the wave-packet group velocity through a set of direction dependent (anisotropic) effective masses, that embody the effect of the crystal on the particle inertia. The effective masses are determined by the band structure through Eq.2.61.

The Eqs.2.121 are especially important because they relate the dynamical behavior of the centroids of the electrons' quantum wave-packet in real and momentum space to the carrier energy and, in particular, to the band structure of the material. As far as we can represent the quantum wave-packet as a classical point charge, these equations describe the dynamics of electrons in crystals under the action of smoothly varying external forces and can thus be used to determine the particle trajectory in phase space. Practical calculation of these trajectories is thoroughly discussed in Chapters 5 and 6.

2.5.3 Carrier scattering by a rapidly fluctuating potential

As pointed out in Section 2.4.1, any term added to the potential $U_C(\mathbf{R})$ given by the periodic arrangement of atoms contributes to a change with time in the coefficients $C_{n\mathbf{K}}(t)$ appearing in the expansion of Eq.2.66. As a result, if a wave-packet is initialized to have the largest components $C_{n\mathbf{K}}(t)$ in a small interval around the wave-vector $\mathbf{K}_0$ in band n (that is, the corresponding particle is in the state defined by the pair $(n, \mathbf{K}_0)$), and if the perturbation is small and slowly varying in space, then after some time interval it will have the largest coefficients in the neighborhood of a slightly different wave-vector $\mathbf{K}'_0$. During such an interval the wave-packet dynamics is described by the semi-classical equations of motion 2.121 derived in Section 2.5.2 and the wave-packet band index n remains constant.

In real crystals, however, the electrons are subject to internal stimuli that change rapidly in space and time on the microscopic scale. If the perturbation fluctuates rapidly over the distance scale of the wave-packet dimensions, then not only the $\mathbf{K}'_0$ can be very different from $\mathbf{K}_0$ but also the band index n may change. In other words, rapidly fluctuating perturbation potentials $U(\mathbf{R}, t)$ added to the periodic crystal potential energy

$U_C(\mathbf{R})$ can cause particle transitions between the state $(n, \mathbf{K}_0)$ and the state $(n', \mathbf{K}_0')$. These transitions are commonly called *scattering events*.

The scattering events have the utmost importance in the study of carrier transport in electron devices, since they tend to restore the equilibrium state when the device is brought out of it by external stimuli. Any realistic transport model should thus account for scattering in at least some region of the device.

As discussed in extensive and more quantitative terms in Chapters 4 and 10, the perturbations causing the transitions can be deterministic or inherently random, they can be stationary or time-dependent. Consequently, in the most general case the scattering should enter the time-dependent Schrödinger equation 2.65 in the form of an operator $\hat{H}_{sc}$ added to the crystal Hamiltonian. The scattering mechanisms described throughout this book, however, with the exception of those discussed in Section 4.4.1, are such that $\hat{H}_{sc}$ reduces to a simple scattering potential $U_{sc}(\mathbf{R}, t)$.

A detailed quantum mechanical treatment of the scattering process is in general very complicated, since it involves solution of the time-dependent Schrödinger equation, Eq.2.65, for the wave-packet representing the particle. A relatively simple rule, the *Fermi golden rule*, can be derived under suitable simplifying assumptions which allow us to express the rate of scattering between two given states, namely the average number of transitions per second occurring from the initial to the final state. Fermi's rule is discussed in several textbooks for a 3D carrier gas [4, 21]. In the following we present the derivation for a stationary scattering potential $U_{sc} = U_{sc}(\mathbf{R})$. Here our objective is to highlight the assumptions behind the Fermi golden rule, its limitations and the support that it provides in development of a complete semi-classical picture of carrier transport. More details on the Fermi golden rule in inversion layers are given in Chapter 4.

2.5.4 The Fermi golden rule

Fermi's rule expresses the probability per unit time that a particle in band n represented by a wave-packet with negligible uncertainty in the wave-vector $\mathbf{K}$ suffers a transition from such an initial state $(n, \mathbf{K})$ to a final state $(n', \mathbf{K}')$. The transition probability per unit time is called *scattering rate*.

We begin by referring to the general form of the time-dependent Schrödinger equation, Eq.2.65, anticipated in Section 2.4.1 and we assume that the perturbation of the crystal Hamiltonian $\hat{H}_C$ is given by $\delta\hat{H} = U_{sc}(\mathbf{R})$, where $U_{sc}(\mathbf{R})$ is a stationary scattering potential. The derivation of Fermi's rule for time-dependent scattering potentials $U_{sc}(\mathbf{R}, t)$ is deferred to Section 4.1.6.

The solution of Eq.2.65 with $U(\mathbf{R}) = U_{sc}(\mathbf{R})$ can be expanded in the superposition of Bloch states as in Eq.2.90, which we rewrite here for convenience

$$\Psi_t(\mathbf{R}, t) = \sum_{n,\mathbf{K}} B_{n\mathbf{K}}(t)\Psi_{n\mathbf{K}}(\mathbf{R}), \tag{2.125}$$

where the $\Psi_{n\mathbf{k}}(\mathbf{R})$ are the stationary part of the eigenfunctions of $\hat{H}_C$, that is the Bloch wave-functions for the band n and the wave-vector $\mathbf{K}$, and we concisely set

$B_{n\mathbf{K}}(t) = C_{n\mathbf{K}}(t) \exp\left(-\mathrm{i}E_{B,n}(\mathbf{K})t/\hbar\right)$. The $\Psi_{n\mathbf{K}}(\mathbf{R})$ and the corresponding energies $E_{B,n}(\mathbf{K})$ satisfy the single-particle Schrödinger equation, Eq.2.24.

By inserting Eq.2.125 into Eq.2.65, projecting on the generic eigenfunction $\Psi_{n'\mathbf{K}'}$, and using the fact that the Bloch functions $\Psi_{n\mathbf{K}}$ are orthogonal and normalized, we readily obtain

$$\frac{\partial B_{n'\mathbf{K}'}(t)}{\partial t} = -\frac{\mathrm{i}}{\hbar}\left\{\sum_{n\mathbf{K}} M_{nn'}(\mathbf{K},\mathbf{K}')B_{n\mathbf{K}}(t) + E_{B,n'}(\mathbf{K}')\, B_{n'\mathbf{K}'}(t)\right\}. \tag{2.126}$$

Equation 2.126 represents a set of coupled differential equations for the coefficients $B_{n\mathbf{K}}(t)$, which in turn give a complete description of the wave-function $\Psi_t(\mathbf{R},t)$. The *matrix elements* of the scattering potential energy term $U_{sc}(\mathbf{R})$ are

$$M_{nn'}(\mathbf{K},\mathbf{K}') = \int_\Omega \Psi^\dagger_{n'\mathbf{K}'}(\mathbf{R})\, U_{sc}(\mathbf{R})\, \Psi_{n\mathbf{K}}(\mathbf{R})\, \mathrm{d}\mathbf{R}, \tag{2.127}$$

where Ω is the normalization volume. Once the matrix elements are known, Eq.2.126 can be solved numerically if the initial values of the coefficients $B_{n\mathbf{K}}(0)$ at time $t = 0$ are known [18].

In particular, if $U_{sc}(\mathbf{R})$ and $M_{nn'}(\mathbf{K},\mathbf{K}')$ are null, Eq.2.126 provides the well known result

$$B_{n\mathbf{K}}(t) = B_{n\mathbf{K}}(0)\, \exp(-\frac{\mathrm{i}}{\hbar}E_{B,n}(\mathbf{K})\, t), \tag{2.128}$$

which corresponds to a perfectly coherent transport where the magnitude of the coefficients $B_{n\mathbf{K}}$ remains constant and only the phase changes with time.

In the search for an analytical solution we assume that at $t = 0$ the wave-packet consists of only one non-null coefficient in Eq.2.125. If we indicate with $(n_0, \mathbf{K}_0)$ the state which is occupied at time $t = 0$, the initial conditions for Eq.2.126 can be written as

$$B_{n'\mathbf{K}'}(0) = \begin{cases} 1 & \text{for} \quad (\mathrm{n}', \mathbf{K}') = (\mathrm{n}_0, \mathbf{K}_0) \\ 0 & \text{otherwise} \end{cases}. \tag{2.129}$$

We now suppose that the scattering is very weak, so that the magnitude of $B_{n_0\mathbf{K}_0}$ does not appreciably reduce during the transient. In such a case $B_{n_0\mathbf{K}_0}$ evolves approximately as predicted by Eq.2.128 and we have

$$B_{n_0\mathbf{K}_0}(t) \approx \exp(-\frac{\mathrm{i}}{\hbar}E_{B,n_0}(\mathbf{K}_0)\, t). \tag{2.130}$$

Furthermore, $|B_{n'\mathbf{K}'}| \ll |B_{n_0\mathbf{K}_0}|$ for all the $(n', \mathbf{K}') \neq (n_0, \mathbf{K}_0)$. Consequently, if the perturbation is weak we can neglect all the terms other than $(n_0, \mathbf{K}_0)$ in the sum over $(n, \mathbf{K})$ of Eq.2.126 and a drastic simplification is achieved. In fact, by using Eq.2.130 for $B_{n_0\mathbf{K}_0}(t)$, for any $(n', \mathbf{K}') \neq (n_0, \mathbf{K}_0)$, Eq.2.126 can be rewritten as

$$\frac{\partial B_{n'\mathbf{K}'}}{\partial t} \approx -\frac{\mathrm{i}}{\hbar} M_{nn'}(\mathbf{K},\mathbf{K}')\, \exp(-\frac{\mathrm{i}}{\hbar}E_{B,n_0}(\mathbf{K}_0)\, t) - \frac{\mathrm{i}E_{B,n'}(\mathbf{K}')}{\hbar} B_{n'\mathbf{K}'}. \tag{2.131}$$

In Eq.2.131 the time evolution of the different $B_{n'\mathbf{K}'}$ is now decoupled and we simply have to solve one linear differential equation for each $B_{n'\mathbf{K}'}$. By taking into account the initial condition $B_{n'\mathbf{K}'}(0) = 0$ given by Eq.2.129, the solution of Eq.2.131 is

$$B_{n'\mathbf{K}'}(t) = -\frac{\mathrm{i}}{\hbar} M_{n_0,n'}(\mathbf{K}_0, \mathbf{K}') \exp\left(-\frac{\mathrm{i}}{\hbar} E_{B,n'}(\mathbf{K}')\, t\right)$$
$$\times \int_0^t \exp\left(-\frac{\mathrm{i}}{\hbar}[E_{B,n_0}(\mathbf{K}_0) - E_{B,n'}(\mathbf{K}')]\, t'\right) \mathrm{d}t'. \qquad (2.132)$$

In view of the assumptions made, the squared magnitude of $B_{n'\mathbf{K}'}(T)$ expresses the probability $P_{n'\mathbf{K}'}(T)$ of finding in the state $(n', \mathbf{K}')$ at time T a carrier with the initial state $(n_0, \mathbf{K}_0)$. A straightforward evaluation of the integral in the r.h.s. of Eq.2.132 gives

$$P_{n'\mathbf{K}'}(T) = |B_{n'\mathbf{K}'}(T)|^2 = \frac{|M_{n_0,n'}(\mathbf{K}_0, \mathbf{K}')|^2}{\hbar^2} \left[\frac{\sin(T\, \Delta E/2\hbar)}{\Delta E/2\hbar}\right]^2, \qquad (2.133)$$

where $\Delta E = E_{B,n_0}(\mathbf{K}_0) - E_{B,n'}(\mathbf{K}')$. Equation 2.133 shows that $P_{n'\mathbf{K}'}$ is an oscillating function of time governed by the quantity $(T\, \Delta E/2\hbar)$.

The net transfer rate $S_{n_0,n'}(\mathbf{K}_0, \mathbf{K}')$ from state $(n_0, \mathbf{K}_0)$ to state $(n', \mathbf{K}')$ may be implicitly defined by writing $P_{n'\mathbf{K}'}$ as

$$P_{n'\mathbf{K}'}(T) = T \cdot S_{n_0,n'}(\mathbf{K}_0, \mathbf{K}'), \qquad (2.134)$$

where $S_{n_0,n'}(\mathbf{K}_0, \mathbf{K}')$ has the physical units of time^{-1}. By using Eqs.2.133 and 2.134 we can write

$$S_{n_0,n'}(\mathbf{K}_0, \mathbf{K}') = \frac{|M_{n_0,n'}(\mathbf{K}_0, \mathbf{K}')|^2}{\hbar^2} F(\Delta E, T), \qquad (2.135)$$

where the function $F(\Delta E, T)$ is defined as

$$F(\Delta E, T) = T \left[\frac{\sin(T\, \Delta E/2\hbar)}{T\, \Delta E/2\hbar}\right]^2. \qquad (2.136)$$

Figure 2.13 shows that, for increasing T values, $F(\Delta E, T)$ becomes more and more peaked around $\Delta E = 0$. Furthermore, it can be shown that the integral of $F(\Delta E, T)$

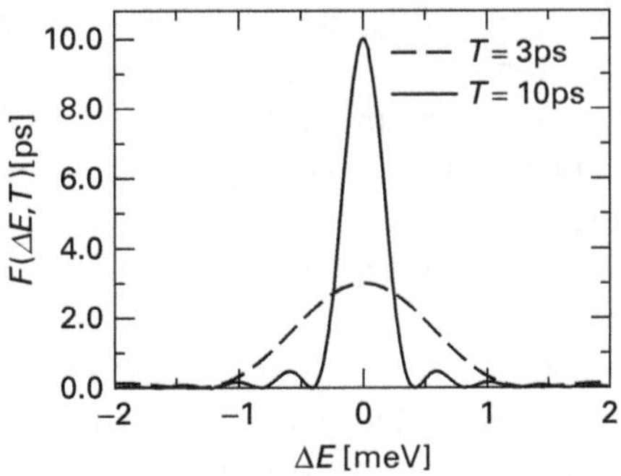

Figure 2.13 Plot of the function $F(\Delta E, T)$ defined in Eq.2.136 for two T values. As T increases the function becomes more peaked around $\Delta E = 0$.

over E is $2\pi\hbar$ independently of T. Hence for very large T values $F(\Delta E, T)$ is proportional to the Dirac function [4]. More precisely, we have

$$\lim_{T\longrightarrow\infty} F(\Delta E, T) = 2\pi\hbar\,\delta(\Delta E). \tag{2.137}$$

Equation 2.137 finally allows us to express the scattering rate, that is the transfer rate produced by the stationary scattering potential $U_{sc}(\mathbf{R})$. By going back to the generic symbol $(n,\mathbf{K})$ for the initial state, the scattering rate can be written as

$$S_{n,n'}(\mathbf{K},\mathbf{K}') = \frac{2\pi}{\hbar}\,|M_{n,n'}(\mathbf{K},\mathbf{K}')|^2\,\delta[E_{B,n}(\mathbf{K}) - E_{B,n'}(\mathbf{K}')]. \tag{2.138}$$

Equation 2.138 is known as the *Fermi golden rule* for a stationary scattering potential. Following arguments such as those above, expressions similar to Eq.2.138 can be derived also for time dependent scattering potentials and even for cases where the perturbation Hamiltonian does not reduce to a simple potential energy term. These generalizations of Eq.2.138 are developed in Chapter 4 with reference to electron and hole inversion layers.

2.5.5 Semi-classical electron transport

The framework for device modeling developed throughout this book is essentially based on the equations of motion (Eq.2.121) and on the Fermi golden rule for calculation of scattering rates (Eq.2.138 and its extensions in Chapter 4). Hence it is very important to clarify the physical meaning of these equations, to state their limits of validity and to understand how a complete picture of carrier transport, including slowly as well as rapidly varying perturbations of the crystal potential, can be built upon them.

We start this discussion from Eq.2.138, observing that the scattering rate between two states is governed by the Dirac function and by the matrix element $M_{n,n'}(\mathbf{K},\mathbf{K}')$, which in turn depends on the scattering potential and on the eigenfunctions $\Psi_{n\mathbf{K}}(\mathbf{R})$. The relation of the matrix element to the scattering potential and wave-functions is analyzed in Chapter 4 for many practically relevant cases. The meaning of the Dirac function instead deserves some immediate discussion.

Due to the term $\delta[E_{B,n}(\mathbf{K}) - E_{B,n'}(\mathbf{K}')]$ in Eq.2.138, scattering can only occur between states conserving energy. The Dirac function derives from the fact that, in the transient produced by the scattering potential $U_{sc}(\mathbf{R})$, we took the limit for a very large time T. Equation 2.136 and Fig.2.13 show that, even by considering only the first lobe of the function $F(\Delta E, T)$, for any finite T value we have non-null transition rates towards all the states with an energy $E_{B,n'}(\mathbf{K}')$ such that $|\Delta E| = |E_{B,n}(\mathbf{K}) - E_{B,n'}(\mathbf{K}')|$ is smaller than $(2\pi\hbar)/T$.

The above considerations clarify that, if we consider a sample carrier undergoing several scattering events, then it is fully justified to use Fermi's rule to describe the scattering rates only as long as the average time between two scattering events is relatively long. More precisely, if we now let T denote the time between two scattering events, then T should be long enough to make $(2\pi\hbar)/T$ practically negligible on the energy scale of the problem at hand. To take a practical example illustrating the order of magnitude of the relevant figures, it is easy to see that if we want to have, e.g., $|\Delta E| < 1$ meV, we need to take T larger than about 4 ps. Clearly, in a regime of very high scattering rates the approximation leading to energy conservation is, strictly speaking, no longer fulfilled.

The ability to calculate the scattering rates given by Fermi's rule makes it possible to include in the transport model those rapidly fluctuating perturbations that cannot be described by Eqs.2.121. In fact, if the rate of transitions is small, then the duration of the particle trajectory under the action of the external forces (the so-called *free-flight*) is large and we can condense the effect of the rapidly fluctuating perturbations in a scattering event that is practically instantaneous on the time scale of free-flight. Consequently the scattering does not change the carrier position but simply causes a sudden modification of its wave-vector and, possibly, of its energy and of the band the carrier belongs to.

It is then legitimate to treat the particle dynamics as an alternating sequence of free-flights with a finite duration and instantaneous collisions. The former describe the action of the external forces on the wave-packet according to the semi-classical equations of motion 2.121; the latter account for the rapid fluctuations of internal potentials and are treated as instantaneous events that transfer the carriers from a state $(n, \mathbf{K})$ to a different state $(n', \mathbf{K}')$ without changing the position in real space. Moreover, if the scattering potential $U_{sc}(\mathbf{R})$ is stationary, then energy conservation is enforced both during the free-flight and during the scattering event. Note that Eq.2.121 assumes the validity of Eq.2.67, which states that all components of the wave-packet belong to the same branch of the dispersion relation. Hence, the band index n is a constant of motion during the free-flight.

The transport model thus developed is semi-classical in the sense that it combines elements of classical physics (namely the idea that both position and momentum can be specified simultaneously for each particle) with a selected number of quantum mechanical ingredients entering Eq.2.121 in the calculation of the total particle energy (and in particular of the kinetic energy), and Eq.2.138 through the matrix elements.

The above discussion underlines that electronic transitions between bands that cause a simultaneous shift of the wave-packet centroid in real space that is large compared to the wave-packet dimensions are not part of the transport picture outlined so far and should be separately introduced in the transport model [22]. This is the case, for instance, for the intra-band tunneling processes which can occur at the source barrier of nanoscale MOSFETs or for the band-to-band tunneling events which are expected to take place at the drain end of the channel of short devices fabricated with low band gap channel materials, and at the source side of tunnel-FETs.

Another physical situation that can be dealt with in the semi-classical model only by resorting to empirical rules occurs when the wave-packet occupies regions of $\mathbf{K}$-space where two branches of $E_{B,n}(\mathbf{K})$ cross each other. Under these circumstances, arbitrarily small and slowly fluctuating perturbations can still produce a change in the band index that is not considered in the semi-classical model outlined so far, as can be seen by using a fully quantistic treatment of the wave-packet dynamics [4, 18, 19].

2.6 Summary

This chapter has outlined the basic concepts behind the semi-classical model of carrier transport in bulk semiconductors with the aim of paving the way to the treatment of carriers in inversion layers, which will be elucidated further in Chapters 3, 4, and 5.

To this purpose we have examined the properties of the solutions of the stationary Schrödinger equation in perfect crystals, namely Bloch waves. We have recognized that the Bloch waves of distinct electron states are unambiguously identified by a band index n and a wave-vector $\mathbf{K}$ belonging to the first Brillouin zone of reciprocal space. The dispersion relation for the Bloch waves is known as the band structure of the crystal and, for a macroscopic sample of semiconductor, it is a continuous multi-valued function of $\mathbf{K}$.

We have then developed the concept of a wave-packet, that is the weighted sum of stationary electron states with possibly time-dependent coefficients. A suitable choice of weights allows us to describe in a simple and straightforward manner quantum entities with limited uncertainty in real and wave-vector space, thus resembling classical point charges. In general, due to the non-linearity of the energy relation, the quantum wave-packet tends to smear out with increasing time. However, we have seen that in sufficiently short time intervals the wave-packet appears to translate almost rigidly with an apparent velocity, the group velocity, given by $\mathbf{V}_g = \nabla_{\mathbf{K}} E_{B,n}(\mathbf{K})/\hbar$. The definition of the group velocity establishes the first of a few important links between the band structure of the material and the transport properties of the charged carriers.

The effective mass approximation allowed us to understand that wave-packets with limited uncertainty in the wave-vector behave similarly to classical particles, in that the crystal momentum $\hbar\mathbf{K}$ obeys the classical equations of motion. The particle inertial mass is replaced by a matrix of effective, direction dependent masses related to the curvature of the bands in the proximity of the extremes. The effective masses embody the effect of the periodic crystal potential on the quasi-free carriers occupying conduction and valence band states in the crystal. They represent the second important element of the semi-classical model which is directly related to the band structure.

The semi-classical equations of motion 2.121 are only valid if all the components of the wave-packet belong to the same branch of the dispersion relation. Hence, the band index n is a constant of motion and it can only change as a consequence of transitions due to rapidly fluctuating scattering potentials. In the presence of the above processes, the reduced expansion in Eq.2.67 is no longer valid (it is restricted to a single branch of $E_{B,n}(\mathbf{K})$), because of transitions of the wave-packet between different bands. Nevertheless, thanks to the Fermi golden rule it is possible to relate the transition probability (scattering rate) to the electron states of the unperturbed crystal through the scattering matrix elements.

At the end of this necessarily condensed journey through the foundations of the semi-classical transport model it appears that the band structure is a crucial ingredient of the overall picture, not only because it defines the available electron states, but also because important properties of electron dynamics can be derived from it. In the next chapters we generalize these concepts to electrons in inversion layers, starting with calculation of the energy relation and with the objective of developing a transport model suitable for a physically sound and comprehensive analysis of modern nano-scale MOS transistors.

References

[1] N. Ashkroft and N. Mermin, *Solid State Physics*. Philadelphia, PA: Sounders College, 1976.

[2] W.A. Harrison, *Solid State Theory*. New York, NY: Dover, 1980.

[3] R. Martin, *Electronic Structure: Basic Theory and Practical Methods*. Cambridge: Cambridge University Press, 2004.

[4] S. Datta, *Quantum Phenomena*. Addison-Wesley Modular Series in Solid State Devices, New York: Addison-Wesley, 1989.

[5] J.R. Chelikowsky and M.L. Cohen, "Nonlocal pseudopotential calculations for the electronic structure of eleven diamond and zinc-blende semiconductors," *Phys. Rev. B*, vol. 14, no. 2, pp. 556–582, 1976.

[6] M.M. Rieger and P. Vogl, "Electronic-band parameters in strained $Si_{1-x}Ge_x$ alloys on $Si_{1-y}Ge_y$ substrates," *Phys. Rev. B*, vol. 48, no. 19, pp. 14276–14287, 1993.

[7] L.R. Saravia and D. Brust, "Spin splitting and the ultraviolet absorption of Ge," *Phys. Rev.*, vol. 176, p. 915, 1968.

[8] E. Kane, "Energy band structure in P-type germanium and silicon," *J. Phys. Chem. Solids.*, vol. 1, pp. 82–99, 1956.

[9] J.M. Hinckley and J. Singh, "Influence of substrate composition and crystallographic orientation on the band structure of pseudomorphic Si-Ge alloy films," *Phys. Rev. B*, vol. 42, no. 6, p. 3546, 1990.

[10] T. Manku and A. Nathan, "Valence energy-band structure for strained group-IV semiconductors," *Journal of Applied Physics*, vol. 73, no. 3, pp. 1205–1213, 1993.

[11] M.V. Fischetti and S.E. Laux, "Band structure, deformation potentials, and carrier mobility in strained Si, Ge, and SiGe alloys," *Journal of Applied Physics*, vol. 80, p. 2234, 1996.

[12] P. Lawaetz, "Valence-band parameters in cubic semiconductors," *Phys. Rev. B*, vol. 4, pp. 3460–3467, Nov 1971.

[13] M. Cardona and F.H. Pollak, "Energy-band structure of germanium and silicon: The $\mathbf{k}\cdot\mathbf{p}$ method," *Phys. Rev.*, vol. 142, pp. 530–543, Feb 1966.

[14] T.B. Bahder, "Eight-band $\mathbf{k}\cdot\mathbf{p}$ model of strained zinc-blende crystals," *Phys. Rev. B*, vol. 41, pp. 11992–12001, Jun 1990.

[15] D. Rideau, M. Feraille, L. Ciampolini, *et al.*, "Strained Si, Ge, and $Si_{1-x}Ge_x$ alloys modeled with a first-principles-optimized full-zone $\mathbf{k}\cdot\mathbf{p}$ method," *Phys. Rev. B*, vol. 74, no. 19, pp. 195208–195228, 2006.

[16] G. Dresselhaus, A.F. Kip, and C. Kittel, "Cyclotron resonance of electrons and holes in silicon and germanium crystals," *Phys. Rev.*, vol. 98, p. 368, 1955.

[17] C. Jacoboni and L. Reggiani, "The Monte Carlo method for the solution of charge transport in semiconductors with applications to covalent materials," *Rev. Mod. Phys.*, vol. 55, pp. 645–705, 1983.

[18] D. Esseni and P. Palestri, "Theory of the motion at the band crossing points in bulk semiconductor crystals and in inversion layers," *Journal of Applied Physics*, vol. 105, no. 5, pp. 053702-1–053702-11, 2009.

[19] J.M. Luttinger and W. Kohn, "Motion of electrons and holes in perturbed fields," *Phys. Rev.*, vol. 97, pp. 869–883, 1955.
[20] J. Singh, *Quantum Mechanics: Fundamentals and Applications to Technology*. New York: Wiley Interscience, 1997.
[21] M. Lundstrom, *Fundamentals of Carrier Transport*. New York: Addison-Wesley, 1990.
[22] K. Hess, *Advanced Theory of Semiconductor Devices*. Piscataway, NJ: IEEE Press, 1999.

3 Quantum confined inversion layers

The working principle of MOS transistors requires formation of an extremely thin layer of carriers at the semiconductor-oxide interface. This is obtained by means of an appropriate gate bias, which produces a narrow minimum of the carrier potential energy at such an interface, resulting in an inversion layer with significant quantization effects.

The carriers in such inversion layers are essentially free to move only in the plane parallel to the silicon-dielectric interface. The allowed energy states stemming from either the conduction or the valence band of the underlying crystal form subbands, inside which the energy depends on a quasi-continuum two-dimensional wave-vector $\mathbf{k}$, so that the carriers are said to form a two-dimensional (2D) gas. The energy dispersion in the 2D subbands is very important to describe the electrostatics and the transport in MOS transistors and it depends both on the confining potential and on the characteristics of the underlying crystal.

This chapter presents the fundamental concepts and the models to determine the energy relation and the carrier densities at equilibrium in the inversion layer of both n-type and p-type MOS transistors.

After introducing the basic concepts related to subband quantization in Section 3.1 by using a pedagogical example, Section 3.2 discusses the application to an electron inversion layer of the effective mass approximation (EMA) approach (see Section 2.4). Then Section 3.3 presents the extension to hole inversion layers of the $\mathbf{k}\cdot\mathbf{p}$ model introduced in Section 2.2.2; a simplified semi-analytical model for the hole energy relation is also discussed. Section 3.4 presents a full-band model for the energy dispersion in inversion layers, which goes beyond the EMA and $\mathbf{k}\cdot\mathbf{p}$ models and can be considered an extension to the 2D carrier gas of the pseudo-potential method presented in Section 2.2.1. In all cases the computational complexity of the resulting eigenvalue problems is discussed.

In Section 3.5 we then discuss how to evaluate the sums over the quasi-continuum wave-vector $\mathbf{k}$ that appear in the calculation of very many physical quantities, as readily illustrated in Section 3.6 which describes how to calculate the carrier densities self-consistently with the electrostatic potential at equilibrium.

The concepts, the models, and the notation introduced in this chapter set the stage for most of the material presented in the following chapters.

3.1 Electrons in a square well

Let us consider the simple problem where an electron is confined by a potential energy $U(z)$ that varies only in the z direction. The stationary Schrödinger equation can be written as

$$\frac{-\hbar^2}{2m_0}\nabla^2_{\mathbf{R}}\Phi_n(\mathbf{R}) + U(z)\,\Phi_n(\mathbf{R}) = E_n\,\Phi_n(\mathbf{R}), \tag{3.1}$$

where m_0 is the rest electron mass while Φ_n and E_n denote the nth eigenfunction and eigenvalue of the system, respectively. $\mathbf{R} = (x, y, z)$ is the 3D position vector.

Since the confining potential does not depend on the in-plane coordinates $\mathbf{r} = (x, y)$, we make the *ansatz* for the form of the eigenfunction

$$\Phi_n(\mathbf{R}) = \xi_n(z)\,\frac{e^{\mathrm{i}\,\mathbf{k}\cdot\mathbf{r}}}{\sqrt{A}}, \tag{3.2}$$

where A is a normalization area in the transport plane (x, y) and $\mathbf{k} = (k_x, k_y)$ is the two-dimensional wave-vector.

The possible values for $\mathbf{k}$ depend on the boundary conditions imposed on the system in the transport plane. If we suppose that A is large and we are not interested in the properties at the edges of the area A, we can apply to $\Phi_n(\mathbf{R})$ periodic boundary conditions with respect to $\mathbf{r}$ by setting

$$e^{-\mathrm{i}\,0.5\,k_x L} = e^{\mathrm{i}\,0.5\,k_x L}, \qquad e^{-\mathrm{i}\,0.5\,k_y L} = e^{\mathrm{i}\,0.5\,k_y L}.$$

Given the conceptual nature of the area A, we have here assumed with no loss of generality that A is a square whose sides are $L = \sqrt{A}$.

Under these circumstances the possible values for $\mathbf{k}$ are

$$k_x = n\,\frac{2\pi}{L} = n\,\frac{2\pi}{\sqrt{A}}, \quad k_y = m\,\frac{2\pi}{L} = m\,\frac{2\pi}{\sqrt{A}}, \qquad n, m = 0, \pm 1, \pm 2\ \cdots \tag{3.3}$$

The periodic boundary conditions imposed on $\Phi_n(\mathbf{R})$ are analogous to the Born–von Karman boundary conditions imposed by Eq.2.28 for a three-dimensional bulk crystal. By substituting Eq.3.2 in Eq.3.1 we obtain the eigenvalue problem for $\xi_n(z)$

$$\frac{-\hbar^2}{2m_0}\frac{\partial^2\xi_n(z)}{\partial z^2} + U(z)\,\xi_n(z) = \varepsilon_n\,\xi_n(z), \tag{3.4}$$

where ε_n is related to the eigenvalue E_n of Eq.3.1 by

$$E_n = \varepsilon_n + \frac{\hbar^2}{2m_0}(k_x^2 + k_y^2) = \varepsilon_n + \frac{\hbar^2}{2m_0}k^2, \tag{3.5}$$

where k is the magnitude of the wave-vector $\mathbf{k}$, namely $k = |\mathbf{k}|$. The wave-function $\xi_n(z)$ is normalized to one in the z direction.

Let us now suppose that the confining potential is a simple square well such as the one sketched in Fig.3.1, where we have conventionally assumed that $U(z)$ is null inside the well. The width of the well and the potential energy barrier height are

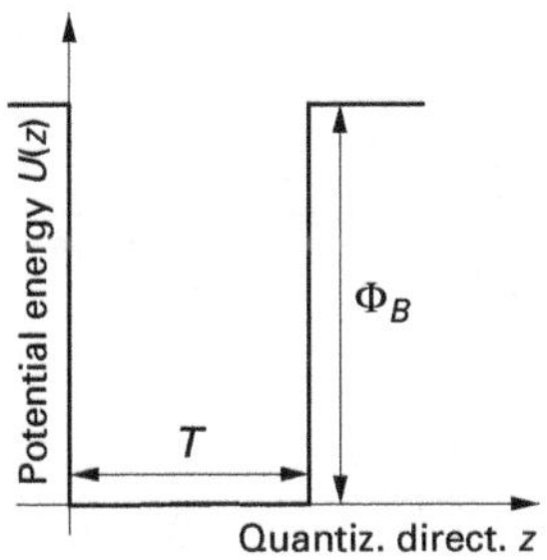

Figure 3.1 Square quantum well: T and Φ_B are respectively the width and the barrier height of the well.

denoted by T and Φ_B, respectively. In such a case Eq.3.4 simply states that inside the well the second derivative of $\xi(z)$ is proportional to $\xi(z)$. The solution will thus be given by an appropriate linear combination of sine and cosine functions, which are unequivocally determined by the boundary conditions. In particular, for a quantum well with an infinitely large energy barrier Φ_B, the wave-function is null at the well boundaries, namely $\xi(0) = \xi(T) = 0$, so that the normalized eigenfunctions of Eq.3.4 are

$$\xi_n(z) = \sqrt{\frac{2}{T}} \sin(k_n z), \qquad k_n = \frac{n\pi}{T}, \qquad n = 1, 2 \cdots \tag{3.6}$$

The corresponding eigenvalues are

$$\varepsilon_n = \frac{\hbar^2 k_n^2}{2m_0} = \frac{\hbar^2 (n\pi)^2}{2m_0 T^2}, \tag{3.7}$$

and the total energy E_n is obtained from Eq.3.5 as

$$E_n = \frac{\hbar^2 (n\pi)^2}{2m_0 T^2} + \frac{\hbar^2}{2m_0} k^2. \tag{3.8}$$

As can be seen, the energy dispersion in Eq.3.8 consists of subbands identified by the index n, whose minima ε_n are spaced according to the inverse square of the well width T and the inverse of the electron mass m_0. Inside each subband the energy has a simple analytical dependence on the wave-vector magnitude $k = |\mathbf{k}|$.

3.2 Electron inversion layers

The carrier transport in nMOS transistors is due to the electrons in the inversion layer of a p-type silicon substrate or silicon film. Thus calculation of the band structure in the inversion layer requires us to tackle a Schrödinger problem that includes both the crystal potential and the confining potential at the silicon-oxide interface. The resulting problem is theoretically complex and computationally very demanding (see also Section 3.4). In this section we illustrate a simple quantization model based on the concept of the equivalent Hamiltonian introduced in Section 2.4; such a model is very widely used for analysis of electron devices.

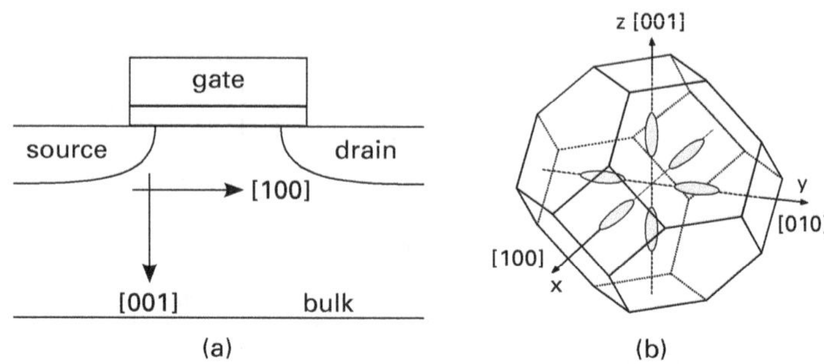

Figure 3.2 (a) A Si (001)/[100] n-MOSFET, that is a silicon MOS transistor realized in a (001) silicon substrate where the source to drain direction is the [100] crystallographic direction; (b) six equivalent Δ minima of the silicon conduction band and the corresponding energy ellipsoids.

3.2.1 Equivalent Hamiltonian for electron inversion layers

In order to introduce the effective mass approximation (EMA) quantization model, let us first consider the (001) silicon n-MOSFET sketched in Fig.3.2.a, where the source to drain direction is [100]. We use throughout the book the notation Si (001)/ [100] to concisely indicate the channel material, interface and transport direction. In this case the quantization, transport and device width directions coincide with the crystallographic axes. Figure 3.2.b shows the six equivalent Δ minima of the silicon conduction band, which are located along the crystallographic axes (see Section 2.3.1 and Fig.2.9).

According to the equivalent Hamiltonian model discussed in Section 2.4.2, for the electrons that occupy states in the bulk silicon close to the conduction band minimum ν, the energies $E_{\nu,n}$ available in the inversion layer are given by

$$E_{\nu,n} = (E'_{\nu,n} + E_{\nu 0}), \tag{3.9}$$

where $E_{\nu 0}$ is the energy value at the conduction band minimum ν, while $E'_{\nu,n}$ and the envelope wave-function $\Phi_{\nu,n}(\mathbf{R})$ are obtained by solving the stationary Schrödinger equation

$$[\hat{E}_{cb}^{(\nu)}(-i\nabla_{\mathbf{R}}) + U(z)]\,\Phi_{\nu,n}(\mathbf{R}) = E'_{\nu,n}\,\Phi_{\nu,n}(\mathbf{R}), \tag{3.10}$$

where $\hat{E}_{cb}^{(\nu)}(-\mathrm{i}\nabla_{\mathbf{R}})$ is the operator satisfying Eq.2.71 and $U(z)$ is the confining potential.

By recalling Eq.2.84 we see that the complete wave-function for a stationary problem can be expressed as

$$\Psi_{\nu,n}(\mathbf{R}) = \Phi_{\nu,n}(\mathbf{R})\,u_{\nu}(\mathbf{R})\,e^{\mathrm{i}\,\mathbf{K}_{\nu}\cdot\mathbf{R}}, \tag{3.11}$$

where $\mathbf{K}_{\nu}$ and $u_{\nu}(\mathbf{R})$ are respectively the value of the wave-vector and the periodic part of the Bloch function at the conduction band minimum ν. The normalization of the complete wave-function $\Psi_{\nu,n}(\mathbf{R})$ is discussed in Section 4.1.4.

3.2.2 Parabolic effective mass approximation

The solution of Eq.3.10 can be either relatively straightforward or very complex depending on the form of the operator $\hat{E}_{cb}^{(\nu)}(-i\nabla_{\mathbf{R}})$, which in turn depends on the form used for the energy dispersion $E_{cb}^{(\nu)}(\mathbf{K})$ in the vicinity of the minimum ν; it should always be remembered that $\mathbf{K}$ here denotes the wave-vector referred to the wave-vector $\mathbf{K}_\nu$ at the minimum. In this respect, we have seen in Section 2.3.1 that, for wave-vectors close enough to a minimum of the conduction band (namely if $\mathbf{K}$ is small enough), the energy dispersion can be approximated by a revolutionary ellipsoid with longitudinal effective mass m_l and transverse effective mass m_t.

If we now consider one of the six equivalent minima shown in Fig.3.2.b, then by recalling Eq.2.86 we can write the corresponding energy dispersion close to the minimum as

$$E_{cb}^{(\nu)}(\mathbf{K}) = \frac{\hbar^2}{2}\left(\frac{K_x^2}{m_x} + \frac{K_y^2}{m_y} + \frac{K_z^2}{m_z}\right), \tag{3.12}$$

where m_x, m_y, and m_z can be either m_l or m_t, as shown in Table 3.1. The corresponding Schrödinger equation derived from Eq.3.10 is

$$\left[-\frac{\hbar^2}{2}\left(\frac{1}{m_x}\frac{\partial^2}{\partial x^2} + \frac{1}{m_y}\frac{\partial^2}{\partial y^2} + \frac{1}{m_z}\frac{\partial^2}{\partial z^2}\right) + U(z)\right]\Phi_{\nu,n}(\mathbf{R}) = E'_{\nu,n}\,\Phi_{\nu,n}(\mathbf{R}), \tag{3.13}$$

where $E'_{\nu,n}$ must be inserted into Eq.3.9 to express the energies $E_{\nu,n}$ available in the inversion layer.

Equation 3.13 has been obtained from the parabolic approximation of the energy dispersion. Thus this quantization model is known as the parabolic EMA model and is very widely used in the analysis of electron devices. We now recall that the quantization direction is z and note that Eq.3.13 is mathematically very similar to Eq.3.1. We thus again use the *ansatz*

$$\Phi_{\nu,n}(\mathbf{R}) = \xi_{\nu,n}(z)\,\frac{e^{\mathrm{i}\,\mathbf{k}\cdot\mathbf{r}}}{\sqrt{A}}, \tag{3.14}$$

and insert $\Phi_{\nu,n}(\mathbf{R})$ into Eq.3.13 to finally express the electron energies $E_{\nu,n}$ defined in Eq.3.9 as

$$E_{\nu,n}(\mathbf{k}) = E_{\nu 0} + \varepsilon_{\nu,n} + \frac{\hbar^2}{2}\left(\frac{k_x^2}{m_x} + \frac{k_y^2}{m_y}\right), \tag{3.15}$$

where $\varepsilon_{\nu,n}$ is the eigenvalue of the equation

$$\frac{-\hbar^2}{2m_z}\frac{\partial^2\xi_{\nu,n}}{\partial z^2} + U(z)\,\xi_{\nu,n}(z) = \varepsilon_{\nu,n}\,\xi_{\nu,n}(z), \tag{3.16}$$

and $|\xi_{\nu,n}(z)|^2$ is normalized to one in the z domain. The effective mass m_z plays the role of the quantization mass , namely the mass that governs the subband splitting through Eq.3.16.

Table 3.1 Valleys relevant for the electron inversion layer in an (001)/[100] n-MOSFET. Note that m_z is the quantization mass, m_x and m_y are the effective masses in the transport plane, while $m_d = \sqrt{m_x m_y}$ is the effective mass for the density of states and μ_ν is the valley degeneracy. The masses are expressed in units of the rest electron mass m_0. As discussed in the text, throughout the book the $\Delta_{0.92}$ valleys are indicated also as Δ_2 or unprimed, and the $\Delta_{0.19}$ valleys are indicated also as Δ_4 or primed.

ν	μ_ν	$m_x[\mathrm{m}_0]$	$m_y[\mathrm{m}_0]$	$m_z[\mathrm{m}_0]$	$m_d[\mathrm{m}_0]$
$\Delta_{0.92}$	2	$m_t = 0.19$	$m_t = 0.19$	$m_l = 0.92$	0.19
$\Delta_{0.19x}$	2	$m_l = 0.92$	$m_t = 0.19$	$m_t = 0.19$	0.42
$\Delta_{0.19y}$	2	$m_t = 0.19$	$m_l = 0.92$	$m_t = 0.19$	0.42

The six Δ minima of the conduction band are degenerate in bulk silicon, hence $E_{\nu 0}$ is the same for all the valleys. Furthermore, as can be seen in Fig.3.2.b and Table 3.1, the six minima can be coupled in three groups of two minima each, which have exactly the same values for the effective masses m_x, m_y, m_z. Equations 3.15 and 3.16 indicate that, according to the EMA model, the two minima in each group yield exactly the same energies in the inversion layer, so that we must attribute a valley multiplicity $\mu_\nu = 2$ to the states described by Eqs.3.15 and 3.16 (besides the further multiplicity given by the electron spin). We shall address the valley multiplicity in more detail in Section 3.5.2.

Table 3.1 summarizes the valleys stemming from the six equivalent Δ minima of silicon in a (001)/[100] n-MOSFET. The $\Delta_{0.92}$ label is used for the valleys with a quantization mass $m_z = 0.92\mathrm{m}_0$, while $\Delta_{0.19x}$ and $\Delta_{0.19y}$ are the valleys with $m_z = 0.19\mathrm{m}_0$ and with respectively m_x or m_y equal to the longitudinal mass $m_l = 0.92\mathrm{m}_0$. As discussed in Section 3.5.2, all the $\Delta_{0.19}$ valleys give the same density of states, thus the $\Delta_{0.19}$ are sometimes considered collectively and assigned a valley multiplicity $\mu_\nu = 4$.

When the different subband minima $\varepsilon_{\nu,n}$ obtained by Eq.3.16 are inserted into Eq.3.15, we readily see that the energy $E_{\nu,n}(\mathbf{k})$ in the inversion layer becomes a multi-value function of the wave-vector $\mathbf{k}$, as already discussed for the energy dispersion of a bulk crystal in Chapter 2. The energy dependence on $\mathbf{k}$ is the same for all the subbands of a given valley and it has a simple analytical expression.

The confining potential $U(z)$ in the depletion region of a bulk MOS transistor can be approximated with a triangular well close to the Si-SiO_2 interface [1, 2]. In a thin film SOI transistor or in a thin fin FinFET, instead, the silicon between the two oxides forms a quantum well that can be approximated with a square well, although the potential drop inside the well can be very significant when the inversion density is large. The solution to Eq.3.16 for a square well has been presented in Section 3.1, and approximate solutions for a triangular well based on the *Airy* functions have already been reported in the seventies [2]. New analytical expressions for the eigenvalues in a non-triangular well have also recently been proposed in [3]. By virtue of the large band discontinuity at the silicon-oxide interface, the Schrödinger equations are typically solved by imposing null values of $\xi(z)$ at the Si-SiO_2 interfaces and, in particular, most of the approximate analytical solutions are derived according to this simplification. In Section 3.2.3 we will see that it is easy to arrange a computer code that solves Eq.3.16 numerically for an arbitrary form of the confining potential energy $U(z)$.

We now note that, for a square well, Eq.3.7 states that the subband minima $\varepsilon_{\nu,n}$ for Eq.3.16 are inversely proportional to m_z. Thus the lowest available states in the inversion layer belong to the valleys labelled $\Delta_{0.92}$ in Table 3.1, which are frequently named *unprimed* in the literature [2, 4, 5]. The $\Delta_{0.19}$ subbands, often referred to as *primed* [2, 4, 5], correspond to $m_z = 0.19$ and result in larger eigenvalues $\varepsilon_{\nu,n}$ with respect to the $\Delta_{0.92}$ subbands. Section 3.6 discusses how the occupation of the electronic states is related to their energy at the equilibrium.

One final remark is necessary about the labels used to identify the valleys in the (100) silicon inversion layer. Since the $\Delta_{0.92}$ are two times degenerate, they will also frequently be referred to as Δ_2 valleys; as discussed above, these are also the lowest, hence the so called unprimed valleys, thus throughout the book we refer to the lowest valleys in the (100) silicon inversion layer by using equivalently $\Delta_{0.92}$, Δ_2 or unprimed. Consistently with this choice, the $\Delta_{0.19}$ will also be denoted Δ_4 or primed valleys. A similar convention for the naming of the valleys will also be used for transistors with different crystal orientations and alternative channel materials (see Chapters 8 and 10), as well as for strained silicon devices (see Chapter 9).

3.2.3 Implementation and computational complexity

In numerical analysis of electronic devices it is not necessary to resort to simplifying assumptions for $U(z)$ to solve Eq.3.16, in fact a numerical solution is straightforward. To this purpose, let us consider a uniformly spaced grid in the quantization direction, where $z_l = l{\cdot}\mathrm{d}z$ are the grid points and $\mathrm{d}z$ is the grid step, as sketched in Fig.3.3. The goal is now to find the eigenvalues ε and the values ξ_l of the unknown functions ξ at the points z_l. In a finite differences approach the first and second derivative of ξ at z_l are approximated as

$$\frac{\partial \xi(z_l)}{\partial z} \simeq \frac{\xi_{l+1} - \xi_{l-1}}{2\mathrm{d}z}, \qquad \frac{\partial^2 \xi(z_l)}{\partial z^2} \simeq \frac{\xi_{l+1} + \xi_{l-1} - 2\xi_l}{\mathrm{d}z^2}. \tag{3.17}$$

Since the Schrödinger equation is a linear eigenvalue differential equation, the finite differences method readily allows us to convert it to a linear algebraic eigenvalue problem, whose eigenvectors are the ξ_l values.

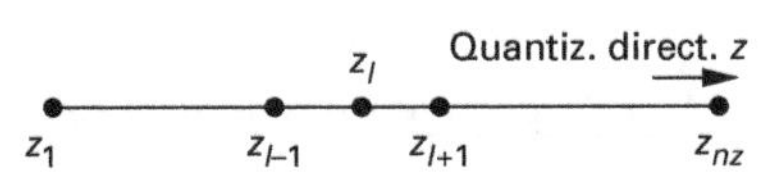

Figure 3.3 Uniformly spaced grid used for a numerical solution of the Schrödinger equation by means of the finite difference method.

More precisely, by introducing the parameter $t_0 = \hbar^2/(2m_z \mathrm{d}z^2)$, the algebraic problem can be written as

$$\begin{pmatrix} \cdot & \cdot & \cdot & \cdot & \cdot & \cdot & \cdot \\ \cdot & -t_0 & 2t_0 + U_{l-1} & -t_0 & 0 & \cdot & \cdot \\ \cdot & 0 & -t_0 & 2t_0 + U_l & -t_0 & 0 & \cdot \\ \cdot & \cdot & 0 & -t_0 & 2t_0 + U_{l+1} & -t_0 & \cdot \\ \cdot & \cdot & \cdot & \cdot & \cdot & \cdot & \cdot \end{pmatrix} \begin{pmatrix} \cdot \\ \xi_{l-1} \\ \xi_l \\ \xi_{l+1} \\ \cdot \end{pmatrix} = \varepsilon \begin{pmatrix} \cdot \\ \xi_{l-1} \\ \xi_l \\ \xi_{l+1} \\ \cdot \end{pmatrix},$$

where $U_l = U(z_l)$ is the potential at the grid points and the matrix of the problem is a tri-diagonal one. If we let n_z denote the number of discretization points, the general form of the matrix elements just shown holds for $2 \le l \le (n_z - 1)$, whereas at $l = 1$ and $l = n_z$ we have to deal with the boundary conditions. Imposing null boundary conditions $\xi_1 = \xi_{n_z} = 0$ is straightforward. In fact, since ξ_1 and ξ_{n_z} are known and null, we can simply solve the problem for the internal $(n_z - 2)$ points

$$\begin{pmatrix} 2t_0 + U_2 & -t_0 & 0 & \cdot & \cdot \\ -t_0 & 2t_0 + U_3 & -t_0 & 0 & \cdot \\ \cdot & \cdot & \cdot & \cdot & \cdot \\ \cdot & 0 & -t_0 & 2t_0 + U_{n_z-2} & -t_0 \\ \cdot & \cdot & 0 & -t_0 & 2t_0 + U_{n_z-1} \end{pmatrix} \begin{pmatrix} \xi_2 \\ \xi_3 \\ \cdot \\ \xi_{n_z-2} \\ \xi_{n_z-1} \end{pmatrix}$$

$$= \varepsilon \begin{pmatrix} \xi_2 \\ \xi_3 \\ \cdot \\ \xi_{n_z-2} \\ \xi_{n_z-1} \end{pmatrix}.$$

The rank of the eigenvalue problem is $(n_z - 2)$.

Thus, according to the EMA quantization model, for a Si (001)/ [100] n-MOS device and a given confining potential $U(z)$, we must numerically solve two eigenvalue problems, for the $\Delta_{0.92}$ and for the $\Delta_{0.19}$ valleys. After the subband minima ε have been numerically calculated, Eq.3.15 provides an analytical energy dependence on the wave-vector $\mathbf{k}$ in the transport plane.

Figure 3.4 shows the eigenvalues and the squared magnitude of the envelope wave-functions for the two lowest subbands in a (001) electron inversion layer. The absolute lowest subband is the lowest subband of the $\Delta_{0.92}$ valley, whereas the second lowest subband is typically the lowest subband of the $\Delta_{0.19}$ valley. As can be seen, the larger quantization mass of the $\Delta_{0.92}$ valley results in wave-functions more closely confined towards the silicon-oxide interface.

3.2.4 Non-parabolic effective mass approximation

Equation 3.10 can be used to calculate the energy dispersion for an electron inversion layer, also with more accurate expressions for the band structure of the bulk crystal than the parabolic approximation discussed in Section 3.2.2.

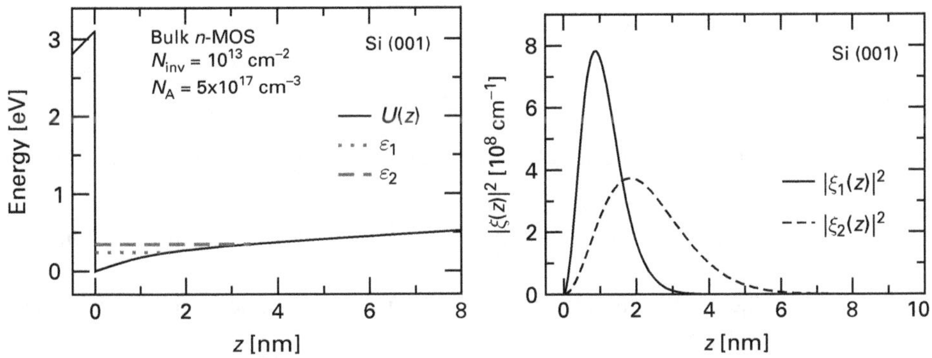

Figure 3.4 Bulk nMOS transistor with channel doping concentration $5{\cdot}10^{17}$cm^{-3}. Left plot shows the confining energy potential $U(z)$ and the ε_1 and ε_2, which are the lowest eigenvalue of the $\Delta_{0.92}$ and $\Delta_{0.19}$ valleys, respectively. Right plot shows the squared magnitude of the corresponding wave-functions $\xi_1(z)$ and $\xi_2(z)$. Results obtained with a self-consistent solution of the Schrödinger–Poisson problem (see Section 3.7), based on the parabolic effective mass approximation.

A relatively simple model for non-parabolic corrections to the energy dispersion of bulk crystals has been introduced with Eq.2.87, and reads

$$E_{cb}^{(\nu)}(\mathbf{K}) = \frac{1}{2\alpha}\left[-1+\sqrt{1+2\alpha\hbar^2\left(\frac{K_x^2}{m_x}+\frac{K_y^2}{m_y}+\frac{K_z^2}{m_z}\right)}\right], \tag{3.18}$$

where the α value typically used for the non-parabolic corrections in the silicon Δ valleys is $\alpha = 0.5$ eV^{-1} [6, 7]. We reiterate that here $\mathbf{K}$ denotes the wave-vector value referred to the $\mathbf{K}_\nu$ at the valley minimum. The corresponding operator to be used in Eq.3.10 is

$$\hat{E}_{cb}^{(\nu)}(-i\nabla_{\mathbf{R}}) = \frac{1}{2\alpha}\left[-1+\sqrt{1-2\alpha\hbar^2\left(\frac{1}{m_x}\frac{\partial^2}{\partial x^2}+\frac{1}{m_y}\frac{\partial^2}{\partial y^2}+\frac{1}{m_z}\frac{\partial^2}{\partial z^2}\right)}\right]. \tag{3.19}$$

By using the *ansatz*

$$\Phi_{\nu,n}(\mathbf{R}) = \xi_{\nu,n,\mathbf{k}}(z)\,\frac{e^{i\,\mathbf{k}\cdot\mathbf{r}}}{\sqrt{A}}, \tag{3.20}$$

the EMA operator for a 2D electron gas can be cast in the form

$$\hat{E}_{cb}^{(\nu)}\left(\mathbf{k},-\mathrm{i}\frac{\partial}{\partial z}\right) = \frac{1}{2\alpha}\left[-1+\sqrt{1+2\alpha\hbar^2\left(\frac{k_x^2}{m_x}+\frac{k_y^2}{m_y}-\frac{1}{m_z}\frac{\partial^2}{\partial z^2}\right)}\right], \tag{3.21}$$

which contains the wave-vector $\mathbf{k}$ in the transport plane as a parameter.

By inserting in Eq.3.10 the operator defined in Eq.3.21 we obtain a non-linear eigenvalue problem. It is thus convenient to introduce a further approximation. To this purpose, for any given $\mathbf{k}$ value, we can expand the energy $E_{cb}^{(\nu)}(\mathbf{K})$ to the second order in k_z and thus obtain an approximate form for the EMA operator that reads

$$\hat{E}_{cb}^{(\nu)}\left(\mathbf{k},-\mathrm{i}\frac{\partial}{\partial z}\right) \approx \frac{1}{2\alpha}\left(-1+\sqrt{1+4\alpha\gamma(\mathbf{k})}\right) - \frac{\hbar^2}{2m_z\sqrt{1+4\alpha\gamma(\mathbf{k})}}\frac{\partial^2}{\partial z^2}, \tag{3.22}$$

where we have introduced the term

$$\gamma(\mathbf{k}) = \frac{\hbar^2}{2}\left(\frac{k_x^2}{m_x} + \frac{k_y^2}{m_y}\right). \tag{3.23}$$

By using the operator defined in Eq.3.22, the eigenvalue problem can be cast in the form

$$\left\{-\frac{\hbar^2}{2m_z\sqrt{1+4\alpha\gamma(\mathbf{k})}}\frac{\partial^2}{\partial z^2} + U(z)\right\}\xi_{\nu,n,\mathbf{k}}(z) = \varepsilon_{\nu,n}(\mathbf{k})\,\xi_{\nu,n,\mathbf{k}}(z), \tag{3.24}$$

and the energy dispersion in the inversion layer, namely the eigenvalues of Eq.3.10, can be finally expressed as

$$E_{\nu,n}(\mathbf{k}) = E_{\nu 0} + \varepsilon_{\nu,n}(\mathbf{k}) + \frac{1}{2\alpha}\left(-1 + \sqrt{1+4\alpha\gamma(\mathbf{k})}\right). \tag{3.25}$$

The main difference between Eq.3.24 and Eq.3.16 is that the eigenvalue problem in Eq.3.24 depends on the wave-vector $\mathbf{k}$, so that it must be solved for any $\mathbf{k}$ value to determine the $\mathbf{k}$ dependent eigenvalues $\varepsilon_{\nu,n}(\mathbf{k})$ and wave-functions $\xi_{\nu,n,\mathbf{k}}$. This fact increases the computational complexity remarkably.

If we neglect the $\mathbf{k}$ dependence of Eq.3.24 and solve it for $\mathbf{k}= 0$, then Eq.3.24 reduces to Eq.3.16, hence the non-parabolic corrections enter the model only through the third term in Eq.3.25. Such an approach has been used frequently to account for the non-parabolic effects in quantization of electron inversion layers [4, 6, 7].

According to such a simplified non-parabolic, effective mass approximation model, the energy dispersion in the valley ν can be written as

$$E_{\nu,n}(\mathbf{k}) = E_{\nu 0} + \varepsilon_{\nu,n} + E_p(\mathbf{k}), \tag{3.26}$$

where $\varepsilon_{\nu,n}$ is obtained from Eq.3.16 and the relation between the kinetic energy E_p and the wave-vector $\mathbf{k}$ is analogous to Eq.2.87, namely

$$\frac{\hbar^2}{2}\left(\frac{k_x^2}{m_x} + \frac{k_y^2}{m_y}\right) = E_p\,(1 + \alpha\, E_p), \tag{3.27}$$

or, equivalently

$$E_p(\mathbf{k}) = \frac{1}{2\alpha}\left(-1 + \sqrt{1 + 2\,\alpha\,\hbar^2\left(\frac{k_x^2}{m_x} + \frac{k_y^2}{m_y}\right)}\right). \tag{3.28}$$

3.3 Hole inversion layers

3.3.1 k·p method in inversion layers

For the hole inversion layers a quantization model can be derived from the $\mathbf{k}\cdot\mathbf{p}$ method described in Section 2.2.2 for a bulk crystal. In particular, Eq.2.57 summarizes the algebraic eigenvalue problem corresponding to the 6×6 $\mathbf{k}\cdot\mathbf{p}$ model, that we rewrite here for convenience as

$$\left[\mathbf{H}_{\mathbf{k}\cdot\mathbf{p}} + \mathbf{H}_{so}\right]\mathbf{C}_{\mathbf{K}} = E\,\mathbf{C}_{\mathbf{K}}, \tag{3.29}$$

where $\mathbf{H}_{\mathbf{k}\cdot\mathbf{p}}$ is the $\mathbf{K}$ dependent matrix defined in Eq.2.58 in terms of the 3×3 matrix $\mathbf{H}^{3\times3}_{\mathbf{k}\cdot\mathbf{p}}$ expressed by Eq.2.55. Note that $\mathbf{H}_{so}$ is the $\mathbf{K}$ independent matrix defined in Eq.2.59 and $\mathbf{C}_{\mathbf{K}}$ is a six-component eigenvector, which, by means of Eq.2.51, unequivocally determines the eigenfunctions $u_{n\mathbf{K}}(\mathbf{R})$ (for each $\mathbf{K}$ value) in terms of the valence band Bloch functions $u_{i\mathbf{0}}(\mathbf{R})$ at the Γ point (with $i \in \{1\uparrow, 2\uparrow, 3\uparrow, 1\downarrow, 2\downarrow, 3\downarrow\}$) (see Section 2.2.2). We are here assuming a (001) hole inversion layer and z is the abscissa along the quantization direction normal to the silicon–dielectric interface.

If we now consider a confining potential $U(z)$ independent of $\mathbf{r} = (x, y)$, the band structure in the hole inversion layer can be determined by the eigenvalue differential problem derived from Eq.3.29 by introducing the operator $\hat{\mathbf{H}}_{\mathbf{k}\cdot\mathbf{p}}$ obtained from the matrix $\mathbf{H}_{\mathbf{k}\cdot\mathbf{p}}$ with the standard prescription $\mathbf{K} \rightarrow -\mathrm{i}\nabla_{\mathbf{R}}$. The set of eigenvalue differential equations reads [8]

$$[\hat{\mathbf{H}}_{\mathbf{k}\cdot\mathbf{p}}(-\mathrm{i}\nabla_{\mathbf{R}}) + \mathbf{H}_{so} + \mathbf{I}\,U(z)]\,\mathbf{\Phi}_{n\mathbf{k}}(\mathbf{R}) = E_n\,\mathbf{\Phi}_{n\mathbf{k}}(\mathbf{R}), \tag{3.30}$$

where $\hat{\mathbf{H}}_{\mathbf{k}\cdot\mathbf{p}}$ is a 6×6 differential operator, $\mathbf{I}$ is the 6×6 identity matrix and $\mathbf{\Phi}_{n\mathbf{k}}(\mathbf{R})$ is a six-component envelope wave-function.

The complete hole wave-functions in the inversion layer can be written as

$$\Psi_{n\mathbf{k}}(\mathbf{R}) = \sum_i \Phi_{n\mathbf{k},i}(\mathbf{R})\,u_{i\mathbf{0}}(\mathbf{R}), \qquad i \in \{1\uparrow, 2\uparrow, 3\uparrow, 1\downarrow, 2\downarrow, 3\downarrow\}, \tag{3.31}$$

where the $u_{i\mathbf{0}}(\mathbf{R})$ are the valence band Bloch functions at the Γ point and the $\Phi_{n\mathbf{k},i}(\mathbf{R})$ is the ith component of the vectorial envelope wave-function $\mathbf{\Phi}_{n\mathbf{k}}(\mathbf{R})$. The expression for the complete wave-function $\mathbf{\Phi}_{n\mathbf{k}}(\mathbf{R})$ is used in Section 4.1.5 for calculation of the scattering matrix elements in hole inversion layers; normalization of $\mathbf{\Phi}_{n\mathbf{k}}(\mathbf{R})$ is also addressed in the same section.

Since $U(z)$ is independent of $\mathbf{r} = (x, y)$, in order to proceed further with Eq.3.30 we make the usual *ansatz*

$$\mathbf{\Phi}_{n\mathbf{k}}(\mathbf{R}) = \boldsymbol{\xi}_{n\mathbf{k}}(z)\,\frac{e^{\mathrm{i}\,\mathbf{k}\cdot\mathbf{r}}}{\sqrt{A}}, \tag{3.32}$$

where A is the normalization area and $\boldsymbol{\xi}_{n\mathbf{k}}(z)$ is a $\mathbf{k}$ dependent six-component function of the abscissa z.

It should now be noted that, according to Eq.3.32, the x derivative of $\mathbf{\Phi}_{n\mathbf{k}}(\mathbf{R})$ can be readily evaluated to give

$$-\mathrm{i}\frac{\partial}{\partial x}\left(\boldsymbol{\xi}_{n\mathbf{k}}(z)\,\frac{e^{\mathrm{i}\,\mathbf{k}\cdot\mathbf{r}}}{\sqrt{A}}\right) = k_x\,\boldsymbol{\xi}_{n\mathbf{k}}(z)\,\frac{e^{\mathrm{i}\,\mathbf{k}\cdot\mathbf{r}}}{\sqrt{A}}, \tag{3.33}$$

and a similar relation holds for the y derivative. Thus, when we insert Eq.3.32 in Eq.3.30 and evaluate the x and y derivatives of the operator $\hat{\mathbf{H}}_{\mathbf{k}\cdot\mathbf{p}}(-\mathrm{i}\nabla_{\mathbf{R}})$ we readily obtain

$$\left[\hat{\mathbf{H}}_{\mathbf{k}\cdot\mathbf{p}}\left(\mathbf{k}, -\mathrm{i}\frac{\partial}{\partial z}\right) + \mathbf{H}_{so} + \mathbf{I}\,U(z)\right]\boldsymbol{\xi}_{n\mathbf{k}}(z) = E_n(\mathbf{k})\,\boldsymbol{\xi}_{n\mathbf{k}}(z), \tag{3.34}$$

where we have now introduced the new operator $\hat{\mathbf{H}}_{\mathbf{k}\cdot\mathbf{p}}(\mathbf{k}, -\mathrm{i}\frac{\partial}{\partial z})$, which depends on the wave-vector $\mathbf{k}$ in the transport plane and is a differential operator only with respect to the abscissa z. According to Eqs.2.55 and 2.58, the elements of $\hat{\mathbf{H}}_{\mathbf{k}\cdot\mathbf{p}}(\mathbf{k}, -\mathrm{i}\frac{\partial}{\partial z})$ are defined in terms of the elements of the operator $\hat{\mathbf{H}}^{3\times3}_{\mathbf{k}\cdot\mathbf{p}}(\mathbf{k}, -\mathrm{i}\frac{\partial}{\partial z})$, which in turn are given by

$$\begin{aligned}
\hat{H}^{3\times3}_{\mathbf{k}\cdot\mathbf{p}(1,1)} &= L\,k_x^2 + M\,k_y^2 - M\,\frac{\partial^2}{\partial z^2}, \\
\hat{H}^{3\times3}_{\mathbf{k}\cdot\mathbf{p}(2,2)} &= L\,k_y^2 + M\,k_x^2 - M\,\frac{\partial^2}{\partial z^2}, \\
\hat{H}^{3\times3}_{\mathbf{k}\cdot\mathbf{p}(3,3)} &= -L\,\frac{\partial^2}{\partial z^2} + M\,k_x^2 + M\,k_y^2, \\
\hat{H}^{3\times3}_{\mathbf{k}\cdot\mathbf{p}(1,2)} &= \hat{H}^{3\times3}_{\mathbf{k}\cdot\mathbf{p}(2,1)} = N k_x k_y, \\
\hat{H}^{3\times3}_{\mathbf{k}\cdot\mathbf{p}(1,3)} &= \hat{H}^{3\times3}_{\mathbf{k}\cdot\mathbf{p}(3,1)} = -\mathrm{i}N k_x\frac{\partial}{\partial z}, \\
\hat{H}^{3\times3}_{\mathbf{k}\cdot\mathbf{p}(2,3)} &= \hat{H}^{3\times3}_{\mathbf{k}\cdot\mathbf{p}(3,2)} = -\mathrm{i}N k_y\frac{\partial}{\partial z}.
\end{aligned} \tag{3.35}$$

The energy values $E_n(\mathbf{k})$ available for the holes in the inversion layer are obtained by solving the set of six eigenvalue differential equations written in Eq.3.34 with a matrix notation.

The differences with respect to the parabolic EMA quantization model for the electron inversion layers are remarkable. In the case of holes, in fact, we are not able to write an equation similar to Eq.3.15 which, in the case of electrons, provides a simple analytical expression for the energy dependence on the wave-vector $\mathbf{k}$. Equation 3.34, instead, contains $\mathbf{k}$ as a parameter of $\hat{\mathbf{H}}_{\mathbf{k}\cdot\mathbf{p}}$, which implies that the eigenvalue problem must be solved for all the $\mathbf{k}$ values of practical interest. The energy dispersion $E_n(\mathbf{k})$ in the hole inversion layer is thus obtained by solving Eq.3.34 for different $\mathbf{k}$ values.

In the $\mathbf{k}\cdot\mathbf{p}$ quantization model even the envelope wave-functions $\xi_{n\mathbf{k}}(z)$ depend on $\mathbf{k}$, hence we do not have one single wave-function for each subband, as for the electrons described with a parabolic EMA model. Furthermore, the wave-functions $\xi_{n\mathbf{k}}(z)$ in general take complex values, as the operators $\hat{\mathbf{H}}_{\mathbf{k}\cdot\mathbf{p}}$ and $\mathbf{H}_{so}$ contain complex terms.

We finally note that the energies in the hole inversion layer stem from a mixing of the three bulk crystal energy branches, namely the heavy, the light, and the split-off band. Hence, although the 2D subbands are sometimes labelled as heavy, light, and split-off subbands, this naming is not fully justified because of the above-mentioned mixing.

3.3.2 Implementation and computational complexity

As already seen in the case of the EMA quantization model for electrons, Eq.3.34 can be solved numerically by resorting to, for example, the finite differences method. We consider again a uniformly spaced grid in the quantization direction with points $z_l = l\,\mathrm{d}z$ and l=1, 2 $\cdots n_z$ and use Eq.3.17 to express the first and second derivative with respect to z for the six components of $\boldsymbol{\xi}(z)$. In this way we obtain an algebraic eigenvalue problem which, for any given **k**, allows us to calculate the eigenvalues E and the values $\boldsymbol{\xi}_l$ of the six-component wave-functions $\boldsymbol{\xi}(z)$ at the points z_l.

The steps necessary to identify the matrix that governs the problem are somewhat tedious but straightforward. The blocks of the matrix corresponding to the generic point l take the form

$$\begin{pmatrix} \cdot & \cdot & \cdot & \cdot & \cdot & \cdot & \cdot \\ \cdot & \mathbf{D}_- & \mathbf{D}_{l-1} & \mathbf{D}_+ & 0 & 0 & \cdot \\ \cdot & 0 & \mathbf{D}_- & \mathbf{D}_l & \mathbf{D}_+ & 0 & \cdot \\ \cdot & 0 & 0 & \mathbf{D}_- & \mathbf{D}_{l+1} & \mathbf{D}_+ & \cdot \\ \cdot & \cdot & \cdot & \cdot & \cdot & \cdot & \cdot \end{pmatrix} \begin{pmatrix} \cdot \\ \boldsymbol{\xi}_{l-1} \\ \boldsymbol{\xi}_l \\ \boldsymbol{\xi}_{l+1} \\ \cdot \end{pmatrix} = E \begin{pmatrix} \cdot \\ \boldsymbol{\xi}_{l-1} \\ \boldsymbol{\xi}_l \\ \boldsymbol{\xi}_{l+1} \\ \cdot \end{pmatrix}, \tag{3.36}$$

where the $\mathbf{D}_-$, $\mathbf{D}_l$ and $\mathbf{D}_+$ are 6×6 matrices defined as follows: $\mathbf{D}_l$ is given by

$$\mathbf{D}_l = \begin{pmatrix} \mathbf{D}' & \mathbf{0} \\ \mathbf{0} & \mathbf{D}' \end{pmatrix} + \mathbf{H}_{so} + \mathbf{I}\, U_l, \tag{3.37}$$

where the elements of the matrix $\mathbf{D}'$ are given by

$$\begin{aligned} D'_{(1,1)} &= L\,{k_x}^2 + M\,{k_y}^2 + \frac{2M}{\mathrm{d}z^2}, \\ D'_{(2,2)} &= M\,{k_x}^2 + L\,{k_y}^2 + \frac{2M}{\mathrm{d}z^2}, \\ D'_{(3,3)} &= M\,{k_x}^2 + M\,{k_y}^2 + \frac{2L}{\mathrm{d}z^2}, \\ D'_{(1,2)} &= D'_{(2,1)} = N k_x k_y, \qquad D'_{(1,3)} = D'_{(3,1)} = D'_{(2,3)} = D'_{(3,2)} = 0, \end{aligned} \tag{3.38}$$

while the matrix $\mathbf{D}_-$ is

$$\mathbf{D}_- = \begin{pmatrix} \mathbf{D}'_- & \mathbf{0} \\ \mathbf{0} & \mathbf{D}'_- \end{pmatrix}, \tag{3.39}$$

with

$$\begin{aligned} D'_{-(1,1)} &= D'_{-(2,2)} = -\frac{M}{\mathrm{d}z^2}, \qquad D'_{-(3,3)} = -\frac{L}{\mathrm{d}z^2}, \\ D'_{-(1,3)} &= D'_{-(3,1)} = \frac{\mathrm{i}N\,k_x}{2\mathrm{d}z}, \qquad D'_{-(2,3)} = D'_{-(3,2)} = \frac{\mathrm{i}N\,k_y}{2\mathrm{d}z}, \\ D'_{-(1,2)} &= D'_{-(2,1)} = 0. \end{aligned} \tag{3.40}$$

We finally have the matrix $\mathbf{D}'_+$, which is the conjugate of matrix $\mathbf{D}'_-$ (not the transpose conjugate). It is clear from the above definitions that the overall matrix is Hermitian, so that the eigenvalues are real numbers. Furthermore, for a spatial grid with n_z points, the eigenvalue problem is described by a $6n_z \times 6n_z$ block-diagonal matrix.

In analysis of the hole band structure in inversion layers it is sometimes very informative to inspect the equi-energy lines, namely the curves in the **k** plane that correspond to a given energy value. Such equi-energy lines can be calculated by rearranging the **k·p** model in order to obtain, for any given energy and direction in the transport plane,

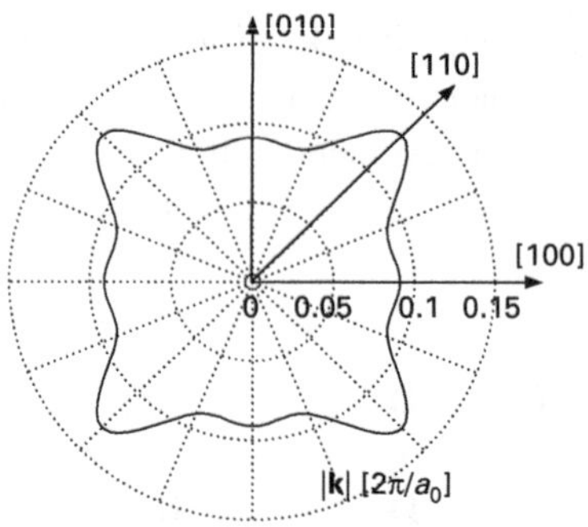

Figure 3.5 Equi-energy curve for the lowest hole subband obtained with the **k·p** method. The energy is 100 meV above the bottom of the subband minimum. Calculations performed for a triangular well with a confining field F_c=1 MV/cm. a_0 is the lattice constant.

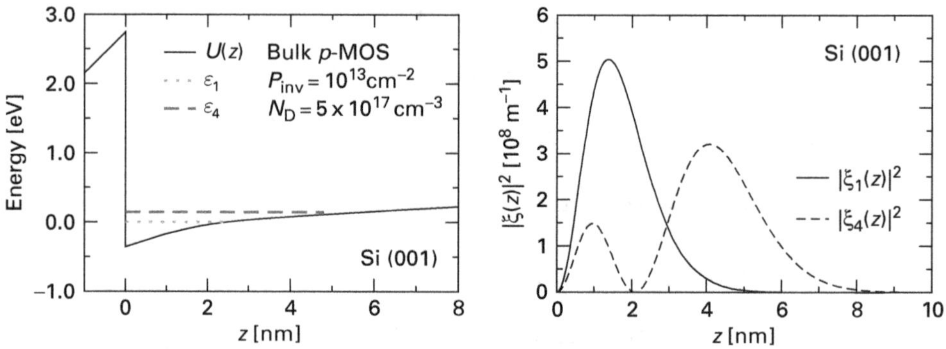

Figure 3.6 Bulk *p*MOS transistor with channel doping concentration $5{\cdot}10^{17}$ cm^{-3}. Left plot shows the confining energy potential $U(z)$ and the corresponding lowest and fourth lowest eigenvalue ε_1 and ε_4. Right plot shows the squared magnitude of the corresponding wave-functions $\boldsymbol{\xi}_1(z)$ and $\boldsymbol{\xi}_4(z)$. Results obtained with a self-consistent solution of the Schrödinger–Poisson problem (see Section 3.7), based on the **k·p** quantization model. The eigenvalues ε_1 and ε_4 correspond to the energy calculated at $\mathbf{k} = 0$.

an eigenvalue problem for the magnitude of the wave-vector. The rearrangement of the equations is not trivial and it is discussed in Appendix C.

Figure 3.5 shows the equi-energy curve of the lowest hole subband obtained with the **k·p** method and according to the procedure explained in Appendix C. For each direction in the **k** plane, the lowest subband is identified as the one resulting in the largest value of the magnitude k of the wave-vector **k**. As can be seen, the energy dispersion is strongly anisotropic. Furthermore, Fig.3.6 shows the eigenvalues and the squared magnitude of the envelope wave-functions for the lowest and the fourth lowest subband in a (001) electron inversion layer. The eigenvalues ε_1 and ε_4 are defined here as the energy calculated with the **k·p** model at $\mathbf{k} = \mathbf{0}$.

As already mentioned in Chapter 2, the hole energies are illustrated throughout the book by using an electron-like convention, so that the ground subband is the one with the lowest minimum.

3.3.3 A semi-analytical model for hole inversion layers

As already shown in Fig.3.5, the hole energy is anisotropic, and, furthermore, for a given direction in the transport plane the energy to $k = |\mathbf{k}|$ relation is also markedly non-parabolic. Thus the development of an analytical model for the hole energy dispersion in an inversion layer is a challenge. However, a semi-analytical model can be very useful in order to provide an insight into the otherwise completely numerical simulation results, and also for development of transport models and multi-subband Monte Carlo simulators, as will become clear in Chapter 6. Thus in this section we introduce a semi-analytical model for the hole energy relation [9].

Such a band structure model aims at describing the energy dispersion of the 2D holes by using three groups $\nu = g_1, g_2, g_3$ of two-dimensional subbands. The choice is suggested by the fact that in the **k·p** results for a (001) inversion layer it is possible to identify quite naturally three families of subbands [9]. Since in the inversion layers the 2D subbands cannot be unequivocally linked to one energy branch of the bulk silicon valence band, we do not label the subband groups as heavy-hole, light-hole, and split-off group, but rather use the generic labels g_1, g_2, g_3.

If we now denote with $\varepsilon_{\nu,n}$ the minimum of the nth subband of the νth group, the total energy in the band (ν,n) is expressed as

$$E_{\nu,n}(\mathbf{k}) = E_{\nu 0} + \varepsilon_{\nu,n} + E_{p,\nu}(\mathbf{k}), \tag{3.41}$$

where $E_{\nu 0}$ denotes an energy reference for the νth group of subbands, whereas $E_p(\mathbf{k})$ indicates the wave-vector dependent kinetic energy in the subband.

Since we assume an electron-like convention for the hole energy dispersion, the energy inside a subband increases above the subband minimum for non-null wave-vectors $\mathbf{k}$. As implied by Eq.3.41, in this model the kinetic energy E_p has the same dependence on $\mathbf{k}$ for all the subbands of a given group ν. Equation 3.41 is thus similar to Eq.3.15 for an electron inversion layer.

In the model, for a given confining potential $U(z)$, the eigenvalues $\varepsilon_{\nu,n}$ and the envelope wave-functions $\xi_{\nu,n}(z)$ for the subband (ν, n) are obtained by solving the Schrödinger equation corresponding to the parabolic EMA model already discussed for the electrons [2]:

$$-\frac{\hbar^2}{2m_z^{(\nu)}}\frac{\mathrm{d}^2\xi_{\nu,n}(z)}{\mathrm{d}z^2} + U(z)\,\xi_{\nu,n}(z) = \varepsilon_{\nu,n}\,\xi_{\nu,n}(z). \tag{3.42}$$

In Eq.3.42 the $m_z^{(\nu)}$ term is an effective quantization mass for the subband group ν. Each eigenvalue provided by Eq.3.42 is assumed to be two times degenerate because of the spin degeneracy. The values of $m_z^{(\nu)}$ can be obtained by best fitting with Eq.3.42 the eigenvalues obtained with the **k·p** model at $\mathbf{k} = \mathbf{0}$ for a set of square well and triangular confining potential energies, which are in fact representative of the actual potential energies in thin body SOI or bulk MOSFETs respectively [10]. The values for $m_z^{(\nu)}$ are given in Table 3.2 for the (001) crystal orientation. Figure 3.7 shows the close agreement between the squared magnitude of the envelope wave-functions obtained

Table 3.2 Band structure parameters for the silicon (001) crystal orientation: $m_z^{(\nu)}$ is the quantization mass and $E_{\nu 0}$ the energy shift of the νth group of subbands with respect to the lowest group $\nu = g_1$. For each group the parameters $a_{\nu,d}$, $b_{\nu,d}$, $c_{\nu,d}$ along the three main directions $d = 0$, $\pi/8$ and $\pi/4$ are listed.

ν	$m_z^{(\nu)}$ [m_0]	$E_{\nu 0}$ [meV]	d	$a_{\nu,d}$	$b_{\nu,d}$ [eV^{-1}]	$c_{\nu,d}$
			0	0.3077	48.0072	0.8258
g_1	0.2687	0	$\pi/8$	0.3695	248.0561	1.1991
			$\pi/4$	0.4071	110.462	0.4109
			0	0.4239	51.442	0.85
g_2	0.2249	0	$\pi/8$	1.8271	20.1326	1.0542
			$\pi/4$	0.3766	71.5079	1.9608
			0	-0.3052	−5.4514	3.74
g_3	0.2085	44	$\pi/8$	-0.4455	−0.5895	3.3847
			$\pi/4$	-0.5866	−89.5203	1.8320

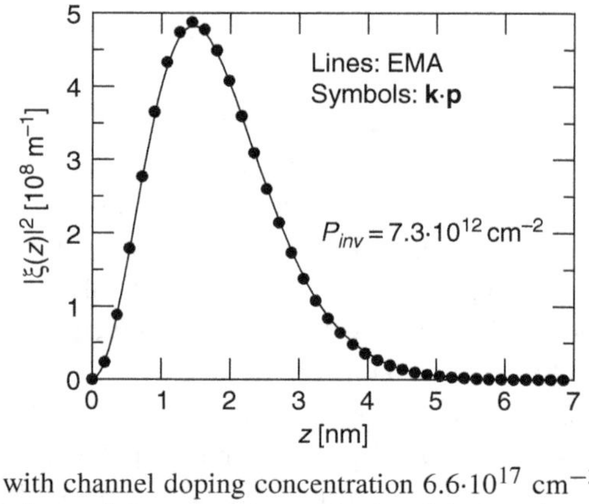

Figure 3.7 Bulk pMOS transistor with channel doping concentration $6.6 \cdot 10^{17}$ cm^{-3} and inversion density $7.3 \cdot 10^{12}$ cm^{-2}. Comparison of the squared magnitude of the envelope wave-functions corresponding to the lowest hole subbands obtained either with the $\mathbf{k}\cdot\mathbf{p}$ or with the EMA Schrödinger equation 3.42. The values of the effective quantization masses $m_z^{(\nu)}$ for the EMA Schrödinger equation are taken from Table 3.2.

either with the $\mathbf{k}\cdot\mathbf{p}$ model or with Eq.3.42. It should be noted that by using Eq.3.42 we are tacitly neglecting the dependence of the wave-function on the wave-vector $\mathbf{k}$; such an approximation is frequently embraced also in the studies based on the $\mathbf{k}\cdot\mathbf{p}$ model [8, 11], even if the $\mathbf{k}$ dependent wave-functions are available in the $\mathbf{k}\cdot\mathbf{p}$ approach.

Below we present the expressions for the kinetic energy $E_p(\mathbf{k})$ employed in the semi-analytical energy model; to this purpose, the wave-vector $\mathbf{k}$ is expressed in polar coordinates, that is by using its magnitude k and the angle θ.

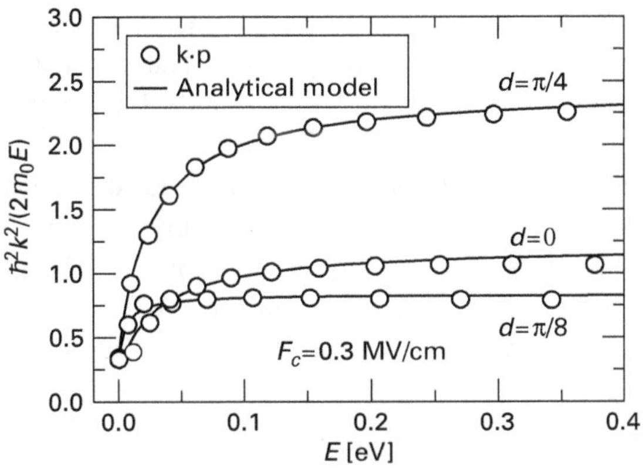

Figure 3.8 Plot of the quantity $(\hbar^2 k^2/2m_0 E)$ versus E obtained with the $\mathbf{k}\cdot\mathbf{p}$ method (symbols) for the three directions $d = 0$, $\pi/8$, and $\pi/4$ of the lowest subband of the group $\nu = g_1$. Each point of the $\mathbf{k}\cdot\mathbf{p}$ results is the average of the two values corresponding to the two different spins. Comparison shown to the best-fitting approximation obtained with the analytical relation expressed in Eq.3.43 (solid lines). Calculations for silicon (001) crystal orientation, triangular well with a confining electric field $F_c = 0.3\,\text{MV/cm}$. *Reprinted with permission from [10]. Copyright 2007 by the Institution of Electrical and Electronics Engineers.*

Energy dispersion along symmetry directions

Figure 3.5 shows that in a (001) hole inversion layer the energy dispersion, because of the crystal symmetries, takes independent values only in the sector $\theta \in [0, \pi/4]$, where $\theta = 0$ and $\theta = \pi/4$ indicate the [100] and [110] crystal directions, respectively. Figure 3.8 reports the unitless quantity $(\hbar^2 k^2/2m_0 E)$ versus E calculated with the $\mathbf{k}\cdot\mathbf{p}$ model (symbols) for the lowest subband in a triangular confining potential. The plots have been taken along the in-plane directions labelled as $d = 0$, $\pi/8$, and $\pi/4$, that correspond to an angle $\theta = 0$, $\pi/8$, and $\pi/4$, respectively.

It should be noted that, when plotted as in Fig.3.8, a parabolic energy dispersion results in a constant value equal to the effective mass (in units of m_0). Hence Fig.3.8 shows that, as already pointed out in [12] for the valence band of bulk silicon, the energy dispersion in the 2D hole subbands is approximately parabolic at relatively high energies; however, a markedly non-parabolic behavior is observed close to the bottom of the subband.

On the basis of above considerations, the model proposed in [9] assumes that, for each subband group ν and along each of the three above-mentioned in-plane directions d, the energy relation can be expressed as

$$\frac{\hbar^2 k_{\nu,d}^2}{2m_0 E_p} = \left(\frac{1}{a_{\nu,d} + b_{\nu,d}\, E_p} + c_{\nu,d} \right)^{-1}, \tag{3.43}$$

where $k_{\nu,d}$ is the magnitude of the wave-vector $\mathbf{k}$ along the direction d, E_p is the kinetic energy inside the subband, while $a_{\nu,d}$, $b_{\nu,d}$, and $c_{\nu,d}$ are the three parameters of the model along the d direction for the bands of group ν. Equation 3.43 reduces to the conventional parabolic model for $b_{\nu,d} = 0$, with an effective mass in the direction

d equal to $m_d^{(\nu)} = (1/a_{\nu,d} + c_{\nu,d})^{-1}$ (in units of m_0). For large values of E_p (and $b_{\nu,d} \neq 0$) the energy tends to a parabolic behavior with an effective mass $c_{\nu,d}^{-1}$, which is consistent with the behavior observed in the **k·p** calculations of Fig.3.8.

The values of the parameters $a_{\nu,d}$, $b_{\nu,d}$, and $c_{\nu,d}$ along the three directions can be calibrated by fitting the **k·p** results for the lowest subband of each group, these results being obtained with a triangular or a square confining potential [9]; more details about the calibration procedure may be found in [9]. The values of the parameters for the (001) crystal orientation are given in Table 3.2, and Fig.3.9 illustrates the corresponding agreement between Eq.3.43 and the **k·p** calculations.

Angular dependence of the energy dispersion

So far we have discussed the energy dispersion employed by the model along the directions d used for fitting of the **k·p** results. In order to derive a model valid for any **k** direction, we need an appropriate function of the angle θ connecting the values along the above d directions.

To this purpose, we note in Fig.3.5 that, for a (001) silicon inversion layer, the magnitude of the wave-vector **k** is, for a given kinetic energy E_p, a periodic function of the angle θ with period $\pi/2$. This feature suggests that, for each subband group ν, we can express the magnitude of **k** as

$$k_\nu(E_p, \theta) = A + B\cos(4\theta) + C\cos(8\theta), \tag{3.44}$$

which represents the expansion of $k_\nu(E_p, \theta)$ in Fourier series up to the second order (where the sine terms have been dropped because $k_\nu(E_p, \theta)$ is an even function of θ). The coefficients A, B, and C are readily determined by imposing the requirement that Eq.3.44 satisfies Eq.3.43 in the three directions $\theta = 0$, $\pi/8$, and $\pi/4$.

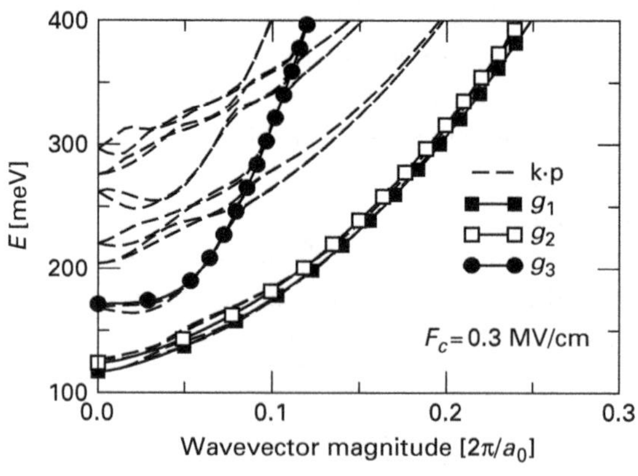

Figure 3.9 Relation between the energy E and the wave-vector magnitude k obtained either with the **k·p** method or with the analytical model. For clarity, only the lowest subband of each group g_1, g_2, and g_3 is shown for the analytical model. Calculations for silicon (001) crystal orientation. Direction $d = 0$. Triangular well with a confining electric field $F_c = 0.3$ MV/cm. *Reprinted with permission from [10]. Copyright 2007 by the Institution of Electrical and Electronics Engineers.*

This leads to the expression

$$\begin{aligned}k_\nu(E_p,\theta) &= \frac{1}{4}\left[k_{\nu,0}(E_p) + 2k_{\nu,\pi/8}(E_p) + k_{\nu,\pi/4}(E_p)\right] \\ &+ \frac{1}{2}\left[k_{\nu,0}(E_p) - k_{\nu,\pi/4}(E_p)\right]\cos(4\theta) \\ &+ \frac{1}{4}\left[k_{\nu,0}(E_p) - 2k_{\nu,\pi/8}(E_p) + k_{\nu,\pi/4}(E_p)\right]\cos(8\theta),\end{aligned} \tag{3.45}$$

that, for a given energy E_p, provides the dependence on θ of the magnitude of the wave-vector **k**. It can be easily verified that $k_\nu(E_p,\theta)$ expressed by Eq.3.45 equals $k_{\nu,0}(E_p)$, $k_{\nu,\pi/8}(E_p)$, and $k_{\nu,\pi/4}(E_p)$ for $\theta = 0$, $\pi/8$, and $\pi/4$, respectively. When Eq.3.43 is used to express $k_{\nu,0}(E_p)$, $k_{\nu,\pi/8}(E_p)$, and $k_{\nu,\pi/4}(E_p)$, then Eq.3.45 essentially provides the function $E_p(\mathbf{k})$ defined in Eq.3.41.

To summarize the main features of the model described in this section, we recall that, for a given crystal orientation and a given group $\nu = g_1, g_2, g_3$ of two-dimensional subbands, the model relies on Eq.3.43 to express the energy dispersion along three directions in the transport plane. The angular dependence of the model is then analytically expressed by Eq.3.45.

3.4 Full-band energy relation and the LCBB method

Both the equivalent Hamiltonian for the electrons and the **k·p** model for the holes are quantization models based on an approximation of the full-band structure of the underlying crystal, which is valid only for wave-vectors close enough to an edge of the silicon conduction or valence band. Some of the methodologies used to calculate the full-band energy dispersion in bulk crystals can be extended to low dimensional systems, such as a 2D carrier gas in an inversion layer or a 1D gas in a nanowire.

The tight-binding method, for example, can be very naturally applied to a low dimensional system; in fact the tight binding develops a bottom-up description of the matter which is very suited for nano-structured devices [13–15].

An alternative to the tight-binding method is the so-called Linear Combination of Bulk Bands (LCBB) approach, which is essentially based on expansion of the unknown wave-function in terms of the Bloch functions of the underlying periodic crystal. If the Bloch functions are determined by using the pseudo-potential method, then the LCBB can be considered to be an extension to low dimensional systems of the pseudo-potential method introduced in Chapter 2.

In formulation of the LCBB model we assume a single material approximation where $\hat{H}_C$ is the Hamiltonian of the underlying periodic crystal (see Eq.2.24), while $U(z)$ is the confining potential superimposed onto the crystalline one. We also assume a (001) silicon inversion layer, so that the z direction is a crystallographic axis.

Under the above assumptions the *unknown* wave-function $\psi(\mathbf{r},z)$ of the quantized system must satisfy the stationary Schrödinger equation

$$[\hat{H}_C + U(z)]\,\psi(\mathbf{r},z) = \varepsilon\,\psi(\mathbf{r},z). \tag{3.46}$$

Given the completeness of the Bloch functions $\Phi_{n\mathbf{k}k_z} = |n\mathbf{k}k_z\rangle$ of the underlying bulk crystal, the wave-function $\psi(\mathbf{r}, z)$ can be expanded as

$$\psi(\mathbf{r}, z) = \sum_{n',(\mathbf{k}',k_z')} A_{n'}(\mathbf{k}', k_z')\, \Phi_{n'\mathbf{k}'k_z'}(\mathbf{r}, z). \tag{3.47}$$

In the above equation $A_{n'}(\mathbf{k}', k_z')$ is the coefficient corresponding to the basis function $\Phi_{n'\mathbf{k}'k_z'}(\mathbf{r}, z)$ and each Bloch function can be written by using Eq.2.25 in terms of the corresponding periodic part $u_{n\mathbf{k}k_z}(\mathbf{r}, z)$ as

$$\Phi_{n'\mathbf{k}'k_z'}(\mathbf{r}, z) = |n'\mathbf{k}'k_z'\rangle = u_{n'\mathbf{k}'k_z'}(\mathbf{r}, z)e^{i\mathbf{k}'\cdot\mathbf{r}}e^{ik_z'z}. \tag{3.48}$$

Since $u_{n\mathbf{k}k_z}(\mathbf{r}, z)$ is periodic over the crystal unit cell, it can be expressed by means of the Fourier series defined in Eq.A.25. In particular we can write

$$u_{n'\mathbf{k}'k_z'}(\mathbf{r}, z) = \sum_{(\mathbf{g},g_z)} B_{n'\mathbf{k}'k_z'}(\mathbf{g}, g_z)\, e^{-i\mathbf{g}\cdot\mathbf{r}}\, e^{-ig_zz}, \tag{3.49}$$

where $B_{n\mathbf{k}k_z}$ are the coefficients

$$B_{n'\mathbf{k}'k_z'}(\mathbf{g}, g_z) = \frac{1}{\Omega_{cell}} \int_{\mathbf{r}}\int_{z} u_{n'\mathbf{k}'k_z'}(\mathbf{r}, z)\, e^{i\mathbf{g}\cdot\mathbf{r}}\, e^{ig_zz}\, d\mathbf{r}\, dz, \tag{3.50}$$

with Ω_{cell} being the volume of the unit cell of the crystal defined in Section 2.1.

The Bloch functions $|n'\mathbf{k}'k_z'\rangle$ are eigenvectors of the Hamiltonian $\hat{H}_C$, thus we have

$$\hat{H}_C\, |n'\mathbf{k}'k_z'\rangle = E_{B,n'}(\mathbf{k}', k_z')\, |n'\mathbf{k}'k_z'\rangle, \tag{3.51}$$

where $E_{B,n}(\mathbf{k}, k_z)$ is the energy value corresponding to the wave-vector $\mathbf{K} = (\mathbf{k}, k_z)$ in the nth band of the full-band energy dispersion. Furthermore the $|n\mathbf{k}k_z\rangle$ form a basis of normalized and orthogonal functions such that

$$\langle n\mathbf{k}k_z|n'\mathbf{k}'k_z'\rangle = \int_{\mathbf{r}} d\mathbf{r} \int_{z} dz\, \Phi^{\dagger}_{n\mathbf{k}k_z}(\mathbf{r}, z)\Phi_{n'\mathbf{k}'k_z'}(\mathbf{r}, z) = \delta_{n,n'}\, \delta_{\mathbf{k},\mathbf{k}'}\, \delta_{k_z,k_z'}. \tag{3.52}$$

If we now substitute Eq.3.47 in the Schrödinger equation 3.46, project Eq.3.46 on the generic state $|n\mathbf{k}k_z\rangle$ and make use of Eqs.3.51 and 3.52, we obtain the eigenvalue problem [16]

$$E_{B,n}(\mathbf{k}, k_z)A_n(\mathbf{k}, k_z) + \sum_{n',(\mathbf{k}',k_z')} \langle n\mathbf{k}k_z|U(z)|n'\mathbf{k}'k_z'\rangle\, A_{n'}(\mathbf{k}', k_z') = \varepsilon A_n(\mathbf{k}, k_z), \tag{3.53}$$

where $\langle n\mathbf{k}k_z|U(z)|n'\mathbf{k}'k_z'\rangle$ is the matrix element of the confining potential $U(z)$

$$\langle n\mathbf{k}k_z|U(z)|n'\mathbf{k}'k_z'\rangle = \int_{\mathbf{r}} d\mathbf{r} \int_{z} dz\, \Phi^{\dagger}_{n\mathbf{k}k_z}\, U(z)\, \Phi_{n'\mathbf{k}'k_z'}, \tag{3.54}$$

ε is the eigenvalue and the $A_n(\mathbf{k}, k_z)$ denote the eigenvectors which allow us to express the unknown wave-function $\psi(\mathbf{r}, z)$ according to Eq.3.47. The index n' in Eq.3.53 runs over the bands of the full-band crystal energy dispersion included in the calculations [17].

In order to proceed towards a practical formulation of the eigenvalue problem, it is necessary to determine the matrix elements $\langle n\mathbf{k}k_z|U(z)|n'\mathbf{k}'k_z'\rangle$ and to clarify the range of $(\mathbf{k}, k_z)$ to be included in the calculation or, in other words, the expansion volume V_{EK} to be used in Eq.3.47.

Given the completeness of the Bloch functions, we can choose as V_{EK} any volume such that

$$\forall(\mathbf{k}, k_z), (\mathbf{k}', k_z') \in V_{EK} : \ (\mathbf{k}' \neq \mathbf{k} + \mathbf{g}) \ \text{or} \ (k_z' \neq k_z + g_z) \ \forall\mathbf{G} = (\mathbf{g}, g_z), \tag{3.55}$$

for any reciprocal lattice vector $\mathbf{G}$=($\mathbf{g}$, g_z). The first Brillouin zone of the 3D crystal is a possible choice but it is not necessarily the most convenient for a 2D system.

In fact, it is demonstrated in [17] that, for a confining potential $U(z)$ which is constant in the $\mathbf{r}$ plane normal to the confining direction z (namely for a 2D carrier gas), it is convenient to choose the expansion volume V_{2D} as

$$\forall(\mathbf{k}, k_z), (\mathbf{k}', k_z') \in V_{2D} : \begin{cases} \mathbf{k}' \neq \mathbf{k} + \mathbf{g} & \forall\mathbf{G}_{nz} = (\mathbf{g} \neq \mathbf{0}, g_z) \\ \quad \text{and} & \\ k_z' \neq k_z + G_z & \forall\mathbf{G}_z = (\mathbf{0}, G_z) \end{cases}, \tag{3.56}$$

where we have separated the lattice vectors $\mathbf{G}$=($\mathbf{g}$, g_z) in two sets. The lattice vectors in the quantization direction $\mathbf{G}_z = (\mathbf{0}, G_z)$ (i.e. those which have a zero in-plane component $\mathbf{g} = \mathbf{0}$) and those that have a non-null in plane component $\mathbf{G}_{nz} = (\mathbf{g} \neq \mathbf{0}, g_z)$. The capital letter G_z for the k_z component of the $\mathbf{G}_z$ vectors is used to remind us that it is the component of a reciprocal lattice vector along k_z (i.e. $|\mathbf{G}_z| = |G_z|$), whereas the symbol g_z is used for the k_z component of any generic $\mathbf{G}$.

Equation 3.56 defines the V_{2D} as a prism with the shape of the base set by the condition on $\mathbf{G}_{nz}$ and height equal to the periodicity interval in the quantization direction, that is equal to the magnitude G_{zm} of the smallest reciprocal lattice vector $\mathbf{G}_z$ along the k_z direction [17].

All the wave-vectors belonging to V_{2D} defined by Eq.3.56 satisfy Eq.3.55, so that the above definition of V_{2D} is perfectly legitimate.

If we now adopt the expansion volume defined in Eq.3.56 and introduce the Fourier transform $U_T(q_z)$ of the confining potential defined according to Eq.A.2

$$U_T(q_z) = \frac{1}{(2\pi)} \int_L U(z) e^{\mathrm{i}q_z z} \mathrm{d}z, \tag{3.57}$$

where L is the normalization length in the z direction, it can be demonstrated that the matrix elements defined in Eq.3.54 take a form such that Eq.3.53 can be rewritten as [17]

$$E_{B,n}(\mathbf{k}, k_z) A_n(\mathbf{k}, k_z) + \frac{2\pi}{L} \sum_{n',k_z'} \Bigg\{ U_T(k_z' - k_z) S^{(n,n')}_{\mathbf{k}k_z\mathbf{k}k_z'}(\mathbf{0}, 0)$$

$$+ \sum_{\mathbf{G}_z=(\mathbf{0},G_z\neq 0)} U_T(k_z' - k_z + G_z) S^{(n,n')}_{\mathbf{k}k_z\mathbf{k}k_z'}(\mathbf{0}, G_z) \Bigg\} A_{n'}(\mathbf{k}, k_z') = \varepsilon(\mathbf{k}) A_n(\mathbf{k}, k_z). \tag{3.58}$$

Equation 3.58 represents a separated eigenvalue problem for each value of the wave-vector **k** in the transport plane. Consistently with the definition of the V_{2D} given in Eq.3.56, k_z must vary in a periodicity interval [17].

The term $S^{(n,n')}_{\mathbf{k}k_z\mathbf{k}k'_z}$ in Eq.3.58 is the overlap integral between the periodic parts of the Bloch functions

$$S^{(n,n')}_{\mathbf{k}k_z\mathbf{k}k'_z}(\mathbf{g}, g_z) = \langle u_{n(\mathbf{k}-\mathbf{g})(k_z-g_z)}(\mathbf{R})|u_{n'\mathbf{k}k'_z}(\mathbf{R})\rangle$$
$$= \int_\Omega [u_{n(\mathbf{k}-\mathbf{G})(k_z-g_z)}(\mathbf{R})]^\dagger\, u_{n'\mathbf{k}k'_z}(\mathbf{R})\, \mathrm{d}\mathbf{R}, \qquad (3.59)$$

where Ω is the normalization volume. In Appendix D it is shown that the overlap integrals can be expressed as

$$S^{(n,n')}_{\mathbf{k}k_z\mathbf{k}'k'_z}(\mathbf{g}, g_z) = \Omega \sum_{(\mathbf{g}',g'_z)} B^\dagger_{n\mathbf{k}k_z}(\mathbf{g}' + \mathbf{g}, g'_z + g_z) B_{n'\mathbf{k}'k'_z}(\mathbf{g}', g'_z) \qquad (3.60)$$

in terms of the Fourier coefficients $B_{n\mathbf{k}k_z}$ of the periodic parts of Bloch functions (see Eq.3.50).

The LCBB approach can be used to calculate the band structure for either an electron or a hole inversion layer by including in Eqs.3.46 and 3.58 the appropriate set of bands of the bulk crystal energy dispersion.

As discussed in [17], the base of the prism V_{2D} defined in Eq.3.56 is the 2D Brillouin zone of the 2D carrier gas, whose shape and extension depends on the quantization direction. Figure 3.10 shows the 2D Brillouin zone for a (100) inversion layer (inferred from the base of the V_{2D} defined in Eq.3.56), as well as the periodicity interval where the k_z must vary for the numerical solution of Eq.3.58.

3.4.1 Implementation and computational complexity

Equation 3.58 represents a separated eigenvalue problem for each value of the in-plane wave-vector **k**. This is reflected in the notation $\varepsilon(\mathbf{k})$ which underlines the fact that each set of eigenvalues obtained by Eq.3.58 can be associated to a single **k** value. Hence, as in the case of the **k·p** model, the band structure must be determined by solving an eigenvalue problem for each **k** wave-vector. The range of the k_z values is set by Eq.3.56 and it is [−1,1] (in units of $2\pi/a_0$) for a (001) silicon inversion layer.

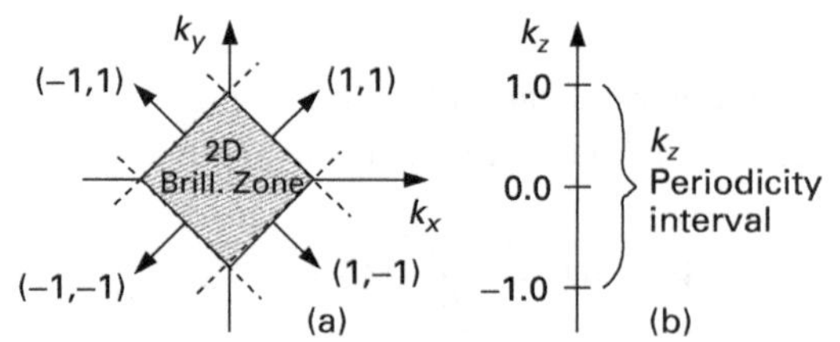

Figure 3.10 (a) 2D first Brillouin zone for a (100) electron inversion layer obtained as the base of the prism defined in Eq.3.56. (b) Periodicity interval for k_z to be used in the numerical solution of Eq.3.58. All vectors are in units of $(2\pi/a_0)$.

As discussed in detail in [17], the electron band structure in an inversion layer can be obtained by solving Eq.3.58 for the two lowest conduction bands of the bulk silicon crystal. More precisely, for each $\mathbf{k}$ in the transport plane, the pseudo-potential method described in Section 2.2.1 can be used to determine both the full-band dispersion $E_{B,n}(\mathbf{k}, k_z)$ and the Fourier components $B_{n\mathbf{k}k_z}$ of the $u_{n\mathbf{k}k_z}$ functions of the underlying 3D crystal (see Eq.3.49). The coefficients $B_{n\mathbf{k}k_z}$ can be used to calculate the overlap integrals $S^{(n,n')}_{\mathbf{k}k_z\mathbf{k}k'_z}(\mathbf{0}, G_z)$ in accordance with Eq.3.60.

It can be verified that the sum over $\mathbf{G}_z$ in the brackets of Eq.3.58 converges to a stable value when at least the two terms corresponding to $\mathbf{G}_z = (\mathbf{0}, \pm\frac{4\pi}{a_0})$ are included. On the contrary, it is by no means a valid approximation to drop the entire sum over $\mathbf{G}_z$ and keep only the first term in the bracket [17].

3.4.2 Calculation results for the LCBB method

It is interesting to note in Eq.3.58 that, if the potential $U(z)$ is slowly varying in the z direction, then the $U_T(k'_z - k_z)$ components become vanishingly small for $k'_z \neq k_z$ and the eigenvalues $\varepsilon(\mathbf{k})$ of Eq.3.58 tend to the energy values $E_{B,n}(\mathbf{k}, k_z)$ of the bulk crystal. This observation suggests that the local minima of the two-dimensional energy dispersion $\varepsilon(\mathbf{k})$ are typically found at the $\mathbf{k}$ values such that the bulk crystal energy dispersion has a minimum at some $(\mathbf{k}, k_z)$ point inside the expansion volume defined in Eq.3.56. This is a useful hint to help us understand, on the basis of the features of the $E_{B,n}(\mathbf{k}, k_z)$, at what $\mathbf{k}$ points the minima of the 2D carrier gas are expected to be located. In the case of a (001) silicon inversion layer, for example, we can readily infer that the energy minima in $\mathbf{k}$ are located at $\mathbf{k} = 0$, at $\mathbf{k} = (0, \pm 0.85)$, and at $\mathbf{k} = (\pm 0.85, 0)$.

Interestingly, however, some minima of the 2D carriers' energy relation do not stem directly from minima of the bulk crystal energy dispersion [18].

In this respect, the LCBB method also allows us to naturally identify the shape of the 2D Brillouin zone for any quantization direction, because the 2D Brillouin zone coincides with the base of the prism defined in Eq.3.56 as the expansion volume V_{2D}. This allows us to identify a priori the 2D Brillouin zone [17].

Figure 3.11 shows the equi-energy contour for the lowest eigenvalue versus the in-plane wave-vector $\mathbf{k}$ for a (001) silicon inversion layer, where the parameters for the pseudo-potential calculations of the bulk silicon energy and overlap integrals were taken from [19]. The square in Fig.3.11 indicates the 2D first Brillouin zone. Similar results can be obtained, even for different crystal orientations and for semiconductors other than silicon.

The LCBB method is computationally demanding and, in particular, self-consistent calculation of the band structure discussed in Section 3.7 requires numerical determination of the density of states (see Section 3.5.1). For this reason, at the time of writing, the LCBB has been used mainly as a benchmark to investigate the limit of applicability of simpler quantization models, such as the EMA for electrons or the $\mathbf{k}{\cdot}\mathbf{p}$ for holes [20, 21]. In this respect, the EMA model was found to be fairly accurate for SOI MOSFETs with silicon and germanium films down to a

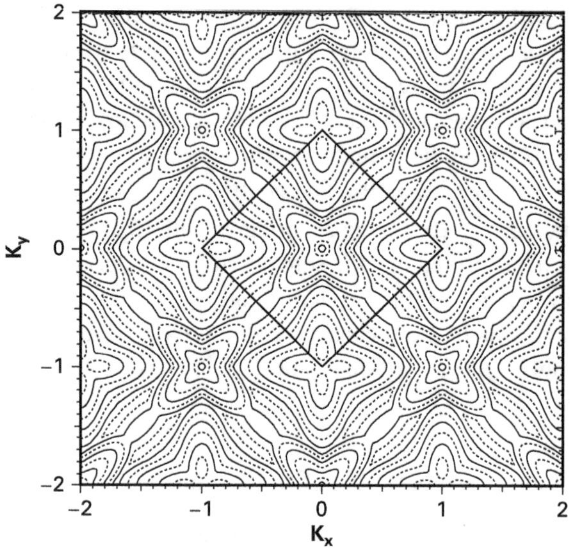

Figure 3.11 Silicon, [001] quantization direction. Contour plots of the energy versus in-plane **k** for the lowest eigenvalue. The energy values are 0.04, 0.3, 0.85, 1.45, 2.05, 2.65 [eV] for the solid lines and 0.12, 0.55, 1.15, 1.75, 2.35 [eV] for the dashed lines. The absolute minimum is in $\mathbf{k} = 0$ and it is approximately two-times degenerate; four more degenerate minima are in $\mathbf{k} = (0, \pm 0.85)$ and $\mathbf{k} = (\pm 0.85, 0)$. *Reprinted, with permission, from [17]. Copyright 2005 by the American Physics Society.*

few nanometers [20]. Qualitatively similar results were obtained for silicon nano-wire transistors by comparing the EMA to the tight-binding quantization model [22].

3.5 Sums and integrals in the k space

In descriptions of transport in electron devices we frequently need to evaluate the sum over the wave-vectors **k** of a function $S(\mathbf{k})$. Such sums are typically converted to integrals in the **k** space, which are in turn calculated either analytically or numerically, depending on the function $S(\mathbf{k})$. This section deals with evaluation of the sums in **k** space and illustrates a few examples of particular practical interest.

Let us consider an inversion layer with an envelope wave-function expressed as

$$\Phi_n(\mathbf{R}) = \xi_{n\mathbf{k}}(z) \frac{e^{i\,\mathbf{k}\cdot\mathbf{r}}}{\sqrt{A}}, \tag{3.61}$$

where A is the normalization area in the transport plane and the index n denotes both the valley and the subband if more valleys exist.

As discussed in Section 3.1, for periodic boundary conditions in the transport plane the possible **k** values are given by Eq.3.3. Thus the area in **k** space to be associated with each **k** value is

$$A_{\mathbf{k}} = \frac{(2\pi)^2}{A}, \tag{3.62}$$

and the density $g(\mathbf{k})$ of the $\mathbf{k}$ states, that is the number of $\mathbf{k}$ states per unit area in *reciprocal* space, is given by

$$g(\mathbf{k}) = \frac{A}{(2\pi)^2}. \tag{3.63}$$

It should be noted that $\mathbf{k}$ denotes the total wave-vector in the transport plane when we are using a full-band description of the 2D carrier gas (such as the one based on the LCBB method introduced in Section 3.4), or even when we are employing a $\mathbf{k}\cdot\mathbf{p}$ quantization model. In the effective mass approximation model, instead, $\mathbf{k}$ is the displacement of the wave-vector from the corresponding value at the energy minimum.

As can be seen from Eq.3.3 and 3.63, the density of available $\mathbf{k}$ states increases with the normalization area A. If A is large enough the $\mathbf{k}$ states are so dense that a sum over $\mathbf{k}$ of a function $S(\mathbf{k})$ can be approximated by an integral. More precisely, according to Eq.3.62 we have

$$\frac{(2\pi)^2}{A} \sum_{\mathbf{k}} S(\mathbf{k}) \approx n_{sp} \int_{\mathbf{k}} S(\mathbf{k})\, \mathrm{d}\mathbf{k}. \tag{3.64}$$

The spin multiplicity factor n_{sp} is 2 if, for each $\mathbf{k}$, two degenerate states with opposite spin exist and, furthermore, we wish to include both of them in the summation; it is 1 otherwise.

In order to clarify the above statement about n_{sp} we consider a few practical examples. The EMA quantization model for the electrons provides two degenerate states with opposite spins, whereas in the $\mathbf{k}\cdot\mathbf{p}$ model for hole inversion layers the spin degeneracy is removed by the spin–orbit interaction. Thus, when the summation involves states calculated with the $\mathbf{k}\cdot\mathbf{p}$ model, n_{sp} cannot but be 1. In the case of the EMA for electrons, instead, n_{sp} is 2 when the summation must include all possible states (as in calculation of the electron density described in Section 3.6). However, n_{sp} is 1 if the summation is used to calculate a scattering rate (see Chapter 4), because the scattering mechanisms considered in this book do not change the spin, hence the state after scattering must have the same spin as the state before scattering.

In all the practical cases the energy model employed in the calculations and the physical meaning of the summation allow us to set the correct value for n_{sp}, as is clarified by many examples discussed throughout the book.

It should be noted finally that the normalization area A is no longer present in the r.h.s. of Eq.3.64. This fact highlights that A is just a conceptual tool, namely a reference area with no physical meaning, which must always disappear in the final calculations.

3.5.1 Density of states

An important quantity for the physical and electrical properties of an inversion layer is the density $g_n(E)$ of states available per unit energy and unit area in real space. By definition, the density $g_n(E)$ of states available in the subband n can be evaluated by counting all the states of the subband available at a given energy E and then dividing by the normalization area A. Thus we have

$$g_n(E) = \frac{1}{A}\sum_{\mathbf{k}} \delta(E_n(\mathbf{k}) - E) = \frac{n_{sp}}{(2\pi)^2}\int_{\mathbf{k}} \delta(E_n(\mathbf{k}) - E)\, d\mathbf{k}, \tag{3.65}$$

where Eq.3.64 has been used to convert the sum to an appropriate integral over **k**. As discussed above, n indicates both the valley and the subband if more valleys exist in the inversion layer.

In order to clarify both the notation and the meaning of Eq.3.65, we note that $E_n(\mathbf{k})$ denotes the energy in the subband n (which depends on the integration variable **k**), whereas E is just a parameter in the integral, namely the energy value at which $g_n(E)$ is evaluated.

The *Dirac* function in Eq.3.65 suggests that for the evaluation of the integral it is convenient to change the integration variables so that one of them is the energy itself; in fact the *Dirac* function reduces the corresponding integral. The details of this procedure depend on the form of the energy dispersion $E_n(\mathbf{k})$.

We also note that in some circumstances we need to evaluate the sum over **k** of a function that depends on **k** only through the energy $E_n(\mathbf{k})$, that is we have $S(\mathbf{k}) = S(E_n(\mathbf{k}))$. In such a case, according to Eq.3.64 we have

$$\begin{aligned}\frac{1}{A}\sum_{\mathbf{k}} S(E_n(\mathbf{k})) &= \frac{1}{A}\sum_{\mathbf{k}}\int_{-\infty}^{+\infty} \delta(E_n(\mathbf{k}) - E)\, S(E)\, dE \\ &= \int_{-\infty}^{+\infty}\left[\frac{1}{A}\sum_{\mathbf{k}} \delta(E_n(\mathbf{k}) - E)\right] S(E)\, dE \\ &= \int_{-\infty}^{+\infty} g_n(E)\, S(E)\, dE,\end{aligned} \tag{3.66}$$

where we have used the definition of $g_n(E)$ given in Eq.3.65. As discussed above, the value of n_{sp} in $g_n(E)$ depends on the quantization model.

3.5.2 Electron inversion layers in the effective mass approximation

Let us consider the elliptic, non-parabolic expression for the electron energy dispersion introduced in Section 3.2.4. The energy $E_{\nu,n}(\mathbf{k})$ is written as

$$E_{\nu,n}(\mathbf{k}) = E_{\nu 0} + \varepsilon_{\nu,n} + E_p(\mathbf{k}), \tag{3.67}$$

where E_p is the kinetic energy inside the subband (ν, n), and it is related to the wave-vector **k** by

$$\frac{\hbar^2 k_x^2}{2m_x} + \frac{\hbar^2 k_y^2}{2m_y} = E_p(1 + \alpha E_p). \tag{3.68}$$

If we now express **k** in polar coordinates as

$$\mathbf{k} = (k_x, k_y) = (k\cos(\theta),\ k\sin(\theta)),$$

then the $\mathbf{k}$ to E_p relation can be written as

$$\frac{\hbar^2 k^2}{2\,m_{xy}(\theta)} = E_p(1+\alpha E_p), \tag{3.69}$$

where we have introduced the θ dependent mass m_{xy},

$$m_{xy}(\theta) = \left[\frac{\cos^2(\theta)}{m_x} + \frac{\sin^2(\theta)}{m_y}\right]^{-1}. \tag{3.70}$$

For a circular subband with effective mass m the m_{xy} is independent of θ and coincides with m; this is the case for the $\Delta_{0.92}$ in (100) inversion layers, as illustrated in Table 3.1.

Equations 3.69 and 3.70 can be used to calculate the contribution $g_{\nu,n}(E)$ to the density of states of the subband (ν,n). To this purpose, we evaluate the integral over $\mathbf{k}$ in the r.h.s. of Eq.3.65 by using polar coordinates and then change the integration variable from $k=|\mathbf{k}|$ to E_p. We thus obtain

$$\begin{aligned} g_{\nu,n}(E) &= \frac{n_{sp}}{(2\pi)^2}\int_0^{2\pi} \mathrm{d}\theta \int_0^{+\infty} k(E_p,\theta)\,\frac{\partial k(E_p,\theta)}{\partial E_p}\,\delta(E_{\nu 0}+\varepsilon_{\nu,n}+E_p-E)\,\mathrm{d}E_p \\ &= H_\upsilon(E_{pE})\,\frac{n_{sp}}{(2\pi)^2}\int_0^{2\pi} k(E_{pE},\theta)\,\frac{\partial k(E_{pE},\theta)}{\partial E_p}\,\mathrm{d}\theta, \end{aligned} \tag{3.71}$$

where $H_\upsilon(x)$ is the step function and E_{pE} is the kinetic energy in the subband (ν,n) corresponding to the total energy E, that is

$$E_{pE} = (E - E_{\nu 0} - \varepsilon_{\nu,n}). \tag{3.72}$$

As can be seen, $g_{\nu,n}(E)$ is null for $E<(E_{\nu 0}+\varepsilon_{\nu,n})$.

We can now complete the calculation of $g_{\nu,n}(E)$ by deriving both sides of Eq.3.69 with respect to E_p to obtain

$$k(E_p,\theta)\,\frac{\partial k(E_p,\theta)}{\partial E_p} = \frac{m_{xy}(\theta)}{\hbar^2}(1+2\alpha\,E_p), \tag{3.73}$$

which can then be inserted in Eq.3.71 to obtain

$$g_{\nu,n}(E) = H_\upsilon(E_{pE})\,\frac{n_{sp}[1+2\alpha\,E_{pE}]}{4\pi^2\hbar^2}\int_0^{2\pi}\left[\frac{\cos^2(\theta)}{m_x}+\frac{\sin^2(\theta)}{m_y}\right]^{-1}\mathrm{d}\theta. \tag{3.74}$$

The integral over θ can be evaluated analytically as $2\pi\sqrt{m_x m_y}$ [23], so that we finally obtain

$$g_{\nu,n}(E) = H_\upsilon(E-E_{\nu 0}-\varepsilon_{\nu,n})\,\frac{n_{sp}\sqrt{m_x m_y}}{2\pi\hbar^2}\,[1+2\alpha\,(E-E_{\nu 0}-\varepsilon_{\nu,n})]. \tag{3.75}$$

If we neglect the non-parabolic corrections by setting $\alpha=0$, we obtain the simpler expression

$$g_{\nu,n}(E) = H_\upsilon(E-E_{\nu 0}-\varepsilon_{\nu,n})\,\frac{n_{sp}\sqrt{m_x\,m_y}}{2\pi\hbar^2}, \tag{3.76}$$

which corresponds to a subband density of states independent of E_p. Because of the above expressions for $g_{\nu,n}(E)$, the subband is said to have an effective mass for the density of states equal to $m_d = \sqrt{m_x m_y}$.

As can be seen in Table 3.1, the subbands of the valleys $\Delta_{0.92}$, $\Delta_{0.19x}$, and $\Delta_{0.19y}$ have a valley multiplicity $\mu_\nu = 2$. Furthermore the $\Delta_{0.19x}$ and $\Delta_{0.19y}$ valleys have the same subband minima and the same $m_d = \sqrt{m_x m_y}$, hence, as far as the density of states is concerned, they can be grouped in a valley with an overall multiplicity $\mu_\nu = 4$.

By introducing the valley multiplicity μ_ν, the total density of states can be expressed as

$$g(E) = \sum_{\nu,n} H_\nu(E - E_{\nu 0} - \varepsilon_{\nu,n}) \frac{n_{sp}\,\mu_\nu\, m_{d,\nu}}{2\pi \hbar^2} [1 + 2\alpha\,(E - E_{\nu 0} - \varepsilon_{\nu,n})]. \tag{3.77}$$

For a (001)/[100] silicon inversion layer the valley multiplicities are those reported in Table 3.1 if we let ν vary in three possible valleys.

As can be seen, in the energy model employed here for the 2D electron gas, we need the two indices (ν,n) to identify a subband in the inversion layer. However, some quantities, such as the density of states effective mass $m_{d,\nu}$, are the same for all the subbands of a given valley.

The procedure used for the density of states of an electron inversion layer can also be used to calculate the sum of a generic function $S(\mathbf{k})$ over all the $\mathbf{k}$ values corresponding to a given energy E. The contribution to the sum corresponding to the subband (ν, n) can be obtained by starting from Eq.3.64 and then evaluating the integral over $\mathbf{k}$ by using the variables E_p and θ. If we now express $S(\mathbf{k})$ as a function of the new integrating variable as $S(E_p, \theta)$, we readily obtain

$$\begin{aligned} &\frac{1}{A}\sum_{\mathbf{k}} S(\mathbf{k})\,\delta(E_{\nu,n}(\mathbf{k}) - E) \\ &= \frac{n_{sp}\,\mu_\nu}{(2\pi)^2} \int_0^{2\pi} \mathrm{d}\theta \int_0^{+\infty} k(E_p,\theta)\,\frac{\partial k(E_p,\theta)}{\partial E_p}\, S(E_p,\theta)\,\delta(E_{\nu 0} + \varepsilon_{\nu,n} + E_p - E)\,\mathrm{d}E_p \\ &= \frac{n_{sp}\,\mu_\nu\, H_\nu(E_{pE})}{(2\pi)^2\,\hbar^2} [1 + 2\alpha\, E_{pE}] \int_0^{2\pi} m_{xy}(\theta) S(E_{pE},\theta)\,\mathrm{d}\theta, \end{aligned} \tag{3.78}$$

with E_{pE} defined in Eq.3.72. A comparison with Eq.3.74 shows that, if the function $S(\mathbf{k})$ is independent of θ, then Eq.3.78 simplifies to

$$\frac{1}{A}\sum_{\mathbf{k}} S(E_p)\,\delta(E_{\nu,n}(\mathbf{k}) - E) = g_{\nu,n}(E)\,S(E - E_{\nu 0} - \varepsilon_{\nu,n}), \tag{3.79}$$

and the sum is simply proportional to the density of states $g_{\nu,n}(E)$ of the subband (ν, n) times the value of the function $S(E_p)$ at the kinetic energy E_{pE} defined in Eq.3.72. The value of n_{sp} is set by the physical meaning of the sum.

By following the same procedure we can also express the sum over $\mathbf{k}$ of a generic function $S(\mathbf{k})$ even if there is no Dirac function selecting a particular energy in the corresponding integral. The contribution to the sum of the subband (ν, n) is

$$\frac{1}{A}\sum_{\mathbf{k}} S(\mathbf{k}) = \frac{n_{sp}\,\mu_\nu}{(2\pi)^2}\int_0^{2\pi} \mathrm{d}\theta \int_0^{+\infty} \mathrm{d}E_p\, k(E_p,\theta)\,\frac{\partial k(E_p,\theta)}{\partial E_p}\, S(E_p,\theta)$$
$$= \frac{n_{sp}\,\mu_\nu}{(2\pi)^2\,\hbar^2}\int_0^{+\infty} \mathrm{d}E_p\,(1+2\alpha\, E_p)\int_0^{2\pi} \mathrm{d}\theta\; m_{xy}(\theta) S(E_p,\theta). \quad (3.80)$$

In particular, if the function $S(\mathbf{k})$ depends on E_p but is independent of θ, then Eq.3.80 simplifies to Eq.3.66. The value of n_{sp} is set by the physical meaning of the sum.

3.5.3 Hole inversion layers with an analytical energy model

If the energy dispersion in the hole inversion layer is expressed as

$$E_{\nu,n}(\mathbf{k}) = E_{\nu 0} + \varepsilon_{\nu,n} + E_p(\mathbf{k}), \quad (3.81)$$

according to the analytical model described in Section 3.3.3, we can then proceed as in the case of the electrons and calculate the integral over $\mathbf{k}$ by changing the integration variables to the kinetic energy E_p and the angle θ.

By doing so we obtain an expression for the density of states of the band (ν, n) formally identical to Eq.3.71:

$$g_{\nu,n}(E) = H_\nu(E_{pE})\,\frac{n_{sp}}{(2\pi)^2}\int_0^{2\pi} \mathrm{d}\theta\; k_\nu(E_{pE},\theta)\,\frac{\partial k_\nu(E_{pE},\theta)}{\partial E_p}, \quad (3.82)$$

where E_{pE} is the kinetic energy in the subband (ν, n) defined in Eq.3.72

Analytical expressions for $k_\nu(E_p,\theta)$ and its derivative with respect to E_p can be obtained by using Eqs.3.43 and 3.45. In fact, from Eq.3.43 we can express the magnitude $k_{\nu,d}$ of the wave-vector $\mathbf{k}$ in the d direction as

$$k_{\nu,d}(E_p) = \frac{\hbar}{2m_0}\sqrt{\frac{E_p(a_{\nu,d} + b_{\nu,d}\, E_p)}{1 + c_{\nu,d}(a_{\nu,d} + b_{\nu,d}\, E_p)}}. \quad (3.83)$$

The above equation can be used to calculate analytically $[\partial k_{\nu,d}(E_p)/\partial E_p]$ for any of the three directions $d = 0, \pi/8, \pi/4$. Then by using Eq.3.45 we can finally write the derivative with respect to E_p of the wave-vector magnitude k_ν in a subband of the group ν as

$$\frac{\partial k_\nu(E_p,\theta)}{\partial E_p} = \frac{\partial k_{\nu,0}(E_p)}{\partial E_p}\left[\frac{1+2\cos(4\theta)+\cos(8\theta)}{4}\right]$$
$$+\frac{\partial k_{\nu,\pi/8}(E_p)}{\partial E_p}\left[\frac{1-\cos(8\theta)}{2}\right]$$
$$+\frac{\partial k_{\nu,\pi/4}(E_p)}{\partial E_p}\left[\frac{1-2\cos(4\theta)+\cos(8\theta)}{4}\right]. \quad (3.84)$$

Equations 3.45 and 3.84 allow us to analytically express the integrand function in Eq.3.82. The integral over the angle θ can be calculated numerically for each E value. It should be noted that, according to the energy model of Section 3.3.3, the dependence of $g_{\nu,n}$ on the kinetic energy is the same for all the subbands of a given group ν. Thus, the density of states versus E_p can be calculated just once and stored in appropriate tables for any possible later use.

Similarly to the electron case, we can also express sums that include a generic function $S(\mathbf{k})$. The contribution given by the subband (ν, n) is

$$\frac{1}{A}\sum_{\mathbf{k}} S(\mathbf{k})\,\delta(E_{\nu,n}(\mathbf{k}) - E)$$
$$= \frac{n_{sp}\,H_\nu(E_{pE})}{(2\pi)^2}\int_0^{2\pi} k_\nu(E_{pE},\theta)\,\frac{\partial k_\nu(E_{pE},\theta)}{\partial E_p}\,S(E_{pE},\theta)\,\mathrm{d}\theta. \qquad (3.85)$$

Furthermore we also have

$$\frac{1}{A}\sum_{\mathbf{k}} S(\mathbf{k}) = \frac{n_{sp}}{(2\pi)^2}\int_0^{2\pi}\mathrm{d}\theta\int_0^{+\infty} k_\nu(E_p,\theta)\frac{\partial k_\nu(E_p,\theta)}{\partial E_p}S(E_p,\theta)\,\mathrm{d}E_p. \qquad (3.86)$$

For the case of a function $S(\mathbf{k})$ independent of θ, the simplified expression given by Eq.3.79 applies also to a hole inversion layer.

3.5.4 Sums and integrals for a numerical energy model

The derivations discussed in Sections 3.5.2 and 3.5.3 are based on the availability of an analytical form for the energy dependence on $\mathbf{k}$. In such a case we can express analytically at least the functions inside the integrals that define the density of states or the sum over $\mathbf{k}$ of a generic function $S(\mathbf{k})$. In the case of electrons with a parabolic EMA quantization model, furthermore, we have been able to evaluate analytically even the final expression for the density of states.

When the energy dispersion is obtained numerically by solving an eigenvalue problem for each $\mathbf{k}$ value, as in the case of the $\mathbf{k}\cdot\mathbf{p}$ or the LCCB models, the integrals over $\mathbf{k}$ cannot but be evaluated numerically by using the energy values at the grid nodes used to calculate the band structure.

In this respect, if we denote with $\mathbf{k}_d$ the $\mathbf{k}$ points used for calculation of the energy dispersion, the sums over $\mathbf{k}$ can be approximated as

$$\frac{1}{A}\sum_{\mathbf{k}} S(\mathbf{k}) \approx \frac{n_{sp}}{(2\pi)^2}\int_{\mathbf{k}} S(\mathbf{k})\,\mathrm{d}\mathbf{k} \approx \frac{n_{sp}}{(2\pi)^2}\sum_{\mathbf{k}_d} A_{\mathbf{k}_d}\,S(\mathbf{k}_d), \qquad (3.87)$$

where $A_{\mathbf{k}_d}$ is the area in the $\mathbf{k}$ space associated with the $\mathbf{k}_d$ point. In particular, if we wish to calculate a sum by considering only the $\mathbf{k}$ states with a given energy E, then the contribution of the subband n can be approximately written as

$$\frac{1}{A}\sum_{\mathbf{k}} \delta(E_n(\mathbf{k}) - E)\,S(\mathbf{k}) \approx \frac{n_{sp}}{\Delta E(2\pi)^2}\sum_{E_n(\mathbf{k}_d)\in(E\mp 0.5\Delta E)} A_{\mathbf{k}_d}\,S(\mathbf{k}_d), \qquad (3.88)$$

where the sum in the r.h.s. of Eq.3.88 is restricted to the $\mathbf{k}_d$ points such that $|E_n(\mathbf{k}_d)-E|<0.5\Delta E$ and ΔE is the energy discretization.

The calculation of the density of states is an important special case of Eq.3.88, where $S(\mathbf{k}_d)$ is simply 1.0. If we wish to express the $g_n(E)$ of a given subband n, we have

$$g_n(E) \approx \frac{n_{sp}}{\Delta E(2\pi)^2} \sum_{E_n(\mathbf{k}_d)\in(E\mp 0.5\Delta E)} A_{\mathbf{k}_d}. \quad (3.89)$$

Here it should be noted that, while A is just a normalization area that never enters either the implementation or the final results, $A_{\mathbf{k}_d}$ is instead the area in the **k** space associated with each point $\mathbf{k}_d$ at which the band structure has been calculated. The more are the points $\mathbf{k}_d$ in the **k** range of practical interest the smaller are the corresponding areas $A_{\mathbf{k}_d}$ and thus the closer is the result of the sum over $\mathbf{k}_d$ to an integral over **k**. Increasing the number of $\mathbf{k}_d$ points, however, proportionally increases the computational effort.

In this respect, the last equality in Eq.3.87 is just the simplest approximation of the integral over **k** when the integrand function is known just in a finite number of points $\mathbf{k}_d$. However, since the energy is continuous versus **k** (which is a parameter in the eigenvalue problem), we can use interpolation methods to approximate the energy values at some **k** points other than the $\mathbf{k}_d$. As an example, Fig.3.12 shows the case where the $\mathbf{k}_d$ points have been arranged as triangles $T_{\mathbf{k}d}$. Interestingly, for each subband the energy values at the vertices of $T_{\mathbf{k}d}$ uniquely identify a linear expression for the energy versus **k** relation inside $T_{\mathbf{k}d}$:

$$E_n = E_{n,0} + a_{n,x}\, k_x + a_{n,y}\, k_y. \quad (3.90)$$

The three constants $E_{n,0}$, $a_{n,x}$, and $a_{n,y}$ are determined by the values of E_n at the vertices of $T_{\mathbf{k}d}$ and the expression is valid only inside a given triangle $T_{\mathbf{k}d}$. The energy gradient with respect to **k**, hence the group velocity, is constant inside $T_{\mathbf{k}d}$. Such properties are also at the basis of the Monte Carlo transport simulations based on the approach of the simplexes (see Section 6.1.8).

If we assume a linear energy dispersion inside the triangles it is then very easy to identify the **k** vectors corresponding to a given energy value. In more general terms the integrals over **k** inside each triangle can be evaluated either analytically or with the standard finite sum approximations. In such a case the sums over **k** for the numerically calculated band structure can be written as

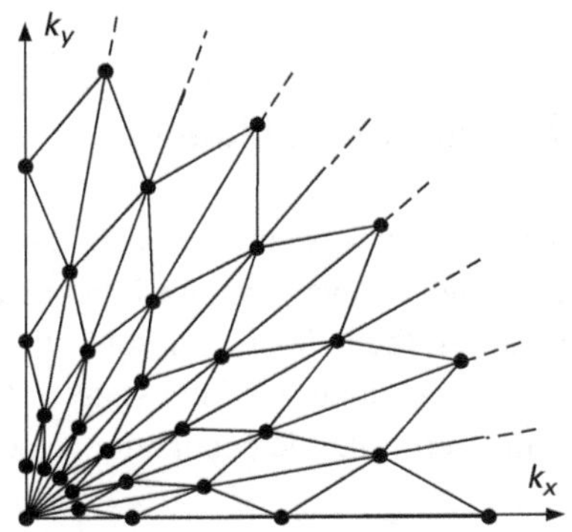

Figure 3.12 Example of discretization points $\mathbf{k}_d$ arranged as triangles $T_{\mathbf{k}d}$ and used for numerical calculation of the band structure of a 2D carrier gas.

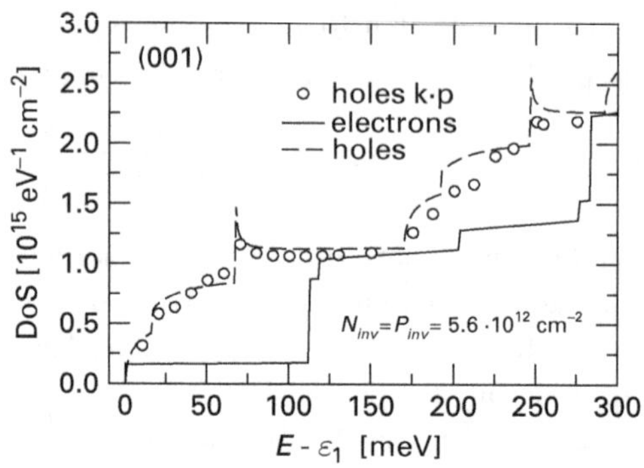

Figure 3.13 Density of states (DoS) versus the energy (referred to the lowest subband minimum ε_1) for either an electron inversion layer (according to Eq.3.77 and for $\alpha = 0.5\,\mathrm{eV}^{-1}$) or for a hole inversion layer. For holes the DoS was numerically calculated either with the energy model of Section 3.3.3 (dashed line) or with the **k**·**p** model (circles). For both the n- and the p-MOS structure the channel doping is $1.5{\cdot}10^{18}$ cm^{-3}.

$$\frac{1}{A}\sum_{\mathbf{k}} S(\mathbf{k}) \approx \frac{n_{sp}}{(2\pi)^2}\sum_{T_{\mathbf{k}d}}\int_{T_{\mathbf{k}d}} S(\mathbf{k})\,d\mathbf{k}. \tag{3.91}$$

Use of a linear approximation of the energy inside the triangles $T_{\mathbf{k}d}$ can help improve the accuracy of the results for a given number of $\mathbf{k}_d$ points.

Figure 3.13 illustrates the density of states (DoS) for either an electron or a hole inversion layer obtained with a self-consistent Schrödinger–Poisson solution for the n- and p-MOSFETs. As can be seen, for a (001) silicon inversion layer and for energies close to the subband minimum, the DoS is much larger for holes than for electrons.

In this and in the previous section we have derived expressions for the conversion of sums in the **k** space to appropriate integrals. Such conversions are used many times in the rest of the book for calculation of carrier spatial densities, of scattering rates and of momentum relaxation times. The detailed treatment illustrated in this section is thus justified by the practical importance of the above derivations.

3.6 Carrier densities at the equilibrium

At the thermodynamic equilibrium the occupation of a state in the inversion layer identified by the band n and the wave-vector **k** depends only on its energy $E_n(\mathbf{k})$; as already mentioned a few times, the index n denotes both the valley and the subband if more valleys exist. In fact the occupation of the state (n,**k**) is simply given by $f_0(E_n(\mathbf{k}))$, with $f_0(E)$ denoting the Fermi–Dirac occupation function

$$f_0(E) = \frac{1}{1+\exp\left(\frac{E-E_F}{K_BT}\right)}, \tag{3.92}$$

where E_F is the Fermi level. The following identity for the energy derivative of the Fermi–Dirac function is frequently useful

$$\frac{\partial f_0(E)}{\partial E} = -\frac{1}{K_B T} f_0(E)(1 - f_0(E)). \tag{3.93}$$

As already pointed out in Chapter 2, in this book we use an electron-like convention also for the hole energy, that is the hole energies tend to increase for increasing magnitudes of **k**. Thus Eq.3.92 provides the occupation of states at equilibrium for both an electron and a hole inversion layer.

If we now recall Eq.3.32 for the general form of the envelope wave-function $\Phi_n(\mathbf{R})$ in the inversion layer, the density per unit volume $d_n(z)$ of carriers in the subband n (either electrons or holes) can be expressed as

$$d_n(z) = \sum_{\mathbf{k}} |\Phi_n(\mathbf{R})|^2 f_0(E_n(\mathbf{k})) = \frac{1}{A} \sum_{\mathbf{k}} |\xi_{n\mathbf{k}}(z)|^2 f_0(E_n(\mathbf{k})). \tag{3.94}$$

When more valleys exist the total carrier density is obtained by summing over all the valleys and the subbands.

According to the Fermi–Dirac function, the occupation of the states is exponentially reduced with increase of the energy $E_n(\mathbf{k})$, hence the sum over **k** always converges to a finite value. When the wave-functions are independent of **k**, Eq.3.94 simplifies to

$$d_n(z) = \frac{1}{A} |\xi_n(z)|^2 \sum_{\mathbf{k}} f_0(E_n(\mathbf{k})) = |\xi_n(z)|^2 \int_{-\infty}^{+\infty} g_n(E) f_0(E) \, \mathrm{d}E. \tag{3.95}$$

Using Eq.3.95, once the wave-functions $\xi_n(z)$ are known, we can calculate the carrier density by evaluating the integral over the energy in Eq.3.95. We can even manage to obtain an analytical expression if $g_n(E)$ has a simple form, as for the electrons in the parabolic EMA approximation. We also note that, since the $\xi_n(z)$ are normalized in the z direction normal to the silicon–oxide interface, the inversion density D_{inv} per unit area of the subband n can be expressed as

$$D_{inv,n} = \int_{-\infty}^{+\infty} d_n(z) \, \mathrm{d}z = \int_{-\infty}^{+\infty} g_n(E) f_0(E) \, \mathrm{d}E. \tag{3.96}$$

The inversion density is an important parameter in the inversion layers, which is closely related to the electric field at the silicon–oxide interface and to the effective field in the inversion layer [24, 25] (see Section 7.1.1).

In the following we denote with $n(z)$ and N_{inv} respectively the carrier concentration and the inversion density in the electron inversion layer of an n-type MOSFET; $p(z)$ and P_{inv} are the equivalent quantities in the hole inversion layer of a p-type transistor.

3.6.1 Electron inversion layers

In the case of a 2D electron gas described according to the EMA approximation discussed in Section 3.2, the wave-function is obtained by solving Eq.3.16 and it is independent of $\mathbf{k}$. By using Eq.3.77 with $n_{sp} = 2$, the electron density per unit volume can be expressed with Eq.3.95 as

$$n(z) = \sum_{\nu,n} |\xi_{\nu,n}(z)|^2 \frac{\mu_\nu m_{d,\nu}}{\pi \hbar^2} \int_{E_{\nu 0}+\varepsilon_{\nu,n}}^{+\infty} [1 + 2\alpha(E - E_{\nu 0} - \varepsilon_{\nu,n})]\, f_0(E)\, \mathrm{d}E, \qquad (3.97)$$

where μ_ν is the valley multiplicity, $m_{d,\nu}$ is the effective mass for the density of states, and $f_0(E)$ is the Fermi–Dirac occupation function defined in Eq.3.92. By introducing the unitless quantities

$$x = \frac{E - E_{\nu 0} - \varepsilon_{\nu,n}}{K_B T}, \qquad \eta_{\nu,n} = \frac{E_F - E_{\nu 0} - \varepsilon_{\nu,n}}{K_B T}, \qquad (3.98)$$

Eq.3.97 can be rewritten as

$$n(z) = \sum_{\nu,n} |\xi_{\nu,n}(z)|^2 \frac{\mu_\nu m_{d,\nu} K_B T}{\pi \hbar^2} \int_0^{+\infty} \left[\frac{1 + (2\alpha K_B T)\, x}{1 + \exp(x - \eta_{\nu,n})} \right] \mathrm{d}x. \qquad (3.99)$$

By recalling the definition of the Fermi integrals given in Appendix A, Eq.3.99 can be finally written as

$$n(z) = \sum_{\nu,n} |\xi_{\nu,n}(z)|^2 \frac{\mu_\nu m_{d,\nu} (K_B T)}{\pi \hbar^2} \left[\ln(1 + \mathrm{e}^{\eta_{\nu,n}}) + 2\alpha (K_B T)\, \mathcal{F}_1(\eta_{\nu,n}) \right], \qquad (3.100)$$

where $\mathcal{F}_1(\eta_{\nu,n})$ denotes the Fermi integral of order $j = 1$. As can be seen, the effect of the non-parabolicity (i.e. a non-null α value) increases the density of states and thus the electron concentration for a given $\eta_{\nu,n}$ value.

There are essentially two elements that govern the electron density provided by a subband (ν,n), namely the position of minimum energy $(E_{\nu 0}+\varepsilon_{\nu,n})$ of the subband with respect to the Fermi level (see Eq.3.98), and the density of states, namely the product $(\mu_\nu m_{d,\nu})$.

The lower the energy minimum and the larger the $(\mu_\nu m_{d,\nu})$ product for a given subband, the larger is its contribution to the total electron density. In a silicon (001) inversion layer the $\Delta_{0.92}$ have lower $\varepsilon_{\nu,n}$ values than the $\Delta_{0.19}$ subbands because of the larger quantization mass; however, the $\Delta_{0.19}$ have a $(\mu_\nu m_{d,\nu})$ product more than four times larger than the $\Delta_{0.92}$ (see Table 3.1).

Figure 3.14 shows the subband energy minima (left) and the subband occupation (right) versus the total inversion density N_{inv} for the $\Delta_{0.92}$ and $\Delta_{0.19}$ subbands in a bulk MOSFET. These results have been obtained by employing the self-consistent solution of the Schrödinger and the Poisson equation explained in Section 3.7. As can be seen, at small inversion density, N_{inv}, the larger DoS of the $\Delta_{0.19}$ subbands partly compensates for the higher values of the subband minimum, so that the fraction of charge in the lowest $\Delta_{0.19}$ subband is comparable to the one in the lowest $\Delta_{0.92}$ subband. However, at large N_{inv} the confining electric field is large and the gap between the lowest $\Delta_{0.19}$ and the lowest $\Delta_{0.92}$ subband increases, hence the $\Delta_{0.92}$ subbands supply most of the carriers.

In this respect, Eq.3.98 points out that the distance between the minima of two subbands should be compared to the thermal energy $K_B T$ to judge the corresponding

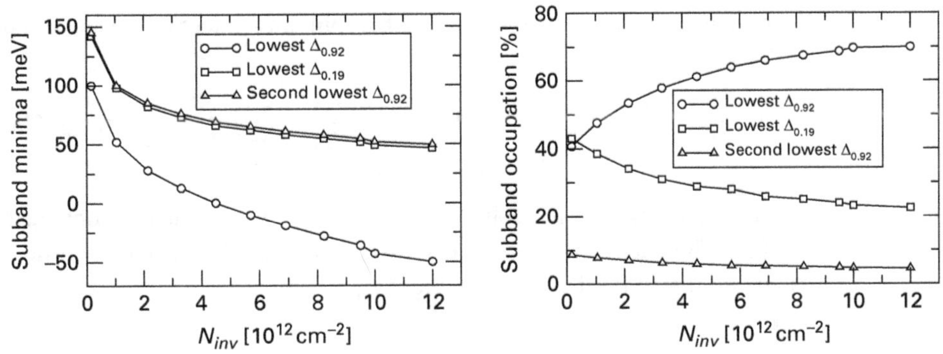

Figure 3.14 Subband minima referred to the Fermi level (left) and relative occupation (right) for the three lowest subbands versus the total inversion density N_{inv} in a (001) silicon n-MOSFET. Note that $\Delta_{0.92}$ and $\Delta_{0.19}$ indicate respectively the unprimed and the prime valley. Results obtained with a self-consistent solution of the Schrödinger–Poisson problem by using the parabolic effective mass approximation. Channel doping concentration $5\cdot 10^{17}$ cm^{-3}.

difference in the subband occupation. More precisely, the lower the temperature the larger the difference in subband occupation for a given gap in the subband minima. As can be seen in Fig.3.14, the energy difference between the lowest $\Delta_{0.19}$ and the lowest $\Delta_{0.92}$ can typically range from very few tens of meV at small N_{inv} to more than 100 meV at large inversion densities. At very low temperatures these differences tend to confine all the electrons within the $\Delta_{0.92}$ subbands and, in particular, in the lowest $\Delta_{0.92}$ subband. When essentially all the carriers belong to the lowest subband the 2D electron gas is said to be in the *quantum limit* [2].

A more detailed analysis of subband occupation and the implications for carrier transport appears in Sections 7.1.2, 7.1.3, and 7.2.2.

3.6.2 Hole inversion layers

If we adopt the semi-analytical model described in Section 3.3.3, then calculation of the hole density is similar to the case of electrons, in fact in this case the envelope wave-function is independent of **k**. We thus obtain:

$$p(z) = \sum_{\nu,n} |\xi_{\nu,n}(z)|^2 \int_{-\infty}^{+\infty} g_{\nu,n}(E)\, f_0(E)\, \mathrm{d}E, \tag{3.101}$$

where the density of states $g_{\nu,n}(E)$ of the subband n in the group ν is given by Eq.3.82. Since $g_{\nu,n}(E)$ is not known in an analytical form, the integral over E in Eq.3.101 must be evaluated numerically.

In the **k**·**p** quantization model, however, we solve an eigenvalue problem for each **k** value, hence the wave-function depends on **k**. If we wish to account for such a **k** dependence of the wave-function in Eq.3.94, then the sum must be evaluated according to the general prescriptions of Section 3.5.4. In particular, if the points $\mathbf{k}_d$ are arranged in triangles, the integral over $\mathbf{k}$ can be split into integrals inside the triangles and we have:

$$p(z) = \frac{1}{A} \sum_{n,\mathbf{k}} |\boldsymbol{\xi}_{n\mathbf{k}}(z)|^2 f_0(E_n(\mathbf{k})) = \frac{1}{(2\pi)^2} \sum_{T_{\mathbf{k}d}} \int_{T_{\mathbf{k}d}} |\boldsymbol{\xi}_{n\mathbf{k}}(z)|^2 f_0(E_n(\mathbf{k}))\, \mathrm{d}\mathbf{k}, \tag{3.102}$$

where $\boldsymbol{\xi}_{n\mathbf{k}}(z)$ is the six components envelope wave-function obtained by solving Eq.3.34. The linear interpolation discussed in Eq.3.90 for the energy inside the triangles can be used also for the wave-functions $\boldsymbol{\xi}_{n\mathbf{k}}(z)$.

3.6.3 Average values for energy and wave-vector at the equilibrium

At the equilibrium the electronic states in the inversion layer are occupied according to the Fermi–Dirac function $f_0(E)$ defined in Eq.3.92, hence it is possible to calculate the average value in the subband of several quantities of practical interest, such as the kinetic energy or the magnitude of the wave-vectors.

At the equilibrium the average value of a $\mathbf{k}$ dependent quantity $S(\mathbf{k})$ in a given subband n may be defined as

$$\langle S(\mathbf{k})\rangle_n = \frac{\frac{1}{A}\sum_{\mathbf{k}} S(\mathbf{k})\, f_0(E_n(\mathbf{k}))}{\frac{1}{A}\sum_{\mathbf{k}} f_0(E_n(\mathbf{k}))} = \frac{\frac{1}{A}\sum_{\mathbf{k}} S(\mathbf{k})\, f_0(E_n(\mathbf{k}))}{D_{inv,n}}, \tag{3.103}$$

where $D_{inv,n}$ is the inversion density in the subband n. The index n denotes both the valley and the subband if more valleys exist. By further averaging over the subbands and the valleys one may obtain the average value of $S(\mathbf{k})$ in the inversion layer.

By resorting to the results of Section 3.5.4, the sums over $\mathbf{k}$ in Eq.3.103 can be converted to integrals over $\mathbf{k}$ and then, if the energy dispersion in the subband $E_n(\mathbf{k})$ is known analytically, to integrals over the energy. If the energy relation is simple, it is thus possible to derive analytical expressions for the average value of some relevant quantities, which are very useful to gain a physical insight.

In this respect, let us consider an electron gas described by a parabolic EMA approximation (i.e. $\alpha = 0$) and with circular bands, namely with $m_x = m_y = m^*$. The energy dispersion is simply

$$E_n(\mathbf{k}) = \frac{\hbar^2 k^2}{2m^*},$$

and it is easy to calculate analytically the average value of both the kinetic energy and the wave-vector magnitude k in any subband n. For the kinetic energy $E_p = (E_n(\mathbf{k}) - E_{\nu 0} - \varepsilon_{\nu,n})$ in the subband n, for example, the numerator of Eq.3.103 can be written by using the results of Section 3.5.4 as

$$\begin{aligned}\langle E_p\rangle_n &= \frac{1}{A}\sum_{\mathbf{k}} (E_n(\mathbf{k}) - E_{\nu 0} - \varepsilon_{\nu,n})\, f_0(E_n(\mathbf{k}))\\ &= \frac{\mu_\nu m_{d,\nu}}{\pi\hbar^2} \int_{E_{\nu 0}+\varepsilon_{\nu,n}}^{+\infty} (E - E_{\nu 0} - \varepsilon_{\nu,n})\, f_0(E)\, \mathrm{d}E,\end{aligned} \tag{3.104}$$

where μ_v and $m_{d,v}$ are respectively the valley multiplicity and the effective mass for the density of states. By introducing the unitless quantities defined in Eq.3.98 and recalling Eq.3.100 to express the inversion density, $N_{inv,n}$, we readily obtain

$$\langle E_p \rangle_n = (K_B T) \left[\frac{\mathcal{F}_1(\eta_n)}{\mathcal{F}_0(\eta_n)} \right] = (K_B T) \left[\frac{\mathcal{F}_1(\eta_n)}{\ln(1 + \mathrm{e}^{\eta_n})} \right], \tag{3.105}$$

where $\mathcal{F}_0(\eta)$ and $\mathcal{F}_1(\eta)$ are the Fermi integrals respectively of order 1 and 0 defined in Appendix A.

For a non-degenerate electron gas, namely for η_n smaller than approximately -3 or -4, the Fermi integrals of any order tend to $\mathrm{e}^{-\eta}$, and the average value for the energy is approximately $(K_B T)$, which is about 26 meV at room temperature. If the electron gas is appreciably degenerate, instead, the ratio in the bracket of Eq.3.105 becomes larger than 1 and the average energy increases above $(K_B T)$; in this case the average energy in the lowest and most populated subband is larger than it is in the higher subbands.

For the average value of the wave-vector magnitude the calculations are similar to those that lead to Eq.3.105, and we obtain

$$\langle k \rangle_n = \frac{\sqrt{2\pi m^* K_B T}}{2\hbar} \left[\frac{\mathcal{F}_{\frac{1}{2}}(\eta_n)}{\ln(1 + \mathrm{e}^{\eta_n})} \right], \tag{3.106}$$

where $\mathcal{F}_{\frac{1}{2}}(\eta_n)$ is the Fermi integral of order 1/2. Thus $\langle k \rangle_n$ is approximately $(\sqrt{2\pi m^* K_B T}/2\hbar)$ for a non-degenerate gas and it increases in a degenerate gas.

In silicon the $\Delta_{0.19}$ subbands have $m^* = 0.19 m_0$ and the corresponding average k value is $\langle k \rangle \simeq 0.318\ \mathrm{nm}^{-1}$ for a non-degenerate electron gas at room temperature. Note that this value is very small compared to $[2\pi/a_0]$, that for silicon is about 11.6 nm^{-1}. Hence we can see that $\langle k \rangle$ is small compared to the extension of the Brillouin zone and to the magnitude of the reciprocal lattice vectors **G** defined in Section 2.1.2. This observation is very relevant for the discussion of inter-valley electronic transitions possibly induced by the scattering mechanisms (see Section 4.1.4).

Figure 3.15 shows numerically calculated $\langle E_p \rangle$ and $\langle k \rangle$ values for an electron inversion layer versus the inversion density, N_{inv}. The upper x-axis shows the η_0 of the lowest subband (see Eq.3.98). As can be seen, for a non-degenerate electron gas $\langle E_p \rangle$ is essentially $K_B T$ and $\langle k \rangle$ is in close agreement with the 0.318 nm^{-1} value calculated a few lines above. Furthermore, both $\langle E_p \rangle$ and $\langle k \rangle$ increase when the electron gas becomes degenerate, as expected from Eqs.3.105 and 3.106.

For a hole inversion gas it is not possible to derive simple expressions for the average kinetic energy and for $\langle k \rangle$, however, $K_B T$ is a reasonable approximation for $\langle E_p \rangle$. In this respect, Fig.3.16 shows the same quantities as in Fig.3.15 for a hole inversion layer and obtained with a self-consistent **k·p** solver. As can be seen, the $\langle E_p \rangle$ is close to $K_B T$ for a non-degenerate gas, whereas $\langle k \rangle$ is significantly larger than for the electrons. Figures 3.15 and 3.16 also show that, for a given inversion density, the η_0 value is significantly smaller for holes than for electrons, so that the effects of the carrier degeneracy are much smaller for the hole 2D gas. This is due to the larger hole density of states

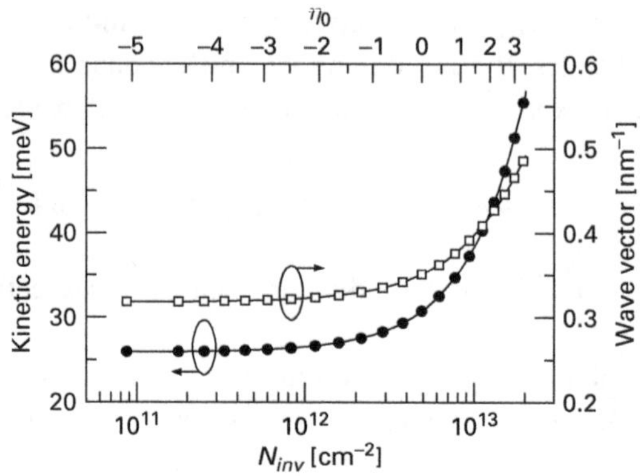

Figure 3.15 Average kinetic energy (left y-axis) and average magnitude k of the wave-vector (right y-axis) in an electron inversion layer versus the electron inversion density, N_{inv}. The upper x-axis shows the η_0 parameter of the lowest subband (see Eq.3.98). Results obtained with a self-consistent Schrödinger–Poisson solver and for a bulk n-MOS with channel doping, $N_A = 5 \cdot 10^{16}$ cm^{-3}, $T = 300\,K$.

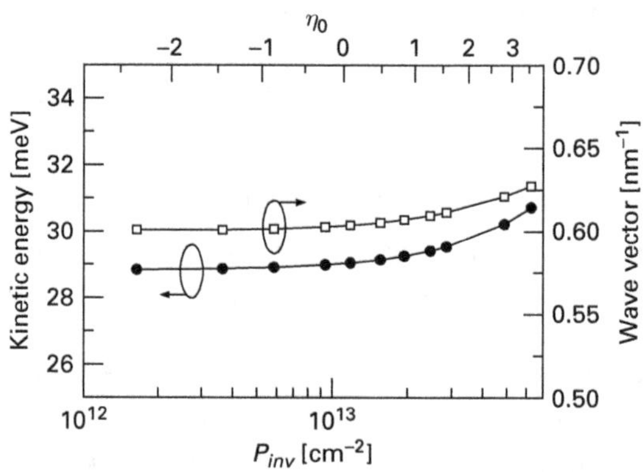

Figure 3.16 Same quantities as in Fig.3.15 shown versus the hole inversion density, P_{inv}. Results obtained with a self-consistent $\mathbf{k}\cdot\mathbf{p}$ solver and for a bulk p-MOS with channel doping, $N_D = 5 \cdot 10^{16}$ cm^{-3}, $T = 300\,K$.

illustrated in Fig.3.13. Such a large density of states makes the degeneracy effects quite modest for holes in the practically relevant range of inversion densities up to about $2{\cdot}10^{13}$ cm^{-2}; we return to this point in Section 7.1.2.

We conclude by noting that the values of $\langle E_p \rangle$ and $\langle k \rangle$ are very relevant for the scattering rates in the inversion layer and we refer to the results of this section several times in subsequent chapters.

3.7 Self-consistent calculation of the electrostatic potential

We have seen how to calculate the energy relation in an inversion layer and then use the density of states to compute the electron and hole concentrations under the assumption of thermal equilibrium. In the solution of the EMA or **k·p** eigenvalue problem necessary to complete the first step of this procedure we have assumed that the potential well $U(z)$ is known. However, the shape of the confining well is controlled by the electrostatic potential profile $\phi(z)$. In fact, in the case of electron inversion layers, $U(z)$ can be defined as:

$$U(z) = -\mathrm{e}\phi(z) - \chi(z), \tag{3.107}$$

where $\chi(z)$ is the position-dependent electron affinity along the quantization direction.

The solution of the eigenvalue problem has thus to be coupled with the solution of the Poisson equation:

$$\frac{\mathrm{d}}{\mathrm{d}z}\left[\epsilon(z)\frac{\mathrm{d}\phi}{\mathrm{d}z}\right] = \mathrm{e}\left[n(z) - p(z) + N_A(z) - N_D(z)\right], \tag{3.108}$$

which links the electrostatic potential $\phi(z)$ to the electron and hole concentration profiles. The dielectric constant ϵ is a material property which is constant inside each device region, so that the Poisson equation becomes

$$\epsilon\frac{\mathrm{d}^2\phi}{\mathrm{d}z^2} = \mathrm{e}\left[n(z) - p(z) + N_A(z) - N_D(z)\right] \tag{3.109}$$

inside each device region; the continuity condition applied at the interfaces between different materials is the conservation of ϕ and $\epsilon(\mathrm{d}\phi/\mathrm{d}z)$.

The calculation of the charge and potential profile obtained by coupling the Schrödinger and Poisson equations is denoted as *self-consistent*. This coupling is necessary since the carrier concentrations $n(z)$ and $p(z)$ in the Poisson equation derive from Eqs.3.100 or 3.101, which contain the subband energies and the wave-functions. These latter are obtained by solving the Schrödinger equation using the potential energy profile $U(z)$ defined in Eq.3.107. However, differently from the classical case, the non-local nature of the Schrödinger equation does not allow us to find an explicit relation between the $\phi(z)$ and the corresponding $n(z)$ and $p(z)$ at the same position. One thus needs to iterate the solution of the Schrödinger and Poisson equations until convergence is reached. This iteration poses stability problems which are examined in Section 3.7.1.

3.7.1 Stability issues

In order to address the stability issues associated with the coupled solving of the Poisson equation and of an independent equation used to compute the charge density, it is convenient to consider a situation much simpler than a 2D inversion layer, namely an uniform n-type semiconductor slab with constant donor doping N_D and dielectric constant ϵ_S and without any external bias. The main results we obtain in such a reference case apply also to more complex situations. We consider a one-dimensional problem in the direction z. In this case, with a suitable choice of the potential reference, the electron concentration profile is simply given by [26]

$$n(z) = N_D \exp\left(\frac{e\phi(z)}{K_B T}\right). \qquad (3.110)$$

In this simple example one could substitute Eq.3.110 into Eq.3.109 and obtain a non-linear differential equation which is easy to solve with the Newton scheme. However, this is not possible in inversion layers, where the non-local nature of the Schrödinger equation does not allow us to find a local relation between the carrier concentrations and the electrostatic potential as Eq.3.110. Thus, in order to emulate the coupling between the Poisson equation 3.109 and the Schrödinger equation in inversion layers, we hereafter study the iterative solution of Eq.3.110 and Eq.3.109. The correct solution of the self-consistent problem is $\phi(z) = 0$ and $n(z) = N_D$ for all z. However, it is easy to show that the iterative scheme (obtained by taking a guess for $\phi(z)$ from Eq.3.109, using it in Eq.3.110 to calculate $n(z)$ and then, by feeding $n(z)$ back into Eq.3.109, to calculate the new guess for $\phi(z)$) is unstable; in fact such a procedure results in solutions that deviate more and more from the correct solution with increasing iteration number.

To demonstrate this point, we consider generic deviations from the exact solutions written as plane waves:

$$\phi^{(k)}(z) = \tilde{\phi}^{(k)} \exp(i\beta z), \qquad (3.111a)$$

$$n^{(k)}(z) - N_D = \tilde{n}^{(k)} \exp(i\beta z), \qquad (3.111b)$$

where (k) refers to the iteration step and β is the wave-vector of the perturbation, which is related to the wave-length λ by $\beta = 2\pi/\lambda$.

By inserting these expressions in Eq.3.109 we get:

$$-\beta^2 \epsilon_S \tilde{\phi}^{(k)} = e\tilde{n}^{(k)}. \qquad (3.112)$$

The charge at the kth iteration is computed using Eq.3.110 from the potential at the iteration $k-1$. We can linearize Eq.3.110, thus readily finding

$$\tilde{n}^{(k)} \simeq N_D \frac{e\tilde{\phi}^{(k-1)}}{K_B T}. \qquad (3.113)$$

Substitution of $\tilde{n}^{(k)}$ in Eq.3.112 yields:

$$\frac{\tilde{\phi}^{(k)}}{\tilde{\phi}^{(k-1)}} = -\frac{e^2 N_D}{\epsilon K_B T \beta^2} = -\frac{1}{L_D^2 \beta^2}, \qquad (3.114)$$

where we have used the definition of the Debye length

$$L_D = \sqrt{\frac{K_B T \epsilon_S}{e^2 N_D}}. \qquad (3.115)$$

We see that perturbations of the correct solution with wave-vector $\beta < 1/L_D$ can grow during the iterative process, since in this case $|\tilde{\phi}^{(k)}/\tilde{\phi}^{(k-1)}| > 1$.

The problem originates from the fact that the charge depends exponentially on the potential, whereas the effect of charge variations on potential variations is linear. It is thus important to damp the variation of the potential profile from the $(k-1)$th to the kth iteration. To this purpose, a widely used technique is the *predictor–corrector* scheme, where we replace Eq.3.109 with the so-called *non-linear Poisson equation*:

$$\epsilon_S \frac{\mathrm{d}^2\phi^{(k)}}{\mathrm{d}z^2} = \mathrm{e}\left\{ n^{(k)} \exp\left[\frac{\mathrm{e}(\phi^{(k)} - \phi^{(k-1)})}{K_B T}\right] - p^{(k)} \exp\left[\frac{\mathrm{e}(\phi^{(k-1)} - \phi^{(k)})}{K_B T}\right] + N_A - N_D \right\}. \tag{3.116}$$

According to Eq.3.116, in the solution of the Poisson equation the changes of $\phi^{(k)}$ with respect to $\phi^{(k-1)}$ are damped by assuming that they result in exponential modifications of the carrier concentrations. Of course, when convergence is reached (i.e. $\phi^{(k)} \simeq \phi^{(k-1)}$), Eq.3.116 is equivalent to Eq.3.109.

The stability of Eq.3.116 when applied to the template uniform slab can be easily demonstrated as follows. We compute from Eq.3.110 the charge at the kth iteration from the potential at the previous iteration as

$$n^{(k)} = N_D \exp\left(\frac{\mathrm{e}\phi^{(k-1)}}{K_B T}\right). \tag{3.117}$$

Substitution into Eq.3.116 and the additional assumptions $N_A = 0$ and $p(z) \simeq 0$ yield:

$$\epsilon_S \frac{\mathrm{d}^2\phi^{(k)}}{\mathrm{d}z^2} = \mathrm{e}N_D \left(\exp\frac{\mathrm{e}\phi^{(k)}}{K_B T} - 1\right). \tag{3.118}$$

The solution of this equation is $\phi^{(k)} = 0$, meaning that any perturbation at the $(k-1)$th iteration is eliminated at the next iteration.

The case of the uniform silicon slab examined so far is not fully representative of the more complex situations found in realistic devices. In particular, in our toy model an explicit local relation exists between the potential $\phi(z)$ and the carrier density $n(z)$ (see Eq.3.110), whereas this is not the case in many practical applications. In an inversion layer, for instance, in order to calculate the charge we need to solve the Schrödinger equation and then use Eq.3.100 for electrons and either Eq.3.101 or Eq.3.102 for holes. Thus the procedure to compute the charge from the potential has to be iterated with the Poisson equation 3.116. This latter equation is non-linear and itself requires iterative methods such as the Newton scheme to be solved.

3.7.2 Electron inversion layers and boundary conditions

In this section we analyze in detail the self-consistent calculation of the potential and charge density profiles in an electron inversion layer by considering different models for the electron charge and different MOS structures. The boundary conditions for the Poisson equation are discussed in context.

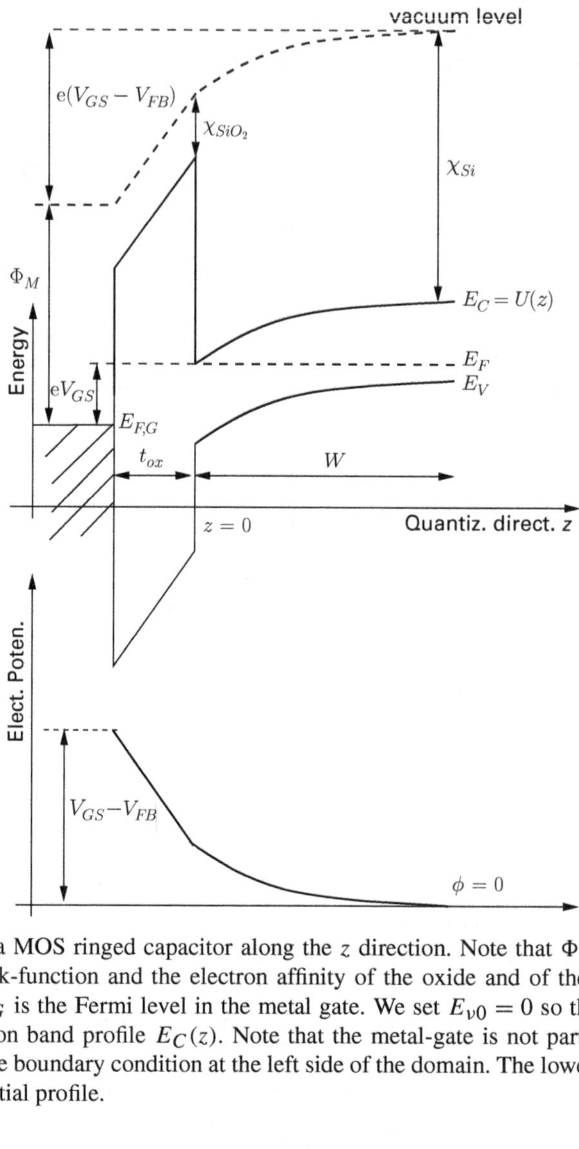

Figure 3.17 Band diagram of a MOS ringed capacitor along the z direction. Note that Φ_M, χ_{SiO_2} and χ_{Si} are the metal work-function and the electron affinity of the oxide and of the silicon substrate, respectively, $E_{F,G}$ is the Fermi level in the metal gate. We set $E_{v0} = 0$ so that $U(z)$ coincides with the conduction band profile $E_C(z)$. Note that the metal-gate is not part of the simulation domain but sets the boundary condition at the left side of the domain. The lower plot sketches the electrostatic potential profile.

Bulk MOSFET with classical charge distribution

Let us consider a bulk MOS ringed capacitor with an acceptor substrate doping $N_A(z)$. The band diagram and potential profile are given in Fig.3.17. We assume an extension W of the substrate region much larger than the depletion region.

Let us consider the self-consistent solution of the potential and charge in the classical (no quantization) and non-degenerate (Maxwell–Boltzmann statistics) case. Under these assumptions the carrier concentrations depend exponentially on the potential $\phi(z)$ [26]. If we set $\phi(W) = 0$ and denote by n_i the intrinsic concentration [26], we have

$$n(z) = n(W) \exp\left(\frac{e\phi(z)}{K_B T}\right) \simeq \frac{n_i^2}{N_A(W)} \exp\left(\frac{e\phi(z)}{K_B T}\right), \tag{3.119a}$$

$$p(z) = p(W)\exp\left(-\frac{e\phi(z)}{K_BT}\right) \simeq N_A(W)\exp\left(-\frac{e\phi(z)}{K_BT}\right), \qquad (3.119b)$$

in the semiconductor material and $n(z) \simeq 0$, $p(z) \simeq 0$ in the dielectric. Direct substitution of Eqs.3.119 into the Poisson equation 3.109 yields a non-linear differential equation in $\phi(z)$:

$$\epsilon_S \frac{d^2\phi}{dz^2} = e\left[n(W)\exp\left(\frac{e\phi(z)}{K_BT}\right) - p(W)\exp\left(-\frac{e\phi(z)}{K_BT}\right) + N_D\right]. \qquad (3.120)$$

This equation can be solved by Newton's method, and there is no need to couple it with any other equation, since the charge is already expressed as a function of the potential. A comment is in order about the boundary conditions for Eq.3.120. The Eqs.3.119 are based on the choice $\phi(W) = 0$. Therefore, as suggested by Fig.3.17, we must impose $\phi(-t_{ox}) = (V_{GS} - V_{FB})$ at the gate, where V_{FB} is the flat-band voltage [26] and V_{GS} is the gate-source voltage, equal to the gate-bulk voltage for a ringed capacitor or for a MOSFET with the $n+$ regions shorted to the bulk. Since we solve Eq.3.120 only in the substrate ($z > 0$), we can use Gauss's law and impose

$$\epsilon_S \frac{d\phi}{dz}(0) = \epsilon_{ox}\frac{V_{GS} - V_{FB} - \phi(0)}{t_{ox}}, \qquad (3.121)$$

where ϵ_S and ϵ_{ox} are the dielectric constants of the substrate and of the dielectric, respectively.

To express the flat-band voltage of the structure in Fig.3.17 we start by observing that [1]

$$e(V_{GS} - V_{FB}) = e[\phi(-t_{ox}) - \phi(W)] = -\chi_{SiO_2} - U(-t_{ox}) - [-\chi_{Si} - U(W)]. \qquad (3.122)$$

In the absence of charge in the oxide, we can write $U(-t_{ox})$ as a function of the Fermi level E_F in the substrate as:

$$U(-t_{ox}) = E_F - eV_{GS} + \Phi_M - \chi_{SiO_2}. \qquad (3.123)$$

Substitution of Eq.3.123 into Eq.3.122 allows us to express the flat band voltage as

$$V_{FB} = \frac{\Phi_M - \chi_{Si} - [U(W) - E_F]}{e}. \qquad (3.124)$$

If the substrate is non-degenerate and $N_A(z)$ is smoothly changing with z, $U(W) - E_F \simeq E_G - K_BT\log[N_V/N_A(W)]$ [26], where E_G is the energy gap of the silicon substrate and N_V is the effective density of states in the valence band.

Bulk MOSFET with quantum mechanical charge distribution

We now consider the case of a quantized inversion layer. As already noted, the expression of the electron charge as a function of the electrostatic potential is not explicit. This requires iteration between the Schrödinger and Poisson equations. In fact, the electron

[1] Use Eq.3.107 and consider that $e(V_{GS} - V_{FB})$ should be equal to the difference in $U(z)$ between the substrate and the SiO_2/metal interface aside from the changes in affinity.

concentration profile $n^{(k)}$ at the kth iteration is given by Eq.3.100, which requires the subband energies and wave-functions obtained from solution of the Schrödinger equation using the potential profile $\phi^{(k-1)}$ calculated with the Poisson equation at step $(k-1)$.

In the case of electron inversion layers, the hole concentration can be expressed as in Eq.3.119b, so that Eq.3.116 can be simplified to

$$\epsilon_S \frac{\mathrm{d}^2\phi^{(k)}}{\mathrm{d}z^2} = \mathrm{e}\left[n^{(k)} \exp\left(\frac{\mathrm{e}(\phi^{(k)}-\phi^{(k-1)})}{K_B T}\right) - p(W)\exp\left(-\frac{\mathrm{e}\phi^{(k)}}{K_B T}\right) + N_A(z)\right], \tag{3.125}$$

where we have again set $\phi(W)=0$. Equation 3.125 must be solved iteratively with the Schrödinger equation until convergence is reached (i.e. the absolute value of the potential updates between two subsequent iterations is below a given tolerance at each grid point).

To solve the Schrödinger equation we need to relate the electrostatic potential $\phi(z)$ and the potential energy $U(z)$. One possibility is to use Eq.3.107. This means that the energy reference is the vacuum level in the substrate, consistently with Fig.3.17. Hence the subband energies $\varepsilon_{v,n}$ obtained from the Schrödinger equation and the Fermi level E_F should be referred to the vacuum level. By recalling the brief discussion after Eq.3.124 we have

$$E_F = -\chi_{Si} - E_G + K_B T \ln\left(\frac{N_V}{N_A(W)}\right). \tag{3.126}$$

Of course, any rigid shift of $U(z)$ and E_F is allowed. In particular, one could set E_F as the energy reference and then express $U(z)$ by subtracting the r.h.s. of Eq.3.126 from Eq.3.107. All these choices are consistent with the boundary condition $\phi(W)=0$, which is convenient in the sense that it simplifies the computation of the $p(z)$ profile. However, any rigid shift of the $\phi(z)$ profile is also allowed as long as it is also included in the definition of $U(z)$ and it is taken into account in computation of $p(z)$. As an example of yet another choice for the energy reference, in many commercial simulators the potential is referred to the intrinsic silicon Fermi level $E_{F,int} = E_G/2 + (K_B T/2)\ln(N_V/N_C)$.

Boundary conditions for an SOI MOSFET

As an example of an alternative choice for the energy reference, we describe below how to determine the boundary conditions for an SOI structure, sketched in Fig.3.18. In this case we use the Fermi level in the silicon film as energy reference, that is we set $E_F = 0$. We furthermore define:

$$U(z) = -\mathrm{e}\phi(z) - [\chi(z) - \chi_{Si}], \tag{3.127}$$

which simplifies the expression of $U(z)$ in the silicon film, where we have $U(z) = -\mathrm{e}\phi(z)$.

In order to find the electrostatic potential values ϕ_{FG} and ϕ_{BG} at the front and back metal gates, we remember that the applied biases V_{FG} and V_{BG} represent the shift of

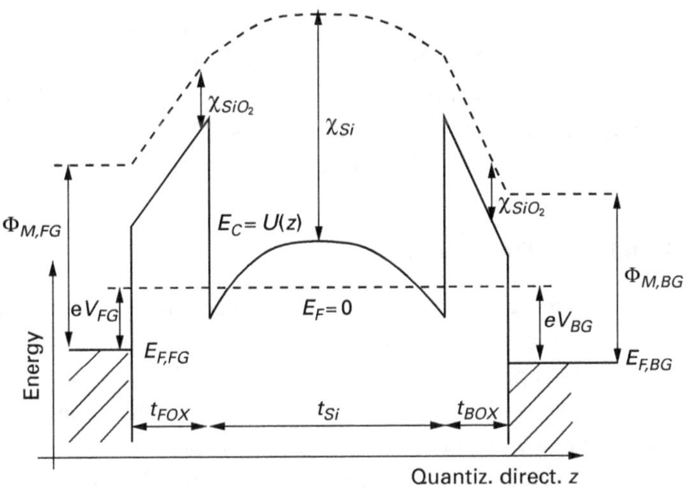

Figure 3.18 Band diagram along the z direction inside an SOI structure. Only the conduction band profile is shown for simplicity. t_{FOX} and t_{BOX} indicate the thickness of the front and back oxide layer, respectively.

the Fermi level in the metal gates ($E_{F,FG}$ and $E_{F,BG}$) with respect to the Fermi level E_F in the substrate. Since we took $E_F = 0$, from Fig.3.18 we see that the value of U at the FG and BG interface (in particular at the SiO_2 side of the metal/SiO_2 interfaces, since $\chi(z)$ is undefined in the metal) is given by

$$U_{FG/BG} = -eV_{FG/BG} + (\Phi_{M,FG/M,BG} - \chi_{SiO_2}), \tag{3.128}$$

where $\Phi_{M,FG}$ and $\Phi_{M,BG}$ are the work-functions of the front and back interface metal gates, respectively.[2]

On the other hand, from Eq.3.127 we find the boundary conditions for the electrostatic potentials ϕ_{FG} and ϕ_{BG} as:

$$\phi_{FG/BG} = V_{FG/BG} - (\Phi_{M,FG/M,BG} - \chi_{Si})/e. \tag{3.129}$$

These boundary conditions can be used also in a MOS transistor biased with a non-null V_{DS}. It must be remembered that now the potential $\phi = 0$ no longer corresponds to a p-type region at equilibrium as in Fig.3.17. Thus, since E_F is null and $E_V = [-e\phi(z) - E_G]$, we need to modify the expression of the hole profile to:

$$p(z) = N_V \exp\left(-\frac{e\phi(z) + E_G}{K_B T}\right). \tag{3.130}$$

The brief overview in this section of self-consistent treatment of MOS capacitors considers a simplified structure with an ideal metal gate and a single dielectric layer.

[2] Considering the front metal/SiO_2 interface in Fig.3.18, U_{FG} can be obtained by starting from the horizontal dashed line denoting E_F, moving down by eV_{FG} to the Fermi level in the metal, then up by Φ_{MFG} (reaching the vacuum level) and then down again by χ_{SiO_2}, thus reaching the $U(z)$ profile (solid line) at the SiO_2 side of the metal/dielectric interface.

There is a vast literature on this topic reporting extensions to gate stacks with multiple dielectric layers [27–30], addition to the quantized levels of a continuum of classical states for $E > U(W)$ (in order to handle bias conditions where the potential well tends to disappear and only few subbands exists) [31–33], treatment of quantum effects in the poly-silicon gate [34, 35] and effects of wave-function penetration in the dielectric [36].

3.7.3 Speed-up of the convergence

The predictor–corrector scheme introduced into Eq.3.125 assumes that the electron charge is modified exponentially by changes in the electrostatic potential, in analogy with what happens in the classical case and with Maxwell–Boltzmann statistics. However, in quantized inversion layers at large (in magnitude) gate bias and strong carrier gas degeneracy, the electron charge dependence on $\phi(z)$ strongly deviates from exponential behavior (as can be seen by inspection of Eq.3.100 where, instead of $\exp(-\eta_{\nu,i})$, we have $\ln[1+\exp(-\eta_{\nu,i})]$). As a result, the damping introduced by the predictor–corrector scheme can significantly slow down the convergence of the self-consistent loop. An effective way to reduce the number of iterations is the scheme proposed in [37]. In this approach the Poisson equation is written as

$$\epsilon_S \frac{d^2\phi^{(k)}}{dz^2} = e\left[n^{(k)}(\eta_{\nu,i}^{(k)}) - p(W)\exp\left(-\frac{e\phi^{(k)}}{K_B T}\right) + N_A\right], \tag{3.131}$$

where, as in Eq.3.125, we assume that the hole concentration can be expressed as $p(z) = p(W)\exp(-e\phi^{(k)}/K_B T)$. The term $n^{(k)}(\eta_{\nu,i}^{(k)})$ is an estimate of the charge given by Eq.3.100, where $\eta_{\nu,i}^{(k)}$ is

$$\eta_{\nu,i}^{(k)}(z) = \frac{E_F - \varepsilon_{\nu,i}^{(k-1)} - E_{\nu 0} + e[\phi^{(k)}(z) - \phi^{(k-1)}(z)]}{K_B T}. \tag{3.132}$$

Differently from the $\eta_{\nu,i}$ in Eq.3.98, the $\eta_{\nu,i}^{(k)}(z)$ in Eq.3.132 depends on z through $[\phi^{(k)}(z) - \phi^{(k-1)}(z)]$.

Since $n^{(k)}(\eta_{\nu,i}^{(k)})$ contains the unknown potential profile $\phi^{(k)}$ then, for the set of subband minima and wave-functions obtained from the Schrödinger equation at the kth iteration, we must substitute into Eq.3.131 the $n^{(k)}(\eta_{\nu,i}^{(k)})$ expression obtained with Eq.3.100 and with the $\eta_{\nu,i}^{(k)}$ in Eq.3.132, and then solve Eq.3.131 with the Newton scheme. The Schrödinger equation and Eq.3.131 must be solved iteratively until an appropriate convergence is reached.

3.8 Summary

This chapter has illustrated the fundamental concepts related to a 2D carrier gas in the inversion layer of an MOS transistor; the models as well as the notation introduced here will be used throughout the rest of the book.

Sections 3.2 and 3.3 discussed how to determine the energy relation for either an electron or a hole inversion layer for a (001) silicon substrate. These models will be naturally extended to arbitrary crystal orientations and to semiconductors alternative to silicon in Chapters 8 and 10, as well as to the case of strained silicon in Chapter 9. Section 3.4 illustrates a full-band quantization approach, whose accuracy and completeness goes beyond the EMA and $\mathbf{k}\cdot\mathbf{p}$ models of Sections 3.2 and 3.3, while Section 3.5 deals in detail with evaluation of the sums over the wave-vector $\mathbf{k}$, that occur in calculation of very many physical quantities in inversion layers.

Section 3.6 explains how, at equilibrium, carrier densities can be directly calculated from the energy relation by using the Fermi–Dirac function for the occupation of the states and appropriate sums and integrals over the $\mathbf{k}$ space.

Section 3.7 clarifies that the confining potential, which determines the energy relation in the inversion layer, depends in turn on the carrier densities, so that a self-consistent determination of the carrier densities and of the electrostatic potential is in general necessary. Such a self-consistent solution completely determines the electrostatics of the MOS device at equilibrium. In out of equilibrium conditions, instead, the electrostatics is inherently coupled to the transport problem through the occupation of states, as discussed in detail in Section 6.2.

The theoretical and practical relevance of the results presented in this chapter is far wider than electrostatics; in fact the energy relation in the inversion layers is a fundamental ingredient also for calculation of carrier velocities, and hence of the currents at device terminals. Furthermore, the prescriptions of Section 3.5.4 concerning evaluation of the sums over the wave-vector $\mathbf{k}$ are used very extensively in the rest of the book for calculation of, for instance, scattering rates (in Chapter 4 and 10), momentum relaxation times (in Section 5.4), and ballistic currents (in Section 5.6).

References

[1] Y. Taur and T. Ning, *Fundamentals of Modern VLSI Devices.* Edinburgh: Cambridge University Press, 1998.

[2] T. Ando, A. Fowler, and F. Stern, "Electronic properties of two-dimensional systems," *Review of Modern Physics*, vol. 54, pp. 437–673, 1982.

[3] M. Ferrier, R. Clerc, G. Ghibaudo, F. Boeuf, and T. Skotnicki, "Analytical model for quantization on strained and unstrained bulk nMOSFET and its impact on quasi-ballistic current," *Solid State Electronics*, vol. 50, no. 1, pp. 69–77, 2006.

[4] M.V. Fischetti and S.E. Laux, "Monte Carlo study of electron transport in silicon inversion layers," *Phys. Rev. B*, vol. 48, pp. 2244–2274, 1993.

[5] D.K. Ferry and S.M. Goodnick, *Transport in Nanostructures.* Cambridge: Cambridge University Press, 1997.

[6] C. Jungemann, A. Edmunds, and W.L. Engl, "Simulation of linear and nonlinear electron transport in homogeneous silicon inversion layers," *Solid State Electronics*, vol. 36, no. 11, pp. 1529–1540, 1993.

[7] L. Lucci, P. Palestri, D. Esseni, L. Bergagnini, and L. Selmi, "Multi-subband Monte-Carlo study of transport, quantization and electron gas degeneration in ultra-thin SOI n-MOSFETs," *IEEE Trans. on Electron Devices*, vol. 54, no. 5, pp. 1156–1164, 2007.

[8] M.V. Fischetti, Z. Ren, P.M. Solomon, M. Yang, and K. Rim, "Six-band k·p calculation of the hole mobility in silicon inversion layers: Dependence on surface orientation, strain, and silicon thickness," *Journal of Applied Physics*, vol. 94, no. 2, pp. 1079–1095, 2003.

[9] M. De Michielis, D. Esseni, and F. Driussi, "Analytical models for the insight into the use of alternative channel materials in ballistic nano-MOSFETs," *IEEE Trans. on Electron Devices*, vol. 54, no. 1, pp. 115–123, 2006.

[10] M. De Michielis, D. Esseni, Y.L. Tsang, *et al.*, "A semianalytical description of the hole band structure in inversion layers for the physically based modeling of pMOS transistors," *IEEE Trans. on Electron Devices*, vol. 54, no. 9, pp. 2164–2173, 2007.

[11] E. Wang, P. Montagne, L. Shifren, *et al.*, "Physics of hole tranport in strained silicon MOSFET inversion layers," *IEEE Trans. on Electron Devices*, vol. 53, no. 8, pp. 1840–1850, 2006.

[12] E.O. Kane, "Energy band structure in p-type germanium and silicon," *Journal of Phys. Chem. Solids*, vol. 1, pp. 82–99, 1956.

[13] A. Di Carlo, "Microscopic theory of nanostructured semiconductor devices: beyond the envelope-function approximation," *Semiconductor Science Technology*, vol. 18, pp. R1–R31, 2003.

[14] G. Klimeck, S.S. Ahmed, H. Bae, *et al.*, "Atomistic simulation of realistically sized nanodevices using NEMO 3D Part I: Models and benchmarks," *IEEE Trans. on Electron Devices*, vol. 54, no. 9, pp. 2079–2089, 2007.

[15] G. Klimeck, S.S. Ahmed, H. Bae, *et al.*, "Atomistic simulation of realistically sized nanodevices using NEMO 3D Part II: Applications," *IEEE Trans. on Electron Devices*, vol. 54, no. 9, pp. 2090–2099, 2007.

[16] J.M. Luttinger and W. Kohn, "Motion of electrons and holes in perturbed fields," *Phys. Rev.*, vol. 97, pp. 869–883, 1955.

[17] D. Esseni and P. Palestri, "Linear combination of bulk bands method for investigating the low-dimensional electron gas in nanostructured devices ," *Phys. Rev. B*, vol. 72, pp. 165342 (1–14), 2005.

[18] D. Esseni and P. Palestri, "Full-band quantization analysis reveals a third valley in (001) silicon inversion layers," *IEEE Electron Device Lett.*, vol. 26, no. 6, pp. 413–415, 2005.

[19] J.R. Chelikowsky and M.L. Cohen, "Nonlocal pseudopotential calculations for the electronic structure of eleven diamond and zinc-blende semiconductors," *Phys. Rev. B*, vol. 14, no. 2, pp. 556–582, 1976.

[20] J.-L.P.J. van der Steen, D. Esseni, P. Palestri, L. Selmi, and R.J.E. Hueting, "Validity of the parabolic Effective Mass Approximation in silicon and germanium n-MOSFETs with different crystal orientations," *IEEE Trans. on Electron Devices*, vol. 54, pp. 1843–1851, Aug. 2007.

[21] V. Sverdlov, D. Esseni, O. Baumgartner, *et al.*, "The linear combination of bulk bands-method for electron and hole subband calculations in strained silicon films and surface layers," in *International Workshop on Computational Electronics*, pp. 49–52, 2009.

[22] J. Wang, A. Rahman, A. Ghosh, G. Klimeck, and M. Lundstrom, "On the validity of the parabolic effective-mass approximation for the I-V calculation of silicon nanowire transistors," *IEEE Trans. on Electron Devices*, vol. 52, no. 7, pp. 1589–1595, 2005.

[23] A. Jeffrey, *Handbook of Mathematical Formulas and Integrals*. San Diego, CA: Academic Press, 2000.

[24] A.G. Sabnis and J.T. Clemens, "Characterization of the electron mobility in the inverted ⟨100⟩ Si surface," in *IEEE IEDM Technical Digest*, pp. 18–21, 1979.

[25] S. Takagi, A. Toriumi, M. Iwase, and H. Tango, "On the universality of inversion layer mobility in Si MOSFET's: Part 1–effects of substrate impurity concentration," *IEEE Trans. on Electron Devices*, vol. 41, pp. 2358–2362, 1994.

[26] S.M. Sze, *Physics of Semiconductor Devices*. New York: Wiley, 1981.

[27] A. Wettstein, A. Schenk, and W. Fichtner, "Simulation of direct tunneling through stacked gate dielectrics by a fully integrated 1D-Schrödinger-Poisson solver," in *Proc.SISPAD*, pp. 243–246, 1999.

[28] B. Govoreanu, P. Blomme, M. Rosmeulen, J. Van Houdt, and K. De Meyer, "A model for tunneling current in multi-layer tunnel dielectrics," *Solid State Electronics*, vol. 47, no. 6, pp. 1045–1053, 2003.

[29] T. Kauerauf, B. Govoreanu, R. Degraeve, G. Groeseneken, and H. Maes, "Scaling CMOS: Finding the gate stack with the lowest leakage current," *Solid State Electronics*, vol. 49, no. 5, pp. 695–701, 2005.

[30] F. Driussi, S. Marcuzzi, P. Palestri, and L. Selmi, "Gate current in stacked dielectrics for advanced FLASH EEPROM cells," in *Proc. European Solid State Device Res. Conf.*, pp. 317–320, 2005.

[31] C. Bowen, C.L. Fernando, G. Klimeck, *et al.*, "Physical oxide thickness extraction and verification using quantum mechanical simulation," in *IEEE IEDM Technical Digest*, pp. 869–872, 1997.

[32] A. Ghetti, A. Hamad, P.J. Silverman, H. Vaidya, and N. Zhao, "Self-consistent simulation of quantization effects and tunneling current in ultra-thin gate oxide MOS devices," in *Proc.SISPAD*, pp. 239–242, 1999.

[33] A. Dalla Serra, A. Abramo, P. Palestri, L. Selmi, and F. Widdershoven, "Closed- and open-boundary models for gate-current calculation in n-MOSFETs," *IEEE Trans. on Electron Devices*, vol. 48, no. 8, pp. 1811–1815, 2001.

[34] A. Pacelli, A.S. Spinelli, and L.M. Perron, "Carrier quantization at flat bands in MOS devices," *IEEE Trans. on Electron Devices*, vol. 46, no. 2, pp. 383–387, 1999.

[35] A.S. Spinelli, A. Pacelli, and A.L. Lacaita, "Polysilicon quantization effects on the electrical properties of MOS transistors," *IEEE Trans. on Electron Devices*, vol. 47, no. 12, pp. 2366–2371, 2000.

[36] S. Mudanai, L.F. Register, A.F. Tasch, and S.K. Banerjee, "Understanding the effects of wave function penetration on the inversion layer capacitance of nMOSFETs," *IEEE Electron Device Lett.*, vol. 22, no. 3, p. 145, 2001.

[37] A. Trellakis, T. Galick, A. Pacelli, and U. Ravaioli, "Iteration scheme for the solution of the two-dimensional Schrödinger–Poisson equations in quantum structures," *Journal of Applied Physics*, vol. 81, no. 12, 1997.

4 Carrier scattering in silicon MOS transistors

In Chapter 3 we discussed in detail determination of the energy levels and wave-functions of a quasi-2D electron or hole gas at equilibrium. Consistent with the time independent Schrödinger equation used in our calculations, the electronic states are stationary states, in the sense that the carriers do not have transitions between the states and the lifetime of the states is infinitely long.

In real physical systems there are perturbations of the Hamiltonian used to calculate the band structure that can be collectively referred to as *scattering mechanisms* and that produce a very large number of electronic transitions per second between the available states.

If the system is driven out of equilibrium by an external stimulus, such as the source to drain electric field responsible for the drain current in a MOSFET, then the scattering mechanisms tend to restore the equilibrium. This latter statement means that, on the one hand, if the stimulus is removed at a given time, then the scattering events govern the transient of the system back to the equilibrium. On the other hand if, say, an electric field tends to accelerate the carriers and transfer energy to the carrier gas, then in a stationary transport condition an equilibrium is reached between the effects of the field and the relaxation of the momentum and energy produced by the scattering events. Such an equilibrium determines the stationary properties of the carrier gas subject to a constant electric field.

This chapter is devoted to the fundamental concepts and to the models for the scattering mechanisms most relevant for MOS transistors.

Section 4.1 presents a formulation of the scattering rates based on the Fermi golden rule (already introduced in Section 2.5.4), which points out that the energy relation and the wave-functions in the inversion layers discussed in Chapter 3 enter the calculation of the scattering rates directly through the scattering matrix elements. In MOS transistors the screening produced by the free carriers in the inversion layers has a large impact on the surface roughness and Coulomb scattering rates; this topic is addressed in detail in Section 4.2.

Section 4.6 is devoted to phonon scattering, after the topic of lattice vibrations has been introduced in Section 4.5. Sections 4.4 and 4.3 describe models for surface roughness and Coulomb scattering.

The expressions for the scattering rates are prominent ingredients of the Boltzmann transport equation discussed in Chapter 5, and they are used several times in the following chapters.

4.1 Theory of the scattering rate calculations

In the semi-classical picture, calculation of the scattering rate between an initial and a final state is governed by the so called Fermi golden rule, that was derived in Section 2.5.4 for a 3D carrier gas. The present section describes in detail the application and physical interpretation of Fermi's rule in the case of a 2D carrier gas in an inversion layer.

4.1.1 The Fermi golden rule in inversion layers

Let us consider an inversion layer where $\Phi_{n\mathbf{k}}(\mathbf{R})$ is the wave-function possibly depending on the wave-vector $\mathbf{k}$; the index n indicates both the valley and the subband if more valleys exist. The $\Phi_{n\mathbf{k}}$ and the corresponding energies $E_n(\mathbf{k})$ satisfy the Schrödinger equation

$$\hat{H}_0\,\Phi_{n\mathbf{k}}(\mathbf{R}) = E_n(\mathbf{k})\,\Phi_{n\mathbf{k}}(\mathbf{R}), \tag{4.1}$$

where $E_n(\mathbf{k})$ is the eigenvalue corresponding to the wave-vector $\mathbf{k}$ in the subband n and $\hat{H}_0$ is the Hamiltonian of the unperturbed system.

Let us now suppose that the perturbation is given by a stationary scattering potential $U_{sc}(\mathbf{R})$. In the presence of $U_{sc}(\mathbf{R})$, the time-dependent Schrödinger equation that governs the evolution of the system reads

$$\mathrm{i}\hbar\,\frac{\partial\psi(\mathbf{R},t)}{\partial t} = [\hat{H}_0 + U_{sc}(\mathbf{R})]\,\psi(\mathbf{R},t), \tag{4.2}$$

and $U_{sc}(\mathbf{R})$ enables transitions from an initial state $(n,\mathbf{k})$ to a final state $(n',\mathbf{k}')$, that is it results in a scattering rate from $(n,\mathbf{k})$ to $(n',\mathbf{k}')$.

The derivation of Fermi's rule in Section 2.5.4 showed that the scattering rate from $(n,\mathbf{k})$ to $(n',\mathbf{k}')$ can be expressed as

$$S_{n,n'}(\mathbf{k},\mathbf{k}') = \frac{2\pi}{\hbar}\,|M_{n,n'}(\mathbf{k},\mathbf{k}')|^2\,\delta[E_n(\mathbf{k}) - E_{n'}(\mathbf{k}')], \tag{4.3}$$

where the $M_{n,n'}(\mathbf{k},\mathbf{k}')$ term denotes the scattering matrix element here defined as

$$M_{n,n'}(\mathbf{k},\mathbf{k}') = \langle\Phi_{n'\mathbf{k}'}|U_{sc}(\mathbf{R})|\Phi_{n\mathbf{k}}\rangle = \int_\Omega \Phi^\dagger_{n'\mathbf{k}'}(\mathbf{R})\,U_{sc}(\mathbf{R})\,\Phi_{n\mathbf{k}}(\mathbf{R})\,\mathrm{d}\mathbf{R}, \tag{4.4}$$

and Ω is the normalization volume of the system. We discuss in detail the calculation and interpretation of the matrix elements $M_{n,n'}(\mathbf{k},\mathbf{k}')$ in the following sections.

The semi-classical device modeling is essentially based on Fermi's rule for calculation of the scattering rates, hence it is very important, on the one hand, to clarify the physical meaning of Eq.4.3 and, on the other hand, to understand the limits of its validity.

According to Eq.4.3 the scattering rate between two states is governed by the corresponding matrix element $M_{n,n'}(\mathbf{k},\mathbf{k}')$, which in turn depends on the scattering potential and on the eigenfunctions $\Phi_{n\mathbf{k}}(\mathbf{R})$ in the inversion layer. In order to underline the role played by the eigenfunctions, in this chapter we have separated discussion of the electron intra-valley transitions (described in Section 4.1.2), from the more complicated cases of electron inter-valley transitions and from the transitions in hole inversion layers described by a **k**·**p** Hamiltonian (discussed respectively in Sections 4.1.4 and 4.1.5).

4.1.2 Intra-valley transitions in electron inversion layers

Let us then consider the transitions between the subbands of a given valley ν. According to the parabolic effective mass approximation described in Section 3.2, the envelope wave-function can be written as

$$\Phi_{n\mathbf{k}}(\mathbf{R}) = \xi_{n\mathbf{k}}(z)\,\frac{e^{i\,\mathbf{kr}}}{\sqrt{A}}, \tag{4.5}$$

where n is the subband index (in this section we drop the valley index ν), and A is the normalization area.

As will be clarified in Section 4.1.4, the scattering matrix elements for electron intra-valley transitions can be calculated by using the envelope wave-functions. Thus we can substitute Eq.4.5 in Eq.4.4 and obtain

$$\begin{aligned} M_{n,n'}(\mathbf{k},\mathbf{k}') &= \frac{1}{A}\int_z \xi^{\dagger}_{n'\mathbf{k}'}(z)\xi_{n\mathbf{k}}(z)\left[\int_A U_{sc}(\mathbf{R})\,e^{i\,(\mathbf{k}-\mathbf{k}')\mathbf{r}}\,d\mathbf{r}\right]dz \\ &= \frac{(2\pi)^2}{A}\int_z \xi^{\dagger}_{n'\mathbf{k}'}(z)\,\xi_{n\mathbf{k}}(z)\,U_{2T}(-\mathbf{q},z)\,dz, \end{aligned} \tag{4.6}$$

where $\mathbf{q}$=($\mathbf{k}'-\mathbf{k}$) is the wave-vector variation produced by the scattering and $U_{2T}(\mathbf{q},z)$ is the Fourier transform defined in Eq.A.17 of the scattering potential $U_{sc}(\mathbf{R})$ with respect to the coordinates $\mathbf{r} = (x, y)$ in the transport plane. Note that $U_{2T}(\mathbf{q},z)$ is thus a hybrid representation of $U_{sc}(\mathbf{r},z)$ because it depends on the wave-vector $\mathbf{q}$ and on the spatial coordinate z in the quantization direction. The presence of the pre-factor $(2\pi)^2$ in the last line of Eq.4.6 stems from the definition of the Fourier transform. Appendix B discusses in more detail the formal implications of the fact that in the square brackets of Eq.4.6 the integration is over a finite area A, as opposed to the entire $\mathbf{r}$ plane.

Equation 4.6 will be used several times in the following sections of the book as the starting point for calculation of the intra-valley scattering rates.

4.1.3 Physical interpretation and validity limits of Fermi's rule

Equation 4.3 provides us with a few neat indications about the possible transitions between electronic states produced by the stationary potential $U_{sc}(\mathbf{R})$. The first constraint imposed by Eq.4.3 through the Dirac function is that, for a given initial state (n,$\mathbf{k}$) with energy $E_n(\mathbf{k})$, transitions are possible only towards states (n',$\mathbf{k}'$) which have the same energy as the initial state, namely we must have $E_{n'}(\mathbf{k}') = E_n(\mathbf{k})$.

Further insight is obtained by considering the expression for the matrix elements. For intra-valley transitions, for example, by substituting Eq.4.6 in Eq.4.3 we see that the possible transitions from the state (n,$\mathbf{k}$) to the state (n',$\mathbf{k}'$) are governed by the spectral component of the scattering potential $U_{sc}(\mathbf{R})$ at the wave-vector $\mathbf{q} = (\mathbf{k}' - \mathbf{k})$. This implies that, in order to have large wave-vector changes $\mathbf{q}$, we need potentials with rapid spatial variations with respect to $\mathbf{r}$.

To be quantitative about this latter point we notice that, for the simple case of an elastic intra-valley transition in an electron gas with parabolic and circular bands, the energy conservation implies that the magnitude k' of $|\mathbf{k}'|$ must be equal to $k = |\mathbf{k}|$. Thus, if we let θ denote the angle between $\mathbf{k}$ and $\mathbf{k}'$, then the magnitude of the wave-vector change $\mathbf{q}$ is simply given by

$$q = 2k\sin(\theta/2). \tag{4.7}$$

Equation 4.7 states that the average q values for elastic intra-valley transitions are related to the average k values in the inversion layer, which have been discussed in Section 3.6.3 and illustrated in Fig.3.15. This observation is important in relation to inter-valley transitions, which, instead, typically imply changes of the wave-vector much larger than the average k values of Fig.3.15 (see Section 4.1.4).

Equation 4.6 also shows that the integral over the quantization direction plays a key role in selecting the allowed transitions and their overall rate; however, the possibility to produce large $\mathbf{k}$ changes depends on the availability of spectral components of the scattering potential at large $\mathbf{q}$ vectors.

Besides the analysis of the matrix element, it is also important to recall that the Dirac function in Eq.4.3 (imposing that the final state (n',$\mathbf{k}'$) has the same energy as the initial state (n,$\mathbf{k}$)), stems from a long time limit in the derivation of Fermi's rule (see Section 2.5.4), which is justified only for weak scattering potentials and low scattering rates.

In this respect, the assumption of a weak scattering regime has been used throughout the derivation of Fermi's rule, in fact Eq.2.130 neglects the finite lifetime that the initial state has because of the scattering events that we wish to describe. By accounting for the finite lifetime of the initial state, the calculation of the scattering rates results in an expression similar to Eq.4.3, but with a Lorentzian function instead of the Dirac function [1, 2]. This implies that, even if we take the long time limit (namely a time much longer than the lifetime of the initial state), the finite lifetime of the initial state results in non-null scattering rates for a *range of possible energies*, as opposed to a precise energy value [1].

The uncertainty in the energy produced by the intense scattering rates is known as *collisional broadening* and it is almost universally neglected in semi-classical treatment of carrier transport. A few studies have been carried out on the possible role played by *collisional broadening* in hot carrier transport, where in fact scattering rates can be as large as $10^{14}s^{-1}$ [3].

4.1.4 Inter-valley transitions in electron inversion layers

This section deals with the inter-valley transitions in an electron inversion layer produced by a stationary perturbation. Hence the derivations are relevant for surface roughness and Coulomb scattering, but they are not directly applicable to phonon assisted inter-valley transitions, which are discussed in Section 4.6.5.

Surface roughness and Coulomb scattering inter-valley transitions are typically neglected in inversion layers because it is stated that the much larger wave-vector exchange with respect to the intra-valley transitions drastically suppresses the matrix elements [4–7]. However, the above argument is much neater for a three-dimensional electron gas than it is for an inversion layer. In fact one may argue that, for example, the two Δ_2 valleys have the same position at $\mathbf{k} = \mathbf{0}$ in the two dimensional Brillouin zone (see Fig.3.11), hence the transitions between these valleys *do not imply* large changes in the in-plane **k**.

The above considerations suggest that a sound theoretical analysis is necessary to understand the physics behind the inter-valley transitions, and, in particular, to assess for what scattering mechanisms they can actually be neglected; this is the purpose of the present section. The mathematical formalism used below is somewhat heavy, however, the final results have a clear physical interpretation. Furthermore, the results will also be used in Section 4.1.5 to derive an expression for the matrix elements of a hole gas described with the $\mathbf{k}\cdot\mathbf{p}$ quantization model.

By recalling Eqs.3.11 and 3.14, the complete wave-function for the electrons in the inversion layer can be expressed as

$$\Psi_{\nu,n}(\mathbf{R}) = \sqrt{L}\, \xi_{\nu n}(z)\, e^{i\,\mathbf{k}\cdot\mathbf{r}}\, u_{\mathbf{K}_\nu}(\mathbf{R})\, e^{i\,\mathbf{K}_\nu\cdot\mathbf{R}}, \tag{4.8}$$

where L is the normalization length in the z direction, while $\mathbf{K}_\nu$ and $u_{\mathbf{K}_\nu}(\mathbf{R})$ are respectively the wave-vector and the periodic part of the Bloch function at the minimum ν of the conduction band.

We now recall that the wave-vector **k** is the displacement from $\mathbf{k}_\nu$, namely from the in-plane component of the wave-vector $\mathbf{K}_\nu$=($\mathbf{k}_\nu$,$\mathrm{k}_{\nu,z}$) at the minimum ν. Furthermore, since throughout this section we consider only the lowest conduction band at each minimum ν of the bulk semiconductor band structure, the index denoting the band of the Bloch function is dropped to simplify the notation.

The $\xi_{\nu n}(z)$ in Eq.4.8 is the envelope wave-function defined in Section 3.2.1 and normalized as

$$\int_L |\xi_{\nu n}(z)|^2 \mathrm{d}z = 1. \tag{4.9}$$

Furthermore the $u_{\mathbf{K}_\nu}(\mathbf{R})$ is normalized to one over the volume Ω (see Eq.2.26). Given the normalization of $\xi_{\nu n}(z)$ and $u_{\mathbf{K}_\nu}(\mathbf{R})$, the pre-factor $\sqrt{L}$ in Eq.4.8 gives the right units to $\Psi_{\nu,n}(\mathbf{R})$ (i.e. cm$^{-3/2}$) in order for it to be normalized over the volume Ω; the correct normalization of $\Psi_{\nu,n}(\mathbf{R})$ is verified below.

The purpose of the derivations presented below is to calculate the matrix element

$$M_{\nu n}^{wn'}(\mathbf{k}, \mathbf{k}') = \langle \Psi_{wn'}(\mathbf{R}) \,|\, U_{sc}(\mathbf{R}) \,|\, \Psi_{\nu n}(\mathbf{R}) \rangle \tag{4.10}$$

between the state (ν,n,**k**) in the valley ν and the state (w,n',**k**$'$) in the valley w by using the wave-functions defined in Eq.4.8. To this end we start by writing the envelope wave-functions $\xi_{vn}(z)$ and $\xi_{wn'}(z)$ by means of the corresponding Fourier transforms $\chi_{vn}(k_z)$ and $\chi_{wn'}(k_z)$ as

$$\xi_{vn}(z) = \int_{-\infty}^{+\infty} \chi_{vn}(k_z) e^{-i k_z z} dk_z, \qquad \xi_{wn'}(z) = \int_{-\infty}^{+\infty} \chi_{wn'}(k'_z) e^{-i k'_z z} dk'_z. \tag{4.11}$$

By substituting Eqs.4.8 and 4.11 in Eq.4.10 we obtain

$$M_{vn}^{wn'}(\mathbf{k}, \mathbf{k}') = L \int_{k'_z} \int_{k_z} \chi^{\dagger}_{wn'}(k'_z) \chi_{vn}(k_z)\, I_{\Omega}\, dk_z\, dk'_z, \tag{4.12}$$

where I_{Ω} is an integral over the normalization volume Ω defined as

$$I_{\Omega} = \int_{\Omega} \left(u_{\mathbf{K}_w}(\mathbf{R})\, e^{i\mathbf{K}_w \cdot \mathbf{R}} \right)^{\dagger} U_{sc}(\mathbf{R})\, e^{i(\mathbf{k}-\mathbf{k}')\cdot \mathbf{r}}\, e^{i(k'_z - k_z)z}\, u_{\mathbf{K}_v}(\mathbf{R})\, e^{i\mathbf{K}_v \cdot \mathbf{R}} d\mathbf{R}, \tag{4.13}$$

and it depends on $\mathbf{K}_v$, $\mathbf{K}_w$, $\mathbf{k}$, $\mathbf{k}'$, k_z, and k'_z.

A convenient expression for the integral I_{Ω} in Eq.4.13 is a key step in the development of the derivations. Appendix D demonstrates that I_{Ω} can be written as

$$I_{\Omega} = \frac{(2\pi)^3}{\Omega} \sum_{\mathbf{G}=(\mathbf{g}, g_z)} S_{w,v}(\mathbf{g}, g_z) U_{3T}(-\Delta\mathbf{k}_{wv} - \mathbf{q} + \mathbf{g}, -\Delta k_{wv,z} + k'_z - k_z + g_z), \tag{4.14}$$

where $\mathbf{G} = (\mathbf{g}, g_z)$ is a reciprocal lattice vector and $U_{3T}(\mathbf{Q})$ is the three-dimensional Fourier transform of the scattering potential $U_{sc}(\mathbf{R})$ defined in Eq.A.15. The symbol $S_{w,v}(\mathbf{g}, g_z)$ in Eq.4.14 denotes the overlap integral between the periodic parts of the Bloch functions

$$S_{w,v}(\mathbf{G}) = \langle u_{(\mathbf{K}_w - \mathbf{G})}(\mathbf{R}) | u_{\mathbf{K}_v}(\mathbf{R}) \rangle = \int_{\Omega} [u_{(\mathbf{K}_w - \mathbf{G})}(\mathbf{R})]^{\dagger}\, u_{\mathbf{K}_v}(\mathbf{R})\, d\mathbf{R} \tag{4.15}$$

already introduced in Eq.3.59. In Eq.4.14 we have also introduced the symbols

$$\mathbf{q} = \mathbf{k}' - \mathbf{k}, \qquad \Delta\mathbf{K}_{wv} = \mathbf{K}_w - \mathbf{K}_v = (\Delta\mathbf{k}_{wv}, \Delta k_{wv,z}), \tag{4.16}$$

where $\Delta\mathbf{K}_{wv}$ is the distance in $\mathbf{K}$ space between the position of the initial valley v and the final valley w, while $\Delta\mathbf{k}_{wv}$ and $\Delta k_{wv,z}$ are the components of $\Delta\mathbf{K}_{wv}$ respectively in the transport plane and in the quantization direction.

By inserting Eq.4.14 in Eq.4.12 we obtain

$$\begin{aligned} M_{vn}^{wn'}(\mathbf{k}, \mathbf{k}') =& \frac{(2\pi)^3}{A} \sum_{\mathbf{G}=(\mathbf{g}, g_z)} S_{w,v}(\mathbf{g}, g_z) \int_{k'_z} \chi^{\dagger}_{wn'}(k'_z) \int_{k_z} \chi_{vn}(k_z) \\ & \times U_{3T}(-\Delta\mathbf{k}_{wv} - \mathbf{q} + \mathbf{g}, k'_z - k_z + g_z - \Delta k_{wv,z})\, dk_z\, dk'_z \end{aligned} \tag{4.17}$$

where A is the normalization area in the transport plane.

In order to cast $M_{vn}^{wn'}(\mathbf{k}, \mathbf{k}')$ in a form that can be directly compared to Eq.4.6 derived for intra-valley transitions by using the envelope the wave-functions, it is necessary to manipulate Eq.4.17 to introduce the hybrid two-dimensional Fourier transform $U_T(\mathbf{q}, z)$ of the scattering potential $U_{sc}(\mathbf{R})$. This can be accomplished by working on the integrals over k_z and k'_z in Eq.4.17. In these integrals the wave-vector components in the transport plane $(-\Delta\mathbf{k}_{wv}-\mathbf{q}+\mathbf{g})$ are only parameters, hence we can momentarily simplify the notation by dropping their indication in the three-dimensional Fourier transform U_{3T}. That is, we write $U_{3T}(v_z)$ rather than $U_{3T}(\mathbf{v}, v_z)$, where $(\mathbf{v}, v_z)$ is a generic three components wave-vector. By doing so we can recognize that the integral over k_z in Eq.4.17 calculates the convolution with respect to k_z between $\chi_{vn}(k_z)$ and $U_{3T}(k'_z - k_z + g_z - \Delta k_{wv,z})$. More precisely, we can write

$$\begin{aligned}\int_{k_z} \chi_{vn}(k_z)\, U_{3T}(k'_z - k_z + g_z - \Delta k_{wv,z})\, \mathrm{d}k_z \\ &= (\chi_{vn} * U_{3T})(k'_z + g_z - \Delta k_{wv,z}) \\ &= \mathcal{F}_z\{\xi_{vn}(z)\, U_{2T}(z)\,\}(k'_z + g_z - \Delta k_{wv,z}) \\ &= \mathcal{F}_z\{\xi_{vn}(z)\, U_{2T}(z)\, \mathrm{e}^{\mathrm{i}(g_z-\Delta k_{wv,z})z}\}(k'_z),\end{aligned} \tag{4.18}$$

where $\mathcal{F}_z\{f(z)\}(k_z)$ denotes the Fourier transform of $f(z)$ with respect to z and calculated in k_z. In the first equality of Eq.4.18 we made use of the property expressed in Eq.A.9 about the Fourier transform of the product of functions; in the second equality we used Eq.A.19 relating the three- and two-dimensional Fourier transforms; in the last we finally exploited Eq.A.7.

We now notice that, by resorting to Eq.A.11, the remaining integral over k'_z in Eq.4.17 can be converted to an integral over the spatial coordinate z. More precisely, by reintroducing in the notation for U_{2T} the dependence on $(-\Delta\mathbf{k}_{wv}-\mathbf{q}+\mathbf{g})$ omitted in Eq.4.18, we can write

$$\begin{aligned}&\int_{k'_z} \chi^{\dagger}_{wn'}(k'_z)\mathcal{F}_z\{\xi_{vn}(z)\, U_{2T}(-\Delta\mathbf{k}_{wv} - \mathbf{q} + \mathbf{g}, z)\, \mathrm{e}^{\mathrm{i}(g_z-\Delta k_{wv,z})z}\}(k'_z)\, \mathrm{d}k'_z \\ &\quad = \frac{1}{2\pi}\int_z \xi^{\dagger}_{wn'}(z)\, \xi_{vn}(z)\, U_{2T}(-\Delta\mathbf{k}_{wv} - \mathbf{q} + \mathbf{g}, z)\, \mathrm{e}^{\mathrm{i}(g_z-\Delta k_{wv,z})z}\mathrm{d}z.\end{aligned} \tag{4.19}$$

By substituting Eq.4.19 in Eq.4.17 we finally obtain

$$\begin{aligned}M^{wn'}_{vn}(\mathbf{k}, \mathbf{k}') &= \frac{(2\pi)^2}{A}\Bigg\{ S_{w,v}(\mathbf{0}, 0)\int_z \xi^{\dagger}_{wn'}(z)\, \xi_{vn}(z)\, U_{2T}(-\Delta\mathbf{k}_{wv} - \mathbf{q}, z) \\ &\quad \times \mathrm{e}^{-\mathrm{i}\Delta k_{wv,z}\cdot z}\mathrm{d}z + \sum_{\mathbf{G}=(\mathbf{g},g_z)\neq\mathbf{0}} S_{w,v}(\mathbf{g}, g_z)\int_z \xi^{\dagger}_{wn'}(z)\, \xi_{vn}(z) \\ &\quad \times U_{2T}(-\Delta\mathbf{k}_{wv} - \mathbf{q} + \mathbf{g}, z)\, \mathrm{e}^{\mathrm{i}(g_z-\Delta k_{wv,z})z}\mathrm{d}z\Bigg\},\end{aligned} \tag{4.20}$$

which represents the expression for the inter-valley matrix element that we wished to derive.

Equation 4.20 can also be used for the case of intra-valley transitions if we set $w = v$; we discuss below the relation with Eq.4.6.

The first term in Eq.4.20 corresponding to $\mathbf{G} = \mathbf{0}$ is called *normal* or N process, whereas the remaining terms corresponding to non-null $\mathbf{G}$ vectors are called *umklapp* or U processes [4]. The overall sum over $\mathbf{G}$ in Eq.4.20 converges rapidly because, by considering the non-null and progressively larger in magnitude $\mathbf{G}$ vectors, both the overlap integrals $S_{w,v}(\mathbf{g}, g_z)$ and the spectral components of the scattering potential tend to vanish. Numerical evaluations show that the sum is frequently dominated by one single term.

Normalization of the wave-function

Before discussing the application of Eq.4.20 to either intra-valley or inter-valley transitions, we consider normalization of the total wave-functions $\Psi_{\nu,n}(\mathbf{R})$ defined in Eq.4.8. To this purpose we must evaluate

$$\langle \Psi_{\nu,n}(\mathbf{R}) | \Psi_{\nu,n}(\mathbf{R}) \rangle = \int_{\Omega} \Psi^{\dagger}_{\nu,n}(\mathbf{R}) \Psi_{\nu,n}(\mathbf{R})\, \mathrm{d}\mathbf{R}. \tag{4.21}$$

All the derivations in going from Eq.4.10 to Eq.4.20 can be used directly to evaluate Eq.4.21 if we specially modify them for $w = \nu$, for $\mathbf{k} = \mathbf{k}'$, and for $U_{sc}(\mathbf{R}) = 1$ (within the normalization volume Ω). In this case we have

$$\Delta \mathbf{K}_{w\nu} = (\mathbf{0}, 0), \qquad \mathbf{q} = \mathbf{k}' - \mathbf{k} = \mathbf{0}, \tag{4.22}$$

and the $U_{sc}(\mathbf{R})$ is such that

$$\frac{(2\pi)^2}{A} U_{2T}(\mathbf{v}, z) = \frac{1}{A} \int_A \mathrm{e}^{\mathrm{i}\,\mathbf{v}\cdot\mathbf{r}}\, \mathrm{d}\mathbf{r} = \delta_{\mathbf{v},\mathbf{0}}, \tag{4.23}$$

where $\mathbf{v}$ is the generic wave-vector in the transport plane, the Kronecker symbol $\delta_{\mathbf{v},\mathbf{0}}$ is 1 if $\mathbf{v} = \mathbf{0}$ and 0 otherwise and the last equality in Eq.4.23 follows directly from Eq.B.8.

Furthermore, for $w = \nu$ the overlap factor $S_{\nu,\nu}(\mathbf{0}, 0)$ coincides with the scalar product of the Bloch function at the minimum ν with itself, hence we have

$$S_{\nu,\nu}(\mathbf{0}, 0) = \int_{\Omega} |u_{\mathbf{K}_\nu}(\mathbf{R})|^2\, \mathrm{d}\mathbf{R} = 1.0, \tag{4.24}$$

as given also in Table 4.1 for the intra-valley transition. By substituting Eqs.4.22, 4.23 and 4.24 in Eq.4.20 we obtain

$$\langle \Psi_{\nu,n}(\mathbf{R}) | \Psi_{\nu,n}(\mathbf{R}) \rangle = \int_z |\xi_{\nu n}(z)|^2 \mathrm{d}z + \sum_{\mathbf{G}=(\mathbf{0},g_z)} S_{\nu,\nu}(\mathbf{0}, g_z) \int_z |\xi_{\nu n}(z)|^2 \mathrm{e}^{\mathrm{i}g_z z} \mathrm{d}z, \tag{4.25}$$

where the sum over $\mathbf{G}$ runs over all the lattice vectors with null in-plane components, namely over the lattice vectors $\mathbf{G} = (\mathbf{0}, \pm 2n)(\frac{2\pi}{a_0})$, n being a positive integer. It can be demonstrated that the above sum takes a real value because, for any given g_z, the term corresponding to $(-g_z)$ implies the conjugation of both $S_{\nu,\nu}(\mathbf{0}, g_z)$ and the integral over z.

The magnitude of the sum over $\mathbf{G}$ in Eq.4.25 is negligible with respect to the first term for two main reasons. First, the magnitude of the overlap integrals $|S_{\nu,\nu}(\mathbf{G})|$ for intra-valley transitions and for $\mathbf{G} \neq \mathbf{0}$ is typically negligible with respect to $S_{\nu,\nu}(\mathbf{0}) = 1.0$ (see the examples in Table 4.1). Furthermore, if the spectral components of the wave-function $|\xi_{\nu n}(z)|^2$ are small at $g_z = \pm 2n\frac{2\pi}{a_0}$, then the integral over z is very small. This is equivalent to requiring that the wave-function is slowly varying on a bulk crystal unit cell, which is a general requirement for the validity of the envelope wave-function approximation (see Section 2.4).

Hence, by neglecting the second term in Eq.4.25, we can conclude that the total wave-functions $\Psi_{\nu,n}(\mathbf{R})$ defined in Eq.4.8 are normalized to 1.0 provided that the envelope wave-function $\xi_{\nu n}(z)$ is in turn normalized according to Eq.4.9.

Application to intra-valley transitions

If Eq.4.20 is used for an intra-valley transition in the valley ν, we have

$$\Delta\mathbf{K}_{w\nu} = (\mathbf{0}, 0), \qquad \mathbf{q} = \mathbf{k}' - \mathbf{k} = \mathbf{0},$$

and $S_{\nu,\nu}(\mathbf{0}, 0) = 1$ (see Eq.4.24), so that the matrix elements simplify to

$$M_{\nu n}^{\nu n'}(\mathbf{k}, \mathbf{k}') = \frac{(2\pi)^2}{A}\Bigg\{\int_z \xi_{\nu n'}^{\dagger}(z)\,\xi_{\nu n}(z)\,U_{2T}(-\mathbf{q}, z)\mathrm{d}z + \sum_{\mathbf{G}=(\mathbf{g},g_z)\neq\mathbf{0}} S_{\nu,\nu}(\mathbf{g}, g_z)\int_z \xi_{\nu n'}^{\dagger}(z)\,\xi_{\nu n}(z)\,U_{2T}(-\mathbf{q}+\mathbf{g}, z)\mathrm{e}^{\mathrm{i}g_z z}\mathrm{d}z\Bigg\}. \tag{4.26}$$

By following similar arguments to those used in the discussion of Eq.4.25, we find that the *umklapp* processes in Eq.4.26 are negligible for intra-valley transitions. This is due to the very small values of both the overlap integrals $S_{\nu,\nu}(\mathbf{g}, g_z)$ (see Table 4.1) and the integrals over z.

By neglecting the *umklapp* processes, Eq.4.26 reduces to Eq.4.6, which demonstrates how *the matrix elements for intra-valley transitions can be calculated by using only the envelope wave-functions.*

Application to inter-valley transitions

According to Eq.4.20, the transitions between valley ν and valley w in the inversion layer are critically influenced by the distance $\Delta\mathbf{K}_{w\nu}$, defined in Eq.4.16, between the position of the corresponding energy minima in the $\mathbf{K}$ space of the bulk crystal.

The transitions between different valleys have in common the fact that the magnitude of the overlap integral $S_{w,\nu}(\mathbf{0}, 0)$ is typically much smaller than the magnitude $|S_{\nu,\nu}(\mathbf{0}, 0)|=1.0$ of the overlap integrals for intra-valley transitions. Aside from this, the main difference between the diverse inter-valley transitions is given by the fact that $\Delta\mathbf{k}_{w\nu}$ (namely the component in the transport plane of the distance $\Delta\mathbf{K}_{w\nu}$ between the valleys defined in Eq.4.16), may be null or non-null.

As an example, let us consider the two Δ_4 valleys in the inversion layers stemming from the bulk silicon minima located at

$$\mathbf{K}_{\nu} = (0.85, 0, 0), \qquad \mathbf{K}_{w} = (-0.85, 0, 0). \tag{4.27}$$

Table 4.1 Magnitude of some overlap integrals $S_{w,\nu}(\mathbf{G})$ for intra- or inter-valley transitions in silicon, calculated by using the pseudo-potential method with parameters taken from [8]. The transitions for $\mathbf{G} \neq \mathbf{0}$ are *umklapp* processes and the corresponding overlap integrals are negligible for the cases illustrated in the table.

	$\lvert S_{w,\nu}(\mathbf{G})\rvert$ $\mathbf{G}=\mathbf{0}$	$\lvert S_{w,\nu}(\mathbf{G})\rvert$ $\mathbf{G}=(2.0,0,0)$	$\lvert S_{w,\nu}(\mathbf{G})\rvert$ $\mathbf{G}=(-2.0,0,0)$
$\mathbf{K}_\nu=(0.85,0,0)$ $\mathbf{K}_w=(0.85,0,0)$	1.0	$3.9\cdot10^{-15}$	$3.9\cdot10^{-15}$
$\mathbf{K}_\nu=(0.85,0,0)$ $\mathbf{K}_w=(-0.85,0,0)$	0.14	$4.8\cdot10^{-14}$	$4.9\cdot10^{-16}$
	$\lvert S_{w,\nu}(\mathbf{G})\rvert$ $\mathbf{G}=\mathbf{0}$	$\lvert S_{w,\nu}(\mathbf{G})\rvert$ $\mathbf{G}=(0,0,2.0)$	$\lvert S_{w,\nu}(\mathbf{G})\rvert$ $\mathbf{G}=(0,0,-2.0)$
$\mathbf{K}_\nu=(0,0,0.85)$ $\mathbf{K}_w=(0,0,-0.85)$	0.14	$2.2\cdot10^{-14}$	$1.5\cdot10^{-16}$

These valleys yield a non-null $\Delta\mathbf{k}_{w\nu}$ and, in particular, we have

$$\Delta\mathbf{k}_{w\nu}=(-1.7,0), \qquad \Delta k_{w\nu,z}=0,$$

where all the wave-vectors are expressed in units of $(2\pi/a_0)$, which for silicon is about $11.6\,\text{nm}^{-1}$.

Table 4.1 shows that $|S_{w,\nu}(\mathbf{0},0)|$ is much smaller than 1.0. Furthermore, the $|U_{2T}(-\Delta\mathbf{k}_{w\nu}-\mathbf{q},z)|$ values for the *normal* processes in Eq.4.20 are expected to be much smaller than $|U_{2T}(-\mathbf{q},z)|$, because $|\Delta\mathbf{k}_{w\nu}|$ is much larger than the average q values for intra-valley transitions expressed by Eq.4.7 (and related to the average k values discussed in Section 3.6.3 and Fig.3.15). As can be seen, the *normal* processes for the inter-valley transitions between two different Δ_4 valleys are much weaker than the corresponding intra-valley transitions.

In order to complete the analysis one may argue that, since the *normal* processes are very weak, the largest term in Eq.4.20 could be *umklapp* processes. In this respect, the dominant *umklapp* process is expected to be the one yielding the minimum value for $|-\Delta\mathbf{k}_{w\nu}-\mathbf{q}+\mathbf{g}|$, hence the maximum value for $|U_{2T}(-\Delta\mathbf{k}_{w\nu}-\mathbf{q}+\mathbf{g},z)|$. This is obtained for the lattice vector $(\mathbf{g},g_z)=(-2.0,0,0)$, and the corresponding term in Eq.4.20 involves the product of $S_{w,\nu}(\mathbf{g},g_z)$ and $U_{2T}[(-0.3\frac{2\pi}{a_0},0)-\mathbf{q},z]$. Table 4.1 shows, however, that in this example the overlap integrals $S_{w,\nu}(\mathbf{g},g_z)$ for $(\mathbf{g},g_z)=(\pm2.0,0,0)$ are negligible, hence the *umklapp* processes are not expected to contribute significantly to the overall inter-valley matrix elements.

An example with quite different characteristics is obtained by considering the inter-valley transitions between the Δ_2 valleys corresponding to the bulk silicon conduction band minima located at

$$\mathbf{K}_\nu=(0,0,0.85), \qquad \mathbf{K}_w=(0,0,-0.85). \tag{4.28}$$

In this case we have

$$\Delta \mathbf{k}_{wv} = (0, 0), \qquad \Delta k_{wv,z} = -1.7,$$

hence, differently from the previous case, we do not have a large exchanged wave-vector in the transport plane. Nevertheless the matrix element is very small compared to an intra-valley transition within one of the Δ_2 valleys.

In fact the *normal* processes in Eq.4.20 consist of $S_{w,v}(\mathbf{0}, 0)$ (whose magnitude is small compared to 1.0, see Table 4.1), and of the integral

$$\int_z \xi^{\dagger}_{wn'}(z)\, \xi_{vn}(z)\, U_{2T}(-\mathbf{q}, z)\, e^{i\left(1.7\frac{2\pi}{a_0}\right)z}\, dz. \tag{4.29}$$

The exponential term in Eq.4.29 oscillates very rapidly in the z direction, hence the above integral is very small if the envelope wave-functions $\xi^{\dagger}_{wn'}(z)$ and $\xi_{vn}(z)$ and the scattering potential are slowly varying on a bulk crystal unit cell. The *normal* process is thus expected to be very small compared to an intra-valley matrix element.

Also in this case one may argue that the dominant process in Eq.4.20 is an *umklapp* process. The largest *umklapp* process is expected to be the one with $(\mathbf{g}, g_z) = (\mathbf{0}, -2.0)$, that involves the integral

$$M^{wn'}_{vn}(\mathbf{k}, \mathbf{k}') = \int_z \xi^{\dagger}_{wn'}(z)\, \xi_{vn}(z)\, U_{2T}(-\mathbf{q}, z)\, e^{-i\left(0.3\frac{2\pi}{a_0}\right)z}\, dz, \tag{4.30}$$

which can actually be larger than the integral in Eq.4.29.

However, Table 4.1 shows that the overlap integral for such an *umklapp* process is negligible, so that the magnitudes of the matrix elements for the inter-valley transitions are expected to be negligible with respect to the corresponding intra-valley transitions.

We note finally that the inter-valley transitions between a Δ_4 and a Δ_2 valley result in $\Delta \mathbf{K}_{wv} = (\Delta \mathbf{k}_{wv}, \Delta k_{wv,z})$ where neither $\Delta \mathbf{k}_{wv}$ nor $\Delta k_{wv,z}$ is null. Hence the arguments illustrated above demonstrate that the matrix elements are small compared to intra-valley transitions.

Based on the above discussion, we can state that the inter-valley transitions in silicon electron inversion layers can be safely neglected if the scattering potentials are slowly varying over a bulk crystal unit cell, or, equivalently, if their spectral components are small for wave-vectors comparable to the extension of the first Brillouin zone. This allows us to legitimately neglect inter-valley transitions in discussion, for example, of the surface roughness and Coulomb scattering (see Sections 4.4 and 4.3). In Section 4.6 we see that the phonon scattering potential, instead, has large spectral components even for wave-vectors at the boundaries of the Brillouin zone, consequently the phonon scattering mechanism can produce significant inter-valley transitions (see Section 4.6.5).

The formalism developed in this section is very useful also for analysis of the matrix elements in a hole inversion layer described with the $\mathbf{k}\cdot\mathbf{p}$ quantization model, as illustrated in the next section.

4.1.5 Hole matrix elements for a k·p Hamiltonian

By recalling Eqs.3.31 and 3.32, the complete wave-function for the holes in an inversion layer can be expressed as

$$\Psi_{n\mathbf{k}}(\mathbf{R}) = \sqrt{L}\sum_i \xi_{n\mathbf{k},i}(z)\,\mathrm{e}^{\mathrm{i}\,\mathbf{k}\cdot\mathbf{r}}\,u_{i\mathbf{0}}(\mathbf{R}), \qquad i \in \{1\uparrow, 2\uparrow, 3\uparrow, 1\downarrow, 2\downarrow, 3\downarrow\}, \tag{4.31}$$

where the $u_{i\mathbf{0}}(\mathbf{R})$ are the periodic parts of the valence band Bloch functions at the Γ point. Since the Γ point is located at $\mathbf{K} = \mathbf{0}$, the $u_{i\mathbf{0}}(\mathbf{R})$ coincide with the Bloch functions themselves. The $\boldsymbol{\xi}_{n\mathbf{k}}$ is the six-component envelope wave-function obtained by solving the **k**·**p** eigenvalue problem given by Eq.3.34; $\xi_{n\mathbf{k},i}(z)$ is the corresponding i component.

A derivation entirely similar to the one presented in the previous section for an electron inversion layer shows that, if the envelope wave-function $\boldsymbol{\xi}_{n\mathbf{k}}$ is normalized over L according to

$$\sum_i \int_L |\xi_{n\mathbf{k},i}(z)|^2 \mathrm{d}z = 1, \qquad i \in \{1\uparrow, 2\uparrow, 3\uparrow, 1\downarrow, 2\downarrow, 3\downarrow\}, \tag{4.32}$$

then the wave-function $\Psi_{n\mathbf{k}}(\mathbf{R})$ in Eq.4.31 is correctly normalized to one over the volume $\Omega{=}AL$.

Starting from Eq.4.31 the matrix element $M_{n,n'}(\mathbf{k}, \mathbf{k}')$ between the state $(n,\mathbf{k})$ and $(n',\mathbf{k}')$ can be written as

$$M_{n,n'}(\mathbf{k}, \mathbf{k}') = \langle \Psi_{n'\mathbf{k}'}(\mathbf{R}) \mid U_{sc}(\mathbf{R}) \mid \Psi_{n\mathbf{k}}(\mathbf{R})\rangle = \sum_{i,j} M_{n,n'}^{(i,j)}(\mathbf{k}, \mathbf{k}'), \tag{4.33}$$

where $M_{n,n'}^{(i,j)}(\mathbf{k}, \mathbf{k}')$ is defined as

$$M_{n,n'}^{(i,j)}(\mathbf{k}, \mathbf{k}'){=}L\,\langle \xi_{n'\mathbf{k}',j}(z)\,\mathrm{e}^{\mathrm{i}\,\mathbf{k}'\cdot\mathbf{r}}\,u_{j\mathbf{0}}(\mathbf{R})|\,U_{sc}(\mathbf{R})\,|\xi_{n\mathbf{k},i}(z)\,\mathrm{e}^{\mathrm{i}\,\mathbf{k}\cdot\mathbf{r}}\,u_{i\mathbf{0}}(\mathbf{R})\rangle. \tag{4.34}$$

The matrix elements $M_{n,n'}^{(i,j)}(\mathbf{k}, \mathbf{k}')$ can be expressed by repeating the derivations going from Eq.4.10 to Eq.4.20 with two simplifications related to the fact that all the hole subbands are located at the Γ point. This implies that on the one hand there is no need to identify the valley by using the symbols ν or w used in Section 4.1.4 (so that the symbol for the valley has been dropped throughout this section), and on the other hand $\Delta\mathbf{K}_{w\nu} = (\Delta\mathbf{k}_{w\nu}, \Delta k_{w\nu,z})$ is null. Consequently Eq.4.20 allows us to write $M_{n,n'}^{(i,j)}(\mathbf{k}, \mathbf{k}')$ as

$$\begin{aligned} M_{n,n'}^{(i,j)}(\mathbf{k}, \mathbf{k}') = \frac{(2\pi)^2}{A}\Big\{ & S_{j,i}(\mathbf{0}, 0) \int_z \xi^{\dagger}_{n'\mathbf{k}',j}(z)\,\xi_{n\mathbf{k},i}(z)\,U_{2T}(-\mathbf{q}, z)\mathrm{d}z \\ & + \sum_{\mathbf{G}=(\mathbf{g},g_z)\neq\mathbf{0}} S_{j,i}(\mathbf{g}, g_z) \int_z \xi^{\dagger}_{n'\mathbf{k}',j}(z)\,\xi_{n\mathbf{k},i}(z)\,U_{2T}(-\,\mathbf{q}+\mathbf{g}, z)\mathrm{e}^{\mathrm{i}g_z z}\mathrm{d}z \Big\}, \end{aligned} \tag{4.35}$$

where $S_{j,i}(\mathbf{g}, g_z)$ is the overlap integral

$$S_{j,i}(\mathbf{g}, g_z) = \langle u_{j(-\mathbf{g},-g_z)}(\mathbf{R}) | u_{i\mathbf{0}}(\mathbf{R})\rangle = \int_{\Omega} [u_{j(-\mathbf{g},-g_z)}(\mathbf{R})]^{\dagger}\, u_{i\mathbf{0}}(\mathbf{R})\,\mathrm{d}\mathbf{R}. \tag{4.36}$$

Since, as discussed above, at the Γ point the $u_{i,\mathbf{0}}$ coincide with the Bloch functions, the overlap integrals for $(\mathbf{g}, g_z) = (\mathbf{0}, 0)$ are simply given by

$$S_{j,i}(\mathbf{0}, 0) = \langle u_{j\mathbf{0}}(\mathbf{R})|u_{i\mathbf{0}}(\mathbf{R})\rangle = \delta_{i,j}, \tag{4.37}$$

because of the orthogonality and normalization of the Bloch functions at the Γ point.

We can now further simplify Eq.4.35 by neglecting the *umklapp* processes because, on the one hand, $|S_{j,i}(\mathbf{g}, g_z)|$ is small with respect to 1.0 for non-null lattice vectors $(\mathbf{g}, g_z)$, and, furthermore, because the integral over z in the second line of Eq.4.35 is very small with respect to the corresponding integral in the *normal* processes. This is in turn due to the fact that, for non-null $\mathbf{g}$ values, $|U_{2T}(-\mathbf{q}+\mathbf{g}, z)|$ is much smaller than $|U_{2T}(-\mathbf{q}, z)|$ because $|-\mathbf{q}+\mathbf{g}|$ is much larger than $q = |\mathbf{q}|$ (see Section 3.6.3 for the average k values in a hole inversion layer). Furthermore, even for lattice vectors with $\mathbf{g} = \mathbf{0}$, the integral over z in the second line of Eq.4.35 is extremely small because of the very rapidly oscillating complex exponential term, which is essentially the same argument used to discuss the integral in the second line of Eq.4.25.

Thus, by neglecting the *umklapp* processes and substituting Eq.4.37 in Eq.4.35 we can write

$$M^{(i,j)}_{n,n'}(\mathbf{k}, \mathbf{k}') \simeq \left[\frac{(2\pi)^2}{A}\int_z \xi^{\dagger}_{n'\mathbf{k}',j}(z)\, \xi_{n\mathbf{k},i}(z)\, U_{2T}(-\mathbf{q}, z)\mathrm{d}z\right]\delta_{i,j}. \tag{4.38}$$

By substituting Eq.4.38 in Eq.4.33 we obtain the desired expression for the matrix elements of a $\mathbf{k}{\cdot}\mathbf{p}$ Hamiltonian:

$$M_{n,n'}(\mathbf{k}, \mathbf{k}') \simeq \frac{(2\pi)^2}{A}\sum_i \int_z \xi^{\dagger}_{n'\mathbf{k}',i}(z)\, \xi_{n\mathbf{k},i}(z)\, U_{2T}(-\mathbf{q}, z)\mathrm{d}z. \tag{4.39}$$

To simplify the notation we can now introduce the scalar product between the six component envelope wave-functions defined as

$$\boldsymbol{\xi}^{\dagger}_{n'\mathbf{k}'}(z) \cdot \boldsymbol{\xi}_{n\mathbf{k}}(z) = \sum_i \xi^{\dagger}_{n'\mathbf{k}',i}(z)\, \xi_{n\mathbf{k},i}(z), \qquad i \in \{1\uparrow, 2\uparrow, 3\uparrow, 1\downarrow, 2\downarrow, 3\downarrow\}. \tag{4.40}$$

By using such a scalar product, we can finally rewrite the matrix element as

$$M_{n,n'}(\mathbf{k}, \mathbf{k}') \simeq \frac{(2\pi)^2}{A}\int_z \boldsymbol{\xi}^{\dagger}_{n'\mathbf{k}'}(z) \cdot \boldsymbol{\xi}_{n\mathbf{k}}(z)\, U_{2T}(-\mathbf{q}, z)\mathrm{d}z. \tag{4.41}$$

Equation 4.41 represents the expression for the matrix elements corresponding to the $\mathbf{k}{\cdot}\mathbf{p}$ Hamiltonian that we use throughout the rest of the book.

4.1.6 A more general formulation of the Fermi golden rule

Equation 4.3 has been derived for a stationary scattering potential $U_{sc}(\mathbf{R})$. This implies that the carriers after a scattering event have the same energy as before the scattering. Some very important scattering mechanisms, however, produce a change in the electron energy and can be represented by a time-dependent scattering potential or, more appropriately, a time dependent Hamiltonian.

Let us first consider the case of a time-dependent potential, which is a natural extension of the stationary potential considered in the previous section. For simplicity we write the time-dependent scattering potential as a simple Fourier component

$$U_{sc}(\mathbf{R}, t) = U_{ab}(\mathbf{R})\,\mathrm{e}^{-\mathrm{i}\omega t} + U_{em}(\mathbf{R})\,\mathrm{e}^{\mathrm{i}\omega t}, \tag{4.42}$$

and we notice that, in order for $U_{sc}(\mathbf{R}, t)$ to be real, the emission scattering potential $U_{em}(\mathbf{R})$ must be the complex conjugate of the absorption scattering potential $U_{ab}(\mathbf{R})$, namely $U_{em}(\mathbf{R}) = U^{\dagger}_{ab}(\mathbf{R})$. We will see in Section 4.6 that the phonon scattering can be approximately described by a scattering potential with the form given in Eq.4.42.

The scattering rate produced by the $U_{sc}(\mathbf{R}, t)$ in Eq.4.42 can be calculated by following a derivation entirely similar to the one that led to Eq.4.3. The final result is

$$\begin{aligned} S_{n,n'}(\mathbf{k}, \mathbf{k}') = {} & \frac{2\pi}{\hbar}\,|M^{(ab)}_{n,n'}(\mathbf{k}, \mathbf{k}')|^2\,\delta[E_n(\mathbf{k}) - E_{n'}(\mathbf{k}') + \hbar\omega] \\ & + \frac{2\pi}{\hbar}\,|M^{(em)}_{n,n'}(\mathbf{k}, \mathbf{k}')|^2\,\delta[E_n(\mathbf{k}) - E_{n'}(\mathbf{k}') - \hbar\omega], \end{aligned} \tag{4.43}$$

where $M^{(em)}_{n,n'}(\mathbf{k}, \mathbf{k}')$ and $M^{(ab)}_{n,n'}(\mathbf{k}, \mathbf{k}')$ denote the emission and the absorption matrix elements defined as in Eq.4.6 and calculated by using $U_{ab}(\mathbf{R})$ and $U_{em}(\mathbf{R})$, respectively.

Equation 4.43 shows that the absorption process results in a final energy $E_{n'}(\mathbf{k}') = [E_n(\mathbf{k}) + \hbar\omega]$, hence the electron has gained an energy $\hbar\omega$, which typically represents the energy of either a photon or a phonon (see Section 4.6 for the concept of phonon). The emission process, instead, forces the electron to lose an energy $\hbar\omega$.

It is easy to realize that not all the perturbations can be described by means of a scattering potential. As a simple example, developed in more detail in Section 4.4, let us consider an electron inversion layer whose subband minima are given by Eq.3.16 according to the effective mass approximation quantization model. We also suppose that the Schrödinger equation has to be solved by accounting for the penetration of the wave-function into the oxide region and we set the silicon–oxide interface at $z = 0$, as sketched in Fig.4.1. In such a case we need to solve the eigenvalue problem

$$\frac{-\hbar^2}{2m_z}\xi(z) + U(z)\,\xi(z) = \varepsilon\,\xi(z) \tag{4.44}$$

both in silicon and in the oxide and the quantization mass m_z is given by

$$m_z = \begin{cases} m_{si} & \text{for} \quad z \geq 0 \\ m_{ox} & \text{for} \quad z < 0 \end{cases}. \tag{4.45}$$

It is understood that $U(z)$ includes the potential energy barrier at the silicon–oxide interface (see Eq.3.107).

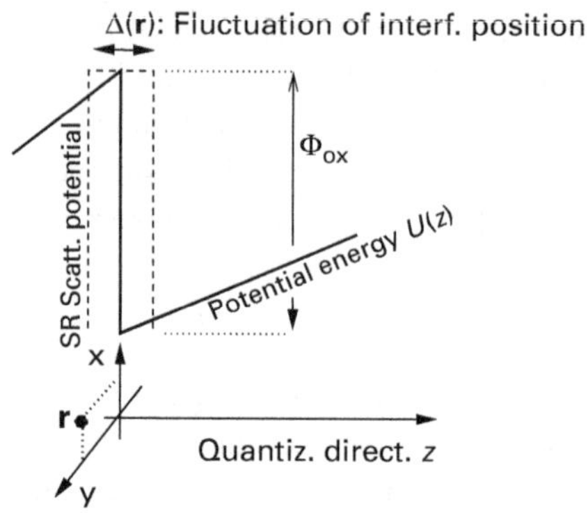

Figure 4.1 The confining potential $U(z)$ for an electron gas and the possible variation $\Delta(\mathbf{r})$ of the position of the silicon–oxide interface with respect to its nominal position at $z = 0$.

In order to solve Eq.4.44 in the entire z domain we impose the continuity conditions at $z = 0$:

$$\xi(0^-) = \xi(0^+), \qquad \frac{1}{m_{ox}}\frac{\partial\xi(0^-)}{\partial z} = \frac{1}{m_{si}}\frac{\partial\xi(0^+)}{\partial z}, \tag{4.46}$$

as well as the boundary conditions for $z\rightarrow\pm\infty$.

We now consider the perturbation of such a physical system produced by a change Δ in the position of the silicon–oxide interface ideally placed at $z = 0$ (see Fig.4.1). In the perturbed case Eq.4.44 has to be solved with a quantization mass

$$m_z = \begin{cases} m_{si} & \text{for} \quad z \geq \Delta \\ m_{ox} & \text{for} \quad z < \Delta \end{cases}. \tag{4.47}$$

Furthermore, the continuity conditions in Eq.4.46 must be imposed at $z = \Delta$ rather than at $z = 0$.

As can be seen, such a perturbation also changes the differential part of the unperturbed Hamiltonian, hence it cannot be represented by a simple scattering potential, in fact we need a scattering Hamiltonian operator $\hat{H}_{sc}$ (as discussed in more detail in Section 4.4).

In the example above the scattering Hamiltonian is stationary, however, in general the scattering Hamiltonian can very well be time dependent; the scattering potential is a particular case of the Hamiltonian formulation. For the Hamiltonian describing the carrier interactions with phonons or photons, for instance, we have an operator $\hat{H}_{ab}$ for the absorption and an operator $\hat{H}_{em}$ for the emission, in other words the total Hamiltonian $\hat{H}$ is

$$\hat{H} = \hat{H}_0 + \hat{H}_{ab} + \hat{H}_{em}, \tag{4.48}$$

where $\hat{H}_0$ is the unperturbed Hamiltonian.

In such a case the scattering rate is expressed by Eq.4.43 and the matrix element is given by

$$M^{(ab)}_{n,n'}(\mathbf{k},\mathbf{k}') = \langle \Phi_{n'\mathbf{k}'}|\hat{H}_{ab}|\Phi_{n\mathbf{k}}\rangle = \int_{\Omega} \Phi^{\dagger}_{n'\mathbf{k}'}(\mathbf{R})\,[\hat{H}_{ab}\,\Phi_{n\mathbf{k}}(\mathbf{R})]\,\mathrm{d}\mathbf{R} \tag{4.49}$$

for absorption, and by

$$M^{(em)}_{n,n'}(\mathbf{k},\mathbf{k}') = \langle \Phi_{n'\mathbf{k}'}|\hat{H}_{em}|\Phi_{n\mathbf{k}}\rangle = \int_{\Omega} \Phi^{\dagger}_{n'\mathbf{k}'}(\mathbf{R})\,[\hat{H}_{em}\,\Phi_{n\mathbf{k}}(\mathbf{R})]\,\mathrm{d}\mathbf{R} \tag{4.50}$$

for emission. As shown in Section 4.6, the absorption and emission matrix elements can be very different.

4.1.7 Total scattering rate

In the previous section we dealt with the scattering rate between a given initial state (n,$\mathbf{k}$) and a given final state (n',$\mathbf{k}'$), that we expressed with Eq.4.3 for a stationary scattering potential and with Eq.4.43 for a time-dependent scattering potential comprising both an emission and an absorption part.

If we now wish to calculate the scattering rate $S_n(\mathbf{k})$ out of the state (n,$\mathbf{k}$) produced by a given scattering mechanism, we need to sum the scattering rates over all the possible final states (n',$\mathbf{k}'$). By using Eq.4.3 we have

$$S_n(\mathbf{k}) = \sum_{n',\mathbf{k}'} S_{n,n'}(\mathbf{k},\mathbf{k}') = \frac{2\pi}{\hbar}\sum_{n',\mathbf{k}'} |M_{n,n'}(\mathbf{k},\mathbf{k}')|^2\,\delta[E_n(\mathbf{k}) - E_{n'}(\mathbf{k}')], \tag{4.51}$$

whereas from Eq.4.43 we obtain

$$\begin{aligned} S_n(\mathbf{k}) &= \frac{2\pi}{\hbar}\sum_{n',\mathbf{k}'} |M^{(ab)}_{n,n'}(\mathbf{k},\mathbf{k}')|^2\,\delta[E_n(\mathbf{k}) - E_{n'}(\mathbf{k}') + \hbar\omega] \\ &+ \frac{2\pi}{\hbar}\sum_{n',\mathbf{k}'} |M^{(em)}_{n,n'}(\mathbf{k},\mathbf{k}')|^2\,\delta[E_n(\mathbf{k}) - E_{n'}(\mathbf{k}') - \hbar\omega]. \end{aligned} \tag{4.52}$$

Evaluation of the scattering rate $S_n(\mathbf{k})$ according to Eqs.4.51 and 4.52 proceeds by converting the sums over $\mathbf{k}'$ to appropriate integrals as discussed in detail in Section 3.5. In particular, it is important to understand that the Dirac function restricts the integrals over $\mathbf{k}'$ to curves corresponding to a specific value of the energy $E_{n'}(\mathbf{k}')$. Thus, if an analytical relation exists between the wave-vector $\mathbf{k}$ and the energy, it is typically convenient to change the integration variables to introduce an integral over the energy, which is readily reduced by the presence of the Dirac function. More detailed examples can be seen in the rest of this chapter.

4.1.8 Elastic and isotropic scattering rates

A scattering mechanism is said to be *elastic* when the change in the carrier energy produced by the scattering is null or practically negligible. The scattering rates can be expressed by means of Eqs.4.3 or 4.51 and the final energy $E_{n'}(\mathbf{k}')$ must be the same as the initial energy $E_n(\mathbf{k})$.

As discussed in Sections 4.4 and 4.3, the surface roughness and the Coulomb scattering are elastic. Section 4.6 explains that some phonon assisted transitions can also be considered approximately elastic, at least at room temperature.

If the matrix element of a scattering mechanism has a very weak, practically negligible dependence on the wave-vector redirection $\mathbf{q} = (\mathbf{k}' - \mathbf{k})$, then the scattering mechanism is said to be *isotropic*. In such a case the wave-vector independent matrix element $M^{(0)}_{n',n}$ can be taken out of the sum over $\mathbf{k}'$ in Eqs.4.51 or 4.52 and the calculation of the scattering rate is drastically simplified. In fact we can now recall Eq.3.79 and evaluate Eq.4.51 as

$$S_n(\mathbf{k}) = \frac{\pi}{\hbar} \sum_{n'} A\,|M^{(0)}_{n,n'}|^2\, g_{n'}(E_n(\mathbf{k})), \tag{4.53}$$

where $g_n(E)$ is the density of states of the subband n' defined in Eq.3.75 (with $n_{sp} = 1$ because the scattering mechanisms considered in this book do not change the spin), and $E_n(\mathbf{k})$ is the energy in the initial state $(n,\mathbf{k})$. The normalization area A always disappears in the final expression for the scattering rates, as can be seen in all the practical cases discussed in this book.

Similarly, by starting from Eq.4.52 we obtain

$$S_n(\mathbf{k}) = \frac{\pi}{\hbar} \sum_{n'} \left[A\,|M^{(ab)}_{n,n'}|^2\, g_{n'}(E_n(\mathbf{k}) + \hbar\omega) + A\,|M^{(em)}_{n,n'}|^2\, g_{n'}(E_n(\mathbf{k}) - \hbar\omega) \right], \tag{4.54}$$

where $M^{(ab)}_{n,n'}$ and $M^{(em)}_{n,n'}$ are the wave-vector independent matrix elements for absorption and emission, respectively.

It is probably useful to remember that $g_{n'}(E)$ is null for any E value below the minimum energy of the subband n' (see Section 3.5.1), thus, for a given $E_n(\mathbf{k})$, the density of states in the r.h.s. of Eqs.4.53 and 4.54 determine the subbands n' that give a non-null contribution to the total scattering rate $S_n(\mathbf{k})$. In particular, for a given energy $E_n(\mathbf{k})$ and final subband n', according to Eq.4.54 the absorption transitions may be possible even if the corresponding emissions are not.

4.2 Static screening produced by the free carriers

A description of the screening effects in a low dimensional carrier gas requires derivations that are both complex and long. Therefore Section 4.2.1 introduces the basic concepts of screening effects considering a much simpler case, namely a 3D electron gas in a uniformly doped bulk silicon sample. The purpose of Section 4.2.1 is to provide a physical insight into screening effects before moving to the quite involved derivations necessary to describe screening in inversion layers.

Throughout this section we consider a static perturbation potential, which is an adequate description for the Coulomb and the surface roughness scattering mechanisms (see Sections 4.3 and 4.4). Dynamic screening is discussed in Section 4.7 in relation to the phonon scattering mechanism.

4.2.1 Basic concepts of screening

Let us consider an n-type silicon region with a constant donor concentration N_D. The electron concentration at a point $\mathbf{R}$ can be simply expressed as [9]

$$n(\mathbf{R}) = n_0 \, \mathrm{e}^{\frac{\mathrm{e}\phi(\mathbf{R})}{K_B T}}, \tag{4.55}$$

where $\phi(\mathbf{R})$ is the electrostatic potential and n_0 is by definition the electron concentration in the reference region where $\phi(\mathbf{R})$ is null.

In this section we are interested in determination of the scattering potential $U_p(\mathbf{R})$ produced by a point charge in the n-silicon sample. To this purpose we write the electrostatic potential as

$$\phi(\mathbf{R}) = \phi_0(\mathbf{R}) + U_p(\mathbf{R}), \tag{4.56}$$

where $\phi_0(\mathbf{R})$ is the potential corresponding to the unperturbed case (namely without the point charge).

The overall potential $\phi(\mathbf{R})$ can be obtained by solving the Poisson equation. If we neglect the hole concentration in the expression for the macroscopic charge density, the Poisson equation can be written in spherical coordinates as

$$\frac{1}{R^2}\frac{\mathrm{d}}{\mathrm{d}R}\left[R^2 \frac{\mathrm{d}\phi(\mathbf{R})}{\mathrm{d}R}\right] = -\frac{\mathrm{e}}{\epsilon_{si}}\,\delta(\mathbf{R}) + \frac{\mathrm{e}}{\epsilon_{si}}(n(\mathbf{R}) - N_D), \tag{4.57}$$

where we have assumed, without any loss of generality, that the point charge is located at the origin of the coordinate system.

The solution of Eq.4.57 *without* the first term in the r.h.s. due to the point charge provides the unperturbed potential $\phi_0(\mathbf{R})$. We readily see that $\phi_0(\mathbf{R}) = 0$ is a valid solution for $n_0 = N_D$. Hence Eq.4.56 gives $\phi(\mathbf{R}) = U_p(\mathbf{R})$ and Eq.4.55 becomes

$$n(\mathbf{R}) = n_0 \, \mathrm{e}^{\frac{\mathrm{e}U_p(\mathbf{R})}{K_B T}}. \tag{4.58}$$

We now suppose that the scattering potential is small enough so that Eq.4.58 can be linearized as

$$n(\mathbf{R}) \simeq n_0 + \delta n = n_0 + \frac{\mathrm{e}\, n_0}{K_B T}\, U_p(\mathbf{R}). \tag{4.59}$$

By substituting $\phi(\mathbf{R}) = U_p(\mathbf{R})$ and Eq.4.59 in Eq.4.57 we finally obtain

$$\frac{1}{R^2}\frac{\mathrm{d}}{\mathrm{d}R}\left[R^2 \frac{\mathrm{d}U_p(\mathbf{R})}{\mathrm{d}R}\right] = -\frac{\mathrm{e}}{\epsilon_{si}}\,\delta(\mathbf{R}) + \frac{U_p(\mathbf{R})}{L_D^2}, \tag{4.60}$$

where L_D is the Debye length defined as

$$L_D = \sqrt{\frac{\epsilon_{si} K_B T}{e^2\, n_0}}. \tag{4.61}$$

The scattering potential $U_p(R)$ produced by a point charge in the n-silicon sample is obtained by solving Eq.4.60 and is given by

$$U_p(R) = \frac{e^2}{4\pi\epsilon_{si}R}\, e^{-\frac{R}{L_D}}. \qquad (4.62)$$

As can be seen, L_D represents a screening length, which is inversely proportional to the square root of the electron density $n = n_0 = N_D$. The larger is the electron density, the smaller is L_D and the stronger is the screening effect. In the limit of infinitely large L_D values, instead, Eq.4.62 simplifies to

$$U_p^{(unscr)}(R) = \frac{e^2}{4\pi\epsilon_{si}R}, \qquad (4.63)$$

and the scattering potential takes the well known unscreened expression.

Equation 4.62 clearly shows that the screening tends to reduce the scattering potential with respect to the unscreened case; hence neglecting the screening in calculation of the scattering rates may result in vast over-estimations of the rates.

4.2.2 Static dielectric function for a 2D carrier gas

Formulation of the screening produced by carriers in the inversion layer is much more complicated than it is for a 3D carrier gas. To begin with we notice that, because of the subband quantization, for a 2D gas it is not possible to express the carrier densities as a local function of the perturbation potential $U_p(\mathbf{r}, z)$ as in Eq.4.58.

As discussed in Appendix E, the charge density $\rho_{ind}(\mathbf{r}, z)$ induced by $U_p(\mathbf{r}, z)$ can be expressed by using perturbation theory as a linear function of all the *unscreened* matrix elements (namely the matrix elements of $U_p(\mathbf{r}, z)$). The potential $V_\rho(\mathbf{r}, z)$ produced by $\rho_{ind}(\mathbf{r}, z)$ is thus similarly dependent on all the unscreened matrix elements. The overall screened perturbation potential is given by $[U_p(\mathbf{r}, z)+V_\rho(\mathbf{r}, z)]$, hence, when we calculate the screened matrix elements (namely the matrix elements of $[U_p(\mathbf{r}, z)+V_\rho(\mathbf{r}, z)]$), we obtain a system of linear equations linking the screened to the unscreened matrix elements. The matrix or tensor governing such a linear system is called the dielectric function; the screened matrix elements must be sought by solving the linear problem [7, 10].

The next two sections discuss separately the formulation of the dielectric function for the case when the envelope wave-function in the inversion layer is either independent of the wave-vector $\mathbf{k}$ or dependent on $\mathbf{k}$. The first case is representative of an electron inversion layer described by the EMA model; the second case applies to the $\mathbf{k}\cdot\mathbf{p}$ model for hole inversion layers. In both cases we consider a static perturbation potential, hence the treatment is directly applicable to the Coulomb and surface roughness scattering mechanisms discussed in Sections 4.4 and 4.3. A discussion of dynamic screening is deferred to Section 4.7, after a description of the phonon scattering mechanisms.

Envelope wave-functions independent of the wave-vector

For an electron inversion layer described with the EMA quantization model or for a hole inversion layer described according to the semi-analytical model of Section 3.3.3, the envelope wave-function $\xi_{\nu,n}(z)$ is independent of $\mathbf{k}$ and we can write

$$\Phi_{\nu,n}(\mathbf{R}) = \frac{e^{i\,\mathbf{k}\cdot\mathbf{r}}}{\sqrt{A}}\,\xi_{\nu,n}(z), \tag{4.64}$$

where n is the subband index and ν is the valley index.

We henceforth denote by $M_{\nu,m,m'}(\mathbf{q})$ the matrix element defined in Eq.4.6 for the static perturbation potential $U_p(\mathbf{r}, z)$. Note that $M_{\nu,m,m'}(\mathbf{q})$ is the unscreened matrix element between the initial state $(\nu,m,\mathbf{k})$ and the final band $(\nu,m',\mathbf{k}')$ (with $\mathbf{q}=(\mathbf{k}'-\mathbf{k})$), hence it corresponds to an intra-valley transition between subbands of the valley ν. As discussed in detail in Section 4.1.4, in silicon inversion layers the electron inter-valley transitions due to Coulomb or surface roughness scattering mechanisms can be reasonably neglected. Hence in the following discussion we consider only intra-valley transitions.

As shown in Appendix E, a perturbative approach allows us to write the charge density $\rho_{ind}(\mathbf{r}, z)$ induced by the perturbation potential $U_p(\mathbf{r}, z)$. In particular, the $\rho_{ind}(\mathbf{r}, z)$ produced by the matrix elements corresponding to a wave-vector variation $\mathbf{q}_p$ is given by

$$\rho_{ind}(\mathbf{r}, z) = e \sum_{w,n,n'} \xi^{\dagger}_{w,n}(z)\,\xi_{w,n'}(z)\,\Pi_{w,n,n'}(\mathbf{q}_p)\,M_{w,n,n'}(\mathbf{q}_p)\,e^{i\mathbf{q}_p\cdot\mathbf{r}} + (c.c.). \tag{4.65}$$

The $\Pi_{w,n,n'}(\mathbf{q})$ is the polarization factor defined in Eq.E.16 as

$$\Pi_{w,n,n'}(\mathbf{q}_p) = \frac{1}{A}\sum_{\mathbf{k}_p} \frac{f_{w,n'}(\mathbf{k}_p+\mathbf{q}_p) - f_{w,n}(\mathbf{k}_p)}{E_{w,n'}(\mathbf{k}_p+\mathbf{q}_p) - E_{w,n}(\mathbf{k}_p)}, \tag{4.66}$$

where $f_{\nu,n}(\mathbf{k})$ is the occupation function of the subband (ν,n) such that

$$\frac{1}{A}\sum_{\mathbf{k}} f_{\nu,n}(\mathbf{k}) = N_{\nu,n}, \tag{4.67}$$

with $N_{\nu,n}$ being the inversion density in the subband (ν,n).

Equation 4.65 expresses ρ_{ind} as a linear function of the unscreened matrix elements $M_{\nu,m,m'}(\mathbf{q})$. Such a charge density ρ_{ind} yields an induced potential $V_\rho(\mathbf{r}$,z) which adds to the unscreened perturbation potential $U_p(\mathbf{r}, z)$. The induced potential $V_\rho(\mathbf{r}$,z) produces in turn the induced matrix elements $M^{(ind)}_{\nu,m,m'}(\mathbf{q})$, so that the overall screened matrix elements are given by

$$M^{(scr)}_{\nu,m,m'}(\mathbf{q}) = M_{\nu,m,m'}(\mathbf{q}) + M^{(ind)}_{\nu,m,m'}(\mathbf{q}). \tag{4.68}$$

The purpose of the derivations in the rest of this section is to find the linear system relating the screened matrix elements $M^{(scr)}_{\nu,m,m'}(\mathbf{q})$ to the unscreened ones $M_{\nu,m,m'}(\mathbf{q})$, which requires elimination of $M^{(ind)}_{\nu,m,m'}(\mathbf{q})$ in Eq.4.68. To this end we first need to calculate the matrix elements $M^{(ind)}_{\nu,m,m'}(\mathbf{q})$ produced by $V_\rho(\mathbf{r}$,z). Recalling Eq.4.6, we start by writing

$$M^{(ind)}_{\nu,m,m'}(\mathbf{q}) = \frac{(2\pi)^2}{A} \int \xi_{\nu,m}(z)\, \xi^\dagger_{\nu,m'}(z)\, V_{\rho,2T}(-\mathbf{q}, z)\, \mathrm{d}z, \tag{4.69}$$

where $V_{\rho,2T}(\mathbf{q}, z)$ is the Fourier transform defined in Eq.A.17 of the potential $V_\rho(\mathbf{r}$,z).

Calculation of $V_{\rho,2T}(\mathbf{q}, z)$ can be achieved by using the results of Section 4.3.1, which describes determination of the potential produced by a point charge. In particular, according to Eq.4.115, the 2D Fourier transform $\Psi_{pc}(\mathbf{q}, z)$ of the potential produced by a point charge located at ($\mathbf{r}_0$,z_0) is given by

$$\Psi_{pc}(\mathbf{q}, z) = \frac{\mathrm{e}^{\mathrm{i}\mathbf{q}\cdot\mathbf{r}_0}}{(2\pi)^2}\, \phi_{pc}(q, z, z_0), \tag{4.70}$$

and explicit expressions for $\phi_{pc}(q, z, z_0)$ are given by Eqs.4.126 and 4.132 for a bulk and an SOI MOS structure, respectively.

Thus, given the linearity of the Poisson equation, the potential $V_{\rho,2T}$ produced by the charge density $\rho_{ind}(\mathbf{r}_0, z_0)$ is given by

$$\frac{1}{A} V_{\rho,2T}(-\mathbf{q}, z) = \frac{1}{A} \int \mathrm{d}\mathbf{r}_0 \int \mathrm{d}z_0\, \rho_{ind}(\mathbf{r}_0, z_0)\, \Psi_{pc}(-\mathbf{q}, z), \tag{4.71}$$

where $\Psi_{pc}(\mathbf{q}, z)$ is defined in Eq.4.70.

The $\rho_{ind}(\mathbf{r}_0, z_0)$ expressed by Eq.4.65 must now be inserted in Eq.4.71. In the r.h.s. of Eq.4.65 we see a first term plus the complex conjugate. In the following calculations we neglect the $(c.c.)$ term in Eq.4.65, in fact at the end of the derivations we can understand that the inclusion of such a term in the calculations is redundant and unnecessary. By proceeding as just explained, Eq.4.71 provides

$$\frac{1}{A} V_{\rho,2T}(-\mathbf{q}, z) = \frac{\mathrm{e}}{(2\pi)^2} \left[\frac{1}{A} \int_A \mathrm{e}^{\mathrm{i}(\mathbf{q}_p - \mathbf{q})\cdot\mathbf{r}_0}\, \mathrm{d}\mathbf{r}_0 \right] \sum_{w,n,n'} \Pi_{w,n,n'}(\mathbf{q}_p)\, M_{w,n,n'}(\mathbf{q}_p)$$
$$\times \int \mathrm{d}z_0\, \xi^\dagger_{w,n}(z_0)\, \xi_{w,n'}(z_0) \phi_{pc}(q, z, z_0). \tag{4.72}$$

As discussed in Appendix B, for the $\mathbf{q}$ and $\mathbf{q}_p$ values defined by Eq.3.3, the integral in square brackets in Eq.4.72 yields $\delta_{\mathbf{q},\mathbf{q}_p}$, hence it is one for $\mathbf{q} = \mathbf{q}_p$ and null otherwise.

By substituting Eq.4.72 in Eq.4.69 we obtain

$$M^{(ind)}_{\nu,m,m'}(\mathbf{q}) = \frac{\mathrm{e}^2}{q(\epsilon_{si} + \epsilon_{ox})} \sum_{w,n,n'} \Pi_{w,n,n'}(\mathbf{q})\; F^{w,n,n'}_{\nu,m,m'}(q)\; M_{w,n,n'}(\mathbf{q}), \tag{4.73}$$

where we have introduced the unitless screening form factor

$$F^{w,n,n'}_{\nu,m,m'}(q) = \int \mathrm{d}z\, \xi_{\nu,m}(z)\, \xi^\dagger_{\nu,m'}(z) \int \mathrm{d}z_0\, \xi^\dagger_{w,n}(z_0)\, \xi_{w,n'}(z_0) \phi_{pcN}(q, z, z_0). \tag{4.74}$$

The symbol $\phi_{pcN}(q, z, z_0)$ denotes the normalized potential

$$\phi_{pcN}(q, z, z_0) = \frac{q(\epsilon_{si} + \epsilon_{ox})}{e}\, \phi_{pc}(q, z, z_0), \tag{4.75}$$

where, as noted above, explicit expressions for $\phi_{pc}(q, z, z_0)$ are given by Eqs.4.126 and 4.132 for a bulk and an SOI MOS structure, respectively.

Equation 4.73 provides a relation between the unscreened matrix elements $M_{w,n,n'}(\mathbf{q})$ and the matrix elements $M^{(ind)}_{v,m,m'}(\mathbf{q})$ induced by the potential $V_\rho(\mathbf{r},z)$. However, the $V_\rho(\mathbf{r},z)$ changes the overall perturbation potential and, in turn, the $M^{(ind)}_{v,m,m'}(\mathbf{q})$. Hence we need to seek a self-consistent solution of the problem by calculating the $M^{(ind)}_{v,m,m'}(\mathbf{q})$ produced by the overall screened perturbation potential. This can be accomplished by using Eq.4.73 to ensure that the screened matrix elements in the r.h.s. of the equation yield matrix elements $M^{(ind)}_{v,m,m'}(\mathbf{q})$ in the l.h.s. equal to $[M^{(scr)}_{v,m,m'}(\mathbf{q}) - M_{v,m,m'}(\mathbf{q})]$.

Mathematically speaking, we can use Eq.4.73 to write

$$M^{(ind)}_{v,m,m'}(\mathbf{q}) = M^{(scr)}_{v,m,m'}(\mathbf{q}) - M_{v,m,m'}(\mathbf{q})$$
$$= \frac{e^2}{q(\epsilon_{si} + \epsilon_{ox})} \sum_{w,n,n'} \Pi_{w,n,n'}(\mathbf{q})\; F^{w,n,n'}_{v,m,m'}(q)\; M^{(scr)}_{w,n,n'}(\mathbf{q}), \qquad (4.76)$$

which provides an implicit definition of the screened matrix elements $M^{(scr)}_{w,n,n'}(\mathbf{q})$. More precisely, Eq.4.76 represents the desired system of linear equations relating the unscreened $M_{v,m,m'}(\mathbf{q})$ to the screened matrix elements $M^{(scr)}_{v,m,m'}(\mathbf{q})$. Equation 4.76 can be cast in the form

$$M_{v,m,m'}(\mathbf{q}) = \sum_{w,n,n'} \epsilon^{w,n,n'}_{v,m,m'}(\mathbf{q})\, M^{(scr)}_{w,n,n'}(\mathbf{q}) \qquad (4.77)$$

by introducing the dielectric function

$$\epsilon^{w,n,n'}_{v,m,m'}(\mathbf{q}) = \delta_{w,v}\,\delta_{n,m}\,\delta_{n',m'} - \frac{e^2}{q(\epsilon_{si} + \epsilon_{ox})} \Pi_{w,n,n'}(\mathbf{q})\; F^{w,n,n'}_{v,m,m'}(q), \qquad (4.78)$$

which is a six component tensor.

As can be seen, according to the dielectric function formulation, determination of the screened matrix elements requires us to solve Eq.4.77 for $M^{(scr)}_{w,n,n'}(\mathbf{q})$. More precisely, Eq.4.77 links all the intra-valley matrix elements corresponding to the wave-vector variation $\mathbf{q}$; hence the calculation must be performed for each $\mathbf{q}$ value.

The effect of the screening is governed by the polarization factors $\Pi_{w,n,n'}(\mathbf{q})$ and by the screening form factors $F^{w,n,n'}_{v,m,m'}(q)$. As discussed in Section 4.2.4, the polarization factors are related to the occupation of the subbands. If the occupation of the subbands and the corresponding carrier density are very small, then the polarization factors tend to zero and Eq.4.77 gives

$$M^{(scr)}_{v,m,m'}(\mathbf{q}) \approx M_{v,m,m'}(\mathbf{q}), \qquad (4.79)$$

hence the screening effect vanishes.

It is finally worth noting that, if we now repeat the derivations from Eq.4.71 to Eq.4.77 by considering the $(c.c.)$ term in Eq.4.65, then we obtain an equation similar to Eq.4.77

but relating $M^{(scr)}_{\nu,m,m'}(-\mathbf{q})$ to $M_{\nu,m,m'}(-\mathbf{q})$, that is the matrix elements corresponding to the wave-vector variation $(-\mathbf{q})$. However, since we know that

$$M^{(scr)}_{\nu,m,m'}(-\mathbf{q}) = [M^{(scr)}_{\nu,m',m}(\mathbf{q})]^{\dagger},$$

the solution of this latter problem is thus redundant with respect to Eq.4.77.

Envelope wave-functions dependent on the wave-vector

When a fully numerical quantization model is used, the envelope wave-functions usually depend on the wave-vector, and Eq.4.64 must be replaced by

$$\Phi_{w,n,\mathbf{k}}(\mathbf{R}) = \xi_{w,n,\mathbf{k}}(z)\,\frac{e^{i\,\mathbf{k}\cdot\mathbf{r}}}{\sqrt{A}}. \tag{4.80}$$

In such a case the matrix element $M_{\nu,m,m'}(\mathbf{k},\mathbf{k}{+}\mathbf{q})$ between the initial state $(\nu,m,\mathbf{k})$ and the final state $(\nu,m',\mathbf{k}{+}\mathbf{q})$ defined in Eq.4.6 depends not only on the wave-vector variation $\mathbf{q}$ but also on the wave-vector $\mathbf{k}$ of the initial state.

Under these circumstances, Appendix E shows that the charge density $\rho_{ind}(\mathbf{r},z)$ produced by the matrix elements corresponding to the wave-vector variation $\mathbf{q}_p$ is given by

$$\rho_{ind}(\mathbf{r},z) = \frac{e}{A}\sum_{w,n,\mathbf{k}_p,n'} \xi^{\dagger}_{w,n,\mathbf{k}_p}(z)\cdot\xi_{w,n',(\mathbf{k}_p+\mathbf{q}_p)}(z)\,M_{w,n,n'}(\mathbf{k}_p,\mathbf{k}_p+\mathbf{q}_p)$$
$$\times\left[\frac{f_{w,n'}(\mathbf{k}_p+\mathbf{q}_p)-f_{w,n}(\mathbf{k}_p)}{E_{w,n'}(\mathbf{k}_p+\mathbf{q}_p)-E_{w,n}(\mathbf{k}_p)}\right] e^{i\mathbf{q}_p\cdot\mathbf{r}} + (c.c.). \tag{4.81}$$

If the wave-functions are vectorial functions, as in the case of the $\mathbf{k}\cdot\mathbf{p}$ model for a hole inversion layer, the dot sign between $\xi^{\dagger}_{w,n,\mathbf{k}_p}$ and $\xi_{w,n',(\mathbf{k}_p+\mathbf{q}_p)}$ in Eq.4.81 indicates the scalar product defined in Eq.4.40.

Starting from Eq.4.81, the derivations of the dielectric function can proceed similarly to those for $\mathbf{k}$ independent wave-functions. In fact the matrix element induced by the potential $V_\rho(\mathbf{r},z)$ produced by $\rho_{ind}(\mathbf{r},z)$ can be expressed as

$$M^{(ind)}_{\nu,m,m'}(\mathbf{k},\mathbf{k}+\mathbf{q}) = \frac{(2\pi)^2}{A}\int \xi_{\nu,m,\mathbf{k}}(z)\cdot\xi^{\dagger}_{\nu,m',(\mathbf{k}+\mathbf{q})}(z)\,V_{\rho,2T}(-\mathbf{q},z)\,dz, \tag{4.82}$$

where $V_{\rho,2T}(\mathbf{q},z)$ is the two-dimensional Fourier transform of $V_\rho(\mathbf{r},z)$, which can be obtained by substituting Eq.4.81 into Eq.4.71.

By following the same derivations that allowed us to go from Eq.4.69 to Eq.4.77, we finally obtain

$$M_{\nu,m,m'}(\mathbf{k},\mathbf{k}+\mathbf{q}) = \frac{1}{A}\sum_{w,n,\mathbf{k}_p,n'} \epsilon^{w,n,n'}_{\nu,m,m'}(\mathbf{k},\mathbf{k}_p,\mathbf{q})\,M^{(scr)}_{w,n,n'}(\mathbf{k}_p,\mathbf{k}_p+\mathbf{q}), \tag{4.83}$$

where we have introduced the dielectric function $\epsilon_{v,m,m'}^{w,n,n'}(\mathbf{k}, \mathbf{k}_p, \mathbf{q})$:

$$\epsilon_{v,m,m'}^{w,n,n'}(\mathbf{k}, \mathbf{k}_p, \mathbf{q}) = \delta_{w,v}\,\delta_{n,m}\,\delta_{n',m'} - \frac{\mathrm{e}^2}{q(\epsilon_{si}+\epsilon_{ox})} \times \left[\frac{f_{w,n'}(\mathbf{k}_p+\mathbf{q}) - f_{w,n}(\mathbf{k}_p)}{E_{w,n'}(\mathbf{k}_p+\mathbf{q}) - E_{w,n}(\mathbf{k}_p)}\right] F_{v,m,m'}^{w,n,n'}(\mathbf{k}, \mathbf{k}_p, \mathbf{q}), \quad (4.84)$$

and the form factor is now given by

$$F_{v,m,m'}^{w,n,n'}(\mathbf{k}, \mathbf{k}_p, \mathbf{q}) = \int \mathrm{d}z\, \xi_{v,m,\mathbf{k}}(z) \cdot \xi_{v,m',(\mathbf{k}+\mathbf{q})}^{\dagger}(z) \times \int \mathrm{d}z_0\, \xi_{w,n,\mathbf{k}_p}^{\dagger}(z_0)\, \xi_{w,n',(\mathbf{k}_p+\mathbf{q})}(z_0)\, \phi_{pcN}(q, z, z_0). \quad (4.85)$$

Thus, in the case of wave-vector dependent wave-functions, calculation of the screened matrix elements can be attained by solving Eq.4.83 for $M_{w,n,n'}^{(scr)}(\mathbf{k},\mathbf{k}+\mathbf{q})$. In practice, using Eq.4.83 is much more computationally demanding than using Eq.4.77, because according to Eq.4.83, for any given wave-vector variation $\mathbf{q}$, the matrix elements corresponding to all the possible initial wave-vectors $\mathbf{k}$ are coupled by a linear problem. Hence, if we consider that $\mathbf{k}$ is a two-dimensional vector, then we can see that, for each $\mathbf{q}$ value, the dielectric function $\epsilon_{v,m,m'}^{w,n,n'}(\mathbf{k}, \mathbf{k}_p, \mathbf{q})$ is a ten component tensor.

The number of components of the dielectric function defined in Eq.4.84 is too large to be practically used. In this respect, it should be noted that, as discussed in Sections 4.3 and 4.4, calculation of the scattering rate according to the $\mathbf{k}{\cdot}\mathbf{p}$ quantization model is often simplified by dropping the wave-vector dependence of the wave-functions, that is by using the wave-functions corresponding to $\mathbf{k} = \mathbf{0}$. For such a simplified use of the $\mathbf{k}{\cdot}\mathbf{p}$ quantization model, the screening can be accounted for according to the dielectric function introduced for $\mathbf{k}$ independent wave-functions.

4.2.3 The scalar dielectric function

Even for wave-vector independent wave-functions, formulation of the screening according to Eq.4.76 implies calculation of a very large number of screening form factors and polarization factors; this can be computationally very demanding and the derivation of approximate formulations is thus of practical interest. In this respect, it is easy to see that several simplifications apply when the wave-vector variation $\mathbf{q}$ is very small, that is in the long wavelength limit.

A bulk MOS structure

We first consider a bulk MOS transistor and note that, according to Eqs.4.75 and 4.126a, the normalized potential $\phi_{pcN}(q, z, z_0)$ tends to 1.0 when q tends to zero. If $\phi_{pcN}(q, z, z_0)$ is approximately one, then the orthogonality and the normalization of the wave-functions imply that the form factors defined in Eq.4.74 tend to 1 for intra-subband transitions (i.e. for $n = n'$ and $m = m'$) and to 0 otherwise. In other words we have

$$\lim_{q\to 0} F^{w,n,n'}_{\nu,m,m'}(q) = \delta_{n,n'}\,\delta_{m,m'}. \tag{4.86}$$

By substituting Eq.4.86 in Eq.4.78 and then in Eq.4.77 we see that simplification of the form-factors makes the screening negligible for the inter-subband transitions, for which we have simply

$$M^{(scr)}_{\nu,m,m'}(\mathbf{q}) \simeq M_{\nu,m,m'}(\mathbf{q}), \qquad m \neq m'. \tag{4.87}$$

The screened matrix elements for the intra-subband transitions, instead, must be determined by using Eq.4.77 in the simplified form

$$M_{\nu,m,m}(\mathbf{q}) = \sum_{w,n} \epsilon^{w,n}_{\nu,m}(\mathbf{q})\, M^{(scr)}_{w,n,n}(\mathbf{q}), \tag{4.88}$$

where we have set $m' = m$ and $n' = n$, so that the dielectric function $\epsilon^{w,n}_{\nu,m}(\mathbf{q})$ is now a four rank tensor, rather than a six rank tensor as in the general formulation of Eq.4.77.

According to Eq.4.78, the dielectric function for the intra-subband transitions can now be written as the matrix

$$\begin{pmatrix} 1+\chi_{1,1,1}(\mathbf{q}) & \cdot & \chi_{1,n_l,n_l}(\mathbf{q}) & \chi_{2,1,1}(\mathbf{q}) & \cdot & \chi_{w_l,n_l,n_l}(\mathbf{q}) \\ \cdot & \cdot & \cdot & \cdot & \cdot & \cdot \\ \chi_{1,1,1}(\mathbf{q}) & \cdot & 1+\chi_{1,n_l,n_l}(\mathbf{q}) & \chi_{2,1,1}(\mathbf{q}) & \cdot & \chi_{w_l,n_l,n_l}(\mathbf{q}) \\ \chi_{1,1,1}(\mathbf{q}) & \cdot & \chi_{1,n_l,n_l}(\mathbf{q}) & 1+\chi_{2,1,1}(\mathbf{q}) & \cdot & \chi_{w_l,n_l,n_l}(\mathbf{q}) \\ \cdot & \cdot & \cdot & \cdot & \cdot & \cdot \\ \chi_{1,1,1}(\mathbf{q}) & \cdot & \chi_{1,n_l,n_l}(\mathbf{q}) & \chi_{2,1,1}(\mathbf{q}) & \cdot & 1+\chi_{w_l,n_l,n_l}(\mathbf{q}) \end{pmatrix}$$

where n_l and w_l denote respectively the number of subbands and the number of valleys used in the calculations, and we have introduced

$$\chi_{w,n,n}(\mathbf{q}) = -\frac{e^2}{q(\epsilon_{si}+\epsilon_{ox})}\Pi_{w,n,n}(\mathbf{q}). \tag{4.89}$$

As can be seen, the dielectric function can be written as the sum of the identity matrix plus a matrix where all the elements of each column are the same. A matrix with this form can be inverted analytically [7], and the screened matrix elements are thus given by

$$M^{(scr)}_{\nu,m,m}(\mathbf{q}) = \frac{M_{\nu,m,m}(\mathbf{q}) + \sum_{w,n\neq m}\chi_{w,n,n}(\mathbf{q})[M_{\nu,m,m}(\mathbf{q}) - M_{w,n,n}(\mathbf{q})]}{\epsilon_D(\mathbf{q})}, \tag{4.90}$$

where $\epsilon_D(\mathbf{q})$ is the determinant of the dielectric matrix $\boldsymbol{\epsilon}$, which takes the form

$$\epsilon_D(\mathbf{q}) = \det(\boldsymbol{\epsilon}) = 1 + \sum_{w,n}\chi_{w,n,n}(\mathbf{q}) = 1 - \sum_{w,n}\frac{e^2}{q(\epsilon_{si}+\epsilon_{ox})}\Pi_{w,n,n}(\mathbf{q}). \tag{4.91}$$

Equation 4.90 provides an explicit expression for the screened matrix elements as a linear combination of the unscreened ones.

In a bulk MOSFET the intra-subband unscreened matrix elements can be quite similar for different subbands, so that the second term in the numerator of Eq.4.90 is frequently negligible with respect to the first. If we adopt this further simplification, then each intra-subband screened matrix element can be simply written as

$$M^{(scr)}_{v,m,m}(\mathbf{q}) = \frac{M_{v,m,m}(\mathbf{q})}{\epsilon_D(\mathbf{q})}. \quad (4.92)$$

Equation 4.92 is a scalar formulation of the screening and the $\epsilon_D(\mathbf{q})$ defined in Eq.4.91 is frequently referred to as the *scalar dielectric function*. In fact, according to Eq.4.92, each intra-subband screened matrix element is simply proportional to the corresponding unscreened matrix element through $\epsilon_D(\mathbf{q})$, that is the screening no longer produces a coupling of the different matrix elements.

By recalling Eq.4.6 for the intra-valley matrix elements, we see that Eq.4.92 is equivalent to saying that the $M^{(scr)}_{v,m,m}(\mathbf{q})$ are obtained from a screened scattering potential $U^{(scr)}_{sc}(\mathbf{R})$, whose two-dimensional Fourier transform is

$$U^{(scr)}_{2T}(\mathbf{q}, z) = \frac{U_{2T}(\mathbf{q}, z)}{\epsilon_D(-\mathbf{q})}. \quad (4.93)$$

Thus, in the scalar formulation, the dielectric function simply describes the $\mathbf{q}$ dependent screening of the scattering potential itself. In particular, knowing the unscreened $U_{2T}(\mathbf{q}, z)$, Eq.4.93 allows one to find the screened scattering potential $U^{(scr)}_{sc}(\mathbf{R})$ by using the appropriate inverse Fourier transform (see Eq.A.14).

Modeling of the screening based on Eqs.4.90 or 4.92 reduces the computational burden remarkably with respect to the full dielectric function defined in Eq.4.78. However, one should be aware that, strictly speaking, the simplified formulation is valid only for very small q values.

Figure 4.2 shows the inverse of the scalar dielectric function $\epsilon_D(q)$ versus q for an electron inversion layer and for different inversion densities N_{inv}. The dependence of $\epsilon_D(q)$ on the $\mathbf{q}$ direction is very small for an electron inversion layer (see Section 4.2.4), thus $\epsilon_D(q)$ is reported as a function of $q = |\mathbf{q}|$. By recalling Eq.4.92, we see that the screening reduces dramatically the scattering matrix elements for small q values. This is because the electrons in the inversion layer can screen much more effectively the

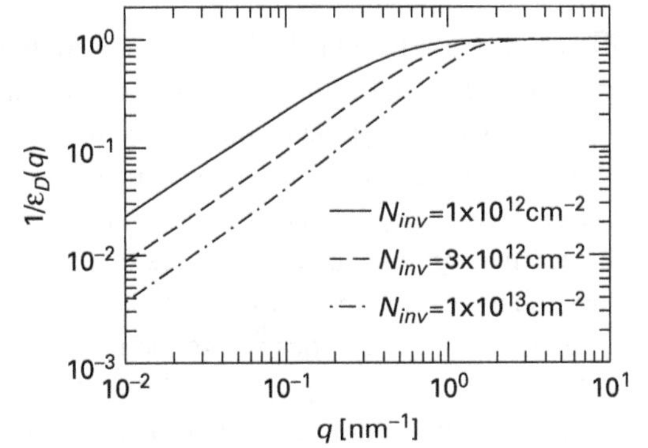

Figure 4.2 Inverse of the scalar dielectric function $\epsilon_D(q)$ defined in Eq.4.91 versus q for three values of the inversion density, N_{inv}. Results obtained with a self-consistent Schrödinger–Poisson solver. Bulk n-MOSFET with channel doping concentration of $3{\cdot}10^{17}$ cm^{-3}.

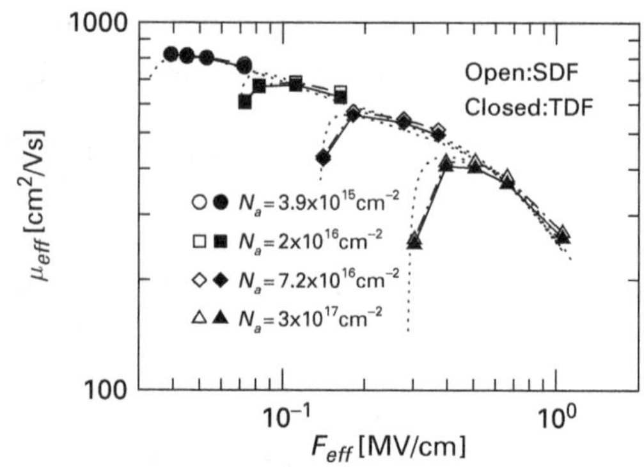

Figure 4.3 Simulated mobility in bulk MOSFETs versus the effective field, F_{eff}, for different channel doping concentrations, N_a, and compared to the experiments from [11]. The screening for Coulomb and surface roughness scattering was accounted for by using either the complete tensorial dielectric function (TDF, closed symbols) or the simplified scalar dielectric function (SDF, open symbols).

spectral components of the scattering potential that are slowly varying in real space (see Eq.4.93).

The scalar dielectric function works quite well in practice, at least for bulk MOSFETs. In this respect, Fig.4.3 shows the simulated mobility of bulk MOS transistors where the screening was accounted for either by means of Eq.4.77 and the full tensorial dielectric function defined in Eq.4.78 (TDF), or by means of Eq.4.92 and the scalar dielectric function defined in Eq.4.91 (SDF). As can be seen, the scalar approximation is in very close agreement with the tensorial dielectric function model in the entire range of channel doping values, N_a, and effective fields, F_{eff}. The simulations are also in good agreement with the experimental data from [11], and, in particular, they reproduce quite well the roll-off of the mobility at small F_{eff} values, which is actually related to screening of the ionized impurities in the channel (see Section 4.3 for the Coulomb scattering mechanism). Section 7.1 illustrates in more detail the definition of F_{eff} and the comparison between mobility simulations and experimental data.

An SOI MOS structure driven in single- or double-gate mode

The scalar screening model summarized by Eq.4.92 has also been used for SOI MOSFETs, and Fig.4.4 shows that the results are again in good agreement with the tensorial dielectric function model when SOI transistors are operated in single-gate (SG) mode.

In double-gate (DG) mode, instead, some of the approximations behind the derivation of Eq.4.92 become largely inaccurate and the scalar dielectric function model fails. This can lead to artifacts and to simulation results inconsistent with experiments, especially at high channel inversion densities where the mobility is limited by surface roughness scattering. In Fig.4.4, for example, we see that the mobility of a thick SOI transistor obtained with the SDF model becomes progressively larger in DG than in SG mode with increasing inversion density (open symbols). This result is an artifact produced by use of the scalar dielectric function for screening of the surface roughness scattering. In fact such behavior is inconsistent with experiment [12–14], and, furthermore, it is not

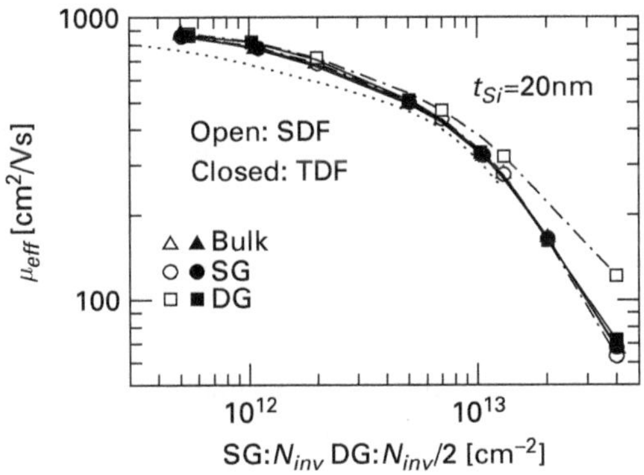

Figure 4.4 Simulated mobility versus inversion density, N_{inv}, for a lightly doped bulk MOSFET and for a thick SOI MOSFET (silicon thickness $t_{si} = 20$ nm) in single-gate (SG) or in double-gate (DG) mode. Screening was accounted for either with the tensorial dielectric function (TDF) or with the scalar dielectric function (SDF). In double-gate mode the simulated mobility of the SOI device deviates remarkably from the mobility of the bulk device at large N_{inv}.

observed when the tensorial dielectric function is used (closed symbols). The reasons for failure of the SDF model for surface roughness scattering in double-gate transistors are discussed in detail in [15].

The reader should be aware that only the tensorial dielectric function is a reliable screening model for double-gate SOI MOSFETs or FinFETs.

4.2.4 Calculation of the polarization factor

In order to obtain the dielectric function for wave-vector independent wave-functions it is necessary to calculate the polarization factor that is defined as

$$\Pi_{\nu,n,n'}(\mathbf{q}) = \frac{1}{A}\sum_{\mathbf{k}} \frac{f_{\nu,n'}(\mathbf{k}+\mathbf{q}) - f_{\nu,n}(\mathbf{k})}{E_{\nu,n'}(\mathbf{k}+\mathbf{q}) - E_{\nu,n}(\mathbf{k})}. \tag{4.94}$$

It is worth noting that, in general terms, calculation of $\Pi_{\nu,n,n'}(\mathbf{q})$ requires a knowledge of the occupation function in the different subbands. The occupation functions are the unknowns of the transport problem based on the Boltzmann transport equation (see Chapter 5), hence, strictly speaking, the $\Pi_{\nu,n,n'}(\mathbf{q})$ should be calculated self-consistently with the transport equations.

For calculation of low field mobility, the deviations of the $f_{\nu,n}(\mathbf{k})$ from the equilibrium Fermi–Dirac occupation $f_0(E)$ are very small and they can be neglected in calculation of the polarization factors. We can thus write

$$f_{\nu,n}(\mathbf{k}) = f_0(E_{\nu,n}(\mathbf{k})). \tag{4.95}$$

Equation 4.95 states that at equilibrium the occupation of the state (ν,n,$\mathbf{k}$) depends only on its energy, $E_{\nu,n}(\mathbf{k})$.

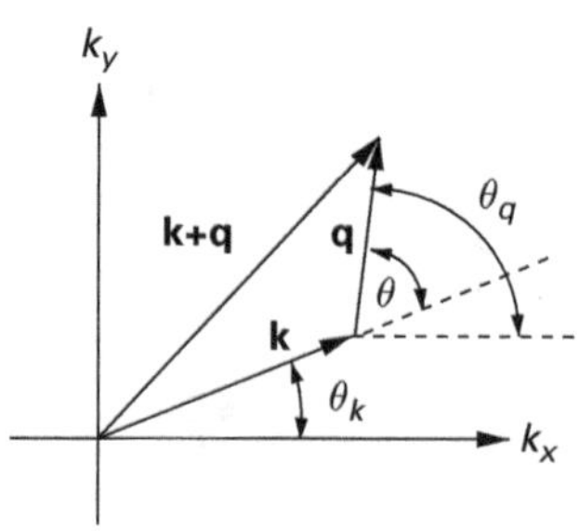

Figure 4.5 Wave-vectors **k** and **q** involved in calculation of the polarization factors and corresponding angles in the two-dimensional wave-vector plane.

As for the wave-vectors involved in the calculations of the $\Pi_{\nu,n,n'}(\mathbf{q})$, by using the angles θ_k, θ_q, and θ defined in Fig.4.5 we can write

$$\mathbf{k}+\mathbf{q} = [k\cos(\theta_k) + q\cos(\theta_q), k\sin(\theta_k) + q\sin(\theta_q)], \tag{4.96a}$$

$$|\mathbf{k}+\mathbf{q}|^2 = k^2 + q^2 + 2kq\cos(\theta). \tag{4.96b}$$

Equation 4.96b shows that the magnitude of (**k**+**q**) depends only on k, q, and on the angle θ between **k** and **q**, whereas the direction of (**k**+**q**) depends on both θ_k and θ_q.

In order to proceed further in the calculations we need to specify a form for the energy dispersion in the inversion layer.

Electrons: parabolic circular bands

In the case of electrons described with a parabolic and circular band model, the energy is given by

$$E_{\nu,n}(\mathbf{k}) = E_{\nu 0} + \varepsilon_{\nu,n} + E_p = E_{\nu 0} + \varepsilon_{\nu,n} + \frac{\hbar^2 k^2}{2m}, \tag{4.97}$$

where m is the effective mass and $E_p = \frac{\hbar^2 k^2}{2m}$ is the kinetic energy in the subband (ν,n).

The sum over **k** in Eq.4.94 can be converted to an integral by resorting to Eq.3.80 with $n_{sp} = 2$. By setting $\alpha = 0$ and $m_x = m_y = m$ we obtain

$$\frac{1}{A}\sum_{\mathbf{k}} \quad \Rightarrow \quad \frac{2\mu_\nu m}{(2\pi)^2\hbar^2}\int_0^{+\infty} \mathrm{d}E_p \int_0^{2\pi} \mathrm{d}\theta_k, \tag{4.98}$$

where μ_ν is the multiplicity of the valley ν.

To proceed further it is thus desirable to express the quantities in Eq.4.94 as a function of the kinetic energy, E_p, and the angle θ_k. This has already been obtained for $E_{\nu,n}(\mathbf{k})$ in Eq.4.97. As for $E_{\nu,n'}(\mathbf{k}+\mathbf{q})$, we can use Eq.4.96a to express the magnitude of (**k**+**q**) and then write

$$\begin{aligned} E_{\nu,n'}(\mathbf{k}+\mathbf{q}) &= E_{\nu,n'}(E_p, \theta, q) = E_{\nu 0} + \varepsilon_{\nu,n'} + E_p + \frac{\hbar^2 q^2}{2m} + \frac{\hbar^2 kq}{m}\cos(\theta) \\ &= E_{\nu 0} + \varepsilon_{\nu,n'} + E_p + \frac{\hbar^2 q^2}{2m} + \hbar q\sqrt{\frac{2E_p}{m}}\cos(\theta), \end{aligned} \tag{4.99}$$

where the notation $E_{\nu,n'}(E_p,\theta,q)$ underlines that in Eq.4.99 $E_{\nu,n'}(\mathbf{k}+\mathbf{q})$ is expressed as a function of E_p, θ, and q.

If we now use Eq.4.98 in Eq.4.94 and change the angular integration variable from θ_k to θ we can finally write

$$\Pi_{\nu,n,n'}(q) = \frac{2\mu_\nu m}{(2\pi)^2\,\hbar^2}\int_0^{+\infty} \mathrm{d}E_p \int_0^{2\pi}\left\{\frac{f_0[E_{\nu,n'}(E_p,\theta,q)] - f_0[E_{\nu0}+\varepsilon_{\nu,n}+E_p]}{E_{\nu,n'}(E_p,\theta,q) - (E_{\nu0}+\varepsilon_{\nu,n}+E_p)}\right\}\mathrm{d}\theta, \tag{4.100}$$

where $E_{\nu,n'}(E_p,\theta,q)$ is given by Eq.4.99.

Equation 4.100 shows that for circular bands the polarization factor depends on the magnitude of $\mathbf{q}$ but not on its direction.

To gain a physical insight into the polarization factors let us now consider an intra-subband polarization factor (i.e. $n=n'$) and take the limit for very small q values. In such a case the terms in the bracket in Eq.4.100 simplify to

$$\lim_{q\to 0}\left\{\frac{f_0[E_{\nu,n}(E_p,\theta,q)] - f_0[E_{\nu0}+\varepsilon_{\nu,n}+E_p]}{E_{\nu,n}(E_p,\theta,q) - (E_{\nu0}+\varepsilon_{\nu,n}+E_p)}\right\} = \frac{\partial f_0(E_{\nu0}+\varepsilon_{\nu,n}+E_p)}{\partial E}, \tag{4.101}$$

and we obtain

$$\lim_{q\to 0}\Pi_{\nu,n,n}(q) = -\frac{\mu_\nu m}{\pi\,\hbar^2} f_0(E_{\nu0}+\varepsilon_{\nu,n}), \tag{4.102}$$

where $f_0(E_{\nu0}+\varepsilon_{\nu,n})$ is the occupation of the energy corresponding to the subband minimum.

Equation 4.102 shows that, physically speaking, a subband produces a screening which is proportional to its occupation. This implies that the largest contribution to the screening is provided by the lowest subbands, whereas the contribution becomes negligible when we move to higher subbands.

Electrons: non-parabolic elliptic bands

For a non-parabolic elliptic energy dispersion, the sum over $\mathbf{k}$ in Eq.4.94 can be converted to an integral according to Eq.3.80 with $n_{sp}=2$, that is

$$\frac{1}{A}\sum_{\mathbf{k}} \quad\Rightarrow\quad \frac{2\mu_\nu}{(2\pi)^2\,\hbar^2}\int_0^{+\infty} \mathrm{d}E_p\,(1+2\,\alpha\,E_p)\int_0^{2\pi}\mathrm{d}\theta_k\; m_{xy}(\theta_k). \tag{4.103}$$

To proceed further we write $E_{\nu,n}(\mathbf{k})$ as

$$E_{\nu,n}(\mathbf{k}) = E_{\nu0}+\varepsilon_{\nu,n}+E_p, \tag{4.104}$$

and then we need to derive a suitable expression for $E_{\nu,n'}(\mathbf{k}+\mathbf{q})$. To this purpose we recall Eqs.3.26 and 3.28 to write

$$E_{\nu,n'}(\mathbf{k}+\mathbf{q}) = E_{\nu0}+\varepsilon_{\nu,n'}+\frac{1}{2\alpha}\left(-1+\sqrt{1+4\,\alpha\,\gamma(\mathbf{k}+\mathbf{q})}\right). \tag{4.105}$$

According to Eq.3.23, the $\gamma(\mathbf{k}+\mathbf{q})$ can be written as

$$\gamma(\mathbf{k}+\mathbf{q}) = \frac{\hbar^2}{2}\left\{\frac{[k\cos(\theta_k)+q\cos(\theta_q)]^2}{m_x} + \frac{[k\sin(\theta_k)+q\sin(\theta_q)]^2}{m_y}\right\}, \tag{4.106}$$

where (**k**+**q**) has been expressed with Eq.4.96a. The term k in Eqs.4.106 and 4.105 can be written in terms of E_p and θ_k by means of Eq.3.69, so that we can finally obtain $E_{\nu,n'}(\mathbf{k}+\mathbf{q})$ as a function of (E_p,θ_k) and (q,θ_q), that is we have

$$E_{\nu,n'}(\mathbf{k}+\mathbf{q}) = E_{\nu,n'}(E_p,\theta_k,q,\theta_q). \tag{4.107}$$

By using Eq.4.103 in Eq.4.94 and then the expressions for $E_{\nu,n}(\mathbf{k})$ and $E_{\nu,n'}(\mathbf{k}+\mathbf{q})$ discussed above, the polarization factor can be written as

$$\Pi_{\nu,n,n'}(q,\theta_q) = \Pi_{\nu,n,n'}(\mathbf{q}) = \frac{2\mu_\nu}{(2\pi)^2\hbar^2}\int_0^{+\infty} dE_p(1+2\alpha E_p)$$
$$\times \int_0^{2\pi} d\theta_k\, m_{xy}(\theta_k)\left\{\frac{f_0[E_{\nu,n'}(E_p,\theta_k,q,\theta_q)] - f_0[E_{\nu 0}+\varepsilon_{\nu,n}+E_p]}{E_{\nu,n'}(E_p,\theta_k,q,\theta_q) - (E_{\nu 0}+\varepsilon_{\nu,n}+E_p)}\right\}. \tag{4.108}$$

Aside from the tedious details of the derivations, Eq.4.108 shows that for an anisotropic energy dispersion the polarization factor depends not only on the magnitude q of **q** but also on θ_q, that is the direction of **q**.

Numerical calculations show that the dependence of $\Pi_{\nu,n,n'}$ on θ_q is typically weak, so that in analysis of electron devices the polarization factor is most frequently considered a function of only $q = |\mathbf{q}|$.

Hole inversion layers

If we adopt the semi-analytical energy model presented in Section 3.3.3, then the wave-functions do not depend on **k** and the dielectric function can be expressed in terms of the polarization factors, which can be calculated by following the procedure explained for the electrons. For any state identified by the group and subband indexes (ν,n) and then either by the wave-vector $\mathbf{k} = (k_x, k_y)$ or, equivalently, by the kinetic energy $E_{p,\nu}$ and the angle θ, we must express (**k**+**q**) and $E_{\nu,n'}$(**k**+**q**) as a function of ($E_{p,\nu}$,θ) and (q,θ_q), exactly as in Eq.4.107.

The polarization factor can then be obtained by converting the sum over **k** in Eq.4.94 to appropriate integrals by using Eq.3.86 with $n_{sp} = 2$

$$\Pi_{\nu,n,n'}(q,\theta_q) = \frac{2}{(2\pi)^2}\int_0^{+\infty} dE_{p,\nu}\int_0^{2\pi} d\theta\; k_\nu(E_\nu,\theta)\frac{\partial k_\nu(E_\nu,\theta)}{\partial E_\nu}$$
$$\times\left\{\frac{f_0[E_{\nu,n'}(E_{p,\nu},\theta,q,\theta_q)] - f_0[E_{\nu 0}+\varepsilon_{\nu,n}+E_{p,\nu}]}{E_{\nu,n'}(E_{p,\nu},\theta,q,\theta_q) - (E_{\nu 0}+\varepsilon_{\nu,n}+E_{p,\nu})}\right\}. \tag{4.109}$$

As can be seen, the anisotropy of the energy dispersion results in a polarization factor that depends not only on $q = |\mathbf{q}|$ but also on the angle θ_q.

If the **k·p** quantization model is used, then, strictly speaking, one should use the formulation of the dielectric function summarized by Eqs.4.83 to 4.85 (with $n_{sp} = 1$). However, if the wave-vector dependence of the wave-functions is neglected by taking, for each subband, the wave-function at $\mathbf{k} = 0$, then we can resort to the dielectric function defined in Eq.4.78. In such a case the polarization factors must be calculated by evaluating numerically the implied sum over $\mathbf{k}$ according to the procedure described in detail in Section 3.5.4. The spin multiplicity for the $\mathbf{k}\cdot\mathbf{p}$ Hamiltonian is always $n_{sp} = 1$.

4.3 Scattering with Coulomb centers

A Coulomb center produces a perturbation potential which can effectively scatter the carriers in a semiconductor device. In an MOS transistor examples of such electrically charged centers are the ionized dopants in the channel and the fixed charges either in localized states at the silicon-oxide interface or in the gate oxide stack.

Calculation of the scattering rates starts from determination of the potential produced by a point charge as a function of its position in the device. Such a stationary scattering potential can then be used to calculate the matrix elements and finally the scattering rates. All these steps are discussed in the following sections.

4.3.1 Potential produced by a point charge

Let us consider a positive electron charge e in a material with dielectric constant ϵ. The potential $\psi_{pc}(\mathbf{r}, z)$ produced by the point charge is obtained by solving the Poisson equation [16–19]

$$\nabla^2_{\mathbf{R}}\psi_{pc}(\mathbf{r}, z) = -\frac{e}{\epsilon}\,\delta(\mathbf{r} - \mathbf{r}_0)\,\delta(z - z_0). \tag{4.110}$$

In Eq.4.110 it is assumed that there are no free charges in the material, so that the point charge is the only term in the r.h.s. of the Poisson equation. Thus the scattering potential obtained from Eq.4.110 is the unscreened potential; the effect of screening is discussed in Section 4.3.3.

As already discussed in Sections 4.1.2 to 4.1.5, for an electron or a hole inversion layer the matrix elements produced by the scattering potential $\psi_{pc}(\mathbf{r}, z)$ are expressed in terms of the hybrid Fourier transform $\Psi_{pc}(\mathbf{q}, z)$ defined in Eq.A.17. Thus, for the Coulomb scattering, the $\Psi_{pc}(\mathbf{q}, z)$ produced by the Coulomb centers is the quantity we need to express. To this purpose, we start by recalling Eq.A.18 and write

$$\psi_{pc}(\mathbf{r}, z) = \int_{\mathbf{q}} \Psi_{pc}(\mathbf{q}, z)\mathrm{e}^{-\mathrm{i}\mathbf{q}\cdot\mathbf{r}}\,\mathrm{d}\mathbf{q}, \tag{4.111}$$

so that

$$\nabla^2_{\mathbf{R}}\psi_{pc}(\mathbf{r}, z) = \int_{\mathbf{q}} \left[\frac{\partial^2}{\partial z^2} - q^2\right]\Psi_{pc}(\mathbf{q}, z)\mathrm{e}^{-\mathrm{i}\mathbf{q}\cdot\mathbf{r}}\,\mathrm{d}\mathbf{q}. \tag{4.112}$$

If we now recall the identity

$$\delta(\mathbf{r} - \mathbf{r}_0) = \frac{1}{(2\pi)^2}\int_{\mathbf{q}} \mathrm{e}^{-\mathrm{i}\mathbf{q}\cdot(\mathbf{r}-\mathbf{r}_0)}\,\mathrm{d}\mathbf{q}, \tag{4.113}$$

then, by introducing Eqs.4.112 and 4.113 into Eq.4.110, this latter can be re-written as

$$\int_{\mathbf{q}} \left[\frac{\partial^2}{\partial z^2} - q^2 \right] \Psi_{pc}(\mathbf{q}, z)\, \mathrm{e}^{-i\mathbf{q}\cdot\mathbf{r}}\, d\mathbf{q} = -\int_{\mathbf{q}} \frac{e}{\epsilon (2\pi)^2}\, \mathrm{e}^{\mathrm{i}\mathbf{q}\cdot\mathbf{r}_0}\, \delta(z - z_0)\, \mathrm{e}^{-\mathrm{i}\mathbf{q}\cdot\mathbf{r}}\, d\mathbf{q}. \quad (4.114)$$

The above equation must be satisfied for any value of $\mathbf{r}$ and $\mathbf{r}_0$, which play the role of parameters in the integrals over $\mathbf{q}$. This implies that the integrand functions must be the same, so that $\Psi_{pc}(\mathbf{q}, z)$ must take the form

$$\Psi_{pc}(\mathbf{q}, z) = \frac{\mathrm{e}^{\mathrm{i}\mathbf{q}\cdot\mathbf{r}_0}}{(2\pi)^2}\, \phi_{pc}(q, z, z_0), \quad (4.115)$$

where $\phi_{pc}(q, z, z_0)$ satisfies the equation

$$\left[\frac{\partial^2}{\partial z^2} - q^2 \right] \phi_{pc}(q, z, z_0) = -\frac{e}{\epsilon}\, \delta(z - z_0). \quad (4.116)$$

As can be verified by direct substitution, the general solution of Eq.4.116 is

$$\phi_{pc}(q, z, z_0) = \frac{e}{2q\epsilon} \mathrm{e}^{-q|z-z_0|} + C_1\, \mathrm{e}^{qz} + C_2\, \mathrm{e}^{-qz}, \quad (4.117)$$

where the two q dependent constants C_1 and C_2 are to be determined according to the boundary conditions associated with Eq.4.116.

One last observation may be useful before moving to a discussion of some cases of practical interest. Let us consider the MOS system schematically illustrated in Fig.4.6 (where the silicon–oxide interface is set at $z = 0$), and a point charge is located at the generic position z_0, that can be either negative or positive. On the basis of the previous analysis, one may think that $\phi_{pc}(q, z, z_0)$ can be expressed by Eq.4.117 only in the region that hosts the point charge, namely either in the oxide or in the silicon for negative or positive z_0 values, respectively. In the second region, instead, one may argue that we should use the expression

$$\phi_{pc}(q, z, z_0) = C_1\, \mathrm{e}^{qz} + C_2\, \mathrm{e}^{-qz}, \quad (4.118)$$

where the term related to the presence of the point charge does not appear.

However, it is easy to understand that Eq.4.117 is completely equivalent to Eq.4.118 in a domain that does not include z_0. In fact, if we suppose that z_0 is negative, then in the silicon region z is always larger than z_0, hence Eq.4.117 simplifies to

$$\phi_{pc}(q, z, z_0) = \frac{e}{2q\epsilon_{si}} \mathrm{e}^{-q(z-z_0)} + C_1\, \mathrm{e}^{qz} + C_2\, \mathrm{e}^{-qz} = C_1\, \mathrm{e}^{qz} + C_2'\, \mathrm{e}^{-qz}, \quad (4.119)$$

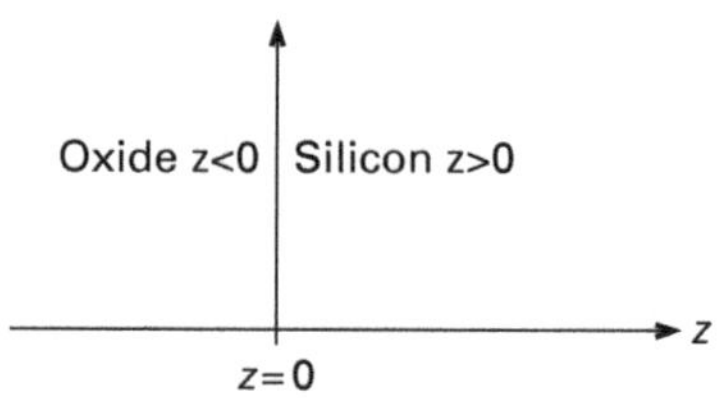

Figure 4.6 Sketch of a bulk-type MOS structure. The silicon–oxide interface is set at $z = 0$.

where the new constant C_2' has been introduced to group the terms with the $[e^{-qz}]$ dependence. Equation 4.119 is identical to Eq.4.118. A similar result can be obtained in the oxide region for a positive z_0 (namely for a point charge located in silicon), provided that ϵ_{ox} is used in place of ϵ_{si}.

The above discussion validates the use of Eq.4.117 to express the potential $\phi_{pc}(q, z, z_0)$ of the point charge in any region of the device (with the corresponding ϵ value), and irrespective of the position of z_0. As clarified by the following examples, this is very convenient because it allows us to determine $\phi_{pc}(q, z, z_0)$ for any position of the point charge by solving one single problem.

Potential in a bulk MOS structure

For a homogeneous silicon sample with an almost infinite volume, the unscreened potential produced by a point charge e is readily obtained by setting $C_1 = C_2 = 0$ in Eq.4.117, so that we have

$$\phi_{pc}(q, z, z_0) = \frac{e}{2q\epsilon_{si}} e^{-q|z-z_0|}, \tag{4.120}$$

where ϵ_{si} is the silicon dielectric constant.

As soon as we have two or more materials with different dielectric constants, calculation of the $\phi_{pc}(q, z, z_0)$ produced by the point charge requires us to take the general solution in the two materials and then determine the unknown constants by imposing appropriate continuity conditions at the interface between the materials.

As a case of prominent practical importance, let us consider an idealized bulk-type MOS structure consisting of an infinite silicon substrate and a similarly infinite oxide region. For convenience of notation the silicon–oxide interface is set at $z = 0$; a sketch of the structure is given in Fig.4.6. One may argue that in a MOS transistor the oxide region is very thin rather than infinitely extended, but typically the finite thickness of the oxide does not play a critical role in determination of the scattering potential: we discuss this point further in Chapter 10.

Equation 4.117 can be used to express $\phi_{pc}(q, z, z_0)$ in silicon (with $C_1 = 0$) and in the oxide (with $C_2 = 0$). In the problem under study we thus have two remaining unknown constants that we denote as A_1 and A_2. More precisely we can write

$$\phi_{pc}(q, z, z_0) = A_1 e^{qz} + \frac{e}{2q\epsilon_{ox}} e^{-q|z-z_0|}, \qquad \text{for } z < 0 \tag{4.121a}$$

$$\phi_{pc}(q, z, z_0) = A_2 e^{-qz} + \frac{e}{2q\epsilon_{si}} e^{-q|z-z_0|}, \qquad \text{for } z > 0 \tag{4.121b}$$

where ϵ_{ox} denotes the oxide dielectric constant.

In order to determine A_1 and A_2 in Eq.4.121 we can impose the continuity of the potential $\psi_{pc}(\mathbf{r}, z)$ and of the z component of the electric displacement field at $z = 0$. The continuity of the z component of the electric displacement field holds for a null density of interface charge at the silicon–oxide interface. Hence such a continuity condition restricts the following calculations to the case $z_0 \neq 0$.

Equations 4.111 and 4.115 show that the continuity of $\psi_{pc}(\mathbf{r}, z)$ at $z = 0$ requires the corresponding continuity of $\phi_{pc}(q, z, z_0)$. From Eq.4.121 we readily obtain

$$A_1 + \frac{e}{2q\epsilon_{ox}} \mathrm{e}^{-q|z_0|} = A_2 + \frac{e}{2q\epsilon_{si}} \mathrm{e}^{-q|z_0|}. \tag{4.122}$$

The continuity of the z component of the electric displacement field at the silicon–oxide interface results in an equation involving the z derivative of the potential $\psi_{pc}(\mathbf{r}, z)$ either in the oxide or in the silicon region. By using Eqs.4.111 and 4.115 to express $\psi_{pc}(\mathbf{r}, z)$, such a continuity condition results in

$$\epsilon_{ox} \frac{\partial \phi_{pc}}{\partial z}(q, z = 0^-, z_0) = \epsilon_{si} \frac{\partial \phi_{pc}}{\partial z}(q, z = 0^+, z_0). \tag{4.123}$$

The z derivative of $\phi_{pc}(q, z)$ at $z = 0^-$ and $z = 0^+$ must be calculated according to Eq.4.121a and Eq.4.121b, respectively, so that we obtain

$$\begin{aligned} \epsilon_{ox} \frac{\partial \phi_{pc}}{\partial z} &= \epsilon_{ox} q A_1 \mathrm{e}^{qz} - \frac{e}{2} \operatorname{sgn}(z - z_0) \mathrm{e}^{-q|z-z_0|}, && \text{for } z < 0 \\ \epsilon_{si} \frac{\partial \phi_{pc}}{\partial z} &= -\epsilon_{si} q A_2 \mathrm{e}^{-qz} - \frac{e}{2} \operatorname{sgn}(z - z_0) \mathrm{e}^{-q|z-z_0|}, && \text{for } z > 0 \end{aligned} \tag{4.124}$$

where sgn(z) denotes the sign function. Hence, by substituting Eq.4.124 in Eq.4.123 we finally obtain

$$\epsilon_{ox}\, q\, A_1 + \epsilon_{si}\, q\, A_2 = 0. \tag{4.125}$$

Equations 4.122 and 4.125 allow us to determine A_1 and A_2 and to finally express $\phi_{pc}(q, z, z_0)$ as

$$\phi_{pc}(q, z, z_0) = \frac{e}{2q\epsilon_{si}} \mathrm{e}^{-q|z-z_0|} + \left(\frac{\epsilon_{si} - \epsilon_{ox}}{\epsilon_{ox} + \epsilon_{si}}\right) \frac{e}{2q\epsilon_{si}} \mathrm{e}^{-q(z+|z_0|)}, \quad \text{for } z > 0 \tag{4.126a}$$

$$\phi_{pc}(q, z, z_0) = \frac{e}{2q\epsilon_{ox}} \mathrm{e}^{-q|z-z_0|} + \left(\frac{\epsilon_{ox} - \epsilon_{si}}{\epsilon_{ox} + \epsilon_{si}}\right) \frac{e}{2q\epsilon_{ox}} \mathrm{e}^{q(z-|z_0|)}. \quad \text{for } z < 0 \tag{4.126b}$$

The expressions for $\phi_{pc}(q, z, z_0)$ in Eq.4.126 can be rearranged when the sign of z_0 is known. For an ionized impurity in the transistor channel z_0 is positive, so that in the silicon region we have

$$\phi_{pc}(q, z, z_0) = \frac{e}{2q\epsilon_{si}} \mathrm{e}^{-q|z-z_0|} + \frac{1}{\epsilon_{si}} \left(\frac{\epsilon_{si} - \epsilon_{ox}}{\epsilon_{si} + \epsilon_{ox}}\right) \frac{e}{2q} \mathrm{e}^{-q|z+z_0|}. \quad \text{for } z > 0 \tag{4.127}$$

A comparison to Eq.4.120 shows that the first term in Eq.4.127 is the potential produced by a point charge at z_0 in a homogeneous silicon substrate. The second term, instead, can be interpreted as the potential produced by an image charge located at $(-z_0)<0$. Such an image charge potential is produced by the difference between the silicon and the oxide dielectric constant; in fact it is null for $\epsilon_{ox} = \epsilon_{si}$.

Charges in the oxide correspond to negative z_0 values and Eq.4.126a gives

$$\phi_{pc}(q, z, z_0) = \frac{e}{2q\overline{\epsilon}} \mathrm{e}^{-q(z-z_0)}, \qquad \text{for } z > 0 \tag{4.128}$$

where $\overline{\epsilon}$ is the average dielectric constant,

$$\overline{\epsilon} = \frac{\epsilon_{si} + \epsilon_{ox}}{2}. \tag{4.129}$$

Charged interface states or fixed charges at the silicon–oxide interface correspond to $z_0 = 0$. As discussed above, Eqs.4.126 have been obtained for $z_0 \neq 0$, in fact we have imposed the continuity of the z component of the electric displacement field at $z = 0$. By repeating the derivations for $z_0 = 0$ and accounting for the corresponding discontinuity of the z component of the electric displacement field at $z = 0$, one obtains a scattering potential in silicon given by

$$\phi_{pc}(q, z, 0) = \frac{e}{2q\overline{\epsilon}} \mathrm{e}^{-qz}. \quad \text{for } z > 0 \tag{4.130}$$

Equation 4.126a reduces to Eq.4.130 by setting $z_0 = 0$, as expected because of the continuity of the potential $\phi_{pc}(q, z, z_0)$ with respect to z_0.

Hence, we conclude that Eq.4.126a allows us to express the potential in the silicon region produced by a point charge at any possible z_0 value.

Potential in an SOI structure

The methodology described in detail in the previous section for a bulk MOS structure can be naturally extended to an SOI MOS transistor. Figure 4.7 shows an SOI structure where T_{SI} is the silicon film thickness and the oxide regions are assumed to be almost infinite. Equation 4.117 can be used in any of the three regions identified in Fig.4.7 (by using the corresponding dielectric constant), so that we can write

$$\begin{aligned}
\phi_{pc}(q, z, z_0) &= A_1 \mathrm{e}^{qz} + \frac{e}{2q\epsilon_{ox}} \mathrm{e}^{-q|z-z_0|}, && \text{for } z < 0 \\
\phi_{pc}(q, z, z_0) &= A_2 \mathrm{e}^{-qz} + A_3 \mathrm{e}^{qz} + \frac{e}{2q\epsilon_{si}} \mathrm{e}^{-q|z-z_0|}, && \text{for } 0 < z < \mathrm{T_{SI}} \\
\phi_{pc}(q, z, z_0) &= A_4 \mathrm{e}^{-qz} + \frac{e}{2q\epsilon_{ox}} \mathrm{e}^{-q|z-z_0|}. && \text{for } z > \mathrm{T_{SI}}
\end{aligned} \tag{4.131}$$

At the silicon–oxide interfaces located at $z = 0$ and $z = T_{SI}$ we must impose the continuity of the potential $\phi_{pc}(q, z, z_0)$ and of the z component of the electric displacement field. We thus obtain four equations that allow us to determine the unknown constants A_1, A_2, A_3, and A_4, and hence the potential $\phi_{pc}(q, z, z_0)$ throughout the oxide–silicon–oxide structure. The calculations are straightforward but quite tedious.

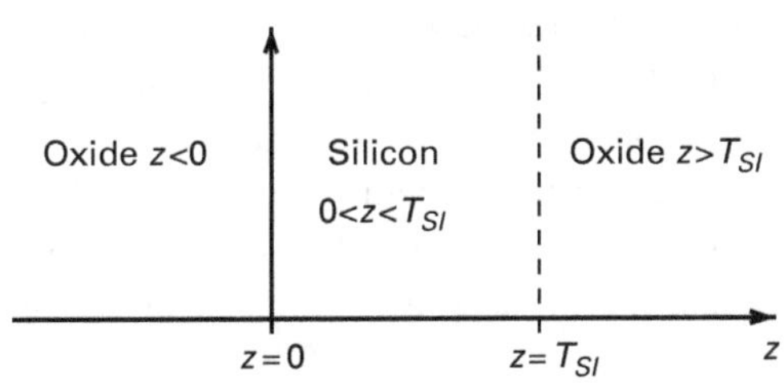

Figure 4.7 Sketch of an SOI-type MOS structure. The silicon–oxide interfaces are set at $z = 0$ and $z = T_{SI}$, where T_{SI} denotes the thickness of the silicon film.

The resulting scattering potential in the silicon region, namely for $0<z<T_{SI}$, can be expressed as

$$\phi_{pc}(q,z,z_0) = \frac{e}{2q\epsilon_{si}}\left[e^{-q|z-z_0|} + C_1\,e^{qz} + C_2\,e^{-qz}\right], \tag{4.132}$$

where the coefficients C_1 and C_2 are given by

$$C_1 = \frac{(\epsilon_{si}-\epsilon_{ox})^2\,e^{-q|z_0|} + (\epsilon_{si}^2-\epsilon_{ox}^2)\,e^{-q(|T_{SI}-z_0|-T_{SI})}}{(\epsilon_{si}+\epsilon_{ox})^2\,e^{2q\,T_{SI}} - (\epsilon_{si}-\epsilon_{ox})^2},$$

$$C_2 = \frac{(\epsilon_{si}-\epsilon_{ox})(C_1+e^{-q|z_0|})}{(\epsilon_{si}+\epsilon_{ox})}. \tag{4.133}$$

The procedure illustrated for a bulk and an SOI structure with infinite oxide regions can be extended to stacks with more dielectric layers, as discussed in Chapter 10.

4.3.2 Scattering matrix elements

Below we go on to discuss calculation of the Coulomb scattering matrix elements. For electron inversion layers we consider only intra-valley transitions for the reasons explained in Section 4.1.4. Furthermore, in order to simplify somewhat the heavy notation, in this sub-section we drop the index indicating the valley. In Section 4.3.3, however, the indication of the valley is reintroduced because it is necessary to discuss effects of the screening in accordance with the approach described in Section 4.2.

The matrix element produced by the Coulomb potential of a point charge located at $(\mathbf{r}_0,z_0)$ can be calculated with Eq.4.6 or Eq.4.41 and by using Eq.4.115 to express the Fourier transform of the scattering potential. We thus obtain

$$M_{n,n'}(\mathbf{k},\mathbf{k}',\mathbf{r}_0,z_0) = \frac{e^{-i\mathbf{q}\cdot\mathbf{r}_0}}{A}\,M^{(0)}_{n,n'}(\mathbf{k},\mathbf{k}',z_0), \tag{4.134}$$

where $\mathbf{q}=(\mathbf{k}'-\mathbf{k})$ and $M^{(0)}_{n,n'}(\mathbf{k},\mathbf{k}',z_0)$ is defined as

$$M^{(0)}_{n,n'}(\mathbf{k},\mathbf{k}',z_0) = \int_z \boldsymbol{\xi}^{\dagger}_{n'\mathbf{k}'}(z)\cdot\boldsymbol{\xi}_{n\mathbf{k}}(z)\,\phi_{pc}(q,z,z_0)\,dz, \tag{4.135}$$

and $\boldsymbol{\xi}_{n\mathbf{k}}(z)$ denotes the envelope wave-function. Determination of the potential $\phi_{pc}(q,z,z_0)$ has been discussed in Section 4.3.1.

The expression for the matrix element given in Eq.4.134 is of general validity and it can be used for both electron and hole inversion layers. In the case of holes described by a **k·p** Hamiltonian, the wave-functions are six component vectors and the dot sign indicates the scalar product defined in Eq.4.40. In the parabolic EMA model for the electrons or in the hole model of Section 3.3.3, instead, the wave-functions are scalar and do not depend on the wave-vector. In these two latter cases the matrix elements simplify to

$$M^{(0)}_{n,n'}(q,z_0) = \int_z \xi^{\dagger}_{n'}(z)\xi_n(z)\,\phi_{pc}(q,z,z_0)\,dz, \tag{4.136}$$

where $M^{(0)}_{n,n'}(q, z_0)$ depends only on the magnitude q of $\mathbf{q} = (\mathbf{k}' - \mathbf{k})$.

To simplify the notation, in the following derivations we drop the indication of the initial and final state, so that the matrix elements in Eqs.4.134 and 4.135 will be indicated as $M(\mathbf{r}_0, z_0)$ and $M^{(0)}(z_0)$, respectively.

Given the linearity of the Poisson equation 4.110 with respect to the charge density, the matrix element corresponding to the scattering potential produced by all the Coulomb centers in the device can be written as

$$M = \frac{1}{A} \sum_{(\mathbf{r}_0, z_0)} M^{(0)}(z_0)\, e^{-i\mathbf{q}\cdot\mathbf{r}_0}, \tag{4.137}$$

where the sum is over the position $(\mathbf{r}_0, z_0)$ of the Coulomb centers.

We now proceed to evaluation of the squared matrix element by temporarily neglecting the screening, which should be accounted for at this stage of the derivations, that is *before* taking the square of the matrix element. We return to discussion of the screening in Section 4.3.3.

Starting from Eq.4.137, the squared *unscreened* matrix element can be expressed as

$$\begin{aligned} |M|^2 &= \frac{1}{A^2} \left(\sum_{(\mathbf{r}_0, z_0)} M^{(0)}(z_0)\, e^{-i\mathbf{q}\cdot\mathbf{r}_0} \right) \cdot \left(\sum_{(\mathbf{r}'_0, z'_0)} [M^{(0)}(z'_0)]^\dagger\, e^{i\mathbf{q}\cdot\mathbf{r}'_0} \right) \\ &= \frac{1}{A^2} \sum_{(\mathbf{r}_0, z_0)} \sum_{(\mathbf{r}'_0, z'_0)} M^{(0)}(z_0)\, [M^{(0)}(z'_0)]^\dagger\, e^{i\mathbf{q}\cdot(\mathbf{r}'_0 - \mathbf{r}_0)}. \end{aligned} \tag{4.138}$$

In the following calculations we separate the Coulomb centers into groups that have essentially the same z_0 value. In other words, we are conceptually grouping the Coulomb centers in two dimensional layers, each of which corresponds to a given z_0 value. This choice is practically convenient because the Coulomb centers, having the same z_0, also result in the same $M^{(0)}(z_0)$. Furthermore, a layer of Coulomb centers is an approximate description of the fixed charges located at the silicon–oxide interface or at interfaces between different dielectrics in the gate stack.

Such a rearrangement of Eq.4.138 yields

$$\begin{aligned} |M|^2 &= \frac{1}{A^2} \sum_{z_0} |M^{(0)}(z_0)|^2 \left(A N_{z0} + \sum_{(\mathbf{r}'_0 \neq \mathbf{r}_0)} e^{i\mathbf{q}\cdot(\mathbf{r}'_0 - \mathbf{r}_0)} \right) \\ &+ \frac{1}{A^2} \sum_{(\mathbf{r}_0, z_0)} \sum_{(\mathbf{r}'_0, z'_0 \neq z_0)} M^{(0)}(z_0)\, [M^{(0)}(z'_0)]^\dagger\, e^{i\mathbf{q}\cdot(\mathbf{r}'_0 - \mathbf{r}_0)}, \end{aligned} \tag{4.139}$$

where the first sum over z_0 corresponds to the terms in Eq.4.138 with $z'_0 = z_0$. This sum includes in turn the terms with $(\mathbf{r}'_0, z'_0) = (\mathbf{r}_0, z_0)$, which simply produce the number $A N_{z0}$ of Coulomb centers in the z_0 layer (N_{z0} being the density of the centers per unit area), and also the cross terms stemming from the centers that have $\mathbf{r}'_0 \neq \mathbf{r}_0$. These latter terms correspond to different Coulomb centers in the same z_0 layer. Finally, the last sum includes the centers belonging to different layers, namely those with $z'_0 \neq z_0$.

In most of the Coulomb scattering studies presented in the literature only the term in Eq.4.139 corresponding to N_{z0} is retained in calculation of the scattering rates. This simplification is justified if the positions of the Coulomb centers are statistically uncorrelated and the area A of the system is very large, so that all the remaining terms in Eq.4.139 tend to vanish.

In this respect, it is very reasonable to assume that no correlation exists between the positions in the $\mathbf{r}$ plane of Coulomb centers belonging to different z_0 layers, so that the last sum of Eq.4.139 is expected to vanish. The problem is more subtle for the Coulomb centers that belong to the same z_0 layer and it is further discussed in Section 4.3.4. For the time being we thus retain only the term with N_{z0} in Eq.4.139 and write

$$|M|^2 \approx \frac{1}{A} \sum_{z_0} N_{z0} \, |M^{(0)}(z_0)|^2. \tag{4.140}$$

The areal density N_{z0} of Coulomb centers in Eq.4.139 may correspond to the Coulomb centers located at a given interface, such as the silicon–oxide interface, or, in the case of the ionized impurities in the transistor channel, it can be viewed as the areal density $[N_{II}(z_0)\mathrm{d}z_0]$ corresponding to the volumetric density $N_{II}(z_0)$ of ionized impurities at $z = z_0$ and in a layer of thickness $\mathrm{d}z_0$.

It is worth pointing out that, in calculation of the Coulomb scattering matrix elements, we should always exclude the transitions corresponding to $\mathbf{q} = (\mathbf{k}' - \mathbf{k}) = \mathbf{0}$. There are several reasons why this is the case and why it is useful to keep this in mind. First of all, for each z, the Fourier component at $\mathbf{q} = \mathbf{0}$ of the scattering potential is proportional to the average value of the potential in the $\mathbf{r}$ plane. This is the macroscopic potential which enters the Poisson equation and it has already been accounted for in the Hamiltonian of the unperturbed case.

To be more explicit we can say that, if we consider the possible fixed charges in the oxide or at the silicon–oxide interface, then the Coulomb potential for $\mathbf{q} = \mathbf{0}$ is the average value of the potential which, for example, is well known to produce shifts of the threshold voltage of the transistor. It is not such a macroscopic potential that scatters the carriers in the channel, but rather the $\mathbf{r}$ dependent deviations of the potential with respect to its average value. Such fluctuations are described by the Fourier components with non-null $\mathbf{q}$ values, and the larger is $q = |\mathbf{q}|$ the more spatially rapid are the potential fluctuations.

Furthermore, we also note that calculation of the matrix elements for $\mathbf{q} = \mathbf{0}$ is also useless. In fact, since the Coulomb scattering is elastic, the scattering events for $\mathbf{q} = (\mathbf{k}' - \mathbf{k}) = \mathbf{0}$ do not change either the energy or the wave-vector, hence they do not affect the carrier transport.

Thus, in all the calculations concerning Coulomb scattering, it is appropriate to assume $\mathbf{q} \neq \mathbf{0}$, which, incidentally, allows us to avoid the singularity at $q = 0$ in the expressions for $\phi_{pc}(q, z, z_0)$ discussed in Section 4.3.1.

By using Eq.4.140 the squared matrix element can be written as

$$|M_{n,n'}(\mathbf{k}, \mathbf{k}')|^2 = \frac{1}{A} \left[\int_0^{z_{max}} |M^{(0)}_{n,n'}(\mathbf{k}, \mathbf{k}', z_0)|^2 N_{II}(z_0)\mathrm{d}z_0 + |M^{(0)}_{n,n'}(\mathbf{k}, \mathbf{k}', 0)|^2 \, N_{SI} \right], \tag{4.141}$$

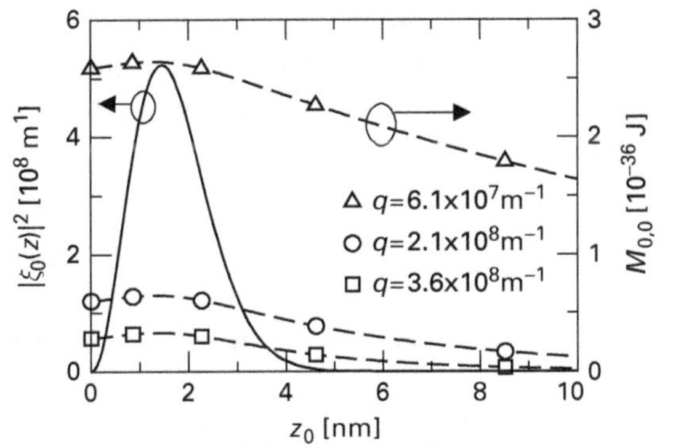

Figure 4.8 Unscreened Coulomb scattering matrix element $M_{0,0}^{(0)}(q, z, z_0)$ for the lowest subband as defined in Eq.4.136 versus the position z_0 of the point charge and for different q values. The squared magnitude $|\xi_0(z)|^2$ of the wave-function is also shown. Bulk n-MOSFET: doping concentration 3×10^{17} cm^{-3} and inversion density 1.6×10^{11} cm^{-2}.

where N_{SI} is the density per unit area of Coulomb centers at the silicon–oxide interface, whereas the first term in Eq.4.141 accounts for the ionized impurities in the transistor channel. The term $M_{n,n'}^{(0)}(\mathbf{k}, \mathbf{k}', z_0)$ is defined in Eq.4.135.

For a bulk MOSFET the upper limit z_{max} of the integration is almost infinite, whereas for an SOI device it is limited to the silicon film. In the case of an SOI transistor, the possible Coulomb centers at the back–oxide interface can be straightforwardly included in Eq.4.141 by adding a term at $z_0 = T_{SI}$.

Regarding the z_0 layers that contribute to the scattering rate the most, let us consider for simplicity an intra-subband transition, namely $n = n'$. In such a case, Eq.4.135 indicates that the Coulomb centers which most effectively scatter the carriers are those located in the region where the squared magnitude $|\xi_{n\mathbf{k}}(z)|^2$ of envelope function is maximum. If $|\xi_{n\mathbf{k}}(z)|^2$ has a single peak, then the z_0 most effective for scattering are those close to such a peak. This is clearly illustrated by Fig.4.8, which shows the unscreened intra-subband matrix element $M_{0,0}^{(0)}(q, z, z_0)$ defined in Eq.4.136 versus the position z_0 of the point charge and for different $q = |\mathbf{q}|$ values. As can be seen, the matrix elements are largest for z_0 close to the maximum of the wave-function magnitude and they decrease for increasing q values.

It is finally worth noticing that, since the envelope wave-functions are confined within a few nanometers from the silicon–oxide interface, the Coulomb centers farther than some tens of nanometers from the silicon–oxide interface produce vanishingly small values of $|M_{n,n'}^{(0)}(\mathbf{k}, \mathbf{k}', z_0)|^2$; hence their contribution to Eq.4.141 is negligible.

4.3.3 Effect of the screening

The screening produced by the carriers in the inversion layer has a strong impact on the Coulomb scattering rate and, in particular, on its dependence on the inversion density and the gate voltage. The treatment of the screening discussed in Section 4.2 can be directly applied to the matrix elements of the Coulomb scattering, but it is important to clarify at what stage of the derivations the screening should be accounted for.

Let us first consider the case of wave-functions independent of the wave-vector **k**. In this case Eq.4.134 can be rewritten as

$$M_{w,n,n'}(\mathbf{q},\mathbf{r}_0,z_0)=\frac{\mathrm{e}^{-\mathrm{i}\mathbf{q}\cdot\mathbf{r}_0}}{A}\,M^{(0)}_{w,n,n'}(\mathbf{q},z_0), \tag{4.142}$$

where $\mathbf{q}=(\mathbf{k}'-\mathbf{k})$ and $M^{(0)}_{w,n,n'}(\mathbf{q},z_0)$ is defined consistently with Eq.4.136 as

$$M^{(0)}_{w,n,n'}(\mathbf{q},z_0)=\int_z \xi^{\dagger}_{n'}(z)\xi_n(z)\,\phi_{pc}(q,z,z_0)\,\mathrm{d}z. \tag{4.143}$$

The symbol w in Eqs.4.142 and 4.143 denotes the valley of the intra-valley matrix elements.

Equation 4.142 expresses the unscreened matrix elements for the scattering potential produced by a point charge located at ($\mathbf{r}_0$,z_0). For any $\mathbf{q}$ value, the corresponding screened matrix elements $M^{(scr)}_{w,n,n'}(\mathbf{q},\mathbf{r}_0,z_0)$ can be obtained by using Eq.4.77. By substituting Eq.4.142 in the l.h.s. of Eq.4.77 we obtain

$$\frac{\mathrm{e}^{-\mathrm{i}\mathbf{q}\cdot\mathbf{r}_0}}{A}\,M^{(0)}_{w,n,n'}(\mathbf{q},z_0)=\sum_{w,n,n'}\epsilon^{w,n,n'}_{v,m,m'}(\mathbf{q})\,M^{(scr)}_{w,n,n'}(\mathbf{q},\mathbf{r}_0,z_0). \tag{4.144}$$

It is now important to understand that $\mathbf{q}$ is a constant parameter in Eq.4.144, which in fact couples the matrix elements corresponding to a given $\mathbf{q}$ and a given position ($\mathbf{r}_0$,z_0) of the point charge. Consequently Eq.4.144 allows us to state that the screened matrix elements can be expressed as

$$M^{(scr)}_{w,n,n'}(\mathbf{q},\mathbf{r}_0,z_0)=\frac{\mathrm{e}^{-\mathrm{i}\mathbf{q}\cdot\mathbf{r}_0}}{A}\,M^{(0,scr)}_{w,n,n'}(\mathbf{q},z_0), \tag{4.145}$$

where the $M^{(0,scr)}_{w,n,n'}(\mathbf{q},z_0)$ are obtained by solving the linear problem

$$M^{(0)}_{w,n,n'}(\mathbf{q},z_0)=\sum_{w,n,n'}\epsilon^{w,n,n'}_{v,m,m'}(\mathbf{q})\,M^{(0,scr)}_{w,n,n'}(\mathbf{q},z_0). \tag{4.146}$$

Since the screened matrix elements $M^{(scr)}_{w,n,n'}(\mathbf{q},\mathbf{r}_0,z_0)$ in Eq.4.145 have exactly the same dependence on $\mathbf{r}_0$ as the unscreened matrix elements in Eq.4.134, all the derivations that allowed us to go from Eq.4.134 to Eq.4.141 hold also in the case of the screened matrix elements. In other words, we can express the squared screened matrix element as

$$|M^{(scr)}_{w,n,n'}(\mathbf{q})|^2=\frac{1}{A}\left[\int_0^{z_{max}}|M^{(0,scr)}_{w,n,n'}(\mathbf{q},z_0)|^2N_{II}(z_0)\mathrm{d}z_0+|M^{(0,scr)}_{w,n,n'}(\mathbf{q},0)|^2\,N_{SI}\right], \tag{4.147}$$

where $M^{(0,scr)}_{w,n,n'}(\mathbf{q},z_0)$ must be calculated by using Eq.4.146.

It is worth noting that the calculation of the screened matrix elements is much simpler if one accepts using the scalar dielectric function as in Eq.4.92. In such a case, in fact, the matrix elements for inter-subband transitions are left unscreened, while each intra-subband screened matrix element can be individually calculated from the corresponding unscreened matrix element. To be more explicit we have

$$M^{(0,scr)}_{w,n,n'}(\mathbf{k}, \mathbf{k}', z_0) = \frac{M^{(0)}_{w,n,n}(\mathbf{q}, z_0)}{\epsilon_D(\mathbf{q})}, \tag{4.148}$$

for intra-subband transitions and $M^{(0,scr)}_{w,n,n'}(\mathbf{k}, \mathbf{k}', z_0) = M^{(0)}_{w,n,n'}(\mathbf{q}, z_0)$ for $n \neq n'$. Even in the case of the scalar dielectric function, we must use Eq.4.147 to calculate the squared screened matrix elements.

In the case of a $\mathbf{k}$ dependent wave-function, as for a hole inversion layer described by the $\mathbf{k}{\cdot}\mathbf{p}$ approach, the screening presents no conceptual differences with respect to the treatment for the $\mathbf{k}$ independent wave-functions. In fact the unscreened matrix elements are given by

$$M_{w,n,n'}(\mathbf{k}, \mathbf{k}', \mathbf{r}_0, z_0) = \frac{\mathrm{e}^{-\mathrm{i}\mathbf{q}\cdot\mathbf{r}_0}}{A} M^{(0)}_{w,n,n'}(\mathbf{k}, \mathbf{k}', z_0), \tag{4.149}$$

and the corresponding screened matrix elements $M^{(scr)}_{w,n,n'}(\mathbf{k}, \mathbf{k}', \mathbf{r}_0, z_0)$ can be obtained by using Eq.4.83. All the matrix elements coupled by Eq.4.83 correspond to the same $\mathbf{q}$, hence the screened matrix elements can be written as

$$M^{(scr)}_{w,n,n'}(\mathbf{k}, \mathbf{k}', \mathbf{r}_0, z_0) = \frac{\mathrm{e}^{-\mathrm{i}\mathbf{q}\cdot\mathbf{r}_0}}{A} M^{(0,scr)}_{w,n,n'}(\mathbf{k}, \mathbf{k}', z_0), \tag{4.150}$$

where the $M^{(0,scr)}_{w,n,n'}(\mathbf{k}, \mathbf{k}', z_0)$ are determined with an equation essentially identical to Eq.4.146, except for the fact that the dielectric function must be the one defined in Eq.4.83. Thus, all the derivations from Eq.4.134 to Eq.4.141 can be used for the screened matrix elements as well.

Such a rigorous treatment of the screening for a $\mathbf{k}$ dependent wave-function is computationally almost prohibitive, because the matrix elements with the same $\mathbf{q}$ value and all the possible initial wave-vector $\mathbf{k}$ are coupled. For this reason, the $\mathbf{k}$ dependence of the wave-functions is typically neglected in the treatment of screening [20, 21].

4.3.4 Small areas and correlation of the Coulomb centers position

It is interesting to discuss in more detail the approximations that allowed us to simplify Eq.4.139 to 4.140. To this purpose let us consider a single layer of point charges, which may represent the Coulomb centers at the silicon–oxide interface. In such a case, Eqs.4.138 and 4.139 reduce to

$$\begin{aligned} |M|^2 &= \frac{1}{A^2}|M^{(0)}(z_0)|^2 \sum_{\mathbf{r}_0', \mathbf{r}_0} \mathrm{e}^{\mathrm{i}\mathbf{q}\cdot(\mathbf{r}_0'-\mathbf{r}_0)} \\ &= \frac{1}{A^2}|M^{(0)}(z_0)|^2 \left(A N_{z0} + \sum_{(\mathbf{r}_0' \neq \mathbf{r}_0)} \mathrm{e}^{\mathrm{i}\mathbf{q}\cdot(\mathbf{r}_0'-\mathbf{r}_0)} \right), \end{aligned} \tag{4.151}$$

which, for uncorrelated Coulomb centers, may be simplified to

$$|M|^2 \approx \frac{1}{A} N_{z0} |M^{(0)}(z_0)|^2. \tag{4.152}$$

Equation 4.152 corresponds to the case of Coulomb centers that individually scatter the carriers, as if the other Coulomb centers did not exist. In fact, it is assumed that the sum over $\mathbf{r}_0' \neq \mathbf{r}_0$ stemming from the presence of the different Coulomb centers is negligible.

Such a simplification holds if the area A of the two dimensional system is very large and if the positions of the Coulomb centers are completely uncorrelated.

Let us first discuss the role played by the area A. We mentioned several times in the previous sections that A is just a normalization area which never enters the final results of the calculations. This is certainly true if A is large enough. In particular, the vanishing of the sum over $\mathbf{r}_0' \neq \mathbf{r}_0$ in Eq.4.151 requires that we have an arbitrarily large number AN_{z0} of Coulomb centers in the system. It is interesting to note that, if this is not the case, then the Coulomb scattering rate of a given device is a characteristic of that particular device. In fact, because of the inherently statistical distribution of the Coulomb centers, nominally identical devices may have different distributions of Coulomb centers and correspondingly different scattering rates. In other words, if the area A is not large enough, then the statistical distribution of the Coulomb centers becomes a source of device to device variability, not only for its electrostatics (hence for the threshold voltage and the short-channel effects), but also for transport properties [22].

The second point that deserves further discussion is the possible correlation between the positions of the Coulomb centers. To clarify this point, let us first consider a toy model consisting of a super-lattice of Coulomb centers [23]. In such a case we have a complete correlation in the positions of the Coulomb centers, which are in fact located at the points $\mathbf{r}_0$ of the direct lattice obtained with the two basis vectors

$$\mathbf{a}_1 = d_c\,\hat{\mathbf{x}}, \qquad \mathbf{a}_2 = d_c\,\hat{\mathbf{y}}, \tag{4.153}$$

where $\hat{\mathbf{x}}$ and $\hat{\mathbf{y}}$ are the unit vectors in the x and y directions. Equation 4.153 describes a simple rectangular lattice where d_c is the nearest neighbor distance. The reciprocal lattice vectors $\mathbf{g}_0$ are obtained from the basis vectors of the reciprocal lattice

$$\mathbf{b}_1 = \frac{2\pi}{d_c}\,\hat{\mathbf{x}}, \qquad \mathbf{b}_2 = \frac{2\pi}{d_c}\,\hat{\mathbf{y}}. \tag{4.154}$$

For such a regular distribution of Coulomb centers it is straightforward to calculate the sum over $\mathbf{r}_0' \neq \mathbf{r}_0$ in Eq.4.151. In fact, since AN_{z0} is the number of Coulomb centers in the z_0 plane, then in the sum over $\mathbf{r}_0' \neq \mathbf{r}_0$ we have $[A^2N_{z0}^2 - AN_{z0}]$ terms (because the terms corresponding $\mathbf{r}_0' = \mathbf{r}_0$ are excluded). Hence the sum evaluates to

$$\sum_{(\mathbf{r}_0' \neq \mathbf{r}_0)} e^{i\mathbf{q}\cdot(\mathbf{r}_0' - \mathbf{r}_0)} \begin{cases} A^2N_{z0}^2 - AN_{z0} & \text{for} \quad \mathbf{q} = \mathbf{g}_0 \\ 0 & \text{otherwise} \end{cases}. \tag{4.155}$$

Equation 4.155 shows that, in the case of a super-lattice of Coulomb centers, the simplification used in Eq.4.151 holds only if $\mathbf{q} = (\mathbf{k}' - \mathbf{k})$ *is not* a reciprocal lattice vector of the super-lattice. If $\mathbf{q}$, instead, can be expressed as

$$\mathbf{q} = n_1\,\frac{2\pi}{d_c}\,\hat{\mathbf{x}} + n_2\,\frac{2\pi}{d_c}\,\hat{\mathbf{y}}, \qquad n_1, n_2 = \pm 1, \pm 2, \pm 3 \cdots$$

then the sum over $\mathbf{r}_0' \neq \mathbf{r}_0$ becomes the dominant term in the matrix element.

The only point of the super-lattice example is to show how significantly a correlation in the position of the Coulomb centers can change the matrix elements. From a practical viewpoint, it is quite reasonable to assume that the positions of the Coulomb centers in a given z_0 plane are uncorrelated as long as the concentration N_{z0} per unit area is relatively small. At large N_{z0}, instead, one may speculate that the repulsion between charges of the same sign makes it unlikely that the Coulomb centers get too close to one another. In other words, the presence of a Coulomb center can reduce the probability of finding other Coulomb centers in its vicinity.

The hard sphere model represents very simplistically such a phenomenological picture by assuming that the positions of the Coulomb centers are random, but to the extent that no two fixed charges exist within a critical radius R_c from each other [23]. The resulting joint probability $P(\mathbf{r}_0, \mathbf{r}_0')$ of finding in a given area A a Coulomb center in $\mathbf{r}_0$ and a second center in $\mathbf{r}_0'$ is given by

$$P(\mathbf{r}_0, \mathbf{r}_0') \begin{cases} \frac{1}{A(A-\pi R_c^2)} & \text{for } |\mathbf{r}_0' - \mathbf{r}_0| \geq \mathrm{R_c} \\ 0 & \text{for } |\mathbf{r}_0' - \mathbf{r}_0| < \mathrm{R_c} \end{cases}. \tag{4.156}$$

When a correlation exists between the positions of the Coulomb centers, evaluation of Eq.4.151 requires introduction of the power spectrum of the density of Coulomb centers in the area A. In fact, if we indicate with $n_c(\mathbf{r})$ the density per unit area of Coulomb centers expressed as

$$n_c(\mathbf{r}) = \sum_{\mathbf{r}_0} \delta(\mathbf{r} - \mathbf{r}_0), \tag{4.157}$$

then the corresponding Fourier transform is

$$n_c(\mathbf{q}) = \frac{1}{(2\pi)^2} \sum_{\mathbf{r}_0} \mathrm{e}^{\mathrm{i}\mathbf{q}\cdot\mathbf{r}_0}, \tag{4.158}$$

and Eq.4.151 can be rewritten as

$$|M|^2 = \frac{(2\pi)^4}{A^2} |M^{(0)}(z_0)|^2 \, n_c(\mathbf{q})\, n_c^{\dagger}(\mathbf{q}) = \frac{(2\pi)^4}{A} |M^{(0)}(z_0)|^2 \, \frac{|n_c(\mathbf{q})|^2}{A}. \tag{4.159}$$

Since $n_c(\mathbf{r})$ is a stochastic process, $(|n_c(\mathbf{q})|^2/A)$ is the corresponding power spectrum (if the area A is large enough) [24]. The power spectrum can be calculated by using the probability density $P(\mathbf{r}_0, \mathbf{r}_0')$ defined in Eq.4.156, so that we can finally express the matrix elements produced by the Coulomb centers located in the z_0 plane as [23]

$$|M|^2 \approx \frac{1}{A} |M^{(0)}(z_0)|^2 \, N_{z0} \left[1 - \frac{2C J_1(qR_c)}{qR_c} \right], \tag{4.160}$$

where $J_1(x)$ is a Bessel function of first order [25], and C is the correlation factor

$$C = \pi R_c^2 N_{z0}, \tag{4.161}$$

defined as the relation between the critical area πR_c^2 and average area N_{z0}^{-1} occupied by a Coulomb center.

The critical radius R_c is an unknown parameter in the hard sphere model, and Eq.4.160 shows that, for a given R_c, the correlation factor C increases with N_{z0}. The term in the brackets of Eq.4.160 represents the correction to Eq.4.152 produced by correlation of the Coulomb centers' positions.

Quite interestingly Eq.4.160 allows us to identify the spatial scale for the correlation effects. In fact the second term in the bracket of Eq.4.160 vanishes when qR_c is much larger than 1. As discussed in Section 4.1.3, for an intra-subband transition assisted by Coulomb scattering the magnitude q of the wave-vector variation is given by $q = 2k\sin(\theta/2)$ (see Eq.4.7), hence we can conclude that for $k = |\mathbf{k}|$ values larger than $1/R_c$ the matrix elements essentially correspond to uncorrelated Coulomb centers. The correlation effects, however, may be significant at longer electron wavelengths (i.e. smaller k values). This result can be intuitively understood by recalling that, in the hard-sphere model, no two Coulomb centers can lie within an area of radius R_c. Consequently, if an electron has a wavelength well below R_c it tends to "see" only one Coulomb center at a time. Only for longer wavelengths can the electrons "feel" the potential produced by more Coulomb centers at the same time, hence the correlation of their positions plays a role.

Since the correlation effects depend on the electron wavelength, their importance may vary with temperature. In fact, in close to equilibrium conditions, Eq.3.106 shows that the average electron wave-vector is reduced for decreasing temperature. Such an increase of the electron wavelength may enhance the correlation effects at low temperatures. As can be seen, the correlation of the Coulomb centers may affect the temperature dependence of the Coulomb-scattering-limited mobility [23].

4.4 Surface roughness scattering

Interaction of carriers with the microscopic asperities of the semiconductor–oxide interface is one of the dominant scattering mechanisms in MOSFETs biased at large gate voltages and inversion densities. In the following we derive the matrix elements and the scattering rate for bulk and SOI n-MOS and p-MOS transistors.

4.4.1 Bulk n-MOSFETs

As already pointed out in Section 4.1.6, the perturbation produced by random fluctuations of the semiconductor–oxide position in an MOS transistor cannot be described simply as a perturbation potential. Instead, in the presence of surface roughness the unperturbed Hamiltonian $\hat{H}_{0,z}$ must be substituted by a perturbed Hamiltonian $\hat{H}_{p,\mathbf{r}z}$, so that $(\hat{H}_{p,\mathbf{r}z} - \hat{H}_{0,z})$ is the operator representing the perturbation of the ideal MOS structure.

We will henceforth consider only intra-valley transitions due to surface roughness scattering for the reasons discussed in detail in Section 4.1.4. Furthermore, the valley index ν will be dropped to simplify the notation, so that the envelope wave-function given by the solution of Eq.3.16 will be indicated by $\xi_n(z)$.

Since we have identified $(\hat{H}_{p,\mathbf{r}z}-\hat{H}_{0,z})$ as the perturbation Hamiltonian, calculation of the matrix elements for an electron inversion layer between the state $\mathbf{k}$ (in the subband n) and the state $\mathbf{k}'$ (in the subband n') starts by writing

$$M_{nn'}(\mathbf{q}) = \int_A \left\{ \int_z \xi_{n'}^{\dagger}(z)[(\hat{H}_{p,\mathbf{r}z} - \hat{H}_{0,z})\, \xi_n(z)]\, \mathrm{d}z \right\} \frac{\exp(-\mathrm{i}\mathbf{q}\cdot\mathbf{r})}{A}\, \mathrm{d}\mathbf{r}, \tag{4.162}$$

where $\mathbf{q} = (\mathbf{k}' - \mathbf{k})$ and A is the normalization area.

The matrix elements given by Eq.4.162 are the unscreened ones. We discuss how to include screening in Section 4.4.3.

In the case of an electron inversion layer described using the parabolic effective mass approximation, the unperturbed and perturbed Hamiltonian can be written as (see also Fig.4.9)

$$\hat{H}_{0,z} = -\frac{\hbar^2}{2}\frac{\mathrm{d}}{\mathrm{d}z}\left(\frac{1}{m_z(z)}\frac{\mathrm{d}}{\mathrm{d}z}\right) - e\phi(z) + \Phi_B H_v(-z), \tag{4.163a}$$

$$\hat{H}_{p,\mathbf{r}z} = -\frac{\hbar^2}{2}\frac{\mathrm{d}}{\mathrm{d}z}\left(\frac{1}{m_z[z-\Delta(\mathbf{r})]}\frac{\mathrm{d}}{\mathrm{d}z}\right) - e\phi(z) + \Phi_B H_v[-z+\Delta(\mathbf{r})], \tag{4.163b}$$

where $\phi(z)$ is the electrostatic potential, Φ_B is the potential energy barrier at the channel/dielectric interface and $H_v(x)$ is the step function. The quantization mass is

$$m_z(z) = \begin{cases} m_{Si} & \text{for} \quad z \geq 0 \\ m_{ox} & \text{for} \quad z < 0 \end{cases}. \tag{4.164}$$

If the quantization mass does not depend on z (that is for $m_{Si} \approx m_{ox}$), and if we assume that $\Delta(\mathbf{r})$ is small enough, the integral over z in Eq.4.162 can be readily evaluated to obtain

$$M_{nn'}(\mathbf{q}) \simeq \xi_n(0)\xi_{n'}^{\dagger}(0)\Phi_B \frac{1}{A}\int_A \Delta(\mathbf{r})\exp(-\mathrm{i}\mathbf{q}\cdot\mathbf{r})\mathrm{d}\mathbf{r}. \tag{4.165}$$

It is evident from Eq.4.165 that one should be careful when analyzing surface roughness scattering by using closed boundary conditions for the Schrödinger equation. In

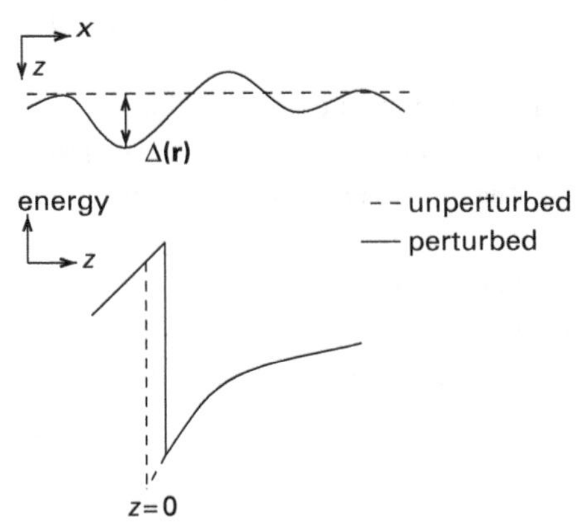

Figure 4.9 Sketch of the surface roughness at the semiconductor–oxide interface of a bulk MOSFET (upper part) and of the potential profile in the direction normal to the Si/SiO_2 interface (lower part).

fact, the so called closed boundary conditions for $\xi_n(z)$ stem from the assumption of an infinitely large potential energy barrier Φ_B at the silicon–oxide interface, which results in a vanishingly small value of the wave-functions at $z = 0$. Under these circumstances the product $[\xi_n(0)\xi_{n'}^{\dagger}(0)\Phi_B]$ is not a well-defined quantity and Eq.4.165 is not a good starting point to determine the matrix elements. However, we see below that by following a different approach it is possible to express the matrix elements $M_{nn'}(\mathbf{q})$ also by using wave-functions obtained by employing closed boundary conditions.

Our goal is now to calculate the matrix elements in accordance with Eqs.4.162 and 4.163 and removing the simplifying assumption $m_{Si} \approx m_{ox}$ used to obtain Eq.4.165. We shall see that if we assume a very large energy barrier at the interface, our results replicate those in [26]. To simplify the calculations, we assume that the wave-functions take real values (hence we have $\xi_n^{\dagger}(z) = \xi_n(z)$), which is actually the case if the Schrödinger equation is solved with closed boundary conditions at the silicon–oxide interface, and it is a good approximation also for open boundary conditions, as long as the Φ_B is large and the electron current through the gate dielectric is negligible.

To proceed in evaluation of the matrix elements we define

$$z' = z - \Delta(\mathbf{r}), \tag{4.166}$$

and, furthermore, we assume that $\Delta(\mathbf{r})$ is small enough that the electrostatic potential can be linearized as $\phi(z) \simeq \phi(z') + \Delta(\mathbf{r})(\mathrm{d}\phi(z')/\mathrm{d}z')$ in the expression for the Hamiltonian operators defined in Eq.4.163. We now note that, with the definition of z', the perturbed Hamiltonian $\hat{H}_{p,\mathbf{r}z}$ in Eq.4.163b can be written as

$$\hat{H}_{p,\mathbf{r}z} = \hat{H}_{0,z'} - e\Delta(\mathbf{r})\frac{\mathrm{d}\phi(z')}{\mathrm{d}z'}, \tag{4.167}$$

where $\hat{H}_{0,z'}$ is the unperturbed Hamiltonian defined in Eq.4.163a but here referred to the abscissa z'. We can now evaluate the matrix elements of $\hat{H}_{p,\mathbf{r}z}$ as

$$\begin{aligned}\langle \xi_{n'}(z)|\hat{H}_{p,\mathbf{r}z}|\xi_n(z)\rangle &= \int \xi_n(z)[\hat{H}_{0,z'}\,\xi_{n'}(z)]\,\mathrm{d}z \\ &\quad - e\Delta(\mathbf{r})\int \xi_n(z)\xi_{n'}(z)\frac{\mathrm{d}\phi(z')}{\mathrm{d}z'}\,\mathrm{d}z.\end{aligned} \tag{4.168}$$

In Eq.4.168 it should be noted that the wave-functions depend on the integration abscissa z, but the Hamiltonian $\hat{H}_{0,z'}$ in the first integral is defined with respect to z'. Also the derivative of the electrostatic potential in the second integral is evaluated at z' and not at z.

Thus, in order to proceed in the derivations, we now expand also the wave-functions as $\xi_n(z) \simeq \xi_n(z') + \Delta(\mathbf{r})\mathrm{d}\xi_n(z')/\mathrm{d}z'$ and then change the integration variable in Eq.4.168 from z to z'. By retaining only the first order terms with respect to $\Delta(\mathbf{r})$ we obtain

$$\begin{aligned}\langle \xi_{n'}(z)|\hat{H}_{p,\mathbf{r}z}|\xi_n(z)\rangle = &\int \xi_n(z')[\hat{H}_{0,z'}\xi_{n'}(z')]\mathrm{d}z' + \Delta(\mathbf{r})\int \xi_n(z')\left[\hat{H}_{0,z'}\frac{\mathrm{d}\xi_{n'}(z')}{\mathrm{d}z'}\right]\mathrm{d}z' \\ &+ \Delta(\mathbf{r})\int \frac{\mathrm{d}\xi_n(z')}{\mathrm{d}z'}[\hat{H}_{0,z'}\,\xi_{n'}(z')]\,\mathrm{d}z' - e\Delta(\mathbf{r})\int \xi_n(z')\xi_{n'}(z')\frac{\mathrm{d}\phi(z')}{\mathrm{d}z'}\,\mathrm{d}z'.\end{aligned} \tag{4.169}$$

It should be noted that in Eq.4.169 all the quantities are now referred to, and expressed in terms of, the same abscissa z', which can thus be renamed z in the following derivations.

It is now convenient to rewrite the second term in the r.h.s. of Eq.4.169 by recalling the identity

$$\int \xi_n(z)\left[\hat{H}_{0,z}\frac{\mathrm{d}\xi_{n'}(z)}{\mathrm{d}z}\right]\mathrm{d}z = \int \frac{\mathrm{d}\xi_{n'}(z)}{\mathrm{d}z}\left[\hat{H}_{0,z}\xi_n(z)\right]\mathrm{d}z,$$

which can be verified by using integration by parts for the differential part of the Hamiltonian $\hat{H}_{0,z}$ defined in Eq.4.163a. Since $\hat{H}_{0,z}\xi_n(z) = \varepsilon_n\xi_n(z)$, Eq.4.169 can be finally rewritten as

$$\begin{aligned}\langle\xi_{n'}(z)|\hat{H}_{p,\mathbf{r}z}|\xi_n(z)\rangle &= \varepsilon_n\delta_{nn'} + \Delta(\mathbf{r})\varepsilon_n\int \xi_n(z)\frac{\mathrm{d}\xi_{n'}(z)}{\mathrm{d}z}\mathrm{d}z \\ &+ \Delta(\mathbf{r})\varepsilon_{n'}\int \xi_{n'}(z)\frac{\mathrm{d}\xi_n(z)}{\mathrm{d}z}\mathrm{d}z - e\Delta(\mathbf{r})\int \xi_n(z)\xi_{n'}(z)\frac{\mathrm{d}\phi(z)}{\mathrm{d}z}\mathrm{d}z.\end{aligned} \tag{4.170}$$

We can now go back to Eq.4.162; in order to obtain $M_{nn'}(\mathbf{q})$ we need to subtract from Eq.4.170 the matrix element of the unperturbed Hamiltonian (which gives $\varepsilon_n\delta_{nn'}$ and thus cancels out with the first term in Eq.4.170), and then evaluate the integral over $\mathbf{r}$. If the wave-functions vanish for $z\rightarrow\pm\infty$, then an integration by parts of the third term in the r.h.s. of Eq.4.170 provides:

$$\int \xi_{n'}(z)\frac{\mathrm{d}\xi_n(z)}{\mathrm{d}z}\mathrm{d}z = -\int \xi_n(z)\frac{\mathrm{d}\xi_{n'}(z)}{\mathrm{d}z'}\mathrm{d}z, \tag{4.171}$$

so that the matrix element can be finally written:

$$M_{nn'}(\mathbf{q}) = \Delta(\mathbf{q})\left[\int \xi_n\left(\frac{-e\mathrm{d}\phi}{\mathrm{d}z}\right)\xi_{n'}\,\mathrm{d}z + (\varepsilon_n - \varepsilon_{n'})\int \xi_n\frac{\mathrm{d}\xi_{n'}}{\mathrm{d}z}\mathrm{d}z\right], \tag{4.172}$$

where $\Delta(\mathbf{q})$ has been implicitly defined as

$$\Delta(\mathbf{q}) = \frac{1}{A}\int_A \Delta(\mathbf{r})\exp(-\mathrm{i}\mathbf{q}\cdot\mathbf{r})\mathrm{d}\mathbf{r}. \tag{4.173}$$

If we denote with $\mathcal{M}_{nn'}$ the term inside the square brackets of Eq.4.172, the unscreened squared matrix element can be written as

$$|M_{nn'}|^2 = |\mathcal{M}_{nn'}|^2\frac{S_R(\mathbf{q})}{A}, \tag{4.174}$$

where the spectrum $S_R(\mathbf{q})$ of the surface roughness is defined as

$$S_R(\mathbf{q}) = \frac{1}{A}\left|\int_A \Delta(\mathbf{r})\exp(-\mathrm{i}\mathbf{q}\cdot\mathbf{r})\mathrm{d}\mathbf{r}\right|^2. \tag{4.175}$$

Recalling the definition of the autocorrelation function $C(\mathbf{r})$,

$$C(\mathbf{r}) = \frac{1}{A}\int_A \Delta(\mathbf{r}')\Delta(\mathbf{r}'+\mathbf{r})\mathrm{d}\mathbf{r}', \tag{4.176}$$

it can be shown that $S_R(\mathbf{q})$ is proportional to the Fourier transform of $C(\mathbf{r})$ (calculated in $-\mathbf{q}$), so that we have [24]

$$S_R(\mathbf{q}) = \int_A C(\mathbf{r}) \exp(-\mathrm{i}\mathbf{q} \cdot \mathbf{r}) \mathrm{d}\mathbf{r}. \tag{4.177}$$

The forms most frequently used in the literature for the SR spectrum $S_R(\mathbf{q})$ are the exponential and the Gaussian [27]. The Gaussian spectrum is expressed as

$$S_R(q) = \pi \Delta_{SR}^2 \lambda_{SR}^2 \exp\left(-\frac{\Delta_{SR}^2 \lambda_{SR}^2}{4}\right), \tag{4.178}$$

where Δ_{SR} and λ_{SR} are the r.m.s. value and the correlation length, respectively. Such a spectrum corresponds to the autocorrelation function

$$C(\mathbf{r}) = \Delta_{SR}^2 \exp\left(-r^2/\lambda_{SR}^2\right). \tag{4.179}$$

The exponential spectrum is instead given by:

$$S_R(q) = \frac{\pi \Delta_{SR}^2 \lambda_{SR}^2}{\left[1 + \frac{q^2 \lambda_{SR}^2}{2}\right]^{3/2}}, \tag{4.180}$$

and the corresponding autocorrelation function is

$$C = \Delta_{SR}^2 \exp\left(-\sqrt{2} r/\lambda_{SR}\right). \tag{4.181}$$

In calculation of the scattering rates the term $1/A$ in Eq.4.174 always cancels out when the sum over the final states is converted into an integral (see Section 4.1.7).

Closed versus open boundary conditions in the Schrödinger equation.

Evaluation of the matrix element using Eq.4.172 requires us to perform integrals over the entire z axis. We demonstrate below that, when the inversion layer is described by the Schrödinger equation with closed boundary conditions, there is an easier way to evaluate the term $\mathcal{M}_{nn'}$.

By proceeding similarly to [26], we evaluate the term

$$\Theta = \int_{z_1}^{z_2} \frac{\mathrm{d}\xi_{n'}(z)}{\mathrm{d}z} [\hat{H}_{0,z} \xi_n(z)] \, \mathrm{d}z + \int_{z_1}^{z_2} \frac{\mathrm{d}\xi_n(z)}{\mathrm{d}z} [\hat{H}_{0,z} \xi_{n'}(z)] \, \mathrm{d}z \tag{4.182}$$

between two generic points z_1 and z_2 (with $z_2 > z_1$) and we assume that m_z does not depend on z in the $[z_1, z_2]$ interval. This is consistent with two possible cases: 1) open boundary conditions for the Schrödinger equation with the same quantization mass in the semiconductor and in the oxide; 2) either closed or open boundary conditions for the Schrödinger problem and $z_1 \geq 0$, so that both z_1 and z_2 belong to the semiconductor.

First of all we compute Θ exploiting the fact that $\xi_n(z)$ and $\xi_{n'}(z)$ are wave-functions of the Schrödinger equation, so that $\hat{H}_{0,z}\xi_n = \varepsilon_n \xi_n$ and $\hat{H}_{0,z}\xi_{n'} = \varepsilon_{n'}\xi_{n'}$. Insertion of these identities in Eq.4.182 and integration by parts yields:

$$\Theta = (\varepsilon_n - \varepsilon_{n'}) \int_{z1}^{z2} \xi_n \frac{\mathrm{d}\xi_{n'}}{\mathrm{d}z} \mathrm{d}z + \varepsilon_{n'} \left[\xi_n \xi_{n'}\right]_{z1}^{z2} . \tag{4.183}$$

We now calculate an equivalent expression for Θ by following a different path, namely we substitute in Eq.4.182 the $\hat{H}_{0,z}$ expression given in Eq.4.163a. Simple manipulations obtained by integrating by parts lead to the new Θ expression:

$$\Theta = -\frac{\hbar^2}{2m_z} \left[\frac{\mathrm{d}\xi_n}{\mathrm{d}z}\frac{\mathrm{d}\xi_{n'}}{\mathrm{d}z}\right]_{z1}^{z2} - \left[e\phi\xi_n\xi_{n'}\right]_{z1}^{z2} + e \int_{z1}^{z2} \xi_{n'} \frac{\mathrm{d}\phi}{\mathrm{d}z} \xi_n \,\mathrm{d}z + \Phi_B \left[\xi_n \xi_{n'}\right]_{z_{\min}}^{z_{\max}} , \tag{4.184}$$

where $z_1 \geq z_2$ and $z_{\min}$, $z_{\max}$ are defined as

$$z_{\min} = \begin{cases} z_1 & \text{if} \quad z_1 < 0 \\ 0 & \text{if} \quad z_1 \geq 0 \end{cases}, \qquad z_{\max} = \begin{cases} z_2 & \text{if} \quad z_2 < 0 \\ 0 & \text{if} \quad z_2 \geq 0 \end{cases}. \tag{4.185}$$

By setting equal the Θ expressions in Eqs.4.183 and 4.184 we obtain [26]

$$\begin{aligned} -\frac{\hbar^2}{2m_z} \left[\frac{\mathrm{d}\xi_n}{\mathrm{d}z}\frac{\mathrm{d}\xi_{n'}}{\mathrm{d}z}\right]_{z1}^{z2} - \left[e\,\phi\,\xi_n\,\xi_{n'}\right]_{z1}^{z2} + e \int_{z1}^{z2} \xi_{n'} \frac{\mathrm{d}\phi}{\mathrm{d}z} \xi_n \,\mathrm{d}z + \Phi_B \left[\xi_n \xi_{n'}\right]_{z_{\min}}^{z_{\max}} \\ = (\varepsilon_n - \varepsilon_{n'}) \int_{z1}^{z2} \xi_n \frac{\mathrm{d}\xi_{n'}}{\mathrm{d}z} \mathrm{d}z + \varepsilon_{n'} \left[\xi_n \xi_{n'}\right]_{z1}^{z2} . \end{aligned} \tag{4.186}$$

Equation 4.186 can be used to rewrite the term in square brackets in Eq.4.172, that is to express the matrix element $\mathcal{M}_{nn'}$ implicitly defined in Eq.4.174. For example, if we consider open boundary conditions (i.e. $z_1 \to -\infty$ and $z_2 \to \infty$) with a constant quantization mass m_z and we assume vanishing values of $\xi_n(z)$, $\xi_{n'}(z)$ and of the corresponding derivatives for $z \to \pm\infty$, then we obtain

$$\mathcal{M}_{nn'} = \Phi_B \, \xi_n(0) \, \xi_{n'}(0), \tag{4.187}$$

which is consistent with Eq.4.165. If, instead, we consider closed boundary conditions for the Schrödinger equation, we have to set $z_1 = 0$, $z_2 = \infty$, and $\xi_n(0) = \xi_{n'}(0) = 0$, and thus obtain:

$$\mathcal{M}_{nn'} = \frac{\hbar^2}{2m_z} \frac{\mathrm{d}\xi_n}{\mathrm{d}z}(0) \frac{\mathrm{d}\xi_{n'}}{\mathrm{d}z}(0), \tag{4.188}$$

which is known as the Prange–Nee term [28].

Additional terms in the surface roughness matrix elements

In calculations of the surface roughness matrix elements we have so far considered only the direct consequences of fluctuations of the semiconductor–oxide interface, namely the change in the position where the discontinuity of the confining potential and of the quantization mass occur. However, the changes in the interface position also induce fluctuations in the inversion charge, which in turn result in a perturbation of the electrostatic potential that contributes to scattering of the carriers [29, 30].

For a bulk MOSFET, the change $\delta n(\mathbf{r}, z)$ in the electron density produced by $\Delta(\mathbf{r})$ can be expressed to the zeroth order as [29]

$$\delta n(\mathbf{r}, z) = n(z - \Delta(\mathbf{r})) - n(z) \simeq \Delta(\mathbf{r}) \frac{\partial n}{\partial z}. \quad (4.189)$$

The potential produced by $\delta n(\mathbf{r}, z)$ can be calculated by using the expressions for the potential of the point charge discussed in Section 4.3.1, with a procedure similar to the one used in Section 4.2.2 to calculate the potential induced by $\rho_{ind}(\mathbf{r}, z)$.

The derivations show that the $\delta n(\mathbf{r}, z)$ lead to a quite complicated, $\mathbf{q}$ dependent contribution to the surface roughness matrix elements that we do not discuss in detail in this book. The interested reader should refer to [30] to find the expressions valid for both bulk and ultra-thin body SOI MOSFETs.

4.4.2 SOI n-MOSFETs

As illustrated by Fig.4.10, the quantization mass in an SOI MOSFET with an ideal semiconductor–oxide interface is given by

$$m_z(z) = \begin{cases} m_{Si} & \text{for} \quad 0 \le z \le T_{si} \\ m_{ox} & \text{otherwise} \end{cases}. \quad (4.190)$$

Given the $m_z(z)$ definition, we can write the unperturbed $\hat{H}_{0,z}$ and the perturbed $\hat{H}_{P,\mathbf{r}z}$ Hamiltonian of the SOI structure as

$$\hat{H}_{0,z} = -\frac{\hbar^2}{2} \frac{\mathrm{d}}{\mathrm{d}z} \left(\frac{1}{m_z(z)} \frac{\mathrm{d}}{\mathrm{d}z} \right) - e\phi(z) + \Phi_B H_v(-z) + \Phi_B H_v(z - T_{Si}), \quad (4.191a)$$

$$\hat{H}_{P,\mathbf{r}z} = -\frac{\hbar^2}{2} \frac{\mathrm{d}}{\mathrm{d}z} \left(\frac{1}{m_z(z')} \frac{\mathrm{d}}{\mathrm{d}z} \right) - e\phi(z)$$
$$+ \Phi_B H_v[-z + \Delta_F(\mathbf{r})] + \Phi_B H_v[z - T_{Si} - \Delta_B(\mathbf{r})]. \quad (4.191b)$$

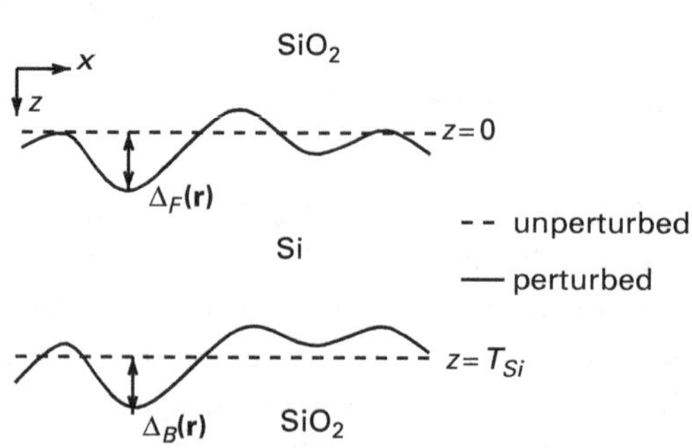

Figure 4.10 Sketch of the surface roughness at the semiconductor–oxide interfaces of a SOI MOSFET.

In order to describe in a compact and convenient form the z dependence of the effective mass m_z in the perturbed case, we henceforth follow [31] and define

$$z' = \frac{T_{Si}}{T_{Si} + \Delta_B(\mathbf{r}) - \Delta_F(\mathbf{r})}[z - \Delta_F(\mathbf{r})], \tag{4.192}$$

which, for $T_{Si} \to \infty$, becomes equivalent to Eq.4.166 used for a bulk MOSFET.

Now we can proceed as in Section 4.4.1 and express the perturbed Hamiltonian $\hat{H}_{P,\mathbf{r}z}$ in terms of the unperturbed Hamiltonian referred to z', namely in terms of $\hat{H}_{0,z'}$. After some tedious manipulations and by keeping only the first order terms in $\Delta_F(\mathbf{r})$ and $\Delta_B(\mathbf{r})$, we obtain:

$$\begin{aligned}\hat{H}_{P,\mathbf{r}z} = \hat{H}_{0,z'} &+ \frac{\Delta_B(\mathbf{r}) - \Delta_F(\mathbf{r})}{T_{Si}} \left(\hbar^2 \frac{\mathrm{d}}{\mathrm{d}z'} \frac{1}{m_z(z')} \frac{\mathrm{d}}{\mathrm{d}z'} - e\, z' \frac{\mathrm{d}\phi(z')}{\mathrm{d}z'} \right) \\ &- e\, \Delta_F(\mathbf{r}) \frac{\mathrm{d}\phi(z')}{\mathrm{d}z'}.\end{aligned} \tag{4.193}$$

In order to calculate the surface roughness matrix elements by using Eqs.4.162 and 4.193 we also need to express the unperturbed wave-functions as a function of z'. To this purpose we first write z in terms of z' according to Eq.4.192, and then expand $\xi_n(z)$ to first order with respect to $\Delta_F(\mathbf{r})$ and $\Delta_B(\mathbf{r})$ as

$$\begin{aligned}\xi_n(z) &= \xi_n \left(z' + z' \frac{\Delta_B(\mathbf{r}) - \Delta_F(\mathbf{r})}{T_{Si}} + \Delta_F(\mathbf{r}) \right) \\ &\simeq \xi_n(z') + \frac{\mathrm{d}\xi_n(z')}{\mathrm{d}z'} \left[\Delta_F(\mathbf{r}) + z' \frac{\Delta_B(\mathbf{r}) - \Delta_F(\mathbf{r})}{T_{Si}} \right].\end{aligned} \tag{4.194}$$

Then, by substituting Eqs.4.193 and 4.194 in Eq.4.162, we can write

$$\begin{aligned}M_{nn'}(\mathbf{q}) = \Delta_F(\mathbf{q}) &\left[-\int \xi_n e \frac{\mathrm{d}\phi}{\mathrm{d}z} \xi_{n'}\, \mathrm{d}z + (\varepsilon_n - \varepsilon_{n'}) \int \xi_n \frac{\mathrm{d}\xi_{n'}}{\mathrm{d}z}\, \mathrm{d}z \right] \\ &+ \frac{\Delta_B(\mathbf{q}) - \Delta_F(\mathbf{q})}{T_{Si}} \int \xi_n \left[\left(\hbar^2 \frac{\mathrm{d}}{\mathrm{d}z} \frac{1}{m_z(z)} \frac{\mathrm{d}}{\mathrm{d}z} - z e \frac{\mathrm{d}\phi}{\mathrm{d}z} \right) \xi_{n'} \right] \mathrm{d}z \\ &+ \frac{\Delta_B(\mathbf{q}) - \Delta_F(\mathbf{q})}{T_{Si}} (\varepsilon_n - \varepsilon_{n'}) \int \xi_n z \frac{\mathrm{d}\xi_{n'}}{\mathrm{d}z}\, \mathrm{d}z,\end{aligned} \tag{4.195}$$

where $\Delta_F(\mathbf{q})$ and $\Delta_B(\mathbf{q})$ are defined as

$$\Delta_{F,B}(\mathbf{q}) = \frac{1}{A} \int_A \Delta_{F,B}(\mathbf{r}) \exp(-i\mathbf{q} \cdot \mathbf{r}) \mathrm{d}\mathbf{r}. \tag{4.196}$$

The first term in Eq.4.195 is the same as in the bulk case (see Eq.4.172), whereas the other two terms are explicitly related to the SOI structure. Equation 4.195 is equivalent to the results in [30].

If we now suppose that the Schrödinger equation in the SOI device was solved by imposing closed boundary conditions, then the integrals in Eq.4.195 are to be evaluated from $z = 0$ to $z = T_{Si}$. We first observe that in such a z interval the two step functions in Eq.4.191a are null, which simplifies the expression of the unperturbed Hamiltonian, $\hat{H}_{0,z}$. Then we continue the derivations with an approach similar to the one used in bulk MOSFETs to derive Eq.4.186, that is we evaluate the term

$$\Theta_B = \int_0^{T_{Si}} z\,\frac{\mathrm{d}\xi_{n'}(z)}{\mathrm{d}z}\,[\hat{H}_{0,z}\,\xi_n(z)]\,\mathrm{d}z + \int_0^{T_{Si}} z\,\frac{\mathrm{d}\xi_n(z)}{\mathrm{d}z}\,[\hat{H}_{0,z}\xi_{n'}(z)]\,\mathrm{d}z \tag{4.197}$$

in two different ways and then equate the resulting expressions. More precisely, we first evaluate Θ_B by exploiting the fact that $\xi_{n'}$ and ξ_n are wave-functions of the unperturbed Schrödinger equation, hence we have $\hat{H}_{0,z}\xi_n = \varepsilon_n\xi_n$ and $\hat{H}_{0,z}\xi_{n'} = \varepsilon_{n'}\xi_{n'}$. Then we express Θ_B by substituting directly in Eq.4.197 the $\hat{H}_{0,z}$ expression given in Eq.4.191a. By equating the two equivalent expressions for Θ_B thus obtained, we can derive the relation

$$-\int_0^{T_{Si}} z\,\xi_n\left(e\frac{\mathrm{d}\phi}{\mathrm{d}z}\right)\xi_{n'}\,\mathrm{d}z + (\varepsilon_n - \varepsilon_{n'})\int_0^{T_{Si}} z\,\xi_n\frac{\mathrm{d}\xi_{n'}}{\mathrm{d}z}\,\mathrm{d}z + \frac{\hbar^2}{m_z}\int_0^{T_{Si}} \xi_n\frac{\mathrm{d}^2\xi_{n'}}{\mathrm{d}z^2}\,\mathrm{d}z = -\frac{\hbar^2}{2m_z}\frac{\mathrm{d}\xi_{n'}(T_{Si})}{\mathrm{d}z}\frac{\mathrm{d}\xi_n(T_{Si})}{\mathrm{d}z}. \tag{4.198}$$

Equation 4.195 can now be rearranged exploiting Eqs.4.198 and 4.186 written for $z_1 = 0$ and $z_2 = T_{SI}$ (in fact Eq.4.186 also holds in an SOI structure if z_1 and z_2 belong to the interval $[0, T_{Si}]$). The scattering matrix element can thus be finally expressed as

$$M_{nn'}(\mathbf{q}) = \Delta_F(\mathbf{q})\frac{\hbar^2}{2m_z}\frac{\mathrm{d}\xi_{n'}(0)}{\mathrm{d}z}\frac{\mathrm{d}\xi_n(0)}{\mathrm{d}z} - \Delta_B(\mathbf{q})\frac{\hbar^2}{2m_z}\frac{\mathrm{d}\xi_{n'}(T_{Si})}{\mathrm{d}z}\frac{\mathrm{d}\xi_n(T_{Si})}{\mathrm{d}z}. \tag{4.199}$$

If the spectra of the two interfaces are uncorrelated, then the unscreened squared matrix element is given by

$$|M_{nn'}(\mathbf{q})|^2 = \frac{S_R^F(\mathbf{q})}{A}\left|\frac{\hbar^2}{2m_z}\frac{\mathrm{d}\xi_{n'}(0)}{\mathrm{d}z}\frac{\mathrm{d}\xi_n(0)}{\mathrm{d}z}\right|^2 + \frac{S_R^B(\mathbf{q})}{A}\left|\frac{\hbar^2}{2m_z}\frac{\mathrm{d}\xi_{n'}(T_{Si})}{\mathrm{d}z}\frac{\mathrm{d}\xi_n(T_{Si})}{\mathrm{d}z}\right|^2, \tag{4.200}$$

where $S_R^F(\mathbf{q})$ and $S_R^B(\mathbf{q})$ are the surface roughness spectrum respectively at the front and at the back interface defined consistently with Eq.4.175.

Surface roughness versus silicon film thickness variations

Let us consider a square well with closed boundary conditions for the Schrödinger equation and in the quantum limit; only the lowest subband is relevant and the corresponding envelope wave-function is $\xi_0(z) = \sqrt{2/T_{Si}}\sin(\pi z/T_{Si})$. The squared matrix element for SR is thus:

$$|M_{00}|^2 = \frac{1}{A}\left|\frac{\hbar^2}{2m_z}\left(\frac{\pi}{T_{Si}}\sqrt{\frac{2}{T_{Si}}}\right)^2\right|^2 [S_R^F(\mathbf{q}) + S_R^B(\mathbf{q})] = \frac{\pi^4\hbar^4}{Am_z^2T_{Si}^6}[S_R^F(\mathbf{q})) + S_R^B(\mathbf{q})]. \tag{4.201}$$

Since the energy of the subband is $\varepsilon_0 = \hbar^2\pi^2/(2m_z T_{Si}^2)$, it follows that:

$$|M_{00}|^2 = \frac{1}{A}\left|\frac{\mathrm{d}\varepsilon_0}{\mathrm{d}T_{Si}}\right|^2 [S_R^F(\mathbf{q}) + S_R^B(\mathbf{q})], \tag{4.202}$$

which is the formula used to evaluate film thickness fluctuations in quantum wells [32].

This means that the effect of the T_{Si} fluctuations is inherently accounted for by the SR scattering if Eq.4.195 or Eq.4.200 is used.

A simple interpretation of Eq.4.202 is that the local film thickness fluctuation $[\Delta_F(\mathbf{r})-\Delta_B(\mathbf{r})]$ produces a perturbation potential $(\mathrm{d}\varepsilon_0/\mathrm{d}T_{Si})[\Delta_F(\mathbf{r}) - \Delta_B(\mathbf{r})]$, which is in turn the origin of the scattering.

4.4.3 Effect of the screening in n-MOSFETs

If we consider a bulk MOSFET and account for the screening by using a scalar dielectric function, then the matrix elements for inter-subband transitions are left unscreened. For intra-subband transitions instead, each matrix element $M_{wnn}^{(scr)}(\mathbf{q})$ can be individually calculated from the corresponding unscreened matrix element as

$$|M_{wnn}^{(scr)}(\mathbf{q})|^2 = \frac{|M_{wnn}|^2(\mathbf{q})}{\epsilon_D^2(\mathbf{q})}, \tag{4.203}$$

where $\epsilon_D^2(\mathbf{q})$ is defined in Eq.4.91, $|M_{wnn}|^2(\mathbf{q})$ is given by Eq.4.174 and we have reintroduced the symbol w to denote the valley in the electron inversion layer.

If we use the full dielectric tensor instead, the procedure is somewhat more complicated. First of all we need to compute the unscreened matrix elements $M_{wnn'}(\mathbf{q})$ for all valleys and subbands. Then Eq.4.77 can be used to compute the screened matrix elements $M_{wnn'}^{(scr)}(\mathbf{q})$, which are proportional to $\Delta(\mathbf{q})$. The squared matrix elements $|M_{wnn'}^{(scr)}(\mathbf{q})|^2$ are thus proportional to the surface roughness spectrum $S_R(\mathbf{q})$, as was the case for the unscreened matrix elements given by Eq.4.174.

Let us now consider an SOI MOS transistor, where the surface roughness is relevant at both interfaces. We assume that surface fluctuations at the front and back interfaces are uncorrelated, so that the squared matrix elements at the two interfaces can be simply summed, as was done in Eq.4.200 for the unscreened matrix elements.

For a scalar dielectric function, Eq.4.200 can be used to calculate the unscreened matrix elements and then the screening is simply introduced dividing by $\epsilon_D^2(\mathbf{q})$, as in Eq.4.203 for bulk transistors.

When the complete dielectric function tensor is used instead, the screening is introduced separately for the scattering matrix elements at the two interfaces. To be more explicit, at the front interface the unscreened matrix elements are

$$M_{wnn'}^{(F)}(\mathbf{q}) = \Delta_F(\mathbf{q})\frac{\hbar^2}{2m_z}\frac{\mathrm{d}\xi_{n'}(0)}{\mathrm{d}z}\frac{\mathrm{d}\xi_n(0)}{\mathrm{d}z}, \tag{4.204}$$

and the corresponding screened matrix elements $M_{wnn'}^{(F,scr)}(\mathbf{q})$ are obtained by using Eq.4.77. The same procedure is used for the screened matrix elements $M_{wnn'}^{(B,scr)}(\mathbf{q})$ at the back interface. By doing so we can see that $M^{(F,scr)}_{wnn'}(\mathbf{q})$ is proportional to $\Delta_F(\mathbf{q})$ while $M^{(B,scr)}_{wnn'}(\mathbf{q})$ is proportional to $\Delta_B(\mathbf{q})$. If we assume that fluctuations at the front and back interface are uncorrelated, then we can sum the squared matrix elements at the two interfaces and finally express the overall screened matrix elements as

$$|M^{(scr)}_{wnn'}(\mathbf{q})|^2 = |M^{(F,scr)}_{wnn'}(\mathbf{q})|^2 + |M^{(B,scr)}_{wnn'}(\mathbf{q})|^2. \tag{4.205}$$

As already discussed in Section 4.2.3, the scalar dielectric function for the surface scattering is *not* a reliable model for SOI MOSFETs in DG mode. This is because some of the approximations behind the derivation of Eq.4.92 fail for a DG SOI MOSFET. This can lead to artifacts and to simulation results inconsistent with experiments, as illustrated in Fig.4.4. More details about the failure of the scalar dielectric function for DG SOI MOSFETs can be found in [15].

4.4.4 Surface roughness in *p*-MOSFETs

If we adopt the semi-analytical energy model described in Section 3.3.3, the derivations of the unscreened matrix elements presented in Sections 4.4.1 and 4.4.2 for an electron inversion layer can be readily extended to a hole inversion layer. Then determination of the scattering rates requires us to evaluate the sums over the final states, which can be calculated by resorting to appropriate integrals in the **k** space as discussed in Section 3.5.3.

In the case of a hole inversion layer described by the **k**·**p** Hamiltonian of Section 3.3.1, we can start the calculation of the scattering matrix elements again by using Eq.4.162. However, now the unperturbed $\hat{\mathbf{H}}_0(z)$ and the perturbed Hamiltonian $\hat{\mathbf{H}}_P(\mathbf{r}, z)$ are 6×6 matrix operators and $\boldsymbol{\xi}_{n\mathbf{k}}(z)$ is a six component wave-function. Furthermore we also assume, consistently with Section 3.3.1, that the **k**·**p** eigenvalue problem is solved by using closed boundary conditions, that is by using a z domain $[0,+\infty]$ and by setting $\boldsymbol{\xi}_{n\mathbf{k}}(0) = \mathbf{0}$. We can now proceed exactly as in Section 4.4.1 and obtain

$$M_{nn'}(\mathbf{k}, \mathbf{k}') = \Delta(\mathbf{q}) \int_0^{+\infty} \left[\boldsymbol{\xi}^{\dagger}_{n',\mathbf{k}'}(z) \cdot \boldsymbol{\xi}_{n,\mathbf{k}}(z) \left(e\frac{\mathrm{d}\phi}{\mathrm{d}z} \right) + E_{n'\mathbf{k}'}\, \boldsymbol{\xi}^{\dagger}_{n'\mathbf{k}'}(z) \cdot \frac{\mathrm{d}\boldsymbol{\xi}_{n,\mathbf{k}}(z)}{\mathrm{d}z} + E_{n,\mathbf{k}} \frac{\mathrm{d}\boldsymbol{\xi}^{\dagger}_{n'\mathbf{k}'}(z)}{\mathrm{d}z} \cdot \boldsymbol{\xi}_{n\mathbf{k}}(z) \right] \mathrm{d}z, \tag{4.206}$$

which is similar to Eq.4.172 except for the different sign of the first term (due to the different sign taken by $\phi(z)$ in the hole with respect to the electron potential energy), and for the fact that the energies $E_{n\mathbf{k}}$ and $E_{n'\mathbf{k}'}$ cannot be grouped. The dot sign denotes the scalar product between the vectorial wave-functions defined in Eq.4.40.

In the rest of this section we no longer explicitly indicate the dot sign, assuming that $\boldsymbol{\xi}^{\dagger}_{n',\mathbf{k}'}$ is a row vector and $\boldsymbol{\xi}_{n,\mathbf{k}}$ is a column vector, so that $(\boldsymbol{\xi}^{\dagger}_{n',\mathbf{k}'}\boldsymbol{\xi}_{n,\mathbf{k}})$ can be considered a standard row-column product.

The fact that the wave-functions depend on **k** complicates the calculation of the scattering rates remarkably. In many cases this dependence is thus neglected and the wave-functions are taken at $\mathbf{k}=\mathbf{0}$ [20, 33, 34]. We henceforth denote with $\boldsymbol{\xi}_n$ and E_n the wave-functions and subband energies at $\mathbf{k}=\mathbf{0}$, which are obtained from the eigenvalue problem $\hat{\mathbf{H}}_\mathbf{0}\,\boldsymbol{\xi}_n = E_n\,\boldsymbol{\xi}_n$, where $\hat{\mathbf{H}}_\mathbf{0}$ is the $\mathbf{k}{\cdot}\mathbf{p}$ Hamiltonian at $\mathbf{k}=\mathbf{0}$ that reads (see Section 3.3.1)

$$\hat{\mathbf{H}}_\mathbf{0} = \begin{pmatrix} -M\frac{\mathrm{d}^2}{\mathrm{d}z^2} & \mathrm{i}\frac{\Delta_{SO}}{3} & 0 & 0 & 0 & -\frac{\Delta_{SO}}{3} \\ -\mathrm{i}\frac{\Delta_{SO}}{3} & -M\frac{\mathrm{d}^2}{\mathrm{d}z^2} & 0 & 0 & 0 & \mathrm{i}\frac{\Delta_{SO}}{3} \\ 0 & 0 & -L\frac{\mathrm{d}^2}{\mathrm{d}z^2} & \frac{\Delta_{SO}}{3} & -\mathrm{i}\frac{\Delta_{SO}}{3} & 0 \\ 0 & 0 & \frac{\Delta_{SO}}{3} & -M\frac{\mathrm{d}^2}{\mathrm{d}z^2} & -\mathrm{i}\frac{\Delta_{SO}}{3} & 0 \\ 0 & 0 & \mathrm{i}\frac{\Delta_{SO}}{3} & \mathrm{i}\frac{\Delta_{SO}}{3} & -M\frac{\mathrm{d}^2}{\mathrm{d}z^2} & 0 \\ -\frac{\Delta_{SO}}{3} & -\mathrm{i}\frac{\Delta_{SO}}{3} & 0 & 0 & 0 & -L\frac{\mathrm{d}^2}{\mathrm{d}z^2} \end{pmatrix} + e\phi(z)\mathbf{I}, \tag{4.207}$$

where the values of L, M, and N are given in Table 2.2. The Hamiltonian in Eq.4.207 can be concisely rewritten as

$$\hat{\mathbf{H}}_0 = \mathbf{A}\frac{\mathrm{d}^2}{\mathrm{d}z^2} + \mathbf{B} + e\phi(z)\mathbf{I}, \tag{4.208}$$

where $\mathbf{A}$ is the diagonal matrix whose elements are $(-M,-M,-L,-M,-M,-L)$ and $\mathbf{B}$ is the remaining part of the matrix in Eq.4.207.

We now recall the eigenvalue problem $\hat{\mathbf{H}}_0\boldsymbol{\xi}_n(z) = E_n\boldsymbol{\xi}_n(z)$ (with the Hamiltonian $\hat{\mathbf{H}}_0$ defined in Eq.4.208), left-multiply both sides of the equation by $(\mathrm{d}\boldsymbol{\xi}^\dagger_{n'}/\mathrm{d}z)$ and thus obtain

$$\frac{\mathrm{d}\boldsymbol{\xi}^\dagger_{n'}}{\mathrm{d}z}\mathbf{A}\frac{\mathrm{d}^2\boldsymbol{\xi}_n}{\mathrm{d}z^2} + \frac{\mathrm{d}\boldsymbol{\xi}^\dagger_{n'}}{\mathrm{d}z}\mathbf{B}\,\boldsymbol{\xi}_n + e\phi(z)\frac{\mathrm{d}\boldsymbol{\xi}^\dagger_{n'}}{\mathrm{d}z}\boldsymbol{\xi}_n = E_n\frac{\mathrm{d}\boldsymbol{\xi}^\dagger_{n'}}{\mathrm{d}z}\boldsymbol{\xi}_n. \tag{4.209}$$

Similarly, we can start from $\hat{\mathbf{H}}_0\boldsymbol{\xi}_{n'}(z) = E_{n'}\boldsymbol{\xi}_{n'}(z)$ and then, by taking the transpose conjugate of the equation and right-multiplying by $(\mathrm{d}\boldsymbol{\xi}_n/\mathrm{d}z)$, we obtain

$$\frac{\mathrm{d}^2\boldsymbol{\xi}^\dagger_{n'}}{\mathrm{d}z^2}\mathbf{A}\frac{\mathrm{d}\boldsymbol{\xi}_n}{\mathrm{d}z} + \boldsymbol{\xi}^\dagger_{n'}\mathbf{B}\frac{\mathrm{d}\boldsymbol{\xi}_n}{\mathrm{d}z} + e\,\phi(z)\boldsymbol{\xi}^\dagger_{n'}\frac{\mathrm{d}\boldsymbol{\xi}_n}{\mathrm{d}z} = E_{n'}\boldsymbol{\xi}^\dagger_{n'}\frac{\mathrm{d}\boldsymbol{\xi}_n}{\mathrm{d}z}, \tag{4.210}$$

where we have exploited the fact that both $\mathbf{A}$ and $\mathbf{B}$ are Hermitian matrices.

By summing Eqs.4.209 and 4.210 and performing some simple re-arrangements we obtain:

$$\frac{\mathrm{d}}{\mathrm{d}z}\left(\frac{\mathrm{d}\boldsymbol{\xi}^\dagger_{n'}}{\mathrm{d}z}\mathbf{A}\frac{\mathrm{d}\boldsymbol{\xi}_n}{\mathrm{d}z}\right) + \frac{\mathrm{d}}{\mathrm{d}z}(\boldsymbol{\xi}^\dagger_{n'}\mathbf{B}\boldsymbol{\xi}_n) + e\,\phi(z)\frac{\mathrm{d}}{\mathrm{d}z}(\boldsymbol{\xi}^\dagger_{n'}\boldsymbol{\xi}_n)$$
$$= E_n\frac{\mathrm{d}\boldsymbol{\xi}^\dagger_{n'}}{\mathrm{d}z}\boldsymbol{\xi}_n + E_{n'}\boldsymbol{\xi}^\dagger_{n'}\frac{\mathrm{d}\boldsymbol{\xi}_n}{\mathrm{d}z}. \tag{4.211}$$

Equation 4.211 can then be integrated over the domain $[0,+\infty]$ to obtain

$$\frac{\mathrm{d}\boldsymbol{\xi}^{\dagger}_{n'}(0)}{\mathrm{d}z}\mathbf{A}\frac{\mathrm{d}\boldsymbol{\xi}_n(0)}{\mathrm{d}z} - e\int_0^{+\infty}\frac{\mathrm{d}\phi}{\mathrm{d}z}\boldsymbol{\xi}^{\dagger}_{n'}\boldsymbol{\xi}_n\mathrm{d}z$$
$$= E_n\int_0^{+\infty}\frac{\mathrm{d}\boldsymbol{\xi}^{\dagger}_{n'}}{\mathrm{d}z}\boldsymbol{\xi}_n\mathrm{d}z + E_{n'}\int_0^{+\infty}\boldsymbol{\xi}^{\dagger}_{n'}\frac{\mathrm{d}\boldsymbol{\xi}_n}{\mathrm{d}z}\mathrm{d}z, \qquad (4.212)$$

where we have integrated by parts the term $[e\phi(z)\mathrm{d}(\boldsymbol{\xi}^{\dagger}_{n'}\boldsymbol{\xi}_n)/\mathrm{d}z]$ in Eq.4.211 and we have used $\boldsymbol{\xi}_n(0) = \boldsymbol{\xi}_{n'}(0) = \mathbf{0}$.

Equation 4.212 allows us to express the integral in the r.h.s. of Eq.4.206 as a function of the wave-function's derivatives at the semiconductor–oxide interface. Hence the matrix element $M_{nn'}$ corresponding to the wave-functions $\boldsymbol{\xi}_n$ and $\boldsymbol{\xi}_{n'}$ calculated at $\mathbf{k} = \mathbf{0}$ can be finally written as

$$M_{nn'}(\mathbf{q}) = -\Delta(\mathbf{q})\left[\frac{\mathrm{d}\boldsymbol{\xi}^{\dagger}_{n'}(0)}{\mathrm{d}z}\mathbf{A}\frac{\mathrm{d}\boldsymbol{\xi}_n(0)}{\mathrm{d}z}\right], \qquad (4.213)$$

where $\mathbf{q} = (\mathbf{k}' - \mathbf{k})$ is the wave-vector change induced by the scattering.

An explicit evaluation of the term in square brackets in Eq.4.213 leads to

$$-\left(\frac{\mathrm{d}\boldsymbol{\xi}^{\dagger}_{n'}(0)}{\mathrm{d}z}\mathbf{A}\frac{\mathrm{d}\boldsymbol{\xi}_n(0)}{\mathrm{d}z}\right)$$
$$= M\left(\frac{\mathrm{d}\xi^{\dagger}_{n',1}}{\mathrm{d}z}\frac{\mathrm{d}\xi_{n,1}}{\mathrm{d}z} + \frac{\mathrm{d}\xi^{\dagger}_{n',2}}{\mathrm{d}z}\frac{\mathrm{d}\xi_{n,2}}{\mathrm{d}z} + \frac{\mathrm{d}\xi^{\dagger}_{n',4}}{\mathrm{d}z}\frac{\mathrm{d}\xi_{n,4}}{\mathrm{d}z} + \frac{\mathrm{d}\xi^{\dagger}_{n',5}}{\mathrm{d}z}\frac{\mathrm{d}\xi_{n,5}}{\mathrm{d}z}\right)_{z=0}$$
$$+ L\left(\frac{\mathrm{d}\xi^{\dagger}_{n',3}}{\mathrm{d}z}\frac{\mathrm{d}\xi_{n,3}}{\mathrm{d}z} + \frac{\mathrm{d}\xi^{\dagger}_{n',6}}{\mathrm{d}z}\frac{\mathrm{d}\xi_{n,6}}{\mathrm{d}z}\right)_{z=0}, \qquad (4.214)$$

which is the expression reported in [33], where $\xi_{n,i}(z)$ denotes the ith component of $\boldsymbol{\xi}_n(z)$. By using Eq.4.213 the squared matrix element is readily found to be proportional to the surface roughness spectrum $S_R(\mathbf{q})$ defined in Eq.4.175.

For a (100) inversion layer the squared matrix elements for the surface roughness scattering obtained with the semi-analytical energy model presented in Section 3.3.3 are in good agreement with the results of the **k·p** approach [35].

SOI *p*-MOSFETs

In the case of a hole inversion layer in an SOI MOSFET, we can essentially follow the approach used for n-type SOI transistors, which is based on the change of coordinates defined by Eq.4.192. The calculation of the matrix elements proceeds similarly to the electron case, with the difference that the scalar terms $\left[-\frac{\hbar^2}{2m_z}\frac{\mathrm{d}^2\xi_n}{\mathrm{d}z^2}\right]$ are replaced by vectorial terms in the form $\left[\mathbf{A}\frac{\mathrm{d}^2\boldsymbol{\xi}_n}{\mathrm{d}z^2}\right]$. After tedious calculations, the matrix elements can be finally written as

$$M_{nn'}(\mathbf{q}) = -\Delta_F(\mathbf{q})\left(\frac{\mathrm{d}\boldsymbol{\xi}^{\dagger}_{n'}(0)}{\mathrm{d}z}\mathbf{A}\frac{\mathrm{d}\boldsymbol{\xi}_n(0)}{\mathrm{d}z}\right) + \Delta_B(\mathbf{q})\left(\frac{\mathrm{d}\boldsymbol{\xi}^{\dagger}_{n'}(T_{Si})}{\mathrm{d}z}\mathbf{A}\frac{\mathrm{d}\boldsymbol{\xi}_n(T_{Si})}{\mathrm{d}z}\right). \qquad (4.215)$$

By using Eq.4.215 and assuming that the roughness at the front and at the back interface is uncorrelated, the squared matrix elements take a form similar to Eq.4.200 and are proportional to the surface roughness spectra $S_R^F(\mathbf{q})$ and $S_R^B(\mathbf{q})$ respectively at the front and back interfaces.

Screening effect in *p*-MOSFETs

If we use the semi-analytical energy model presented in Section 3.3.3, then the wave-functions do not depend on the wave-vector and screening can be introduced by following the procedure explained in Section 4.4.3 for the electrons. The dielectric function should be calculated consistently with the hole energy model [35].

If the $\mathbf{k}\cdot\mathbf{p}$ quantization model is used instead, one should use the formulation of the dielectric function summarized by Eqs.4.83 to 4.85 (with $n_{sp} = 1$); this is computationally very demanding. However, if the $\mathbf{k}$ dependence of the wave-functions is neglected (as was the case for the matrix elements described in this section), then we can resort to the dielectric function defined in Eq.4.78. In such a case, the polarization factors must be calculated by evaluating numerically the sum over $\mathbf{k}$ according to the procedure described in detail in Section 3.5.4; the spin multiplicity is $n_{sp} = 1$.

4.5 Vibrations of the crystal lattice

At any finite temperature the atoms in the crystal lattice oscillate with respect to their nominal positions. Such vibrations of the lattice produce perturbations of the potential corresponding to the ideal lattice which are a very important source of carrier scattering, at least at not too low operating temperatures.

Lattice vibrations are discussed in detail in several textbooks of solid state physics [36, 37], in fact they are important not only for electronic transport but also for the thermal as well as the optical properties of the crystals. Thus, we henceforth limit ourselves to introducing concepts about lattice vibrations that are most important for the operation of electron devices and, in particular, for understanding electron–phonon scattering in MOS transistors.

4.5.1 Classical model for the lattice vibrations

The concept of the phonon itself stems from quantum mechanical analysis of the lattice vibrations, however, a discussion based on the classical Newton's laws of vibrations in a macroscopic mechanical system is very instructive in providing an insight into the main features of the phonon modes.

In this respect, let us consider a one-dimensional chain of atoms. Since the equilibrium positions of the atoms correspond to a minimum of the potential energy of the system, then expansion of the energy in Taylor series with respect to the atomic displacements has a quadratic leading term. The corresponding forces exerted on the atoms when they are displaced from their equilibrium positions are thus proportional to

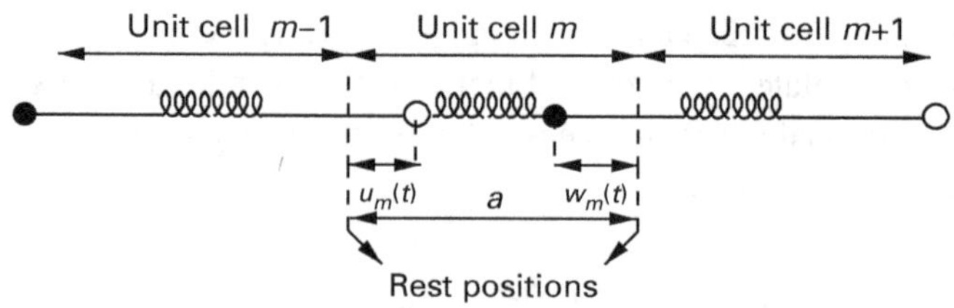

Figure 4.11 One dimensional mechanical model for atomic vibrations in a lattice. Interatomic forces are represented by springs. a is the length of the unit cell at rest.

the displacements. Hence a chain of point masses connected by springs, as illustrated in Fig.4.11, can be considered a mechanical system phenomenologically similar to a chain of atoms.

In particular, Fig.4.11 illustrates a linear chain of atoms where the unit cell consists of two atoms; $u_m(t)$ and $w_m(t)$ indicate the displacements of the two atoms in the unit cell m, and are much smaller than the length a of the unit cell.

By letting M denote the mass of the atoms and C the constant of the springs, the displacement $u_m(t)$ of the left atom in the unit cell m is governed by Newton's equation

$$M\,\frac{\partial^2 u_m}{\partial t^2} = C\,[w_{m-1}(t) - u_m(t)] + C\,[w_m(t) - u_m(t)], \tag{4.216}$$

and a similar equation can be written for the displacement $w_m(t)$. Hence $u_m(t)$ and $w_m(t)$ are linked by the equations

$$M\,\frac{\partial^2 u_m}{\partial t^2} = C\,[w_{m-1}(t) + w_m(t) - 2\,u_m(t)], \tag{4.217a}$$

$$M\,\frac{\partial^2 w_m}{\partial t^2} = C\,[u_m(t) + u_{m+1}(t) - 2\,w_m(t)]. \tag{4.217b}$$

The vibrations of a system such as the one in Fig.4.11 can be conveniently studied by using the so called normal modes. In particular, since we are interested in the steady state sinusoidal modes of the system, we express the unknown atomic displacements as

$$u_m(t) = \tilde{u}_q\,\mathrm{e}^{\mathrm{i}(qma-\omega t)} + \tilde{u}_q^{\dagger}\mathrm{e}^{-\mathrm{i}(qma-\omega t)} = 2|\tilde{u}_q|\cos[qma - \omega t + \phi_{uq}], \tag{4.218a}$$

$$w_m(t) = \tilde{w}_q\mathrm{e}^{\mathrm{i}(qma-\omega t)} + \tilde{w}_q^{\dagger}\mathrm{e}^{-\mathrm{i}(qma-\omega t)} = 2|\tilde{w}_q|\cos[qma - \omega t + \phi_{wq}], \tag{4.218b}$$

where $\tilde{u}_q$ and $\tilde{w}_q$ are the complex numbers representing the vibration modes for the two atoms (ϕ_{uq} and ϕ_{wq} being the corresponding phases), while q and ω are the wave-vector and the frequency of the mode.

Let us first note that the modes defined in Eq.4.218 are such that, if we take two q values that differ by $(2\pi/a)$, then we obtain exactly the same mode. This implies that the values of the wave-vector q can be restricted to a range $(2\pi/a)$ wide: the range $[-\pi/a, \pi/a]$ is the typical choice and it can be considered the first Brillouin zone of the atomic displacements (see Section 2.1.2). Furthermore, if we are interested in the vibrations of a chain with a virtually infinite length, we can impose periodic boundary conditions on the atomic displacements in Eq.4.218. To this purpose, if we denote by $L = N_u a$ the length of the atomic chain (where N_u is the number of unit cells), then the periodic boundary conditions impose $e^{iqL} = 1$, namely $q = n(2\pi/L)$ with n being an integer. Thus the length L of the chain sets the spacing $(2\pi/L)$ of the allowed wave-vectors in the $[-\pi/a, \pi/a]$ interval.

If we now substitute Eq.4.218 in, for instance, Eq.4.217a, we obtain

$$-\omega^2 M\tilde{u}_q e^{i(qma-\omega t)} + (c.c.) = C[\tilde{w}_q e^{-iqa} + \tilde{w}_q - 2\tilde{u}_q]e^{i(qma-\omega t)} + (c.c.). \tag{4.219}$$

In order for Eq.4.219 to be verified at any time t, that is at any value of the argument of the exponential term, the complex numbers multiplying the exponentials must be the same. In other words we must have

$$-\omega^2 M\,\tilde{u}_q = C\,[\tilde{w}_q\,e^{-iqa} + \tilde{w}_q - 2\,\tilde{u}_q], \tag{4.220}$$

which is an algebraic equation linking the complex numbers $\tilde{u}_q$ and $\tilde{w}_q$ representative of the two lattice displacements inside the unit cell.

By substituting Eq.4.218 in Eq.4.217b we obtain a second equation similar to Eq.4.220, so that the two complex numbers $\tilde{u}_q$ and $\tilde{w}_q$ can be finally determined by solving the algebraic linear problem

$$(2C - \omega^2 M)\,\tilde{u}_q - C(1 + e^{-iqa})\,\tilde{w}_q = 0, \tag{4.221a}$$

$$-C(1 + e^{iqa})\,\tilde{u}_q + (2\,C - \omega^2 M)\,\tilde{w}_q = 0. \tag{4.221b}$$

Since Eq.4.221 is a homogeneous linear system, non-null values for $\tilde{u}_q$ and $\tilde{w}_q$ are possible only if the determinant of the corresponding matrix is null. By setting such a determinant to zero we obtain the secular equation for the vibrational modes, which provides, for each wave-vector q, the possible ω_q values. The equation reads

$$(2\,C - \omega_q^2 M) = \pm C\sqrt{2[1 + \cos(qa)]}, \tag{4.222}$$

and the two possible frequency values ω_{q1} and ω_{q2} are

$$\omega_{q1,q2} = \sqrt{\frac{1}{M}\left[2C \mp C\sqrt{2(1 + \cos(qa))}\right]}, \tag{4.223}$$

where the upper and lower sign correspond respectively to ω_{q1} and ω_{q2}. The linear model for the atomic displacements described by Eq.4.221 does not provide any information about the magnitude of $\tilde{w}_q$ and $\tilde{u}_q$. However, when Eq.4.222 is satisfied, we can use Eq.4.221b to express the relation between $\tilde{w}_q$ and $\tilde{u}_q$ and obtain

$$\tilde{w}_q = \pm\frac{1 + e^{iqa}}{\sqrt{2(1 + \cos(qa))}}\,\tilde{u}_q, \qquad \text{for } (qa) \neq \pi \tag{4.224}$$

where we have expressed $(2\,C - \omega_q^2 M)$ by means of Eq.4.222 and the upper and the lower sign correspond respectively to $\tilde{w}_{q1}, \tilde{u}_{q1}$ and to $\tilde{w}_{q2}, \tilde{u}_{q2}$. Equation 4.224 shows that $|\tilde{w}_q| = |\tilde{u}_q|$ for both the solutions of Eq.4.221, hence $\tilde{w}_q$ differs from $\tilde{u}_q$ just for a q dependent phase shift.

Equation 4.223 describes two branches for the atomic displacements illustrated in Fig.4.12. The branch with the upper sign is referred to as acoustic and the other as

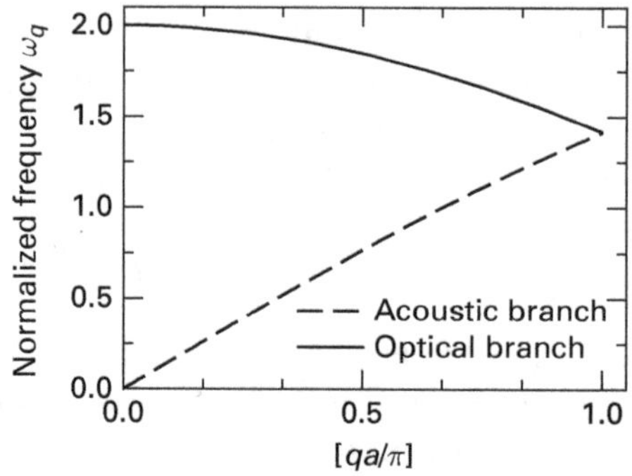

Figure 4.12 Frequency versus the wave-vector relation given by Eq.4.223 for a one-dimensional chain of point masses. The frequency is normalized to $\sqrt{C/M}$ and the wave-vector to (π/a). The acoustic and optical branches are shown.

optical. The differences between the two branches are particularly evident when we consider very small q values, that is in the so-called *long wavelength limit.* In such a case, in fact, Eqs.4.223 and 4.224 yield

$$\lim_{q\to 0}\begin{cases}\omega_{q1}=0\\ \omega_{q2}=\sqrt{\frac{4C}{M}}\end{cases}, \qquad \lim_{q\to 0}\begin{cases}\tilde{w}_{q1}=\tilde{u}_{q1}\\ \tilde{w}_{q2}=-\tilde{u}_{q2}\end{cases}. \tag{4.225}$$

Hence, in the long wavelength limit, for the acoustic branch the energy $\hbar\omega_{q1}$ is null and the displacement $\tilde{w}_{q1}$ of the second atom in the unit cell is perfectly in phase with the displacement $\tilde{u}_{q1}$ of the first atom. Since for very small q values the atoms of different cells are also displaced in phase (because e^{imqa} in Eq.4.218 tends to 1), then for vanishing q values the acoustic mode corresponds to a rigid translation of the crystal that understandably results in a null vibration frequency.

For the optical branch, Eq.4.225 states instead that, for $q\to 0$, the energy $\hbar\omega_{q2}$ is non-null and the phase of $\tilde{w}_{q2}$ is opposite with respect to $\tilde{u}_{q2}$, namely the two atoms in the unit cell move in opposite directions.

The kinetic energy associated with $\tilde{u}_q$ is readily given by Eq.4.218a as

$$E_{kin}=\frac{M}{2}\left(\frac{\partial u}{\partial t}\right)^2=2M\omega_q^2|\tilde{u}_q|^2\sin^2(qma-\omega_q t+\phi_{uq}). \tag{4.226}$$

Furthermore, in a harmonic oscillator the total energy oscillates between the kinetic and the potential energy, so that we can express the total energy as the maximum value $2M\omega^2|\tilde{u}_q|^2$ of the kinetic energy. Since $|\tilde{w}_q|$ is equal to $|\tilde{u}_q|$, the total energy related to the atomic displacements is given by

$$E_q=2NM\omega_q^2|\tilde{u}_q|^2=2\rho\Omega\omega_q^2|\tilde{u}_q|^2, \tag{4.227}$$

where N is the total number of atoms in the volume Ω (twice as many as the unit cells for our example with two atoms per unit cell), and ρ is the mass density of the crystal, so that (NM) is equal to $(\rho\Omega)$.

4.5.2 Quantization of the lattice vibrations

The chain of point masses illustrated in Fig.4.11 is a macroscopic and one dimensional representation of the atomic vibrations in a crystal. When dealing with the lattice Hamiltonian that describes the atom vibrations, if we assume that the atomic displacements are small, then we can expand the potential energy in Taylor series about the equilibrium positions of the atoms. The first order term in the Taylor series is null for the definition itself of the equilibrium positions, so that the leading term is a quadratic function of the atomic displacements.

This approach turns the lattice vibrations into a set of harmonic oscillators. As discussed in detail in most textbooks of quantum mechanics [38–40], the quantum theory of such harmonic oscillators yields a quantization of the possible energy values.

More precisely, for the mode identified by the wave-vector q and the corresponding frequency ω_q, the energy is quantized in units of $\hbar\omega_q$ according to

$$E_q = \hbar\omega_q \left(n_q + \frac{1}{2}\right), \tag{4.228}$$

where n_q is a positive integer (i.e. $n_q = 0, 1, 2, \cdots$).

Equation 4.228 provides a phenomenological way to introduce the concept of the phonon. In fact, since the energy is quantized in terms of $\hbar\omega_q$, Eq.4.228 can be visualized as the energy of a group of n_q virtual particles having an energy $\hbar\omega_q$ each. Such particles are called *phonons* and n_q is the number of phonons in the phonon mode.

We can link the classical to the quantum treatment of lattice vibrations by equating the corresponding expressions for the energy given by Eq.4.227 and Eq.4.228, respectively. We can thus express the amplitude $|\tilde{u}_q|$ of the lattice wave in terms of the corresponding phonon number

$$|\tilde{u}_q|^2 = \frac{\hbar}{2\rho\Omega\omega_q}\left(n_q + \frac{1}{2}\right). \tag{4.229}$$

By recalling Eq.4.218a we can now express the atomic displacements as

$$u_q(x,t) = \sqrt{\frac{\hbar}{2\rho\Omega\omega_q}}\left[a_q\, \mathrm{e}^{\mathrm{i}(qx-\omega_q t)} + a_q^{\dagger}\mathrm{e}^{-\mathrm{i}(qx-\omega_q t)}\right], \tag{4.230}$$

where the squared magnitude of the complex number a_q is

$$|a_q|^2 = \left(n_q + \frac{1}{2}\right). \tag{4.231}$$

In Eq.4.230 the atomic displacements are expressed as a function of a continuum abscissa $x = (ma)$ which identifies the equilibrium positions of the atoms.

The results obtained with a simple diatomic one-dimensional model may be summarized by saying that, for a unit cell with two atoms, we have an acoustic and an optical mode which exhibit remarkably different relations between the frequency ω_q and the wave-vector q.

If we now move to a three-dimensional crystal with a diatomic unit cell, as is the case for Si, Ge, and GaAs (see Section 2.1.1), then we obtain six possible phonon branches described by a three-dimensional wave-vector **Q**. Three branches are acoustic and three are optical phonon modes. The three branches of each mode correspond to different polarizations, namely to different directions of the atomic displacements with respect to the wave-vector **Q**. The modes that have the atomic displacements aligned with **Q** are referred to as longitudinal; these are the only possible modes in a one dimensional lattice. In a three-dimensional solid, however, for any given **Q** we also have two possible transverse modes corresponding to atomic displacements normal to the **Q** direction.

The wave of the atomic displacements can then be written as

$$\mathbf{u}_{\nu,\mathbf{Q}}(\mathbf{R},t) = \hat{\mathbf{e}}_\nu \sqrt{\frac{\hbar}{2\rho\Omega\omega_{\nu,\mathbf{Q}}}} \left[a_{\nu,\mathbf{Q}} \mathrm{e}^{\mathrm{i}(\mathbf{Q}\cdot\mathbf{R}-\omega_{\nu,\mathbf{Q}}t)} + a^\dagger_{\nu,\mathbf{Q}} \mathrm{e}^{-\mathrm{i}(\mathbf{Q}\cdot\mathbf{R}-\omega_{\nu,\mathbf{Q}}t)} \right], \tag{4.232}$$

where ν indicates the polarization of the phonon mode and $\hat{\mathbf{e}}_\nu$ is the polarization vector, which is aligned to **Q** for the longitudinal modes and normal to **Q** for the transverse modes. The magnitude of $a_{\nu,\mathbf{Q}}$ is related to the phonon number $n_{\nu,\mathbf{Q}}$ by Eq.4.231 as

$$|a_{\nu,\mathbf{Q}}|^2 = \left(n_{\nu,\mathbf{Q}} + \frac{1}{2} \right). \tag{4.233}$$

The energy of the phonon of a given mode is linked to its wave-vector by the phonon energy dispersion

$$E_\nu(\mathbf{Q}) = \hbar\,\omega_{\nu,\mathbf{Q}}, \tag{4.234}$$

and there is a perfect analogy with the electron energy described as a function of the wave-vector **K**. It is worth noting that Eq.4.234 expresses the energy of a single phonon of mode ν; this should not be confused with the energy defined by Eq.4.228, which is instead the total energy carried by the entire mode.

All the properties of the electron energy dispersion discussed in Section 2.1.2 can be extended to the phonon energy dispersion. In particular, the phonon wave-vector **Q** can be restricted to the first Brillouin zone of the reciprocal lattice, in fact the energy $\hbar\omega_{\nu,\mathbf{Q}}$ is periodically repeated outside the first Brillouin zone.

For a given semiconductor crystal the energy dispersion of the acoustic and optical phonon branches is a property of the material, exactly as the electron dispersion in the valence or in the conduction band is. Figure 4.13 shows the energy dispersion for silicon for a wave-vector **Q** along the [100] direction as reported in [41, 42]. The phonon branches are actually six because the transverse branches are doubly degenerate.

As can be seen, for acoustic phonons the energy is approximately proportional to the magnitude Q of the wave-vector for relatively small Q values, whereas the energy of the optical modes is only weakly dependent on Q. Thus, for modeling of electron devices, the phonon energy dispersions are very frequently approximated as [4–6, 43, 44]

$$\text{acoustic phonons}: \quad \hbar\omega_{\nu,\mathbf{Q}} \approx \hbar v_s Q, \tag{4.235a}$$

$$\text{optical phonons}: \quad \hbar\omega_{\nu,\mathbf{Q}} \approx \hbar\omega_0. \tag{4.235b}$$

In particular, Eq.4.235a is valid only for small Q values, namely in the so-called long wavelength limit. Figure 4.13 shows that the sound velocity v_s is somewhat different

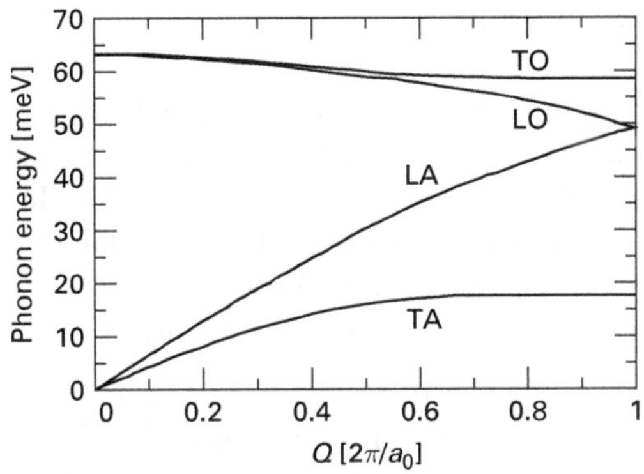

Figure 4.13 Acoustic and optical phonon energy dispersion in silicon versus the magnitude Q of the wave-vector $\mathbf{Q}$ along the [100] direction. After Long [41] and Brockhouse [42]. The symbols TO, LO, LA, TA denote, respectively, transverse optical, longitudinal optical, longitudinal acoustic and transverse acoustic branches. The transverse branches are doubly degenerate, so that the phonon branches are six in all. The sound velocity v_s is approximately $4.7{\cdot}10^5$ cm/s and $9.2{\cdot}10^5$ cm/s for the transverse and longitudinal acoustic modes, respectively.

for the transverse and the longitudinal acoustic modes, in fact it is approximately $4.7{\cdot}10^5$ cm/s and $9.2{\cdot}10^5$ cm/s for the transverse and the longitudinal case, respectively.

Before moving to a description of scattering processes we must clarify how to determine the number of phonons. Differently from electrons, phonons are Bose particles and do not obey Pauli's exclusion principle. Consequently, any number of phonons can occupy a state $(\nu,\mathbf{Q})$. The average number of phonons at equilibrium is given by Bose–Einstein statistics

$$n_{\nu,\mathbf{Q}} = \frac{1}{\exp\left(\frac{\hbar\omega_{\nu,\mathbf{Q}}}{K_B T}\right) - 1}, \tag{4.236}$$

and it depends only on the phonon energy $\hbar\omega_{\nu,\mathbf{Q}}$ and on the lattice temperature.

Depending on whether $\hbar\omega_{\nu,\mathbf{Q}}$ is large or small compared to $K_B T$, the average phonon number can be much smaller or much larger than 1. In particular, the energy of the acoustic phonons tends to 0 for small $\mathbf{Q}$ values and, when $\hbar\omega_{\nu,\mathbf{Q}}$ is much smaller than $K_B T$, the phonon number can be approximated as

$$n_{\nu,\mathbf{Q}} \approx \frac{K_B T}{\hbar\omega_{\nu,\mathbf{Q}}}. \tag{4.237}$$

Equation 4.237 is known as the *energy equipartition approximation*, and is very often used for calculation of acoustic phonon scattering rates (see Section 4.6).

Equations 4.236 and 4.237 show that, if we use the approximate phonon energies given in Eq.4.235, then the phonon number of the optical modes is approximately independent of Q, while the number of the acoustic phonons is inversely proportional to Q.

4.6 Phonon scattering

Equation 4.232 expresses the atomic displacement in the crystal lattice, where the magnitude of $a_{\nu,\mathbf{Q}}$ is related to the phonon number by Eq.4.233 and the phonon number is in turn related to the phonon energy by Eq.4.236. In order to proceed further towards the phonon scattering rates we need to understand how atomic displacements are converted to scattering potentials. The scattering potentials can then be used to calculate the scattering matrix elements and scattering rates as discussed in Section 4.1.

Before proceeding with the derivations, an important remark about the notation may be useful. In Section 4.1.2, the symbol $\mathbf{q} = (\mathbf{k}' - \mathbf{k})$ has been defined as the wave-vector variation produced by the scattering. However, in this section $\mathbf{q}$ denotes the component in the transport plane of the phonon wave-vector $\mathbf{Q} = (\mathbf{q}, q_z)$. *The wave-vector change* $(\mathbf{k}'-\mathbf{k})$ *and the phonon wave-vector* $\mathbf{q}$ *are conceptually different quantities that should not be confused.* In fact, we will see in Section 4.6.2 that for the phonon scattering $(\mathbf{k}'-\mathbf{k})$ can be either $(+\mathbf{q})$ or $(-\mathbf{q})$ respectively for phonon absorption and phonon emission. A more rigorous notation would probably require introduction of the symbol $\mathbf{Q}_{ph} = (\mathbf{q}_{ph}, q_{z,ph})$ for the phonon wave-vector. However this latter notation is ponderous and unusual, in fact $\mathbf{Q} = (\mathbf{q}, q_z)$ is a standard symbol for the phonon wave-vector, which we have also adopted in this book. We hope that this explanatory remark can help avoid any confusion between the phonon wave-vector and the wave-vector change produced by phonon scattering.

4.6.1 Deformation potentials and scattering potentials

The acoustic phonon modes are closely related to mechanical strain in the crystal, which is discussed in detail in Chapter 9. In fact, propagation of an acoustic wave corresponds to an $\mathbf{R}$ and time dependent strain of the crystal, which in turn alters the electron band structure. As sketched in Fig.4.14, the local strain is associated with an energy shift of the crystal electronic bands, whereas possible changes of the band curvatures, that is of

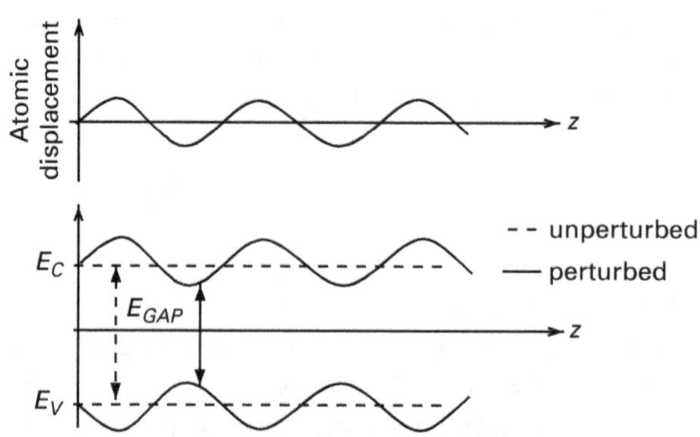

Figure 4.14 Sketch of the spatial fluctuations of the conduction and valence bands due to lattice vibrations. After Nag [48].

the effective masses, are typically neglected in calculation of the scattering rates. Such an energy shift is what we can interpret as the phonon scattering potential.

If we now consider the **R** and time dependent displacement of the unit cells described by Eq.4.232, the corresponding strain in the crystal is defined as

$$\varepsilon_{ij}(\mathbf{R},t) = \frac{1}{2}\left(\frac{\partial u_i}{\partial x_j} + \frac{\partial u_j}{\partial x_i}\right), \qquad i,j = 1,2,3 \tag{4.238}$$

where the indication of the mode (ν,**Q**) has been dropped to simplify the notation and x_1, x_2, x_3 are the three spatial coordinates (i.e. $\mathbf{R} = (x_1, x_2, x_3)$).

Equation 4.238 states that the strain is associated with the spatial variations of the displacements, that is with the differential displacements. In fact, for the acoustic modes the atoms inside a unit cell tend to move in the same direction, hence only the spatial variations of the displacements produce a strain.

As discussed in Section 9.2.2, the strain is a symmetric 3×3 matrix with only six independent components and, by substituting Eq.4.232 in Eq.4.238, we can express the strain components as

$$\varepsilon_{ij}(\mathbf{R},t) = \frac{i}{2}(e_i Q_j + e_j Q_i)\sqrt{\frac{\hbar}{2\rho\Omega\omega_{\nu,\mathbf{Q}}}}\, a_{\nu,\mathbf{Q}}\, \mathrm{e}^{\mathrm{i}(\mathbf{Q}\cdot\mathbf{R}-\omega_{\nu,\mathbf{Q}}t)} + (c.c.), \tag{4.239}$$

where e_i and Q_i are the components along x_i of the polarization vector $\hat{\mathbf{e}}_\nu$ and of the wave-vector **Q** of the phonon mode (ν,**Q**).

The basic idea behind the deformation potential model for phonon scattering is that the scattering potential for the carriers close to a given band edge is equal to $\delta E(\mathbf{R},t)$, which is defined as the shift of the band edge produced by a uniform strain equal to the local strain at the point **R** and at the time t.

The scattering potential $U_{ph}(\mathbf{R},t)$ is thus expressed in terms of the local strain by means of the 3×3 matrix Ξ_{ij} of the deformation potentials [45]

$$U_{ph}(\mathbf{R},t) = \delta E(\mathbf{R},t) = \sum_{i,j=1}^{3} \Xi_{ij}\,\varepsilon_{ij}(\mathbf{R},t), \qquad i,j = 1,2,3 \tag{4.240}$$

where ε_{ij} is given by Eq.4.239. Each band edge has its own set of deformation potentials and the matrix of the Ξ_{ij} is symmetric, so that it has at most six independent components; the crystal symmetries at the band edges typically reduce further the number of independent Ξ_{ij} components (see Section 4.6.3).

The deformation potentials of the different band edges are properties of the crystal and, in principle, they can be determined with the numerical methods used for calculation of the band structure, such as the pseudo-potentials approach. In practice, however, the deformation potentials have frequently been determined as fitting parameters to reproduce electrical as well as optical measurements (see [46] and references therein). Admittedly, the values extracted by different authors exhibit a significant spread, as can be seen in Table III of [46].

In addition to the strain due to the acoustic modes, the optical modes also produce a strain and a corresponding scattering potential. For optical lattice vibrations, atoms of the same unit cell tend to be displaced in opposite directions, hence the scattering potential is assumed to be directly proportional to the atomic displacements, rather than to their spatial derivative as for the acoustic modes [45, 47]. For a given band edge the scattering potential due to optical phonons can be expressed as

$$U_{ph}(\mathbf{R},t) = \delta E(\mathbf{R},t) = D_{op}\sqrt{\frac{\hbar}{2\rho\Omega\omega_{\nu,\mathbf{Q}}}}a_{\nu,\mathbf{Q}}\mathrm{e}^{\mathrm{i}(\mathbf{Q}\cdot\mathbf{R}-\omega_{\nu,\mathbf{Q}}t)} + (c.c.), \tag{4.241}$$

where D_{op} is a scalar optical deformation potential.

4.6.2 General formulation of the phonon matrix elements

By using Eqs.4.239 to 4.241 it is possible to write a quite general formulation of the phonon scattering matrix elements in inversion layers, which is then used in Sections 4.6.3 to 4.6.6 to discuss the different cases of practical interest. First note that, for a given phonon mode (ν,$\mathbf{Q}$), Eqs.4.240 and 4.241 can be cast in the form

$$U_{ph}(\mathbf{R},t) = D(\mathbf{Q})\sqrt{\frac{\hbar}{2\rho\Omega\omega_{\nu,\mathbf{Q}}}}a_{\nu,\mathbf{Q}}\,\mathrm{e}^{\mathrm{i}(\mathbf{Q}\cdot\mathbf{R}-\omega_{\nu,\mathbf{Q}}t)} + (c.c.), \tag{4.242}$$

where $D(\mathbf{Q})$ is

$$D(\mathbf{Q}) = \frac{i}{2}\sum_{i,j=1}^{3}\Xi_{ij}\,(e_i Q_j + e_j Q_i), \qquad \text{acoustic phonons} \tag{4.243a}$$

$$D(\mathbf{Q}) = D_{op}, \qquad \text{optical phonons} \tag{4.243b}$$

and is thus $\mathbf{Q}$ dependent for acoustic phonons.

Equation 4.242 provides the time dependent scattering potential produced by the phonon mode (ν,$\mathbf{Q}$), which can be used to calculate the matrix elements and scattering rate according to Fermi's rule described in Section 4.1.

We start by calculating the matrix element of the first term of U_{ph} in Eq.4.242, which corresponds to phonon absorption. According to Eq.4.6 or Eq.4.41 we have

$$M^{(ab)}_{n,n'}(\mathbf{k},\mathbf{k}') = \frac{D(\mathbf{Q})}{A}\sqrt{\frac{\hbar}{2\rho\Omega\omega_{\nu,\mathbf{Q}}}}a_{\nu,\mathbf{Q}}\int_z \xi^{\dagger}_{n'\mathbf{k}'}(z)\cdot\xi_{n\mathbf{k}}(z)\,\mathrm{e}^{\mathrm{i}q_z z}\mathrm{d}z \times\int_A \mathrm{e}^{\mathrm{i}\,(\mathbf{k}-\mathbf{k}'+\mathbf{q})\cdot\mathbf{r}}\,\mathbf{dr}, \tag{4.244}$$

where the total phonon wave-vector $\mathbf{Q} = (\mathbf{q}, q_z)$ has been split into the components $\mathbf{q}$ and q_z in the transport plane and quantization direction, respectively. For the vectorial wave-functions of a $\mathbf{k}\cdot\mathbf{p}$ Hamiltonian the dot sign in Eq.4.244 denotes the scalar product defined in Eq.4.40. Since we have used Eq.4.6, then, in the case of an electron inversion layer, the results of the present section are directly applicable only to intra-valley transitions.

As discussed in Appendix B, when the $\mathbf{q}$ and $\mathbf{k}$ values are given by Eq.3.3 the integral over $\mathbf{r}$ in Eq.4.244 gives

$$\frac{1}{A}\int_A e^{i(\mathbf{k}-\mathbf{k}'+\mathbf{q})\cdot\mathbf{r}}\, d\mathbf{r} = \delta_{\mathbf{k}',(\mathbf{k}+\mathbf{q})}, \tag{4.245}$$

where the Kronecker $\delta_{\mathbf{k}',(\mathbf{k}+\mathbf{q})}$ is one for $\mathbf{q} = (\mathbf{k}' - \mathbf{k})$ and zero otherwise.

We now introduce the $\mathbf{k}$ space form factor

$$G_{n'\mathbf{k}',n\mathbf{k}}(q_z) = \frac{1}{2\pi}\int_z \xi^{\dagger}_{n'\mathbf{k}'}(z)\cdot\xi_{n\mathbf{k}}(z)\, e^{iq_z z} dz = F_z\{\xi^{\dagger}_{n'\mathbf{k}'}\cdot\xi_{n\mathbf{k}}\}, \tag{4.246}$$

which is the Fourier transform with respect to z of the function $(\xi^{\dagger}_{n'\mathbf{k}'}\cdot\xi_{n\mathbf{k}})$, and accordingly rewrite the matrix element as

$$M^{(ab)}_{n,n'}(\mathbf{k},\mathbf{k}') = (2\pi)\,\delta_{\mathbf{k}',(\mathbf{k}+\mathbf{q})}\, D(\mathbf{Q})\sqrt{\frac{\hbar}{2\rho\Omega\omega_{\nu,\mathbf{Q}}}}\, a_{\nu,\mathbf{Q}}\, G_{n'\mathbf{k}',n\mathbf{k}}(q_z). \tag{4.247}$$

We now note that the total matrix element produced by the lattice vibrations is obtained by summing the contribution of Eq.4.247 over all the phonon modes (ν,$\mathbf{Q}$), where ν identifies the phonon branch and the $\mathbf{Q} = (\mathbf{q}, q_z)$ values must vary inside the first Brillouin zone. Differently from the phonons representing a macroscopic acoustic wave propagating in the crystal, the thermal phonons have no phase correlation, so that, by invoking a random phase approximation, the overall squared matrix element is simply obtained by summing the squared matrix elements of all the modes [1].

For a given phonon branch ν, we thus need to sum the squared matrix elements given by Eq.4.247 over all the wave-vectors $\mathbf{Q}$. The Kronecker delta $\delta_{\mathbf{k}',(\mathbf{k}+\mathbf{q})}$ selects only the term with $\mathbf{q} = (\mathbf{k}' - \mathbf{k})$ in the summation over $\mathbf{q}$. If we then use the standard prescription

$$\sum_{q_z} = \frac{L}{2\pi}\int_{q_z}$$

to convert the summation over q_z to an integral, the squared matrix element for a given mode ν can finally be written as

$$|M^{(ab)}_{n,n'}(\mathbf{k},\mathbf{k}')|^2 = \delta_{\mathbf{k}',(\mathbf{k}+\mathbf{q})}\frac{\pi\hbar}{\rho A}\int_{q_z}\frac{|D(\mathbf{Q})|^2}{\omega_{\nu,\mathbf{Q}}}\,|a_{\nu,\mathbf{Q}}|^2\,|G_{n'\mathbf{k}',n\mathbf{k}}(q_z)|^2\, dq_z, \tag{4.248}$$

where $\mathbf{Q} = (\mathbf{q}, q_z)$ is the phonon wave-vector and the symbol $M^{(ab)}_{n,n'}$ indicates that the matrix element produced by the first term in Eq.4.242 corresponds to the absorption of a phonon. Note that A is the normalization area such that $\Omega = LA$.

The squared matrix elements for the second term of the $U_{ph}(\mathbf{R}, t)$, denoted by (c.c.) in Eq.4.242, are obtained with a derivation entirely similar to the one that led to Eq.4.248. The final result is

$$|M^{(em)}_{n,n'}(\mathbf{k},\mathbf{k}')|^2 = \delta_{\mathbf{k}',(\mathbf{k}-\mathbf{q})}\frac{\pi\hbar}{\rho A}\int_{q_z}\frac{|D(\mathbf{Q})|^2}{\omega_{\nu,\mathbf{Q}}}\,|a^{\dagger}_{\nu,\mathbf{Q}}|^2\,|G_{n'\mathbf{k}',n\mathbf{k}}(-q_z)|^2\, dq_z, \tag{4.249}$$

and it corresponds to emission of a phonon.

Equations 4.248 and 4.249 deserve several comments. The first point to note is that, if we express the squared magnitude of $a_{\nu,\mathbf{Q}}$ according to Eq.4.231, then both absorption and emission are proportional to $(n_{\nu,\mathbf{Q}}+0.5)$, where $n_{\nu,\mathbf{Q}}$ is the phonon number for the mode $(\nu,\mathbf{Q})$ given by Eq.4.236.

This result is not fully correct and it stems from the fact that, for the electron–phonon interaction, we have used a scattering potential where the number of phonons has been introduced phenomenologically, that is by equating the energy of the lattice vibrations obtained by using either a classical or a quantum mechanical model (see Section 4.5.2). A rigorous quantum treatment of the electron–phonon interaction requires use of a phonon Hamiltonian (rather than just a phonon scattering potential), which must be written in terms of creation and annihilation operators; for more details about this point the interested reader should refer to [1, 45]. We limit ourselves here to giving the final result consisting in the fact that the squared matrix element for phonon absorption is proportional to $n_{\nu,\mathbf{Q}}$, whereas the squared matrix element for phonon emission is proportional to $(n_{\nu,\mathbf{Q}}+1)$.

This leads us to rewrite Eqs.4.248 and 4.249 as

$$|M^{(ab)}_{n,n'}(\mathbf{k},\mathbf{k}')|^2 = \delta_{\mathbf{k}',(\mathbf{k}+\mathbf{q})} \frac{\pi\hbar}{\rho A} \int_{q_z} \frac{|D(\mathbf{Q})|^2}{\omega_{\nu,\mathbf{Q}}} n_{\nu,\mathbf{Q}}\, |G_{n'\mathbf{k}',n\mathbf{k}}(q_z)|^2\, dq_z, \tag{4.250}$$

and

$$|M^{(em)}_{n,n'}(\mathbf{k},\mathbf{k}')|^2 = \delta_{\mathbf{k}',(\mathbf{k}-\mathbf{q})} \frac{\pi\hbar}{\rho A} \int_{q_z} \frac{|D(\mathbf{Q})|^2}{\omega_{\nu,\mathbf{Q}}} (n_{\nu,\mathbf{Q}}+1)\, |G_{n'\mathbf{k}',n\mathbf{k}}(-q_z)|^2\, dq_z, \tag{4.251}$$

where it should be noted that the phonon number $n_{\nu,\mathbf{Q}}$ in general depends on $\mathbf{Q} = (\mathbf{q}, q_z)$ through Bose–Einstein statistics, Eq.4.236.

Starting from Eqs.4.250 and 4.251, in the following sub-sections of Section 4.6 we discuss the formulation of the phonon scattering rates for the cases of most practical interest in silicon inversion layers.

4.6.3 Electron intra-valley scattering by acoustic phonons

According to the deformation potential theory, scattering by acoustic phonons is closely related to calculation of band edge shifts in the presence of strain. Herring and Vogt studied the deformation potentials Ξ_{ij} for the Γ, the Λ, and the Δ minima of the conduction band in detail; Table 4.2 gives the set of deformation potentials proposed in [49]. For the Λ and Δ valleys the deformation potentials Ξ_{ij} are always defined in the principal coordinate system of the ellipsoids that describe the energy dispersion close to the conduction band minima.

As can be seen in Table 4.2, for a non-degenerate minimum of the conduction band at the Γ point, the matrix of the deformation potentials reduces to the scalar $\Xi_{d\Gamma}$. In such a case, representative of the lowest conduction band for GaAs, Eqs. 4.239 and 4.243a show that the shear strain components (namely the ε_{ij} components with $i \neq j$) do not

Table 4.2 Deformation potentials for the conduction band minima for the Γ, Δ, and Λ valleys. In each case the components of the deformation potentials are related to a system of axes that coincide with the principal axes of the ellipsoidal or spheroidal minima; in particular the x_1 axis is a $\langle 100\rangle$ direction for the Γ and Δ valleys and a $\langle 111\rangle$ direction for the Λ valleys.

	Γ Valley	Δ Valleys	Λ Valleys
Ξ_{11}	$\Xi_{d\Gamma}$	$\Xi_{dX}+\Xi_{uX}$	$\Xi_{dL}+\frac{1}{3}\,\Xi_{uL}$
Ξ_{22}	$\Xi_{d\Gamma}$	Ξ_{dX}	$\Xi_{dL}+\frac{1}{3}\,\Xi_{uL}$
Ξ_{33}	$\Xi_{d\Gamma}$	Ξ_{dX}	$\Xi_{dL}+\frac{1}{3}\,\Xi_{uL}$
Ξ_{23}	0	0	$\frac{1}{3}\,\Xi_{uL}$
Ξ_{31}	0	0	$\frac{1}{3}\,\Xi_{uL}$
Ξ_{12}	0	0	$\frac{1}{3}\,\Xi_{uL}$

contribute to the scattering potential. Furthermore, Eqs.4.242 and 4.243a can be used to express $U_{ph}(\mathbf{R},t)$ as

$$U_{ph}(\mathbf{R},t)=i\;\Xi_{d\Gamma}\,\hat{\mathbf{e}}_\nu\cdot\mathbf{Q}\sqrt{\frac{\hbar}{2\rho\Omega\omega_{\nu,\mathbf{Q}}}}\;a_{\nu,\mathbf{Q}}e^{\mathrm{i}\mathbf{Q}\cdot\mathbf{R}-\omega_{\nu,\mathbf{Q}}t}+(c.c.). \tag{4.252}$$

The scalar product between the polarization vector $\hat{\mathbf{e}}_\nu$ and $\mathbf{Q}$ in Eq.4.252 indicates that the transverse modes result in a null scattering potential for a conduction band minimum at the Γ point. In other words, only the longitudinal modes can produce scattering, in which case Eq.4.252 yields

$$U_{ph}(\mathbf{R},t)=i\;\Xi_{d\Gamma}\;Q\sqrt{\frac{\hbar}{2\rho\Omega\omega_{\nu,\mathbf{Q}}}}\;a_{\nu,\mathbf{Q}}e^{\mathrm{i}\mathbf{Q}\cdot\mathbf{R}-\omega_{\nu,\mathbf{Q}}t}+(c.c.), \tag{4.253}$$

because $\hat{\mathbf{e}}_\nu$ is aligned with $\mathbf{Q}$.

According to Table 4.2, the stress induced energy shifts of the Δ valleys located along the $\langle 100\rangle$ directions can be expressed as

$$\frac{1}{2}\sum_{i,j=1}^{3}\Xi_{ij}\,(e_iQ_j+e_jQ_i)=\Xi_{dX}\,\hat{\mathbf{e}}_\nu\cdot\mathbf{Q}+\Xi_{uX}e_1Q_1, \tag{4.254}$$

where x_1 is the direction of the longitudinal axis of the energy ellipsoids (namely a $\langle 100\rangle$ direction), and Ξ_{dX} and Ξ_{uX} are the so called dilation and uniaxial shear deformation potentials, respectively.

Equation 4.254 shows that at the Δ minima the transverse modes (i.e. those with $\hat{\mathbf{e}}_\nu\cdot\mathbf{Q}=0$) can also produce a non-null matrix element. Herring and Vogt showed that the term $D(\mathbf{Q})$ entering the squared matrix element for the acoustic phonon scattering can be expressed as [49]

$$D(\mathbf{Q})=Q\,[\Xi_{dX}+\Xi_{uX}\cos^2(\theta_{\mathbf{Q}})], \tag{4.255}$$

for the longitudinal modes and

$$D(\mathbf{Q})=Q\;\Xi_{uX}\sin(\theta_{\mathbf{Q}})\cos(\theta_{\mathbf{Q}}), \tag{4.256}$$

for the transverse modes, where $\theta_{\mathbf{Q}}$ is the angle between the wave-vector $\mathbf{Q}$ and the longitudinal axis of the valley.

The insertion of Eqs.4.255 and 4.256 in the general expression for the matrix elements derived in the previous section leads to a quite complicated formulation, where the matrix element depends not only on the magnitude but also on the direction of the wave-vector $\mathbf{Q} = (\mathbf{q}, q_z)$. Several approximations are typically introduced to obtain the formulations of the acoustic phonon scattering rates most frequently used in electron device modeling.

Elastic and energy equipartition approximation

A first step towards simplification of the matrix elements for acoustic intra-valley transitions is an analysis of the phonon energy which is exchanged with the electrons in a scattering event. To this purpose we first note that, in calculations of phonon scattering rates obtained by inserting Eqs.4.250 and 4.251 in Eq.4.52, the Kronecker symbol $\delta_{\mathbf{k}',(\mathbf{k}\pm\mathbf{q})}$ enforces conservation of the component in the transport plane of the electron momentum, while the Dirac function imposes energy conservation. This can be mathematically expressed as

$$\hbar\mathbf{k}' = \hbar\mathbf{k} \pm \hbar\mathbf{q} \tag{4.257}$$

for the momentum, and as

$$E_{w,n'}(\mathbf{k}') = E_{w,n}(\mathbf{k}) \pm \hbar\omega_{\nu,\mathbf{Q}} \tag{4.258}$$

for the energy, where the upper sign is for phonon absorption and the lower for phonon emission. The valley w to which the subbands n, n' belong has been explicitly indicated in Eq.4.258. Let us now consider a parabolic, circular energy dispersion

$$E_{w,n}(\mathbf{k}) = E_{w0} + \varepsilon_n + \frac{\hbar^2 k^2}{2m}, \tag{4.259}$$

and take the scalar product of Eq.4.257 with itself, which provides

$$(\hbar k')^2 = (\hbar k)^2 + (\hbar q)^2 \pm 2\hbar^2 kq\cos(\theta_{kq}), \tag{4.260}$$

where θ_{kq} is the angle between $\mathbf{k}$ and $\mathbf{q}$.

If we now consider an intra-subband transition (i.e. $n = n'$), then by substituting Eq.4.259 and Eq.4.260 in Eq.4.258 we obtain an equation that accounts for both the energy and the crystal momentum conservation

$$q = \mp 2k\left[\cos(\theta_{kq}) - \frac{\omega_{\nu,\mathbf{Q}}/q}{(\hbar k/m)}\right]. \tag{4.261}$$

By using Eq.4.235a for the frequency $\omega_{\nu,\mathbf{Q}}$ of the acoustic phonons and noting that $(\hbar k/m)$ is the electron velocity, v_{el}, Eq.4.261 can be rewritten

$$q = \mp 2k\left[\cos(\theta_{kq}) - \frac{v_s}{v_{el}}\frac{Q}{q}\right], \tag{4.262}$$

where $Q=\sqrt{q^2+q_z^2}$ is the magnitude of the phonon wave-vector.

Since the magnitude q of the **Q** component in the transport plane must be positive, Eq.4.262 identifies the allowed angles θ_{kq} between **k** and **q** for phonon absorption and emission. In particular we see that, in order for phonon emission to be possible (lower sign), the electron must have a large enough velocity $v_{el} > v_s Q/q$.

The maximum q value corresponds to phonon absorption (upper sign) with a backward redirection of the electron, that is for $\cos(\theta_{kq}) = -1$. In such a case we have

$$q_{max} = 2k \left[1 + \frac{v_s}{v_{el}} \frac{Q}{q} \right]. \tag{4.263}$$

The sound velocity in silicon, v_s, is 4.7 and $9.2 \cdot 10^5$ cm/s for the transverse and the longitudinal acoustic phonon modes, respectively (see Fig.4.13), hence the electron velocity at the thermal energy is typically much larger than v_s. For example, the unprimed subbands with a transport effective mass $m = 0.19 m_0$ have a velocity at the thermal energy $v_{th} = \sqrt{2K_B T/m} \simeq 2.2 \cdot 10^7$ cm/s at room temperature.

Since v_{el} is typically much larger than v_s, then according to Eq.4.263 the maximum q value is approximately $2k$. The phonon energy $\hbar\omega_{\nu,\mathbf{Q}}$ for $q = 2k$ and the corresponding electron kinetic energy, E_{el}, are

$$\hbar\omega_{\nu,\mathbf{Q}} = \hbar v_s Q = 2\hbar\, v_s \sqrt{k^2 + q_z^2/4}, \qquad E_{el} = \frac{\hbar k}{2}\, v_{el}. \tag{4.264}$$

Equation 4.264 suggests that $\hbar\omega_{\nu,\mathbf{Q}}$ is typically small compared to E_{el}, because v_s is much smaller than v_{el}. By considering again the example of the unprimed subbands, at the thermal energy we have $E_{el} = K_B T \simeq 26$ meV and Eq.4.264 provides $\hbar\omega_{\nu,\mathbf{Q}} \simeq$ 4meV for $q_z = 0$. As discussed above, this is the maximum possible value for $\hbar\omega_{\nu,\mathbf{Q}}$, and, statistically speaking, most of the scattering events will involve phonons with an energy well below 4 meV. On the basis of these arguments, for intra-valley acoustic phonon scattering the phonon energy is often neglected with respect to the electron energy and the scattering is thus considered as approximately elastic.

It should be noted, however, that the range of q_z values important for the overall scattering rate is set by Eqs.4.250 and 4.251, and the stronger are the effects of quantum confinement the larger are the q_z values, as clarified in Section 4.6.3. The phonon energy $\hbar\omega_{\nu,\mathbf{Q}}$ increases with q_z, so that the elastic approximation may become questionable in the case of very strong quantization.

If we now adopt the elastic approximation, then we can consistently assume that $\hbar\omega_{\nu,\mathbf{Q}}$ is small compared to $K_B T$ and express the number of phonons by using the energy equipartition approximation Eq.4.237. By doing so, for the longitudinal modes in Δ valleys we obtain

$$\frac{|D(\mathbf{Q})|^2}{\omega_{\nu,\mathbf{Q}}} n_{\nu,\mathbf{Q}} \simeq \frac{K_B T}{\hbar v_s^2} \left[\Xi_{dX} + \Xi_{uX} \cos^2(\theta_{\mathbf{Q}})\right]^2, \tag{4.265}$$

where Eq.4.255 has been used to express $|D(\mathbf{Q})|$.

Isotropic acoustic phonon approximation

Equation 4.265 expresses the first two terms that enter the integral over q_z in Eq.4.250 and it shows that, thanks to the elastic and energy equipartition approximation, such terms depend on q_z only through the direction of the $\mathbf{Q}$ vector, that is through the angle $\theta_{\mathbf{Q}}$. Such a dependence on $\theta_{\mathbf{Q}}$ stems from the non-scalar nature of the deformation potentials, which makes the energy shifts at the minimum of a Δ valley depend on the $\mathbf{Q}$ direction, as can be seen in Eq.4.254. Such an anisotropy of the problem is very often neglected in analysis of electron devices, by introducing an effective value of the deformation potential independent of $\theta_{\mathbf{Q}}$; this is known as the isotropic acoustic phonon approximation.

The transverse acoustic modes result in an equation similar to Eq.4.265, but with the $\theta_{\mathbf{Q}}$ dependence given by Eq.4.256. If we neglect such a $\theta_{\mathbf{Q}}$ dependence, the longitudinal and transverse modes can be collectively described by using an effective deformation potential, D_{ac}, for the intra-valley acoustic phonons and an average value for the sound velocity, v_s, which is somewhat different for transverse and longitudinal modes (see Fig.4.13).

By exploiting the elastic energy equipartition and the isotropic approximation, Eq.4.250 allows us to write the matrix element for the intra-valley acoustic phonon absorption as

$$|M^{(ab)}_{w,n,n'}(\mathbf{k},\mathbf{k}')|^2 = \delta_{\mathbf{k}',(\mathbf{k}+\mathbf{q})} \frac{\pi K_B T D_{ac}^2}{\rho A v_s^2} \int_{q_z} |G^{(w)}_{n',n}(q_z)|^2 \, \mathrm{d}q_z, \tag{4.266}$$

where $G^{(w)}_{n',n}(q_z)$ is defined, consistently with Eq.4.246, as

$$G^{(w)}_{n',n}(q_z) = \frac{1}{2\pi} \int_z \xi^{\dagger}_{w,n'}(z)\xi_{w,n}(z)\, \mathrm{e}^{\mathrm{i}q_z z}\mathrm{d}z = \mathcal{F}_z\{\xi^{\dagger}_{w,n'}(z)\xi_{w,n}(z)\}. \tag{4.267}$$

As can be seen, $G^{(w)}_{n',n}$ is the Fourier transform of the function $[\xi^{\dagger}_{w,n'}\xi_{w,n}]$, where w indicates the valley to which the subbands n and n' belong, and, according to the EMA quantization model, the electron envelope wave-functions are independent of the wave-vector $\mathbf{k}$ in the transport plane.

By recalling Eq.A.12, we can then introduce the form factor in real space:

$$F^{(w)}_{n',n} = 2\pi \int_{q_z} |G^{(w)}_{n',n}(q_z)|^2 \, \mathrm{d}q_z = \int_z |\xi^{\dagger}_{w,n'}(z)\xi_{w,n}(z)|^2 \mathrm{d}z, \tag{4.268}$$

and finally rewrite Eq.4.266 as

$$|M^{(ab)}_{w,n,n'}(\mathbf{k},\mathbf{k}')|^2 = \delta_{\mathbf{k}',(\mathbf{k}+\mathbf{q})} \frac{K_B T D_{ac}^2}{2\rho\, A\, v_s^2} F^{(w)}_{n',n}. \tag{4.269}$$

By starting from Eq.4.251, we can obtain an expression for the matrix element of the phonon emission very similar to Eq.4.269. In this respect, we notice that in the emission matrix element we have a proportionality to $(n_{\nu,\mathbf{Q}}+1)$ rather than to $n_{\nu,\mathbf{Q}}$, but the energy equipartition approximation implies that

$$(n_{\nu,\mathbf{Q}} + 1) \simeq n_{\nu,\mathbf{Q}} \simeq \frac{K_B T}{\hbar Q\, v_s}, \tag{4.270}$$

so that for the phonon emission we obtain

$$|M^{(em)}_{w,n,n'}(\mathbf{k},\mathbf{k}')|^2 = \delta_{\mathbf{k}',(\mathbf{k}-\mathbf{q})} \frac{K_B T D^2_{ac}}{2\rho\, A\, v_s^2} F^{(w)}_{n',n}. \tag{4.271}$$

It should be noted that in Eqs.4.269 and 4.271 the Kronecker symbols $\delta_{\mathbf{k}',(\mathbf{k}\pm\mathbf{q})}$ serve only to select the correct $\mathbf{q}$ and $\mathbf{Q}$ values for the phonon wave-vector corresponding respectively to the absorption and the emission process. A knowledge of $\mathbf{Q}$ is necessary to perform the integration over q_z in the general formulation of the phonon matrix elements given in Eqs.4.250 and 4.251, however, in Eqs.4.269 and 4.271 there is no longer a dependence on $\mathbf{Q}$, hence the Kronecker symbols are dropped in the rest of this section. The scattering rate can now be readily obtained from Eq.4.43 as

$$S^{(w)}_{n,n'}(\mathbf{k},\mathbf{k}') = \frac{\pi\, K_B T D^2_{ac}}{\rho\, A\, \hbar\, v_s^2} F^{(w)}_{n',n}\, \delta[E_n(\mathbf{k}) - E_{n'}(\mathbf{k}')] + \frac{\pi\, K_B T D^2_{ac}}{\rho\, A\, \hbar\, v_s^2} F^{(w)}_{n',n}\, \delta[E_n(\mathbf{k}) - E_{n'}(\mathbf{k}')], \tag{4.272}$$

where n and n' are subbands of the valley w, namely two unprimed Δ_2 subbands or two primed Δ_4 subbands.

As a result of the approximations adopted to derive Eq.4.272, the scattering rates for acoustic phonon absorption and emission are the same. Hence Eq.4.272 can finally be written as

$$S^{(w)}_{n,n'}(\mathbf{k},\mathbf{k}') = \frac{2\pi\, K_B T D^2_{ac}}{\rho\, A\, \hbar\, v_s^2} F^{(w)}_{n',n}\, \delta[E_n(\mathbf{k}) - E_{n'}(\mathbf{k}')]. \tag{4.273}$$

In this formulation the intra-valley acoustic phonon scattering is thus elastic and isotropic, so that it is straightforward to perform the integration over the final states prescribed by Eq.4.52 to obtain the overall scattering rate $S_n(\mathbf{k})$ of the initial state (n,$\mathbf{k}$). By recalling Eq.4.54 we obtain

$$S^{(w)}_{n}(\mathbf{k}) = S^{(w)}_{n}(E_n(\mathbf{k})) = \frac{2\pi\, K_B T D^2_{ac}}{\rho\, \hbar\, v_s^2} \sum_{n'} F^{(w)}_{n',n}\, g_{w,n'}(E_n(\mathbf{k})), \tag{4.274}$$

where n, n' are subbands of the valley w and $g_{w,n'}(E)$ is the electron density of states defined in Eq.3.75 for $n_{sp} = 1$ (because the scattering does not change the spin).

As can be seen, the scattering rate depends on $\mathbf{k}$ only through the energy of the initial state $E = E_{w,n}(\mathbf{k})$, which is a typical property of isotropic scattering mechanisms.

Physical interpretation of the form factors

Equation 4.274 shows that the electron intra-valley transitions assisted by acoustic phonons are governed by the form factors $F^{(w)}_{n',n}$ given by Eq.4.268, which are in turn defined in terms of the $\mathbf{k}$ space form factors $G^{(w)}_{n',n}(q_z)$ defined in Eq.4.246. In the rest of this sub-section we drop the indication of the valley w to simplify the notation.

For an electron inversion layer the wave-functions $\xi_n(z)$ are scalar and independent of $\mathbf{k}$ and the definition of $G_{n',n}(q_z)$ provides a physically transparent interpretation of its role in the electron–phonon interaction. In fact, according to Eq.A.10, Eq.4.246 can be rewritten as

$$G_{n',n}(q_z) = \mathcal{F}\{\xi_{n'}^{\dagger}(z)\xi_n(z)\} = \int_{-\infty}^{\infty} \chi_n(q'_z)\,\chi_{n'}^{\dagger}(q'_z - q_z)\,\mathrm{d}q'_z, \tag{4.275}$$

where $\chi_n(q_z)$ is the Fourier transform of $\xi_n(z)$.

As schematically illustrated in Fig.4.15, the q_z enforces a shift between $\chi_n(q'_z)$ and $\chi_{n'}^{\dagger}(q'_z)$ in the convolution integral of Eq.4.275. Hence, for a given χ_n and $\chi_{n'}^{\dagger}$, there is a q_z range that corresponds to the largest values of the integral over q'_z. Furthermore, for large enough q_z values the integral over q'_z tends to vanish.

The above arguments explain that the $G_{n',n}(q_z)$ is essentially a measure of how effectively a phonon with wave-vector $\mathbf{Q} = (\mathbf{q}, q_z)$ can couple with the electron gas. More precisely, while the Kronecker symbols $\delta_{\mathbf{k}',(\mathbf{k}\pm\mathbf{q})}$ affirm the conservation of the electron wave-vector component in the transport plane, so that $\mathbf{q}$ is unequivocally determined by $\mathbf{k}$ and $\mathbf{k}'$, the quantum confinement in the z direction implies that the phonons that produce significant scattering have a range of possible q_z values, as opposed to a single q_z value.

Thus the $\mathbf{k}$ space form factors $|G_{n',n}(q_z)|$ express the effectiveness of the phonon coupling for each q_z value. Since for the acoustic intra-valley phonons in the elastic, isotropic, energy equipartition approximation, the $|G_{n',n}(q_z)|$ are the only terms which depend on q_z in the integrals of Eqs.4.250 and 4.251, under these approximations it becomes convenient to introduce the real space form factors $F_{n',n}$ defined in Eq.4.268.

For intra-subband transitions the form-factors $F_{n,n}$ can be related to the inverse of the effective width of the squared magnitude $|\xi_n(z)|^2$ of the wave-function, which means that the stronger is the confinement the larger are the $F_{n,n}$ values. Practically speaking, thin silicon films in SOI transistors and large confining electric fields (in the direction normal to the transport plane) result in large form factors and phonon scattering rates.

These latter considerations are consistent with the interpretation of the form factors in the q_z space. In fact, the narrower the confinement in real space, the broader is the Fourier transform $\chi_n(q_z)$ in the q_z space, which results in a larger number of phonons able to produce an appreciable scattering (see Fig.4.15).

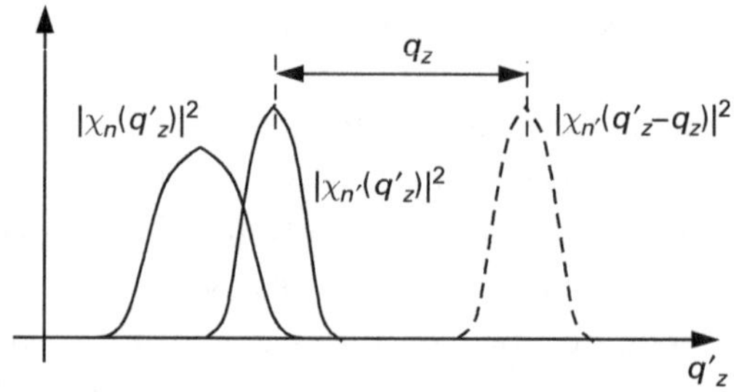

Figure 4.15 The role played by q_z in determination of the $\mathbf{k}$ space form factors defined in Eqs.4.246 and 4.275.

Beyond the elastic equipartition energy approximation

The elastic, the energy equipartition and the isotropic approximations for intra-valley acoustic phonon scattering have been very widely used for analysis of electron devices and represent quite reliable simplifications, at least for not too low operating temperatures [4–6, 44, 50–53].

At low temperatures the electron thermal energy $K_B T$ may become comparable to the phonon energy, so that neither the elastic nor the energy equipartition approximation is very accurate. Furthermore, from the viewpoint of Monte Carlo simulations, there is one more reason why one should not neglect the energy of the acoustic phonons in the corresponding scattering events. In fact, as will be clarified in Chapter 6, in order to obtain the correct steady-state electron distribution in a low field transport simulation, we need a scattering mechanism able to exchange an arbitrarily small energy with the electron gas; such a scattering mechanism is intra-valley acoustic phonon scattering.

Of course, even optical phonons can exchange energy with the electron gas, however, the corresponding energies are essentially fixed values which can be large compared to the electron thermal energy (see Fig.4.13). In the case of relatively large temperatures or high electric fields the electrons may gain relatively high energies and the optical phonons may be sufficient to effectively enforce the steady-state electron distribution. At very low temperatures and at small electric fields, the inelasticity of the acoustic phonons plays a key role instead.

In an inelastic formulation of acoustic phonon scattering, calculation of the matrix elements according to Eqs.4.250 and 4.251 must be performed by properly accounting for the q_z dependence of all the terms inside the integral. To be more specific, the value of $\mathbf{q}$ is given by $\pm(\mathbf{k}'-\mathbf{k})$ for absorption and emission, respectively, and the magnitude of the total phonon wave-vector is $Q = \sqrt{q^2 + q_z^2}$. The phonon energy $\hbar\omega_{\nu,\mathbf{Q}}$ and the phonon number $n_{\nu,\mathbf{Q}}$ can then be readily expressed by using respectively Eq.4.235a and 4.236. The value of $|D(\mathbf{Q})|$ can finally be expressed by means of Eqs.4.255 and 4.256 (if we wish to account for the dependence on the $\mathbf{Q}$ direction of the phonon matrix element), or through the isotropic approximation already discussed in the previous section. In any case, the integrals over q_z in Eqs.4.250 and 4.251 must be performed numerically.

We finally note that in the inelastic treatment of acoustic phonon scattering the absorption and emission must be considered separately. Once $|M^{(ab)}_{w,n,n'}(\mathbf{k},\mathbf{k}')|^2$ and $|M^{(em)}_{w,n,n'}(\mathbf{k},\mathbf{k}')|^2$ have been calculated with Eqs.4.250 and 4.251, the scattering rates must be expressed by means of Eq.4.52. In practice, the sums over $\mathbf{k}'$ involving the Dirac functions can be evaluated according to Eq.3.78 (with $n_{sp} = 1$), that is by resorting to appropriate integrals over the angle of the wave-vector $\mathbf{k}'$.

4.6.4 Electron intra-valley scattering by optical phonons

For optical phonon intra-valley scattering transitions the phonon energy is practically constant, that is $\hbar\omega_{\nu,\mathbf{Q}} = \hbar\omega_0$, and, furthermore, $|D(\mathbf{Q})| = D_{op}$ is independent of $\mathbf{Q}$, hence of q_z. The only term that depends on q_z in Eqs.4.250 and 4.251 is thus the $G^{(w)}_{n',n}(q_z)$ defined in Eq.4.267, so that the derivation of the matrix element and scattering rate is very similar to the one illustrated in detail in Section 4.6.3. In particular, it is convenient to introduce the form factor $F^{(w)}_{n',n}$, defined in Eq.4.268. Then, by using the $F^{(w)}_{n',n}$ definition and substituting $\hbar\omega_{\nu,\mathbf{Q}} = \hbar\omega_0$ and $|D(\mathbf{Q})| = D_{op}$ in Eqs.4.250 and 4.251, the matrix element for optical phonon absorption becomes

$$|M^{(ab)}_{w,n,n'}(\mathbf{k},\mathbf{k}')|^2 = \delta_{\mathbf{k}',(\mathbf{k}+\mathbf{q})}\,\frac{\hbar D^2_{op}}{2\omega_0\,\rho\,A}\,F^{(w)}_{n',n}\,n_{op}, \tag{4.276}$$

whereas for the corresponding emission process we have

$$|M^{(em)}_{w,n,n'}(\mathbf{k},\mathbf{k}')|^2 = \delta_{\mathbf{k}',(\mathbf{k}-\mathbf{q})}\,\frac{\hbar D^2_{op}}{2\omega_0\,\rho\,A}\,F^{(w)}_{n',n}\,(n_{op}+1), \tag{4.277}$$

where n_{op} is the phonon number given by Eq.4.236 with $\hbar\omega_{\nu,\mathbf{Q}} = \hbar\omega_0$.

The scattering rate from the state $(n,\mathbf{k})$ to the state $(n',\mathbf{k}')$ in the valley w is given by Eq.4.43 as

$$\begin{aligned} S^{(w)}_{n,n'}(\mathbf{k},\mathbf{k}') = {} & \frac{\pi D^2_{op}}{\omega_0\,\rho\,A}\,F^{(w)}_{n',n}\,n_{op}\;\delta[E_{w,n}(\mathbf{k}) - E_{w,n'}(\mathbf{k}') + \hbar\omega_0] \\ & + \frac{\pi D^2_{op}}{\omega_0\,\rho\,A}\,F^{(w)}_{n',n}\,(n_{op}+1)\;\delta[E_{w,n}(\mathbf{k}) - E_{w,n'}(\mathbf{k}') - \hbar\omega_0]. \end{aligned} \tag{4.278}$$

Furthermore, since the resulting formulation of the optical phonon scattering rate is isotropic, the overall scattering rate for a given initial state is obtained from Eq.4.54 as

$$S^{(w)}_{n,n'}(\mathbf{k},\mathbf{k}') = \frac{\pi D^2_{op}}{\omega_0\,\rho}\sum_{n'} F^{(w)}_{n',n}\left[n_{op} + \frac{1}{2} \mp \frac{1}{2}\right] g_{w,n'}(E_{w,n}(\mathbf{k}) \pm \hbar\omega_0), \tag{4.279}$$

where $g_{w,n}(E)$ is the electron density of states defined in Eq.3.75 with $n_{sp} = 1$ (because the scattering does not change the spin).

Equation 4.279 is written with the frequently used concise notation that indicates with the upper and lower signs the expressions corresponding to phonon absorption and phonon emission, respectively. As can be seen, the scattering rate depends only on the energy $E_{w,n}(\mathbf{k})$ of the initial state; furthermore, for a given $E_{w,n}(\mathbf{k})$ and a given final subband n', the $g_{w,n'}(E)$ imposes the constraint that the scattering processes are possible only if the final energy $(E_{w,n}(\mathbf{k}) \pm \hbar\omega_0)$ is larger than the minimum of the n' subband. Thus, optical phonon emissions are prohibited for electrons with an energy smaller than $(E_{\nu 0} + \varepsilon_{n'} + \hbar\omega_0)$, whereas the same electrons may very well experience phonon absorptions. In this respect it should be noted that, according to Eq.4.236, the phonon occupation number n_{op} can be much smaller than 1 if the phonon energy $\hbar\omega_0$ is larger than the thermal energy. Consequently the rate of phonon emission in Eq.4.279 may be much larger than the rate for phonon absorption (for the electron energies at which both processes are allowed).

It is worth mentioning that the intra-valley transitions assisted by optical phonons may be prohibited for the valleys located at the high symmetry points or along the high symmetry directions of the bulk crystal Brillouin zone. On the basis of group theory considerations, the selection rules for optical phonon-assisted intra-valley transitions have been studied for the most relevant valleys of common semiconductors and they are discussed in Section 4.6.7.

4.6.5 Electron inter-valley phonon scattering

Electron transitions between subbands belonging to different valleys can be assisted by both acoustic and optical phonons. It is important to understand that in such inter-valley transitions the phonon wave-vector $\mathbf{Q}$ is essentially set by the distance of the valleys in the first Brillouin zone, in fact the displacements of the electron wave-vectors from the valley minima are typically small compared to the distance between the valleys.

To be more precise about the electron and phonon wave-vectors involved in inter-valley transitions, we note that in Eqs.4.250 and 4.251 the Kronecker symbols set the condition $\mathbf{q} = \pm(\mathbf{k}' - \mathbf{k})$ stemming from the form of the envelope wave-function given in Eq.4.5. Thus the $\mathbf{k}$ values in Eqs.4.250 and 4.251 are the wave-vector displacements from the valley minima.

A more detailed treatment of the electron–phonon scattering interaction should consider the complete wave-function in the electron inversion layer Eq.4.8 (see also Section 3.2.1) as well as the phonon Hamiltonian [1, 4, 45], rather than just a phonon scattering potential. For a bulk crystal the intra and inter-valley phonon scattering processes are described in several textbooks of solid state physics and review papers [4, 36, 37, 45], where it is shown that the general condition to identify the $\mathbf{Q}$ of a given phonon assisted transition is

$$\mathbf{Q} = \pm(\mathbf{K}' - \mathbf{K}) + \mathbf{G}, \tag{4.280}$$

where $\mathbf{K}$ and $\mathbf{K}'$ are the wave-vectors for a 3D electron gas and $\mathbf{G}$ is a reciprocal lattice vector (as defined in Section 2.1.2). The upper and lower sign in Eq.4.280 corresponds to absorption and emission, respectively.

We now recall that, given the crystal periodicity, the phonon wave-vector $\mathbf{Q}$ may always be taken inside the first Brillouin zone. Equation 4.280 is the relation which allows us to identify the $\mathbf{G}$ vectors possibly necessary in order for $\mathbf{Q}$ to belong to the first Brillouin zone. Consistently with Section 4.1.4, the processes corresponding to a non-null $\mathbf{G}$ vector are denoted *umklapp* processes [4, 45].

As a practical example, let us consider the transitions between the six equivalent Δ minima of silicon. For symmetry reasons there are only two different such inter-valley transitions, namely the g-type processes between two Δ valleys along the same $\langle 100 \rangle$ direction and the f-type processes between Δ valleys along different $\langle 100 \rangle$ directions. The f-type and g-type processes are illustrated in Fig.4.16.

If we choose $\mathbf{K}$ and $\mathbf{K}'$ coincident with the valley minima, we may take

$$\mathbf{K} = (0.85, 0, 0), \qquad \mathbf{K}' = (-0.85, 0, 0), \tag{4.281}$$

as an example of a g-type phonon transition (the wave-vector components are written in units of $(2\pi/a_0)$), and

$$\mathbf{K} = (0, 0.85, 0), \qquad \mathbf{K}' = (0, 0, 0.85), \tag{4.282}$$

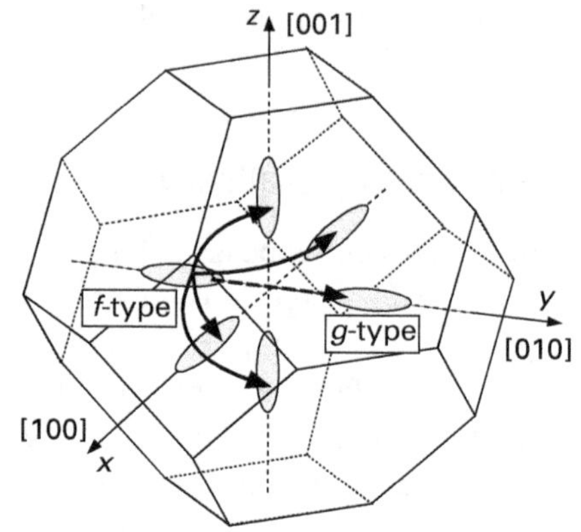

Figure 4.16 g-type and f-type phonon assisted transitions involving the Δ valleys of silicon located along the $\langle 100 \rangle$ directions.

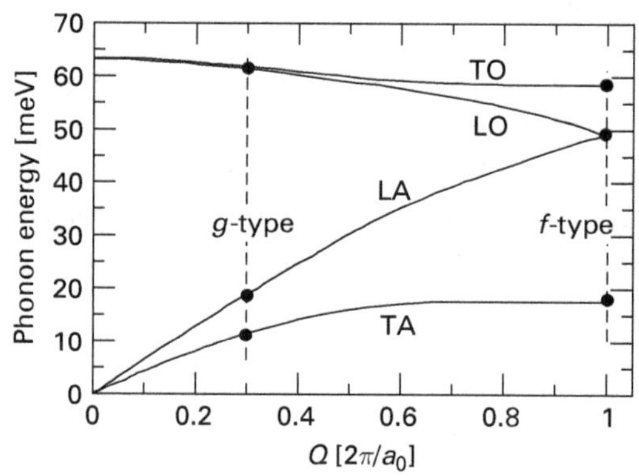

Figure 4.17 Energy versus wave-vector relation for acoustic and optical phonons and for a $\mathbf{Q}$ along the [100] direction; same curves as in Fig.4.13; after Long [41] and Brockhouse [42]. The values of $Q = |\mathbf{Q}_g| = 0.3$ and $Q = |\mathbf{Q}_f| \simeq 1.0$ identify the energies of the g-type and f-type processes corresponding to the different phonon branches. As can be seen, we obtain three different energy values for both the g-type and the f-type processes.

as an example of an f-type process.

By recalling Section 2.1.2 we know that, in order for $\mathbf{Q}$ to belong to the first Brillouin zone, it must fulfill the condition

$$\begin{cases} |Q_x|, \ |Q_y|, \ |Q_z| < 1 \\ \\ |Q_x| + |Q_y| + |Q_z| < 1.5 \end{cases} . \tag{4.283}$$

Thus it is easy to verify that the above examples of g-type and f-type transition are *umklapp* processes [4]. In fact, in the case of phonon absorption, for the g-type process Eqs.4.280 and 4.281 provide

$$\mathbf{Q}_g = (0.3, 0, 0) \quad \text{with} \quad \mathbf{G} = (2, 0, 0), \tag{4.284}$$

whereas for the f-type process we have

$$\mathbf{Q}_f = (1.0, 0.15, -0.15) \quad \text{with} \quad \mathbf{G} = (1, 1, -1). \tag{4.285}$$

It can be easily recognized that the corresponding phonon emissions are also *umklapp* processes.

All the possible g-type and f-type transitions require $\mathbf{Q}_g$ and $\mathbf{Q}_f$ lattice vectors that are obtained from Eqs.4.284 and 4.285 by simply changing either the position or the sign of the non-null $\mathbf{Q}$ components. In particular, the magnitudes are $|\mathbf{Q}_g| = 0.3$ and $|\mathbf{Q}_f| = 1.022$ in all cases.

In a silicon inversion layer the g-type inter-valley transitions occur between the subbands of two equivalent valleys, namely valleys that lie along the same $\langle 100 \rangle$ direction in the bulk silicon, whereas the f-type transitions occur between subbands of non-equivalent valleys. The $\mathbf{Q}$ values of the phonons that assist the inter-valley transitions are not exactly the $\mathbf{Q}_g$ and $\mathbf{Q}_f$ discussed above, both because the electrons have non-null $\mathbf{k}$ values in the transport plane (which represent displacements from the position of the valley minima), and because the quantization removes the conservation of the wave-vector component in the z direction, so that phonons *with a range of* q_z *values* can effectively couple with the electron gas (see Section 4.6.3 and Fig.4.15). However, the relative deviations of $\mathbf{Q}$ from $\mathbf{Q}_g$ and $\mathbf{Q}_f$ are small, so that the inter-valley phonon scattering transitions are typically described by assuming that $\mathbf{Q}$ is approximately fixed by the distance between the valleys in the first Brillouin zone.

If we suppose that for the g-type or the f-type transitions the corresponding phonon wave-vector is approximately fixed, then the phonon energy is correspondingly constant and the matrix elements may be expressed exactly as in the case of the intra-valley optical phonons described in the previous section. For an electron in the subband n of a ν type valley, the scattering rate towards the n' subband of a w type valley produced by the p phonon type is given by

$$S_{\nu,n}^{w,n'}(E_{\nu,n}(\mathbf{k})) = \frac{\pi D_p^2}{\omega_p \rho} \sum_{w \neq \nu, n'} \mu_{w,\nu}^{(p)} F_{\nu,n}^{w,n'} \times \left[n_{op}(\hbar\omega_p) + \frac{1}{2} \mp \frac{1}{2} \right] g_{w,n'}(E_{\nu,n}(\mathbf{k}) \pm \hbar\omega_p), \tag{4.286}$$

where $F_{\nu,n}^{w,n'}$ is the form factor:

$$F_{\nu,n}^{w,n'} = \int_z |\xi_{w,n'}^{\dagger}(z)\xi_{\nu,n}(z)|^2 \mathrm{d}z. \tag{4.287}$$

In Eq.4.286 $\hbar\omega_p$ and D_p are the phonon energy and the deformation potential of the p phonon type and $n_{op}(\hbar\omega_p)$ is the phonon number given by Eq.4.236; $g_{w,n'}(E)$ is the electron density of states of a w type valley as defined in Eq.3.75 (with $n_{sp} = 1$). Furthermore, as can be seen in Fig.4.16, the inter-valley transitions can occur towards more equivalent final valleys, hence $\mu_{w,\nu}^{(p)}$ denotes the multiplicity of such final valleys for the p phonon type, which depends on the type ν and w of the initial and final valley.

Table 4.3 Energy and deformation potentials for the three f-type and three g-type inter-valley phonons in silicon (see Fig.4.17). Numerical values from [4, 5].

Phonon process	Energy [meV]	Deform. potent. [10^8 eV/cm]
g-type, TA	12	0.5
g-type, LA	18.5	0.8
g-type, LO	61.2	11
f-type, TA	19	0.3
f-type, LA	47.4	2.0
f-type, TO	59	2.0

Table 4.4 f-type and g-type inter-valley phonon processes between the Δ_2 and Δ_4 valleys of a (100) silicon inversion layer and corresponding multiplicity $\mu_{w,\nu}$ of the final subbands (see Eq.4.286).

	Initial valley $\nu = \Delta_2$	Initial valley $\nu = \Delta_4$
Final valley $w = \Delta_2$	g-type, $\mu_{w,\nu} = 1$	f-type, $\mu_{w,\nu} = 2$
Final valley $w = \Delta_4$	f-type, $\mu_{w,\nu} = 4$	g-type, $\mu_{w,\nu} = 1$ f-type, $\mu_{w,\nu} = 2$

For a silicon (001) inversion layer, to which the analysis of the present chapter is restricted, the phonon assisted transitions between Δ_2 unprimed subbands can be only of g-type and have $\mu_{w,\nu}^{(p)} = 1$, whereas the transitions between Δ_4 primed subbands can be either g-type ($\mu_{w,\nu}^{(p)} = 1$) or f-type ($\mu_{w,\nu}^{(p)} = 2$). Only f-type transitions are possible from Δ_2 to Δ_4 subbands ($\mu_{w,\nu}^{(p)} = 4$) as well as from Δ_4 to Δ_2 subbands ($\mu_{w,\nu}^{(p)} = 2$) [5]. The multiplicities $\mu_{w,\nu}^{(p)}$ are summarized in Table 4.4.

One last important point concerning inter-valley phonon assisted transitions concerns the number of g-type and f-type processes that we must account for. To answer this question, we inspect Fig.4.17 showing the phonon spectra for a $\mathbf{Q}$ vector aligned with the [100] direction; the transverse branches are doubly degenerate so that overall we actually have six branches in the figure. According to Eqs.4.284 and 4.285 the $\mathbf{Q}_g$ vectors for g-type processes are aligned with the $\langle 100 \rangle$ direction, while the $\mathbf{Q}_f$ vectors for f-type processes form a small angle (approximately 12°) with respect to the $\langle 100 \rangle$ direction. Figure 4.17 can thus be used to identify the energy values for both g-type and f-type processes. In particular, the formulation of the scattering rate given in Eq.4.286 suggests grouping in a single phonon mode all the modes that have the same energy $\hbar\omega_p$, which enters Eq.4.286 both directly in the pre-factor and through the phonon number $n_{op}(\hbar\omega_p)$.

Figure 4.17 shows that for $Q = |\mathbf{Q}_g| = 0.3$ the TO and LO modes give essentially the same energy. Similarly the LO and LA modes have the same energy for

$Q = |\mathbf{Q}_f| \simeq 1.0$. This suggests that we can describe the inter-valley transitions in silicon by using three g-type processes and three f-type processes with the energy values identified in Fig.4.17.

The above discussion explains the model for inter-valley phonon transitions in (001) inversion layers that was originally proposed in [4], and was then very widely used in analysis of electron devices [4–6, 44, 50–53]. The corresponding parameters for the f-type and g-type phonon modes are reported in Table 4.3 and the valley multiplicities $\mu_{w,\nu}^{(p)}$ for the different valley types are summarized in Table 4.4.

In order to clarify the practical use of Eq.4.286 for calculation of the scattering rates due to inter-valley phonons we emphasize that, for a given initial subband n of a ν type valley, the scattering rate is obtained by considering all the possible final subbands n' for all the w type valleys. For each (ν, w) couple, Table 4.4 provides the multiplicity $\mu_{w,\nu}^{(p)}$ of the process and Table 4.3 indicates the corresponding phonon energy and deformation potential. The energy $\hbar\omega_p$ finally allows us to calculate the phonon number n_{op} from Eq.4.236, as well as the density of states $g_{w,n'}$ in the final subband at the energy $[E_{\nu,n}(\mathbf{k}) \pm \hbar\omega_p]$ (upper and lower sign for phonon absorption and emission, respectively).

4.6.6 Hole phonon scattering

Phonon scattering for holes is theoretically more complicated than for electrons, not only because the hole energy dispersion is non-parabolic and anisotropic, but also because the phonon matrix elements are anisotropic, that is they depend on the direction of the phonon wave-vector $\mathbf{Q}$ [45].

For acoustic phonon scattering, difficulties are encountered right at the beginning of the derivations, because the strain-induced energy shifts of the three quasi-degenerate branches of the valence band at the Γ point cannot be simply expressed with a scalar deformation potential. In fact, as discussed in more detail in Section 9.4.1, we need three deformation potentials (denoted by l, m, and n in Section 9.4.1), to describe the shifts of the bulk silicon valence band edges as a function of the strain [54].

In this respect, comprehensive studies have been devoted to hole phonon scattering for bulk semiconductors in an attempt to fully account for the anisotropy of both the phonon matrix elements and the energy dispersion [55–58]. The physical picture in the inversion layers is further complicated by the quantization [33, 59], so that it is widely recognized that, for the purposes of electron device analysis, an approximate isotropic treatment of hole phonon scattering is necessary, which essentially leads to use of a scalar deformation potential. Admittedly it is not easy to estimate a priori the accuracy of such a simplification [33].

Besides the isotropic approximation, for acoustic phonons we can also invoke the elastic and energy equipartition approximations, which are justified for holes to the same extent as they are for electrons. Having adopted the above approximations, the derivations and final expressions for hole acoustic and optical phonon scattering become very similar to those for electrons.

Before moving to the expressions for the scattering rates, a comment about the notation and definition of the hole subbands may be useful. As already discussed in

Section 3.3.1, in the **k**·**p** model the subbands are labelled with a single index n and the envelope wave-functions $\boldsymbol{\xi}_{n\mathbf{k}}(z)$ are six component, **k** dependent functions. In the semi-analytical energy model of Section 3.3.3, instead, the subbands are labelled with two indexes (a group and a subband index), and the envelope wave-functions are scalar and independent of **k**. In the rest of this section we use a notation consistent with the **k**·**p** picture. However, the results directly apply also to the hole energy model of Section 3.3.3 by assuming that, in this latter case, the symbol $\boldsymbol{\xi}_{n\mathbf{k}}(z)$ denotes a scalar wave-function and the index n actually identifies both the group and the subband.

For a scalar deformation potential D_{ac} and by adopting the elastic and energy equipartition approximation, the acoustic phonon scattering rate between the states (n,**k**) and (n',**k**$'$) is

$$S_{n,n'}(\mathbf{k},\mathbf{k}') = \frac{2\pi\, K_B T\, D_{ac}^2}{\rho\, A\, \hbar\, v_s^2}\, F_{n'\mathbf{k}',n\mathbf{k}}\, \delta[E_n(\mathbf{k}) - E_{n'}(\mathbf{k}')], \tag{4.288}$$

which differs from Eq.4.273 because the form factors

$$F_{n'\mathbf{k}',n\mathbf{k}} = 2\pi \int_{q_z} |G_{n'\mathbf{k}',n\mathbf{k}}(q_z)|^2\, dq_z = \int_z |\boldsymbol{\xi}^{\dagger}_{n'\mathbf{k}'}(z) \cdot \boldsymbol{\xi}_{n\mathbf{k}}(z)|^2 dz \tag{4.289}$$

may depend on **k** and **k**$'$ for a hole inversion layer.

More precisely, the form factors depend on **k**, **k**$'$ if we employ a **k**·**p** quantization model, in which case the dot in the wave-functions product of Eq.4.289 denotes the scalar product between the wave-functions defined in Eq.4.40. For the hole energy model of Section 3.3.3, instead, the dot in Eq.4.289 is the conventional product between scalars, and the wave-functions are independent of **k**.

The presence of **k** and **k**$'$ dependent form factors $F_{n'\mathbf{k}',n\mathbf{k}}$ in Eq.4.288 makes the scattering rate non-isotropic. In such a case, in order to calculate the scattering rate $S_n(\mathbf{k})$ in the state (n,**k**), the $F_{n'\mathbf{k}',n\mathbf{k}}$ must be included in the integration over the final states and one cannot use Eq.4.54 to obtain a scattering rate simply proportional to the density of final states.

By assuming that the dependence of $F_{n'\mathbf{k}',n\mathbf{k}}$ on **k** and **k**$'$ is relatively weak, in the literature such a dependence has very frequently been neglected, even in transport studies based on the **k**·**p** model [20, 21, 33, 60], so that the form factors have been taken as those corresponding to $\mathbf{k}=\mathbf{k}'=\mathbf{0}$, namely

$$F_{n'\mathbf{0},n\mathbf{0}} = 2\pi \int_{q_z} |G_{n'\mathbf{0},n\mathbf{0}}(q_z)|^2\, dq_z = \int_z |\boldsymbol{\xi}^{\dagger}_{n'\mathbf{0}}(z) \cdot \boldsymbol{\xi}_{n\mathbf{0}}(z)|^2 dz. \tag{4.290}$$

By substituting $F_{n'\mathbf{0},n\mathbf{0}}$ in Eq.4.288 we obtain an isotropic scattering rate $S_{n,n'}(\mathbf{k},\mathbf{k}')$, as in the case of electrons. The scattering rate $S_n(\mathbf{k})$ in the state (n,**k**) thus becomes

$$S_n(\mathbf{k}) = S_n(E_n(\mathbf{k})) = \frac{2\pi\, K_B T\, D_{ac}^2}{\rho\, \hbar\, v_s^2} \sum_{n'} F_{n'\mathbf{0},n\mathbf{0}}\, g_{n'}(E_n(\mathbf{k})), \tag{4.291}$$

where $g_n(E)$ is the density of states defined in Eq.3.89 or Eq.3.82 for either the **k**·**p** or the semi-analytical hole energy model, respectively, in both cases with $n_{sp} = 1$. We reiterate that, in the case of the semi-analytical energy model of Section 3.3.3, the index

n denotes both the group and the subband, hence the sum in the r.h.s. of Eq.4.291 is a sum over all the groups and all the subbands.

If we use the wave-vector independent form factors $F_{n'\mathbf{0},n\mathbf{0}}$, the hole scattering rate produced by the optical phonons also takes an expression very similar to the electron case:

$$S_n(\mathbf{k}) = S_n(E_n(\mathbf{k})) = \frac{\pi D_{op}^2}{\omega_0 \rho} \sum_{n'} F_{n'\mathbf{0},n\mathbf{0}} \left[n_{op} + \frac{1}{2} \mp \frac{1}{2} \right] g_{n'}(E_n(\mathbf{k}) \pm \hbar\omega_0). \tag{4.292}$$

All the transitions for the hole gas considered in this book relate to subbands located at the Γ point in the 2D Brillouin zone. Hence for holes we do not need to consider inter-valley transitions.

4.6.7 Selection rules for phonon scattering

The treatment of phonon scattering discussed in the previous sections is based on the envelope wave-function approximation, hence it does not take into account the Bloch states of the underlying crystal. The properties of the Bloch wave-functions at the conduction band minima or valence maxima located at the high symmetry points or along the high symmetry directions of the bulk crystal Brillouin zone can forbid some of the intra-valley transitions assisted by optical phonons. A rigorous justification of such selection rules is provided by group theory and is beyond the scope of this book [47, 61].

The selection rules most relevant for transport in MOS transistors can be summarized as follows [4]. For cubic semiconductors the intra-valley electron transitions assisted by optical phonons are forbidden for the conduction band minima located at the Γ point (e.g. the Γ valley of gallium arsenide or germanium) or along the $\langle 100 \rangle$ directions (e.g. the Δ valleys of silicon), whereas they are *not* forbidden for the minima located along the $\langle 111 \rangle$ directions (e.g. the Λ valleys of germanium) [47, 61]. In contrast, no restrictions apply to intra-valley optical phonon transitions for holes at the Γ point [47, 54].

It should be mentioned that, strictly speaking, the selection rules have been derived for the high-symmetry points at the edges of the conduction or valence bands, whereas the initial and final states for the carriers do not coincide exactly with such symmetry points. However, continuity arguments suggest that phonon matrix elements which are null at the symmetry points should remain small for states close to such symmetry points. In practice, the selection rules have been accounted for in the most widely accepted and used models for carrier transport in electron devices [4–6, 44, 50–53].

Finally, it is appropriate to mention that the discussion of phonon scattering in this chapter is based on a zero order treatment in the phonon wave-vector; in particular, the selection rules that we have described in this section apply to such a zero order treatment. The matrix elements for phonon optical and inter-valley phonons to first order in the phonon wave-vector were calculated by Ferry [62]. The processes corresponding to first order treatment may be significant when the zero order transitions are forbidden by the symmetry arguments.

Before concluding the discussion of phonon scattering mechanisms, one last important topic should be addressed, namely the possible effects of screening on the phonon matrix elements and scattering rate. This necessitates extending the screening theory presented in Section 4.2 to the case of a time dependent perturbation potential, which is the purpose of Section 4.7.

4.7 Screening of a time-dependent perturbation potential

The theory of screening presented in Section 4.2 was restricted to a static perturbation potential and it is appropriate, for instance, for Coulomb or surface roughness scattering. For a time-dependent scattering potential, however, one cannot neglect the fact that the response of the carriers in the inversion layer *is not instantaneous*, so that the dielectric function introduced in Section 4.2 must be extended in order to make it both wave-vector and ω dependent, where ω is the angular frequency of the perturbation.

Before discussing the dynamic dielectric function for a 2D carrier gas (see Section 4.7.1), we introduce the basic concepts by considering a bulk material. In a bulk semiconductor we have a dielectric response which typically consists of several terms: the contribution of the valence electrons (resulting in the high frequency dielectric constant $\epsilon_{(\infty)}$), a possible contribution of the lattice polarization (in polar semiconductors), and finally the dielectric response of the free carriers in the valence or conduction band.

In a bulk polar semiconductor the dielectric function can thus be written [45, 63]

$$\epsilon(\mathbf{Q}, \omega) = \epsilon_{(\infty)} + \frac{\epsilon_{(0)} - \epsilon_{(\infty)}}{1 - \left(\frac{\omega}{\omega_{TO}}\right)^2} + \epsilon_{LH}(\mathbf{Q}, \omega), \tag{4.293}$$

where the second term is due to the lattice polarization and ω_{TO} is the frequency of the dominant polar phonon in the material. The $\epsilon_{LH}(\mathbf{Q}, \omega)$ is the so-called *Lindhart dielectric function* which describes the $\mathbf{Q}$ and ω response of the free carriers and is given by [45, 63]

$$\epsilon_{LH}(\mathbf{Q}, \omega) = \frac{e^2}{Q^2 \Omega} \sum_{\mathbf{K}} \frac{f(\mathbf{K}+\mathbf{Q}) - f(\mathbf{K})}{E(\mathbf{K}+\mathbf{Q}) - E(\mathbf{K}) - \hbar\omega - \mathrm{i}\alpha\hbar}, \tag{4.294}$$

where Ω is the normalization volume, α is a small and positive parameter that can be taken arbitrarily small (see also the discussion of Eq.4.295 below), and $f(\mathbf{K})$ is the occupation function of the state $\mathbf{K}$; a single band of the bulk crystal energy relation is considered in Eq.4.294. In purely covalent semiconductors, such as silicon and germanium, there is no contribution to the dielectric constant due to lattice polarization, whereas in polar semiconductors the second term in Eq.4.293 can increase the static dielectric constant $\epsilon_{(0)}$ well above the high-frequency value $\epsilon_{(\infty)}$ (see Section 10.7 and Table 10.7 for GaAs); this is true also in many so-called high-κ dielectrics (see Section 10.2 and Table 10.1), whose static dielectric constant $\epsilon_{(0)}$ is high just because of a strong lattice polarization.

The frequency dependence of the first two terms in Eq.4.293 has a clear physical interpretation. When the perturbation potential has a frequency ω well below ω_{TO} both the core electrons and the lattice vibrations contribute to the overall dielectric constant $\epsilon_{(0)}$. For ω much larger than ω_{TO}, instead, the lattice vibrations cannot follow the perturbation potential and the dielectric constant reduces to $\epsilon_{(\infty)}$.

The Lindhart dielectric function $\epsilon_{LH}(\mathbf{Q}, \omega)$ gives an additional $\mathbf{Q}$ and ω dependent contribution to the overall $\epsilon(\mathbf{Q}, \omega)$. We discuss such a contribution due to the free carriers and for a 2D carrier gas in the next section.

4.7.1 Dynamic dielectric function for a 2D carrier gas

Let us consider the time dependent scattering potential

$$U_{sc}(\mathbf{R}, t) = U_{ab}(\mathbf{R})\, \mathrm{e}^{-\mathrm{i}\omega t}\, e^{\alpha\hbar t} + U_{em}(\mathbf{R})\, \mathrm{e}^{\mathrm{i}\omega t}\, e^{\alpha\hbar t}, \tag{4.295}$$

where $U_{em}(\mathbf{R})$ must be the complex conjugate of $U_{ab}(\mathbf{R})$ in order for $U_{sc}(\mathbf{R}, t)$ to be real. The small and positive parameter α describes a very slow "turning on" of the potential $U_{sc}(\mathbf{R}, t)$, namely it guarantees that $U_{sc}(\mathbf{R}, t)$ is negligible for large and negative t values [45, 63].

By following an approach similar to the one presented in Section 4.2 and Appendix E for the static dielectric function, it is possible first to express the charge $\rho_{ind}(\mathbf{r}, z, t)$ induced by $U_{sc}(\mathbf{R}, t)$ by using time-dependent perturbation theory [40], then to calculate the overall scattering potential and the screened matrix elements. By doing so we obtain a linear relation between the unscreened and screened scattering matrix elements for both absorption and emission. By assuming that the wave-functions in the inversion layer are independent of the wave-vector $\mathbf{k}$, the final formulation of the problem is the same as in Eq.4.77, except for the important difference that the dielectric function $\epsilon^{w,n,n'}_{\nu,m,m'}(\mathbf{q}, \omega)$ is now $\mathbf{q}$ and ω dependent.

More precisely, $\epsilon^{w,n,n'}_{\nu,m,m'}(\mathbf{q}, \omega)$ is still expressed by Eq.4.78, however, the polarization factor is now ω-dependent and, for the absorption matrix elements, it reads

$$\Pi_{w,n,n'}(\mathbf{q}, \omega) = \frac{1}{A} \sum_{\mathbf{k}} \frac{f_{w,n}(\mathbf{k}+\mathbf{q}) - f_{w,n}(\mathbf{k})}{E_{w,n'}(\mathbf{k}+\mathbf{q}) - E_{w,n}(\mathbf{k}) - \hbar\omega - \mathrm{i}\alpha\hbar}. \tag{4.296}$$

Equation 4.296 shows that the parameter α introduced in Eq.4.295 avoids a possible singularity in the calculation of the polarization factors, and it can typically be taken as vanishingly small in practical calculations. For the emission matrix elements, the polarization factors have the $\hbar\omega$ term with a plus sign in the denominator of Eq.4.296.

As discussed in detail in Section 4.2, Eq.4.77 allows us to calculate the screened matrix elements, provided that $\epsilon^{w,n,n'}_{\nu,m,m'}(\mathbf{q}, \omega)$ is not singular. The singularities of the tensorial dielectric function $\epsilon^{w,n,n'}_{\nu,m,m'}(\mathbf{q}, \omega)$ correspond to the zeros of the scalar dielectric function in a bulk system, namely the zeros of the Lindhart dielectric function $\epsilon_{LH}(\mathbf{Q}, \omega)$ defined in Eq.4.294 [45, 63]. In fact, for the ($\mathbf{q}$,ω) values that result in a singular $\epsilon^{w,n,n'}_{\nu,m,m'}(\mathbf{q}, \omega)$, the carrier gas in the inversion layer has spontaneous oscillations that exist even in the absence of an external perturbation potential. These are the plasma oscillations of the 2D carrier gas and may be represented by the virtual particles called *plasmons*, which are analogous to the phonons used to describe the lattice vibrations [7].

The energy relation $\hbar\omega(\mathbf{q})$ for the plasmons of the 2D carrier gas can be obtained by searching for the zeros of the determinant of the tensorial dielectric function $\epsilon^{w,n,n'}_{\nu,m,m'}(\mathbf{q},\omega)$; this is in general very complicated. Thus, henceforth we provide an analytical expression for the plasma frequencies by resorting to the scalar dielectric function introduced in Section 4.2.3 and in the limit of small q values.

The $\mathbf{q}$ and ω dependent scalar dielectric function $\epsilon_D(\mathbf{q},\omega)$ for the intra-subband transitions is expressed by replacing the $\Pi_{w,n,n}(\mathbf{q})$ in Eq.4.91 with the intra-subband polarization factors $\Pi_{w,n,n}(\mathbf{q},\omega)$ given by Eq.4.296. The purpose of our derivation is now to calculate the ω values at which the $\epsilon_D(\mathbf{q},\omega)$ is null. In order to achieve this, we first rewrite the polarization factors by setting $\mathbf{k}' = (\mathbf{k}{+}\mathbf{q})$ in the term of Eq.4.296 containing $f_{w,n}(\mathbf{k}{+}\mathbf{q})$, so that we have

$$\Pi_{w,n,n}(\mathbf{q},\omega) = \frac{1}{A}\left\{\sum_{\mathbf{k}'} \frac{f_{w,n}(\mathbf{k}')}{E_{w,n}(\mathbf{k}') - E_{w,n}(\mathbf{k}'-\mathbf{q}) - \hbar\omega - \mathrm{i}\alpha\hbar} - \sum_{\mathbf{k}} \frac{f_{w,n}(\mathbf{k})}{E_{w,n}(\mathbf{k}+\mathbf{q}) - E_{w,n}(\mathbf{k}) - \hbar\omega - \mathrm{i}\alpha\hbar}\right\}. \tag{4.297}$$

By changing the symbol $\mathbf{k}'$ of the first sum to $\mathbf{k}$, we can now group the two sums in a single sum over $\mathbf{k}$, where the argument of the sum reads

$$\frac{f_{w,n}(\mathbf{k})[E_{w,n}(\mathbf{k}+\mathbf{q}) + E_{w,n}(\mathbf{k}-\mathbf{q}) - 2E_{w,n}(\mathbf{k})]}{[E_{w,n}(\mathbf{k}+\mathbf{q}) - E_{w,n}(\mathbf{k}) - \hbar\omega - \mathrm{i}\alpha\hbar][E_{w,n}(\mathbf{k}) - E_{w,n}(\mathbf{k}-\mathbf{q}) - \hbar\omega - \mathrm{i}\alpha\hbar]}.$$

For circular parabolic bands with a transport mass m_w (w being the valley index), in the numerator of the above equation we have

$$E_{w,n}(\mathbf{k}+\mathbf{q}) + E_{w,n}(\mathbf{k}-\mathbf{q}) - 2E_{w,n}(\mathbf{k}) = \frac{\hbar^2 q^2}{m_w}, \tag{4.298}$$

while, for very small q values and in the limit of a vanishing α, the denominator tends to $(\hbar\omega)^2$. Under these circumstances the polarization factor can be approximately expressed as

$$\Pi_{w,n,n}(\mathbf{q},\omega) \approx \frac{q^2}{m_w\,\omega^2}\left[\frac{1}{A}\sum_{\mathbf{k}} f_{w,n}(\mathbf{k})\right] = \frac{q^2\,N_{w,n}}{m_w\,\omega^2}, \tag{4.299}$$

where we have used Eq.4.67 to rewrite the sum over $\mathbf{k}$ in the squared bracket, with $N_{w,n}$ being the inversion density in the subband (w,n). By substituting Eq.4.299 in Eq.4.91 the dynamic dielectric function can be written

$$\epsilon_D(q,\omega) = 1 - \frac{e^2 q}{\omega^2(\epsilon_{si}+\epsilon_{ox})}\sum_{w,n}\frac{N_{w,n}}{m_w}, \tag{4.300}$$

and the plasma angular frequencies ω_p that make the $\epsilon_D(\mathbf{q},\omega)$ null are thus given by

$$\omega_p = \sqrt{\frac{e^2 q}{(\epsilon_{si}+\epsilon_{ox})}\sum_{w,n}\frac{N_{w,n}}{m_w}}. \tag{4.301}$$

Equation 4.301 suggests that the plasma frequencies of the 2D carrier gas increase with the inversion density and tend to zero when q tends to zero. This latter feature is different with respect to the plasma frequency of a 3D electron gas, which, for large enough electron volumetric densities n, is $\mathbf{Q}$ independent and reads [45, 63]

$$\omega_{p,3D} = \sqrt{\frac{e^2\, n}{m\, \epsilon_{si}}}, \tag{4.302}$$

where m is the effective electron mass. We return to this difference in Section 4.7.2.

Figure 4.18 shows the inverse of the magnitude of the dielectric function versus $\hbar\omega$ numerically calculated by using the eigenvalues and wave-functions obtained with a Schrödinger–Poisson solver; the same device structure was used to calculate the static dielectric function shown in Fig.4.2. For small enough angular frequencies we observe an ω-independent value of $[1/\epsilon_D(q,\omega)]$ which corresponds to the static dielectric function. Furthermore, the plasma frequencies ω_p (see Eq.4.301) are easily recognized as singularities of $[1/|\epsilon_D(q,\omega)|]$ and their values are in fairly good agreement with Eq.4.301. As can be seen, for a scattering potential with an ω well below ω_p the static dielectric model is a good approximation, while for frequencies well above ω_p the screening effect vanishes and $\epsilon_D(q,\omega)$ tends to 1. This high-frequency behavior corresponds to so-called dynamic de-screening and, physically speaking, is due to the fact that the carriers in the inversion layer cannot follow the rapid variations of the scattering potential.

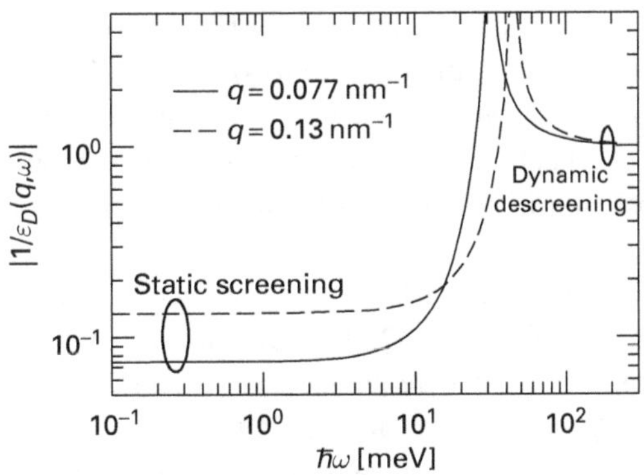

Figure 4.18 Inverse of the magnitude of the dynamic dielectric function versus $\hbar\omega$ for two different q values and calculated by using a self-consistent Schrödinger–Poisson solver. Bulk n-MOSFET with channel doping concentration of $3{\cdot}10^{17}$ cm^{-3} (same device structure as in Fig.4.2); inversion density N_{inv}=$3{\cdot}10^{12}$ cm^{-2}. The ω-independent $[1/\epsilon_D(q,\omega)]$ values at small frequencies are consistent with the static dielectric function reported in Fig.4.2. The plasma frequencies correspond to singularities of $[1/\epsilon_D(q,\omega)]$.

For $\hbar\omega$ values close to plasma frequencies the magnitude of $[1/|\epsilon_D(q,\omega)|$ is larger than 1 and this corresponds to an *anti-screening* effect, such that the magnitude of the scattering potential is amplified rather than reduced by screening. This behavior is not unexpected because the ω_p are the self-resonance frequencies of the 2D carrier gas, hence a stimulus with a frequency close to ω_p is amplified by the dielectric response of the carrier gas.

4.7.2 Screening for phonon scattering

For non-polar semiconductors, such as silicon and germanium, the ω dependence of the overall dielectric response essentially stems from the screening produced by the free charges. In an MOS transistor this results in the $\mathbf{q}$ and ω dependent dielectric function described in the previous section.

Thus, the screening of the inversion layer on a phonon scattering potential with energy $\hbar\omega(\mathbf{Q})$ and wave-vector $\mathbf{Q} = (\mathbf{q}, q_z)$ depends both on $Q = |\mathbf{Q}|$ and on how large is $\omega(\mathbf{Q})$ with respect to the plasma frequencies ω_p of the 2D carrier gas (see Fig.4.18). The ω_p values depend in turn on q and on the inversion density (see Eq.4.301).

For inter-valley phonon assisted transitions in electron inversion layers, the very large $\mathbf{Q}$ values make the screening very ineffective; furthermore, the phonon energy for the dominant phonon mode, namely the third g-type process in Table 4.3, is about 61 meV, hence it is large with respect to ω_p at relatively small inversion densities, N_{inv}. Thus screening is typically neglected for inter-valley phonons [5]. The intra-valley optical phonons, that in silicon are relevant only for hole inversion layers (see Section 4.6.7), have small $\mathbf{Q}$ values but energies around 60 meV (see Fig.4.17), which are quite large compared to the ω_p at small N_{inv}, so that the intra-valley optical phonon transitions can also be safely left unscreened.

The situation is much more complicated for intra-valley acoustic phonons. In fact, for such phonons both $\mathbf{q}$ and $\hbar\omega(\mathbf{Q})$ are small, so that the arguments in favor of, or against, the application of screening are more subtle. Below we illustrate only a few relevant points about screening for intra-valley acoustic phonons; the interested reader should refer to [5] (and the references therein) for a more detailed discussion of this topic.

First we note that in bulk semiconductors the intra-valley acoustic phonons have an energy $\hbar\omega(\mathbf{Q}) = \hbar v_s Q$ that goes to zero with Q (see Section 4.5.2), while the corresponding plasma frequency $\hbar\omega_{p,3D}$ obtained from Eq.4.302 does not. Thus, for vanishing Q values, we can expect to have $\hbar\omega(\mathbf{Q}) \ll \hbar\omega_{p,3D}$, which explains why in some studies the acoustic intra-valley phonons have been screened by using the static dielectric function [64].

It is interesting to note that the situation is quite different in an inversion layer. In fact, by recalling the results of Section 4.6.2 and, in particular, the $\mathbf{k}$ space phonon form factors defined in Eq.4.246, we see that the carriers in the inversion layer are scattered by phonons with $Q = \sqrt{q^2 + q_z^2}$, so that Q is non-null even when the wave-vector $\mathbf{q}$

in the transport plane goes to zero. The plasma energies $\hbar\omega_p$ of the inversion layer, instead, go to zero for vanishing q values (see Eq.4.301). Hence one may argue that for very small q values $\hbar\omega(\mathbf{Q})$ is much larger than $(\hbar\omega_p)$, hence the screening becomes ineffective (see Fig.4.18). More complex arguments can be raised about the screening for acoustic intra-valley phonons [5], however, practically speaking, in silicon inversion layers the scattering with acoustic intra-valley phonons has finally been left unscreened in essentially all the studies of which the authors are aware.

4.8 Summary

Chapter 4 is a broad chapter, ranging from a description of Fermi's golden rule in inversion layers (universally used to calculate scattering rates in semi-classical transport modeling), to formulation of scattering rates for Coulomb scattering in Section 4.3, surface roughness scattering in Section 4.4 as well as acoustic and optical phonons in Section 4.6. Furthermore, Sections 4.2 and 4.7 addressed the practically relevant and theoretically complex topic of the screening effect produced by the carriers in the inversion layer, respectively for a static and a time-dependent scattering potential.

The Fermi golden rule clarified that the total scattering rates in MOS transistors are directly affected by the band structure and wave-functions discussed in Chapter 3. Fermi's rule also showed that the scattering potential enters the matrix elements through a hybrid Fourier transform with respect to the position $\mathbf{r}$ in the transport plane, hence large wave-vector variations $\mathbf{q} = (\mathbf{k}' - \mathbf{k})$ can be produced only by scattering potentials rapidly varying in the real space.

Explicit expressions for scattering rates were given both for electrons in n-MOSFETs, described with the effective mass approximation, and for holes in p-MOSFETs, according either to the $\mathbf{k}\cdot\mathbf{p}$ model described in Section 3.3.1 or to the semi-analytical energy model of Section 3.3.3.

This chapter focused on theoretical formulation of the scattering rates. Chapter 7 provides an insight into the relative importance of scattering mechanisms for low field mobility in long channel MOSFETs and for the on-current in nanoscale transistors.

The scattering mechanisms described in this chapter are those considered the most important for carrier transport in silicon transistors with an SiO_2 gate dielectric. More scattering mechanisms related to alternative channel materials and high-κ gate dielectrics are illustrated in Chapter 10. The device structure may also suggest introduction of corrections or integrations to the formulation of scattering mechanisms reported in this chapter.

As a first example, we note that in very thin silicon films use of the bulk phonon modes for carrier-phonon scattering may be questioned, because the lattice vibrations in the silicon film are influenced by the surrounding gate oxide material. This is expected to be the case especially for the long wavelength acoustic phonon modes that govern phonon assisted intra-valley transitions (see Sections 4.6.3 and 4.6.6). Calculation of the phonon spectra and treatment of the carrier-phonon interaction in nano-structured quantum wells is a topic addressed in detail in [65], and the effect of acoustic phonon confinement on the mobility of ultra-thin SOI transistors has recently been investigated in [66, 67]. The impact of acoustic phonon quantization on the hole mobility is found to be essentially negligible [67], whereas the simulated electron mobility is somewhat reduced by the phonon confinement [66].

A second example is related to surface roughness scattering in very thin gate oxide transistors. In fact, it has been argued that in ultra-thin oxide devices the carriers in the inversion layer may be scattered not only by the roughness of the semiconductor–oxide interface (discussed in Section 4.4), but also by the asperities at the oxide–polysilicon interface, that is the oxide interface at the gate side. Such a mechanism has been designated remote surface roughness scattering, and some simulation studies suggested that it may contribute to degradation of the carrier mobility in ultra-thin oxide MOSFETs [68, 69].

In this context, we also remind readers that, based on both experimental results and numerical simulations, some authors have argued that the remote Coulomb scattering due to ionized dopants in the gate polysilicon may appreciably degrade the channel mobility in ultra-thin oxide MOSFETs [70, 71]. This effect was debated for some years because, on the one hand, an accurate extraction of effective mobility for very thin gate oxides is hampered by possible gate doping penetration in the channel and by gate leakage that complicates the I_{DS} and CV measurements [72–74]. On the other hand, a theoretical evaluation of the scattering produced by polysilicon Coulomb centers should account for the screening effect due to the free carriers in the polysilicon, which is additive and possibly dominant with respect to the screening of the carriers in the inversion layers [53]. The effect of the polysilicon screening on remote Coulomb scattering has been approached in [53] by resorting to a model different from the dielectric function discussed in Sections 4.2.2 and based on the perturbative treatment originally proposed in [16]. Such an approach naturally lends itself to account also for the screening effect produced by the free carriers in the polysilicon. Incidentally, the appropriate inclusion of the polysilicon screening in the mobility calculations helped reconcile the simulations with experiments and relegated the scattering with remote Coulomb centers in the polysilicon to an effect of only modest practical importance [53, 75].

We conclude with the hope that the detailed and somewhat pedagogical treatment of scattering theory set out in this chapter provides the reader with a solid background also for an understanding of scattering mechanisms different from those explicitly discussed in the previous sections.

References

[1] S. Datta, *Quantum Phenomena.* Addison-Wesley Modular Series in Solid State Devices, 1989.

[2] D.K. Ferry, *Quantum Mechanics — An Introduction for Device Physicists and Electrical Engineers.* Institute of Physics publishing, 2001.

[3] J. Bude, K. Hess, and G.J. Iafrate, "Impact ionization in semiconductors: effects of high electric fields and high scattering rates," *Phys. Rev. B*, vol. 45, pp. 10958–10964, 1992.

[4] C. Jacoboni and L. Reggiani, "The Monte Carlo method for the solution of charge transport in semiconductors with applications to covalent materials," *Rev. Mod. Phys.*, vol. 55, pp. 645–705, 1983.

[5] M.V. Fischetti and S.E. Laux, "Monte Carlo study of electron transport in silicon inversion layers," *Phys. Rev. B*, vol. 48, pp. 2244–2274, Jul 1993.

[6] C. Jungemann, A. Edmunds, and W.L. Engl, "Simulation of linear and nonlinear electron transport in homogeneous silicon inversion layers," *Solid State Electronics*, vol. 36, no. 11, pp. 1529–1540, 1993.

[7] D.K. Ferry and S.M. Goodnick, *Transport in Nanostructures*. Cambridge: Cambridge University Press, 1997.

[8] J.R. Chelikowsky and M.L. Cohen, "Nonlocal pseudopotential calculations for the electronic structure of eleven diamond and zinc-blende semiconductors," *Phys. Rev. B*, vol. 14, no. 2, pp. 556–582, 1976.

[9] E.H. Nicollian and J.R. Brews, *MOS Physics and Technology*. New York: Wiley, 1983.

[10] E.D. Siggia and P.C. Kwok, "Properties of electrons in semiconductor inversion layers with many occupied electric subbands. I. Screening and impurity scattering," *Phys. Rev. B*, vol. 2, no. 4, pp. 1024–1036, 1970.

[11] S. Takagi, A. Toriumi, M. Iwase, and H. Tango, "On the universality of inversion layer mobility in Si MOSFET's: Part I-Effects of substrate impurity concentration," *IEEE Trans. on Electron Devices*, vol. 41, pp. 2357–2362, Dec 1994.

[12] D. Esseni, M. Mastrapasqua, C. Fiegna, *et al.*, "An experimental study of low field electron mobility in double-gate, ultra-thin SOI MOSFETs," in *IEEE IEDM Technical Digest*, pp. 445–448, 2001.

[13] D. Esseni, M. Mastrapasqua, G.K. Celler, *et al.*, "An experimental study of mobility enhancement in ultra-thin SOI transistors operated in double-gate mode ," *IEEE Trans. on Electron Devices*, vol. 50, no. 3, pp. 802–808, 2003.

[14] K. Uchida, H. Watanabe, A. Kinoshita, *et al.*, "Experimental study on carrier transport mechanisms in ultrathin-body SOI n- and p-MOSFETs with SOI thickness less than 5nm," in *IEEE IEDM Technical Digest*, pp. 47–50, 2002.

[15] P. Toniutti, D. Esseni, P. Palestri, and L. Selmi, "Failure of the scalar dielectric function approach for the screening modelling in double-gate SOI MOSFETs," *IEEE Trans. on Electron Devices*, vol. 57, 2010.

[16] F. Stern and W.E. Howard, "Properties of semiconductor surface inversion layers in the electric quantum limit ," *Phys. Rev.*, vol. 163, no. 3, pp. 816–835, 1967.

[17] K. Hess, "Impurity and phonon scattering in layered structures," *Applied Physics Letters*, vol. 35, no. 7, pp. 484–486, 1979.

[18] K. Yokoyama and K. Hess, "Monte Carlo study of electronic transport in $Al_{1-x}Ga_xAs$/GaAs single-well heterostructures," *Phys. Rev. B*, vol. 33, no. 8, pp. 5595–5606, 1986.

[19] F. Gamiz, J.A. Lopez-Villanueva, J.A. Jimenez-Tejada, I. Melchor, and A. Palma, " A comprehensive model for Coulomb scattering in inversion layers ," *Journal of Applied Physics*, vol. 75, no. 2, pp. 924–934, 1994.

[20] L. Donetti, F. Gamiz, A. Godoy, and N. Rodriguez, "Fully self-consistent $\mathbf{k}\cdot\mathbf{p}$ solver and Monte Carlo simulator for hole inversion layers," in *Proc. European Solid State Device Res. Conf.*, pp. 254–257, 2008.

[21] A.T. Pham, C. Jungemann, and B. Meinerzhagen, "Modeling of hole inversion layer mobility in unstrained and uniaxially strained Si on arbitrarily oriented substrates," *Proc. European Solid State Device Res. Conf.*, pp. 390–393, September 2007.

[22] C. Alexander, G. Roy, and A. Asenov, "Random-dopant-induced drain current variation in nano-MOSFETs: A three-dimensional self-consistent Monte Carlo simulation study using ab initio ionized impurity scattering," *IEEE Trans. on Electron Devices*, vol. 55, no. 11, pp. 3251–3258, 2008.

[23] T.H. Ning and C.T. Sah, "Theory of scattering in a nondegenerate-semiconductor-surface inversion layer by surface-oxide charges," *Phys. Rev. B*, vol. 6, no. 12, pp. 4605–4613, 1972.

[24] A. Papoulis, *Probability, Random Variables and Stochastic Processes*. 3rd ed. New York: Mc Graw-Hill, 1991.

[25] A. Jeffrey, *Handbook of Mathematical Formulas and Integrals*. San Diego, CA: Academic Press, 2000.

[26] D. Esseni, "On the modeling of surface roughness limited mobility in SOI MOSFETs and its correlation to the transistor effective field ," *IEEE Trans. on Electron Devices*, vol. 51, no. 3, pp. 394–401, 2004.

[27] S.M. Goodnick, D.K. Ferry, C.W. Wilmsen, *et al.*, "Surface roughness at the Si(100)-SiO_2 interface," *Phys. Rev. B*, vol. 32, pp. 8171–8186, 1985.

[28] R.E. Prange and T.-W. Nee, "Quantum spectroscopy of the low-field oscillations in the surface impedance," *Phys. Rev. B*, vol. 168, pp. 779–786, 1968.

[29] T. Ando, "Screening effect and quantum transport in a silicon inversion layer in strong magnetic field," *Journal of the Physical Society of Japan*, vol. 43, no. 5, pp. 1616–1626, 1977.

[30] S. Jin, M.V. Fischetti, and T.W. Tang, "Modeling of surface-roughness scattering in ultrathin-body SOI MOSFETs," *IEEE Trans. on Electron Devices*, vol. 54, no. 9, pp. 2191–2203, 2007.

[31] T. Ishihara, K. Uchida, J. Koga, and S. Takagi, "Unified roughness scattering model incorporating scattering component induced by thickness fluctuations in silicon-on-insulator metal-oxide-semiconductor field-effect transistor," *Japanese Journal of Applied Physics*, vol. 45, no. 4B, pp. 3125–3132, 2006.

[32] A. Gold, "Electronic transport properties of a two-dimensional electron gas in a silicon quantum-well structure at low temperature ," *Phys. Rev. B*, vol. 35, pp. 723–733, 1987.

[33] M.V. Fischetti, Z. Ren, P.M. Solomon, M. Yang, and K. Rim, "Six-band $\mathbf{k}\cdot\mathbf{p}$ calculation of the hole mobility in silicon inversion layers: Dependence on surface orientation, strain, and silicon thickness," *Journal of Applied Physics*, vol. 94, no. 2, pp. 1079–1095, 2003.

[34] A.T. Pham, C. Jungemann, and B. Meinerzhagen, "Physics-based modeling of hole inversion-layer mobility in strained-SiGe-on-insulator," *IEEE Trans. on Electron Devices*, vol. 54, no. 9, pp. 2174–2182, 2007.

[35] M. De Michielis, D. Esseni, P. Palestri, and L. Selmi, "Semiclassical modeling of quasi-ballistic hole transport in nanoscale pMOSFETs based on a multi-subband Monte Carlo approach," *IEEE Trans. on Electron Devices*, vol. 56, no. 9, pp. 2081–2091, 2009.

[36] N. Ashkroft and N. Mermin, *Solid State Physics*. Philadelphia, PA: Sounders College, 1976.

[37] C. Kittel, *Introduction to Solid State Physics*. New York: John Wiley and Sons, 1996.

[38] A. Messiah, *Quantum Mechanics*. New York: Dover Publications, 1999.

[39] A.T. Fromhold, *Quantum Mechanics for Applied Physics and Engineering*. New York: Dover Publications Inc., 1981.

[40] J. Singh, *Quantum Mechanics: Fundamentals and Applications to Technology*. New York: Wiley Interscience, 1997.

[41] D. Long, "Scattering of conduction electron by lattice vibrations in silicon," *Phys. Rev.*, vol. 120, no. 6, pp. 2024–2032, 1960.

[42] B.N. Brockhouse, "Lattice vibrations in silicon and germanium," *Phys. Rev. Lett.*, vol. 2, no. 6, pp. 256–258, 1959.

[43] M.V. Fischetti and S.E. Laux, "Monte Carlo analysis of electron transport in small semiconductor devices including band-structure and space-charge effects," *Phys. Rev. B*, vol. 38, p. 9721, 1988.

[44] L. Lucci, P. Palestri, D. Esseni, L. Bergagnini, and L. Selmi, "Multisubband Monte Carlo study of transport, quantization, and electron-gas degeneration in ultrathin SOI n-MOSFETs," *IEEE Trans. on Electron Devices*, vol. 54, no. 5, pp. 1156–1164, May 2007.

[45] B.K. Ridley, *Quantum Processes in Semiconductors*. Oxford: Oxford Science Publication, 1993.

[46] M.V. Fischetti and S.E. Laux, "Band structure, deformation potentials, and carrier mobility in strained Si, Ge, and SiGe alloys," *Journal of Applied Physics*, vol. 80, p. 2234, 1996.

[47] W.A. Harrison, "Scattering of electrons by lattice vibrations in nonpolar crystals," *Phys. Rev.*, vol. 104, no. 5, pp. 1281–1289, 1956.

[48] B.R. Nag, *Electron Transport in Compound Semiconductors*. New York: Springer-Verlag, 1980.

[49] C. Herring and E. Vogt, "Transport and deformation-potential theory for many-valley semiconductors with anisotropic scattering," *Phys. Rev.*, vol. 101, p. 944, 1956.

[50] F. Gamiz, J.A. Lopez-Villanueva, J.B. Roldan, J.E. Carceller, and P. Cartujo, "Monte Carlo simulation of electron transoprt properties in extremely thin SOI MOSFET's," *IEEE Trans. on Electron Devices*, vol. 45, no. 5, pp. 1122–1126, 1998.

[51] M. Shoji and S. Horiguchi, "Electronic structures and phonon-limited electron mobility of double-gate silicon-on-insulator Si inversion layers," *Journal of Applied Physics*, vol. 85, no. 5, pp. 2722–2731, 1999.

[52] D. Esseni, A. Abramo, L. Selmi, and E. Sangiorgi, "Physically based modeling of low field electron mobility in ultrathin single- and double-gate SOI n-MOSFETs," *IEEE Trans. on Electron Devices*, vol. 50, no. 12, pp. 2445–2455, 2003.

[53] D. Esseni and A. Abramo, "Modeling of electron mobility degradation by remote Coulomb scattering in ultrathin oxide MOSFETs," *IEEE Trans. on Electron Devices*, vol. 50, pp. 1665–1674, July 2003.

[54] G.L. Bir and G.E. Pikus, *Symmetry and Strain Induced Effects in Semiconductors*. New York: Wiley, 1974.

[55] F.L. Madarasz and F. Szmulowicz, "Transition rates for acoustic-phono-hole scattering in p-type silicon with nonparabolic bands," *Phys. Rev. B*, vol. 24, no. 8, pp. 4611–4622, 1981.

[56] F. Szmulowicz and F.L. Madarasz, "Angular dependence of hole-acoustic-phonon transition rates in silicon," *Phys. Rev. B*, vol. 26, no. 4, pp. 2101–2112, 1982.

[57] F. Szmulowicz, "Calculation of optical- and acoustic-phonon-limited conductivity and Hall mobilities for p-type silicon and germanium," *Phys. Rev. B*, vol. 28, no. 10, pp. 5943–5963, 1983.

[58] F. Szmulowicz, "Calculation of the mobility and the Hall factor for doped p-type silicon," *Phys. Rev. B*, vol. 34, no. 6, pp. 4031–4047, 1986.

[59] R. Oberhuber, G. Zandler, and P. Vogl, "Subband structure and mobility of two-dimensional holes in strained Si/SiGe MOSFETs," *Phys. Rev. B*, vol. 58, no. 15, pp. 9941–9948, 1998.

[60] E. Wang, P. Montagne, L. Shifren, *et al.*, "Physics of hole tranport in strained silicon MOSFET inversion layers," *IEEE Trans. on Electron Devices*, vol. 53, no. 8, pp. 1840–1850, 2006.

[61] M. Lax and J.J. Hopfield, "Selection rules connecting different points in the Brillouin zone," *Phys. Rev. B*, vol. 124, no. 1, pp. 115–123, 1961.

[62] D.K. Ferry, "First order optical and intervalley scattering in semiconductors," *Phys. Rev. B*, vol. 14, no. 4, pp. 1605–1609, 1976.

[63] D.K. Ferry, *Semiconductors*. New York: Macmillan, 1991.

[64] M.J. Ziman, *Electrons and Phonons*. Oxford: Clarendon, 1960.

[65] M.A. Stroscio and M. Dutta, *Phonons in Nanostructures*. Cambridge: Cambridge University Press, 2001.

[66] L. Donetti, F. Gamiz, J.B. Roldan, and A. Godoy, "Acoustic phonon confinement in silicon nanolayers: Effect on electron mobility," *Journal of Applied Physics*, vol. 100, pp. 0137001–7, 2006.

[67] L. Donetti, F. Gamiz, N. Rodriguez, and A. Godoy, "Hole mobility in ultrathin double-gate SOI devices: The effect of acoustic phonon confinement," *IEEE Electron Device Lett.*, vol. 30, no. 12, pp. 1338–1340, 2009.

[68] F. Gamiz and J.B. Roldan, "Scattering of electrons in silicon inversion layers by remote surface roughness," *Journal of Applied Physics*, vol. 94, pp. 392–399, 2003.

[69] B. Ghosh, X.-F. Fan, L.F. Register, and S.K. Banarjee, "Monte Carlo study of remote Coulomb and remote surface roughness scattering in nanoscale Ge PMOSFETs with ultrathin high-κ dielectrics," in *Proc.SISPAD*, pp. 170–172, 2006.

[70] M.S. Krishnan, Y.C. Yeo, Q. Lu, *et al.*, "Remote charge scattering in MOSFETs with ultra-thin gate dielectrics ," in *IEEE IEDM Technical Digest*, pp. 571–574, 1998.

[71] N. Yang, W.K. Henson, J.R. Hauser, and J.J. Wortman, "Estimation of the effects of remote charge scattering on electron mobility of n-MOSFET's with ultrathin gate oxides," *IEEE Trans. on Electron Devices*, vol. 47, no. 2, pp. 440–446, 2000.

[72] W.K. Henson, K.Z. Ahmed, J.R. Hauser, *et al.*, "Estimating oxide thickness of tunnel oxide down to 1.4 nm using conventional capacitance-voltage measurements on MOS capacitors," *IEEE Electron Device Lett.*, vol. 20, no. 4, pp. 179–181, 1999.

[73] S. Takagi and M. Takayanagi, "Experimental evidence of inversion-layer mobility lowering in ultrathin gate oxide metal-oxide-semiconductor field-effect-transistors with direct tunneling current," *Japanese Journal of Applied Physics*, vol. 41, no. 12, pp. 2348–2352, 2002.

[74] P. Palestri, D. Esseni, L. Selmi, G. Guegan, and E. Sangiorgi, "A methodology to extract the channel current of permeable gate oxide MOSFETs ," *IEEE Trans. on Electron Devices*, vol. 50, no. 5, pp. 1314–1321, 2003.

[75] L. Lucci, D. Esseni, J. Loo, *et al.*, "Quantitative assessment of mobility degradation by remote Coulomb scattering in ultra-thin oxide MOSFETs: Measurements and simulations ," in *IEEE IEDM Technical Digest*, pp. 463–466, December 2003.

5 The Boltzmann transport equation

In this chapter we introduce the Boltzmann transport equation (BTE), which is the basis for the semi-classical description of carrier transport in electron devices. We begin with a brief reminder of the key assumptions behind the formulation of the BTE for a free carrier gas. Since this topic is extensively discussed in many textbooks [1, 2], we go on to focus on use of the BTE for the description of transport in inversion layers.

In particular, Section 5.4 explains how the carrier mobility in inversion layers can be computed solving the BTE in the Momentum Relaxation Time approximation, once the scattering rates introduced in Chapter 4 are known.

Section 5.5 reviews the methodology to solve the BTE in the limiting cases of near-equilibrium transport through the derivation of balance equations and of the widely used Drift-Diffusion model.

At the end of the chapter, Section 5.6 overviews the modeling of the far from equilibrium ballistic transport and Section 5.7 illustrates the quasi-ballistic transport regime. Expressions for the MOSFET on-current are derived in all cases from Drift-Diffusion to purely ballistic transport. These equations become useful in Chapter 7 to interpret the results of numerical simulations.

5.1 The BTE for the free-carrier gas

As discussed in Section 2.5, the dynamics of electrons in crystals can be described in terms of classical point charges provided that the extension of the wave-packet in real and momentum space is assumed to be negligible. In this case, the particle is described by a wave-packet with a position set by the centroid **R** of the square modulus of the envelope function and by the wave-vector **K** which is the centroid of the wave-packet in the **K**-space. The evolution of the wave-packet centroid in real and wave-vector space is governed by Eq.2.121.

In this section we aim to derive an equation describing in statistical terms the evolution of an ensemble of charged carriers, each moving according to Eq.2.121. This equation will be essentially the same as the Boltzmann Transport Equation which describes the evolution in real space, momentum space and time of a statistical distribution of classical particles.

We start with the case of a free electron gas and then extend the results to a free hole gas.

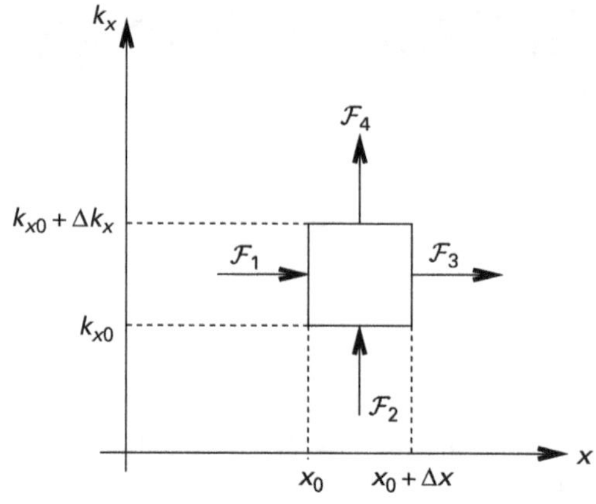

Figure 5.1 A control volume in phase-space.

5.1.1 The BTE for electrons

We obtain the BTE by imposing carrier conservation in phase-space, where for electrons in crystals we consider (**R**,**K**) as phase-space for the reasons explained above. The derivation is very close to the one in [1], except that here we consider the wave-vector instead of the carrier momentum.

A projection of the phase-space in the (x,k_x) plane is shown in Fig.5.1. At this stage we proceed as if the phase-space were two-dimensional; we will later extend the results to the 6-dimensional space (**R**,**K**). Furthermore, we neglect scattering. We can compute the variation over a time interval Δt of the number of particles N_P inside the control volume as:

$$\frac{\Delta N_P}{\Delta t} = \mathcal{F}_1 + \mathcal{F}_2 - \mathcal{F}_3 - \mathcal{F}_4. \tag{5.1}$$

We can write

$$N_P \simeq G(x_0, k_{x0}, t)\Delta k_x \Delta x, \tag{5.2}$$

where $G(x_0, k_{x0}, t)$ is the distribution of the particles in the phase-space, that is the number of particles per unit phase-space volume at time t. The fluxes $\mathcal{F}_1$, $\mathcal{F}_2$, $\mathcal{F}_3$, and $\mathcal{F}_4$ can be expressed as

$$\mathcal{F}_1 \simeq G(x_0, k_{x0}, t)\Delta k_x \left.\frac{\mathrm{d}x}{\mathrm{d}t}\right|_{x_0,k_{x0}}, \tag{5.3a}$$

$$\mathcal{F}_2 \simeq G(x_0, k_{x0}, t)\Delta x \left.\frac{\mathrm{d}k_x}{\mathrm{d}t}\right|_{x_0,k_{x0}}, \tag{5.3b}$$

$$\mathcal{F}_3 \simeq G(x_0 + \Delta x, k_{x0}, t)\Delta k_x \left.\frac{\mathrm{d}x}{\mathrm{d}t}\right|_{x_0+\Delta x,k_{x0}}, \tag{5.3c}$$

$$\mathcal{F}_4 \simeq G(x_0, k_{x0} + \Delta k_x, t)\Delta x \left.\frac{\mathrm{d}k_x}{\mathrm{d}t}\right|_{x_0,k_{x0}+\Delta k_x}. \tag{5.3d}$$

For vanishingly small Δt, Δx, and Δk_x, substitution of Eqs.5.3 and 5.2 into Eq.5.1 gives

$$\frac{\partial G}{\partial t} + \frac{\partial G}{\partial x}\frac{\mathrm{d}x}{\mathrm{d}t} + \frac{\partial G}{\partial k_x}\frac{\mathrm{d}k_x}{\mathrm{d}t} = 0. \tag{5.4}$$

A simple generalization considering the other dimensions of the phase-space also yields

$$\frac{\partial G}{\partial t} + \nabla_{\mathbf{R}} G \cdot \frac{\mathrm{d}\mathbf{R}}{\mathrm{d}t} + \nabla_{\mathbf{K}} G \cdot \frac{\mathrm{d}\mathbf{K}}{\mathrm{d}t} = 0. \tag{5.5}$$

The l.h.s. of Eq.5.5 is the total time derivative $\mathrm{d}G/\mathrm{d}t$ of the distribution G.

Equation 5.5 is linear in G. Since the density of states in the phase-space does not depend on either $\mathbf{R}$ or $\mathbf{K}$ (Section 2.1.4 and Eq.2.37), we can re-write the equation simply substituting G with the occupation function f, that is the probability of finding the state $(\mathbf{R}, \mathbf{K})$ occupied by a particle. We thus obtain

$$\frac{\mathrm{d}f}{\mathrm{d}t} = \frac{\partial f}{\partial t} + \nabla_{\mathbf{K}} f \cdot \frac{\mathrm{d}\mathbf{K}}{\mathrm{d}t} + \nabla_{\mathbf{R}} f \cdot \frac{\mathrm{d}\mathbf{R}}{\mathrm{d}t} = 0. \tag{5.6}$$

Consistently with the semi-classical picture of motion, we then use Eq.2.121 to express $\mathrm{d}\mathbf{R}/\mathrm{d}t$ and $\mathrm{d}\mathbf{K}/\mathrm{d}t$. For a wave-packet belonging to the nth branch of the energy relation $E_{B,n}(\mathbf{R}, \mathbf{K})$ with total energy $E(\mathbf{R}, \mathbf{K}) = E_{B,n}(\mathbf{R}, \mathbf{K}) + U(\mathbf{R})$:

$$\frac{\mathrm{d}\mathbf{R}}{\mathrm{d}t} = \mathbf{V}_{g,n} = \frac{1}{\hbar}\nabla_{\mathbf{K}} E(\mathbf{R}, \mathbf{K}), \tag{5.7}$$

where $\mathbf{V}_{g,n}(\mathbf{K})$ is the group velocity. Furthermore

$$\hbar\frac{\mathrm{d}\mathbf{K}}{\mathrm{d}t} = \mathbb{F} = -\frac{1}{\hbar}\nabla_{\mathbf{R}} E(\mathbf{R}, \mathbf{K}), \tag{5.8}$$

where, in the absence of magnetic fields and of band structure discontinuities, $\mathbb{F} = -e\mathbf{F}$ and $\mathbf{F}$ is the electric field. Consistently with the formulation of the semi-classical transport model in Section 2.5, in the following we consider for simplicity only a single band and drop the index n.

In the presence of scattering events perturbing the free-flights, we have $(\mathrm{d}f/\mathrm{d}t) \neq 0$, because the carriers can jump from one ballistic trajectory to a new one. By regarding scattering as a process taking place on a very short time scale and one that does not change the particle position, we can define a scattering-out flux $S_{out}(\mathbf{R}, \mathbf{K}, t)$ from the phase-space element centered in $(\mathbf{R}, \mathbf{K})$ as the number of carriers in $\mathbf{R}$ moving from state $\mathbf{K}$ to any state $\mathbf{K}'$ per unit time:

$$S_{out}(\mathbf{R}, \mathbf{K}, t) = f(\mathbf{R}, \mathbf{K}, t) \sum_{\mathbf{K}'} S(\mathbf{K}, \mathbf{K}')[1 - f(\mathbf{R}, \mathbf{K}', t)], \tag{5.9}$$

where $\mathbf{K}'$ is the wave-vector of the final state and $S(\mathbf{K}, \mathbf{K}')$ is the scattering rate from state $\mathbf{K}$ to state $\mathbf{K}'$ in units of s^{-1}. A methodology and an expression to compute $S(\mathbf{K}, \mathbf{K}')$ have been given in Section 2.5.4 and Eq.2.138. Note that, depending on the type of scattering mechanism and on the device, $S(\mathbf{K}, \mathbf{K}')$ may be an explicit function of $\mathbf{R}$. As in the rest of the book, summations over $\mathbf{K}$ are transformed into integrals as discussed in Section 3.5 for the 2D gas. The transformation for the 3D free carrier gas is expressed as

$$\sum_{\mathbf{K}} = \frac{n_{sp}\,\Omega}{(2\pi)^3}\int_{\Omega} \mathbf{dK}. \tag{5.10}$$

When summing over the final states, the normalization volume Ω cancels out in the product with $S(\mathbf{K}, \mathbf{K}')$, similarly to what happens to the normalization area in the 2D gas (see for example how the area term appearing in $S^{(w)}_{n,n'}(\mathbf{k}, \mathbf{k}')$ in Eq.4.273 disappears when computing the total scattering rate $S^{(w)}_n(\mathbf{k})$ in Eq.4.274).

By analogy with the $S_{out}(\mathbf{R}, \mathbf{K}, t)$ flux, we can define a scattering-in flux:

$$S_{in}(\mathbf{R}, \mathbf{K}, t) = [1 - f(\mathbf{R}, \mathbf{K}, t)] \sum_{\mathbf{K}'} S(\mathbf{K}', \mathbf{K}) f(\mathbf{R}, \mathbf{K}', t). \tag{5.11}$$

The total time derivative for the occupation function is then given by

$$\frac{\mathrm{d}f}{\mathrm{d}t} = S_{in} - S_{out}, \tag{5.12}$$

and the BTE thus reads:

$$\frac{\partial f}{\partial t} - \nabla_{\mathbf{K}} f \cdot \mathbf{F}\frac{e}{\hbar} + \nabla_{\mathbf{R}} f \cdot \mathbf{V}_g = S_{in} - S_{out}. \tag{5.13}$$

It is important to remember that this expression does not include phenomena such as Shockley–Read–Hall (SRH) generation–recombination or generation by band-to-band tunneling and impact ionization.[1]

The solution of the coupled BTE and Poisson equations with appropriate boundary conditions provides a self-consistent distribution function $f(\mathbf{R}, \mathbf{K}, t)$ which represents a complete solution of the transport and electrostatic problem. In fact, once $f(\mathbf{R}, \mathbf{K}, t)$ is known, all the internal quantities of interest in the device can be easily derived. For instance, the free electron concentration is straightforwardly computed as

$$n(\mathbf{R}, t) = \frac{1}{\Omega} \sum_{\mathbf{K}} f(\mathbf{R}, \mathbf{K}, t), \tag{5.14}$$

where the summation extends to all states in the first Brillouin zone and the normalization volume Ω cancels out when converting the summation into an integral (see Eq.5.10). The current density, in contrast, is given by:

$$\mathbf{J}(\mathbf{R}, t) = -\frac{e}{\Omega} \sum_{\mathbf{K}} \mathbf{V}_g(\mathbf{K}) f(\mathbf{R}, \mathbf{K}, t). \tag{5.15}$$

It is worth remembering that, when the full band structure of the material is considered, electron states are identified not only by the $\mathbf{K}$ vector but also by the band index n.

[1] This simplification is expected to be a good one if the focus is on nanoscale devices biased at low drain-source voltage $V_{DS} \le E_G/e$ (where E_G is the energy gap), so that generation by impact ionization is negligible, and the dimensions of the devices are so small that the net SRH recombination rate is also small. Neglecting band-to-band tunneling may imply large errors in the off-current, I_{OFF}, of devices with low-bandgap channel materials, but it is not expected to cause major errors in the on-current, I_{ON}.

Each band has its own distribution function $f_n(\mathbf{R}, \mathbf{K}, t)$ and one BTE, such as that of Eq.5.13 can then be written for each band. Since, according to the semi-classical model, the carriers cannot change band during ballistic flights, the different BTEs are coupled by inter-band scattering. Therefore the S_{out} and S_{in} fluxes of Eqs.5.9 and 5.11 should be calculated by summing also on the final bands.

As already mentioned, the assumption that the carriers do not change band during ballistic flights becomes unreasonable if we have crossing points between the bands. Unfortunately, such points exist in all the most common semiconductors. In silicon, for instance, the lowest conduction band crosses the second lowest one at the X point. The band energy at this crossing point is only $\approx 130meV$ above the conduction band minimum; therefore the carriers can easily populate the states at the crossing point even in the presence of moderate driving fields. The behavior of carriers at the crossing points goes beyond the semi-classical model [3] and needs to be tackled by specific rules, as discussed in Section 6.1.8.

5.1.2 The BTE for holes

So far we have derived the BTE considering electrons in the conduction band. In this section we summarize the main concepts related to hole transport and derive the BTE for holes.

First of all, we note that the summation in Eq.5.15 that defines the current density $\mathbf{J}$ should in principle extend to the electron states in all the bands of the crystal. However, it is easy to show that the contribution to $\mathbf{J}$ of a band which is completely filled with electrons is zero. In fact, since $\mathbf{V}_g(\mathbf{K})$ is an odd function of $\mathbf{K}$ (Eq.2.96), for every state $\mathbf{K}$ in a given band with group velocity $\mathbf{V}_g$ there is a corresponding state with velocity $[-\mathbf{V}_g]$ so that the total current density for a completely filled band is

$$\mathbf{J}_{\text{filled band}} = -\frac{e}{\Omega}\sum_{\mathbf{K}} \mathbf{V}_g(\mathbf{K}) = 0. \tag{5.16}$$

Therefore, on a macroscopic scale, current can flow only if the band is partially filled.

We can evaluate the contribution of a generic band to the current density using Eqs.5.15 and 5.16:

$$\begin{aligned}
\mathbf{J}(\mathbf{R}, t) &= -\frac{e}{\Omega}\sum_{\mathbf{K}} \mathbf{V}_g(\mathbf{K}) f(\mathbf{R}, \mathbf{K}, t) \\
&= \frac{e}{\Omega}\sum_{\mathbf{K}} \mathbf{V}_g(\mathbf{K}) - \frac{e}{\Omega}\sum_{\mathbf{K}} \mathbf{V}_g(\mathbf{K}) f(\mathbf{R}, \mathbf{K}, t) \\
&= \frac{e}{\Omega}\sum_{\mathbf{K}} \mathbf{V}_g(\mathbf{K}) \left[1 - f(\mathbf{R}, \mathbf{K}, t)\right] = \frac{e}{\Omega}\sum_{\mathbf{K}} \mathbf{V}_g(\mathbf{K}) f_h(\mathbf{R}, \mathbf{K}, t),
\end{aligned} \tag{5.17}$$

where the sums run over all $\mathbf{K}$ states in the first Brillouin zone and we have implicitly defined $f_h = (1 - f)$ as the probability for a state to be empty of electrons, i.e. occupied by holes. Note that the last expression of the current density $\mathbf{J}(\mathbf{R}, t)$ in Eq.5.17 can be interpreted as the current carried by a pseudo-particle with positive charge.

The choice of considering f_h instead of f is especially convenient for the valence band, where the empty states are fewer than the full, thus leading to more efficient calculations.

To derive the BTE and the equations of motion for the empty valence band states, we note first that the BTE was derived in Eqs.5.1 to 5.13 for the occupation probability f. Since $f = (1 - f_h)$, Eq.5.12 can be equivalently written as

$$\frac{\mathrm{d}f_h}{\mathrm{d}t} = -S_{in} + S_{out}, \tag{5.18}$$

where S_{in} and S_{out} are the scattering-in and scattering-out fluxes defined in Eq.5.11 and Eq.5.9, respectively. We now note that, according to Eq.5.11, S_{in} contains terms in the form $[1 - f(\mathbf{R}, \mathbf{K}, t)]f(\mathbf{R}, \mathbf{K}', t)$, thus in the form $f_h(\mathbf{R}, \mathbf{K}, t)[1 - f_h(\mathbf{R}, \mathbf{K}', t)]$. By comparison to Eq.5.9 we see that the scattering-in flux for electron states is indeed a scattering-out flux for holes. Similarly, the scattering-out flux for electron states is the scattering-in flux for hole states, so that we have $S_{in} = S^h_{out}$ and $S_{out} = S^h_{in}$. Thus the BTE for f_h reads:

$$\frac{\mathrm{d}f_h}{\mathrm{d}t} = S^h_{in} - S^h_{out}, \tag{5.19}$$

and it has exactly the same form as the one for the occupation function of the electron states (Eq.5.12).

In order to understand if and how we can interpret, for the valence band, the current associated with f_h (Eq.5.17) as due to pseudo-particles with positive charge, and then link the term $\mathrm{d}f_h/\mathrm{d}t$ in Eq.5.19 with the electric field and the band structure, we follow essentially the same reasoning as in [4]. As a first step, we start by observing that in the semi-classical model the electrons move according to Eq.2.121, where the total energy used to compute $\mathrm{d}\mathbf{R}/\mathrm{d}t$ and $\mathrm{d}\mathbf{K}/\mathrm{d}t$ is $E(\mathbf{R}, \mathbf{K}) = E_{B,n}(\mathbf{K}) + U(\mathbf{R})$, with $U(\mathbf{R}) = -e\phi(\mathbf{R})$ being the electron potential energy.

Equation 2.121 states that the group velocity of the valence band electrons is

$$\mathbf{V}_g(\mathbf{K}) = \frac{1}{\hbar}\nabla_{\mathbf{K}} E(\mathbf{R}, \mathbf{K}) = \frac{1}{\hbar}\nabla_{\mathbf{K}} E_{B,n}(\mathbf{K}), \tag{5.20}$$

and the wave-vector dynamics is

$$\frac{\mathrm{d}\mathbf{K}}{\mathrm{d}t} = -\frac{1}{\hbar}\nabla_{\mathbf{R}} E(\mathbf{R}, \mathbf{K}) = -\frac{1}{\hbar}\nabla_{\mathbf{R}} U(\mathbf{R}) = -\frac{e\mathbf{F}}{\hbar}, \tag{5.21}$$

where $\mathbf{F} = -e\nabla_{\mathbf{R}}\phi(\mathbf{R})$ is the electric field. Let us now define the pseudo-particle associated with the empty states (the *hole*) as having

$$\mathbf{K}_h = -\mathbf{K}, \qquad E_h(\mathbf{R}, \mathbf{K}_h) = -E(\mathbf{R}, -\mathbf{K}) = -E_{B,n}(-\mathbf{K}) - U(\mathbf{R}). \tag{5.22}$$

We now note that the definitions in Eq.5.22 yield:

$$\mathbf{V}_{g,h}(\mathbf{K}_h) = \frac{1}{\hbar}\nabla_{\mathbf{K}_h} E_h(\mathbf{R}, \mathbf{K}_h) = -\frac{1}{\hbar}\nabla_{(-\mathbf{K})} E_{B,n}(-\mathbf{K}) = \mathbf{V}_g(\mathbf{K}), \tag{5.23}$$

hence the group velocity of the empty states has the same expression as that of the filled electron states. In order to bring out the implications of the assumptions and definitions introduced so far, we consider a valence band at the Γ point with a simple spherical parabolic energy dispersion. The energy is given by

$$E(\mathbf{K}) = -\frac{\hbar^2 K^2}{2|m|} + U(\mathbf{R}), \tag{5.24}$$

where, consistently with Eq.2.61, we have introduced $m = (1/\hbar^2)\partial^2 E(K)/\partial K^2$. Note that, since the curvature of the band is negative, for the sake of clarity a minus sign and the absolute value of m appear in Eq.5.24. Substituting Eq.5.24 in Eq.5.22 we get:

$$E_h(\mathbf{K}_h) = \frac{\hbar^2 K_h^2}{2m_h} - U(\mathbf{R}), \tag{5.25}$$

where $m_h = (1/\hbar^2)\partial^2 E_h(K_h)/\partial K_h^2$ is now positive. Simple manipulations show that

$$\hbar \mathbf{V}_{g,h} = \nabla_{\mathbf{K}_h} E_h(\mathbf{K}_h) = \hbar^2 \mathbf{K}_h / m_h = -\hbar^2 \mathbf{K}/|m| = \nabla_{\mathbf{K}} E(\mathbf{K}) = \hbar \mathbf{V}_g. \tag{5.26}$$

We thus see that for a spherical parabolic energy relation, Eq.5.23 is consistent with a positive effective mass for holes and it is also consistent with Eq.5.17, where we have simply substituted f with $[1 - f_h]$ when computing the current, taking the same $\mathbf{V}_g$.

As for the driving force, by combining Eq.5.21 and Eq.5.22 we find:

$$\frac{\mathrm{d}\mathbf{K}_h}{\mathrm{d}t} = \frac{e\mathbf{F}}{\hbar}, \tag{5.27}$$

which implies that the driving force appearing in the BTE for holes should be that of positively charged particles, which is again consistent with Eq.5.17.

In summary, the equations above show that, if we define the hole energy axis as pointing in the opposite direction with respect to the electron energy axis and we invert the direction of the wave-vector (i.e. $E_h(\mathbf{R}, \mathbf{K}_h) = -E(\mathbf{R}, -\mathbf{K})$), the pseudo-particle corresponding to an empty electron state behaves as if having the usual expression for the group velocity, a positive effective mass (when dealing with spherical bands) and a driving force consistent with a positive charge.

The above discussion allows us to state that the holes in the valence band obey a BTE that reads

$$\frac{\partial f_h}{\partial t} + \nabla_{\mathbf{K}_h} f_h \cdot \mathbf{F}\frac{e}{\hbar} + \nabla_{\mathbf{R}} f_h \cdot \mathbf{V}_g = S_{in}^h - S_{out}^h, \tag{5.28}$$

which has exactly the same form as Eq.5.13, except for the sign of the second term of the l.h.s.

Note that for a complete description of transport, a BTE for electrons in the conduction band (Eq.5.13, with the associated current given by Eq.5.15) and a BTE for holes in the valence band (Eq.5.28, with the associated current given by Eq.5.17) has to be solved. However, since the MOS transistor is a unipolar device and impact ionization can generally be neglected in nanoscale transistors, far from equilibrium phenomena typically affect at most one type of carrier. Therefore the solution of the BTE for the carriers forming the inversion layer is usually sufficient; the transport of the second carrier type can be either neglected or treated with simple Drift-Diffusion equations (see Section 5.5).

We reiterate that, as already mentioned in the previous chapters, in this book we adopt an electron like convention for the hole energy consistent with Eq.5.22 and use the same expression for the electron and hole group velocity consistently with Eq.5.23. Consequently the hole wave-vector will be opposite with respect to the $\mathbf{K}$ of the electron state. Everywhere in the book, except in the present Section 5.1.2, the suffix h is dropped and we simply use $\mathbf{K}$, $\mathbf{V}_g$, and f to indicate the wave-vector, the group velocity, and the occupation probability, respectively, for both electrons and holes.

5.2 The BTE in inversion layers

5.2.1 Real and wave-vector space in a 2D carrier gas

Before discussing the BTE for the inversion layer, we need to define the appropriate phase-space. The task is not obvious and the discussion that follows is partly semi-empirical.

So far, in Chapters 3 and 4 we have considered situations where the device structure and the electrostatic potential are uniform in the transport plane. This condition is representative for instance of a large area MOS transistor with small V_{DS}, but not of a nanoscale MOSFET with $V_{DS} = V_{DD}$. In the uniform case, it has been possible to write a Schrödinger equation in the single real space variable z, having assumed a plane wave dependence of the envelope wave-function with respect to the position $\mathbf{r}$ in the transport plane. When, instead, we consider a nanoscale MOSFET, we have to solve the 2D (or 3D) Schrödinger equation. This is what is done by the so-called *full-quantum* simulation approaches [5]. In the semi-classical picture, instead, we proceed in a somewhat empirical way, resorting to solution of the 1D Schrödinger equation along the quantization direction z at different in-plane positions, and then using the solutions at different $\mathbf{r}$ to build an appropriate real and wave-vector space.

The procedure to generate the phase-space can be understood with the help of Fig.5.2. Assuming that the potential energy profile $U(\mathbf{r}, z)$ is slowly varying with $\mathbf{r}$, the one-dimensional Schrödinger equation is solved along z at each position $\mathbf{r}$:

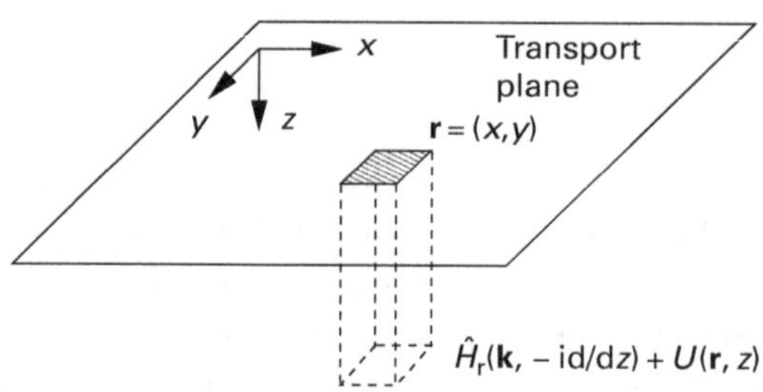

Figure 5.2 The semi-classical picture applied to inversion layers.

$$\left[\hat{H}_{\mathbf{r}}\left(\mathbf{k}, -\mathrm{i}\frac{\mathrm{d}}{\mathrm{d}z}\right) + U(\mathbf{r}, z)\right] \Phi_{n,\mathbf{k},\mathbf{r}}(z) = E_{n,\mathbf{k},\mathbf{r}} \Phi_{n,\mathbf{k},\mathbf{r}}(z), \tag{5.29}$$

where $\hat{H}_{\mathbf{r}}$ is the equivalent Hamiltonian of the bulk crystal (where the crystal potential does not explicitly appear), $U(\mathbf{r}, z)$ is the confining potential energy (which incorporates the $\mathbf{r}$ dependence imposed by the external biases) and $\Phi_{n,\mathbf{k},\mathbf{r}}(z)$ is the envelope wave-function. Following the assumption that $U(\mathbf{r}, z)$ is slowly varying with $\mathbf{r}$, in the region surrounding the position $\mathbf{r}$ we have $\Phi_{n,\mathbf{k},\mathbf{r}}(z) \simeq \exp(\mathrm{i}\mathbf{r} \cdot \mathbf{k})\xi_{n,\mathbf{k},\mathbf{r}}(z)$ (where the normalization prefactor has been omitted).

In the case of electrons described with the EMA we have $\hat{H}_{\mathbf{r}} = [\hat{E}_{CV}^{(\nu)}(-\mathrm{i}\nabla_{\mathbf{r}}, -\mathrm{i}d/dz) + E_{\nu 0}]$ and Eq.5.29 is equivalent to Eq.3.10 where the energy $E'_n = [E_n - E_{\nu 0}]$ is substituted by E_n, which is the total energy as defined in Eq.3.9. The operator $[-\mathrm{i}\nabla_{\mathbf{r}}]$ applied to the term $\exp(\mathrm{i}\mathbf{r} \cdot \mathbf{k})$ in $\Phi_{n,\mathbf{k},\mathbf{r}}(z)$ produces the $\mathbf{k}$ dependent operator $\hat{H}_{\mathbf{r}}(\mathbf{k}, -\mathrm{i}d/dz)$ which defines an eigenvalue problem whose envelope wave-function is $\xi_{n,\mathbf{k},\mathbf{r}}(z)$. In the case of holes described with the $\mathbf{k}\cdot\mathbf{p}$ approach we have $\hat{H}_{\mathbf{r}} = [\hat{\mathbf{H}}_{\mathbf{k}\cdot\mathbf{p}}(\mathbf{k}, -\mathrm{i}d/dz) + \mathbf{H}_{so}]$ and Eq.5.29 becomes equivalent to Eq.3.34, except that now the $\Phi_{n,\mathbf{k},\mathbf{r}}(z)$ are six component envelope wave-functions. As can be seen, $\hat{H}_{\mathbf{r}}$ may depend on $\mathbf{r}$ only if the material properties change with $\mathbf{r}$.

It is clear that the semi-classical transport approach neglects any quantum-mechanical effect in the transport plane, because no Schrödinger equation is solved in the $\mathbf{r}$ plane.

In Eq.5.29 $\mathbf{r}$ and $\mathbf{k}$ are parameters and the eigenvalue problem must be solved for any $\mathbf{r}$ and $\mathbf{k}$ value, unless some simplifications apply. In this sense, $\mathbf{r}$ and $\mathbf{k}$ define the real and wave-vector space for the inversion layer described according to the semi-classical picture. The solutions obtained at different $\mathbf{r}$ and $\mathbf{k}$ points are then assembled to form a continuum four-dimensional phase-space ($\mathbf{r}$,$\mathbf{k}$). Consistently with such a definition of the phase-space, in the following we denote $E_{n,\mathbf{k},\mathbf{r}}$ as $E_n(\mathbf{k}, \mathbf{r})$ and $\Phi_{n,\mathbf{k},\mathbf{r}}(z)$ as $\Phi_{n,\mathbf{k}}(\mathbf{r}, z)$, making explicit the point that $\mathbf{r}$ and $\mathbf{k}$ are treated as continuous variables.

In closing this section, we underline that we should have introduced two indexes for each subband: one for the valley, the other for the different subbands within that valley. However, to simplify the notation we adopt henceforth a single index running on both the valleys and the subbands.

5.2.2 The BTE without collisions

The previous considerations make it clear that the real and wave-vector spaces for a 2D carrier gas are two-dimensional and are identified by the two component vectors $\mathbf{r}$ and $\mathbf{k}$. The state of a particle is then identified by the wave-vector $\mathbf{k}$, the position $\mathbf{r}$ and the subband index n (see Chapter 3). Each subband is characterized by its own occupation function $f_n(\mathbf{r}, \mathbf{k}, t)$, which represents the probability of finding, at time t, a particle in the in-plane position $\mathbf{r}$, in the subband n and with in-plane wave-vector $\mathbf{k}$. The probability of finding the particle at different positions in the direction z normal to the transport plane is described by the envelope wave-function $\Phi_{n,\mathbf{k}}(\mathbf{r}, z)$ obtained solving Eq.5.29. Consistently, if $\Phi_{n,\mathbf{k}}(\mathbf{r}, z)$ is properly normalized, the term $f_n(\mathbf{r}, \mathbf{k}, t)$

$|\Phi_{n,\mathbf{k}}(\mathbf{r}, z)|^2 \mathrm{d}z = f_n(\mathbf{r}, \mathbf{k}, t)|\xi_{n,\mathbf{k}}(\mathbf{r}, z)|^2 \mathrm{d}z$ is the probability of finding a particle with in-plane wave-vector $\mathbf{k}$ belonging to subband n at the position $(\mathbf{r}, z)$.

Since in semi-classical dynamics the carriers cannot change subband in the absence of scattering, the transport in an inversion layer is described by an equation similar to Eq.5.6 for each subband:

$$\frac{\mathrm{d}f_n}{\mathrm{d}t} = \frac{\partial f_n}{\partial t} + \nabla_{\mathbf{k}} f_n \cdot \frac{\mathrm{d}\mathbf{k}}{\mathrm{d}t} + \nabla_{\mathbf{r}} f_n \cdot \frac{\mathrm{d}\mathbf{r}}{\mathrm{d}t} = 0. \tag{5.30}$$

These equations form a system of independent BTEs.

5.2.3 Driving force

In the absence of scattering, the total energy of a particle (E_n, defined as in Eq.3.9 for electrons described with the EMA) is conserved during motion:

$$\frac{\mathrm{d}E_n(\mathbf{r}, \mathbf{k})}{\mathrm{d}t} = \nabla_{\mathbf{k}} E_n \cdot \frac{\mathrm{d}\mathbf{k}}{\mathrm{d}t} + \nabla_{\mathbf{r}} E_n \cdot \frac{\mathrm{d}\mathbf{r}}{\mathrm{d}t} = 0. \tag{5.31}$$

In the most general case, E_n depends on $\mathbf{k}$ and $\mathbf{r}$ and it is provided by the solution of the Schrödinger equation 5.29.

The properties of the wave-packets discussed in Section 2.5 allow us to write:

$$\frac{\mathrm{d}\mathbf{r}}{\mathrm{d}t} = \mathbf{v}_g = \frac{1}{\hbar} \nabla_{\mathbf{k}} E_n(\mathbf{r}, \mathbf{k}), \tag{5.32}$$

which makes Eq.2.95 specific to the 2D case of particles in inversion layers. Equation 5.31 then yields:

$$\frac{\mathrm{d}\mathbf{k}}{\mathrm{d}t} = -\frac{1}{\hbar} \nabla_{\mathbf{r}} E_n(\mathbf{r}, \mathbf{k}). \tag{5.33}$$

Equations 5.32 and 5.33 are equivalent to Eq.2.121 for the free electron gas.

As seen in Chapter 3, in many cases (including electrons described by the EMA and holes described by the analytical band model in Section 3.3.3) it is possible to separate the $\mathbf{k}$ and $\mathbf{r}$ dependences of E_n and write (see Eq.3.15):

$$E_n(\mathbf{r}, \mathbf{k}) = E_{\nu 0}(\mathbf{r}) + \varepsilon_n(\mathbf{r}) + E_p(\mathbf{k}), \tag{5.34}$$

where $\varepsilon_n(\mathbf{r})$ is the energy at the bottom of the nth subband, obtained by solving the Schrödinger equation[2] in the z direction at the position $\mathbf{r}$, E_p is the in-plane energy and $E_{\nu 0}(\mathbf{r})$ the energy at the bottom of the ν valley. In such cases we have

$$\frac{\mathrm{d}\mathbf{r}}{\mathrm{d}t} = \mathbf{v}_g = \frac{1}{\hbar} \nabla_{\mathbf{k}} E_p(\mathbf{k}), \tag{5.35}$$

and

$$\frac{\mathrm{d}\mathbf{k}}{\mathrm{d}t} = -\frac{1}{\hbar} \nabla_{\mathbf{r}} \left[E_{\nu 0}(\mathbf{r}) + \varepsilon_n(\mathbf{r}) \right]. \tag{5.36}$$

[2] Equation 3.16 for electrons described with the EMA approach, Eq.3.42 for holes described by the analytical model in Section 3.3.3.

Comparing Eq.5.36 with Eq.5.8 shows that, in the semi-classical transport approach, the driving force for particle motion in the transport plane is the gradient of the subband energy:

$$\mathbb{F} = -\nabla_{\mathbf{r}}[E_{\nu 0}(\mathbf{r}) + \varepsilon_n(\mathbf{r})]. \tag{5.37}$$

We show explicitly in Chapters 9 and 10 that $E_{\nu 0}$ is sensitive to strain and to the channel material, so that $E_{\nu 0}$ becomes a function of $\mathbf{r}$ if non-uniform strain or materials are present in the device.

The use of Eq.5.36 is justified only if the $\mathbf{r}$ and $\mathbf{k}$ dependence of the total energy can be separated. In hole inversion layers described with the $\mathbf{k}{\cdot}\mathbf{p}$ method (Section 3.3.1), this separation does not apply and we should thus use the more general expressions Eqs.5.32 and 5.33. In this case, closed form expressions for the driving force can be found only in simple situations, as in the example below.

Example 5.1: Long channel MOSFET at low drain-source bias.
This situation is typical for the experimental setup employed to measure the low-field mobility described in Section 7.1.1. The shape of the confining potential energy well is essentially the same in all points of the transport plane, because $U(\mathbf{r}, z)$ changes very slowly with $\mathbf{r}$ and thus $|\nabla_{\mathbf{r}} U| \ll |\mathrm{d}U/\mathrm{d}z|$. There is only a rigid shift of the confining well in the field direction, hence the minimum of each subband is shifted together with the channel potential. In this case, the driving field is the same for all the subbands and it is equal to the electric field in the transport direction. Consequently

$$\frac{\mathrm{d}k_x}{\mathrm{d}t} = \pm\frac{1}{\hbar} e F_x \simeq \frac{e}{\hbar}\frac{|V_{DS}|}{L}, \tag{5.38}$$

where we have used the sign $-$ and $+$ for electrons and holes, respectively, V_{DS} is the drain voltage and L the channel length.

In the following part of the section we assume that the driving force is given by Eq.5.37 and we consider electron inversion layers described with the EMA. In the general case, the change of $\varepsilon_n(\mathbf{r})$ with position cannot be straightforwardly related to the source-drain potential or the surface potential. The shape of the potential well changes with x and the reason for that is simple: at different positions along the channel we may have different inversion and depletion charges that, through the coupling between carrier transport and electrostatics, result in different shapes of the potential well.

Example 5.2: *n*-MOSFET on a thin uniform SOI layer.
If we assume an ultra-thin body SOI device as in Fig.5.3, the electrostatic potential profile can be approximately written as $\phi(x, z) \simeq \phi_x(x) + \phi_{qm}(z)$. Also in this case

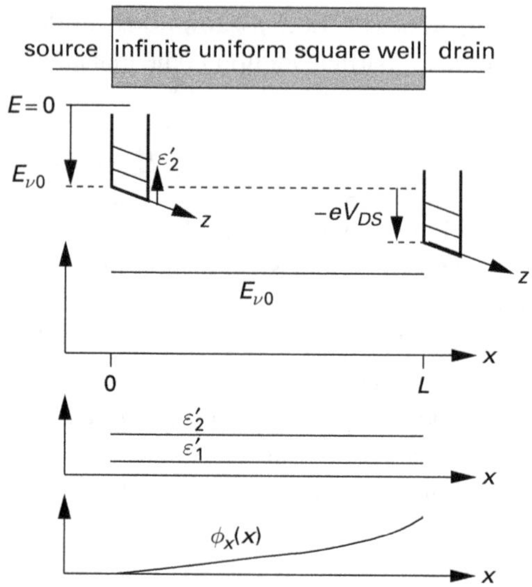

Figure 5.3 A MOSFET in a thin uniform SOI layer. We set as reference energy the vacuum level, so that $E_{\nu 0}$ is the same for all the six valleys of silicon and it is equal to $-\chi_{Si}$. For simplicity we assume a box confining well.

the vertical profile is the same regardless of the position along the channel. Separation of the z and x variables in the Schrödinger equation gives

$$\varepsilon_n(x) = \varepsilon'_n - e\phi_x(x), \tag{5.39}$$

where ε'_n are the subband energy levels obtained by solving the one-dimensional Schrödinger equation, Eq.3.16, at the position where $\phi_x = 0$ and by using $U(x, z) = -e\phi_{qm}(z)$ as confining potential (i.e. differently from what was done in Section 3.7, where U was defined in Eq.3.107 including the electron affinity, here χ enters the $E_{\nu 0}$ term). We can thus write

$$E_n = E_{\nu 0} + \varepsilon'_n + E_p(\mathbf{k}) - e\phi_x(x). \tag{5.40}$$

This result is sketched in Fig.5.3, where $E_{\nu 0}$ and ε'_n are constant along the channel, since we assume that the material properties do not change along x. As a result

$$\frac{\mathrm{d}k_x}{\mathrm{d}t} = \frac{e}{\hbar}\frac{\mathrm{d}\phi_x}{\mathrm{d}x}. \tag{5.41}$$

Example 5.3: *n*-MOSFET on a thin non-uniform SOI layer.
Let us now consider a thin SOI film with a highly non-uniform potential profile in the x direction and non-uniform SOI film thickness, as sketched in Fig.5.4. Compared to the previous example, since the thickness of the well changes with position, we can write $\phi(x, z) = \phi_x(x) + \phi_{qm}(x, z)$, where $\phi_{qm}(x, z)$ is assumed to be a box profile with a width dependent on the x position. Furthermore, we assume that the properties

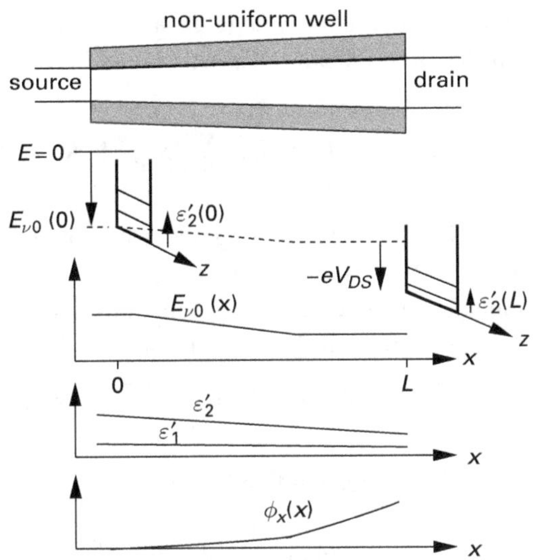

Figure 5.4 A MOSFET in a thin SOI layer with a non-uniform thickness and composition. Note that a change of the channel material may not only result in a modification of $E_{\nu 0}$, as shown, but also of the ε_n' profiles because of possible changes in the quantization mass.

that determine $E_{\nu 0}$ change along the channel. This could be the case, for instance, in a device with non-uniform strain profile along the channel, as will be apparent after reading Chapter 9. Referring to Eqs.5.34 and 5.39 we now see that all variables depend on the position x, so that the x derivative of Eq.5.40 gives:

$$\frac{\mathrm{d}k_x}{\mathrm{d}t} = -\frac{1}{\hbar}\left(\frac{\mathrm{d}E_{\nu 0}}{\mathrm{d}x} + \frac{\mathrm{d}\varepsilon_n'}{\mathrm{d}x} - e\frac{\mathrm{d}\phi_x}{\mathrm{d}x}\right). \tag{5.42}$$

It is apparent from the discussion and the examples above that in the most general case all the terms in Eq.5.34 contribute to the driving force because of:

1. the change of the electrostatic potential from source to drain;
2. the change of the shape of the potential well at different positions along the channel;
3. the change of the band structure of the underlying bulk material due to changes in the strain conditions or in the semiconductor composition along the channel.

All these effects are accounted for by the solution of the Schrödinger equation, Eq.5.29, with an appropriate choice of the potential energy profile $U(\mathbf{r}, z)$. For example, for electrons described by the EMA model we must solve Eq.3.16 with $U(z)$ replaced by $U_\nu(\mathbf{r}, z) = -e\phi(\mathbf{r}, z) - E_{\nu 0}$ (see also Eq.2.83).

5.2.4 Scattering

The inclusion of scattering in the BTE for inversion layers follows the same reasoning as that outlined in Section 5.1.1 for the free-carrier gas. The scattering integral out of the

nth subband (i.e. the rate of change of the occupation probability due to the scattering events from state $\mathbf{k}$ in subband n) must account for inter-subband transitions (Chapter 4) and reads:

$$S_{out,n} = f_n(\mathbf{r}, \mathbf{k}, t) \sum_{\mathbf{k}', n'} S_{n,n'}(\mathbf{k}, \mathbf{k}', \mathbf{r})[1 - f_{n'}(\mathbf{r}, \mathbf{k}', t)], \tag{5.43}$$

where $S_{n,n'}(\mathbf{k}, \mathbf{k}', \mathbf{r})$ is the scattering rate from state $\mathbf{k}$ in subband n to state $\mathbf{k}'$ in subband n' at location $\mathbf{r}$. Note that $S_{n,n'}(\mathbf{k}, \mathbf{k}', \mathbf{r})$ depends on position because the scattering matrix elements depend on the wave-functions at position $\mathbf{r}$ (Eq.4.4).
The scattering integral entering the nth subband is:

$$S_{in,n} = [1 - f_n(\mathbf{r}, \mathbf{k}, t)] \sum_{\mathbf{k}', n'} f_{n'}(\mathbf{r}, \mathbf{k}', t) S_{n',n}(\mathbf{k}', \mathbf{k}, \mathbf{r}). \tag{5.44}$$

The BTE for the nth subband finally reads:

$$\frac{\partial f_n}{\partial t} - \nabla_{\mathbf{k}} f_n \cdot \frac{1}{\hbar} \nabla_{\mathbf{r}} E_n + \nabla_{\mathbf{r}} f_n \cdot \mathbf{v}_g = S_{in,n} - S_{out,n}. \tag{5.45}$$

5.2.5 Macroscopic quantities

Equation 5.45 represents a system of coupled BTEs. Any macroscopic quantity of interest can be readily calculated once the f_n are known. For example, the carrier concentration is:

$$n(\mathbf{r}, z, t) = \frac{1}{A} \sum_n \sum_{\mathbf{k}} |\Phi_{n,\mathbf{k}}(\mathbf{r}, z)|^2 f_n(\mathbf{r}, \mathbf{k}, t), \tag{5.46}$$

where $|\Phi_{n,\mathbf{k}}(\mathbf{r}, z)|^2$ is assumed to be normalized when integrated over z at given $\mathbf{r}$. The normalization area A cancels when the summation over $\mathbf{k}$ is transformed into an integral.

Concerning the current, let us assume that the device is translationally invariant in the y direction. The current in the x direction per unit width in the y direction is

$$\frac{I_x(x, t)}{W} = \pm \frac{e}{A} \sum_n \sum_{\mathbf{k}} v_{gx}(\mathbf{k}, x) f_n(x, \mathbf{k}, t), \tag{5.47}$$

where the $+$ sign is used for holes and the $-$ for electrons. It is understood that in the EMA approach $v_{gx}(\mathbf{k}, x)$ takes different values in the different valleys, and that it may depend on x if the channel material changes along x. For holes described with the $\mathbf{k}{\cdot}\mathbf{p}$ approach, v_{gx} depends on x just because the in-plane dispersion relationship depends on the shape of the potential well, which changes with x.

5.2.6 Detailed balance at equilibrium

The terms $(1 - f_n)$ and $(1 - f_{n'})$ in the collision integral (Eqs.5.43, 5.44) make the BTEs of the subbands a coupled system of *non-linear* integro-differential equations.

It would be tempting to neglect these terms to reduce the system to a linear problem. We see in Chapter 7, however, that carrier degeneracy is quite strong in the inversion layer of modern MOSFETs, so that enforcement of Pauli's exclusion principle cannot be avoided. To underline the importance of Pauli's exclusion principle, we demonstrate here that the $(1 - f_n)$ and $(1 - f_{n'})$ terms *must be* included in order to recover the Fermi–Dirac expression of the statistical distribution of the carrier gas at equilibrium. The derivation also provides an example of the so-called detailed balance which is a powerful methodology to verify the correct implementation of complex scattering models.

Indeed, at equilibrium the number of transitions per unit time between the state $(n, \mathbf{k})$ and the state $(n', \mathbf{k}')$ must balance the number of opposite transitions between $(n', \mathbf{k}')$ and $(n, \mathbf{k})$. By neglecting the $(1 - f_n)$ and $(1 - f_{n'})$ terms, this means:

$$f_n(\mathbf{r}, \mathbf{k}) S_{n,n'}(\mathbf{k}, \mathbf{k}', \mathbf{r}) = f_{n'}(\mathbf{r}, \mathbf{k}') S_{n',n}(\mathbf{k}', \mathbf{k}, \mathbf{r}). \tag{5.48}$$

If the initial and final states have the same energy, $E_n(\mathbf{k}) = E_{n'}(\mathbf{k}')$ (that is, they are connected by elastic transitions such as, for instance, elastic acoustic phonons, surface roughness, ionized impurity scattering), it is evident from the treatment given in Chapter 4 that $S_{n,n'}(\mathbf{k}, \mathbf{k}', \mathbf{r}) = S_{n',n}(\mathbf{k}', \mathbf{k}, \mathbf{r})$. Hence we have $f_n(\mathbf{r}, \mathbf{k}) = f_{n'}(\mathbf{r}, \mathbf{k}')$, meaning that, provided the states have the same total energy, the occupation probability is independent of the subband and $\mathbf{k}$ vector. Therefore we can write $f_n(\mathbf{r}, \mathbf{k}) = f(\mathbf{r}, E_n)$.

If the initial and final states have different total energies they are coupled by inelastic scattering such as, for instance, the intervalley phonons described in Section 4.6.5. In this latter case, if we assume $E_{n'}(\mathbf{k}') = E' > E = E_n(\mathbf{k})$ the transitions between states $(n, \mathbf{k})$ and $(n', \mathbf{k}')$ are due to phonon absorption, whereas the ones between $(n', \mathbf{k}')$ and $(n, \mathbf{k})$ to phonon emission. The phonon energy is $\hbar\omega_{ph} = (E' - E)$. Since the scattering rates for phonon emission and absorption only differ by the term $(1 + n_{ph})$ or n_{ph} (see Eq.4.286), neglecting the $(1 - f)$ terms the net balance gives:

$$f(\mathbf{r}, E) n_{ph} = f(\mathbf{r}, E')(1 + n_{ph}). \tag{5.49}$$

Upon substitution of the phonon occupation probability

$$n_{ph} = \frac{1}{\exp\left[\hbar\omega_{ph}/(K_B T)\right] - 1} \tag{5.50}$$

one finds:

$$f(\mathbf{r}, E') = f(\mathbf{r}, E) \exp\left(\frac{E - E'}{K_B T}\right). \tag{5.51}$$

Since this equation is valid for all energies, including the lowest one, $E = E_{min}$, we can define $A(\mathbf{r}) = f(\mathbf{r}, E_{min}) \exp(E_{min}/K_B T)$ and write:

$$f(\mathbf{r}, E') = A(\mathbf{r}) \exp\left(-\frac{E'}{K_B T}\right), \tag{5.52}$$

that depends neither on the subband nor on the $\mathbf{k}$ but only on the total energy. We can thus write:

$$f_n(\mathbf{r},\mathbf{k}) = A(\mathbf{r})\exp\left(-\frac{E_n(\mathbf{r},\mathbf{k})}{K_BT}\right). \tag{5.53}$$

Now, it is easy to show that the constant A is actually independent of $\mathbf{r}$. In fact, substitution of Eq.5.53 into the multi-subband BTE (Eq.5.45) and the consideration that at equilibrium $\partial f_n/\partial t = 0$ and $S_{out,n} = S_{in,n}$ allow us to write:

$$\begin{aligned}
&\nabla_{\mathbf{r}} f_n \cdot \nabla_{\mathbf{k}} E_p - \nabla_{\mathbf{k}} f_n \cdot \nabla_{\mathbf{r}} \varepsilon_n \\
&\quad = \nabla_{\mathbf{r}} A \cdot \nabla_{\mathbf{k}} E_p \exp\left[-\frac{E_p-\varepsilon_n}{K_BT}\right] + A\exp\left[-\frac{E_p-\varepsilon_n}{K_BT}\right]\left(-\frac{\nabla_{\mathbf{r}}\varepsilon_n}{K_BT}\right)\cdot\nabla_{\mathbf{k}} E_p \\
&\quad\quad - A\exp\left[-\frac{E_p-\varepsilon_n}{K_BT}\right]\left(-\frac{\nabla_{\mathbf{k}} E_p}{K_BT}\right)\cdot\nabla_{\mathbf{r}}\varepsilon_n = 0,
\end{aligned} \tag{5.54}$$

where we have assumed without loss of generality $E_{\nu 0} = 0$, hence $E_n = \varepsilon_n(\mathbf{r}) + E_p(\mathbf{k})$. It is easy to see that Eq.5.54 gives $\nabla_{\mathbf{r}} A = 0$ since the second and the third terms on the l.h.s. cancel out. We may choose to set $A = \exp(-E_F/K_BT)$, and thus rewrite Eq.5.53 as:

$$f_n(\mathbf{r},\mathbf{k}) = \exp\left(-\frac{E_n(\mathbf{r},\mathbf{k}) - E_F}{K_BT}\right), \tag{5.55}$$

that is the well-known Maxwell–Boltzmann function with Fermi energy E_F.

If instead we consider the $(1 - f_n)$ and $(1 - f_{n'})$ terms, the balance between the transition from $(n,\mathbf{k})$ and $(n',\mathbf{k}')$ and vice versa gives:

$$f(\mathbf{r},E)n_{ph}[1 - f(\mathbf{r},E')] = f(\mathbf{r},E')(1+n_{ph})[1 - f(\mathbf{r},E)], \tag{5.56}$$

that becomes

$$\frac{1}{f(\mathbf{r},E')} - 1 = \left(\frac{1}{f(\mathbf{r},E)} - 1\right)\exp\left(\frac{E'-E}{K_BT}\right), \tag{5.57}$$

which is valid for all energies E and E'. Setting the minimum energy equal to zero, we consider $E' = 0$ and define $A(\mathbf{r}) = f(\mathbf{r},0)^{-1} - 1$, so that

$$\frac{1}{f(\mathbf{r},E)} - 1 = A(\mathbf{r})\exp\left(\frac{E}{K_BT}\right). \tag{5.58}$$

Since it can be demonstrated that also in this case A is constant, we can use again the position $A = \exp(-E_F/K_BT)$, thus obtaining:

$$f_n(\mathbf{r},\mathbf{k}) = \frac{1}{1+\exp\left(\frac{E_n(\mathbf{r},\mathbf{k})-E_F}{K_BT}\right)}, \tag{5.59}$$

which is the well-known expression for the Fermi–Dirac distribution.

5.3 The BTE for one-dimensional systems

A short note is in order with respect to extension of the semi-classical approach based on the multi-subband BTE to 1D systems such as, e.g. nanowires. The difference with respect to Eq.5.45 is that now the transport space and the k-space are one-dimensional, i.e. we consider particle propagation only in the transport direction x, the carrier distribution in subband n reads $f_n(x, k_x, t)$, and the probability density of finding the carrier along the other two dimensions in real space is described by the solution of the two-dimensional Schrödinger equation in the plane normal to the transport direction.

Again, we have a BTE for each subband that reads:

$$\frac{\partial f_n}{\partial t} - \frac{\partial f_n}{\partial k_x}\frac{1}{\hbar}\frac{\partial E_n(x,k_x)}{\partial x} + \frac{\partial f_n}{\partial x}v_g = S_{in,n} - S_{out,n}, \tag{5.60}$$

where the expression for the collision integral is the same as for the 2D gas case. The BTE for 1D systems can be efficiently solved by numerical techniques without resorting to the MC method. The reader is referred to [6, 7] for a more extensive treatment of the BTE in 1D systems.

5.4 Momentum relaxation time approximation

The Momentum Relaxation Time (MRT) approximation is a very convenient approach to determining transport parameters such as the carrier mobility and diffusion coefficients for both bulk semiconductors and low dimensional systems [1, 8–13]. We discuss in detail below the MRT approach for inversion layers.

The MRT is an approximate solution of the BTE valid for small displacements from equilibrium and for a uniform transport condition, hence a case where the occupation functions and all the macroscopic quantities are independent of the position $\mathbf{r}$ in the transport plane. In particular, if we suppose that the carrier gas is subject to a small electric field $\mathbf{F} = F_\alpha \hat{\mathbf{e}}_\alpha$ (where $\hat{\mathbf{e}}_\alpha$ is the unit vector in direction $\alpha = x$ or y), then the basic assumption behind the MRT is that the scattering integral for the subband i in the r.h.s. of Eq.5.45 can be written as

$$S_{in} - S_{out} = -\frac{f_i(\mathbf{k}) - f_0(E_i(\mathbf{k}))}{\tau_{i,\alpha}(\mathbf{k})} = -\frac{\delta f_i(\mathbf{k})}{\tau_{i,\alpha}(\mathbf{k})}, \tag{5.61}$$

where $f_0(E)$ is the equilibrium Fermi–Dirac occupation function defined in Eq.3.92.

In the following derivations the index i denotes both the valley and the subband when more valleys are relevant for the problem.

Equation 5.61 is an implicit definition of the non-null momentum relaxation time $\tau_{i,\alpha}(\mathbf{k})$ of the subband i for an electric field in the α direction, where $E_i(\mathbf{k})$ is the energy corresponding to $\mathbf{k}$ in the subband i. The deviation $\delta f_i(\mathbf{k})$ of the occupation function $f_i(\mathbf{k})$ with respect to $f_0(E_i(\mathbf{k}))$ is also implicitly defined as

$$\delta f_i(\mathbf{k}) = f_i(\mathbf{k}) - f_0(E_i(\mathbf{k})). \tag{5.62}$$

In order to clarify the meaning of $\tau_{i,\alpha}(\mathbf{k})$ we observe that, if at a given time the electric field $\mathbf{F}$ disappears, then Eq.5.45 becomes

$$\frac{\partial f_i(\mathbf{k})}{\partial t} = -\frac{f_i(\mathbf{k}) - f_0(E_i(\mathbf{k}))}{\tau_{i,\alpha}(\mathbf{k})}. \tag{5.63}$$

Thus $\tau_{i,\alpha}(\mathbf{k})$ is the time constant governing the decay of the occupation function of the state $(i,\mathbf{k})$ to its equilibrium value.

Since the occupation $f_i(\mathbf{k})$ produced by the stimulus $\mathbf{F}$ yields a non-null average value of the crystal momentum $\hbar\mathbf{k}$ and of the carrier velocity, then $\tau_{i,\alpha}(\mathbf{k})$ can be considered the time scales over which the crystal momentum is relaxed by the scattering mechanisms; each state $(i,\mathbf{k})$ has its own value of momentum relaxation time.

If, instead of considering the transients of the occupation function, we now focus on the stationary, spatially uniform problem corresponding to a weak electric field $\mathbf{F}$, then the $(\nabla_{\mathbf{r}} E_i)$ in Eq.5.45 is proportional to $\mathbf{F}$ and the equation can be rewritten as

$$\frac{\mp e\mathbf{F} \cdot \nabla_{\mathbf{k}} f_i(\mathbf{k})}{\hbar} = -\frac{\delta f_i(\mathbf{k})}{\tau_{i,\alpha}(\mathbf{k})}, \tag{5.64}$$

where the minus and plus signs are for electrons and holes, respectively.
We can now write the gradient of $f_i(\mathbf{k}) = \delta f_i(\mathbf{k}) + f_0(E_i(\mathbf{k}))$ with respect to $\mathbf{k}$ as

$$\nabla_{\mathbf{k}} f_i(\mathbf{k}) = \frac{\partial f_0(E_i(\mathbf{k}))}{\partial E} \nabla_{\mathbf{k}} E_i(\mathbf{k}) + \nabla_{\mathbf{k}}\, \delta f_i(\mathbf{k}), \tag{5.65}$$

and substitute it in Eq.5.64. By dropping the scalar product of $\mathbf{F}$ to $\nabla_{\mathbf{k}} \delta f_i(\mathbf{k})$ (in the spirit of a first order approximation with respect to a small field $\mathbf{F}$), Eq.5.64 yields

$$\begin{aligned} \delta f_i(\mathbf{k}) &= \pm e\, \tau_{i,\alpha}(\mathbf{k})\, \mathbf{F} \cdot \mathbf{v}_i(\mathbf{k})\, \frac{\partial f_0(E_i(\mathbf{k}))}{\partial E} \\ &= \pm e \tau_{i,\alpha}(\mathbf{k})\; F_\alpha\; v_{i,\alpha}(\mathbf{k})\, \frac{\partial f_0(E_i(\mathbf{k}))}{\partial E}, \end{aligned} \tag{5.66}$$

where $\mathbf{v}_i(\mathbf{k})$ and $v_{i,\alpha}(\mathbf{k})$ denote the group velocity in the subband i and its component in the direction $\alpha = x$ or y.

Equation 5.66 validates the statement that the MRT approximation is an approximate solution of the BTE, in fact, if $\tau_{i,\alpha}(\mathbf{k})$ is known, then Eq.5.66 uniquely determines $f_i(\mathbf{k}) = f_0(E_i(\mathbf{k})) + \delta f_i(\mathbf{k})$. Consequently, all the macroscopic quantities can be expressed as functions of the momentum relaxation times $\tau_{i,\alpha}(\mathbf{k})$ (see Eq.5.102 for the current density).

The procedure for calculating mobility by using the MRT consists of two steps: the first is determination of the $\tau_{i,\alpha}(\mathbf{k})$ for all the relevant subbands and scattering mechanisms; the second is calculation of the mobility starting from Eq.5.66. The two steps are described in the following sections.

5.4.1 Calculation of the momentum relaxation time

As already noted, the basic assumption behind the MRT approach is writing of the scattering integral for the ith subband $(S_{in,i} - S_{out,i})$ according to Eq.5.61, that is in a form which is simply proportional to the deviation $\delta f_i(\mathbf{k})$ of the occupation function in the ith subband with respect to $f_0(E_i(\mathbf{k}))$; the proportionality factor is the inverse of the momentum relaxation time $\tau_{i,\alpha}(\mathbf{k})$. We now write the scattering integral for the ith subband $(S_{in,i}-S_{out,i})$ by using Eqs.5.43 and 5.44 for a stationary and uniform system, where the occupation functions do not depend on either the position $\mathbf{r}$ or the time t, so that we have

$$S_{in,i} - S_{out,i} = \sum_{j,\mathbf{k}'} f_j(\mathbf{k}')S_{j,i}(\mathbf{k}',\mathbf{k})[1 - f_i(\mathbf{k})] - f_i(\mathbf{k}) \sum_{j,\mathbf{k}'} S_{i,j}(\mathbf{k},\mathbf{k}')[1 - f_j(\mathbf{k}')]. \tag{5.67}$$

The general procedure for calculation of the momentum relaxation time $\tau_{i,\alpha}(\mathbf{k})$ can be clarified by employing a manipulation of Eq.5.67 able to express $(S_{in,i}-S_{out,i})$ as a quantity proportional to $\delta f_i(\mathbf{k})$. When such a task has been accomplished, the $\tau_{i,\alpha}(\mathbf{k})$ can be readily identified in the proportionality factor between $(S_{in,i}-S_{out,i})$ and $\delta f_i(\mathbf{k})$.

There are two ingredients to be used in the derivations. The first is the detailed balance stating that, at equilibrium, the flux from state $(i,\mathbf{k})$ to state $(j,\mathbf{k}')$ must equal the flux from $(j,\mathbf{k}')$ to $(i,\mathbf{k})$. Since the Fermi occupation function $f_0(E_i(\mathbf{k}))$ depends on $\mathbf{k}$ only through the energy $E_i(\mathbf{k})$, application of the detailed balance to Eq.5.67 implies

$$f_0(E')S_{j,i}(\mathbf{k}',\mathbf{k})[1 - f_0(E)] = f_0(E)\, S_{i,j}(\mathbf{k},\mathbf{k}')[1 - f_0(E')], \tag{5.68}$$

where, for convenience of notation, we have introduced the symbol E to denote the energy $E_i(\mathbf{k})$ corresponding to the wave-vector $\mathbf{k}$ in the subband i and the symbol E' to denote the energy $E_j(\mathbf{k}')$ corresponding to $\mathbf{k}'$ in the subband j.

The second ingredient is writing of the $\delta f_i(\mathbf{k})$ and $\delta f_j(\mathbf{k}')$ in terms of the momentum relaxation time itself. By recalling Eq.5.66 we have

$$\delta f_i(\mathbf{k}) = \pm e\tau_{i,\alpha}(\mathbf{k})\, F_\alpha\, v_{i,\alpha}(\mathbf{k}) \frac{\partial f_0(E)}{\partial E}, \tag{5.69a}$$

$$\delta f_j(\mathbf{k}') = \pm e\tau_{j,\alpha}(\mathbf{k}')\, F_\alpha\, v_{j,\alpha}(\mathbf{k}') \frac{\partial f_0(E')}{\partial E}, \tag{5.69b}$$

where α is the direction x or y of the electric field.

General formulation for the momentum relaxation time

We start by substituting in Eq.5.67 the $f_i(\mathbf{k})$ and $f_j(\mathbf{k}')$ expressed as in Eq.5.62; all the terms that contain only the Fermi occupation function $f_0(E)$ give a null contribution to the scattering integral, because of the detailed balance in Eq.5.68. Then, by retaining only the first order terms with respect to $\delta f_i(\mathbf{k})$ and $\delta f_j(\mathbf{k}')$, we obtain

$$\begin{aligned} S_{in,i} - S_{out,i} = &\sum_{j,\mathbf{k}'} S_{j,i}(\mathbf{k}',\mathbf{k})\, [\delta f_j(\mathbf{k}')(1 - f_0(E)) - f_0(E')\delta f_i(\mathbf{k})] \\ &- \sum_{j,\mathbf{k}'} S_{i,j}(\mathbf{k},\mathbf{k}')[\delta f_i(\mathbf{k})(1 - f_0(E')) - f_0(E)\delta f_j(\mathbf{k}')]. \end{aligned} \tag{5.70}$$

We now use Eq.5.68 to write the scattering rate $S_{j,i}(\mathbf{k}', \mathbf{k})$ as

$$S_{j,i}(\mathbf{k}', \mathbf{k}) = S_{i,j}(\mathbf{k}, \mathbf{k}') \frac{f_0(E)(1 - f_0(E'))}{f_0(E')(1 - f_0(E))}, \tag{5.71}$$

and use Eq.5.71 to eliminate $S_{j,i}(\mathbf{k}', \mathbf{k})$ in the first sum in the r.h.s. of Eq.5.70. By so doing we obtain

$$S_{in,i} - S_{out,i} = \sum_{j,\mathbf{k}'} S_{i,j}(\mathbf{k}, \mathbf{k}') \left[\delta f_j(\mathbf{k}') \frac{f_0(E)}{f_0(E')} - \delta f_i(\mathbf{k}) \frac{1 - f_0(E')}{1 - f_0(E)} \right]. \tag{5.72}$$

In order to proceed further we consider the states $(i,\mathbf{k})$ with a non-null $\delta f_i(\mathbf{k})$. In fact, according to Eq.5.69, the states having a null $\delta f_i(\mathbf{k})$ have a null velocity $v_{i,\alpha}(\mathbf{k})$, hence they do not contribute to the current and the mobility. Thus, for states corresponding to $\delta f_i(\mathbf{k}) \neq 0$, we rewrite Eq.5.72 as

$$S_{in,i} - S_{out,i} = -\delta f_i(\mathbf{k}) \sum_{j,\mathbf{k}'} S_{i,j}(\mathbf{k}, \mathbf{k}') \left[\frac{1 - f_0(E')}{1 - f_0(E)} - \frac{f_0(E)}{f_0(E')} \frac{\delta f_j(\mathbf{k}')}{\delta f_i(\mathbf{k})} \right]. \tag{5.73}$$

We can now use Eqs.3.93 and 5.69 to write

$$\frac{\delta f_j(\mathbf{k}')}{\delta f_i(\mathbf{k})} = \frac{\tau_{j,\alpha}(\mathbf{k}')}{\tau_{i,\alpha}(\mathbf{k})} \frac{v_{j,\alpha}(\mathbf{k}')}{v_{i,\alpha}(\mathbf{k})} \frac{\frac{\partial f_0(E')}{\partial E}}{\frac{\partial f_0(E)}{\partial E}} = \frac{\tau_{j,\alpha}(\mathbf{k}')}{\tau_{i,\alpha}(\mathbf{k})} \frac{v_{j,\alpha}(\mathbf{k}')}{v_{i,\alpha}(\mathbf{k})} \frac{f_0(E')(1 - f_0(E'))}{f_0(E)(1 - f_0(E))}. \tag{5.74}$$

By substituting Eq.5.74 in Eq.5.73 we have

$$S_{in,i} - S_{out,i} = -\delta f_i(\mathbf{k}) \sum_{j,\mathbf{k}'} S_{i,j}(\mathbf{k}, \mathbf{k}') \left[\frac{1 - f_0(E')}{1 - f_0(E)} \right] \left[1 - \frac{\tau_{j,\alpha}(\mathbf{k}')}{\tau_{i,\alpha}(\mathbf{k})} \frac{v_{j,\alpha}(\mathbf{k}')}{v_{i,\alpha}(\mathbf{k})} \right], \tag{5.75}$$

and a comparison to Eq.5.61 allows us finally to obtain:

$$\frac{1}{\tau_{i,\alpha}(\mathbf{k})} = \sum_{j,\mathbf{k}'} S_{i,j}(\mathbf{k}, \mathbf{k}') \left[\frac{1 - f_0(E')}{1 - f_0(E)} \right] \left[1 - \frac{\tau_{j,\alpha}(\mathbf{k}')}{\tau_{i,\alpha}(\mathbf{k})} \frac{v_{j,\alpha}(\mathbf{k}')}{v_{i,\alpha}(\mathbf{k})} \right], \tag{5.76}$$

where $E = E_i(\mathbf{k})$ and $E' = E_j(\mathbf{k}')$ and α can be either x or y. Equation 5.76 does not provide an explicit expression for $\tau_{i,\alpha}(\mathbf{k})$ because the momentum relaxation time is present also in the r.h.s. of the equation. In fact, Eq.5.76 couples the momentum relaxation times of all the subbands in the inversion layer, thus it is an implicit definition of all the relaxation times.

In order to understand more precisely the procedure to calculate the $\tau_{i,\alpha}(\mathbf{k})$ by using Eq.5.76, we write the rate of a generic inelastic scattering mechanism by using Eq.4.43 as

$$S_{i,j}(\mathbf{k}, \mathbf{k}') = \frac{2\pi}{\hbar} |M_{i,j}(\mathbf{k}, \mathbf{k}')|^2 \delta[E' - E - \Delta E(\mathbf{q})], \tag{5.77}$$

where $E = E_i(\mathbf{k})$ and $E' = E_j(\mathbf{k}')$ are the initial and final energy, respectively. The energy $\Delta E(\mathbf{q}) = \pm\hbar\omega(\mathbf{q})$ is written here as the energy of a phonon and may depend on the phonon wave-vector $\mathbf{q}$; the wave-vector variation enforced by the scattering is

$(\mathbf{k}' - \mathbf{k}) = \pm\mathbf{q}$ (see Section 4.6.2). The plus and minus sign correspond to phonon absorption and emission, respectively.

It is now important to recall that, for a given initial state $(i,\mathbf{k})$ with energy $\mathrm{E} = E_i(\mathbf{k})$ and a given $\Delta E(\mathbf{q})$ (set by the scattering mechanism), Eq.5.77 restricts the sum in Eq.5.76 only to those final states $(j,\mathbf{k}')$ that have an energy $E' = (E+\Delta E)$. More precisely, if we convert the sum over $\mathbf{k}'$ to an appropriate integral according to Eq.3.64 (with $n_{sp} = 1$ because the scattering does not change the spin), and express $S_{i,j}(\mathbf{k}.\mathbf{k}')$ with Eq.5.77, then Eq.5.76 can be re-written as

$$\frac{1}{\tau_{i,\alpha}(\mathbf{k})} = \frac{1}{2\pi\hbar}\sum_j \int_{\mathbf{k}'} A|M_{i,j}(\mathbf{k},\mathbf{k}')|^2 \left[\frac{1-f_0(E')}{1-f_0(E)}\right] \times \left[1 - \frac{\tau_{j,\alpha}(\mathbf{k}')}{\tau_{i,\alpha}(\mathbf{k})}\frac{v_{j,\alpha}(\mathbf{k}')}{v_{i,\alpha}(\mathbf{k})}\right] \delta[E' - E - \Delta E(\mathbf{q})]\,\mathrm{d}\mathbf{k}', \quad (5.78)$$

where the delta function restricts the integrals in the $\mathbf{k}'$ plane to the curves corresponding to $E_j(\mathbf{k}') = (E_i(\mathbf{k}) + \Delta E)$. The normalization area A always cancels in the product to the squared matrix elements $|M_{i,j}(\mathbf{k},\mathbf{k}')|^2$.

As can be seen, the implicit definition of the $\tau_{i,\alpha}(\mathbf{k})$ obtained by using Eqs.5.76 and 5.78 yields a system of integral equations which couple the $\tau_{i,\alpha}(\mathbf{k})$ of all the subbands.

Although an exact solution of Eq.5.78 may be sought by using numerical methods, such a general formulation precludes us from writing an explicit expression for the $\tau_{i,\alpha}(\mathbf{k})$ and, in turn, for the mobility. Some simplifications of the general formulation are possible depending on the features of the scattering mechanisms and of the energy model; we discuss them in the following two sub-sections.

Isotropic scattering mechanisms

Let us consider an isotropic scattering mechanism such that both the matrix elements $M_{i,j}$ and the exchanged energy ΔE are independent of $\mathbf{k}$ and $\mathbf{k}'$. The corresponding scattering rate can be written as

$$S_{i,j}(E) = \frac{2\pi}{\hbar}|M_{i,j}|^2\,\delta[E' - E - \Delta E], \quad (5.79)$$

thus the scattering rate and the momentum relaxation time depend only on the energy E of the initial state and we can write the relaxation time as $\tau_{i,\alpha}(E)$.

Under these circumstances, the sum over $\mathbf{k}'$ that includes $\delta f_j(\mathbf{k}')$ in Eq.5.72 vanishes for any j value. In fact, let us write the sum as

$$\sum_{\mathbf{k}'} S_{i,j}(\mathbf{k},\mathbf{k}')\,\delta f_j(\mathbf{k}')\frac{f_0(E)}{f_0(E')} = \frac{2\pi}{\hbar}|M_{i,j}|^2\sum_{\mathbf{k}'}\delta f_j(\mathbf{k}')\frac{f_0(E)}{f_0(E')}\delta[E' - E - \Delta E]$$
$$= -\frac{2\pi}{\hbar}|M_{i,j}|^2 F_\alpha \sum_{\mathbf{k}'}\tau_{j,\alpha}(E')\,v_{j,\alpha}(\mathbf{k}')\frac{f_0(E)}{f_0(E')}\frac{\partial f_0(E')}{\partial E}\delta[E' - E - \Delta E], \quad (5.80)$$

where we have used Eq.5.79 to express the scattering rate $S_{i,j}(\mathbf{k},\mathbf{k}')$ and Eq.5.69b to write $\delta f_j(\mathbf{k}')$ (for a $\tau_{j,\alpha}(E')$ dependent only on the energy E'). For any given energy E of the initial state, the above sum over all the final states $\mathbf{k}'$ having an energy

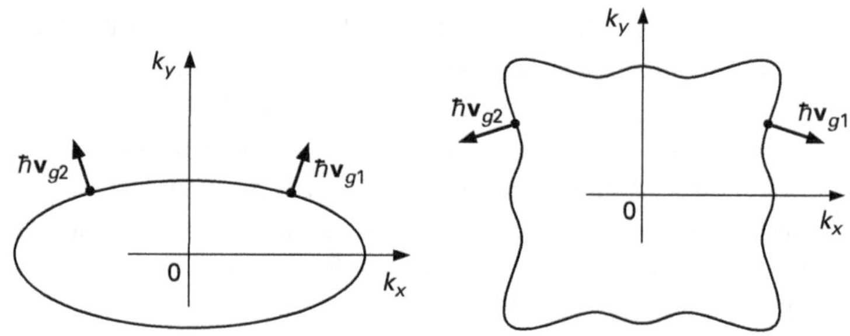

Figure 5.5 Equi-energy contours for the parabolic, elliptic bands of the electrons and for the warped bands of the holes. In both cases, for any **k** point belonging to an equi-energy curve, it is always possible to find a second point of the same equi-energy curve with the opposite value for the product $(v_{g,x} \cdot v_{g,y})$ of the two components of the group velocity.

$E' = E + \Delta E$ is zero, because, for any possible $\mathbf{k}'$ value, the symmetries of the equi-energy contours sketched in Fig.5.5 imply that a second wave-vector exists with an opposite group velocity $v_{j,\alpha}(\mathbf{k}')$.

Since the sum in Eq.5.80 is null for an isotropic scattering, then the corresponding term in Eq.5.76 containing $\tau_{j,\alpha}$ is also null and we obtain the explicit expression for the relaxation time

$$\frac{1}{\tau_i(E)} = \frac{2\pi}{\hbar} \sum_{j,\mathbf{k}'} |M_{i,j}|^2 \left[\frac{1 - f_0(E')}{1 - f_0(E)}\right] \delta[E' - E - \Delta E], \tag{5.81}$$

where we have dropped the subscript α with respect to Eq.5.76 because the relaxation time is now independent of the direction α of the electric field.
By using Eq.3.79 we can now express the relaxation time as

$$\frac{1}{\tau_i(E)} = \frac{\pi}{\hbar} \sum_{j} A|M_{i,j}|^2\, g_j(E + \Delta E) \left[\frac{1 - f_0(E + \Delta E)}{1 - f_0(E)}\right], \tag{5.82}$$

where E $= E_i(\mathbf{k})$, E$' = E_j(\mathbf{k}')$ and $g_j(E)$ is the density of states of the final subbands j (with $n_{sp} = 1$).

Hence, for an isotropic scattering mechanism, the inverse of the momentum relaxation time of a state (i,**k**) is essentially the same as the corresponding scattering rate, except for the term in squared brackets involving the occupation of the initial and final states. Clearly, for an elastic and isotropic scattering mechanism the term in brackets of Eq.5.82 disappears and the inverse of the relaxation time coincides with the rate of scattering.

This last result is not surprising if we recall that $\tau_{i,\alpha}$ is the time necessary for the scattering events to dissipate an initially non-null average value of the crystal momentum $\hbar\mathbf{k}$. When a scattering mechanism is isotropic, it produces a randomization of the momentum because the rate of scattering is the same towards all the final states (j,$\mathbf{k}'$) with the right energy. Hence, for such a randomizing scattering mechanism, the rate of momentum dissipation is the same as the rate of scattering itself. In general, however, the rate of momentum dissipation is smaller than the scattering rate.

Anisotropic and elastic scattering mechanisms

For anisotropic and inelastic scattering mechanisms it is not possible to obtain a formulation of the relaxation time simpler than Eq.5.76 other than by resorting to approximations. The key difficulty with Eq.5.76 is the presence of the relaxation time in the r.h.s. of the equation. In this respect we note that, in general, the relaxation times $\tau_{i,\alpha}$ and $\tau_{j,\alpha}$ correspond to different subbands and to different energies (namely $E = E_i(\mathbf{k})$ for $\tau_{i,\alpha}$ and $E' = E_j(\mathbf{k}) = (E + \Delta E)$ for $\tau_{j,\alpha}$). Consequently the values of $\tau_{i,\alpha}$ and $\tau_{j,\alpha}$ in the ratio in the r.h.s. of Eq.5.76 can be very different and there are no easily defensible simplifications.

We thus restrict our analysis to elastic scattering mechanisms and intra-subband transitions, that is $j = i$ and $E' = E$. In this case, in the r.h.s. of Eq.5.76 we have the ratio $[\tau_{i,\alpha}(\mathbf{k}')/\tau_{i,\alpha}(\mathbf{k})]$ between the relaxation times of the subband i corresponding to the same energy E but to two different wave-vectors. If we neglect the dependence on the $\mathbf{k}$ direction of $\tau_{i,\alpha}(\mathbf{k})$ and introduce the approximation

$$\frac{\tau_{i,\alpha}(\mathbf{k}')}{\tau_{i,\alpha}(\mathbf{k})} \approx 1, \tag{5.83}$$

we obtain the explicit expression for the relaxation time

$$\frac{1}{\tau_{i,\alpha}(\mathbf{k})} = \sum_{\mathbf{k}'} S_{i,i}(\mathbf{k}, \mathbf{k}') \left[1 - \frac{v_{i,\alpha}(\mathbf{k}')}{v_{i,\alpha}(\mathbf{k})} \right], \tag{5.84}$$

where the ratio of the Fermi occupation functions has been dropped for $E' = E$.

Here we should note that Eq.5.83 is not inconsistent with the fact that the scattering mechanism is anisotropic. In fact, as long as the scattering matrix elements depend only on the magnitude q of the wave-vector variation $\mathbf{q}$ (as is approximately the case for all the scattering mechanisms discussed in Chapter 4), then the possible dependence of $\tau_{i,\alpha}(\mathbf{k})$ on the $\mathbf{k}$ direction stems only from the anisotropy of the energy relation. Thus the condition set by Eq.5.83 is satisfied if the anisotropy of the energy relation is weak enough.

By using Eq.5.77 we can finally rewrite Eq.5.84 as

$$\frac{1}{\tau_{i,\alpha}(\mathbf{k})} = \frac{2\pi}{\hbar} \sum_{\mathbf{k}'} |M_{i,i}(\mathbf{k}, \mathbf{k}')|^2 \left[1 - \frac{v_{i,\alpha}(\mathbf{k}')}{v_{i,\alpha}(\mathbf{k})} \right] \delta(E_j(\mathbf{k}') - E_i(\mathbf{k})), \tag{5.85}$$

where E $= E_i(\mathbf{k})$ and E$' = E_j(\mathbf{k}')$.

Equation 5.85 indicates that, for an anisotropic scattering mechanism, the inverse of the momentum relaxation time is typically smaller than the scattering rate. In fact, the transitions that produce a small change in the α component of the group velocity give only a small contribution to Eq.5.85.

5.4.2 Momentum relaxation time for an electron inversion layer

The momentum relaxation time for the isotropic scattering mechanisms is given by Eq.5.82, which is an expression similar to the scattering rate. Hence, since the formulations of both the acoustic and the optical phonon scattering mechanisms presented in Sections 4.6.3 and 4.6.5 correspond to an isotropic treatment, then Eqs.4.274 and 4.286 can be readily modified to express the momentum relaxation time.

In particular, for an electron with an energy $E = E_{\nu,i}(\mathbf{k})$ (ν and i denoting the valley and the subband, respectively), the momentum relaxation time $\tau_{\nu,i}$ for the intra-valley transitions assisted by acoustic phonons depends only on E and is given by

$$\frac{1}{\tau_{\nu,i}(\mathbf{k})} = \frac{1}{\tau_{\nu,i}(E)} = \frac{2\pi K_B T D_{ac}^2}{\rho \hbar v_s^2} \sum_j F_{j,i}^{(\nu)} g_{\nu,j}(E), \tag{5.86}$$

where $F_{j,i}^{(\nu)}$ is the phonon form factor defined in Eq.4.268 and $g_{\nu,j}(E)$ is the density of states of the subband (ν, j) defined in Eq.3.75 (for $n_{sp} = 1$) and evaluated at the energy $E = E_{\nu,i}(\mathbf{k})$.

For inter-valley phonons, instead, we have

$$\frac{1}{\tau_{\nu,i}(E)} = \frac{\pi D_p^2}{\omega_p \rho} \sum_{\nu' \neq \nu, j} \mu_{w,\nu}^{(p)} F_{\nu,i}^{(\nu',j)} \left[n_{op}(\hbar\omega_p) + \frac{1}{2} \mp \frac{1}{2} \right] \times \left[\frac{1 - f_0(E \pm \hbar\omega_m)}{1 - f_0(E)} \right] g_{\nu',j}(E \pm \hbar\omega_p), \tag{5.87}$$

where the meaning of the symbols is the same as in Eq.4.286, $E = E_{\nu,i}(\mathbf{k})$ and the plus and minus sign correspond to phonon absorption and emission, respectively.

Figure 5.6 shows the MRT for the lowest subband in an electron inversion layer calculated for acoustic and optical phonon scattering. The discontinuities in the plot arise at the energies corresponding to the minima of the higher subbands, as explained by the presence of the density of states of the final subband in Eqs.5.86 and 5.87.

For an anisotropic elastic scattering mechanism, such as Coulomb or surface roughness scattering (see Sections 4.3 and 4.4), we can start from Eq.5.85 to express the relaxation time produced by the intra-subband transitions. By introducing the kinetic energy $E_p = (E - E_{\nu 0} - \varepsilon_{\nu,i})$ and using Eq.3.78 with $n_{sp} = 1$ we obtain

$$\frac{1}{\tau_{i,\alpha}(E,\theta)} = \frac{H_v(E_p)}{2\pi\hbar^3} [1 + 2\alpha E_p] \times \int_0^{2\pi} d\theta' \, m_{xy}(\theta') \, |M_{i,i}(\mathbf{q})|^2 \left[1 - \frac{v_{i,\alpha}(E_p,\theta')}{v_{i,\alpha}(E_p,\theta)} \right], \tag{5.88}$$

where θ and θ' are the angles formed respectively by $\mathbf{k}$ and $\mathbf{k}'$ with the direction α of the electric field and $H_v(x)$ is the step function. Since the wave-functions are independent of $\mathbf{k}$, then the matrix element $M_{i,i}(\mathbf{q})$ depends on $\mathbf{q} = (\mathbf{k}' - \mathbf{k})$, rather than separately on $\mathbf{k}$ and $\mathbf{k}'$.

In order to obtain a more explicit form for the relaxation time $\tau_{i,\alpha}(E,\theta)$ we need an expression for the group velocity components $v_{i,x}(E_p,\theta)$ and $v_{i,y}(E_p,\theta)$ in terms of the kinetic energy E_p and of the angle θ. Recalling the energy relation discussed in Section 3.2.4 for the non-parabolic effective mass approximation, we see that, for a given valley, the relation between the kinetic energy E_p and the wave-vector $\mathbf{k}$ is the same for all the subbands (see Eqs.3.27 and 3.28), and so the group velocity components

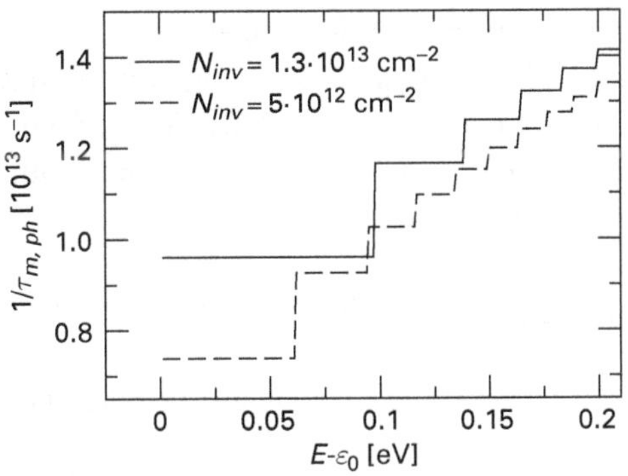

Figure 5.6 Inverse of the momentum relaxation time of the lowest electron subband for acoustic and optical phonon scattering. Bulk nMOS transistor with channel doping concentration $N_A = 1.6{\cdot}10^{16}\text{cm}^{-3}$. Parabolic effective mass approximation.

are expressed from now on by dropping the subband index i. The v_x is readily obtained by first deriving both sides of Eq.3.27 with respect to k_x to obtain

$$v_x = \frac{1}{\hbar}\frac{\partial E_p}{\partial k_x} = \frac{1}{1+2\alpha E_p}\frac{\hbar\, k_x}{m_x}, \tag{5.89}$$

and then by writing k_x as

$$k_x = k\,\cos(\theta) = \frac{\cos(\theta)}{\hbar}\sqrt{2m_{xy}(\theta)E_p(1+\alpha E_p)}, \tag{5.90}$$

where the magnitude k of $\mathbf{k}$ has been expressed by using Eq.3.69 and α is the non-parabolicity factor (see Section 3.2.4).
By doing so we obtain

$$v_x = \frac{\sqrt{2m_{xy}(\theta)E_p(1+\alpha E_p)}}{m_x(1+2\alpha E_p)}\cos(\theta), \tag{5.91}$$

and we can similarly derive

$$v_y = \frac{\sqrt{2m_{xy}(\theta)E_p(1+\alpha E_p)}}{m_y(1+2\alpha E_p)}\sin(\theta). \tag{5.92}$$

We now substitute in Eq.5.88 the $v_{i,x}$ according to Eq.5.91 and obtain

$$\frac{1}{\tau_{i,x}(E,\theta)} = \frac{H_v(E_p)}{2\pi\,\hbar^3}\,[1+2\,\alpha\,E_p)] \times \int_0^{2\pi} \mathrm{d}\theta'\; m_{xy}(\theta')\,|M_{i,i}(\mathbf{q})|^2\left[1-\frac{\cos(\theta')}{\cos(\theta)}\right]. \tag{5.93}$$

The expression for $\tau_{i,y}(E,\theta)$ is the same except for the ratio of cosines in the last term which becomes a ratio of sines. The squared magnitude of the wave-vector change $\mathbf{q} = (\mathbf{k}' - \mathbf{k})$ is given by

$$q^2 = k^2 + k'^2 - 2k\,k'\cos(\theta' - \theta). \tag{5.94}$$

In the case of circular bands with an effective mass m the relaxation time $\tau_i(E)$ depends on the energy E but not on the direction of the initial wave-vector $\mathbf{k}$ and, furthermore, it is independent of the direction α of the electric field. In such a simple case the relaxation time can be calculated by choosing a $\mathbf{k}$ aligned with the electric field, namely $\theta = 0$, so that Eq.5.93 simplifies to

$$\frac{1}{\tau_i(E)} = H_v(E - E_{\nu 0} - \varepsilon_{\nu,i}) \frac{m\,[1 + 2\alpha(E - E_{\nu 0} - \varepsilon_{\nu,i})]}{2\pi\hbar^3} \times \int_0^{2\pi} \mathrm{d}\theta'\, |M_{i,i}(\mathbf{q})|^2\, [1 - \cos(\theta')], \tag{5.95}$$

where we have used $E_p = (E - E_{\nu 0} - \varepsilon_{\nu,i})$. It should be noted that, since $\mathbf{k}$ is aligned to the field, then θ' in Eq.5.95 is the angle between the initial wave-vector $\mathbf{k}$ and the final wave-vector $\mathbf{k}'$.

Equations 5.93 and 5.95 highlight once again that the inverse of the momentum relaxation time can be smaller than the scattering rate. In fact, the transitions that produce small wave-vector changes, namely small θ' angles and small q values, result in only modest contributions to the integral over θ' in Eq.5.95.

Because of its simplicity, Eq.5.95 has frequently been used in the literature, sometimes also for inter-subband transitions, even if it actually applies only to intra-subband transitions. For circular bands, however, the momentum relaxation time is independent of the direction of $\mathbf{k}$ and Eq.5.83 is certainly fulfilled. Under these circumstances, it is possible to calculate the momentum relaxation time in the presence of both intra-subband and inter-subband transitions [10].

The matrix elements for surface roughness and Coulomb scattering to be used in calculations of momentum relaxation time have been discussed in detail in Sections 4.4 and 4.3, respectively. In the same sections it has also been emphasised that the effect of the screening produced by the electrons in the inversion layer is a very important physical ingredient for a credible treatment of such scattering mechanisms.

Figure 5.7 shows the momentum relaxation time for surface roughness scattering for two large inversion densities, N_{inv}, where this scattering mechanism is dominant for the

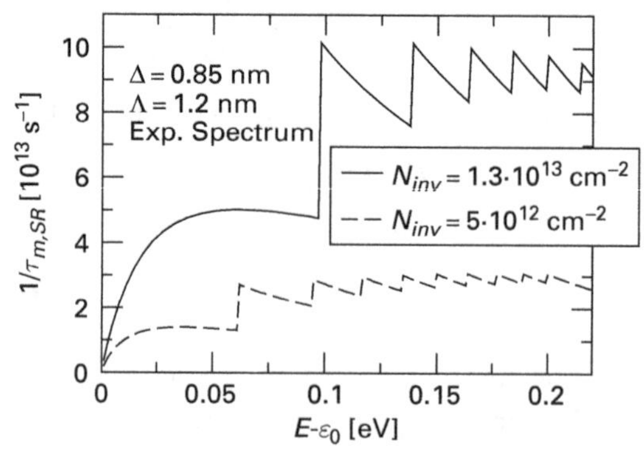

Figure 5.7 Inverse of the momentum relaxation time of the lowest electron subband for surface roughness scattering. Same device as in Fig.5.6. Parabolic effective mass approximation.

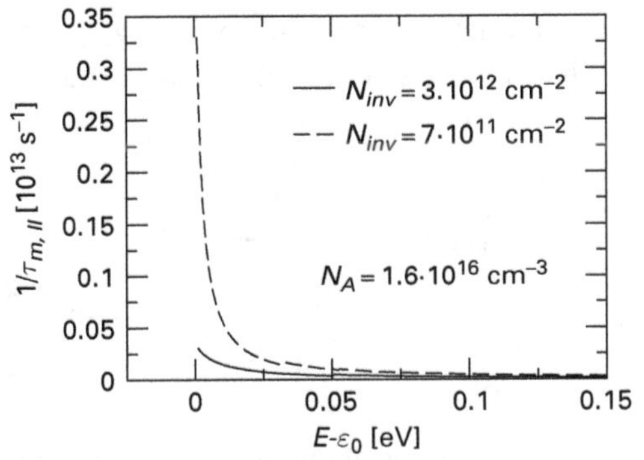

Figure 5.8 Inverse of the momentum relaxation time of the lowest electron subband for Coulomb scattering. Same device as in Fig.5.6. Parabolic effective mass approximation.

overall mobility. The inverse of the MRT increases with N_{inv} because the unscreened matrix elements for surface roughness scattering are proportional to the derivatives of the wave-functions at the silicon–oxide interface, which increases with the quantum confinement. The inverse of the MRT becomes very small, hence the scattering rates are small, at low energies. This is due to the strong screening effect for small values of the magnitude q of the wave-vector variation $\mathbf{q} = (\mathbf{k}' - \mathbf{k})$, hence for small energy values.

Figure 5.8 finally illustrates the MRT for Coulomb scattering and for two relatively small inversion densities. As can be seen, by increasing the inversion density the inverse of the MRT is reduced, hence the mobility is increased. This effect is essentially due to the screening produced by the electrons in the inversion layer, which increases with N_{inv}.

As a final remark, we note that Figs.5.6, 5.7, and 5.8 allow us to estimate the scattering rates in inversion layers, which are in fact of the same order of magnitude as the inverse of the momentum relaxation times. In particular, Fig.5.7 shows that at large inversion densities the surface roughness scattering rate can exceed 10^{13} s^{-1}. By recalling Fig.2.13 and the related discussion in Section 2.5.5, we find that scattering rates around 10^{13} s^{-1} imply a large uncertainty $\Delta E \simeq 40$ meV in the energy after scattering. This suggests that the energy conservation enforced by Fermi's rule may be significantly relaxed in the inversion layers of MOS transistors at the largest vertical electric fields.

5.4.3 Momentum relaxation time for a hole inversion layer

If we use the the semi-analytical hole energy model described in Section 3.3.3, then calculation of the momentum relaxation time for a hole inversion layer is very similar to the electron case. In particular, for a hole with an energy E in the band (ν, i), the momentum relaxation time due to acoustic phonon scattering can be written as

$$\frac{1}{\tau_{\nu,i}(E)} = \frac{2\pi K_B T D_{ac}^2}{\rho \hbar v_s^2} \sum_{\nu',j} F_{j,i}^{(\nu)} g_{\nu,j}(E), \tag{5.96}$$

where $F_{j,i}^{(\nu)}$ is the phonon form factor and $g_{\nu,j}(E)$ is the density of states defined in Eq.3.82 (with $n_{sp} = 1$). For optical phonon scattering, instead, we have

$$\frac{1}{\tau_{\nu,i}(E)} = \frac{\pi D_p^2}{\omega_p \rho} \sum_{\nu',j} \mu_{w,\nu}^{(p)} F_{\nu,i}^{\nu',j} \left[n_{op}(\hbar\omega_p) + \frac{1}{2} \mp \frac{1}{2} \right]$$
$$\times \left[\frac{1 - f_0(E \pm \hbar\omega_m)}{1 - f_0(E)} \right] g_{\nu',j}(E \pm \hbar\omega_p). \tag{5.97}$$

A more detailed discussion about the intra and inter-group phonon assisted transitions is developed in [14], which illustrates a systematic comparison of the phonon form factors obtained with either the energy model of Section 3.3.3 or the **k**·**p** model.

For anisotropic and elastic scattering mechanisms, such as surface roughness and Coulomb scattering, integration of the matrix elements can be obtained according to the general prescriptions of Section 3.5. For a hole in the subband (ν, i) with an energy E and a wave-vector $\mathbf{k}$ forming an angle θ with the direction k_x, the intra-subband momentum relaxation time can be written as [12]

$$\frac{1}{\tau_{\nu,i}(E,\theta)} = \frac{H_\nu(E_p)}{2\pi\hbar} \int_0^{2\pi} k_\nu(E_p, \theta + \beta) \frac{\mathrm{d}k_\nu(E_p, \theta + \beta)}{\mathrm{d}E_p}$$
$$\times |M_{i,i}^{(\nu)}(\mathbf{q})|^2 (1 - \cos\beta)\, \mathrm{d}\beta, \tag{5.98}$$

where $E_p = (E - E_{\nu 0} - \varepsilon_{\nu,i})$ is the kinetic energy and $M_{i,i}^{(\nu)}(\mathbf{q})$ is the intra-subband matrix element.

If the hole inversion layer is described by the **k**·**p** model, all the calculations related to the momentum relaxation time cannot but be performed numerically. Strictly speaking, the wave-vector dependence of the envelope wave-functions makes all the scattering mechanisms anisotropic, hence, in particular, phonon scattering becomes inelastic and anisotropic, which makes calculation of the momentum relaxation time problematic.

As already discussed in Section 4.6.6, the dependence of the wave-functions on **k** is frequently neglected, even in transport studies based on the **k**·**p** model [11–13, 15, 16], and the wave-functions used for the calculations are those corresponding to $\mathbf{k} = \mathbf{0}$. By doing so, the expressions for the momentum relaxation time due to phonon scattering become formally similar to Eqs.5.96 and 5.97, where, however, the density of states for the **k**·**p** model must be calculated numerically as explained in Section 3.5.4 (with $n_{sp} = 1$).

The momentum relaxation time for surface roughness and Coulomb scattering must be calculated numerically even if we accept using the wave-functions for $\mathbf{k} = \mathbf{0}$. In particular, given the **k** wave-vectors used to solve the **k**·**p** problem, the momentum relaxation time due to the intra-subband transitions of an anisotropic scattering mechanism can be obtained by evaluating the sums over $\mathbf{k}'$ in Eq.5.85 as explained in Section 3.5.4, that is by converting them into appropriate integrals by means of Eq.3.87 or Eq.3.91.

5.4.4 Calculation of mobility

By using Eq.5.66, the occupation function of the subband i can be expressed as

$$f_i(\mathbf{k}) = f_0(E_i(\mathbf{k})) \pm e\,\tau_{i,\alpha}(\mathbf{k})\,F_\alpha\,v_{i,\alpha}(\mathbf{k})\,\frac{\partial f_0(E_i(\mathbf{k}))}{\partial E}, \tag{5.99}$$

and, since the equilibrium $f_0(E)$ yields a null current, the current is proportional to $\delta f_i(\mathbf{k})$, hence to the electric field F_α.

Irrespective of the direction of the field (i.e. $\alpha = x$ or y), the distribution function $f_i(\mathbf{k})$ in Eq.5.99 can in general produce a non-null current along both the x and y directions. Since the momentum relaxation time $\tau_{i,\alpha}(\mathbf{k})$ can in turn depend on the direction of $\mathbf{F}$, we conclude that for a 2D carrier gas the mobility of the subband i is a 2×2 matrix. More precisely, assuming a linear relation between the current and the field (which is at the basis of the MRT approach), we can write

$$\begin{aligned} J_{i,x} &= (eN_i)(\mu_{i,xx}\,F_x + \mu_{i,xy}\,F_y), \\ J_{i,y} &= (eN_i)(\mu_{i,yx}\,F_x + \mu_{i,yy}\,F_y), \end{aligned} \tag{5.100}$$

where $J_{i,x}$ and $J_{i,y}$ denote the currents per unit width already defined in Eq.5.47 and N_i is the inversion density in the subband i. Equation 5.100 is the implicit definition of the four mobility components $\mu_{i,\beta\alpha}$ (with $\alpha, \beta = x$ or y).
Consistently with Eq.5.47, the component $J_{i,\beta}$ of the current per unit width in the subband i can be written as [9, 17]

$$J_{i,\beta} = \frac{\mp e}{A}\sum_{\mathbf{k}} v_{i,\beta}(\mathbf{k})\,f_i(\mathbf{k}) = \frac{\mp e}{A}\sum_{\mathbf{k}} v_{i,\beta}(\mathbf{k})\delta f_i(\mathbf{k}). \tag{5.101}$$

By substituting Eq.5.66 in Eq.5.101 we obtain

$$J_{i,\beta} = \frac{e^2 F_\alpha}{A}\sum_{\mathbf{k}} v_{i,\beta}(\mathbf{k})\,v_{i,\alpha}(\mathbf{k})\,\tau_{i,\alpha}(\mathbf{k})\left|\frac{\partial f_0(E_i(\mathbf{k}))}{\partial E}\right|, \tag{5.102}$$

where a minus sign has been absorbed in the negative E derivative of $f_0(E)$, thus leading to the corresponding absolute value.

Hence, recalling Eq.5.100, we can express the generic element of the mobility matrix as

$$\mu_{i,\beta\alpha} = \frac{J_{i,\beta}}{e\,N_i\,F_\alpha} = \frac{e}{N_i\,A}\sum_{\mathbf{k}} v_{i,\beta}(\mathbf{k})\,v_{i,\alpha}(\mathbf{k})\,\tau_{i,\alpha}(\mathbf{k})\left|\frac{\partial f_0(E_i(\mathbf{k}))}{\partial E}\right|, \tag{5.103}$$

where α and β can be either x or y. Equation 5.103 is also known as the Kubo–Greenwood formula for mobility calculation according to the MRT approximation.

When the momentum relaxation time $\tau_{i,\alpha}(\mathbf{k})$ of the subband i is known, Eq.5.103 allows us to calculate the mobility simply by calculating the sum over $\mathbf{k}$, which must be converted to an appropriate integral by using the general prescriptions discussed in Section 3.5. Practically speaking, the complexity of the calculation depends on the models employed for energy dispersion in the inversion layer.

It is worth noting that, if the momentum relaxation time depends on **k** only through the energy $E_i(\mathbf{k})$ (i.e. $\tau_{i,\alpha}(E) = \tau_{i,\alpha}(E_i(\mathbf{k}))$), then the sum over **k** in Eq.5.103 typically tends to vanish for $\alpha \neq \beta$, hence $\mu_{i,\beta\alpha}$ is zero for $\alpha \neq \beta$. In fact, under these circumstances, the sum over **k** in Eq.5.103 can be written as

$$\int_{-\infty}^{\infty} \mathrm{d}E \, \tau_{i,\alpha}(E) \left| \frac{\partial f_0(E)}{\partial E} \right| \sum_{\mathbf{k}} v_{i,\beta}(\mathbf{k}) \, v_{i,\alpha}(\mathbf{k}) \, \delta(E_i(\mathbf{k}) - E), \tag{5.104}$$

that is, it can be evaluated along equi-energy curves that correspond to constant $\tau_{i,\alpha}$ values. The typical symmetries of the equi-energy contours sketched in Fig.5.5 imply that, for any given point belonging to an equi-energy curve and featuring a group velocity $\mathbf{v}_{g1}$, a second point exists with velocity $\mathbf{v}_{g2}$ such that the product of the x and y velocity components results in $(v_{g2,x} \, v_{g2,y}) = (-v_{g1,x} \, v_{g1,y})$. Consequently, in Eq.5.104 the sum over **k** along an equi-energy curve vanishes.

Such a property is of very general validity for holes, because Γ is a symmetry point in the 2D Brillouin zone. The above considerations hold also for electrons when we consider the parabolic or non-parabolic effective mass approximation models discussed in Sections 3.2.2 and 3.2.4.

We finally note that, at not too low temperatures, many subbands may carry a significant charge in the inversion layer and the total current density is thus given by the sum of the contribution of the different subbands. Under these circumstances the average or effective mobility $\mu_{\beta\alpha}$ in the inversion layer can be written as

$$\mu_{\beta\alpha} = \sum_i \frac{N_i}{N_{inv}} \mu_{i,\beta\alpha}, \qquad N_{inv} = \sum_i N_i, \tag{5.105}$$

where N_{inv} denotes the overall inversion density.

In very many circumstances the mobility is regarded as a scalar quantity. For example, when we consider the low field mobility measured in a long channel MOSFET the electric field is directed from source to drain and so is the drain current. In this case, if we denote with x the channel length direction, the relevant mobility element for the transistor channel is μ_{xx}, which is simply referred to as the effective channel mobility of the transistor (see Section 7.1).

5.4.5 Mobility for an electron inversion layer

In order to evaluate the mobility in an electron inversion layer by using Eq.5.103, we can evaluate the sum over **k** by resorting to Eq.3.80 for an elliptic, non-parabolic energy dispersion. The effective masses, which are characteristic parameters of each valley ν (see Table 3.1), affect the carrier velocities and the overall mobility calculation through Eq.3.80. Nevertheless, in order to simplify the notation, in the rest of this section we drop the valley index ν from the symbols for the effective masses, the carrier velocities, and the subband mobilities.

We thus use Eq.3.80 with $n_{sp} = 2$ (because the states with both spins contribute to the current), and write the mobility element in Eq.5.103 as

$$\mu_{i,xy} = \frac{e}{N_i\,\pi^2\hbar^2}\int_0^{+\infty} \frac{E_p(1+\alpha\,E_p)}{1+2\alpha\,E_p}\left|\frac{\partial f_0(E)}{\partial E}\right| \times \int_0^{2\pi} \frac{m_{xy}^2(\theta)}{m_x m_y}\cos(\theta)\,\sin(\theta)\,\tau_{i,y}(E_p,\theta)\,\mathrm{d}\theta\,\mathrm{d}E_p, \tag{5.106}$$

where the energy E in the Fermi occupation function $f_0(E)$ is given by $E = (E_{v0}+\varepsilon_i+E_p)$ and $v_x(E_p,\theta)$ and $v_y(E_p,\theta)$ have been expressed by using Eq.5.91 and 5.92, respectively. The $\mu_{i,yx}$ has the same expression as $\mu_{i,xy}$ except that $\tau_{i,y}$ is replaced by $\tau_{i,x}$.

Here we should note that in Eq.5.103 the sum over $\mathbf{k}$ and the definition of the subband inversion density, N_i, must be consistent as far as a possible valley multiplicity is concerned. In this respect, Eq.5.106 has been obtained by setting the valley multiplicity to 1 in Eq.3.80, consequently the N_i in Eq.5.106 must be the subband inversion density evaluated *without accounting* for the valley multiplicity.

The remaining mobility components are

$$\mu_{i,xx} = \frac{e}{N_i\,\pi^2\hbar^2}\int_0^{+\infty} \frac{E_p(1+\alpha\,E_p)}{1+2\alpha\,E_p}\left|\frac{\partial f_0(E)}{\partial E}\right| \times \int_0^{2\pi} \frac{m_{xy}^2(\theta)}{m_x^2}\cos^2(\theta)\,\tau_{i,x}(E_p,\theta)\,\mathrm{d}\theta\,\mathrm{d}E_p, \tag{5.107}$$

$$\mu_{i,yy} = \frac{e}{N_i\,\pi^2\hbar^2}\int_0^{+\infty} \frac{E_p(1+\alpha\,E_p)}{1+2\alpha\,E_p}\left|\frac{\partial f_0(E)}{\partial E}\right| \times \int_0^{2\pi} \frac{m_{xy}^2(\theta)}{m_y^2}\sin^2(\theta)\,\tau_{i,y}(E_p,\theta)\,\mathrm{d}\theta\,\mathrm{d}E_p, \tag{5.108}$$

where the energy E is given by $E = (E_{v0}+\varepsilon_i+E_p)$. It is worth noting that, if the momentum relaxation time is independent of θ (namely of the $\mathbf{k}$ direction), then the integral over θ in Eq.5.106 vanishes (because of the symmetry properties of the sine and cosine functions), thus $\mu_{i,xy}$ and $\mu_{i,yx}$ are null, as anticipated in the discussion of Eq.5.103.

In particular, the relaxation time is certainly independent of θ as well as of the direction of the electric field for circular bands, namely for $m_x = m_y = m$, and we can simply denote the relaxation time by $\tau_i(E_p)$. In such a case the mobility is a scalar quantity, in fact $\mu_{i,xx} = \mu_{i,yy}$ and $\mu_{i,xy} = \mu_{i,yx} = 0$, and Eqs.5.107 and 5.108 reduce to

$$\mu_{i,xx} = \mu_{i,yy} = \frac{e}{N_i\,\pi\hbar^2}\int_0^{+\infty} \mathrm{d}E_p\,\frac{E_p(1+\alpha\,E_p)}{1+2\alpha\,E_p}\,\tau_i(E_p)\left|\frac{\partial f_0(E)}{\partial E}\right|. \tag{5.109}$$

For parabolic bands, Eq.5.109 further simplifies to

$$\mu_{i,xx} = \mu_{i,yy} = \frac{e}{N_i\,\pi\hbar^2}\int_0^{+\infty} \mathrm{d}E_p\,E_p\,\tau_i(E_p)\left|\frac{\partial f_0(E)}{\partial E}\right|. \tag{5.110}$$

In order to gain some insight into the values of momentum relaxation time that critically affect electron mobility, we recall Eq.3.93 and write

$$\frac{\partial f_0(E_{\nu 0}+\varepsilon_i+E_p)}{\partial E}=\frac{-1}{K_BT}f_0(E_{\nu 0}+\varepsilon_i+E_p)[1-f_0(E_{\nu 0}+\varepsilon_i+E_p)], \quad (5.111)$$

where $f_0(E)$ is defined in Eq.3.92. When the electron gas is non-degenerate, namely $(E_{\nu 0}+\varepsilon_i-E_F)$ is larger than a few (K_BT), then $f_0(E)$ is much smaller than 1.0 and decays with energy as a Boltzmann function, hence we have

$$\frac{\partial f_0(E_{\nu 0}+\varepsilon_i+E_p)}{\partial E}\approx\frac{-1}{K_BT}\exp\left[\frac{-(E_{\nu 0}+\varepsilon_i+E_p-E_F)}{K_BT}\right]. \quad (5.112)$$

Equation 5.112 indicates that, for a non-degenerate electron gas, the values of momentum relaxation time $\tau_i(E_p)$ have an exponentially decreasing weight in calculation of the mobility with increasing kinetic energy E_p inside the subband. Thus the mobility is essentially determined by the $\tau_i(E_p)$ values in an energy range of very few (K_BT) above the subband minimum.

If, instead, the subband minimum $(E_{\nu 0}+\varepsilon_i)$ is lower than E_F, then the electron gas is degenerate and Eq.5.111 indicates that the derivative of f_0 is practically negligible except in an energy range of only a few K_BT around the Fermi level. This implies that, in the mobility calculation with Eqs.5.106 to 5.110, the most influential relaxation times are those for energies close to E_F. Furthermore, since the energy interval is just a few (K_BT), then the smaller the temperature T the more peaked is the f_0 derivative at $E=E_F$, as can be seen in Fig.5.9. At low temperatures only electrons with an energy very close to E_F yield a non-null current; under these circumstances it is sometimes said that only the Fermi electrons contribute to mobility.

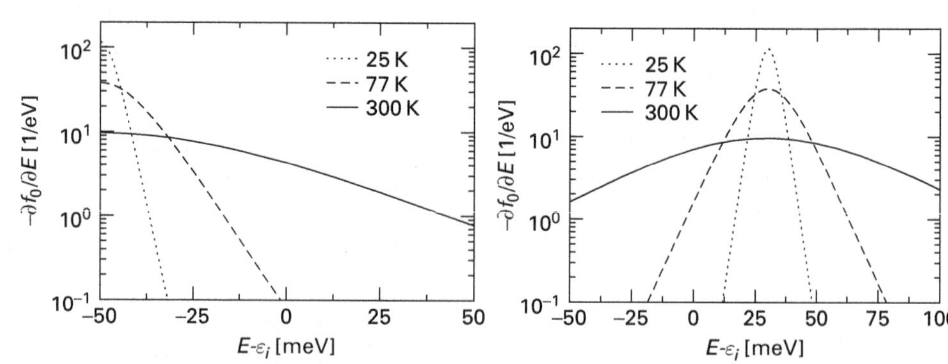

Figure 5.9 The derivative of the Fermi function for different temperatures and versus the energy E referred to the minimum of the subband. (a) A non-degenerate case with $(E_F-\epsilon_i)=-50meV$; (b) A degenerate case with $(E_F-\epsilon_i)=30meV$. In case (b) the derivative is strongly peaked for E equal to the Fermi level.

5.4.6 Mobility for a hole inversion layer

Calculation of the mobility for a hole inversion layer according to Eq.5.103 obviously depends on the energy model. For the semi-analytical hole energy model described in Section 3.3.3, Eq.5.103 can be converted to appropriate integrals over the parallel energy E_p and the angle θ. Similarly to the previous section, we henceforth drop the index ν denoting the subband group.

Thus, by using Eq.3.86 with $n_{sp} = 2$, the mobility component $\mu_{i,\beta\alpha}$ of the subband i can be written as [18]

$$\mu_{i,\beta\alpha} = \frac{2}{(2\pi)^2} \frac{e}{K_B T\, P_i} \int_0^{\infty} \mathrm{d}E_p \int_0^{2\pi} \mathrm{d}\theta\; k(E_p,\theta) \frac{\mathrm{d}k(E_p,\theta)}{\mathrm{d}E_p}$$
$$\times\, v_\beta(E_p,\theta)\; v_\alpha(E_p,\theta)\; \tau_\alpha^{(i)}(E,\theta)\; f_0(E)\,(1 - f_0(E)), \qquad (5.113)$$

where α and β can be either x or y and the argument of the Fermi function $f_0(E)$ is the energy $E = (E_{\nu 0}+\varepsilon_{\nu,i}+E_p)$.

The components $v_x^{(\nu)}$ and $v_y^{(\nu)}$ of the group velocity can be obtained from the relation $k_\nu(E_p,\theta)$ given in Eq.3.45 and read

$$v_x^{(\nu)}(E_p,\theta) = \frac{1}{\hbar}\frac{\partial E_p}{\partial k_x} = \frac{1}{\hbar}\left(\frac{\partial k_\nu}{\partial E_p}\right)^{-1}\left(\cos\theta + \frac{1}{k_\nu}\frac{\partial k_\nu}{\partial \theta}\sin\theta\right),$$
$$v_y^{(\nu)}(E_p,\theta) = \frac{1}{\hbar}\frac{\partial E_p}{\partial k_y} = \frac{1}{\hbar}\left(\frac{\partial k_\nu}{\partial E_p}\right)^{-1}\left(\sin\theta - \frac{1}{k_\nu}\frac{\partial k_\nu}{\partial \theta}\cos\theta\right), \qquad (5.114)$$

where we have used the derivative of an implicit function to express $(\partial E_p/\partial\theta)$ by means of the E_p and θ derivatives of $k_\nu(E_p,\theta)$. The effective mobility in the channel of a MOSFET is obtained with Eq.5.113 by setting $\alpha = \beta = x$.

If the hole inversion layer is described by the $\mathbf{k}\cdot\mathbf{p}$ model, all the calculations related to mobility must be performed numerically. More precisely, calculation of the mobility with Eq.5.103 requires evaluation of the sums over $\mathbf{k}$ as explained in Section 3.5.4, namely by using Eq.3.87 or Eq.3.91.

5.4.7 Multiple scattering mechanisms and Matthiessen's rule

An important point concerning application of the MRT approximation for mobility calculations is related to the presence of multiple scattering mechanisms. In fact, so far we have expressed the MRT of a given subband by implicitly accounting for a single scattering mechanism and its corresponding matrix elements. If such a mechanism is, for example, acoustic phonon scattering, then the mobility obtained by using the MRT approach is the so called acoustic phonon limited mobility. Actually, an advantage of the MRT approach is the insight gained by analyzing the mobility components while considering the scattering mechanisms one at a time.

It is now important to understand that, if more uncorrelated scattering mechanisms are relevant, then the total scattering rate to be used for calculation of the MRT is given by the sum of the scattering rates. This is, for example, the approach employed in the Monte Carlo simulations accounting for several scattering mechanisms (see Chapter 6). When using the momentum MRT approach for mobility calculations, however, it is common practice to calculate the momentum relaxation times $\tau_i^{(s)}(\mathbf{k})$ of each scattering mechanisms, s (the subscript α indicating the direction of the electric field has been dropped to simplify the notation), and then express the total relaxation time $\tau_i(\mathbf{k})$ as

$$\frac{1}{\tau_i(\mathbf{k})} = \sum_s \frac{1}{\tau_i^{(s)}(\mathbf{k})}, \tag{5.115}$$

where the sum is taken over all the relevant scattering mechanisms, s.

The relaxation time $\tau_i(\mathbf{k})$ calculated with Eq.5.115 is in general different from the one obtained by first summing the rates of the different scattering mechanisms and then calculating the relaxation time according to the procedure described in the previous sub-sections. Only if all the scattering mechanisms are elastic and isotropic does Eq.5.115 provide the same $\tau_i(\mathbf{k})$ values as those obtained by summing the scattering rates *before* calculating the relaxation times.

By using Eq.5.115 as an approximate starting point, the composition rule of the MRT has been extended also to mobility values, thus leading to the well known *Matthiessen's rule*. More precisely, if we now denote with $\mu_i^{(s)}$ the mobility limited by the scattering mechanism s in the subband i, Matthiessen's rule states that the overall mobility μ_i in the subband i can be expressed as

$$\frac{1}{\mu_i} = \sum_s \frac{1}{\mu_i^{(s)}}. \tag{5.116}$$

If we now recall that the mobility is related to a weighted average of the momentum relaxation time (see Eq.5.103 or the simplified expression in Eq.5.110), then it is clear that Eq.5.116 cannot be considered a consequence of Eq.5.115 unless the $\tau_i^{(s)}$ of the different scattering mechanisms have the same $\mathbf{k}$ or energy dependence. Unfortunately this is not the case (as illustrated also in Figs.5.6 to 5.8), thus Eq.5.116 is in general not valid.

The above reasoning can be taken one step further, and Eq.5.116 can be extended to the average or effective mobility μ in the inversion layer by writing

$$\frac{1}{\mu} = \sum_s \frac{1}{\mu^{(s)}}, \tag{5.117}$$

where $\mu^{(s)}$ is the effective mobility in the inversion layer that we would have if s were the only scattering mechanism. Equation 5.117 is the formulation of Matthiessen's rule most frequently quoted and used.

However, also in this case, we note that Eq.5.117 is by no means an exact consequence of Eq.5.116. In fact, recalling Eq.5.105, we can express μ and $\mu^{(s)}$ as

$$\mu = \sum_i \frac{N_i}{N_{inv}} \mu_i, \qquad \mu^{(s)} = \sum_i \frac{N_i}{N_{inv}} \mu_i^{(s)}. \tag{5.118}$$

By using Eq.5.116 in Eq.5.118 to express μ as a function of the $\mu_i^{(s)}$ and then substituting the μ and $\mu^{(s)}$ expressions is Eq.5.117, we can easily see that, in order for Eq.5.117 to be consistent with Eq.5.116, we would need the equality

$$\left[\sum_i \frac{N_i}{N_{inv}} \left(\sum_s \frac{1}{\mu_i^{(s)}}\right)^{-1}\right]^{-1} = \sum_s \left(\sum_i \frac{N_i}{N_{inv}} \mu_i^{(s)}\right)^{-1} \tag{5.119}$$

to be verified. Unfortunately, Eq.5.119 is not verified except for the case when one single band is occupied, namely unless we have $N_1 \approx N_{inv}$ and $N_i \approx 0$ for $i \neq 1$.

The above discussion shows that Eq.5.116 is in general not verified (because the MRT of the different scattering mechanisms have a different **k** or energy dependence) and, furthermore, Eq.5.117 does not stem from Eq.5.116, unless we assume a quantum limit inversion layer where only the lowest subband is occupied.

Clearly, Eq.5.117 lacks a sound theoretical basis; nevertheless, it has been and is still widely used to extract from experiments the mobility components limited by the different scattering mechanisms. The possible inaccuracies related to Matthiessen's rule were pointed out a long time ago [19, 20] and critically reconsidered in recent papers [21–23]. The reader should thus be aware that use of Matthiessen's rule in Eq.5.117 is not a quantitatively reliable procedure to separate the mobility components limited by the different scattering mechanisms.

5.5 Models based on the balance equations of the BTE

In the previous sections we have seen how the BTE can be linearized in the low-field regime and solved in uniform transport conditions. The exact solution of the BTE for the general case of non-uniform structures under arbitrarily large electric fields is definitely more complicated and requires sophisticated numerical techniques such as, for instance, the Monte Carlo method described in Chapter 6.

Approximate solutions of the BTE can be found by using the so-called moment's method, that is a set of balance equations derived from the BTE itself by performing appropriate integrals and averages over **K**. A thorough discussion and derivation of balance equations is outside the scope of this book and can be found in [2, 24]. However, given the widespread use of balance equations in TCAD tools, a short discussion of their derivation is necessary to highlight the limitations with respect to the exact solution of the BTE described in Chapter 6. In particular, in the following we show how the well-known Drift–Diffusion model can be derived from the BTE.

5.5.1 Drift–Diffusion model

For the sake of simplicity, we consider a free-electron gas (3D **K** space) and a homogeneous semiconductor with a single minimum of the conduction band, described with a parabolic and isotropic energy relation. Furthermore, we assume a one-dimensional problem in real space and express the collision term in the r.h.s. of Eq.5.13 with an energy independent relaxation time τ. Equation 5.13 becomes:

$$\frac{\partial f}{\partial t} - \frac{\partial f}{\partial k_x} F_x \frac{e}{\hbar} + \frac{\partial f}{\partial x} v_{gx} = -\frac{f - f_0}{\tau}, \tag{5.120}$$

where f_0 is the so-called *local equilibrium distribution*, that is a Maxwell–Boltzmann function with a local Fermi level defined such that the sum of f_0 over $\mathbf{K}$ multiplied by $2/\Omega$ gives the local electron concentration.

We obtain a first balance equation by computing the zeroth order moment of the BTE, that is by summing both sides of Eq.5.120 over $\mathbf{K}$:

$$\frac{2}{\Omega} \sum_{\mathbf{K}} \left[\frac{\partial f}{\partial t} - \frac{\partial f}{\partial k_x} F_x \frac{e}{\hbar} + \frac{\partial f}{\partial x} v_{gx} \right] = -\frac{2}{\Omega} \sum_{\mathbf{K}} \frac{f - f_0}{\tau}. \tag{5.121}$$

Equation 5.14 and the definition of f_0 imply that the summation over the r.h.s. of Eq.5.121 is null and that the first term on the l.h.s. yields $\partial n / \partial t$. The second term on the l.h.s of Eq.5.121, instead, can be computed as follows:

$$\begin{aligned} \frac{2}{\Omega} \sum_{\mathbf{K}} \frac{\partial f}{\partial k_x} &= \frac{1}{4\pi^3} \int_{-\infty}^{+\infty} \int_{-\infty}^{+\infty} \int_{-\infty}^{+\infty} \frac{\partial f}{\partial k_x} \mathrm{d}k_x \mathrm{d}k_y \mathrm{d}k_z \\ &= \frac{1}{4\pi^3} \int_{-\infty}^{+\infty} \int_{-\infty}^{+\infty} \left[f(k_x \to +\infty) - f(k_x \to -\infty) \right] \mathrm{d}k_y \mathrm{d}k_z = 0. \end{aligned} \tag{5.122}$$

From the definition of J_x (Eq.5.15) and recalling that in a homogeneous material v_{gx} does not depend on x, we see that the third term on the l.h.s of Eq.5.121 is $[-(1/e)\partial J_x / \partial x]$. We thus obtain the well-known continuity equation:

$$\frac{\partial n}{\partial t} - \frac{1}{e} \frac{\partial J_x}{\partial x} = 0, \tag{5.123}$$

without the generation–recombination terms, since we have not included these processes in the BTE.

A second balance equation can be found from the first-order moment of the BTE with respect to v_{gx}:

$$-\frac{2e}{\Omega} \sum_{\mathbf{K}} \left[\frac{\partial f}{\partial t} - \frac{\partial f}{\partial k_x} F_x \frac{e}{\hbar} + \frac{\partial f}{\partial x} v_{gx} \right] v_{gx} = -\frac{2e}{\Omega} \sum_{\mathbf{K}} \left[-\frac{f - f_0}{\tau} \right] v_{gx}. \tag{5.124}$$

From the definition of the electron current density J_x (Eq.5.15) we see that the first term on the l.h.s. becomes $\partial J_x / \partial t$. On the other hand, since the contribution of f_0 to the current is null, the r.h.s. gives $[-J_x / \tau]$.

Considering now the second term on the l.h.s. of Eq.5.124, we can convert the sum into an integral and then integrate by parts and obtain

$$\frac{e^2 F_x}{4\pi^3\hbar}\int_{-\infty}^{+\infty}\int_{-\infty}^{+\infty}\int_{-\infty}^{+\infty}\frac{\partial f}{\partial k_x}\frac{\hbar k_x}{m_x}\,\mathrm{d}k_x\mathrm{d}k_y\mathrm{d}k_z$$
$$= -\frac{e^2 F_x}{4\pi^3 m_x}\int_{-\infty}^{+\infty}\int_{-\infty}^{+\infty}\int_{-\infty}^{+\infty} f\,\mathrm{d}k_x\mathrm{d}k_y\mathrm{d}k_z = -\frac{e^2 F_x n}{m_x}, \tag{5.125}$$

where we have replaced v_{gx} with $\hbar k_x/m_x$ since we assume parabolic bands. The third term on the l.h.s. of Eq.5.124,

$$\mathcal{V}^2 = \frac{-2e}{\Omega}\frac{\partial}{\partial x}\sum_{\mathbf{K}} f v_{gx}^2, \tag{5.126}$$

cannot be simply related to macroscopic quantities such as n and J_x. The evaluation of $\mathcal{V}^2$ creates a link to the second-order moment of the BTE. However, if we try to compute such a second-order moment we find a summation of $f v_{gx}^3$ over $\mathbf{K}$ space, whose evaluation would require calculation of the third order moment of the BTE. It is clear, then, that a chain of additional equations and higher order momenta is generated by this method. A closure relation is thus necessary to solve the system of equations. To this purpose, in the Drift–Diffusion model we assume the semiconductor is non-degenerate and we replace f with the Maxwell–Boltzmann local equilibrium distribution f_0, solely for calculation of $\mathcal{V}^2$ in Eq.5.126. We thus evaluate:

$$\mathcal{V}^2 = -\frac{e^2}{4\pi^3}\frac{\partial}{\partial x}\int_{-\infty}^{+\infty}\int_{-\infty}^{+\infty}\int_{-\infty}^{+\infty} f_0 v_{gx}^2\mathrm{d}k_x\mathrm{d}k_y\mathrm{d}k_z$$
$$= -\frac{e^2}{4\pi^3}\frac{\partial}{\partial x}\int_{-\infty}^{+\infty}\int_{-\infty}^{+\infty}\int_{-\infty}^{+\infty} f_0\mathrm{d}k_x\mathrm{d}k_y\mathrm{d}k_z$$
$$\times\frac{\int_{-\infty}^{+\infty}\int_{-\infty}^{+\infty}\int_{-\infty}^{+\infty} f_0 v_{gx}^2\mathrm{d}k_x\mathrm{d}k_y\mathrm{d}k_z}{\int_{-\infty}^{+\infty}\int_{-\infty}^{+\infty}\int_{-\infty}^{+\infty} f_0\mathrm{d}k_x\mathrm{d}k_y\mathrm{d}k_z}. \tag{5.127}$$

The last term on the r.h.s. of Eq.5.127 is the average of the squared velocity and can be written as:

$$\langle v_{gx}^2\rangle = \frac{\int \mathrm{d}k_x \exp\left(-\frac{\hbar^2 k_x^2}{2m_x K_B T}\right)\frac{\hbar^2 k_x^2}{m_x^2}}{\int \mathrm{d}k_x \exp\left(-\frac{\hbar^2 k_x^2}{2m_x K_B T}\right)}, \tag{5.128}$$

where the integrals over k_y and k_z as well as the terms $\exp[E_F(x)/(K_B T)]$ in f_0 have been canceled because they appear in both the numerator and the denominator. Since $\langle v_{gx}^2\rangle$ does not depend on x and, by definition, the integral of f_0 over $\mathbf{K}$ evaluates to $n(x)$, we can write:

$$\mathcal{V}^2 = -e\frac{\partial n}{\partial x}\langle v_{gx}^2\rangle. \tag{5.129}$$

The denominator of Eq.5.128 is a Gauss integral and evaluates to $\sqrt{m_x K_B T 2\pi}/\hbar$. In order to perform the integral at the numerator, we change the integration variable from k_x to $\alpha = \hbar^2 k_x^2/(2m_x K_B T)$, and then from α to $\rho = \sqrt{2\alpha}$, obtaining $(K_B T)^{3/2}\hbar\sqrt{2\pi/m_x}$. We have thus found:

$$\langle v_{gx}^2 \rangle = \frac{K_B T}{m_x}. \tag{5.130}$$

The first moment of the BTE then reads:

$$\frac{\partial J_x}{\partial t} - \frac{e^2 F_x}{m_x} n - e \frac{\partial n}{\partial x} \frac{K_B T}{m_x} = -\frac{J_x}{\tau}. \tag{5.131}$$

The first term can be neglected based on simple reasoning: the partial derivative of J_x over time accounts for the change of J_x due to time-varying external potentials, thus when compared to J_x/τ, $\partial J_x/\partial t$ is significantly smaller, since the momentum relaxation time τ is usually orders of magnitude shorter than the period of the applied signals. Furthermore, by remembering that $\mu = e\tau/m_x$ is the low-field mobility, we obtain:

$$J_x = en\mu F_x + eV_{th}\mu \frac{\partial n}{\partial x}, \tag{5.132}$$

where $V_{th} = K_B T/e$ is the thermal voltage.

Equations 5.123 and 5.132 together with the corresponding equations for holes and the Poisson equation form the well-known Drift–Diffusion model of semiconductors which is widely used for the analysis and design of electron devices.

5.5.2 Analytical models for the MOSFET drain current

Beside being the model implemented in many commercial device simulators, the Drift–Diffusion model is also the basis of many analytical and compact models for MOS devices. In particular, from Eq.5.132, under a proper set of approximations on the device electrostatics (namely the charge sheet and the gradual channel approximations) and on the relative importance of the drift and diffusion components of J_x, one may derive the well-known square law for the drain current of long channel MOSFETs [25]:

$$I_D = \frac{W}{L} \mu_{eff} C_{G,eff} \left[(V_{GS} - V_T) V_{DS}^* - \frac{{V_{DS}^*}^2}{2} \right], \tag{5.133}$$

where W and L are the effective device width and length, respectively, μ_{eff} is the effective electron mobility (which is assumed to be independent of the lateral electric field), $C_{G,eff}$ the effective gate capacitance per unit area, V_T the threshold voltage, and the voltage V_{DS}^* is given by:

$$V_{DS}^* = \begin{cases} V_{DS} & \text{if} \quad V_{DS} < V_{GS} - V_T \\ V_{GS} - V_T & \text{if} \quad V_{DS} \geq V_{GS} - V_T \end{cases}, \tag{5.134}$$

where $(V_{GS} - V_T)$ is the saturation voltage $V_{DS,sat}$. The voltages V_{GS} and V_{DS} are the *intrinsic* gate-source and drain-source voltages.

Equation 5.133 predicts that the drain current increases as $1/L$ in a constant voltage scaling scenario (see Table 1.1). However, Eq.5.133 is by no means applicable to short channel devices where the large value of the lateral field causes a departure of the transport regime from the near equilibrium conditions embedded in Eq.5.132.

A simple extension of the MOSFET model useful for short channel devices is obtained by introducing the saturation of the carrier velocity at high electric fields according to the well known expression [25, 26]:

$$v_x = \frac{\mu_{eff} F_x}{\left[1 + \left(\mu_{eff} F_x / v_{sat}\right)^{1/\beta}\right]^{\beta}}, \tag{5.135}$$

where v_x denotes the x-component of the average carrier velocity v. By setting $\beta = 1$ and under essentially the same assumptions that led to Eq.5.133, it is easy to obtain [25]

$$I_D = \frac{W}{L + \mu_{eff} V_{DS}^* / v_{sat}} \mu_{eff} C_{G,eff} \left[(V_{GS} - V_T)\, V_{DS}^* - \frac{{V_{DS}^*}^2}{2} \right], \tag{5.136}$$

where:

$$V_{DS}^* = \begin{cases} V_{DS} & \text{if} \quad V_{DS} < V_{DS,sat} \\ V_{DS,sat} & \text{if} \quad V_{DS} \geq V_{DS,sat} \end{cases}. \tag{5.137}$$

The saturation voltage is given by:

$$V_{DS,sat} = \frac{\sqrt{1 + 2\mu_{eff}(V_{GS} - V_T)/(L v_{sat})} - 1}{\mu_{eff}/(L v_{sat})}. \tag{5.138}$$

The dependence of I_D on the effective channel length L predicted by Eq.5.136 is definitely weaker than in Eq.5.133. In particular, at high V_{DS} and for $L \to 0$:

$$I_D \simeq W C_G (V_{GS} - V_T) v_{sat}, \tag{5.139}$$

which no longer depends on L.

Equation 5.139 shows that, if we account for velocity saturation, in the frame of the Drift–Diffusion model the current of a short channel MOSFET is limited by the saturation velocity v_{sat}, that is by the maximum average velocity of the carriers in uniform high field transport conditions. In bulk silicon $v_{sat} \simeq 10^7$cm/s for electrons and holes [26]. In MOSFETs, instead, v_{sat} is expected to depend on the inversion charge [27, 28] and the device architecture [29], as also briefly discussed in Section 7.2.1. Most important, however, is to emphasize that in short nanoscale MOSFETs the physical conditions are very different from those of uniform transport, so that the validity of expressions such as Eq.5.139 or Eq.5.133 is questionable.

To overcome the limitations of these models, in the next sections we derive an expression for I_D in the far from equilibrium ballistic and quasi-ballistic transport regimes, where future MOSFETs are expected to work and where the assumptions behind the Drift–Diffusion model are no longer valid. Interestingly enough, we see in Section 5.7 that a compact expression for I_D in the quasi-ballistic regime can be derived, which is very similar to Eq.5.136. Comparison of these models helps us to explain why expressions of I_D based on the Drift–Diffusion approximations are still used for the analysis and design of ultra-scaled CMOS transistors and circuits.

5.6 The ballistic transport regime

The solution of the BTE is drastically simplified when scattering is neglected. This regime of transport is usually named *ballistic* [30]. In the following we discuss ballistic transport in the case of a 2D inversion layer. As in other parts of the book, we assume translational invariance in the device width direction y and we set z as the quantization direction.

In the absence of scattering, the BTEs for the different subbands are decoupled from one another (Section 5.2.4) and can be solved separately. The contribution of the distributions f_i to the total inversion charge density N_{inv} and to the total current I_x are then summed.

We begin the derivation of the ballistic transport model noting that in steady state conditions Eq.5.45 becomes:

$$\frac{\partial f_i}{\partial x} v_{gx} - \frac{\partial f_i}{\partial k_x} \frac{1}{\hbar} \frac{\mathrm{d}E_i}{\mathrm{d}x} = 0, \tag{5.140}$$

where i is the subband index. For the sake of simplicity, we now assume that the in-plane energy can be separated into a term E_x that only depends on k_x and a term E_y that only depends on k_y, namely we rewrite Eq.5.34 as:

$$E_i(x, \mathbf{k}) = E_x(k_x) + E_y(k_y) + \varepsilon_i(x) + E_{\nu 0}, \tag{5.141}$$

and we set $E_{\nu 0} = 0$. This case is representative of, for example, elliptical/parabolic subbands oriented along the transport direction, where

$$E_i(x, \mathbf{k}) = \frac{\hbar^2 k_x^2}{2m_x} + \frac{\hbar^2 k_y^2}{2m_y} + \varepsilon_i(x). \tag{5.142}$$

Following [31], we define the total energy in the x direction as

$$w = E_x(k_x) + \varepsilon_i(x), \tag{5.143}$$

and we attempt to rewrite the BTE by considering a new auxiliary distribution F_i, that is a function of (x, w, k_y) instead of (x, k_x, k_y). The definition of the symbols is given in graphical form in Fig.5.10.

In general, the relationship between k_x and w is not one-to-one, as is immediately understood by recalling, for example, the case of elliptical subbands where w depends on k_x^2 and is thus invariant to sign exchange of k_x. Therefore we need to define two separate auxiliary distributions: $F_i^+(x, w, k_y)$ for the states with $k_x > 0$ (i.e. moving left to right) and $F_i^-(x, w, k_y)$ for those with $k_x < 0$ (moving right to left). We can formally define F_i^+ and F_i^- by writing

$$F_i^+(x, w, k_y) = f_i(x, k_x(w - \varepsilon_i(x)), k_y) \text{ for } \quad k_x > 0, \tag{5.144a}$$

$$F_i^-(x, w, k_y) = f_i(x, k_x(w - \varepsilon_i(x)), k_y) \text{ for } \quad k_x < 0, \tag{5.144b}$$

where w and k_y can be seen as parameters of $F_i^{\pm}$, while k_x is a function of $[w - \varepsilon_i(x)]$.

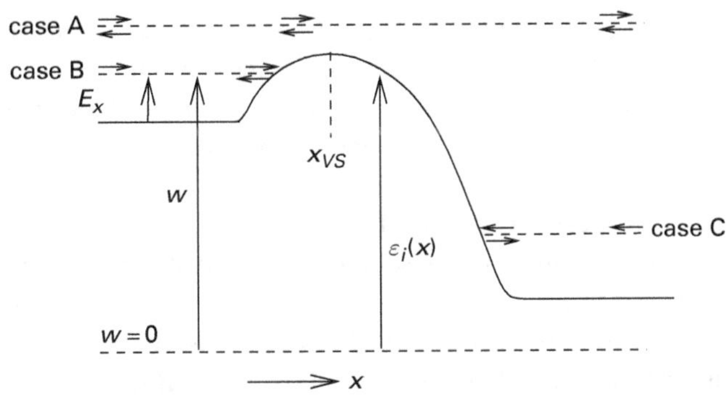

Figure 5.10 Representation of the subband energy profile along the channel of a MOSFET and definition of the energies w, E_x, and ε_i. In case (A) the electron states with $v_{gx}>0$ are filled according to the distribution F_S at the source, whereas those with $v_{gx}<0$ are filled according to the distribution F_D at the drain. In case (B) all states are filled according to F_S, due to the presence of the classical turning point at the left of the x_{VS}. Similarly, in case (C) all states are filled according to F_D.

In order to derive a BTE in $F_i^{\pm}$, we compute the total derivative of $F_i^{\pm}$ with respect to x:

$$\frac{\mathrm{d}F_i^{\pm}}{\mathrm{d}x} = \frac{\partial f_i}{\partial x} + \frac{\partial f_i}{\partial k_x}\frac{\mathrm{d}k_x}{\mathrm{d}x}. \tag{5.145}$$

Since we consider w as a fixed parameter of $F_i^{\pm}$, we must impose $\mathrm{d}w/\mathrm{d}x = 0$ when evaluating $\mathrm{d}F_i^{\pm}/\mathrm{d}x$, that implies $\mathrm{d}E_x/\mathrm{d}x = -\mathrm{d}\varepsilon_i/\mathrm{d}x$. From Eq.5.143 we thus have:

$$\frac{\mathrm{d}k_x(w-\varepsilon_i(x))}{\mathrm{d}x} = \frac{\mathrm{d}k_x}{\mathrm{d}\varepsilon_i}\frac{\mathrm{d}\varepsilon_i}{\mathrm{d}x} = -\frac{\mathrm{d}k_x}{\mathrm{d}E_x}\frac{\mathrm{d}\varepsilon_i}{\mathrm{d}x}. \tag{5.146}$$

However, since $(1/\hbar)\mathrm{d}E_x/\mathrm{d}k_x$ is the group velocity, substitution of Eqs.5.140 and 5.146 into Eq.5.145 gives

$$\frac{\mathrm{d}F_i^{\pm}}{\mathrm{d}x}(x, w, k_y) = 0, \tag{5.147}$$

which is the BTE for the distribution functions $F_i^{\pm}$.

Equation 5.147 underlines an important property that we use in the following section, namely that for a given k_y, the auxiliary distributions $F_i^{\pm}(x, w, k_y)$ are constant along the transport direction x.

5.6.1 Carrier distribution in a ballistic MOSFET

From Eq.5.147 we see that, if we know the distributions $F_i^{\pm}$ at the boundary of the device, then we also know them at any other point inside the device. Since $F_i^{\pm}$ are equivalent to the carrier distributions f_i, we would then possess a complete solution of the ballistic transport problem. In other words, Eq.5.147 defines ballistic transport as a boundary value problem. Indeed, F_i^{+} and F_i^{-} can be expressed in terms of F_S (the distribution of carriers with $v_{gx} > 0$ at the source) and F_D (the distribution of carriers with $v_{gx} < 0$ at the drain) by the following reasoning.

If we consider a typical subband profile along the channel of a MOSFET such as the one sketched in Fig.5.10, and denote with x_{VS} the position of maximum ε_i (the so called *virtual source*), from Eq.5.147 we see that $F_i^+ = F_S$ and $F_i^- = F_D$, only for $w>\varepsilon_i(x_{VS})$ (case A in Fig.5.10). In fact, Eq.5.147 fails at the classical turning points and the cases with $w<\varepsilon_i(x_{VS})$ must be handled separately with special care. In fact, in the ballistic limit f_i is discontinuous for $k_x = 0$, so that at the turning points, where E_x becomes zero, the term $\mathrm{d}f_i/\mathrm{d}k_x$ becomes singular. In the presence of a classical turning point for particles moving with $k_x>0$ (see case B in Fig.5.10), the distribution F_i^+ must become null at the right of the turning point, whereas the flux of particles represented by F_i^+ becomes a flux F_i^- at the left of the turning point. In other words, for $w<\varepsilon_i(x_{VS})$ we have $F_i^+ = F_i^- = F_S$ at the left of the turning point. Similar arguments apply to particles moving with $k_x < 0$ (see case C in Fig.5.10).

The considerations above yield the following relations:

$$\begin{array}{ll} \text{for} \quad w > \varepsilon_i(x_{VS}) \left\{ \begin{array}{l} F_i^+ = F_S \\ F_i^- = F_D \end{array} \right. & \text{(case A)} \\ \text{for} \quad w \leq \varepsilon_i(x_{VS}) \left\{ \begin{array}{lll} F_i^+ = F_i^- = F_S & \text{if} \quad x < x_{VS} & \text{(case B)} \\ F_i^+ = F_i^- = F_D & \text{if} \quad x > x_{VS} & \text{(case C)} \end{array} \right. . \end{array} \tag{5.148}$$

It is worth recalling that since $F_i^{\pm}(w, k_y, x)$ is just a different way to write $f_i(x, k_x, k_y)$ and we know how to relate carrier density and current density to $f_i(x, k_x, k_y)$ (Eqs.5.46 and 5.47), it is easy to convert $F_i^{\pm}(w, k_y, x)$ to $f_i(x, k_x, k_y)$ and then perform all the integrals using $f_i(x, k_x, k_y)$. To this purpose, $f_i(x, k_x, k_y)$ is easily derived from $F_i^+(w, k_y, x)$ and $F_i^-(w, k_y, x)$ (for the states with $k_x>0$ and $k_x<0$, respectively) by substitution of the expression relating w to k_x (Eq.5.143).

If, instead, concentrations and currents are derived directly from $F_i^{\pm}(w, k_y, x)$, attention has to be paid when performing the integrals over w. For instance, the carrier density at the position x is not just the integral over w of $F_i^{\pm}(w, k_y, x)$, but instead reads:

$$\begin{aligned} N(x) = \sum_i \frac{1}{2\pi^2} \int_{-\infty}^{+\infty} \Bigg[\int_{-\infty}^{+\infty} F_i^+(w, k_y, x) \frac{\sqrt{m_x}\mathrm{d}w}{\hbar\sqrt{2[w - \epsilon_i(x)]}} \\ + \int_{-\infty}^{+\infty} F_i^-(w, k_y, x) \frac{\sqrt{m_x}\mathrm{d}w}{\hbar\sqrt{2[w - \epsilon_i(x)]}} \Bigg] \mathrm{d}k_y. \end{aligned} \tag{5.149}$$

The considerations about the conservation of $F_i^{\pm}(w)$ along x are useful to derive the shape of the distribution function in the **k**-plane as well. In particular, with the help of Fig.5.11, we can derive the evolution of the equi-occupation lines in the **k** plane when moving from the VS into the channel. For simplicity we consider circular subbands, set $E_{v0} = 0$ and $\varepsilon_i(x_{VS}) = 0$ and assume that the V_{DS} is large, so that only states with $k_x>0$ are occupied at the VS (and at all other points of interest in the channel) according to the equilibrium distribution at the source contact. Therefore an equi-occupation line at the VS is also an equi-energy line. We denote by E_i the energy of the chosen contour.

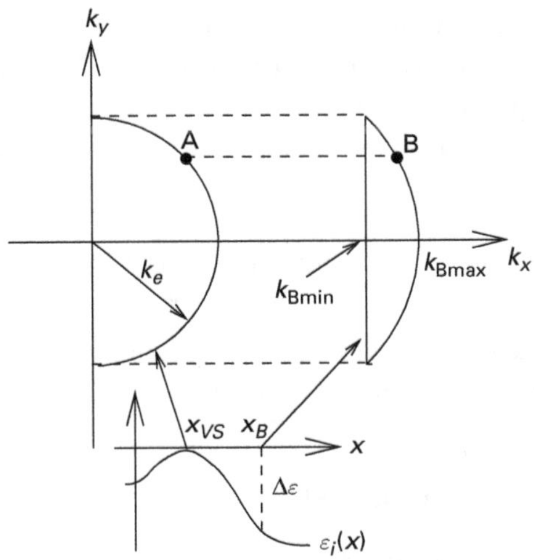

Figure 5.11 The evolution of an equi-occupation curve in the **k**-plane going from the VS toward the drain under ballistic transport at high V_{DS}.

For a circular subband with $m_x = m_y = m$, an equi-energy line in the **k**-plane is a circle with radius $k_e = \sqrt{2mE_i}/\hbar$. Obviously, since we do not have particles with $k_x<0$ at the VS, only the hemi-circle with $k_x>0$ belongs to the contour, as sketched in Fig.5.11.

We now see how a generic **k**-plane point A on the equi-occupation and equi-energy curve at the VS moves to point B on the equi-occupation curve at a position x_B, where the subband energy has dropped by $\Delta\varepsilon$ with respect to the value at the VS. First of all, A and B must have the same k_y, since the forces are aligned with x and thus do not modify k_y. In order to have the same occupation, points A and B should have the same w, but since k_y is also the same, so is the total energy (Eqs.5.142 and 5.143). Thus, in any position along the channel the constant occupation contours are also constant energy contours. To determine the shape of the equi-energy curve at position x_B we consider that, since the total energy is the same at points A and B:

$$E_i = \frac{\hbar^2(k_{x,A}^2 + k_{y,A}^2)}{2m} = \frac{\hbar^2(k_{x,B}^2 + k_{y,B}^2)}{2m} - \Delta\varepsilon, \tag{5.150}$$

thus the equi-energy curve at x_B is an arc with radius $k_{\text{Bmax}} = \sqrt{2m(E_i + \Delta\varepsilon)}/\hbar$. Since transport in the x direction does not modify k_y, the maximum k_y occupied at position x_B is $k_e = \sqrt{2mE_i}/\hbar$. As a result, the minimum k_x is $k_{\text{Bmin}} = \sqrt{2m\Delta\varepsilon}/\hbar$ and the aperture of the arc is

$$\Theta(E_i, x) = 2\arccos[\sqrt{\Delta\varepsilon/(E_i + \Delta\varepsilon)}]. \tag{5.151}$$

From the previous considerations we can also derive the energy distribution function $f_i(E_i, x)$. At a generic position x, compared to the distribution at the VS, only a fraction θ/π of the hemi-circle is occupied at the point x_B. Therefore we have

$$f_i(E_i, x) = f_i(E_i, x_{VS})\frac{\Theta(E_i, x)}{\pi}H_v(E_i), \tag{5.152}$$

where $\Theta(E_i, x)$ depends on x through $\Delta\varepsilon$. The step function $H_v(E_i)$ accounts for the fact that the states with energy below the top of the barrier at the VS cannot be populated if the transport is ballistic.

From $f_i(E_i, x)$ we can compute the concentration of forward moving electrons in the subband as

$$N_i^+(x) = \frac{1}{2} \int f_i(E_i, x) g(E_i) \mathrm{d}E_i, \tag{5.153}$$

where the density of states $g(E_i)$ is given by Eq.3.76, and the factor $1/2$ comes from the fact that states with k_x<0 are empty.

5.6.2 Ballistic current in a MOSFET

We now use Eq.5.148 to calculate the steady-state drain current I_{BL} of a MOSFET under ballistic transport conditions. The results are generalized to quasi-ballistic transport in Section 5.7. The current can be computed at any position because it is independent of x, but the calculation is easier if carried out at the virtual source. In this respect, we have seen in Chapter 3 and we see again in Chapter 7 that the carrier gas in MOSFET devices can be strongly degenerated, thus demanding the use of Fermi–Dirac statistics in the source and drain. However, for the sake of simplicity we assume here a Maxwell–Boltzmann distribution in the source and drain and elliptical subbands as in Eq.5.142. We extend the main results to Fermi–Dirac statistics in Section 5.6.4. A complete treatment of ballistic transport under Fermi–Dirac statistics can be found in [32].

The distributions at the source and drain are:

$$f_i(S, k_x > 0, k_y) = \exp\left(-\frac{\frac{\hbar^2 k_x^2}{2m_x} + \frac{\hbar^2 k_y^2}{2m_y} + \varepsilon_i(S) - E_{FS}}{K_B T}\right), \tag{5.154a}$$

$$f_i(D, k_x < 0, k_y) = \exp\left(-\frac{\frac{\hbar^2 k_x^2}{2m_x} + \frac{\hbar^2 k_y^2}{2m_y} + \varepsilon_i(D) - E_{FD}}{K_B T}\right), \tag{5.154b}$$

where $f_i(S, k_x, k_y)$ for k_x<0 and $f_i(D, k_x, k_y)$ for k_x>0 are a priori unknown and the index i comprises both valleys and subbands within each valley. Substitution of Eqs.5.142 and 5.143 in Eqs.5.154 gives:

$$F_S(w, k_y) = \exp\left(-\frac{w + \frac{\hbar^2 k_y^2}{2m_y} - E_{FS}}{K_B T}\right), \tag{5.155a}$$

$$F_D(w, k_y) = \exp\left(-\frac{w + \frac{\hbar^2 k_y^2}{2m_y} - E_{FD}}{K_B T}\right). \tag{5.155b}$$

Since the VS corresponds to case A in Fig.5.10 for any possible w value, from Eq.5.148 we derive:

$$F_i^+(x_{VS}, w, k_y) = F_S(w, k_y), \tag{5.156a}$$

$$F_i^-(x_{VS}, w, k_y) = F_D(w, k_y). \tag{5.156b}$$

Using again f_i instead of $F_i^{\pm}$:

$$f_i(x_{VS}, k_x > 0, k_y) = \exp\left(-\frac{\frac{\hbar^2 k_x^2}{2m_x} + \frac{\hbar^2 k_y^2}{2m_y} + \varepsilon_i(x_{VS}) - E_{FS}}{K_B T}\right), \tag{5.157a}$$

$$f_i(x_{VS}, k_x < 0, k_y) = \exp\left(-\frac{\frac{\hbar^2 k_x^2}{2m_x} + \frac{\hbar^2 k_y^2}{2m_y} + \varepsilon_i(x_{VS}) - E_{FD}}{K_B T}\right). \tag{5.157b}$$

From Eq.5.47, transforming the sums in integrals according to the standard prescriptions, we can write the current per unit width for the ith subband as

$$\frac{I_i}{W} = \frac{e n_{sp}}{(2\pi)^2}\int_{-\infty}^{+\infty}\int_{-\infty}^{+\infty}\frac{\hbar k_x}{m_x} f_i \,\mathrm{d}k_x\,\mathrm{d}k_y, \tag{5.158}$$

where f_i is the total distribution obtained combining Eqs.5.157a and 5.157b and $n_{sp} = 2$ is the spin degeneracy. We readily get

$$\frac{I_i}{W} = \frac{e}{2\pi^2}\exp\left[-\frac{\varepsilon_i(x_{VS})}{K_B T}\right]\left[\exp\left(\frac{E_{FS}}{K_B T}\right) - \exp\left(\frac{E_{FD}}{K_B T}\right)\right]$$
$$\times \int_{-\infty}^{+\infty}\exp\left(-\frac{\hbar^2 k_y^2}{2m_y K_B T}\right)\mathrm{d}k_y \int_0^{+\infty}\frac{\hbar k_x}{m_x}\exp\left(-\frac{\hbar^2 k_x^2}{2m_x K_B T}\right)\mathrm{d}k_x. \tag{5.159}$$

The integral in k_y can be transformed into the Gauss integral by defining $\zeta = \hbar k_y/\sqrt{m_y K_B T}$, whereas the integral in k_x is transformed into the integral of an exponential function by defining $\xi = \hbar^2 k_x^2/(2m_x K_B T)$. We thus obtain:

$$\frac{I_i}{W} = \frac{e(K_B T)^{3/2}\sqrt{m_y}}{\sqrt{2}\pi^{3/2}\hbar}\exp\left(-\frac{\varepsilon_i(x_{VS})}{K_B T}\right)\left[\exp\left(\frac{E_{FS}}{K_B T}\right) - \exp\left(\frac{E_{FD}}{K_B T}\right)\right]. \tag{5.160}$$

Equation 5.160 can be rewritten as the physically transparent difference between the carrier flux $\mathcal{F}_i^+$ injected into the channel from the source and the one coming from the drain $\mathcal{F}_i^-$:

$$\frac{I_i}{W} = e(\mathcal{F}_i^+ - \mathcal{F}_i^-), \tag{5.161}$$

with

$$\mathcal{F}_i^{\pm} = \frac{(K_B T)^{3/2}\sqrt{m_y}}{\sqrt{2}\pi^{3/2}\hbar}\exp\left(-\frac{\varepsilon_i(x_{VS}) - E_{FS,D}}{K_B T}\right), \tag{5.162}$$

where E_{FS} is used to compute $\mathcal{F}_i^+$ and E_{FD} is used to compute $\mathcal{F}_i^-$.

In order to take one step further and cast Eq.5.161 in a form suited for direct comparison with Eq.5.139, it is useful, at this stage, to evaluate the density of carriers in subband i with $k_x>0$ at the VS (N_i^+) and the corresponding density of carriers with $k_x<0$ (N_i^-). We thus proceed as in Section 3.6 but perform the sum over $\mathbf{k}$ separating the states with $k_x>0$ and $k_x<0$:

$$N_i^+ = \frac{1}{A} \sum_{\mathbf{k},k_x>0} f_i(x_{VS}, \mathbf{k}), \tag{5.163a}$$

$$N_i^- = \frac{1}{A} \sum_{\mathbf{k},k_x<0} f_i(x_{VS}, \mathbf{k}). \tag{5.163b}$$

We readily see that

$$N_i^{\pm} = \frac{1}{2\pi^2} \exp\left(-\frac{\varepsilon_i(x_{VS}) - E_{FS,D}}{K_B T}\right) \times \int_{-\infty}^{+\infty} \exp\left(-\frac{\hbar^2 k_y^2}{2m_y K_B T}\right) dk_y \int_0^{+\infty} \exp\left(-\frac{\hbar^2 k_x^2}{2m_x K_B T}\right) dk_x, \tag{5.164}$$

where E_{FS} is used to compute N_i^+ and E_{FD} is used to compute N_i^-. Both integrals in Eq.5.164 can be easily transformed into the Gauss integral. It is then straightforward to see that:

$$\frac{\mathcal{F}_i^+}{N_i^+} = \frac{\mathcal{F}_i^-}{N_i^-} = \sqrt{\frac{2K_B T}{\pi m_x}} = \upsilon_{th,i}, \tag{5.165}$$

where $\upsilon_{th,i}$ is the so called *thermal velocity*, which is the average velocity of all carriers moving only with positive or only with negative group velocity for a Maxwell–Boltzmann distribution and assuming parabolic bands. We denote the thermal velocity as $\upsilon_{th,i}$ because subbands belonging to different valleys can have different values of m_x and the index i spans both valleys and subbands.

5.6.3 Compact formulas for the ballistic current

The N_i^+ and N_i^- are sensitive to the structure, the electrostatic potential, and transport conditions of the entire MOS device. The correct evaluation of the ballistic current thus requires a full solution of the 2D Poisson equation in the (x, z) directions coupled with the one-dimensional Schrödinger equation solved at each position x and with the electron states occupied according to Eq.5.148. However, since in the VS region the lateral electric field is small compared to the confining field, we can determine the charge at the VS in strong inversion conditions from simple 1D electrostatics, with the self-consistent solution of the one-dimensional Schrödinger and Poisson equations, as described in Section 3.7, with the important difference that now only the states with positive υ_{gx} are populated at high V_{DS}.

In more detail, if we assume Maxwell–Boltzmann statistics, the formulas for the charge at $V_{DS}=0$ must be multiplied by a factor $[1+\exp(-V_{DS}/V_{th})]/2$ where

$V_{th} = (K_B T/e)$ is the thermal voltage, because states with negative v_{gx} are populated by a distribution with a Fermi level $E_{FD} = E_{FS} - eV_{DS}$, where V_{DS} is the intrinsic source-drain voltage.

To obtain results in a compact form, we set:

$$e(N^+ + N^-) \simeq C_{G,eff}(V_{GS} - V_T), \tag{5.166}$$

where V_{GS} is the intrinsic gate-source voltage, V_T is the threshold voltage, $C_{G,eff}$ is an effective gate capacitance and

$$N^+ = \sum_i N_i^+, \qquad N^- = \sum_i N_i^-. \tag{5.167}$$

Since Eq.5.164 gives $N_i^- = N_i^+ \exp(-V_{DS}/V_{th})$, we obtain:

$$I_{BL} = \sum_i I_i = W v^+ C_{G,eff}(V_{GS} - V_T) \frac{1 - \exp(-V_{DS}/V_{th})}{1 + \exp(-V_{DS}/V_{th})}, \tag{5.168}$$

where

$$v^+ = \frac{\sum_i N_i^+ v_{th,i}}{\sum_i N_i^+} \tag{5.169}$$

is the average velocity at the VS of the carriers moving with positive group velocity.

Compared to the well-known equation for long channel MOSFETs (Eq.5.133), we see that the ballistic current does not depend on the gate length and that the dependence on $(V_{GS} - V_T)$ is linear instead of quadratic. For V_{DS} much larger than V_{th}, Eq.5.168 simplifies to

$$I_{BL} = W C_{G,eff}(V_{GS} - V_T) v^+, \tag{5.170}$$

which is very similar to Eq.5.139, derived from the Drift–Diffusion model under the assumption of velocity saturation across the whole channel. The difference between the expressions is that the average thermal velocity v^+ replaces the saturation velocity v_{sat}.

From the point of view of the numerical values, we observe that in bulk unstrained silicon these two quantities are very similar, since $v^+ = 1.2 \times 10^7$cm/s (considering the dominant 2-fold valleys), whereas $v_{sat} = 10^7$cm/s. From the point of view of device physics, however, it is clear that the meaning of Eq.5.139 and Eq.5.170 is totally different. While Eq.5.139 embodies the approximation of carrier distributions close to equilibrium (for the closure condition Eq.5.130) and the concept of velocity saturation due to intense scattering, in Eq.5.170 the upper limit to the carrier velocity is set in a purely ballistic transport regime and it is the result of the limited average velocity of the carriers supplied by the source reservoir to the channel.

It is worth pointing out at this stage that Eqs.5.168 and 5.170 do not contain the low-field mobility parameters, and rightly so, since the concept of mobility is related to the effect of scattering in the low field transport regime (see Section 5.4), whereas no scatterings occur in a ballistic device. Nevertheless, if V_{DS} is low compared to V_{th}, we can rewrite Eq.5.168 as:

$$I_D \simeq \frac{W}{L} \mu_{bl} C_{G,eff} (V_{GS} - V_T) V_{DS}, \tag{5.171}$$

which resembles the expression for the current of long channel devices (Eq.5.133) biased in the linear region at low V_{DS}, except that the mobility is replaced by the so-called *ballistic mobility*

$$\mu_{bal} = \frac{L v^+}{2 V_{th}}. \tag{5.172}$$

Clearly, the ballistic mobility has nothing to do with the momentum relaxation times and with the scattering. It is just an *apparent* mobility. Since the ballistic current tends to be independent of L, whereas the conventional MOSFET model of Eq.5.133 predicts an inverse scaling with L, an apparent mobility proportional to L is necessary to reconcile the two expressions.

Equation 5.172 indicates that we should expect a reduction of the apparent mobility determined by conventional parameter extraction techniques with the scaling of L. Indeed it has been observed in [33, 34] that the experimental mobility reduces as the gate length is reduced. However, this reduction appears to be even larger than predicted by simulations accounting for finite velocity at the source [35]. Additional scattering mechanisms related to the fabrication of very short channel devices have been proposed as being responsible for the discrepancy [34]

The concept of ballistic mobility suggests that in ballistic transport conditions the linear drain current is also expected to become insensitive to L.

5.6.4 Injection velocity and subband engineering

The model expressed by Eq.5.170 predicts that the drain current at high V_{DS} is proportional to v^+, that is the average velocity at the point where carriers are injected in the channel (namely the VS). At high V_{DS} all carriers move with a positive group velocity at the VS, since the population of carriers coming from the drain is negligible. The velocity v^+ is sometimes referred to as the *injection velocity*.

In order to maximize the I_D one could increase the current by increasing the term $W C_{G,eff}(V_{GS} - V_T)$, but this would result in an increase of the gate capacitance and would leave the CV/I delay unaffected at first order. Increasing the v^+ is thus preferable.

Under Maxwell–Boltzmann statistics, Eqs.5.165 and 5.169 tell us that we can increase v^+ by populating subbands with the lowest m_x. In the (001)-silicon inversion layers considered so far this means increasing the population of the 2-fold valleys featuring $m_x = 0.19 m_0$. Techniques aimed at increasing v^+ by optimizing the effective masses and the occupation probability of the subbands are known as *subband engineering* [36].

As a relevant example of the possibilities opened by subband engineering to improve CMOS performance, we consider ultra-thin-body (UTB) SOI devices. In fact, due to the higher quantization mass of the 2-fold valleys compared to the 4-fold ($0.916 m_0$ instead of $0.19 m_0$), a thin quantization well increases the energy splitting between 2-fold and

4-fold valleys, thus increasing the relative population of the former with respect to the latter. However, we see in Chapter 7 that the increase in v^+ provided by UTB-SOI devices is counteracted by the enhanced scattering rate related to the roughness of the interfaces, so that the optimum trade-off between the two is not so trivial to find. An alternative technique to increase the population of the two-fold valleys is the use of strained silicon in the channel, as we see in Chapter 9.

Another important aspect of subband engineering is the effect of the degeneracy of the electron gas. Following the same approach as in Section 5.6.2 we can derive an expression for the injection velocity for the subband i in the case of Fermi–Dirac statistics, because it is easy to see that the distribution of positively directed carriers at the VS follows a Fermi–Dirac distribution with the source Fermi level [32]. Assuming for simplicity a circular subband with effective mass m we can write

$$f_i(x_{VS}, k_x > 0, k_y) = \left[\exp\left(\frac{\hbar^2(k_x^2 + k_y^2)/(2m) + \varepsilon_i - E_{FS}}{k_B T}\right) + 1\right]^{-1}. \tag{5.173}$$

We can thus compute the injection velocity as in Eq.5.165. Using polar coordinates:

$$v_{inj,i} = \frac{\int_0^{+\infty} \int_{-\pi/2}^{+\pi/2} f_i(x_{VS}, k) \frac{\hbar k \cos\theta}{m} \mathrm{d}\theta k \mathrm{d}k}{\int_0^{+\infty} \int_{-\pi/2}^{+\pi/2} \mathrm{d}\theta f_i(x_{VS}, k) k \mathrm{d}k}, \tag{5.174}$$

where $k = \sqrt{k_x^2 + k_y^2}$ and we have written the group velocity as $v_{gx} = \hbar k_x/m = \hbar k \cos\theta/m$. By integrating over θ and changing the variable of integration from k to $E_p = \hbar^2 k^2/(2m)$, one finds:

$$v_{inj,i} = \frac{2\hbar}{\pi m} \frac{\int_0^{+\infty} f_i(E_p)\sqrt{E_p}\mathrm{d}E_p}{\int_0^{+\infty} f_i(E_p)\mathrm{d}E_p}. \tag{5.175}$$

By defining $x = E_p/(K_B T)$ and $\eta_i = (E_{FS} - \varepsilon_i)/(K_B T)$ we get:

$$v_{inj,i} = \frac{2\sqrt{2K_B T}}{\pi\sqrt{m}} \frac{\int_0^{+\infty} \frac{\sqrt{x}}{\exp(x-\eta_i)+1}\mathrm{d}x}{\int_0^{+\infty} \frac{1}{\exp(x-\eta_i)+1}\mathrm{d}x} = v_{th,i} \frac{\mathcal{F}_{1/2}(\eta_i)}{\mathcal{F}_0(\eta_i)}, \tag{5.176}$$

where $\mathcal{F}_0$ and $\mathcal{F}_{1/2}$ are the Fermi integrals of order 0 and 1/2 (see Appendix A). The average velocity at the VS is then given by:

$$v^+ = \frac{\sum_i N_i^+ v_{inj,i}}{\sum_i N_i^+}. \tag{5.177}$$

Figure 5.12 shows that $v_{inj,i}$ can become significantly larger than $v_{th,i}$ for a degenerate carrier gas. This condition is well met also in silicon inversion layers at high V_{GS}, as we see in Chapter 7.

Increasing η_i means that to reach a given inversion charge it is necessary to push the subband minimum ε_i well below the source Fermi level E_{FS}. We return to this point in Section 10.5.1, where we also illustrate the expression for the ballistic current in inversion layers with arbitrary orientation.

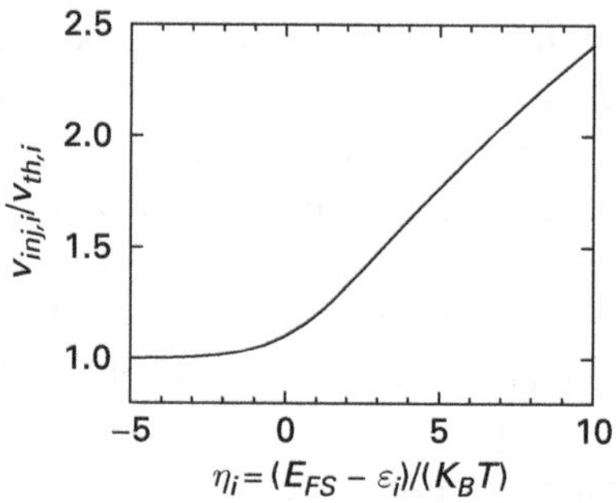

Figure 5.12 Subband injection velocity under Fermi–Dirac statistics normalized to the corresponding thermal velocity, as a function of $\eta_i = (E_{FS} - \varepsilon_i)/(K_B T)$ (Eq.5.176).

5.7 The quasi-ballistic transport regime

Simulations based on exact solutions of the BTE using the Monte Carlo method (such as those reported in Chapter 7 and in [37–39]) as well as experimental evidence [40, 41] suggest that nanoscale MOSFETs are likely to operate still quite far from the ballistic limit. The transport regime appears to lie somewhere in between the purely ballistic and the diffusive regimes. While we can represent the ballistic current with analytical models such as those described in Section 5.6, the diffusive regime is well described by the Drift–Diffusion model implemented in many commercial device simulators (see Section 5.5).

The so-called *quasi-ballistic* transport regime taking place in nanoscale MOSFETs, instead, is a complex intermediate condition, where the self-consistent feedback between the charge, the potential, and the reduced carrier velocity can only be understood with the help of exact solutions of the BTE without a-priori assumptions. The Monte Carlo method illustrated in Chapter 6 can provide full solutions of the BTE and, therefore, the quantities at the device terminals (currents), as well as the internal ones (carrier density, velocity, subband population, etc). Many examples of multi-subband Monte Carlo results on scaled devices are reported in Chapter 7.

5.7.1 Compact formulas for the quasi-ballistic current

Here we describe the widely used analytical quasi-ballistic MOSFET model proposed by Lundstrom *et al.* [42, 43]. Since it is very difficult to develop approximate solutions of the BTE valid in the quasi-ballistic transport regime, to obtain a compact model for the drain current one needs to introduce many simplifications to the problem. For these reasons the correctness and the accuracy of the model are still debated. The interested reader is referred to [44] and [45] for a detailed description of the arguments in favor of or against the validity of the model.

Here we note that Lundstrom's model is explicitly not intended for accurate I/V calculations, but, instead, to provide basic insight into the operation of nanoscale MOSFETs [44]. The model provides a vision of the operation of nano-transistors useful

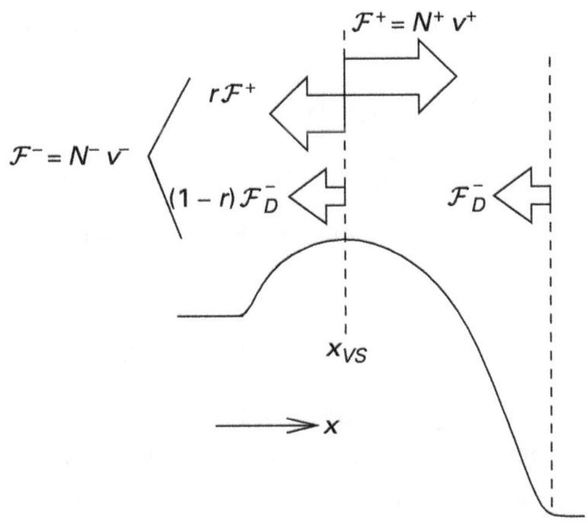

Figure 5.13 A typical subband profile along the channel of a MOSFET operating in the quasi-ballistic transport regime and identification of the main carrier fluxes.

for interpreting the results of simulations based on numerical solution of the BTE and for identifying some of the key quantities for an understanding of device behavior. We follow this in Chapter 7.

For the sake of simplicity we restrict the derivations below to the case of Maxwell–Boltzmann statistics in the source and drain, i.e. neglecting the Pauli exclusion principle. In fact, the extension of the model to Fermi–Dirac statistics is definitely not trivial and still an open field of research.

As in the ballistic transport model of Section 5.6, the treatment of quasi-ballistic transport focuses on the virtual source. With the help of Fig.5.13, we can identify three fluxes of carriers at the VS.

First of all, we have a flux $\mathcal{F}^+$ of electrons injected into the channel from the source reservoir. The model assumes that the transport *uphill* from the source to the top of the barrier does not modify the distribution of the carriers moving with positive group velocity, so that the flux $\mathcal{F}^+$ can be computed as in the ballistic case:

$$\mathcal{F}^+ = N^+ v^+, \tag{5.178}$$

where the meaning of the symbols N^+ and v^+ is the same as in Section 5.6.3. In particular, v^+ is the average velocity of the flux of carriers injected into the channel and it is given by Eq.5.169. If all carriers are in the same subband, v^+ becomes equal to the thermal velocity of that subband and reflects its transport mass (Eq.5.165). In the following, we assume operation in the quantum limit so that only one subband is populated. The index i then becomes useless and will thus be dropped in the notation.

In the ballistic limit and considering a high V_{DS}, the flux of electrons arriving at the VS from the drain is negligible and $\mathcal{F}^+$ determines the drain current. To develop the quasi-ballistic drain current model, instead, we have to consider that a fraction r of the injected flux suffers scattering events in the channel and it is eventually re-absorbed by the source, thus creating a flux $\mathcal{F}^- = N^- v^-$ (see Fig.5.13) that subtracts from F^+. As visible in Fig.5.13, in principle also a fraction of the carriers injected from the drain is able to reach the source, but the assumption of a large V_{DS} allows us to neglect this term at the moment.

The drain current is then given by:

$$I_D = eW(\mathcal{F}^+ - \mathcal{F}^-) = eW\mathcal{F}^+(1-r). \tag{5.179}$$

The term r is called *reflection* or *back-scattering coefficient*, and it is implicitly defined as

$$r = \frac{\mathcal{F}^-}{\mathcal{F}^+}. \tag{5.180}$$

To obtain a closed form expression for I_D the model embraces a number of approximations. First of all, it assumes that the average velocities v^+ and v^- appearing in the fluxes $\mathcal{F}^+$ and $\mathcal{F}^-$ respectively, are equal. Since we are assuming Maxwell–Boltzmann statistics and a single subband, we have $v^+ = v^- = v_{th} = \sqrt{2K_BT/(\pi m)}$. This assumption implies $N^- = N^+ r$ so that $N_{inv} = N^+ + N^- = N^+(1+r)$. It is further assumed that the device electrostatics at the VS is essentially 1D, meaning that the inversion charge density at the VS can be calculated by a self-consistent solution of the one-dimensional Schrödinger and Poisson equations along z at $x = x_{VS}$. Therefore N_{inv} at the VS is directly controlled by the gate capacitance and by the gate voltage overdrive. In other words, N_{inv} at the VS is independent of V_{DS} and of the transport regime, i.e. the charge at the VS under quasi-ballistic transport is the same as in the case of purely ballistic transport.

The assumptions above have been checked by means of Monte Carlo and Non-Equilibrium Green's Function simulations by several authors [38, 45–47] In particular, it has been found that v^- can be significantly smaller than v^+ ($v^- \simeq 0.6v^+$ in [38]), but the impact of the assumption $v^- \simeq v^+$ on the final expression of I_D is not dramatic in the cases of practical interest where r is small. Furthermore, it has been verified that the inversion charge density at the VS is essentially the same with and without scattering, as long as the device is well-tempered and insensitive to DIBL. Otherwise, the different velocity and charge profiles along the channel affect the VS region through the 2D electrostatics [38].

If we substitute $eN_{inv} = C_{G,eff}(V_{GS} - V_T)$ into Eq.5.179 and set $v^+ = v^- = v_{th}$, we obtain

$$I_D = WC_{G,eff}(V_{GS} - V_T)v_{th}\left(\frac{1-r}{1+r}\right) = I_{BL}\left(\frac{1-r}{1+r}\right), \tag{5.181}$$

which is the desired result.

From Eq.5.181 we see that the effect of scattering on I_D is two-fold: the current is reduced compared to the ballistic limit because part of the injected electrons are back-scattered to the source (the $(1-r)$ term at the numerator), but also because for the same total charge (which is set by V_{GS}) the carrier density available to support the positively directed flux $\mathcal{F}^+$ (that is N^+) is smaller compared to the ballistic case, that is $N^+ < C_{G,eff}(V_{GS} - V_T)/e$, since, due to back-scattering, N^- is not zero. This effect is embedded in the $(1+r)$ term in the denominator of Eq.5.181 and is illustrated in Chapter 7.

It is worth noting that Monte Carlo simulation of MOSFETs with L down to 14 nm [38] have shown that the ratio I_D/I_{BL} is fairly close to $(1-r)/(1+r)$, if r is extracted as $(\mathcal{F}^-/\mathcal{F}^+)$ from Monte Carlo simulations including scattering and if I_{BL} is estimated by simulations of the same structure performed switching off scattering in the channel.

One of the assumptions at the basis of Eq.5.181 is that the source region behaves as a perfect reservoir of carriers at thermal equilibrium up to the VS point. This is equivalent to assuming that the region on the left side of the VS is essentially non-influential on the current. The assumption that the population of carriers with positive v_{gx} is at equilibrium with the source contact may become questionable in conditions where *source starvation* limits the amount of available carriers [48] or if long range Coulomb interactions [49] produce significant deviations of the carrier distribution function at the VS with respect to the form given in Eq.5.157.

Equation 5.181 can be readily extended to any V_{DS}. With the help of Fig.5.13, we see that the flux $\mathcal{F}^-$ consists of the back-scattered flux $(\mathcal{F}^+ r)$ plus the flux injected by the drain reservoir and transmitted through the channel back to the source. Since this flux also suffers scattering events in the channel region, it takes the expression $\mathcal{F}_D^-(1-r)$, where the flux $\mathcal{F}_D^-$ accounts only for those electrons able to surmount the VS. Since the Fermi level in the drain is shifted by $[-eV_{DS}]$ with respect to the Fermi level in the source, we get:

$$\mathcal{F}_D^- = \mathcal{F}^+ \exp(-V_{DS}/V_{th}), \tag{5.182}$$

and then

$$I_D = eW\left[\mathcal{F}^+(1-r) - (1-r)\mathcal{F}^+ \exp(-V_{DS}/V_{th})\right], \tag{5.183}$$

which finally gives:

$$I_D = WC_{G,eff}(V_{GS}-V_T)v_{th}\frac{(1-r)-(1-r)\exp(-V_{DS}/V_{th})}{(1+r)+(1-r)\exp(-V_{DS}/V_{th})}. \tag{5.184}$$

It is useful to note that for $r=0$ Eq.5.184 reduces to Eq.5.168, whereas for $V_{DS} \gg V_{th}$ we recover Eq.5.181.

5.7.2 Back-scattering coefficient

By itself Eq.5.184 is not useful for estimating the drain current of a nanoscale MOSFET unless a model to relate the reflection coefficient r to the bias and to the rate of scattering is available. This is probably the most critical aspect of the model when attempts are made to make quantitative use of it. For a vanishingly small lateral field (i.e. very low V_{DS}), the model assumes that [43]:

$$r = \frac{L}{L+\lambda_\mu}, \tag{5.185}$$

where λ_μ is a suitable mean-free-path. Equation 5.185 is straightforwardly derived by considering a situation in the absence of any lateral electric field where carriers belonging to the forward and backward directed fluxes ($\mathcal{F}^+$ and $\mathcal{F}^-$ respectively) change their direction when moving (on average) over a distance λ_μ [50].

The flux theory [51–54] relates the mean-free-path λ_μ to the low-field mobility through

$$\lambda_\mu = \frac{2\mu_{eff} V_{th}}{v_{th}}. \tag{5.186}$$

The accuracy of Eqs.5.185 and 5.186 in reproducing the reflection coefficient r in the limit of zero field in the lateral direction has been analyzed by using Monte Carlo simulations in [55]. It has been found that these equations indeed reproduce the L-dependence of r over a wide range of channel lengths and for values of the scattering rate corresponding to a wide range of μ_{eff} values.

The expression for r at high V_{DS} proposed in [43] is:

$$r = \frac{L_{kT}}{L_{kT} + \lambda_\mu}, \tag{5.187}$$

and it is identical to the low-field expression except for the substitution of the channel length with the so-called *KT-layer* length, L_{kT}, which is the distance from the VS where the subband profile drops by $K_B T$. Since $K_B T$ is much smaller than V_{DS}, the *KT*-layer is only a small fraction of the channel.

The main idea behind Eq.5.187 is that scattering events along the channel generate a thermal distribution of back-scattered carriers, so that the $\mathcal{F}^-$ flux has a Maxwell–Boltzmann distribution. The back-scattered electrons then diffuse to the source and only those that suffered scattering events within L_{kT} still possess enough energy to surmount the VS barrier. In fact, it is demonstrated in [56] that Eq.5.187 can be derived from the BTE by assuming a constant (over energy and over space) mean-free-path and a thermal distribution for the fluxes $\mathcal{F}^+$ and $\mathcal{F}^-$. The same analysis yields an expression for r which is valid at arbitrary V_{DS} values and reduces to Eq.5.185 and to Eq.5.187 in the limit of small and large V_{DS} respectively:

$$r = \frac{L_{kT}\left[1 - \exp(-L/L_{kT})\right]}{L_{kT}\left[1 - \exp(-L/L_{kT})\right] + \lambda_\mu}. \tag{5.188}$$

Monte Carlo simulations have shown that indeed the critical length over which scattering events contribute to r is longer than L_{kT} [37, 38, 57]. We return to this point in Section 7.3.1.

Another important consideration related to Eq.5.187 is that the low-field mobility μ_{eff} appears to be relevant in determining the current drive of short channel MOSFETs at high V_{DS}: the larger is μ_{eff}, the closer the device works to the ballistic limit. This is a non-trivial result with important consequences for device design and optimization.

A systematic analysis of the validity of Eq.5.187 has been carried out in [55] by comparing it with the results of Monte Carlo simulations: the functional form of the dependence of r on L_{kT} and λ_μ appears to be correct; however, if one of the two quantities is set according to its definition or to Eq.5.186, then the other quantity must be used as an adjustable parameter. In practice, either a mean-free-path shorter than λ_μ or a critical length longer than L_{kT} has to be taken. The main reason for this inconsistency is that the forward and backward fluxes are far from being thermalized even inside the KT-layer [55]. Models accounting for the non-thermal nature of the fluxes $\mathcal{F}^+$ and $\mathcal{F}^-$ have been proposed in [57–59].

5.7.3 Critical analysis of the quasi-ballistic model

One of the advantages of Eqs.5.185 and 5.187 is that they can be inserted into Eq.5.184, so that, by using Eq.5.186 to relate λ_μ to μ_{eff}, one obtains expressions for I_D as a function of V_{GS} and V_{DS} that can be directly compared to the ones from the Drift–Diffusion and ballistic models. In fact, for $L \ll \lambda_\mu$, r tends to zero and Eq.5.184 recovers the expected form of Eq.5.168 for the ballistic regime. On the other hand, for long channel devices (i.e. $L \gg \lambda_\mu$) we recover the expression of the drain current in the Drift–Diffusion regime (Eq.5.133). For V_{DS} much smaller than V_{th} this can be easily obtained by assuming $\exp(-V_{DS}/V_{th}) \simeq 1 - V_{DS}/V_{th}$ in Eq.5.184 and then writing (from Eq.5.185) $(1-r) = \lambda_\mu/(L+\lambda_\mu) \simeq \lambda_\mu/L$. These steps give:

$$I_D \simeq W C_{G,eff}(V_{GS} - V_T) v_{th} \frac{\lambda_\mu V_{DS}}{2 L V_{th}}, \tag{5.189}$$

which, upon substitution of λ_μ from Eq.5.186 yields an expression equal to the low V_{DS} limit of Eq.5.133. The fact that Eq.5.184 recovers Eq.5.133 for long channel MOSFETs in saturation is much less trivial. In particular, it is not obvious how to relate L_{kT} to the bias. However, a simple way to proceed exists if we assume a priori that for long channels the quasi-ballistic and Drift–Diffusion models provide similar results. We can thus obtain the potential profile from the Drift–Diffusion model and use it to find L_{kT}. In the Drift–Diffusion model, under the assumptions leading to Eq.5.133, the drain current is given by [25]

$$I_D = \mu_{eff} W C_{G,eff} \left[V_{GS} - V_T - \phi_s(x)\right] \frac{d\phi_s}{dx}, \tag{5.190}$$

where ϕ_s is the surface potential profile. By equating Eq.5.190 and Eq.5.133, we obtain a differential equation in ϕ_s. The solution in saturation is

$$\phi_s(x) = (V_{GS} - V_T)\left(1 - \sqrt{1 - \frac{x}{L}}\right). \tag{5.191}$$

If $(V_{GS} - V_T) \gg V_{th}$, we can extract

$$L_{kT} = L \frac{V_{GS} - V_T}{2 V_{th}}. \tag{5.192}$$

If we substitute this expression in Eq.5.187 and then in Eq.5.184, we obtain an expression equal to the limit of Eq.5.133 in saturation (i.e. $V_{DS}^* = V_{GS} - V_T$).

It is interesting to note that the model expressed by Eq.5.184 has the potential to link the long channel Drift–Diffusion transport regime and the ballistic transport limit. A significant limitation, however, is that it is not self-consistent with the potential. This is apparent from the observation that the expression for r does not assume any specific shape of the potential energy profile along the channel, although in [56] it has been found that Eq.5.187 holds only for a linear profile. Moreover, the impact of the transport in the channel on the shape of the potential profile is not part of the model itself. This is a significant difference with respect to the analytical models of Eqs.5.133 and 5.136, where the potential and charge profile along the channel as well as the drain current are computed in a self-consistent way through the Pao–Sah integral [25].

Self-consistent Monte Carlo simulations [38] point out that the potential profile close to the VS, hence the value of L_{kT}, is affected by the amount of back-scattering in the channel. We illustrate this aspect in more detail in Chapter 7. Since the potential affects the expressions for r and L_{kT}, it is not straightforward to relate L_{kT} and r with the effective channel length L and the bias voltages (V_{GS} and V_{DS}). To overcome this difficulty, we can follow an empirical approach and define an effective drain voltage V_{DS}^{*} such that

$$L_{kT} = L\frac{V_{th}}{V_{DS}^{*}}. \tag{5.193}$$

Equation 5.193 implies that at a first approximation the average field in the KT-layer is the same as in the portion of the channel up to the pinch off point.

It is instructive to note that, if we assume that in saturation $V_{DS}^{*} = V_{DS,sat}$ is given by Eq.5.138 (i.e. is the same expression as in the Drift–Diffusion model with velocity saturation) and that L is short enough so that $V_{DS,sat} \ll (V_{GS} - V_T)$, then Eqs.5.181, 5.186, and 5.187 can be rearranged to write

$$I_D = \frac{W}{L + \mu_0 V_{DS}^{*}/v_{th}} \mu_{eff} C_{G,eff} \left(V_{GS} - V_T\right) V_{DS}^{*}, \tag{5.194}$$

which is the same as the low V_{DS}^{*} limit of Eq.5.136 but with v_{th} replacing v_{sat}. Since in silicon the numerical values of v_{th} and v_{sat} at room temperature are quite close to each other, it comes as no surprise that the drain current predicted by Drift–Diffusion based simulators does not dramatically differ from the one given by the full solution of the BTE, even for gate lengths in the decananometric range where transport is far from the quasi-equilibrium conditions assumed by the Drift–Diffusion model and described in Section 5.5.

Note that in the Drift–Diffusion models the upper limit of the velocity (that is v_{sat}) is typically reached in the high-field drain region. In the quasi-ballistic transport model of [42, 43], instead, the limiting factor lies in the velocity at the virtual source (v_{th}). At the drain side v can significantly exceed v_{th}, but as far as the drain does not affect the charge at the VS (i.e. the device is well tempered and robust against DIBL), the drain region is almost non-influential on the current.

In fact, probably the most important insight given by the Lundstrom model is that in saturation I_D of nanoscale transistors is limited by a small region close to the injection barrier. Since v_{th} is very similar to v_{sat} in unstrained bulk silicon, models based on the Drift–Diffusion or the quasi-ballistic theory provide essentially the same predictions for the drain current. On the other hand, short channel devices may often suffer from poor electrostatic integrity, so that the source barrier is influenced by the charge at the drain as well. Since in the Drift–Diffusion approach the velocity at the drain is limited to v_{sat}, whereas higher velocity values are allowed for in the quasi-ballistic models, extra charge piles up at the drain in Drift–Diffusion models, which counteracts injection from the source reservoir into the channel, thus lowering the drain current with respect to the quasi-ballistic transport case. For this reason, the agreement between the Drift–Diffusion approach and the Monte Carlo simulation of the I_D curves is improved by increasing v_{sat} to values larger than its real value (that is the one measured in the uniform transport conditions achieved in long channel resistive gate transistors) [60].

5.8 Summary

In this chapter we have illustrated the Boltzmann transport equation in inversion layers described as a 2D carrier gas, that is a set of integro-differential equations (one for each subband) coupled by the inter-subband scattering terms (Eq.5.45). This set of equations describes the evolution of semi-classical particles in the transport plane, whereas their distribution in the quantization direction normal to the transport plane is obtained by solution of the Schrödinger equation. All the effects related to quantization in the vertical direction (subband splitting, modulation of the transport properties induced by the bias and by the use of SOI structures) are thus taken into account by this approach.

The exact solution of the set of Boltzmann transport equations requires complex numerical techniques, such as the Monte Carlo method described in the next chapter. However, limiting cases like low-field uniform transport and ballistic transport allow for substantial simplifications of the problem at hand. In particular, in the case of low-field uniform transport, the momentum relaxation time approximation can be used to determine the carrier mobility from the knowledge of the subband structure and of the scattering rates, as described in Section 5.4. When, instead, scattering is neglected, we obtain an upper bound for the drain current, which is limited by the finite carrier velocity at the injection point, as discussed in Section 5.6.

Analysis of short channel devices requires the exact solution of the BTE, but some basic insight can be gained by following the model for quasi-ballistic transport reviewed in Section 5.7. It has been shown that this model provides a link between the diffusive quasi-equilibrium transport taking place in long channel devices and the ballistic transport regime. However, the resulting equations are quite close to the ones obtained with the Drift–Diffusion model employed in many commercial TCAD tools (reviewed in Section 5.5), mainly because the numerical values of the limiting velocities in the two models (the saturation velocity at the drain in the Drift–Diffusion model and the thermal velocity in the quasi-ballistic model) are very close in unstrained bulk silicon.

References

[1] M. Lundstrom, *Fundamentals of Carrier Transport.* New York: Addison Wesley, 1990.

[2] C. Jungemann and B. Meinerzhagen, *Hierarchical Device Simulation: The Monte Carlo Perspective.* Wien, New York: Springer, 2003.

[3] D. Esseni and P. Palestri, "Theory of the motion at the band crossing points in bulk semiconductor crystals and in inversion layers," *Journal of Applied Physics*, vol. 105, no. 5, pp. 053702–1–053702–11, 2009.

[4] K. Hess, *Advanced Theory of Semiconductor Devices.* New York: Wiley-IEEE Press, 1999.

[5] S. Datta, *Quantum Transport – Atom to Transistor.* Cambridge: Cambridge University Press, 2005.

[6] M. Lenzi, E. Gnani, S. Reggiani, *et al.*, "A deterministic solution of the Boltzmann transport equation for a one-dimensional electron gas in silicon nanowires," in *Proc. Int. Conf. on Ultimate Integration on Silicon (ULIS)*, (Leuven, Belgium), pp. 47–50, 15–16 March 2007.

[7] M. Lenzi, P. Palestri, E. Gnani, *et al.*, "Investigation of the transport properties of silicon nanowires using deterministic and Monte Carlo approaches to the solution of the Boltzmann transport equation," *IEEE Trans. on Electron Devices*, vol. 55, no. 8, pp. 2086–2096, 2008.

[8] D. K. Ferry and S. M. Goodnick, *Transport in Nanostructures.* Cambridge: Cambridge University Press, 1997.

[9] M.V. Fischetti, "Long-range Coulomb interactions in small Si devices. Part II: Effective electron mobility in thin-oxide structures," *Journal of Applied Physics*, vol. 89, no. 2, p. 1232, 2001.

[10] D. Esseni and A. Abramo, "Modeling of electron mobility degradation by remote Coulomb scattering in ultra-thin oxide MOSFETs," *IEEE Trans. on Electron Devices*, vol. 50, no. 7, 2003.

[11] A.T. Pham, C. Jungemann, and B. Meinerzhagen, "Physics-based modeling of hole inversion-layer mobility in strained-SiGe-on-insulator," *IEEE Trans. on Electron Devices*, vol. 54, no. 9, pp. 2174–2182, 2007.

[12] M. De Michielis, D. Esseni, Y.L. Tsang, *et al.*, "A semianalytical description of the hole band structure in inversion layers for the physically based modeling of pMOS transistors," *IEEE Trans. on Electron Devices*, vol. 54, no. 9, pp. 2164–2173, 2007.

[13] L. Donetti, F. Gamiz, A. Godoy, and N. Rodriguez, "Fully self-consistent $\mathbf{k}\cdot\mathbf{p}$ solver and Monte Carlo simulator for hole inversion layers," in *Proc. European Solid State Device Res. Conf.*, pp. 254–257, 2008.

[14] M. De Michielis, D. Esseni, P. Palestri, and L. Selmi, "Semiclassical modeling of quasi-ballistic hole transport in nanoscale pMOSFETs based on a multi-subband Monte Carlo approach," *IEEE Trans. on Electron Devices*, vol. 56, no. 9, pp. 2081–2091, 2009.

[15] M.V. Fischetti, Z. Ren, P.M. Solomon, M. Yang, and K. Rim, "Six-band $\mathbf{k}\cdot\mathbf{p}$ calculation of the hole mobility in silicon inversion layers: Dependence on surface orientation, strain, and silicon thickness," *Journal of Applied Physics*, vol. 94, no. 2, pp. 1079–1095, 2003.

[16] E. Wang, P. Montagne, L. Shifren, *et al.*, "Physics of hole tranport in strained silicon MOSFET inversion layers," *IEEE Trans. on Electron Devices*, vol. 53, no. 8, pp. 1840–1850, 2006.

[17] D. K. Ferry and S. M. Goodnick, *Transport in Nanostructures.* Cambridge: Cambridge University Press, 1997.

[18] M. De Michielis, D. Esseni, and F. Driussi, "Analytical models for the insight into the use of alternative channel materials in ballistic nano-MOSFETs," *IEEE Trans. on Electron Devices*, vol. 54, no. 1, pp. 115–123, 2006.

[19] F. Stern, "Calculated temperature dependence of mobility in silicon inversion layers," *Phys. Rev. Lett.*, vol. 44, no. 22, pp. 1469–1472, 1980.

[20] Y. Takeda and T. Pearsall, "Failure of Matthiessen's rule in the calculation of carrier mobility and alloy scattering effects in $Ga_{0.47}In_{0.53}As$," *Electronics Letters*, vol. 17, no. 16, pp. 573–574, 1981.

[21] M.V. Fischetti, F. Gamiz, and W. Hansch, "On the enhanced electron mobility in strained-silicon inversion layers," *Journal of Applied Physics*, vol. 92, no. 12, pp. 7320–7324, 2002.

[22] O. Weber and S. Takagi, "Experimental examination and physical understanding of the Coulomb scattering mobility in strained-Si nMOSFETs," *IEEE Trans. on Electron Devices*, vol. 55, no. 9, pp. 2386–2396, 2008.

[23] F. Driussi and D. Esseni, "Simulation study of Coulomb mobility in strained silicon," *IEEE Trans. on Electron Devices*, vol. 56, no. 9, pp. 2052–2059, 2009.

[24] S. Selberherr, *Analysis and Simulation of Semiconductor Devices*. Wien, New York: Springer Verlag, 1984.

[25] Y. Taur and T. Ning, *Fundamentals of Modern VLSI Devices*. Edinburgh: Cambridge University Press, 1998.

[26] S.M. Sze, *Physics of Semiconductor Devices*. New York: Wiley, 1981.

[27] J.A. Cooper and D.F. Nelson, "High-field drift velocity of electrons at the $Si–SiO_2$ interface as determined by a time-of-flight technique," *Journal of Applied Physics*, vol. 54, pp. 1445–1456, 1983.

[28] A. Modelli and S. Manzini, "High-field drift velocity of electrons in silicon inversion layers," *Solid State Electronics*, vol. 21, pp. 99–104, 1988.

[29] L. Lucci, P. Palestri, D. Esseni, and L. Selmi, "Modeling the uniform transport in thin film SOI MOSFETs with a Monte-Carlo simulator for the 2D electron gas," *Solid State Electronics*, vol. 49, no. 9, pp. 1529–1535, 2005.

[30] K. Natori, "Ballistic metal-oxide-semiconductor field effect transistor," *Journal of Applied Physics*, vol. 76, no. 8, pp. 4879–4890, 1994.

[31] H.U. Baranger and J.W. Wilkins, "Ballistic structure in the electron distribution function of small semiconducting structures: General features and specific trends," *Phys. Rev. B*, vol. 36, no. 3, pp. 1487–1502, 1987.

[32] A. Rahman, J. Guo, S. Datta, and M.S. Lundstrom, "Theory of ballistic nanotransistors," *IEEE Trans. on Electron Devices*, vol. 50, no. 9, pp. 1853–1863, 2003.

[33] M.S. Shur, "Low ballistic mobility in submicron HEMTs," *IEEE Electron Device Lett.*, vol. 23, no. 9, pp. 511–513, 2002.

[34] A. Cros, K. Romanjek, D. Fleury, *et al.*, "Unexpected mobility degradation for very short devices: A new challange for CMOS scaling," in *IEEE IEDM Technical Digest*, pp. 663–666, 2006.

[35] M. Zilli, D. Esseni, P. Palestri, and L. Selmi, "On the apparent mobility in nanometric n-MOSFETs," *IEEE Electron Device Lett.*, vol. 28, no. 11, pp. 1036–1039, 2007.

[36] S. Takagi, "Re-examination of subband structure engineering in ultra-short channel MOSFETs under ballistic carrier transport," in *VLSI Symposium*, p. 115, 2003.

[37] J. Saint Martin, A. Bournel, and P. Dollfus, "On the ballistic transport in nanometer-scaled DG MOSFETs," *IEEE Trans. on Electron Devices*, vol. 51, no. 7, pp. 1148–1155, 2004.

[38] P. Palestri, D. Esseni, S. Eminente, *et al.*, "Understanding quasi-ballistic transport in nano-MOSFETs. Part I: Scattering in the channel and in the drain," *IEEE Trans. on Electron Devices*, vol. 52, no. 12, pp. 2727–2735, 2005.

[39] S. Eminente, D. Esseni, P. Palestri, *et al.*, "Understanding quasi-ballistic transport in nano-MOSFETs. Part II: Technology scaling along the ITRS roadmap," *IEEE Trans. on Electron Devices*, vol. 52, no. 12, pp. 2736–2743, 2005.

[40] A. Lochtefeld and D.A. Antoniadis, "On experimental determination of carrier velocity in deeply scaled NMOS: How close to the thermal limit?," *IEEE Electron Device Lett.*, vol. 22, no. 2, pp. 95–97, 2001.

[41] M.-J. Chen, H.-T. Huang, K.-C. Huang, *et al.*, "Temperature dependent channel backscattering coefficients in nanoscale MOSFETs," in *IEEE IEDM Technical Digest*, pp. 39–42, 2002.

[42] M. Lundstrom, "Elementary scattering theory of the Si MOSFET," *IEEE Electron Device Lett.*, vol. 18, pp. 361–363, July 1997.

[43] M. Lundstrom and Z. Ren, "Essential physics of carrier transport in nanoscale MOSFETs," *IEEE Trans. on Electron Devices*, vol. 49, no. 1, pp. 133–141, 2002.

[44] C. Jeong, D.A. Antoniadis, and M.S. Lundstrom, "On backscattering and mobility in nanoscale silicon MOSFETs," *IEEE Trans. on Electron Devices*, vol. 56, no. 11, p. 2762, 2009.

[45] M. Fischetti, S. Jin, T.-W. Tang, *et al.*, "Scaling MOSFETs to 10 nm: Coulomb effects, source starvation, and virtual source model," *Journal of Computational Electronics*, vol. 8, no. 2, pp. 60–77, 2009.

[46] A. Rahman and M.S. Lundstrom, "A compact scattering model for the nanoscale double-gate MOSFET," *IEEE Trans. on Electron Devices*, vol. 49, no. 3, pp. 481–489, 2002.

[47] M.-J. Chen, S.-G. Yan, R.-T. Chen, *et al.*, "Temperature-oriented experiment and simulation as corroborating evidence of MOSFET backscattering theory," *IEEE Electron Device Lett.*, vol. 28, pp. 177–179, 2007.

[48] M.V. Fischetti, L. Wang, B. Yu, *et al.*, "Simulation of electron transport in high-mobility MOSFETs: Density of states bottleneck and source starvation," in *IEEE IEDM Technical Digest*, pp. 109–112, 2007.

[49] K. Nakanishi, T. Uechi, and N. Sano, "Self-consistent Monte Carlo device simulations under nano-scale device structures: Role of Coulomb interaction, degeneracy, and boundary condition," in *IEEE IEDM Technical Digest*, pp. 79–82, 2009.

[50] M. Lundstrom and J. Guo, *Nanoscale Transistor: Device Physics, Modeling and Simulation*. New York: Springer, 2006.

[51] J.P. McKelvey, R.L. Longini, and T.P. Brody, "Alternative approach to the solution of added carrier transport problems in semiconductors," *Phys. Rev.*, vol. 123, no. 1, pp. 51–57, 1961.

[52] W. Schockley, "Diffusion and drift of minority carrier in semiconductors for comparable capture and scattering mean free paths," *Phys. Rev.*, vol. 125, no. 5, pp. 1570–1576, 1962.

[53] J.P. McKelvey and J.C. Balogh, "Flux methods for the analysis of transport problems in semiconductors in the presence of electric fields," *Phys. Rev.*, vol. 137, no. 5A, pp. A1555–A1561, 1965.

[54] E.F. Pulver and J.P. McKelvey, "Flux methods for transport problems in solids with nonconstant electric fields," *Phys. Rev.*, vol. 149, no. 2, pp. 617–623, 1966.

[55] P. Palestri, R. Clerc, D. Esseni, L. Lucci, and L. Selmi, "Multi-subband-Monte-Carlo investigation of the mean free path and of the kT layer in degenerated quasi-ballistic nanoMOSFETs," in *IEEE IEDM Technical Digest*, pp. 945–948, 2006.

[56] R. Clerc, P. Palestri, and L. Selmi, "On the physical understanding of the kT-layer concept in quasi-ballistic regime of transport in nanoscale devices," *IEEE Trans. on Electron Devices*, vol. 53, no. 7, pp. 1634–1640, 2006.

[57] E. Fuchs, P. Dollfus, G. Le Carval, *et al.*, "A new backscattering model giving a description of the quasi-ballistic transport in nano-MOSFET," *IEEE Trans. on Electron Devices*, vol. 52, no. 10, pp. 2280–2288, 2005.

[58] R. Clerc, P. Palestri, L. Selmi, and G. Ghibaudo, "Back-scattering in quasi-ballistic nanoMOSFETs: The role of non-thermal carrier distributions," in *Proc. Int. Conf. on Ultimate Integration on Silicon (ULIS)*, pp. 125–128, 2008.

[59] J.-L.P.J. van der Steen, P. Palestri, D. Esseni, and R.J.E. Hueting, "A new model for the backscatter coefficient in nanoscale MOSFETs," in *Proc. European Solid State Device Res. Conf.*, pp. 234–237, 2010.

[60] J.D. Bude, "MOSFET modeling into the ballistic regime," in *Proc.SISPAD*, pp. 23–26, 2000.

6 The Monte Carlo method for the Boltzmann transport equation

The Boltzmann Transport Equation (BTE) is an integro-differential equation where the unknown occupation function f depends on up to seven independent variables for a 3D carrier gas and five for a 2D carrier gas: the real and momentum space coordinates and the time. The equation is also non-linear due to the $(1 - f)$ terms in the collision integral (Eq.5.9). Symmetries in the simulated device structure can reduce the number of variables in real-space, but the problem still remains very hard to solve with standard numerical methods [1].

In the previous chapter we examined ways to solve the BTE under various simplifying assumptions: absence of scattering (that is, ballistic transport), momentum-relaxation-time approximation for near equilibrium conduction in a low and uniform electric field, solutions via the moments of the BTE as in the Drift–Diffusion model.

Modern MOS transistors operate in a regime of transport where the number of scattering events suffered by the carriers traveling along the channel is largely reduced compared to long-channel devices, leading to quasi-ballistic transport, as described in Section 5.7. In this situation the distribution function is quite different from an equilibrium distribution and the simulation approaches relying on the moments of the BTE become inadequate [2, 3], thus demanding an exact solution of the BTE.

The Monte Carlo (MC) method is a powerful technique for solving exactly the semi-classical BTE without a-priori assumptions on the carrier distribution function [4–7]. The method is statistical and based on simulation of the motion of sample particles in real and momentum space.

The Monte Carlo method has found widespread use since the late 1970s for studying carrier transport [4], but practical Monte Carlo device simulations became feasible about one decade later [8, 9]. In the last 20 years several numerical and physical improvements to Monte Carlo device simulation techniques have been introduced, e.g. numerical description of the full band dispersion relationship [8], efficient numerical schemes to implement the consistency with the Poisson equation [9], simulation of "hot electron" phenomena (impact ionization, injection in the gate dielectric, etc.) [10].

In this chapter we review the use of the MC method to solve the forms of BTE corresponding to different carrier gas dimensionality in real and momentum space. In particular, we examine the case of a free-electron gas (Section 6.1), which is characteristic of devices with bulk conduction (such as BJTs), and the multi-subband method for the 2D carrier gas (2Deg) in the MOS inversion layer (Section 6.3).

The coupling of the Monte Carlo BTE solver with the Poisson equation is described in Section 6.2 from a practical standpoint aimed at presenting the different implementations and several related issues.

6.1 Basics of the MC method for a free-electron-gas

The MC method [4, 11] achieves the solution of the BTE by simulating the motion of sample particles in phase space either one by one (as in the single particle MC) or as an ensemble. Electrons and holes are represented as point charges that move in all the free directions inside the device. Consistently with the picture of the semi-classical model provided in Section 2.5.5 and depicted in Fig.6.1, the motion is an alternate sequence of free-flights (where the carriers move ballistically according to the generalized Newton's law expressed by Eq.2.121) and scattering events that are instantaneous in time and localized in space. In the bulk semiconductor case considered in this section, the state of the particle is identified by the 3D vectors **R** and **K**. The dimensionality of the **R** space can be reduced if the device structure features spatial symmetries.

The flowchart of a typical *single-particle* MC solver is shown in Fig.6.2. The figure highlights all the steps that are inherently stochastic because the algorithm includes decisions to be taken on the basis of randomly generated numbers. While the computation of the free-flight trajectories is deterministic (step 2), the scattering is a stochastic process described by scattering probabilities, namely the probability per unit time for a carrier in a given state to jump to a different state. The total scattering rate enters the calculation by determining the duration of the free-flight (step 1). The scattering probabilities are computed considering the band structure of the materials inside the device following the Fermi golden rule (see Sections 2.5.4, 4.1.1, 4.1.6). When the free-flight is interrupted by a scattering event, statistical data such as particle position and velocity are gathered (step 3). Then, in order to determine the state after scattering, that is the initial condition of the next free-flight, the scattering mechanism responsible for the interruption of the free-flight is chosen reflecting the relative importance of all the mechanisms. The main blocks of the flowchart are described in the following sections. Although the flowchart refers to a single-particle MC solver, essentially the same steps are necessary in the ensemble MC case, as discussed in Section 6.1.5.

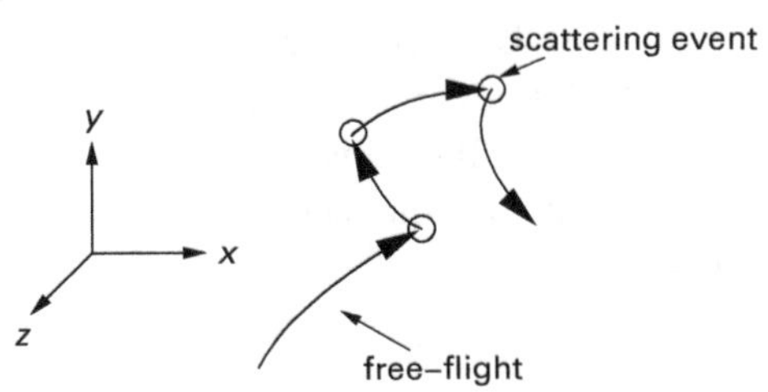

Figure 6.1 Motion of a sample particle as a sequence of free-flights and scattering events.

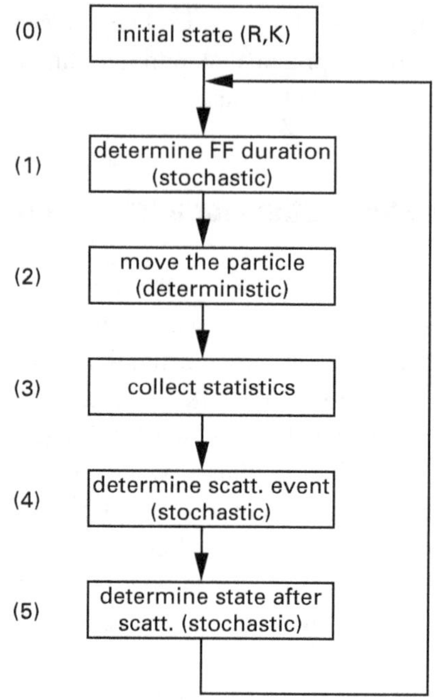

Figure 6.2 Typical flowchart of a single particle Monte Carlo solver of the BTE.

6.1.1 Particle dynamics

The most relevant model ingredient to trace the particle trajectory in phase space during a free-flight is the band structure of the semiconductor material, i.e. the relation $E_{B,n}(\mathbf{K})$ between the carrier wave-vector and the kinetic energy for the nth branch of the dispersion relation. The sum of $E_{B,n}(\mathbf{K})$ and the potential energy gives the total energy $E(\mathbf{R}, \mathbf{K})$ (Eq.2.122). The electron[1] dynamics during the free-flight is described by (see Eq.2.121)

$$\hbar \frac{\mathrm{d}\mathbf{K}}{\mathrm{d}t} = -\nabla_{\mathbf{R}} E(\mathbf{R}, \mathbf{K}), \tag{6.1a}$$

$$\hbar \frac{\mathrm{d}\mathbf{R}}{\mathrm{d}t} = \nabla_{\mathbf{K}} E(\mathbf{R}, \mathbf{K}), \tag{6.1b}$$

where the first equation is the law of motion for packets of Bloch waves in a crystal and closely resembles Newton's law of motion; the second one is the definition of group velocity for a wave-packet (Section 2.5.1).

Regardless of the MC algorithm adopted (single particle or ensemble), the integration of Eq.6.1 during the free-flight time (from $t = 0$ to $t = t_{FF}$) is not trivial. An analytical solution exists only in very simple conditions that we begin illustrating by means of a simple example.

[1] We consider electrons except when otherwise specified. However, most of the equations reported in this chapter can be readily extended to holes.

Example 6.1: Constant electric field and parabolic bands

Assuming parabolic bands and a constant electric field F_x along the x direction, the wave-vector and real space position at the end of the free-flight ($t = t_{FF}$) are given by:

$$k_x^{(f)} = k_x^{(i)} - \frac{eF_x}{\hbar} t_{FF}, \tag{6.2a}$$

$$k_y^{(f)} = k_y^{(i)}, \tag{6.2b}$$

$$k_z^{(f)} = k_z^{(i)}, \tag{6.2c}$$

$$x^{(f)} = x^{(i)} + \frac{\hbar k_x^{(i)}}{m_x} t_{FF} - \frac{eF_x}{2m_x} t_{FF}^2, \tag{6.2d}$$

$$y^{(f)} = y^{(i)} + \frac{\hbar k_y^{(i)}}{m_y} t_{FF}, \tag{6.2e}$$

$$z^{(f)} = z^{(i)} + \frac{\hbar k_z^{(i)}}{m_z} t_{FF}, \tag{6.2f}$$

where the superscripts (i) and (f) stand for the initial ($t = 0$) and final ($t = t_{FF}$) states, respectively.

In the general case of a non-uniform electric field many integration schemes have been proposed for Eq.6.1. Below we report two cases, considering 1D structures in real-space and 3D transport in **K** space and assuming parabolic bands.

The first order integrator scheme results in:

$$k_x^{(f)} = k_x^{(i)} - \frac{eF_x(x^{(i)})}{\hbar} t_{FF}, \tag{6.3a}$$

$$x^{(f)} = x^{(i)} + \frac{\hbar k_x^{(i)}}{m_x} t_{FF}, \tag{6.3b}$$

where $F_x(x)$ is the position-dependent electric field.

The second order Runge–Kutta scheme, instead, results in:

$$k_x^{(f)} = k_x^{(i)} - \frac{e}{2\hbar}\left[F_x(x^{(i)}) + F_x\left(x^{(i)} + \frac{\hbar k_x^{(i)}}{m_x} t_{FF}\right)\right] t_{FF}, \tag{6.4a}$$

$$x^{(f)} = x^{(i)} + \frac{\hbar k_x^{(i)}}{m_x} t_{FF} - \frac{eF_x(x^{(i)})}{2m_x} t_{FF}^2. \tag{6.4b}$$

An alternative approach consists in assuming a constant electric field inside each element of the spatial mesh, then using Eq.6.2 and interrupting the free-flight at the crossing of a boundary between two elements. The crossing point is found from geometrical considerations, since the trajectory of the particle is known from Eq.6.2. Once the crossing point is found, we can invert Eq.6.2 to obtain the time t_A elapsed until the particle hit the interface. The trajectory of the residual portion of the free-flight with duration $[t_{FF} - t_A]$ is computed with the electric field of the new mesh element.

Stability issues in the integration of motion

The round-off errors accumulated in the integration of Eq.6.1 can lead to unstable solutions and erroneous results. It is worth introducing a general approach to model the numerical behavior of the integration algorithm. To this purpose we express the initial (i) and final (f) value of any variable (position and wave-vector) as

$$a^{i,f} = A^{i,f} + \xi^{i,f}, \tag{6.5}$$

where $A^{i,f}$ is the exact value of the variable (that is the solution of the problem at hand without any round-off error), and $\xi^{i,f}$ is the error. Substitution into the integration schemes, that is, for instance, Eqs.6.3 or 6.4, and subsequent linearization gives:

$$\boldsymbol{\xi}^f = \mathbf{T}\boldsymbol{\xi}^i \tag{6.6}$$

where $\mathbf{T}$ is the matrix that transforms the vector of the initial errors $\boldsymbol{\xi}^i$ into the vector $\boldsymbol{\xi}^f$ of the errors at the end of the integration step. We have thus cast the error propagation problem in the usual form for linear discrete-time systems. In such a system stability is analyzed by using the Z transform and imposing the condition that the eigenvalues of the matrix $\mathbf{T}$ fall within the unit circle in the complex plane [12, 13]. We thus write

$$\mathbf{T}\boldsymbol{\xi}^i = Z\boldsymbol{\xi}^i, \tag{6.7}$$

where Z is a scalar complex number. Stability requires $|Z|<1$.

Example 6.2: First-order integrator.

Following the general method, the relations between the initial and final values of x and k_x (Eq.6.3) become:

$$K_x^{(f)} + \xi_k^{(f)} = K_x^{(i)} + \xi_k^{(i)} - \frac{eF_x(X^{(i)} + \xi_x^{(i)})}{\hbar} t_{FF}, \tag{6.8a}$$

$$X^{(f)} + \xi_x^{(f)} = X^{(i)} + \xi_x^{(i)} + \frac{\hbar(K_x^{(i)} + \xi_k^{(i)})}{m_x} t_{FF}, \tag{6.8b}$$

where $F_x(X^{(i)} + \xi_x^{(i)})$ is the electric field at the initial point $X^{(i)}$ affected by the round-off error $\xi_x^{(i)}$. Equations similar to Eq.6.8 hold for the X and K_x variables in the absence of round-off errors. Linearization yields:

$$F_x(X^{(i)} + \xi_x^{(i)}) \simeq F_x(X^{(i)}) + \frac{\mathrm{d}F_x}{\mathrm{d}x}\xi_x^{(i)} = F_x(X^{(i)}) - \frac{en}{\epsilon_S}\xi_x^{(i)}, \tag{6.9}$$

where n is the free-charge concentration and we have used the Poisson equation to express $\mathrm{d}F_x/\mathrm{d}x$, assuming that the free charge is given only by the electrons. Equation 6.6 thus becomes:

$$\begin{pmatrix} \xi_k^{(f)} \\ \xi_x^{(f)} \end{pmatrix} = \begin{bmatrix} 1 & \frac{e^2 n t_{FF}}{\epsilon_S \hbar} \\ \frac{\hbar t_{FF}}{m_x} & 1 \end{bmatrix} \begin{pmatrix} \xi_k^{(i)} \\ \xi_x^{(i)} \end{pmatrix} = Z \begin{pmatrix} \xi_k^{(i)} \\ \xi_x^{(i)} \end{pmatrix}. \tag{6.10}$$

The eigenvalues Z are thus obtained as the eigenvalues of Eq.6.10 and are:

$$Z = 1 \pm \mathrm{i}\omega_p t_{FF}, \tag{6.11}$$

where

$$\omega_p = e\sqrt{\frac{n}{\epsilon_S m_x}} \tag{6.12}$$

is the plasma frequency. We readily see that the scheme is unstable ($|Z|$ is always larger than 1) but, as long as $\omega_p t_{FF}$ is much smaller than 1, the round-off error remains small for simulation times corresponding to the evaluation of the particle dynamics over many free-flights.

The methodology illustrated above and described in Example 6.2 for the first order integrator is applicable also to the second-order Runge–Kutta integration scheme which is often adopted in MC device simulations. The resulting eigenvalues in the Z domain are

$$Z = 1 - \frac{1}{2}\omega_p^2 t_{FF}^2 \pm \mathrm{i}\omega_p t_{FF}, \tag{6.13}$$

which correspond once again to instability (being $|Z| = \sqrt{1 + (\omega_p t_{FF})^4/4} > 1$). The growth rate of the round-off error is slower compared to the first-order integrator and almost always guarantees that the round-off error remains small for simulation times in the order of many pico-seconds or more.

Results concerning other less popular integration schemes can be found in [14]. In general, while it is possible to find schemes that are stable in some ranges of the $\omega_p t_{FF}$ product, one should remember that in practice what matters most is not the stability of the integration scheme but the growth rate of the round-off error. The reader is referred to [14] for a thorough discussion of this point.

According to common practice it appears that essentially all schemes deriving from the first order integrator and the second order Runge–Kutta algorithm do not introduce significant numerical errors for realistic values of the free-flight duration.

6.1.2 Carrier scattering and state after scattering

The stochastic selection of the free-flight duration, of the type of scattering, and of the state after scattering can be performed exploiting the random number generation routines of modern programming languages. However, appropriate algorithms are necessary to reproduce correctly the probability density distribution of the variables identifying the state after scattering.

We describe below two commonly used algorithms known as the direct technique and the rejection technique to generate random numbers according to given distributions.

Generation of random numbers using the direct technique

Given a number r uniformly distributed between 0 and 1, the random continuous variable x with probability density $f(x)$ between the values a and b can be obtained as the solution of the equation

$$\frac{\int_a^x f(x')\mathrm{d}x'}{\int_a^b f(x')\mathrm{d}x'} = r. \tag{6.14}$$

If we consider, instead, a discrete variable with values x_i and associated probabilities P_i (with the index i running from 1 to N and $\sum_i P_i = 1$), the index i and the abscissa x_i are determined according to the inequality:

$$\sum_{j=1}^{i-1} P_j \le r < \sum_{j=1}^{i} P_j. \tag{6.15}$$

Generation of random numbers using the rejection technique

In this technique, a stochastic variable x between a and b with probability density $f(x)$ is generated according to the following procedure [4]:

1. generate a random number r_1 uniformly distributed between 0 and 1,
2. compute $x_1 = a + (b - a)r_1$,
3. generate a random number r_2 uniformly distributed between 0 and 1,
4. compute $f_1 = r_2 \max\{f(x)\}$,
5. if $f_1 > f(x_1)$ go back to point 1; otherwise $x = x_1$.

Determination of the free-flight duration

The Fermi golden rule provides us with the transition probability per unit time for the mth scattering mechanism $S_m(\mathbf{K}, \mathbf{K}')$ (Section 2.5.4). In order to determine the free-flight duration, we have to consider the total scattering rate out of the state $\mathbf{K}$:

$$S_{tot}(\mathbf{K}) = \sum_m \sum_{\mathbf{K}'} S_m(\mathbf{K}, \mathbf{K}'). \tag{6.16}$$

Since $\mathbf{K}$ varies during the free-flight, the probability distribution function of the random variable t_{FF} is [4]:

$$P(t_{FF}) = S_{tot}(\mathbf{K}(t_{FF})) \exp\left(-\int_0^{t_{FF}} S_{tot}(\mathbf{K}(t'))\mathrm{d}t'\right). \tag{6.17}$$

The determination of t_{FF} is very complex because it requires solution of the integral equation 6.17. Different approaches have been proposed to circumvent the numerical solution of Eq.6.17. The simplest is based on the concept of *self-scattering*. In the self-scattering algorithm, $S_{tot}(\mathbf{K})$ is substituted by its upper bound

$$\Gamma = \max\{S_{tot}(\mathbf{K})\}. \tag{6.18}$$

If we insert Eq.6.18 into Eq.6.17, the argument of the integral becomes constant and the probability distribution of t_{FF} becomes

$$P(t_{FF}) = \Gamma \exp(-\Gamma t_{FF}). \tag{6.19}$$

By using Eq.6.14 we see that we can generate t_{FF} as:

$$t_{FF} = -\frac{\ln r_1}{\Gamma}, \tag{6.20}$$

where r_1 is a random number uniformly distributed between 0 and 1. At the end of the free-flight, scattering is rejected (i.e. we compute a new free-flight without modifying the particle momentum) if $S_{tot}(\mathbf{K}(t_{FF})) < r_2\Gamma$, where r_2 is uniformly distributed between 0 and 1.

The main disadvantage of this technique is its inefficiency; in the energy range where $S_{tot}(\mathbf{K})$ is much lower than Γ the method yields many short free-flights instead of a single long free-flight, thus multiplying the number of time consuming checkpoints in the routine. This problem is mitigated by use of the variable gamma scheme [15].

Determination of the scattering mechanisms

Regardless of the algorithm to determine t_{FF}, at the end of the free-flight we have to identify the scattering mechanism that interrupted the flight and the corresponding state after scattering.

In order to accomplish this task, we compute the relative contribution of the mth mechanism to the total scattering rate:

$$P_m(\mathbf{K}) = \frac{1}{S_{tot}(\mathbf{K})} \sum_{\mathbf{K}'} S_m(\mathbf{K}, \mathbf{K}'). \tag{6.21}$$

We can interpret Eq.6.21 as the probability that the free-flight has been stopped by mechanism m.[2] By using the direct technique, we can then select the jth mechanism if

$$\sum_{m=1}^{j-1} P_m(\mathbf{K}) < r_3 < \sum_{m=1}^{j} P_m(\mathbf{K}), \tag{6.22}$$

where r_3 is yet another random number uniformly distributed between 0 and 1.

Determination of the state after scattering

Assuming that the free-flight was interrupted by the jth scattering mechanism, the probability distribution of the final state $\mathbf{K}'$, for a given initial state $\mathbf{K}$, is

$$P_j(\mathbf{K}', \mathbf{K}) = \frac{S_j(\mathbf{K}, \mathbf{K}')}{\sum_{\mathbf{K}'} S_j(\mathbf{K}, \mathbf{K}')}. \tag{6.23}$$

If we sort the allowed states-after-scattering from $\mathbf{K}_1$ to $\mathbf{K}_M$, we can use the direct technique and select the state $\mathbf{K}_n$ as the one which satisfies the inequality:

$$\sum_{i=1}^{n-1} P_j(\mathbf{K}_i, \mathbf{K}) \le r_4 < \sum_{i=1}^{n} P_j(\mathbf{K}_i, \mathbf{K}), \tag{6.24}$$

where r_4 is uniformly distributed between 0 and 1.

[2] To be consistent with the notation used when describing the particle dynamics during free-flight, we should denote $\mathbf{K}$ as $\mathbf{K}^{(f)}$; however, in order to simplify the following equation we drop the superscript (f).

Example 6.3: Phonon scattering in spherical bands.

The transition rate for phonon scattering in bulk materials can be found in many textbooks [11] and has an expression similar to Eqs.4.274 and 4.286 for inversion layers. For the purpose of this discussion it can be written as

$$S_{ph}(\mathbf{K},\mathbf{K}') = C\,\delta(E(\mathbf{K}) - E(\mathbf{K}') \pm E_{ph}), \tag{6.25}$$

where C is a $\mathbf{K}$-independent term and $E_{ph} = \hbar\omega_{ph}$. Since $E = \hbar^2K^2/(2m)$, it is convenient to evaluate the sum over $\mathbf{K}'$ and then the corresponding integral by using spherical coordinates:

$$\sum_{\mathbf{K}'} S_{ph}(\mathbf{K},\mathbf{K}') = \frac{C}{8\pi^3}\int_0^{2\pi} \mathrm{d}\phi \int_0^{\pi} \mathrm{d}\theta \int_0^{\infty} \mathrm{d}K'\delta(E(\mathbf{K}) - E(\mathbf{K}') \pm E_{ph})K'^2 \sin(\theta). \tag{6.26}$$

It is thus straightforward to determine the probability function as:

$$P(K',\theta,\phi) = \frac{\sin(\theta)}{2}\frac{1}{2\pi}\delta(K^2 - K'^2 \pm 2mE_{ph}/\hbar^2). \tag{6.27}$$

The magnitude of $\mathbf{K}'$ is thus imposed by energy conservation ($E(\mathbf{K}') = E(\mathbf{K}) \pm E_{ph}$). Note that phonon emission and absorption are described as two separate scattering mechanisms, so that we know a priori if we should use the plus or the minus sign in Eq.6.27. Since $P(K',\theta,\phi)$ does not contain ϕ, the probability distribution of the azimuth angle is uniform. We can thus determine ϕ as

$$\phi = 2\pi\, r_5, \tag{6.28}$$

where r_5 is yet another random number uniformly distributed between 0 and 1.

Regarding the polar angle θ, the direct technique gives:

$$\frac{1}{2}\int_0^{\theta} \sin(\theta')\mathrm{d}\theta' = r_6, \tag{6.29}$$

hence, solving the integral in Eq.6.29, the angle θ is selected as:

$$\theta = \arccos(1 - 2r_6), \tag{6.30}$$

where r_6 is also a random number uniformly distributed between 0 and 1.

Example 6.4: Phonon scattering in ellipsoidal bands.

In this case we can use the Vogt–Herring transformations [16] to map the ellipsoidal band structure in an equivalent spherical band. We define an effective isotropic final state $\mathbf{K}^*$ as:

$$E = \frac{\hbar^2K_x^2}{2m_x} + \frac{\hbar^2K_y^2}{2m_y} + \frac{\hbar^2K_z^2}{2m_z} = \frac{\hbar^2}{2m}|\mathbf{K}^*|^2, \tag{6.31}$$

where $m = (m_x m_y m_z)^{1/3}$. The final state $\mathbf{K}^*$ is obtained as explained in Example 6.3. Then, the actual final state is given by:

$$K'_x = K^*_x \sqrt{m_x/m^*}$$
$$K'_y = K^*_y \sqrt{m_y/m^*}$$
$$K'_z = K^*_z \sqrt{m_z/m^*}. \tag{6.32}$$

Example 6.5: Anisotropic and elastic scattering.

Following the Fermi golden rule for a stationary perturbation potential (Eq.2.138) and considering that the matrix element depends on the modulus Q of the exchanged wave-vector, we write [11]

$$S(\mathbf{K}, \mathbf{K}') = M(K, Q)\,\delta(E(K) - E(K')), \tag{6.33}$$

where $K = |\mathbf{K}|$ and $K' = |\mathbf{K}'|$. Since in this case $K = K'$, the exchanged wave-vector is $Q = 2K \sin(\theta/2)$, where θ is the angle between $\mathbf{K}$ and $\mathbf{K}'$. The angle θ can be found by solving

$$\frac{\int_0^\theta M(K, 2K \sin(\theta'/2))d\theta'}{\int_0^\pi M(K, 2K \sin(\theta'/2))d\theta'} = r_7, \tag{6.34}$$

where r_7 is a random number uniformly distributed between 0 and 1. In a restricted number of cases, Eq.6.34 can be solved analytically. In other situations we need to tabulate the numerical evaluation of the left-hand-side for a given set of θ values and lookup in the table when evaluating the state after scattering. As for the azimuthal angle ϕ, we observe that the $\mathbf{K}'$ vectors satisfying energy conservation and forming an angle θ with the initial state $\mathbf{K}$ lie on a surface of the cone having $\mathbf{K}$ as axis (Fig. 6.3). The azimuthal angle ϕ is thus selected randomly according to Eq.6.28.

Considering as an example ionized impurities scattering, $M(Q)$ is proportional to $(1 + Q^2/\beta_S^2)^{-2}$, where β_S is the inverse screening length and Q is the magnitude of the exchanged wave-vector [11]. By evaluating the integral in Eq.6.34 we obtain:

$$\cos(\theta) = 1 - \frac{2r_7}{1 + (1 - r_7)K^2/\beta_S^2}. \tag{6.35}$$

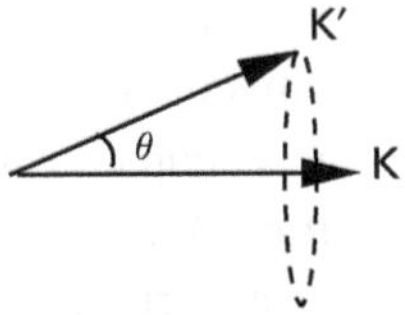

Figure 6.3 The wave-vectors involved in an elastic and anisotropic scattering event. $\mathbf{K}$ is the initial state, $\mathbf{K}'$ the final state. All possible $\mathbf{K}'$ vectors form a cone with $\mathbf{K}$ as axis.

The determination of the final state for anisotropic scattering mechanisms in ellipsoidal bands is particularly difficult. Strictly speaking, the Vogt–Herring transformations used in Example 6.4 for isotropic scattering mechanisms cannot be used, since the exchanged wave-vector Q is not conserved by the transformation. Nevertheless, in order to simplify the implementation, many MC simulators use the Vogt–Herring transformations for anisotropic scattering in ellipsoidal bands.

It is often the case that anisotropic scattering mechanisms favor transitions with a small Q (i.e. a small angle θ between the initial and final $\mathbf{K}$) [11] so that significant simulation time is spent in computing events with a weak influence on the particle momentum. A possible approach to limit the computational burden is to replace the scattering rate with the momentum relaxation rate [17]:

$$S_m^*(\mathbf{K}) = \sum_{\mathbf{K}'} (1 - \cos\theta) S_m(\mathbf{K}, \mathbf{K}'). \tag{6.36}$$

The state after scattering is then selected randomly following Example 6.3, i.e. $\cos\theta = (1 - 2r_6)$.

This approach is justified in low-field regions where the carrier distribution is almost isotropic in $\mathbf{K}$ space. It has been shown [17] that in these circumstances the method gives the same low-field mobility as the rigorous approach; this is not the case, however, in regions of high field and for $\mathbf{K}$-space distributions significantly elongated in one direction.

Carrier degeneracy

The prescriptions given so far to determine the free-flight time and the state after scattering neglect the $(1 - f)$ terms in the collision integral of the BTE (Eqs.5.9, 5.11), i.e. the Pauli exclusion principle. Consistently they refer to the total scattering rate $S_{tot}(\mathbf{K}) = \sum_{m,\mathbf{K}'} S(\mathbf{K}, \mathbf{K}')$. However, as seen in Chapter 7, carrier degeneracy often plays a crucial role in nanoscale MOSFETs and cannot be neglected.

An effective and simple way to account for the occupation probability of the state after scattering [18] is to consider the total scattering rate without the $(1 - f)$ terms, as in the non-degenerate case, and then, once the state after scattering has been found, reject the scattering event if the state is occupied. In other words the state of the particle is left as at the end of the free-flight if $f(\mathbf{R}^{(f)}, \mathbf{K}') > r_8$, where r_8 is uniformly distributed between 0 and 1, $f(\mathbf{R}^{(f)}, \mathbf{K}')$ is the occupation of the final state and $\mathbf{R}^{(f)}$ is the particle position at the end of the free-flight.

An improved version of this method has been proposed in [19]. Beside rejecting scattering events based on the occupation of the final state, if $f(\mathbf{R}^{(f)}, \mathbf{K}') > 1$ is found, not only the scattering from $\mathbf{K}$ to $\mathbf{K}'$ is rejected, but an inverse transition from $\mathbf{K}'$ to $\mathbf{K}$ is enforced, namely, a particle is moved from the $\mathbf{K}$-space bin around $\mathbf{K}'$ to the bin around $\mathbf{K}$ with probability $[f(\mathbf{R}^{(f)}, \mathbf{K}') - 1]$. With this approach the regions with $f(\mathbf{R}^{(f)}, \mathbf{K}') > 1$ are more efficiently depopulated and rapidly brought again to the condition $f \leq 1$.

6.1.3 Boundary conditions

As already mentioned, the BTE is an integro-differential equation in the unknown carrier distribution $f(\mathbf{R}, \mathbf{K}, t)$ and the MC method solves the BTE by tracing the motion of sample particles in the phase space. Imposing the boundary conditions on the BTE in the frame of a Monte Carlo solver means defining rules to handle the events that occur when the particles reach the boundaries of the simulation domain or the interfaces between regions with different physical properties. In the following we illustrate some of these rules and report examples of how to select the most suitable boundary conditions in some specific cases.

Among the most common boundary conditions are those of *reflecting* and *absorbing* boundaries (see Fig.6.4): in the first case particles hitting the boundary are reflected, while in the second case they are removed from the simulation domain. Reflecting boundaries are often placed at the Si/SiO_2 interface in MOSFETs. In order to mimic surface roughness scattering under the free-carrier gas approximation, it is common to assume that a fraction of the carriers hitting the boundary at the Si/SiO_2 interface are not reflected specularly, but are instead randomized in direction, still imposing conservation of the carrier energy [20].

Other useful boundary conditions are the *injecting* contacts, which steadily inject carriers with a given momentum distribution inside the simulation domain. If N_p particles with charge equal to the elementary charge e enter the simulation domain at the beginning of each time step with duration Δt, the average particle flux injected by the contact is:

$$\mathcal{F}_{inj} = \frac{N_p}{\Delta t}. \tag{6.37}$$

The *injecting* contacts can behave either as *absorbing* or *reflecting* for the carriers in the simulation domain. Only in the latter case

$$I_C = I_{inj} = \pm e\mathcal{F}_{inj} \tag{6.38}$$

is the actual current at the contact (where the signs $+$ and $-$ are used for electrons and holes, respectively and $\mathcal{F}_{inj}$ is given by Eq.6.37); otherwise we have a flux $\mathcal{F}_{abs}$ that is the number of particles absorbed by the contact per unit time and the current at the contact is

$$I_C = \pm e(\mathcal{F}_{inj} - \mathcal{F}_{abs}). \tag{6.39}$$

Regarding the distribution, assuming equilibrium Maxwell–Boltzmann statistics, the probability of injecting particles with a given state is proportional to $V_{gx}(\mathbf{K})$

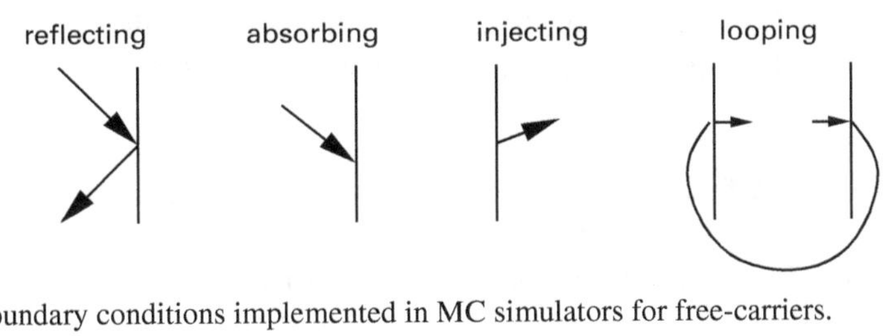

Figure 6.4 Different boundary conditions implemented in MC simulators for free-carriers.

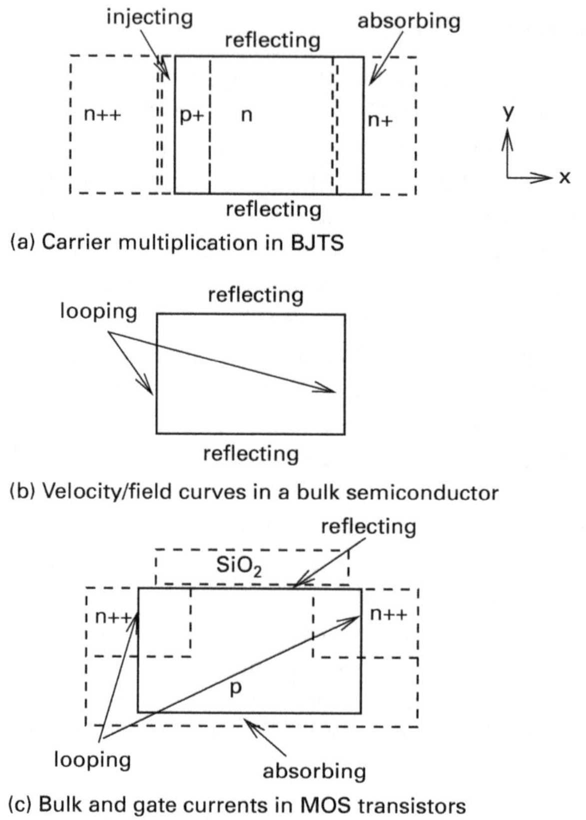

Figure 6.5 Some examples of the practical use of boundary conditions. The simulation domain is indicated by the solid rectangle.

$\exp[-E(\mathbf{K})/k_BT]$, where V_{gx} is the group velocity in the direction normal to the interface. However, injecting particles according to an equilibrium distribution may be inconsistent with the out-of-equilibrium regime inside the device. We come back to this issue when addressing ohmic contacts (Section 6.1.4).

Another kind of useful boundary condition is *looping* contacts, which consist of a pair of coupled interfaces such that carriers trying to exit the simulation domain from one side re-enter the domain at the twin side.

Selection of the most appropriate boundary conditions is not trivial and in general depends on the device structure as well as on the specific physical problem that the simulation is aimed to investigate. The choice of incorrect boundary conditions can produce undesired or unphysical artifacts in the carrier distribution. In the following we give a few examples of the choice of boundary conditions commonly used in device simulation with the Monte Carlo method (see Fig.6.5).

Example 6.6: Carrier multiplication in bipolar transistors.
Let us consider the problem of simulating carrier multiplication in the base-collector depletion region of a bipolar transistor at large base-collector voltages (see Fig.6.5a).

Assuming that the emitter area is large, only the direction perpendicular to the wafer plane (that we denote here as x) is relevant for transport. We can then rely on a 1D simulation in real space. If we use a 2D simulator, we can still enforce a 1D behavior of the physical system by placing reflecting interfaces at the top and bottom of the structure as sketched in Fig.6.5a. For the purpose of computing the multiplication factor (i.e. number of e–h pairs generated in the base-collector space-charge region) we can simply place an injecting contact with a unitary current $I_{inj} = -1$A at the edge of the neutral base (i.e. at the top of the potential energy profile). An absorbing contact is placed in the neutral collector region. The multiplication factor is then easily estimated as I_{coll}/I_{inj}, where I_{coll} is the collector current [21].

Example 6.7: Velocity–field curves in a bulk semiconductor.
Drift velocity versus electric field curves of bulk materials are rigorously defined in uniform electric field situations, where the carrier distribution is unequivocally related to the field. These simulations are instrumental to the calibration of the scattering models (e.g. the coupling constant for acoustic phonons [4], the parameters for ionized impurity scattering, etc.). In principle, the electric field being uniform and the carrier distribution independent of position, we do not need to trace the position of the carriers, hence the simulation can be zero dimensional in real space. Alternatively, a 2D simulator in real space with reflecting boundaries at the interfaces parallel to the electric field and looping contacts at the boundaries normal to the transport direction perfectly mimics an infinitely long device (see Fig.6.5b) provided that enough simulation time is given to the simulation to reach the steady state. We come back to the simulation of velocity–field curves in Section 7.2.1.

Example 6.8: Hot carrier currents in MOSFETs.
Monte Carlo simulations of a 3D carrier gas have been extensively used to evaluate hot-carrier effects in scaled MOSFETs and non-volatile memory cells [22–26]. Hot carrier currents such as the substrate current I_B and the gate current I_G often vary by orders of magnitude in the bias range of practical interest. In several cases the most relevant information is not the absolute value of I_B and I_G but rather the ratios of I_B and I_G to the drain current I_D. The estimate of the drain current is thus not critical and one can perform *frozen-field* simulations, where the electric field profile is taken from Drift–Diffusion or even better from hydrodynamic simulations and then kept fixed during the particle simulation. It has been shown in [27] that use of a self-consistent electric field profile has a small effect on the calculated substrate current, at least as long as (I_B/I_D is smaller than about 0.1).

During the simulation, looping boundary conditions can be imposed at the source and drain (see Fig.6.5c). Compared to the previous example, where the carriers hitting one of the looping boundaries were re-entered at the twin side with the same momentum, now particles are re-injected according to an equilibrium distribution. By doing so, memory of the acquisition of kinetic energy and heating of the carrier population in the channel is canceled and the desired equilibrium distribution in the source is enforced even if the drain region is too short to guarantee the complete thermalization of the particles. By following this algorithm the charge in the channel remains the same imposed during the initialization phase (usually the one provided by the Drift–Diffusion or hydrodynamic simulator) and the current flow is sustained by the looping contacts. Since I_G is in most cases negligible compared to I_D, we can neglect charge loss through the Si/SiO_2 interface and impose a reflecting boundary at the top. The gate dielectric is thus excluded from the simulation domain.

6.1.4 Ohmic contacts

A very delicate issue in MC simulations is the correct definition of ohmic boundary conditions. Various techniques to implement ohmic contacts are reviewed and compared in [28]. Here we focus on the use of carrier reservoirs, that is of neutral regions where the concentration of free carriers compensates exactly the charge of the active dopants. The exchange of particles with the reservoir is sketched in Fig.6.6. We see that particles trying to enter the reservoir from the outside are removed from the simulation domain (plot a). If a particle exits the reservoir to enter the simulation domain, another carrier with the same wave-vector is injected at the other side of the reservoir (plot b). When a particle exits the reservoir from the opposite side with respect to the simulation domain (plot c), it is moved to the other side. The electric field in the reservoir is null, so that scattering enforces an equilibrium Maxwell–Boltzmann or a Fermi–Dirac occupation function, depending on whether or not the rejection of the scattering events based on the $(1 - f)$ term is implemented (as explained in Section 6.1.2).

The main limitation with the use of reservoirs to implement ohmic contacts is that the carrier flux injected by the reservoir into the simulation domain is upper limited by the product between the carrier concentration and the thermal velocity (Eq.5.165), regardless of the actual current flowing in the device. In short channel MOSFETs, the drain current can be very large and the equilibrium reservoir may become unable to provide the simulation domain with an adequate amount of charge per unit time, thus

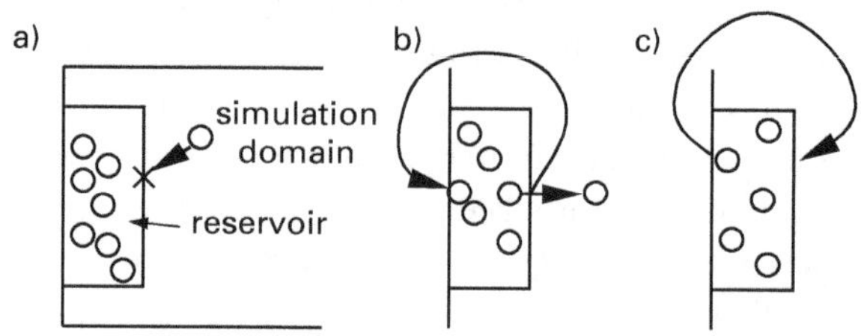

Figure 6.6 Ohmic contacts implemented as carrier reservoirs. (a) particles trying to enter the reservoir are removed from the simulation. (b) particles moving from the reservoir to the device are duplicated. (c) particles trying to exit the reservoir and the simulation domain are re-injected in the reservoir.

resulting in a depletion of the free charge in the source region. This situation would result in an unphysical *source starvation* enforced by inadequate boundary conditions. The case is particularly critical when considering thin film SOI devices where the charge per unit area in the source reservoir, essentially given by the product of the doping concentration times the silicon film thickness, is rather low.

A way to overcome this limitation is to set a non-null electric field inside the reservoir. For example in [30] it is proposed to employ a feed-back loop, where the electric field in the contact region is adjusted to guarantee charge neutrality in the source and drain extensions. Another possibility is to inject a displaced distribution of carriers with a non-zero average velocity [28], somehow accounting for out-of-equilibrium conditions in the reservoir.

However, the problem of finding an efficient and always correct implementation of ohmic contacts is still open. In nanoscale devices the modeling of the source reservoir is critical because effects such as source starvation [29] can impact significantly the drain current in strong inversion.

6.1.5 Gathering of the statistics

In order to infer the carrier distribution function from the motion of the particles, statistical data concerning their position in **R** and **K** space must be collected periodically. From this viewpoint, we can define two main categories of MC simulators: *single-particle* and *ensemble*. Figure 6.7 explains their main characteristics.

In the single-particle algorithm only one particle is simulated at a time. The carrier motion is evaluated from the point of injection into the simulation domain (for example at the source of a MOSFET) until the particle exits the simulation domain because either it is back-scattered to the source or it is collected by the drain. Then, new particles are considered until the desired degree of statistical convergence is reached. In the ensemble MC, instead, many particles are simulated at the same time.

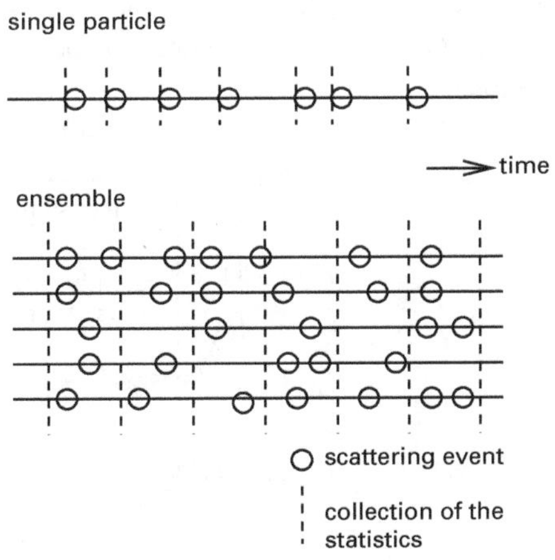

Figure 6.7 The time evolution of a single particle and of an ensemble MC.

A fair comparison between the two algorithms is not trivial. Roughly speaking we may say that, as long as we consider stationary conditions, the same physical configuration (device structure, bias, etc.) can be accurately calculated with both methods. The ensemble MC is, however, needed when considering non-stationary situations, that is the simulation of current transients, since the single particle approach is inherently stationary.

In the single particle MC the gathering of the statistics on the position and the wave-vector of the carriers occurs at the end of the free-flight just before scattering is computed (see again Fig.6.7). In the ensemble MC case, instead, statistics are collected for all the particles at synchronous times $t_i = i\Delta t$. At such time instants, the free-flights must be momentarily interrupted.

Focusing on the ensemble MC, the observable quantity X is estimated as:

$$\bar{X}(t_i) = \frac{1}{N_P} \sum_{p=1}^{N_P} X_p(t_i), \tag{6.40}$$

where N_P is the number of particles in the population contributing to the estimation of X. For example, if X is the velocity at the position $\mathbf{R}$, N_P counts only the particles in the spatial bin around the grid point $\mathbf{R}$.

Figure 6.8 shows, as an example, the time evolution of the average electron velocity obtained from ensemble MC simulations with a uniform electric field. At $t = 0$ particles were initialized in the simulation domain according to an equilibrium Maxwell–Boltzmann distribution. The initial average velocity is thus zero. When the simulation begins and the electrons are accelerated by the electric field, their average velocity increases. As the time elapses, scattering events balance the effect of the electric field and the average velocity reaches a stable value, with an uncertainty produced by statistical noise.

Generalizing the results in Fig.6.8, we can thus divide the time evolution of any quantity X (as for instance, the velocity represented in Fig.6.8) into a *transient* phase, where that quantity changes during time due to the evolution from the initial state to the steady-state value, and a *stationary* phase, where the accuracy in determination of X is limited by statistical noise.

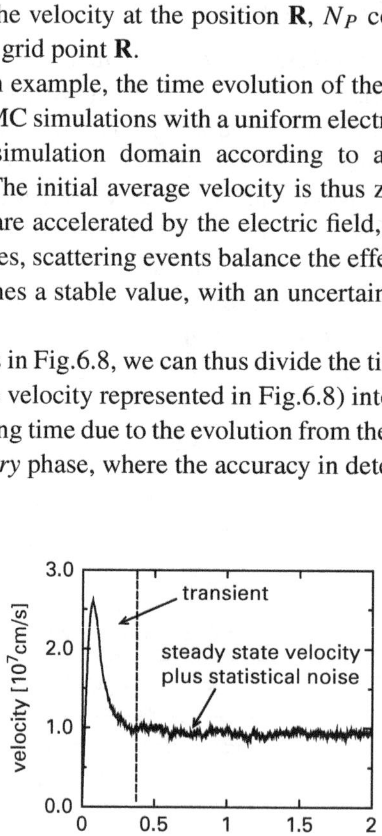

Figure 6.8 Average velocity versus time in a bulk silicon slab with a uniform $F_x = 100$kV/cm.

If our interest is in the static value of X, the duration of the simulation should be sufficiently larger than the transient phase. In fact, the simulation results are not physically meaningful during the transient because they are still affected by the initial conditions used for the carrier distribution. Furthermore, a convergence criterion is needed in order to decide how long the simulation of the stationary phase should last. In principle we should continue the simulation until the error drops below a given tolerance. In stationary cases the variance of the estimate can be reduced by averaging over the time instants t_i after the end of the initial transient:

$$\tilde{X} = \frac{1}{N_{stats}} \sum_{i=N_{tran}+1}^{N_{tran}+N_{stats}} \bar{X}(t_i), \tag{6.41}$$

where N_{stats} is the number of averaged time steps in the stationary phase of the simulation and N_{tran} the number of steps of the initial transient. If all time steps were uncorrelated, the variance of the time averaged expectation value would be

$$\sigma_{\tilde{X}}^2 \simeq \frac{\sigma_{\bar{X}}^2}{N_{stats}} = \frac{\Delta t}{T_{stats}} \sigma_{\bar{X}}^2, \tag{6.42}$$

where $T_{stats} = \Delta t N_{stats}$, and

$$\sigma_{\bar{X}}^2 = \sigma_{\bar{X}}^2(t_i) = \frac{1}{N_P} \sum_{p=1}^{N_P} (X_p(t_i) - \bar{X})^2 \tag{6.43}$$

should not depend on t_i.

Equations 6.42 and 6.43 tell us that the variance $\sigma_{\bar{X}}$ (and thus $\sigma_{\tilde{X}}^2$) can be reduced by increasing the number of particles N_P. Furthermore, $\sigma_{\tilde{X}}^2$ can be reduced by increasing the number of time steps N_{stats}.

On the other hand, if the X values at the different time steps (t_i) are correlated, Eq.6.42 should be replaced by [31]

$$\sigma_{\tilde{X}}^2 \simeq \frac{\max\{\Delta t, 2\tau_C\}}{T_{stats}} \sigma_{\bar{X}}^2 \tag{6.44}$$

where τ_C is the correlation time. This means that in order to reduce statistical noise, we still have to increase the number of time steps (as suggested by Eq.6.42), but being aware that this is an effective measure only if Δt is much larger than τ_C.

6.1.6 Enhancement of the statistics

In many practical cases it is necessary to increase the number of particles in specific regions of the phase-space in order to reduce statistical noise. In fact, considering for instance a MOSFET, most of the carriers would be allocated in the source and drain, because their spatial distribution follows the carrier concentration. Furthermore, most of the particles would possess a low kinetic energy, because they stay in regions (the source and drain) close to thermal equilibrium. Given the total number of particles, if the charge concentration changes by orders of magnitude inside the structure, then there is the risk of finding not even a single particle in some regions of the device. This is for example the case for a MOSFET biased in the subthreshold region.

Another common situation where statistics may be insufficient in some portion of the phase space occurs when we need to compute high energy phenomena such as impact ionization or the gate current in a MOSFET, which require stable estimation of the carrier distribution function for a kinetic energy as high as a few eV.

Many techniques to enhance the statistics in specific portions of the phase space have been proposed [32–35]. They share the common goal of equalizing (or, in general, of controlling) the number of particles in the elements of the phase-space relevant for the estimator of interest.

A simple scheme for statistical enhancement of single particle Monte Carlo simulations is proposed in [32]. Once a particle enters the phase space region of interest, its position in the phase space is stored. Then, N trajectories originating from the same phase space point are simulated, but each of them is given a weight $1/N$ in collection of the statistical averages.

The idea of weighting the particles is exploited also for the statistical enhancement of ensemble Monte Carlo simulations. Denoting the statistical weights w_p, the averages are then computed as:

$$\bar{X}(t_i) = \frac{\sum_{p=1}^{N_P} X_p(t_i) w_p}{\sum_{p=1}^{N_P} w_p}. \tag{6.45}$$

The essence of the statistics enhancement techniques lies in the algorithm to distribute the particles in the bins of the phase space. Many different schemes have been proposed for this task, depending on the application. A good compromise is to have approximately the same number of particles in all the bins [36]. To achieve this objective the simulation algorithm periodically takes snapshots of the particle distribution, deletes particles in elements which are over-populated and duplicates particles in elements which are under-populated. An example of such an algorithm is given below.

Example 6.9: Algorithm for statistical enhancement enforcing a constant number of particles in each element of the phase space.

Assume N is the desired number of particles in each element and N_{old} is the actual number found in a given element of the phase space when collecting the statistics. The N_{old} particles are removed and then N particles are generated with states randomly chosen among those of the N_{old} particles according to their statistical weight. More precisely, to the nth particle of the new ensemble (n going from 1 to N) is assigned the state of the ith particle of the old ensemble (i going from 1 to N_{old}), if the random number r_n uniformly distributed between 0 and 1 satisfies the inequality

$$\frac{\sum_{p=1}^{i-1} w_p}{\sum_{p=1}^{N_{old}} w_p} \le r_n < \frac{\sum_{p=1}^{i} w_p}{\sum_{p=1}^{N_{old}} w_p}. \tag{6.46}$$

A single particle can be selected many times but each particle is given a new free-flight time; it is then very unlikely that the new trajectories are the same. All the new particles have the same weight, a choice which minimizes the variance of the ensemble [36]:

$$w_{new} = \frac{1}{N} \sum_{p=1}^{N_{old}} w_p. \tag{6.47}$$

This approach cannot populate phase-space elements where $N_{old} = 0$, however, it sets to N the number of particles in all the phase space elements with $N_{old} > 0$. In the elements with $N_{old} > N$ the particles will have a larger statistical weight with respect to elements with $N_{old} < N$. After the equalization to N of the number of particles per element, the carriers move along their trajectories and the system will deviate from the condition where exactly N particles lie in each element of the phase-space. Thus, the algorithm has to be periodically repeated during the simulation.

For example, in [33] it is proposed to redistribute the particles in a given bin only when $N_{old} > s_A N$ or $N_{old} < s_A^{-1} N$ or when $(\sum_{p=1}^{N_{old}} w_p^2)/(\sum_{p=1}^{N_{old}} w_p) > (s_B/N_{old})$, that is when N_{old} significantly deviates from the target value N or when the spread between the statistical weight of the various particles in the bin becomes too large. The parameters s_A and s_B can be adjusted by the user and are typically not too far from unity.

Note that the algorithm above is in fact not limited to cases where we aim at the same number of particles, N, in each bin, but to any target distribution. In fact, each bin is treated independently and can have its own N.

6.1.7 Estimation of the current at the terminals

Estimating the current at the contacts simply as the number of particles reaching the contact in unit time is a natural choice but it is very inefficient and noisy for steady state simulations, and it is even wrong when simulating the transient response of a terminal current to voltage waveforms in the time domain. In fact, according to Ramo's theorem [37, 38], all carriers moving inside the device induce charges and thus current at the terminals.

The current at the ith terminal in the presence of the time-varying potentials $V_j(t)$ at the contacts is given by [39]

$$I_i(t) = -\sum_n Q_n \mathbf{V}_n(t) \cdot \nabla f_i + \sum_j \frac{\mathrm{d}V_j}{\mathrm{d}t} \int_{\mathbf{S}_i} \epsilon_i \nabla f_j \cdot \mathrm{d}\mathbf{S}, \tag{6.48}$$

where the index n runs over all the particles in the simulation domain, Q_n is the charge of the particle and $\mathbf{V}_n$ its velocity, the index j runs over all the device contacts; $f_i(\mathbf{R})$ (and f_j as well) are the solutions of the Laplace equation with a boundary condition equal to 1 on the ith (jth) terminal and 0 on the other terminals. The integral is performed over the surface $\mathbf{S}_i$ of the ith terminal and ϵ_i is the dielectric constant at the terminal.

The estimator of Eq.6.48 provides the current as an average where all particles in the simulation domain contribute to the result, thus reducing appreciably the statistical noise with respect to simple counting of the particles hitting the contact.

An alternative approach to estimating the current is proposed in [40]. The current integrated over the contact surface is the sum of the conduction and displacement currents. The former is just the sum of the contributions of all particles at the contact, whereas the latter is estimated as the time derivative of the electric field, multiplied by the dielectric constant.

In case we are interested in calculating very small currents due, for example, to hot carrier effects, it is useful to develop ad hoc estimators such as those discussed in [24, 41, 42]. In particular, for the gate current I_G the simplest estimator counts the number of carriers impinging on the Si/SiO$_2$ interface during a unit time multiplied by their injection probability, whereas for the substrate current I_B it counts the number of e–h pairs generated by impact ionization events during a unit time. An alternative approach for I_B is to calculate the product of the impact ionization scattering rate times the carrier distribution. By following this choice, the whole distribution function, and hence all particles in the simulation, contribute to the estimator. Consequently, the statistical noise is much reduced compared with the case where only the generated e–h pairs are counted.

6.1.8 Full band Monte Carlo

The numerical calculation of the free-flight trajectory in Section 6.1.1 has considered an analytical dispersion relationship such as, for instance, the parabolic and anisotropic energy relation for the conduction band (Eq.2.60). These analytical models are valid only close to the valley minima, as shown in Fig.2.10. When considering high energy transport phenomena in far from equilibrium conditions it is often mandatory to use an accurate numerical description of the $E_{B,n}(\mathbf{K})$ relationship up to high energy, such as the one given by the empirical-pseudo-potential method (Section 2.2.1) or the $\mathbf{k}{\cdot}\mathbf{p}$ method (Section 2.2.2). MC simulators using such a description of the bands are called *Full-Band Monte Carlo* (FBMC) [8]. Many excellent papers and books are devoted to FBMC programs [43–46]. In this book we briefly discuss only the problems associated with the handling of a numerical description of the bands, since they might be of interest also for the treatment of inversion layer dispersion relations such as those provided by the LCBB method (Section 3.4).

The main difference between the FBMC and the MC programs based on analytical bands is the fact that the full band relationship is known only for a finite set of $\mathbf{K}$ points and requires interpolation in the $\mathbf{K}$-space between such points. This aspect significantly complicates evaluation of the state of the particle at the end of the free-flight.

Several techniques have been developed over the last 20 years to manage numerical energy bands [8, 10, 44, 46, 47]. As an example, we describe here the simplex Monte Carlo [47], where the real and reciprocal space are divided into regions (simplexes) where the potential energy versus position and kinetic energy versus $\mathbf{K}$ are

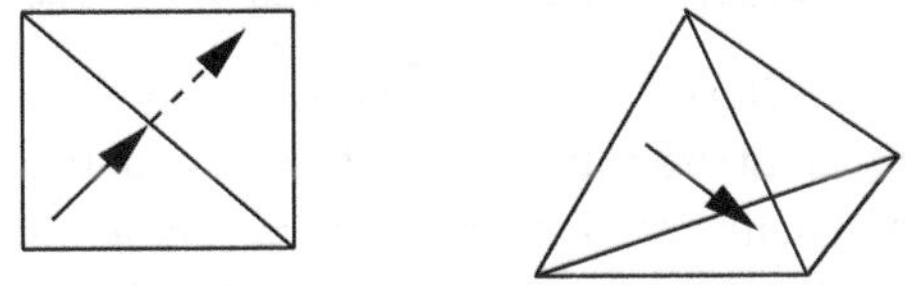

Figure 6.9 A free-flight in a simplex Monte Carlo. The sample free-flight (indicated by arrows) maintains the carrier inside the simplex (tetrahedron) in the 3D **K** space, but it crosses the border between two simplexes (triangles) in the 2D real space.

approximated with a linear relation. If the simulator is 2D in real space and 3D in **K**-space, the simplexes in real space are triangles. Once the potential at each vertex is known, the electric field inside the triangle is a constant two component vector. The simplexes in **K** space, instead, are tetrahedra. The gradient of $E_{B,n}(\mathbf{K})$ (i.e. the group velocity) is thus a constant three component vector inside the tetrahedron. As a result, as long as the particle remains inside a simplex during free-flight, it is very simple to evaluate the final position and velocity. When the particle exits the simplex, free-flight is interrupted and continued in the neighboring simplex (see Fig.6.9).

Another significant difference of the FBMC with respect to MC solvers with analytical bands is the evaluation of the state after scattering. While in the latter case we convert the sum in **K** over the final states into an integral, and thus identify the final state determining for instance the azimuth and polar angles by means of random numbers (see Examples 6.3 and 6.5), in the FBMC we have to use Eq.6.24 directly to select the element in the phase space corresponding to the final state. The interested reader can refer to [44] where an efficient approach based on a linear interpolation within tetrahedra is used in determination of the state after scattering.

It is worth noting that a numerical description of the $E_{B,n}(\mathbf{K})$ relationship may be found critically inaccurate in the vicinity of the band minima because of the abrupt sign change of the group velocity, which complicates calculation of low-field properties such as the mobility. To overcome this tedious problem an analytical description of the bands matched to the numerical one is often used close to the minima, even in the FBMC codes [48].

Another important aspect is how to handle the band index during free-flight. In fact, as discussed in Section 5.1, the semi-classical model assumes that the band index does not change during free-flight, but this may become problematic and questionable in FBMC. As an example, in [8] it is proposed that when a particle reaches a crossing point, it remains in the branch of the energy dispersion that conserves the group velocity. In the case of the X point in silicon, an electron in the lowest conduction band should then be moved to the second conduction band.

The FBMC approach has so far been applied essentially only to descriptions of the carriers as a free gas. In fact, we have seen in Chapter 3 that a full band description of the inversion layer as a 2D carrier gas (such as the one provided by the LCBB method, Section 3.4) is still very demanding from a computational point of view, so that inversion layers are frequently described with analytical models for the conduction and valence bands.

6.1.9 Quantum corrections to free carrier gas MC models

Several methods have been proposed to extend conventional MC BTE solvers for the free electron gas in order to account for the quantization effects at least in the direction normal to the channel/dielectric interface. These models are mostly based on quantum corrections to the electrostatics and transport physics [49–53].

The basic idea behind the adoption of quantum corrections to the potential is to replace the semi-classical potential energy with an effective (or quantum) potential and to modify the force moving the free carriers in such a way that the charge concentration becomes similar to the one predicted by quantum mechanics. The main qualitative features of a quantum-corrected potential energy profile at the Si/SiO_2 interface of a MOSFET are sketched in Fig.6.10. The rise of the potential energy at the interface generates an electric field repelling electrons and thus resulting in an electron concentration qualitatively similar to the one obtained from the wave-functions in the 2D electron gas (see for example the right plot in Fig.3.4), while the increase of the bottom energy of the well reduces the charge and mimics the reduction of the density of states caused by quantization of the energy levels. These qualitative effects are common to many approaches.

As an example, in the effective potential method the classical potential energy is smoothed by a convolution with a Gaussian function to resemble the spatial distribution of the electron wave-packet [49]:

$$U_{\text{corr}}(x,z) = \int\int U(x',z') \frac{1}{2\pi\sigma_x\sigma_z} \exp\left(-\frac{(x-x')^2}{2\sigma_x^2} - \frac{(z-z')^2}{2\sigma_z^2}\right) \mathrm{d}x'\mathrm{d}z', \quad (6.49)$$

where the classical potential energy $U(x,z)$ is assumed to be invariant with y and includes the electrostatic potential and the affinity:

$$U(x,z) = -e\phi - \chi. \quad (6.50)$$

Selection of the smoothing parameters σ_x and σ_z is not trivial since there is no easy way to relate them to the results of the Schrödinger equation. In [54] σ_z has been adjusted to reproduce the inversion charge profiles as obtained from self-consistent solutions of the Schrödinger and Poisson equations, and an optimum $\sigma_z = 0.5$nm was found.

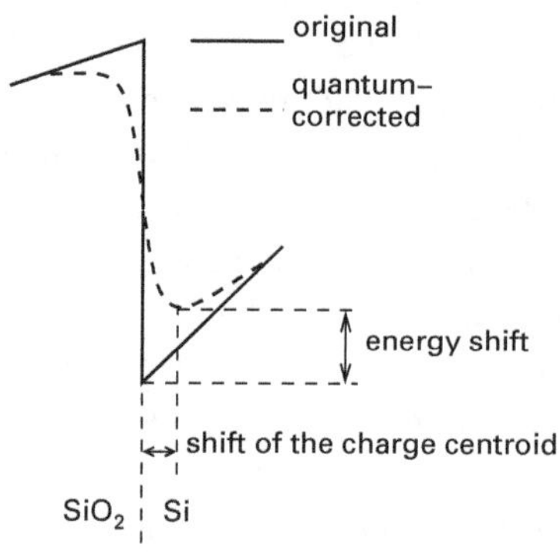

Figure 6.10 Effect of quantum corrections on the potential energy profile at the Si/SiO_2 interface inside a MOSFET.

Note that, since the smoothing is applied to both the quantization z and the transport x directions, the effective potential approach, beside mimicking the main effects related to quantization in the vertical direction (as sketched in Fig.6.10) also accounts, at first order, for quantum effects in the transport direction. In particular the smoothing reduces the height of the potential energy barrier at the source side of the channel, mimicking the effect of source to drain tunneling.

In [50] an additional potential similar to the density-gradient correction [55, 56] and thus proportional to the derivative of the charge concentration $n(x, z)$ is added to the classical potential energy:

$$U_{\mathrm{corr}}(x, z) = U(x, z) - \frac{\hbar^2}{12m_x}\frac{\partial^2 \ln(n)}{\partial x^2} - \frac{\hbar^2}{12m_z}\frac{\partial^2 \ln(n)}{\partial z^2}. \tag{6.51}$$

Instead, Ref.[51] solves the Schrödinger equation in the vertical direction and introduces an additional term to the classical potential energy to force the charge distribution of the MC simulation for the free electron gas to reproduce that of the 2D electron gas.

However, it is clear that these approaches account only for the effect of quantization on the device electrostatics, since effects such as subband splitting and modulation of the scattering rates are not captured. Consequently ad hoc models for the mobility parameters, for example, should complement the use of such corrections in TCAD.

In [52, 53] different corrections are employed for each minimum of the conduction band, to include, at first order, the effect of valley splitting on transport.

It is worth mentioning also that, since the modified potential U_{corr} tends to repel carriers from the Si/SiO_2 interface, the modeling of surface roughness as a specular/diffusive reflection of the impinging carriers [20] does not work any more. Surface roughness scattering must thus be included as an additional scattering mechanism related to the average confining field as proposed, for instance, in [57, 58].

The most complete approach to exactly taking into account the effects of quantization in the vertical direction on both the electrostatic and the transport, however, is the multi-subband Monte Carlo method (MSMC) that we describe in Section 6.3.

6.2 Coupling with the Poisson equation

If there are good reasons to believe that the initial guess of the potential profile used to calculate the force and solve the BTE is far from the real one, it is not wise to maintain the field frozen during the simulation, but it is necessary to couple the solution of the BTE with that of the Poisson equation (PE) [59].

A self-consistent simulation scheme is obtained by partitioning the Monte-Carlo transport simulation into steps of duration Δt and solving the PE at the end of each MC step.

The self-consistent calculation of the electrostatic potential and carrier density has a strong influence on the solution of the BTE in terms of both accuracy and efficiency [8, 60–62]. In the specific case of MOS devices, we note that the current depends exponentially on the energy barrier between source and channel because of the thermionic emission process. More specifically, the inversion density at the virtual source for carriers moving from the source into the channel (N^+ in Section 5.7) depends exponentially on such an energy barrier. As long as the band structure used by the MC is consistent with the one used in the simulator that provided the initial guess of the potential (for instance a simulator based on the Drift–Diffusion model, DD), then N^+ at the VS is similar in both simulators.

However, due to different transport models, the back-scattering implicitly or explicitly considered by the two simulators is different. As a result, the total charge at the VS is different and this results in two different potential profiles in the DD and MC models. This effect is not captured if the MC is run in frozen-field mode; in particular if, as expected, back-scattering is smaller in the MC than in the DD simulation, then a lower charge at the VS is found in the MC than in the DD model if the same potential profile is used. Self-consistency, however, tends to equalize the charge density at the VS roughly to the value $C_{G,eff}(V_{GS} - V_T)$;[3] in the MC this means increasing the charge injected from the source into the channel to compensate for the lower back-scattering.

From the above discussion it is clear that self-consistency is mandatory when evaluating the current drive of nanoscale MOSFETs. In the following sections we illustrate the main issues related to the coupling between MC and PE.

6.2.1 Poisson equation: linear and non-linear solution schemes

The Poisson equation (PE) can be solved with either a linear or a non-linear iteration scheme [9, 59, 63–66], which are the extension to the non-equilibrium case of the schemes described in Section 3.7.

In the linear scheme, the charge computed by the MC solution of the BTE is assumed to be fixed during the solution of the PE, leading to

$$\nabla \cdot \epsilon \nabla \phi^{(k+1)} = -e\left(p^{(k)} - n^{(k)} + N_D - N_A\right), \tag{6.52}$$

where $n^{(k)}$ and $p^{(k)}$ are the electron and hole concentrations at the end of the kth MC iteration that are used to update the potential profile from $\phi^{(k)}$ to $\phi^{(k+1)}$.

The non-linear scheme [9], instead, assumes that the charge exponentially depends on the variation of the potential between two successive solutions of the PE. Thus, the PE for the non-linear scheme becomes:

$$\begin{aligned}\nabla \cdot \epsilon \nabla \phi^{(k+1)} = &- e\Big\{p^{(k)} \exp\left[e(\phi^{(k)} - \phi^{(k+1)})/kT\right] \\ &- n^{(k)} \exp\left[e(\phi^{(k+1)} - \phi^{(k)})/kT\right] + N_D - N_A\Big\}.\end{aligned} \tag{6.53}$$

When the convergence is reached, the two schemes provide the same potential profile. In fact, in this case we have $\phi^{(k+1)} = \phi^{(k)}$ and Eq. 6.53 reduces to Eq. 6.52.

The advantages and disadvantages of the two schemes are mainly related to the stability of the PE/MC coupling and are discussed in Section 6.2.5.

[3] In this discussion we assume for simplicity that $C_{G,eff}$ and V_T have similar values in the MC and in the DD models.

6.2.2 Boundary conditions

The boundary conditions for the electrostatic potential ϕ usually implemented in self-consistent MC simulations are the following. The value of the potential is fixed at the ohmic contacts (e.g. source, drain, gate, and substrate in the case of a bulk MOSFET). Elsewhere the derivative of the potential is set to zero.

These boundary conditions differ from the ones employed in the physics of mesoscopic systems, such as for instance those used in the non-equilibrium Green's function formalism. In this latter case the derivative of the potential is assumed to be zero also at the source and drain contacts [67]. In this case the conduction band profile in the source and drain regions is free to move with respect to the Fermi level and charge neutrality is enforced by the coupled solution of carrier transport and electrostatics that is the solution of the Poisson Equation. In the MC case, instead, charge neutrality in the source and drain regions is not guaranteed, depending on the algorithm used to implement ohmic contacts (see Section 6.1.4). This is one of the reasons why a direct comparison of MC and NEGF simulations of nanoscale devices is often difficult to accomplish [68].

6.2.3 Charge and force assignment

The PE is a partial differential equation commonly solved by the finite differences or the finite elements methods over a discrete mesh. The particles' trajectories, instead, are calculated by the MC loop and the position of the particles inside the simulation domain is not linked to the mesh points. However, since we need to compute the charge at the mesh nodes in order to solve the PE, we have to associate the charge of each particle with one or more mesh points. This process is called *charge assignment* and can be implemented in different ways. Two common schemes are illustrated in Fig.6.11 and refer to a rectangular mesh, as the one employed for the solution of the PE with the finite differences method.

In the Nearest-Grid-Point (NGP) scheme the charge of a given particle is associated with the nearest grid node. In the Cloud-In-Cell (CIC) scheme, instead, the charge is distributed to the four vertices of the rectangle containing the particle. In this latter case, the nodes receive a fraction of the charge proportional to the distance from the particle. More precisely, if we denote with $x_{1,2,3,4}$ and $z_{1,2,3,4}$ the coordinates of the nodes (with $z_1 = z_2$, $z_3 = z_4$, $x_1 = x_3$ and $x_2 = x_4$, see Fig.6.11) and with (x_Q, z_Q) the position of the particle with charge Q, the fractional charges assigned to the nodes are

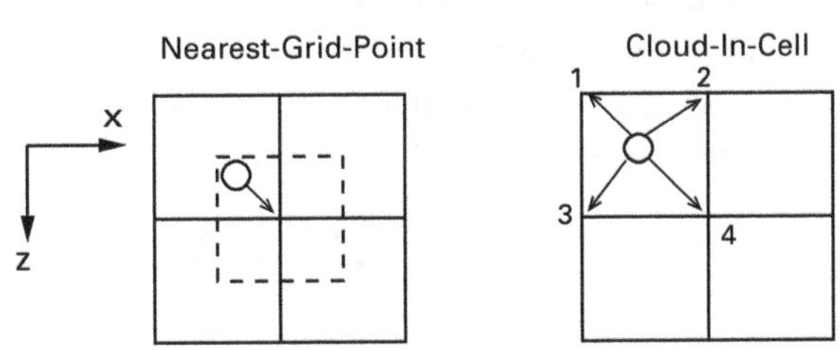

Figure 6.11 Nearest-Grid-Point and Cloud-In-Cell schemes for charge assignment.

$$Q_1 = Q\left(\frac{x_2 - x_Q}{x_2 - x1}\right)\left(\frac{z_3 - z_Q}{z_3 - z_1}\right), \tag{6.54a}$$

$$Q_2 = Q\left(\frac{x_Q - x_1}{x_2 - x1}\right)\left(\frac{z_3 - z_Q}{z_3 - z_1}\right), \tag{6.54b}$$

$$Q_3 = Q\left(\frac{x_2 - x_Q}{x_2 - x1}\right)\left(\frac{z_1 - z_Q}{z_3 - z_1}\right), \tag{6.54c}$$

$$Q_4 = Q\left(\frac{x_Q - x_1}{x_2 - x1}\right)\left(\frac{z_1 - z_Q}{z_3 - z_1}\right). \tag{6.54d}$$

As can be easily verified, the sum of Q_1 to Q_4 equals Q.

Since the PE is solved on a discretized mesh, the electrostatic potential is known only at the mesh points. Due to the use of rectangular elements in the mesh, we cannot define an electric field inside each element but, instead, we must define an electric field at each node. For example, for the node with coordinates (x_i, z_j):

$$F_x(i, j) = -\frac{\frac{\phi_{i+1,j}-\phi_{i,j}}{x_{i+1}-x_i} - \frac{\phi_{i,j}-\phi_{i-1,j}}{x_i-x_{i-1}}}{(x_{i+1} - x_{i-1})/2}, \tag{6.55a}$$

$$F_z(i, j) = -\frac{\frac{\phi_{i,j+1}-\phi_{i,j}}{z_{i+1}-z_i} - \frac{\phi_{i,j}-\phi_{i,j-1}}{z_i-z_{i-1}}}{(z_{i+1} - z_{i-1})/2}, \tag{6.55b}$$

where $\phi_{i,j}$ is the potential at the node (x_i, z_j).

Particles move in-between the mesh points and one must decide which value of the electric field should be used to evaluate the free-flights. This problem is known as *force-assignment*. Also in this case we can use different strategies to define the force on the particle. Similarly to the charge assignment problem, we can use a NGP scheme, where the particles are moved using the electric field at the nearest mesh point. In the CIC scheme, instead, the force is an average of the electric field at the vertices of the rectangle containing the particle. The average is weighted on the basis of the distance of the particle from each vertex.

It is important to note that the same scheme must be used for force and charge assignment, otherwise particles may produce a force on themselves (*self-force*) leading to systematic errors as exemplified below for uniform grids [59].

Example 6.10: NGP assignment.

Let us assume a one-dimensional problem where a single sheet of charge (Q in Coulomb/m^2) between node 0 and node 1 is closer to node 0 than to node 1 (Fig.6.12). The NGP charge assignment algorithm implies that the whole charge is attributed to node number 0. Using finite differences we write the Poisson equation as:

$$\frac{\phi_1 + \phi_{-1} - 2\phi_0}{\Delta x^2} = -\frac{Q}{\epsilon \Delta x}, \tag{6.56a}$$

$$\phi_2 + \phi_0 - 2\phi_1 = 0, \tag{6.56b}$$

$$\phi_0 + \phi_{-2} - 2\phi_{-1} = 0. \tag{6.56c}$$

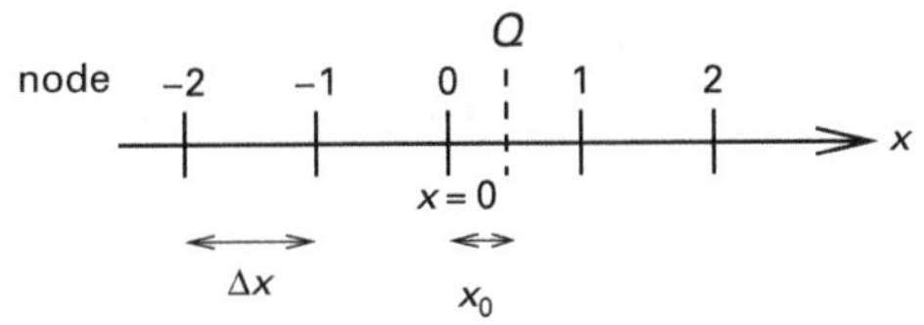

Figure 6.12 A simple 1D case useful to describe self-forces.

We can set $\phi_0 = 0$ and assume that the simulation domain extends indefinitely, so that symmetry imposes

$$\phi_i = \phi_{-i}. \tag{6.57}$$

From Eq.6.56a we readily find:

$$\phi_1 = -\frac{Q}{2\epsilon}\Delta x. \tag{6.58}$$

The electric field at the nodes close to the charge is thus:

$$F_0 = \frac{\phi_{-1} - \phi_1}{2\Delta x} = 0, \tag{6.59a}$$

$$F_1 = \frac{\phi_0 - \phi_2}{2\Delta x} = -\frac{2\phi_1}{2\Delta x} = \frac{Q}{2\epsilon}. \tag{6.59b}$$

If the force assignment scheme is the NGP, hence consistent with the charge assignment, the electric field used to move the particle is taken at the node 0 and it is null, as it should be, since a carrier does not produce a force on itself. If, instead, the CIC scheme is used for force assignment, the field used to move the particle is:

$$F_Q = F_1 \frac{x_Q}{\Delta x} + F_0 \left(1 - \frac{x_Q}{\Delta x}\right) \neq 0, \tag{6.60}$$

which is not null. The unphysical force $-eF_Q$ is the *self-force*.

Example 6.11: CIC assignment.
Let us adopt the CIC scheme for charge assignment in the simple one-dimensional device of Fig. 6.12. We now have:

$$\frac{\phi_1 + \phi_{-1} - 2\phi_0}{\Delta x^2} = -\frac{Q(1 - x_Q/\Delta x)}{\epsilon \Delta x}, \tag{6.61a}$$

$$\frac{\phi_2 + \phi_0 - 2\phi_1}{\Delta x^2} = -\frac{Q x_Q}{\epsilon \Delta x^2}. \tag{6.61b}$$

It is easy to show that

$$F_0 = -\frac{Q}{2\epsilon}\left(\frac{x_Q}{\Delta x}\right), \tag{6.62a}$$

$$F_1 = +\frac{Q}{2\epsilon}\left(1 - \frac{x_Q}{\Delta x}\right). \tag{6.62b}$$

In this case, when the force assignment scheme is NGP, we move the charge with F_0, which is not null, and we thus have self-forces. When, instead, the CIC scheme is used also for the force assignment, we have

$$F_Q = F_1 \frac{x_Q}{\Delta x} + F_0 \left(1 - \frac{x_Q}{\Delta x}\right) = 0. \tag{6.63}$$

The situation is much more complicated than depicted in Examples 6.10 and 6.11 if the grid is non-uniform [59]. An assignment scheme which is robust in terms of immunity to self-forces is the Nearest Element Center (NEC) and can be described briefly as follows[59]. All vertices of the element containing the charge Q receive $Q/4$, while the force on the charge is just the bare arithmetic average (without weighting) of the forces at the four vertices of the element.

6.2.4 Self-consistency and Coulomb interactions

An important topic related to the coupled solution of the BTE and Poisson equations regards the long- and short-range spherically symmetric Coulomb interactions among carriers.

In principle a very fine three-dimensional mesh in real space and a frequent solution of the Poisson equation could account for all these interactions between carriers. However, this approach would require an intolerably large number of grid points. Practical methods rely on the Poisson equation to account for the long-range interactions, whereas the short-range are accounted for by computing explicitly the Coulombic force on each carrier. Care should be taken to avoid double counting, since part of the short-range interaction is already included in the electrostatic potential provided by the Poisson equation [69–72].

An alternative way is to add the short and long range carrier–carrier interactions, usually called carrier–carrier and carrier–plasmon, respectively, as additional scattering mechanisms. This approach is also suited for non-self-consistent models based on two-dimensional grids in real space [73, 74].

A technique to include the carrier–plasmon interaction in a two-dimensional simulation mesh has been proposed in [8, 75]; self-consistent simulations are performed by properly selecting the grid spacing, the time step and the number of particles in each grid element in order to induce plasma oscillations, whereas short range interactions are treated as an additional scattering mechanism.

6.2.5 Stability

The coupling between MC and PE can lead to unstable simulations related to sampling and discretization. One cause for instability is that the electric field is kept constant during the MC simulation between two subsequent solutions of the PE and, furthermore, the electric field is computed at grid nodes or elements (i.e. it is a discrete space

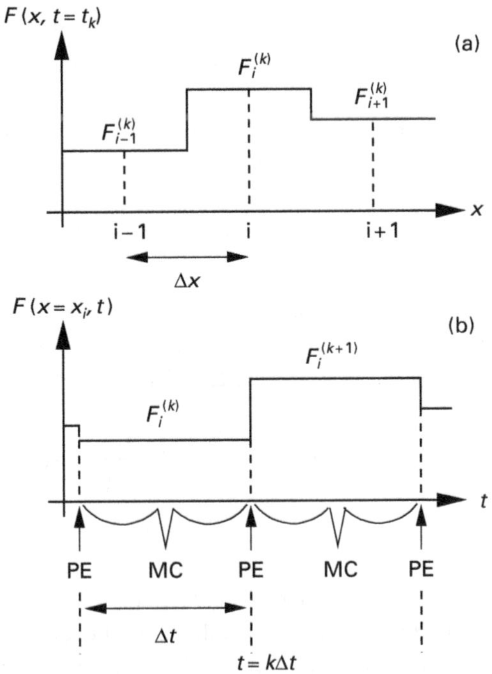

Figure 6.13 The electric field profile as a function of position x at fixed time (plot a) and as a function of time at fixed position x (plot b). *Reprinted with permission from [63]. Copyright 2006 by the Institute of Electrical and Electronics Engineers.*

variable), whereas the MC transport is continuous in space, in the sense that the particle position is not restricted to the grid points (see Fig. 6.13). The time-step Δt between two subsequent solutions of the PE and the grid spacing must be chosen with care to prevent divergent iterations. Another cause is the particle granularity, which is especially critical in the highly doped regions, where the charge assigned to each particle can be very large, so that fluctuations of the velocity and/or the concentration of the carriers induce correspondingly large fluctuations in the electric field [63].

Unfortunately, criteria for choosing the time-step and the grid spacing are available only for very simple physical situations [63, 64, 76, 77]: uniform semiconductor samples, energy-independent scattering rate, uniform mesh. Despite their limitations, stability models for the MC/PE coupling can provide useful guidelines for the simulation of realistic devices with non-uniform grids.

In the following we describe a simple model for the time stability of the linear solution scheme of the Poisson equation discussed in Section 6.2.1 and we summarize the main results about the MC–PE stability and the effects of the grid spacing Δx and of the time-step Δt between two PE solutions [63, 64].

We begin by assuming a uniform one-dimensional structure whose net doping N_{dop} is compensated by a carrier concentration $n(x)$ that deviates from the equilibrium value $n_0 = N_{dop}$ due to plasma oscillations. For a 3D carrier gas and a spherical parabolic band structure the concentration n_0 sets the Debye length (see Section 4.2.1)

$$L_d = \sqrt{\frac{K_B T \epsilon}{e^2 n_0}} \tag{6.64}$$

and the plasma frequency (see Section 4.7.1 and Eq.4.302)

$$\omega_p = \sqrt{\frac{e^2 n_0}{\epsilon m}}, \tag{6.65}$$

where m is the effective mass.

To model the MC transport algorithm, we consider a linearized version of the first two moments of the BTE (i.e. of Eqs.5.123 and 5.131):

$$\frac{\partial v}{\partial t} = -\frac{eF}{m} - \nu_C\, v, \tag{6.66a}$$

$$\frac{\partial n}{\partial t} + n_0 \frac{\partial v}{\partial x} = 0, \tag{6.66b}$$

where $n(x,t)$, $v(x,t)$, and $F(x,t)$ here denote the displacements with respect to the corresponding equilibrium values of the electron density, the electron velocity and the electric field, respectively. The scattering rate ν_C is assumed to be independent of the energy. Furthermore, in deriving Eq.6.66 we have neglected the contribution stemming from the linearization of the diffusion term in Eq.5.131, which greatly simplifies the following derivation.

The simplest model for the stability of the MC–PE loop [76] neglects space discretization. The time-varying variables are thus written as:

$$n(x,t) = \tilde{n}(t) \exp(\mathrm{i}\beta x), \tag{6.67a}$$

$$v(x,t) = \tilde{v}(t) \exp(\mathrm{i}\beta x), \tag{6.67b}$$

$$F(x,t) = \tilde{F}(t) \exp(\mathrm{i}\beta x). \tag{6.67c}$$

The spatial derivatives in Eq.6.66 yield a term $(\mathrm{i}\beta)\exp(\mathrm{i}\beta x)$. If we take $t = 0$ as the time at which we have solved the last PE equation, the evolution of $\tilde{n}$ and $\tilde{v}$ from $t = 0$ to $t = \Delta t$ is given by the solution of Eq.6.66:

$$\tilde{v}(t) = \tilde{v}(0) e^{-\nu_C t} - \frac{e\tilde{F}(0)}{m\nu_C}(1 - e^{-\nu_C t}), \tag{6.68a}$$

$$\tilde{n}(t) = \tilde{n}(0) - \frac{n_0 \mathrm{i}\beta}{\nu_C}\tilde{v}(0)(1 - e^{-\nu_C t}) + \frac{n_0 \mathrm{i}\beta e \tilde{F}(0)}{m\nu_C}\left(t - \frac{1 - e^{-\nu_C t}}{\nu_C}\right). \tag{6.68b}$$

The electric field $\tilde{F}(0)$ is kept constant in the time interval and is given by the Poisson equation that, according to Eq.6.67, reads:

$$\mathrm{i}\beta \tilde{F}(0) = -\frac{e\tilde{n}(0)}{\epsilon_S}. \tag{6.69}$$

Upon substitution of $\tilde{F}(0)$ from Eq.6.69, Eq.6.68 can be cast in the form of a linear system

$$\begin{pmatrix} \tilde{v}(\Delta t) \\ \tilde{n}(\Delta t) \end{pmatrix} = \begin{pmatrix} A_{11} & A_{12} \\ A_{21} & A_{22} \end{pmatrix} \begin{pmatrix} \tilde{v}(0) \\ \tilde{n}(0) \end{pmatrix}, \tag{6.70}$$

where $A_{11}, A_{12}, A_{21}, A_{22}$ can be expressed in terms of n_0, ν_C, Δt, ω_p, and β. Following the established methodology to investigate stability in the Z domain [12, 13], we assume that the time evolution takes the form

$$\begin{pmatrix} \tilde{v}(\Delta t) \\ \tilde{n}(\Delta t) \end{pmatrix} = Z \begin{pmatrix} \tilde{v}(0) \\ \tilde{n}(0) \end{pmatrix}, \tag{6.71}$$

where Z is a scalar complex number. By substituting Eq.6.71 in Eq.6.70 we obtain an eigenvalue problem such that Z can be calculated as [76]:

$$(Z-1)^2 + (Z-1)\left(1 - \delta + \frac{\alpha - (1-\delta)}{\lambda^2}\right) + \frac{(1-\delta)\alpha}{\lambda^2} = 0, \tag{6.72}$$

where $\delta = e^{-\nu_C \Delta t}$, $\alpha = \nu_C \Delta t$, and $\lambda = \nu_C/\omega_p$. If the solutions of Eq.6.72 yield complex numbers with $|Z|>1$, for some combinations of the simulation parameters $(\nu_C, \Delta t, n_0)$ then the corresponding simulations are unstable. It is clear that the only independent parameters in Eq.6.72 are $\alpha = \nu_C \Delta t$ and ν_C/ω_p. To analyze the stability of the MC/PE coupling we thus have to map stable and unstable configurations on the $\nu_C \Delta t$–ν_C/ω_p plane represented in Fig. 6.14. The border between stable and unstable configurations has been computed considering the values of $\nu_C \Delta t$ and ν_C/ω_p giving $|Z| = 1$ and is indicated by the dashed line in Fig. 6.14.a. We see that for the linear scheme to be stable short Δt values are necessary when ω_p is large (i.e. when the doping is high). For instance, when simulating MOSFETs with S/D doping of approximately 10^{20} cm^{-3}, Δt is in the range of fractions of femtoseconds.

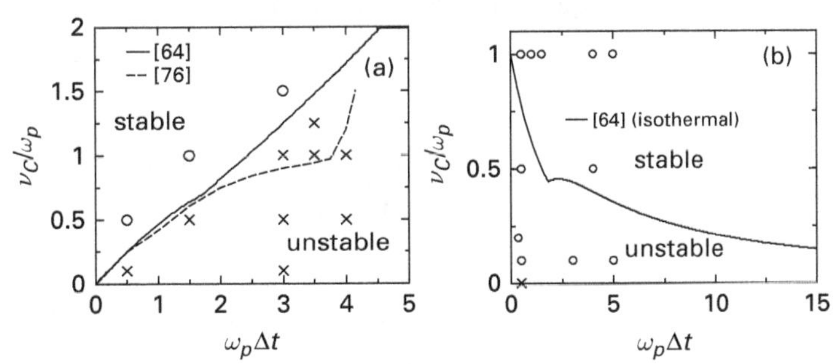

Figure 6.14 Stability regions of the self-consistent MC/PE loop for the linear (a) and non-linear (b) solution scheme of the Poisson equation (PE) as evaluated with the model in [64], considering the effect of the finite time-step Δt between two solutions of the PE. Results of MC simulations of uniform silicon slabs are indicated by circles (stable simulations) and crosses (unstable ones). In the unstable simulations the average carrier energy unphysically grows during the simulation time. *Reprinted with permission from [64]. Copyright 2006 by the Institute of Electrical and Electronics Engineers.*

The stability model has been extended in [64] by considering the additional terms in the linearized BTE (Eq.6.66) which take into account the effects of carrier diffusion. The new result is represented by the solid line in Fig. 6.14.a, that deviates only slightly from the model of [76].

The model has also been reformulated for the non-linear solution scheme of the PE [64]. In this case, different results are found depending on the assumptions behind the small-signal expansion of the diffusion term of the BTE, which can be either isothermal (the average kinetic energy $(3/2)K_BT$ is equally distributed between the three spatial directions) or adiabatic (the kinetic energy $(3/2)K_BT$ is assigned to the direction of the perturbation). Results for the former case are shown in Fig. 6.14.b. We see that stability requires long Δt intervals. However, the calculation of the stability limit is more conservative than real cases (symbols in Fig. 6.14.b), that are stable also for shorter Δt than predicted by the model.

The analysis of the effect of Δx is considerably more complicated than the analysis concerning Δt, and the interested reader should refer to [63] for a detailed discussion of the topic. Assuming again a uniform 1D structure and energy independent scattering rates, the main findings can be summarized as follows:

- in the linear scheme illustrated in Fig. 6.15, stability requires $\Delta x < \pi L_d$, except when very strong scattering rates relax this constraint [63]. In MOSFETs with S/D doping of approximately 10^{20} cm^{-3}, this means a grid spacing of about 1nm.
- in the non-linear scheme, the requirements on Δx for small Δt are the same as for the linear scheme. However, as seen in Fig. 6.14.b, the non-linear scheme offers a possibility of using long Δt, in which case any Δx is acceptable for stable simulations [63]. This is indeed the case for many MC simulators implementing the non-linear solution of the Poisson equation (for instance, the single particle approach of [65, 66]). Usually these solvers iterate between the stationary solution of the BTE obtained with a frozen field profile with Δt long enough to gather good statistics and the non-linear Poisson equation, until the desired convergence is achieved.

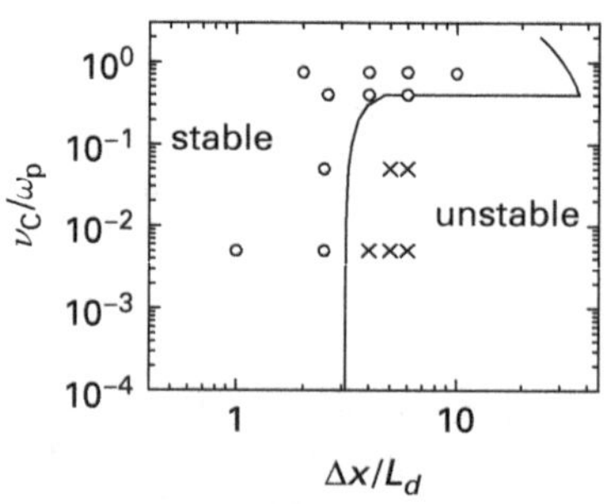

Figure 6.15 Stability regions for the linear PE scheme as evaluated from the model in [63], considering the effect of finite grid size Δx. Results of MC simulations are indicated by circles (stable simulations) and crosses (unstable). *Reprinted with permission from [63]. Copyright 2006 by the Institute of Electrical and Electronics Engineers.*

By considering the influence of Δt and Δx on the stability, it is thus evident that the non-linear coupling scheme provides many advantages. Counter to this, the non-linear scheme is not suited for simulating transients, since self-consistency between MC and PE is achieved only after the transient has expired.

6.3 The multi-subband Monte Carlo method

So far we have focused on the MC method as a means of solving the BTE for the free carrier gas and we have discussed issues of general interest in the implementation of the boundary conditions, the contacts, and the self-consistent loop.

Compared to the case where electrons and holes behave as a free-carrier gas, the strong size and bias induced quantization of modern MOS devices has many important consequences (see Chapter 3):

- the device electrostatics changes, because of the displacement of the charge from the Si/SiO_2 interface;
- because of different effective masses in the quantization direction, an energy splitting is produced among some of the conduction or valence band extrema that are degenerate in the bulk crystal;
- the scattering rates are affected by the device geometry and by the applied bias because they depend on the subband structure and on the corresponding wave-functions (as discussed in Chapter 4).

Quantum corrections to the potential energy such as those presented in Section 6.1.9 can mimic the main consequences of quantization on the electrostatics, but fail to represent accurately the corresponding impact on the scattering rates and the carrier degeneracy.

The self-consistent multi-subband Monte Carlo (MSMC) approach provides a self-consistent solution of the BTEs describing the transport in the electron subbands (Section 5.2) [78–84] and of the Poisson equation governing the device electrostatics.

6.3.1 Flowchart of the self-consistent MSMC method

Consistent with the discussion in Section 5.2.1, in an MSMC simulator the state of a carrier is identified by the state variable $(\mathbf{r}, \mathbf{k}, i)$. The $\mathbf{r}$ is a 2D vector indicating the particle position in the transport plane normal to the semiconductor/dielectric interface, but in many practical implementations reduces to the position x along the channel, because translational invariance is assumed in the device width direction. The $\mathbf{k}$ is the 2D wave vector in the transport plane. If more valleys exist in the inversion layer, the subband index i may be split into a valley index ν and a subband index n identifying the subband in the νth valley. The position of the particle in the quantization direction z is essentially described by the solution of the Schrödinger equation (see Section 5.2.1). In fact the square modulus of the envelope wave-function $|\psi_{\nu,n,x}(z)|^2$ represents the

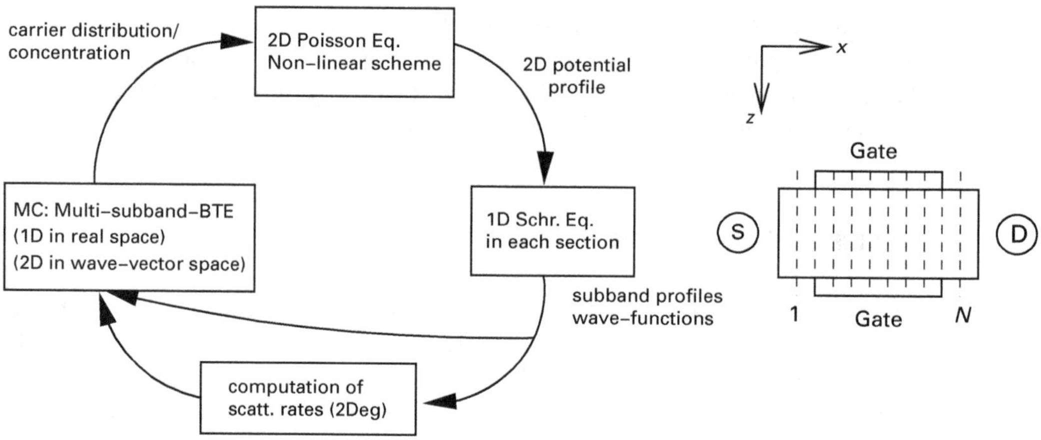

Figure 6.16 Flowchart of an MSMC simulator. The device is partitioned into N sections, as indicated in the sketch.

probability that a particle in the valley ν, subband n, with longitudinal position x, will be found at the vertical position z.

A typical flowchart of a self-consistent MSMC numerical model is shown in Fig. 6.16. The device cross-section is partitioned into N sections along the channel direction x, and the Schrödinger equation in the z direction is solved in all the slices; as discussed in Section 5.2, the derivative of the subband minima with respect to x provides the driving force $\mathbb{F}_{\nu,n,x}$. We consider a scalar driving force since we assume that the problem is translationally invariant in the width direction. Consequently the MC transport solver is 1D in real space. The envelope wave-functions $\psi_{\nu,n,x}(z)$ are then used to compute, at each section, the scattering rates based on the theory of the 2D carrier gas described in Chapter 4. The MC transport procedure yields the occupation probability $f_{\nu,n,x}(\mathbf{k})$ at each section, and the latter is also used to reject scattering events with unavailable final states according to Pauli exclusion principle, as described in Section 6.1.2. After each MC step, the electron concentration $n(x, z)$ is computed as

$$n(x, z) = \frac{2}{A} \sum_{\nu} \sum_{n} \sum_{\mathbf{k}} f_{\nu,n,x}(\mathbf{k}) |\psi_{\nu,n,x}(z)|^2, \qquad (6.73)$$

and then used to calculate the new guess for the potential profile $\phi(x, z)$ by solving the non-linear Poisson equation. The potential $\phi(x, z)$ at each section is then fed into the Schrödinger equation solver, thus starting a new MC iteration. The loop is repeated until convergence is reached according to appropriate criteria.

The non-linear scheme for solution of the Poisson equation (Section 6.2.1) is at present the preferred choice for MSMC models, because, since the scattering rates must be computed after each solution of the PE, this time consuming procedure encourages use of long Δt intervals, that push the linear scheme to an unstable region (see Section 6.2.5).

The flowchart in Fig.6.16 illustrates the whole procedure for the case where the objective is to compute the drain current in the short and wide double gate MOSFET in the sketch. If the goal were to compute the low-field mobility in a long channel device, instead, we could assume that the lateral electric field is constant along x and set it to a low enough value, such as $F_x = 1\text{kV/cm}$. Since there would be no need to know the position of the particles, the MC transport simulation could reduce to zero dimensions (0D) in real space and 2D in **k**-space. In this case, in order to self-consistently calculate the profile of the potential energy well in the vertical direction z, we would need to solve the Poisson equation only in one single device slice.

6.3.2 Free-flight, state after scattering and boundary conditions

The evaluation of the free-flight trajectories follows the algorithms described in Section 6.1.1. The only difference with respect to the 3D case is the number of components of the state vectors **r** and **k** in real and **k**-space. Consistently with the semi-classical model, during free-flight the subband index is not modified. The implications of this latter point have been discussed in Chapter 5; more details can be found in [85].

Regarding scattering, differently from the 3D case, we now have to account for inter-subband scattering that modifies the subband index. We then have to add a new step in the procedure of Fig.6.2 and Section 6.1.2. In particular, after finding the scattering mechanism responsible for the transition, we first determine the final state subband and then the final state **k**. This implies the need to store, for each section x and for each state **k** in subband i, the probability for each scattering mechanism to scatter the particle to state $\mathbf{k}'$ in subband j. An example of determination of the state after scattering is given in the next section.

As to the boundary conditions of the MC transport part, ohmic contacts are necessary at the source and drain for the simulation of drain current in MOS transistors. These contacts can be implemented as discussed in Section 6.1.4 and also sketched in Fig.6.6, aside from the fact that particles can only move in the lateral x direction.

An important point related to boundary conditions is the fact that the potential well confining the carriers in the S/D regions is usually not as narrow as in the channel. Moreover, carrier–carrier and carrier–impurity scattering drastically shorten the mean-free-path in the S/D regions. Thus the existence of significant quantization effects in the source and drain becomes questionable. One approach to taking these effects into consideration is to model the carriers as a 2D gas in the inversion layer and as a 3D gas in the S/D. To do so, one should find ways to match the state of the particle at the boundary between the channel and the S/D regions. The state of a carrier in the source is identified by two additional variables, namely the z position and the wave-vector component k_z, which are not state variables for the 2D carrier gas in the channel. Possible algorithms to handle the event of carriers moving from regions with negligible quantization to regions of strong quantum confinement and vice versa can be found in [78].

Since a 2D carrier gas with very many subbands closely packed in energy approaches the behavior of a 3D gas, some MSMC simulators [86] also use the 2D electron gas description in the source and drain, by employing a large number of subbands.

6.3.3 Multi-subband Monte Carlo transport for electrons

The overall structure of a MSMC transport model for electrons follows the one described in previous sections. However, many simulation steps can be simplified thanks to the fact that the minima of the bulk crystal conduction band can be described by ellipsoidal constant energy surfaces (Eq.3.15).

First of all, the energy and the wave-function of the subbands can be evaluated for $\mathbf{k} = \mathbf{0}$ and used also for $\mathbf{k} \neq \mathbf{0}$, since all $\mathbf{k}$ dependence of the energy is analytical. Secondly, the constant energy lines for the different subbands are either circles (2-fold subbands) or ellipses (4-fold subbands), a situation that enormously simplifies the calculation and the storage of the scattering rates, as well as the determination of the state after scattering.

Considering for example the phonon scattering mechanisms, the total scattering rate depends only on the energy of the initial state and not on the $\mathbf{k}$ direction; the final energy is given by the energy conservation, and the direction of the final wave-vector is random. Hence, for the circular subbands the angle associated with the final $\mathbf{k}$ is random, whereas for the elliptical ones we can assign a random angle and then apply the Vogt–Herring transformations.

With regard to evaluation of the electron dynamics, that is the free-flight trajectory, it is important to note that inside a spatial bin the driving force is constant. Assuming a parabolic $E(\mathbf{k})$ relation, integration of the equation of motion becomes trivial (see Section 6.1.1). When the particle moves to another element, the free-flight is interrupted at the boundary between the two elements and a new free-flight using the driving force in the new element is evaluated.

6.3.4 Multi-subband Monte Carlo transport for holes

An MSMC transport solver for hole inversion layers is more complicated than its electron counterpart. If the hole inversion layer is described by the $\mathbf{k}{\cdot}\mathbf{p}$ theory developed in Section 3.3.1, we need to solve the Schrödinger equation for many points in the $\mathbf{k}$-plane. Furthermore, the wave-functions depend on $\mathbf{k}$ thus complicating remarkably the evaluation of the scattering rates, so that most of the transport studies devoted to hole inversion layers compute the scattering rates by considering the wave-functions only at $\mathbf{k} = \mathbf{0}$ [87, 88].

Since the $\mathbf{k}{\cdot}\mathbf{p}$ energy relation in the transport plane is not analytic, the evaluation of the free-flight should be done as in the case of the full band MC described in Section 6.1.8. For example, in [89] the simplex method has been used. Of course, while the simplex in the 3D $\mathbf{K}$-space of the free-carrier gas is a tetrahedron, in the inversion layer it is a triangle, since $\mathbf{k}$ is a two component vector.

The above-mentioned complications can be mitigated if an analytical relationship is used for the hole energy relation. To illustrate the simplifications allowed for by such an approach, we briefly examine the structure of the MSMC described in [90, 91], which is in fact based on the analytical hole band model of Section 3.3.3. Since in this model quantization in the vertical direction is treated by solving the Schrödinger equation under the effective mass approximation at $\mathbf{k} = \mathbf{0}$, the determination of the subband energy $\varepsilon_i(x)$, of the driving field $d\varepsilon_i/dx$, and of the scattering rates can follow similar procedures to those developed in Section 6.3.3 for the electron case. The only important difference lies in the evaluation of the free-flights, due to the specificity of the energy relation in the transport plane.

In fact the in-plane energy dispersion of Section 3.3.3 does not allow us to analytically integrate the equation of motion

$$x(t) = x_0 + \int_0^t v_{gx}(\mathbf{k}(t'))dt'. \tag{6.74}$$

To avoid the time consuming numerical integration of Eq.6.74, since we are interested only in the position $x(t)$ in the direction of the driving force $\mathbb{F}_{\nu,i,x}$, we can invoke energy conservation and write

$$x(t) = x_0 + \frac{E(k_x(t), k_{y0}) - E(k_{x0}, k_{y0})}{\mathbb{F}_{\nu,i,x}}, \tag{6.75}$$

where k_{x0} is the k_x value at the beginning of the free-flight while k_{y0} denotes the y component of $\mathbf{k}$, which is constant during the free-flight because the force is aligned in the x direction. Since $k_x(t)$ can easily be calculated during the free-flight as

$$k_x(t) = k_{x0} + \mathbb{F}_{\nu,i,x}t/\hbar, \tag{6.76}$$

then the energy $E(k_x(t), k_{y0})$ and the position $x(t)$ are determined by the analytical band model (see again Section 3.3.3). Hence Eqs.6.75 and 6.76 allow us to calculate the hole trajectories inside a section of the device where the force is constant. As already mentioned, when the carrier moves to the adjacent element, the free-flight is stopped and then continued in the next element using the local driving force. More details about the identification of the crossing points (in real and $\mathbf{k}$ space) are given in [91] and are summarized in Fig.6.17.

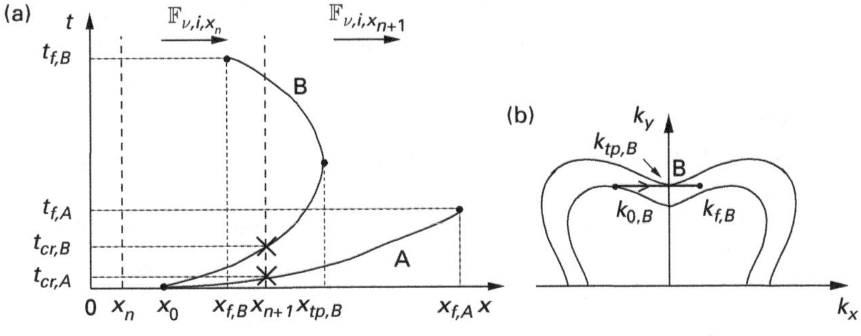

Figure 6.17 a) Sample trajectories in the real space of two holes A and B. The driving forces in two adjacent sections are denoted as $\mathbb{F}_{\nu,i,x_n}$ and $\mathbb{F}_{\nu,i,x_{n+1}}$. b) Sample trajectory of hole B in $\mathbf{k}$-space due to the force $\mathbb{F}_{\nu,i,x_n}$ during free-flight. $k_{0,B}$ and $k_{f,B}$ are the initial and final wave-vectors. Also the equienergy curves corresponding to the $k_{0,B}$ and to the turning point wavevector $k_{tp,B}$ are shown. *Reprinted with permission from [91]. Copyright 2009 by the Institute of Electrical and Electronics Engineers.*

In plot (a), we see two sample trajectories A and B in real space during free-flights with duration t_{fA} and t_{fB}, respectively. The driving forces in two adjacent sections are denoted as $\mathbb{F}_{\nu,i,x_n}$ and $\mathbb{F}_{\nu,i,x_{n+1}}$. The carriers start from the same initial position x_0. The trajectories imply the crossing of the section boundary x_{n+1} at the time $t_{cr,A}$ and $t_{cr,B}$. For the sample hole A at $t > t_{cr,A}$ the motion is governed by the force $\mathbb{F}_{\nu,i,x_{n+1}}$ of the $n+1$ section. In this case, the occurrence of a crossing of the section boundary can be easily identified by checking that the final abscissa $x_{f,A}$ lies in the section $n+1$. The trajectory of particle B is more complicated because if $x_{f,B}$ is checked only for $t = t_{f,B}$, then the section crossing at $t = t_{cr,B}$ is missed. This happens because of the turning point taking place at $x = x_{tp,B}$. In such a case, in order to find the time $t_{cr,B}$ of the crossing, the position $x_{tp,B}$ of the turning point must be identified first.

In [91] it has been shown that the procedure based on energy conservation provides the same results as time consuming integration of the equation of motion, Eq.6.74.

6.4 Summary

In this chapter we have seen how to solve the Boltzmann Transport Equation with a statistical approach, namely the Monte Carlo method. In particular, we have analyzed the main building blocks of Monte Carlo transport simulators for the free-electron gas as well as for the 2D carrier gas.

The most critical and complex parts of the algorithm are the computation of the free-flights according to the chosen band model (Section 6.1.1) and the selection of the state after scattering consistent with the matrix elements of the collision mechanisms (Section 6.1.2). Other critical aspects are the selection of proper boundary conditions (Section 6.1.3), in particular when dealing with ohmic contacts (Section 6.1.4), and the stability of the coupled MC/PE loop (Section 6.2.5).

The multi-subband Monte Carlo (MSMC) approach is a very powerful modeling framework for nanoscale MOSFETs. Firstly, it allows for a description of the carrier gas in the presence of strong size- and bias-induced quantization (in the direction normal to transport) and of far-from-equilibrium transport, including many scattering mechanisms such as lattice vibrations, surface roughness and ionized impurities (described in Chapter 4), as well as polar phonons (in III-V materials), remote phonons and remote Coulomb scattering (in devices with high-κ gate dielectrics) that are described in Chapter 10.

However, the MSMC approach does not account for possible effects of quantization in the transport direction (e.g. electron interference phenomena and source-to-drain-tunneling). Reference [92] has shown, by solving the 1D Schrödinger equation for a coupled electron–phonon many-body system, that at least down to $L_G = 10$nm the effects of coherent transport are only modest. To assess quantitatively the relevance of quantization in the transport direction one should rely on solution of the Wigner equation , which goes beyond the semi-classical modeling provided by the MSMC. Interestingly, the Wigner equation can be solved with the Monte Carlo method [93–96], also in a multi-subband framework, where a set of Wigner equations is considered [96]. The subband profiles are computed as in the Boltzmann transport case, by solving the 1D Schrödinger equation in each section normal to the transport direction. To a first order approximation, scattering can be treated exactly as in the MSMC. In [97], Wigner Monte Carlo simulations of an $L_G = 6$nm device have shown that quantum effects in the transport direction are negligible in strong inversion conditions, whereas scattering still plays a role even in these ultra-short devices. These results support the conclusion that accounting for scattering is somehow more relevant than accounting for quantum transport effects and that the semi-classical multi-subband approach described in this book is more than adequate even in ultimately scaled devices. Furthermore, effects of quantization in the transport direction can be empirically introduced in the MSMC framework by applying a procedure similar to the effective potential approach described in Section 6.1.9 for the free-carrier gas case: the subband profiles along the channel are smoothed with a Gaussian function; the smoothing reduces the height of the potential energy barrier at the virtual source of the device, mimicking source-to-drain-tunneling [68].

As a final remark, we should mention that deterministic approaches to solving the BTE (also, recently, in the multi-subband case) have been presented [1, 98–101] as alternatives to the Monte Carlo method. The rationale behind these new solution methods is that the precision of the MC results defined as the variance of the estimators (Section 6.1.5) scales only as $1/\sqrt{N_P N_S}$, where N_P is the number of particles and N_S the number of time steps. This means that enhancing the precision may often result in a long computation time. A direct comparison between deterministic solutions and the MC method is not trivial. In [102] an attempt is made to carry out such a comparison for the case of silicon nanowires with quasi-1D electron gas. An excellent agreement was found between MSMC and the determinist solutions of the BTE, consistent with the fact that the same equation is solved in both cases. With respect to the simulation time, the deterministic solver is faster than the MC only when a few subbands are considered, whereas the simulation time in the MSMC only weakly depends on the number of subbands and it becomes advantageous for large diameter nanowires. The deterministic solution is also more efficient for bias points in the MOSFET subthreshold region of operation.

References

[1] K. Banoo, M. Lundstrom, and R.K. Smith, "Direct solution of the Boltzmann transport equation in nanoscale Si devices ," in *Proc.SISPAD*, pp. 50–53, 2000.

[2] C. Jungemann, T. Grasser, B. Neinhüs, and B. Meinerzhagen, "Failure of moments-based transport models in nanoscale devices near equilibrium," *IEEE Trans. on Electron Devices*, vol. 52, no. 11, pp. 2404–2408, 2005.

[3] T. Grasser, T.-W. Tang, H. Kosina, and S. Selberherr, "A review of hydrodynamic and energy-transport models for semiconductor device simulation," *Proc. IEEE*, vol. 91, no. 2, pp. 251–274, 2003.

[4] C. Jacoboni and L. Reggiani, "The Monte Carlo method for the solution of charge transport in semiconductors with applications to covalent materials," *Rev. Mod. Phys.*, vol. 55, pp. 645–705, 1983.

[5] R.W. Hockney and J.P. Eastwood, *Computer Simulation Using Particles*. New York: McGraw Hill, 1981.

[6] C. Jungemann, *Efficient Full-Band Monte Carlo Simulation for Device Engineering*. PhD thesis, University of Bremen, 2001.

[7] K. Tomizawa, *Numerical Simulation of Submicron Semiconductor Devices*. Boston: Artech House, 1993.

[8] M.V. Fischetti and S.E. Laux, "Monte Carlo analysis of electron transport in small semiconductor devices including band-structure and space-charge effects," *Phys. Rev. B*, vol. 38, pp. 9721–9745, 1988.

[9] F. Venturi, R.K. Smith, E. Sangiorgi, M.R. Pinto, and B. Riccò, "A general purpose device simulator coupling Poisson and Monte Carlo transport with applications to deep submicron MOSFETs," *IEEE Trans. on Computer-Aided Design*, vol. CAD-8, p. 360, 1989.

[10] C. Jungemann, S. Keith, M. Bartels, and B. Meinerzhagen, "Efficient full-band Monte Carlo simulation of silicon devices," *IEICE Transactions on Electronics*, vol. E82-C, pp. 870–879, 1999.

[11] M. Lundstrom, *Fundamentals of Carrier Transport*. New York: Addison Wesley, 1990.

[12] K. Åström and B. Wittenmark, *Computer-Controlled Systems: Theory and Design*. Upper Saddle River, NJ: Prentice-Hall, 1996.

[13] A. Oppenheim and R. Schafer, *Discrete-Time Signal Processing*. Upper Saddle River, NJ: Prentice-Hall, 2009.

[14] S.E. Laux and M.V. Fischetti, "Numerical aspects and implementation of the *DAMOCLES* Monte Carlo device simulation program," in *Monte Carlo Device Simulation: Full Band and Beyond* (K. Hess, ed.), ch. 1. Norwel, MA: Kluwer, 1991.

[15] E. Sangiorgi, F. Venturi, and B. Riccò, "Mos^2, an efficient Monte Carlo simulator for MOS devices," *IEEE Trans. on Computer-Aided Design*, vol. CAD-7, no. 2, pp. 259–271, 1988.

[16] C. Herring and E. Vogt, "Transport and deformation-potential theory for many-valley semiconductors with anisotropic scattering," *Phys. Rev.*, vol. 101, p. 944, 1956.

[17] H. Kosina, "A method to reduce small-angle scattering in Monte Carlo device analysis," *IEEE Trans. on Electron Devices*, vol. 46, no. 6, pp. 1196–1200, 1999.

[18] P. Lugli and D. K. Ferry, "Degeneracy in the ensemble Monte Carlo method for high field transport in semiconductors," *IEEE Trans. on Electron Devices*, vol. 32, no. 11, pp. 2431–2437, 1985.

[19] P. Borowik and J.-L. Thobel, "Improved Monte Carlo method for the study of electron transport in degenerate semiconductors," *Journal of Applied Physics*, vol. 84, no. 7, pp. 3706–3709, 1998.

[20] E. Sangiorgi and M.R. Pinto, "A semi-empirical model of surface roughness scattering for Monte Carlo simulation of silicon n-MOSFETs," *IEEE Trans. on Electron Devices*, vol. 39, pp. 356–361, 1992.

[21] P. Palestri, A. Pacelli, M. Mastrapasqua, and J.D. Bude, "Monte Carlo simulation of impact ionization in SiGe HBTs," *IEEE Electron Device Lett.*, vol. 22, no. 11, pp. 533–535, 2001.

[22] M.V. Fischetti, S. Laux, and E. Crabbè, "Understanding hot electron transport in silicon devices: Is there a shortcut?," *Journal of Applied Physics*, vol. 78, no. 2, p. 1058, 1995.

[23] J.D. Bude and M. Mastrapasqua, "Impact ionization and distribution functions in sub-micron nMOSFET technologies," *IEEE Electron Device Lett.*, p. 439, 1995.

[24] A. Ghetti, L. Selmi, and R. Bez, "Low voltage hot electrons and soft-programming lifetime prediction in non-volatile memory cells," *IEEE Trans. on Electron Devices*, vol. 46, no. 4, p. 696, 1999.

[25] D. Esseni, L. Selmi, A. Ghetti, and E. Sangiorgi, "Injection efficiency of CHISEL gate currents in short MOS devices:Physical mechanisms, device implications and sensitivity to technological parameters," *IEEE Trans. on Electron Devices*, vol. 47, no. 11, pp. 2194–2200, 2000.

[26] J.D. Bude, M.R. Pinto, and R.K. Smith, "Monte Carlo simulation of CHISEL flash memory cell," *IEEE Trans. on Electron Devices*, vol. 47, no. 10, p. 1873, 2000.

[27] C. Jungemann, S. Yamaguchi, and H. Goto, "On the accuracy and efficiency of substrate current calculations for sub-μm nMOSFETs," *IEEE Electron Device Lett.*, p. 464, 1996.

[28] T. Gonzales and D. Pardo, "Physical models of ohmic contact for Monte Carlo device simulation," *Solid State Electronics*, vol. 39, p. 555, 1996.

[29] M.V. Fischetti, L. Wang, B. Yu, *et al.*, "Simulation of electron transport in high-mobility MOSFETs: Density of states bottleneck and source starvation," in *IEEE IEDM Technical Digest*, pp. 109–112, 2007.

[30] I. Riolino, M. Braccioli, L. Lucci, *et al.*, "Monte Carlo simulation of decananometric nMOSFETS: Multi-subband vs. 3D-electron gas with quantum corrections," *Solid State Electronics*, vol. 51, pp. 1558–1564, 2007.

[31] C. Jungemann, S. Yamaguchi, and H. Goto, "Convergence estimation for stationary ensemble Monte Carlo simulations," in *Proc.SISPAD*, p. 209, 1997.

[32] A. Phillips and P.J. Price, "Monte Carlo calculations of hot electron energy tails," *Applied Physics Letters*, vol. 30, pp. 528–530, 1977.

[33] C. Jungemann, S. Decker, R. Thoma, W.L. Engl, and H. Goto, "Phase space multiple refresh: a general purpose statistical enhancement technique for MC device simulation," in *Proc.SISPAD*, pp. 65–66, 1996.

[34] A. Pacelli and U. Ravaioli, "Analysis of variance-reduction schemes for ensemble Monte Carlo simulation of semiconductor devices," *Solid State Electronics*, vol. 41, no. 4, pp. 599–605, 1997.

[35] M.G. Gray, T.E. Booth, J.T. Kwan, and C.M. Snell, "A multicomb variance reduction scheme for Monte Carlo semiconductor simulators," *IEEE Trans. on Electron Devices*, vol. 45, no. 4, pp. 918–923, 1998.

[36] C. Jungemann, S. Decker, R. Thoma, W.L. Engl, and H.Goto, "Phase space multiple refresh: a general purpose statistical enhancement technique for MC device simulation," *IEEE J. Tech. Comp. Aided Design*, no. 2, 1997.

[37] S. Ramo, "Currents induced by electron motion," *Proc.IRE*, p. 584, 1939.

[38] W. Shockley, "Currents to conductors induced by a moving point charge," *Journal of Applied Physics*, vol. 9, p. 635, 1938.

[39] H. Kim, H. Min, T. Tang, and Y. Park, "An extended proof of the Ramo-Shockley theorem," *Solid State Electronics*, vol. 34, no. 11, p. 1251, 1992.

[40] P. Graf, S. Keith, and B. Meinerzhagen, "Evaluation of solenoidal and statistically enhanced total current densities in multi-particle Monte Carlo device simulators," in *Proc.SISPAD*, p. 221, 1997.

[41] J.M. Higman, K. Hess, C.G. Hwang, and R.W. Dutton, "Coupled Monte Carlo–Drift–Diffusion analysis of hot-electron effects in MOSFET's," *IEEE Trans. on Electron Devices*, vol. 36, p. 930, 1989.

[42] C. Jungemann, S. Yamaguchi, and H. Goto, "Efficient full band Monte Carlo hot carrier simulation for silicon devices," in *Proc. European Solid State Device Res. Conf.*, pp. 821–824, 1996.

[43] K. Hess, ed., *Monte Carlo Device Simulation: Full Band and Beyond.* Norwel, MA: Kluwer, 1991.

[44] T. Kunikiyo, M. Takenaka, Y. Kamakura, *et al.*, "Monte Carlo simulation of anisotropic electron transport in silicon including full band structure and anisotropic impact-ionization model ," *Journal of Applied Physics*, vol. 75, no. 1, pp. 297–312, 1994.

[45] C. Jungemann, *Efficient Full-Band Monte Carlo Simulation for Device Engineering.* Berlin: Springer-Verlag, 2001.

[46] F.M. Bufler, A. Schenk, and W. Fichtner, "Efficient Monte Carlo device modeling," *IEEE Trans. on Electron Devices*, vol. 47, pp. 1891–1897, 2000.

[47] J. Bude and R.K. Smith, "Phase-space simplex Monte Carlo for semiconductor transport," *Semiconductor Science Technology*, vol. 9, p. 840, 1994.

[48] F. Venturi and A. Ghetti, "Assessment of accuracy limitations of full band Monte Carlo device simulation," in *Proc.SISPAD*, p. 343, 1997.

[49] D.K. Ferry, R. Akis, and D. Vasileska, "Quantum effects in MOSFETs: Use of an effective potential in 3D Monte Carlo simulation of ultra-short channel devices," in *IEEE IEDM Technical Digest*, pp. 287–290, 2000.

[50] H. Tsuchiya, B. Fischer, and K. Hess, "A full-band Monte Carlo model for silicon nanoscale devices with a quantum mechanical correction of the potential," in *IEEE IEDM Technical Digest*, pp. 283–286, 2000.

[51] B. Winstead and U. Ravaioli, "A quantum correction based on Schrödinger equation applied to Monte Carlo device simulation ," *IEEE Trans. on Electron Devices*, vol. 50, pp. 440–446, 2003.

[52] X.F. Fan, X. Wang, B. Winstead, *et al.*, "MC simulation of strained-Si MOSFET with full-band structure and quantum correction," *IEEE Trans. on Electron Devices*, vol. 51, no. 6, pp. 962–970, 2004.

[53] C. Sampedro-Matarin, F. Gamiz, A. Godoy, and F.J. Gracia Ruiz, "The multivalley effective conduction band-edge method for Monte Carlo simulation of nanoscale structures," *IEEE Trans. on Electron Devices*, vol. 53, pp. 2703–2710, 2006.

[54] P. Palestri, D. Esseni, A. Abramo, R. Clerc, and L. Selmi, "Carrier quantization in SOI MOSFETs using an effective potential based Monte-Carlo tool," in *Proc. European Solid State Device Res. Conf.*, pp. 407–410, 2003.

[55] M.G. Ancona and G.J. Iafrate, "Quantum correction to the equation of state of an electron gas in a semiconductor," *Phys. Rev. B*, vol. 39, no. 13, pp. 9536–9540, 1989.

[56] A. Wettstein, A. Schenk, and W. Fichtner, "Quantum device-simulation with the density-gradient model on unstructured grids," *IEEE Trans. on Electron Devices*, vol. 48, no. 2, pp. 279–283, 2001.

[57] S.M. Ramey and D.K. Ferry, "Implementation of surface roughness scattering in Monte Carlo modeling of thin SOI MOSFETs using the effective potential," *IEEE Trans. on Nanotechnology*, vol. 2, no. 2, pp. 110–114, 2003.

[58] P. Palestri, S. Eminente, D. Esseni, *et al.*, "An improved semiclassical Monte-Carlo approach for nano-scale MOSFET simulation," *Solid State Electronics*, vol. 49, no. 5, pp. 727–732, 2005.

[59] S.E. Laux, "On particle-mesh coupling in Monte Carlo semiconductor device simulation," *IBM Research Report*, 1995.

[60] C. Jungemann and B. Meinerzhagen, "Analysis of the stochastic error of stationary Monte Carlo device simulation," *IEEE Trans. on Electron Devices*, vol. 48, no. 5, pp. 985–992, 2001.

[61] C. Jungemann and B. Meinerzhagen, "On the applicability of non-self-consistent Monte Carlo device simulations," *IEEE Trans. on Electron Devices*, vol. 49, no. 6, pp. 1072–1074, 2002.

[62] P. Palestri, D. Esseni, S. Eminente, *et al.*, "Understanding quasi-ballistic transport in nano-MOSFETs. Part I: Scattering in the channel and in the drain," *IEEE Trans. on Electron Devices*, vol. 52, no. 12, pp. 2727–2735, 2005.

[63] P. Palestri, N. Barin, D. Esseni, and C. Fiegna, "Stability of self-consistent Monte Carlo simulations: Effects of the grid size and of the coupling scheme," *IEEE Trans. on Electron Devices*, vol. 53, no. 6, pp. 1433–1442, 2006.

[64] P. Palestri, N. Barin, D. Esseni, and C. Fiegna, "Revised stability analysis of the non-linear Poisson scheme in self-consistent Monte-Carlo device simulations," *IEEE Trans. on Electron Devices*, vol. 53, no. 6, pp. 1443–1451, 2006.

[65] F.M. Bufler, C. Zechner, A. Schenk, and W. Fichtner, "Self-consistent single-particle simulation," in *Proc.SISPAD*, pp. 159–162, 2002.

[66] F.M. Bufler, C. Zechner, A. Schenk, and W. Fichtner, "Single-particle approach to self-consistent Monte Carlo device simulation," *IEICE Transactions on Electronics*, vol. E86-C, pp. 308–313, 2003.

[67] Z. Ren, R. Venugopal, S. Goasguen, S. Datta, and M.S. Lundstrom, "nanomos 2.5: A two-dimensional simulator for quantum transport in double-gate MOSFETS," *IEEE Trans. on Electron Devices*, vol. 50, no. 9, pp. 1914–1925, 2003.

[68] P. Palestri, L. Lucci, S. D. Tos, D. Esseni, and L. Selmi, "An improved empirical approach to introduce quantization effects in the transport direction in multi-subband Monte Carlo simulations," *Semiconductor Science and Technology*, vol. 25, p. 055011, 2010.

[69] W.J. Gross, D. Vasileska, and D.K. Ferry, "A novel approach for introducing the electron–electron and the electron–impurity interactions in particle-based simulations," *IEEE Electron Device Lett.*, vol. 20, pp. 463–465, 1999.

[70] C.J. Wordelman and U. Ravaioli, "Integration of particle-particle-particle-mesh algorithm with the ensemble Monte Carlo method for the simulation of ultra-small semiconductor devices," *IEEE Trans. on Electron Devices*, vol. 47, no. 2, pp. 410–416, 2000.

[71] W.J. Gross, D. Vasileska, and D.K. Ferry, "Ultra-small MOSFETs: The importance of the full Coulomb interaction on device characteristics," *IEEE Trans. on Electron Devices*, vol. 47, no. 10, pp. 1831–1837, 2000.

[72] K. Nakanishi, T. Uechi, and N. Sano, "Self-consistent Monte Carlo device simulations under nano-scale device structures: Role of Coulomb interaction, degeneracy, and boundary condition," in *IEEE IEDM Technical Digest*, pp. 79–82, 2009.

[73] P. Lugli and D.K. Ferry, "Investigation of plasmon-induced losses in quasi-ballistic transport," *IEEE Electron Device Lett.*, vol. EDL-6, no. 1, pp. 25–27, 1985.

[74] A. Abramo, R. Brunetti, C. Jacoboni, F. Venturi, and E. Sangiorgi, "A multiband Monte Carlo approach to Coulomb interaction for device analysis," *Journal of Applied Physics*, vol. 76, p. 5786, 1994.

[75] M.V. Fischetti and S.E. Laux, " Long-range Coulomb interactions in small Si devices. Part I: Performance and reliability," *Journal of Applied Physics*, vol. 89, no. 2, pp. 1205–1231, 2001.

[76] P.W. Rambo and J. Denavit, "Time stability of Monte Carlo device simulation," *IEEE Trans. on Computer-Aided Design*, vol. 12, p. 1734, 1993.

[77] A. Ghetti, X. Wang, F. Venturi, and F.A. Leon, "Stability issues in self-consistent Monte Carlo–poisson simulations," in *Simulation of Semiconductor Devices and Process*, v.6, H.Ryssel and P.Pichler (eds.), pp. 388–391. New York: Springer-Verlag, 1995.

[78] M.V. Fischetti and S.E. Laux, "Monte Carlo study of electron transport in silicon inversion layers," *Phys. Rev. B*, vol. 48, pp. 2244–2274, 1993.

[79] C. Jungemann, A. Edmunds, and W.L. Engl, "Simulation of linear and nonlinear electron transport in homogeneous silicon inversion layers," *Solid State Electronics*, vol. 36, no. 11, pp. 1529–1540, 1993.

[80] F. Gamiz, J.A. Lopez-Villanueva, J.B. Roldan, J.E. Carceller, and P. Cartujo, "Monte Carlo simulation of electron transport properties in extremely thin SOI MOSFET's," *IEEE Trans. on Electron Devices*, vol. 45, no. 5, pp. 1122–1126, 1998.

[81] S.C. Williams, K.W. Kim, and W.C. Holton, "Ensemble Monte Carlo study of channel quantization in a 25nm n-MOSFET," *IEEE Trans. on Electron Devices*, vol. 47, no. 10, pp. 1864–1872, 2000.

[82] T. Ezaki, P. Werner, and M. Hane, "Self-consistent quantum mechanical Monte Carlo MOSFET device simulation," *Journal of Computational Electronics*, vol. 2, no. 2–4, pp. 97–103, 2003.

[83] L. Lucci, P. Palestri, D. Esseni, and L. Selmi, "Multi-subband Monte Carlo modeling of nano-MOSFETs with strong vertical quantization and electron gas degeneration," in *IEEE IEDM Technical Digest*, pp. 631–634, 2005.

[84] J. Saint-Martin, A. Bournel, F. Monsef, C. Chassat, and P. Dollfus, "Multi sub-band Monte Carlo simulation of an ultra-thin double gate MOSFET with 2D electron gas ," *Semiconductor Science and Technology*, vol. 21, pp. L29–L31, 2006.

[85] D. Esseni and P. Palestri, "Theory of the motion at the band crossing points in bulk semiconductor crystals and in inversion layers," *Journal of Applied Physics*, vol. 105, no. 5, pp. 053702–1–053702–11, 2009.

[86] L. Lucci, P. Palestri, D. Esseni, L. Bergagnini, and L. Selmi, "Multi-subband Monte-Carlo study of transport, quantization and electron gas degeneration in ultra-thin SOI n-MOSFETs," *IEEE Trans. on Electron Devices*, vol. 54, no. 5, pp. 1156–1164, 2007.

[87] M.V. Fischetti, Z. Ren, P.M. Solomon, M. Yang, and K. Rim, "Six-Band $k \cdot p$ calculation of the hole mobility in silicon inversion layers: Dependence on surface orientation, strain, and silicon thickness," *Journal of Applied Physics*, vol. 94, pp. 1079–1095, July 2003.

[88] A.T. Pham, C. Jungemann, and B. Meinerzhagen, "Modeling of hole inversion layer mobility in unstrained and uniaxially strained Si on arbitrarily oriented substrates," in *Proc. European Solid State Device Res. Conf.*, pp. 390–393, 2007.

[89] L. Donetti, F. Gamiz, A. Godoy, and N. Rodriguez, "Fully self-consistent $\mathbf{k} \cdot \mathbf{p}$ solver and Monte Carlo simulator for hole inversion layers," in *Proc. European Solid State Device Res. Conf.*, pp. 254–257, 2008.

[90] M. De Michielis, D. Esseni, P. Palestri, and L. Selmi, "A new multi subband Monte Carlo simulator for nano p-MOSFETs," in *Proc. Int. Conf. on Ultimate Integration on Silicon (ULIS)*, pp. 67–70, 2008.

[91] M. De Michielis, D. Esseni, P. Palestri, and L. Selmi, "Semi-classical modelling of quasi-ballistic hole transport in nanoscale p-MOSFETs based on a multi-subband Monte Carlo approach," *IEEE Trans. on Electron Devices*, vol. 56, no. 9, pp. 2081–2091, 2009.

[92] W. Chen, L. Register, and S.K. Banerjee, "Simulation of quantum effects along the channel of ultrascaled Si-based MOSFETs," *IEEE Trans. on Electron Devices*, vol. 49, no. 4, pp. 652–657, 2002.

[93] L. Shifren, C. Ringhofer, and D.K. Ferry, "A Wigner function-based quantum ensemble Monte Carlo study of a resonant tunneling diode," *IEEE Trans. on Electron Devices*, vol. 50, no. 3, pp. 769–773, 2003.

[94] V. Sverdlov, A. Gehring, H. Kosina, and S. Selberherr, "Quantum transport in ultra-scaled double-gate MOSFETs: A Wigner function-based Monte Carlo approach," *Solid State Electronics*, vol. 49, pp. 1510–1515, 2005.

[95] D. Querlioz, P. Dollfus, V.-N. Do, A. Bournel, and V. Nguyen, "An improved Wigner Monte Carlo technique for the self-consistent simulation of RTDs," *Journal of Computational Electronics*, vol. 5, pp. 443–446, 2006.

[96] D. Querlioz, J. Saint-Martin, K. Huet, *et al.*, "On the ability of particle Monte Carlo technique to include quantum effects in nano-MOSFET simulation," *IEEE Trans. on Electron Devices*, vol. 54, no. 9, pp. 2232–2242, 2007.

[97] P. Dollfus, D. Querlioz, J. Saint-Martin, V.-N. Do, and A. Bournel, "Wigner Monte Carlo approach to quantum transport in nanodevices," in *Proc. SISPAD*, pp. 277–280, 2008.

[98] D. Ventura, A. Gnudi, G. Baccarani, and F. Odeh, "Multidimensional spherical harmonics expansion of Boltzmann equation for transport in semicondutors," *Appl. Math. Lett.*, vol. 5, pp. 85–90, 1992.

[99] A. Gnudi, D. Ventura, G. Baccarani, and F. Odeh, "Two-dimensional MOSFET simulation by means of a multidimensional spherical harmonics expansion of the Boltzmann transport equation," *Solid State Electronics*, vol. 36, pp. 575–581, 1993.

[100] C. Jungemann, A.T. Pham, B. Meinerzhagen, C. Ringhofer, and M. Bollhofer, "Stable discretization of the Boltzmann equation based on spherical harmonics, box integration, and a maximum entropy dissipation principle," *Journal of Applied Physics*, vol. 100, no. 2, pp. 024502–024502–13, 2006.

[101] A.T. Pham, C. Jungemann, and B. Meinerzhagen, "Deterministic multisubband device simulations for strained double gate PMOSFETs including magnetotransport," in *IEEE IEDM Technical Digest*, pp. 895–898, 2008.

[102] M. Lenzi, P. Palestri, E. Gnani, *et al.*, "Investigation of the transport properties of silicon nanowires using deterministic and Monte Carlo approaches to the solution of the Boltzmann transport equation," *IEEE Trans. on Electron Devices*, vol. 55, no. 8, pp. 2086–2096, 2008.

7 Simulation of bulk and SOI silicon MOSFETs

The semi-classical transport model for inversion layers developed in the previous chapters finds a natural field of application in the analysis of advanced nano-scale MOSFETs.

In this chapter we illustrate the ability of this model to describe low and high field transport in unstrained (001) silicon. To this purpose, extensive simulations are presented for the low field effective mobility of long channel devices and for the on-current of short bulk and SOI transistors. The first half of the chapter describes how the multiple and complex dependencies of the mobility on the bias, channel doping, silicon film thickness and temperature relate to the physical ingredients of the inversion layer transport model, namely the discrete energy levels, the occupation probability and the scattering rates in the subbands.

The second half of the chapter covers high field transport in uniform silicon slabs and in short channel MOSFETs. The quasi-ballistic transport model outlined in Chapter 5 guides the interpretation of the transistor simulations.

The results proposed in this chapter set a reference for the analysis of more complex cases of interest for advanced CMOS technologies. In particular, the impact on carrier transport of technology boosters such as crystal orientation, strain and alternative materials are analyzed in Chapters 8, 9, and 10, respectively.

7.1 Low field transport

7.1.1 Measurement and representation of mobility data

Low field transport in uniformly doped bulk cubic semiconductors such as unstrained crystalline silicon is characterized by a scalar conductivity mobility, which is defined as the ratio between the carrier drift velocity ($\mathbf{V}$) and the electric field ($\mathbf{F}$) in the limit of a vanishingly small driving field:

$$\mu = \lim_{|\mathbf{F}|\to 0} \frac{|\mathbf{V}|}{|\mathbf{F}|}. \tag{7.1}$$

The carrier mobility and the velocity versus field curves of the most common bulk semiconductors have been measured by different authors using, e.g., time of flight techniques [1, 2], and are reported in many semiconductor device physics textbooks and review

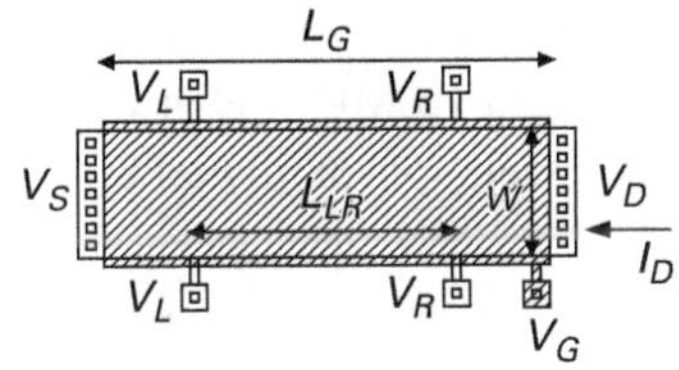

Figure 7.1 Test structure for reliable effective mobility measurements with the split CV method. The pairs of V_L and V_R terminals are highly doped diffusions of the same type as the source and drain and provide lateral contacts to the inversion layer to measure the intrinsic voltage drop $V_{DS,int} = V_R - V_L$ free of series resistance effects.

papers [3–5]. In undoped silicon, low field mobility values of 1420 cm^2/Vs for electrons and 470 cm^2/Vs for holes are found at room temperature. A remarkable reduction of the mobility is observed at substrate doping concentrations exceeding 10^{18} cm^{-3} [6] due to ionized impurity and carrier–carrier scattering. The mobility decreases for increasing temperature mostly because of the increase of the phonon scattering rate.

When considering low field transport in inversion layers, the carrier mobility is most frequently characterized by means of the split CV measurement technique [7–9]. Figure 7.1 shows a typical test structure for accurate inversion layer mobility measurements [10]. It consists of a long channel MOSFET equipped with direct contacts to the inversion layer located close to the source and drain [11]. The drain current I_D and the intrinsic drain-source voltage $V_{RL} = (V_R - V_L)$ are measured for a low external drain voltage V_{DS}. The inversion charge per unit area ($Q_{inv} = -eN_{inv}$ and eP_{inv} for n-MOS and p-MOS devices, respectively) is obtained by integration of the differential gate-source capacitance curve $C_{GS} = |\mathrm{d}Q_G/\mathrm{d}V_{GS}|$ measured by the split CV technique. The effective mobility is then calculated as:

$$\mu_{eff} = \frac{L_{LR}}{W} \frac{I_D}{V_{RL}|Q_{inv}|}. \tag{7.2}$$

If the device has a relatively large area and the layout is carefully designed, the parasitic edge capacitance is negligible, while the parasitic series resistance is de-embedded from the measurements thanks to the V_L and V_R contacts. Accurate values of V_{RL} and Q_{inv} are thus obtained.

The main limitation of the split CV technique is that, since L_G is large and the gate dielectric may be thin, the parasitic gate leakage current ($I_G \propto WL_G$) can offset the small value of the drain current at low drain voltage ($I_G/I_D \propto L_G^2$ [12, 13]), generating non-negligible errors on I_D and μ_{eff}. Indeed, gate leakage ultimately limits the bias range where reliable determination of μ_{eff} can be achieved [14]. Moreover, if the device is too long, the measured C_{GS} is affected by distributed RC effects and the frequency of the small AC signal must be proportionally reduced during the measurements to avoid systematic errors. In these situations, suitable corrections to the experimental data can help to reconstruct more accurate values of I_D/V_{RL} and Q_{inv} [9, 14, 15].

An alternative technique to characterize low field transport resorts to the geometrical magnetoresistive effect [7, 16, 17], by which the resistivity of a semiconductor increases with an externally applied magnetic induction field (B) normal to the transport plane.

If the device geometry is such that the Hall voltage is effectively shorted out (e.g., in relatively short and large MOSFETs with $W/L \gtrsim 5$ [7]), the current density in the transport direction is well approximated by

$$J_x = \sigma_{xx} F_x = \frac{\sigma_0}{1 + \mu_{mr}^2 B^2} F_x, \tag{7.3}$$

where μ_{mr} is the magnetoresistance mobility and $\sigma_0 = eN_{inv}\mu_{xx}$. Straightforward calculations based on the charge sheet approximation [5] lead us to express the change of the device resistance $R = [V_{DS}/I_D]$ produced by the magnetic induction field B as

$$\frac{R(B) - R(0)}{R(0)} = 1 + \mu_{mr,eff}^2 B^2, \tag{7.4}$$

where $\mu_{mr,eff}$ is the effective magnetoresistance mobility of the inversion layer. The $\mu_{mr,eff}$ is thus easily extracted from the relative change of the device resistance induced by the magnetic induction field.

The geometrical condition on W/L makes magnetoresistance measurements applicable to short devices; the regime of small inversion charge values can be easily explored too. However, the parasitic source and drain resistances have a large impact on the current of short devices and they also depend on the magnetic induction field B. A reliable de-embedding of the series resistances is not straightforward [17].

Energy dependent scattering rates cause a discrepancy between the effective mobility μ_{eff} (Eq.7.2) and the effective magnetoresistance mobility $\mu_{mr,eff}$ (Eq.7.4). The relation between the two mobilities is not trivial and has been investigated with the MRT approach and with multi-subband Monte Carlo simulations [7, 18–20]. In this respect we note that the quantum energy levels and wave-functions are essentially not affected by a magnetic induction field B perpendicular to the transport plane for the B magnitudes typically used in the magnetoresistance experiments (up to approximately 10 Tesla [17]). Therefore, the effect of the magnetic induction field is introduced in the multi-subband Monte Carlo transport model simply by modifying the expression for the driving force of particle motion (Eqs.2.89b and 5.8) by adding the Lorenz's force term

$$\mathbb{F} = \pm e(\mathbf{F} + \mathbf{v}_g \times \mathbf{B}). \tag{7.5}$$

Studies of the inversion layer magnetoresistance based on multi-subband Monte Carlo simulations [20–22] suggest that $\mu_{mr,eff}$ tends to μ_{eff} at low temperature while it can be from 20% to 30% higher than μ_{eff} at room temperature. Since the effective mobility determined by electrical measurements on inversion layers is the physical parameter governing low field transport in the practical situations of interest for electronic devices, in the rest of this chapter we restrict our analysis uniquely to μ_{eff} as extracted by means of Eq.7.2.

The inversion layer effective mobility of unstrained silicon MOSFETs measured by the split CV method is significantly smaller than the bulk silicon mobility and depends on the gate voltage, the oxide thickness, the quality of the interface, the substrate doping, the substrate bias and the temperature. A major step in understanding the inversion layer mobility was achieved in [23], where it was observed that μ_{eff} values extracted from devices fabricated with different processes lie on a unique curve if plotted as a function of the so-called *effective field*. With reference to an n-MOSFET, the effective field is

$$F_{eff} = -\frac{1}{\epsilon_{Si}} \left(Q_B + \eta Q_{inv} \right), \tag{7.6}$$

where Q_B and Q_{inv} are the (negative) bulk and inversion charge per unit area, respectively, while $\eta = 1/2$. The effective field was introduced as an estimate of the average component of the field in the quantization direction

$$F_{ave} = \frac{\int_0^{z_0} n(z) F_z(z) \mathrm{d}z}{\int_0^{z_0} n(z) \mathrm{d}z} = -\frac{1}{N_{inv}} \int_0^{z_0} n(z) \left(\frac{\mathrm{d}\phi(z)}{\mathrm{d}z} \right) \mathrm{d}z, \tag{7.7}$$

where z_0 should be taken in a region of the semiconductor substrate where either $n(z)$ or $F_z(z)$ have decayed to zero. In fact, still with reference to the n-MOSFET case, integration along the z direction of the one dimensional Poisson equation gives:

$$F_{ave} = -\frac{\epsilon_{Si}}{2Q_{inv}} \left[F_z^2(z_0) - F_z^2(0) \right] - \frac{eN_A}{Q_{inv}} \left[\phi(z_0) - \phi(0) \right]. \tag{7.8}$$

In a bulk MOSFET and in a symmetric double gate SOI MOSFET we can always set $F_z(z_0) = 0$ by taking z_0 at the bottom edge of the depletion region or at $z_0 = T_{Si}/2$, respectively. Consequently, Gauss's law gives $F_z(0) = -(Q_B + Q_{inv})/\epsilon_{Si}$ where in the double gate case Q_B and Q_{inv} refer to one channel. If we now assume that the depletion charge in the thin interfacial region occupied by the inversion layer is negligible, then we can express $[\phi(z_0) - \phi(0)]$ in terms of Q_B through the relations $Q_B \simeq -eN_A z_d$ where the depletion depth z_d is given by $z_d \simeq \sqrt{2\epsilon_{Si}[\phi(0) - \phi(z_0)]/eN_A}$. Straightforward calculations lead to

$$F_{ave} = -\frac{1}{\epsilon_{Si}} \left(Q_B + \frac{Q_{inv}}{2} \right), \tag{7.9}$$

that is, F_{ave} coincides with the F_{eff} in Eq.7.6 calculated for $\eta = 1/2$.

The relation between F_{eff} and N_{inv} in bulk and in undoped thin body SOI n-MOSFETs is exemplified by the curves with filled symbols in Fig.7.2. We observe that for electrons in undoped channel SOI MOSFETs (right plot) $F_{eff} \simeq F_{ave}$, as expected from the calculation above. In bulk n-MOSFETs with high channel doping instead (left plot), F_{eff} deviates from F_{ave} because the depth of the inversion layer is a non-negligible fraction of the depletion depth.

For the purpose of plotting the experimental mobility data, Q_{inv} is determined from the measured C_{GS} versus V_{GS}, while the bulk charge Q_B is either calculated by integration of the differential gate-bulk capacitance curve ($C_{GB} = |\mathrm{d}Q_G/\mathrm{d}V_{GB}|$) or estimated by means of the analytical expression $|Q_B| = eN_{dop}z_d$ (where z_d is the depletion depth and $N_{dop} = N_A$ in the n-MOSFET case), or derived by more accurate numerical simulations.

Measurements of the inversion layer effective mobility such as those in [23] have been extended to holes in [24] and later carefully repeated on a wide set of advanced devices in [25]. These latter measurements are shown in Fig.7.3, where appropriate

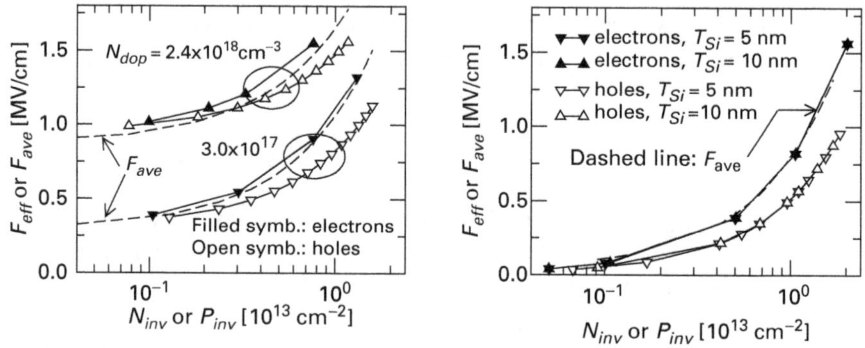

Figure 7.2 Effective field F_{eff} (symbols, Eq.7.6) and average field F_{ave} (dashed line, Eq.7.7) as a function of inversion charge density N_{inv} or P_{inv} for bulk MOSFETs with different substrate doping N_{dop} (left) and for undoped single gate SOI MOSFETs with different silicon film thickness T_{Si} (right). Filled symbols for electrons, open symbols for holes.

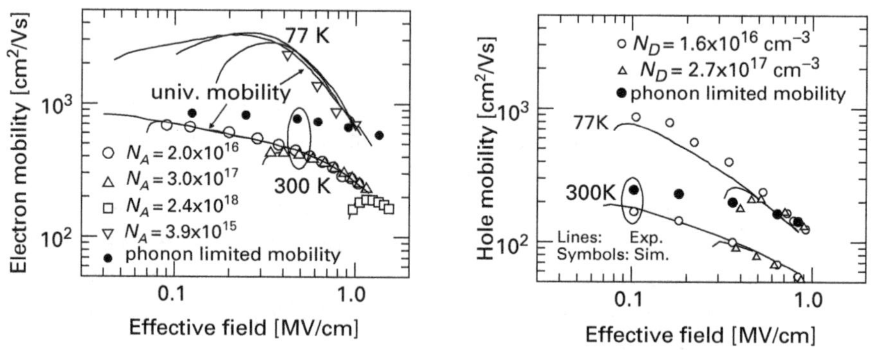

Figure 7.3 Lines: measured electron (left) and hole (right) μ_{eff} as a function of F_{eff}. Data from [25]. Symbols: MSMC simulations with the EMA model for electrons (Section 3.2.2) and with the analytical band model for holes (Section 3.3.3). The scattering model parameters are those in Table 7.1.

sign exchanges have been made to maintain F_{eff} and F_{ave} as positive quantities in the p-MOSFET case also.

The apparently universal behavior of inversion layer mobility as found in [23] is confirmed over a broad range of the effective field, substrate doping and temperature. In fact, the experimental μ_{eff}–F_{eff} curves lie on top of one another and seem to define a universal upper limit for inversion layer mobility in unstrained silicon which is commonly referred to as the *universal mobility* curve.

Quite interestingly, however, the universal behavior of the hole mobility is achieved for $\eta = 1/3$. It is clear from the derivation in the previous section that if $\eta \neq 1/2$ the effective field and the average field over the inversion layer are different, as also visible from the curves with open symbols in Fig.7.2.

The original idea behind the introduction of the effective field as the independent variable to represent the mobility data is that, for a given F_{eff}, the shape of the confining potential energy well is roughly fixed [23]. Therefore, by comparing the μ_{eff} of different samples at a given F_{eff}, the effect of the wave-function shape on the scattering rates is approximately compensated; any residual discrepancy in the mobility of the examined samples should be due to the substrate doping, to the quality of the interface, or to different scattering mechanisms [23]. Since μ_{eff} is by no means universal if plotted versus quantities such as V_G, N_{inv}, or the field at the interface $F_z(0)$, the universal mobility curve versus the effective field F_{eff} defined in Eq.7.6 brings indirect evidence of the reproducible quality of the fabricated Si–SiO_2 interfaces and of the importance of quantum mechanical effects on the carrier transport in inversion layers.

However, it should be remarked that while in an undoped n-MOSFET the effective field with $\eta = 1/2$ essentially coincides with the average field, in Eq.7.7 (see the right plot in Fig.7.2), different and theoretically not yet justified values of η are necessary to obtain a universal F_{eff}–μ_{eff} relation for holes (Fig.7.3 and [24]) and for crystal orientations other than (001), as discussed in more detail in Chapter 8. The effective field should thus be seen as an *empirical* parameter, useful to achieve a simple description of the inversion layer mobility, rather than a physically based parameter. The possibility of identifying an effective field that may result in a universal mobility curve in strained silicon and in other semiconductor materials is still under investigation [26–28].

What is most relevant from an applicative point of view is the value of the inversion layer mobility at a given inversion carrier density N_{inv} or P_{inv}. Some ambiguity in the representation of the μ_{eff}–N_{inv} curve may arise for planar double gate devices when it is not clear if the reported density refers to the total inversion density in the silicon film or to the inversion density "per gate". A similar ambiguity may arise for multiple gate and gate-all-around architectures, since the definition of the gate periphery is not always straightforward in these devices [29], so that it is sometimes more practical to normalize the current "per finger" or "per wire". Unless otherwise stated, in this book we use the inversion density "per gate" when reporting the mobility data.

7.1.2 Low field mobility in bulk devices

Having clarified the meaning and the limitations behind the concept of effective field, we can now analyze the data in Fig.7.3. First, we observe that for a given substrate doping three separate regions can be identified on the experimental curves: in the central region a modest degradation of the mobility is found upon increase of the effective field. At low F_{eff} (that is at low Q_{inv} values corresponding to the near threshold, weak inversion region) μ_{eff} is an increasing function of F_{eff}. At high F_{eff} values, instead, μ_{eff} degrades rapidly upon further increase of the F_{eff}.

Figure 7.3 also shows the results of MSMC calculations which are in good agreement with the experiments over a broad range of F_{eff}, temperature and substrate doping. A unique set of model parameters, reported in Table 7.1, has been used for these simulations. The parameters governing the numerics of the calculations (number of particles, energy bins, etc.) instead, have been separately optimized at 300 and 77 K.

It is worth noting that the calibration of the electron and hole phonon and surface roughness scattering models for inversion layers does not yield a unique set of model parameters, and open issues still exist related to their values [30]. For instance, the

Table 7.1 Model parameters used for the simulations reported in this chapter.

		Electrons			
Phonon	intra-v. acoustic	$\hbar\omega$=0	[meV]	D_{ac}=13.0	[eV]
	inter-v. g-type, TA	$\hbar\omega_{op}$=12	[meV]	D_{op}=0.5 $\cdot 10^8$	[eV/cm]
	inter-v. g-type, LA	$\hbar\omega_{op}$=18.5	[meV]	D_{op}=0.8 $\cdot 10^8$	[eV/cm]
	inter-v. g-type, LO	$\hbar\omega_{op}$=61.2	[meV]	D_{op}=11 $\cdot 10^8$	[eV/cm]
	inter-v. f-type, TA	$\hbar\omega_{op}$=19	[meV]	D_{op}=0.3 $\cdot 10^8$	[eV/cm]
	inter-v. f-type, LA	$\hbar\omega_{op}$=47.4	[meV]	D_{op}=2.0 $\cdot 10^8$	[eV/cm]
	inter-v. f-type, TO	$\hbar\omega_{op}$=59	[meV]	D_{op}=2.0 $\cdot 10^8$	[eV/cm]
Roughness	gauss spectrum	λ_{SR}=1.0	[nm]	Δ_{SR}=0.62	[nm]
		Holes			
Phonon	intra-v. acoustic	$\hbar\omega$=0	[meV]	D_{ac}=5.2	[eV]
	inter-v. optical	$\hbar\omega_{op}$=61.2	[meV]	D_{op}=11.5 $\cdot 10^8$	[eV/cm]
Roughness	exp. spectrum	λ_{SR}=2.6	[nm]	Δ_{SR}=0.5	[nm]

deformation potential $D_{ac} = 13\,\text{eV}$ for intra-valley acoustic phonon scattering needed to reproduce the experiments for electron inversion layers deviates remarkably from the $D_{ac} = 9\,\text{eV}$ value reported for bulk silicon [3, 31, 32].

Another aspect concerns the surface roughness scattering model parameters. Changes in the Δ_{SR} value cause rigid shifts of the μ_{eff}–F_{eff} curves while changes of λ_{SR} modify the slope of these curves as well. For a given crystal orientation and temperature, reasonable agreement with experiments can be achieved with both the Gaussian and the exponential roughness spectrum [33] discussed in Section 4.4.1. The issue of what is the most credible surface roughness spectrum has recently been addressed experimentally [34, 35]. The choice of the surface roughness spectrum and the calibration of the corresponding scattering model is especially relevant for the I_{ON} of modern nano-scale devices and is discussed again in Section 9.5 with reference to the simulation of strained silicon MOSFETs.

With regard to the methodology for the mobility calculations, Fig.7.4 compares simulated electron and hole mobilities computed with the momentum relaxation time approximation (Section 5.4.4) and with the multi-subband Monte Carlo method (Section 6.3) considering phonon and surface roughness scattering mechanisms. An excellent mutual agreement is typically observed between these two calculation techniques, which underlines the validity of the MRT approximation in the limit of low field transport. The small adjustment of the D_{ac} in the MRT calculations compensates the lack of inter-subband transitions in the hole surface roughness scattering and the slightly different energy dispersion models for electrons (parabolic and non-parabolic).

Momentum relaxation time calculations and multi-subband Monte Carlo simulations provide an effective means to interpret the experimental mobility curves of Fig.7.3 in terms of their constituent microscopic ingredients. These ingredients are the subband structure of the inversion layer, the relative population of the subbands, the scattering rates and the mobility in each subband (Section 5.4.4).

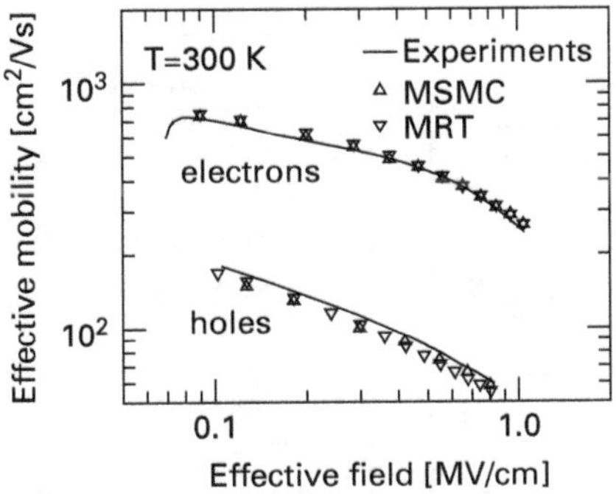

Figure 7.4 Comparison between effective mobility of unstrained bulk n-MOS ($N_A = 2.0 \cdot 10^{16}$ cm^{-3}) and p-MOS ($N_D = 1.6 \cdot 10^{16}$ cm^{-3}) transistors calculated with the MRT approximation (∇) and with a MSMC solution of the BTE (Δ). For the sake of a fair comparison between the models, n-MOSFET simulations assume parabolic bands. Model parameters are the same as in Table 7.1 except for $D_{ac} = 14.6$ eV and $D_{ac} = 5.6$ eV in the MRT calculations for electrons and holes, respectively.

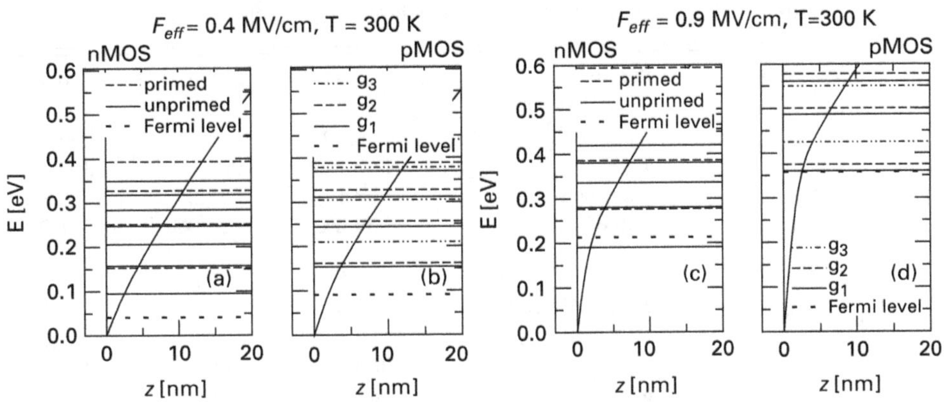

Figure 7.5 Subband structure of bulk n-MOS and p-MOS devices at $F_{eff} = 0.4$ ((a) and (b)) and 0.9 MV/cm ((c) and (d)). Due to the different value of η for electrons and holes the inversion charge density is not the same for n-MOS and p-MOS. $N_A = 3.0{\cdot}10^{17}$ cm^{-3} for n-MOS and $N_D = 2.7{\cdot}10^{17}$ cm^{-3} for p-MOS as in Fig.7.3. $T = 300$ K. The energy reference $E = 0$ is at the bottom of the bulk silicon band edge at the interface with the SiO_2.

Figures 7.5 and 7.6 illustrate the first of these ingredients, namely, the quantized energy levels in the electron and hole inversion layer of a few bulk n-MOS and p-MOS devices selected among those of Fig.7.3 at a few values of the effective field, inversion charge density and temperature. The simulations refer to the EMA quantization model for electrons and to the analytical valence band model of Section 3.3.3 for holes. As already noted in Section 3.6.1, we observe the presence of a significant splitting of the energy levels. The subband energy with respect to the extreme of the bulk silicon band edges and the separation in energy between the subbands increase with increase of the effective field (hence, of the inversion charge density) because of the narrowing of the potential energy well. Due to the different quantization effective masses, distinct

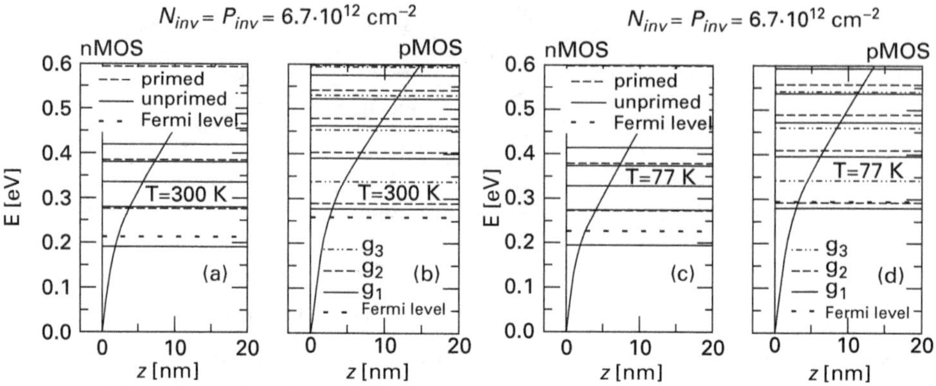

Figure 7.6 Subband structure of the same n-MOS and p-MOS devices as in Fig.7.5 at equal inversion charge density $N_{inv} = P_{inv} = 6.7 \times 10^{12}$ cm^{-2}. $F_{eff} = 0.9$ MV/cm for n-MOS and $F_{eff} = 0.63$ MV/cm for p-MOS. $T = 300$ K ((a) and (b)) and $T = 77$ K ((c) and (d)).

Table 7.2 Relative occupation of the lowest subbands for the devices and bias conditions in Figs.7.5 and 7.6. $\eta = 1/2$ for electrons and $\eta = 1/3$ for holes, respectively.

n-MOS	Figure	7.5.a	7.5.c	7.6.a	7.6.c
	F_{eff} [MV/cm]	0.4	0.9	0.9	0.89
	N_{inv} [10^{12} cm^{-2}]	0.79	6.7	6.7	6.7
	T [K]	300	300	300	77
	unprimed	62.1%	73.8%	73.8%	99.7%
	primed	30.2%	21.0%	21.0%	0.28%
p-MOS	Figure	7.5.b	7.5.d	7.6.b	7.6.d
	F_{eff} [MV/cm]	0.4	0.9	0.64	0.63
	P_{inv} [10^{12} cm^{-2}]	2.5	12.0	6.7	6.7
	T [K]	300	300	300	77
	g_1	61.2%	66.0%	64.5%	78.8%
	g_2	30.3%	28.0%	28.8%	21.2%
	g_3	5.3%	4.96%	5.13%	0.02%

ladders of subbands are formed by the different valleys, as explained in Section 3.2. The separation between the subbands is often larger than the room temperature thermal energy $K_B T$. Consequently, the occupation of the subbands in the close-to-equilibrium conditions that are typical of low field transport rapidly becomes vanishingly small for increasing energy and only the lowest subbands are occupied. Table 7.2 gives the occupation for the different groups of subbands, indicating that, especially for electrons, a non-negligible redistribution of the carriers takes place between the subbands for increasing F_{eff}, hence N_{inv}. It is apparent from Figs.7.5 and 7.6 that at large N_{inv} values the lowest subbands of the n-MOS devices lie below the Fermi energy, which implies that the electron inversion layer is heavily degenerate. At a given fixed inversion charge density, degeneracy is less evident in the hole inversion layer because of the larger density of states (see Fig.3.13).

For a given N_{inv} or P_{inv} and at low temperature the energy levels are almost the same as for $T = 300$ K, but the degeneracy effects are more pronounced because of the reduced value of the thermal energy $K_B T$ (see Fig.7.6.c and .d compared to Fig.7.6.a and .b). In many situations only the lowest subband is populated at low temperature, a condition which is commonly referenced as the *quantum limit* [36]. This situation is nearly reached in the n-MOS device of Fig.7.6 at 77 K, where the relative occupation of the lowest unprimed subband approaches 100% (see the rightmost column in Table.7.2).

In order to make one step further in the interpretation of the experimental mobility curves based on the MSMC simulations, we observe that the phonon-limited mobility μ_{ph} (filled symbols in Fig.7.3) has a weaker dependence on the effective field than the experimental effective mobility μ_{eff}. At inversion charge densities in the order of 10^{13} cm^{-2} (roughly corresponding to $F_{eff} \approx 1$ MV/cm) μ_{ph} is much larger than μ_{eff}, indicating that another scattering mechanism, namely surface roughness scattering, has become dominant.

To illustrate this aspect in detail, we examine in Fig.7.7 the phonon, surface roughness and Coulomb scattering limited mobility of one of the n-MOSFETs in Fig.7.3. Note that MSMC simulations cannot provide surface roughness or Coulomb scattering limited mobilities separately, because in the absence of dissipative scattering mechanisms the simulations do not converge to the correct carrier distributions. Therefore, to compute these mobilities we rely on MRT calculations. In view of the results in Fig.7.4, resorting to the MRT methodology for the mobility calculations does not infringe the validity of the analysis.

The MRT calculations of Fig.7.7 indicate that the modest reduction of the phonon limited mobility with increasing F_{eff} observed in Fig.7.3 is explained by the increase of the form factors for phonon scattering defined in Section 4.6.3 and Eq.4.268 and directly shown in the right plot of Fig.7.7.

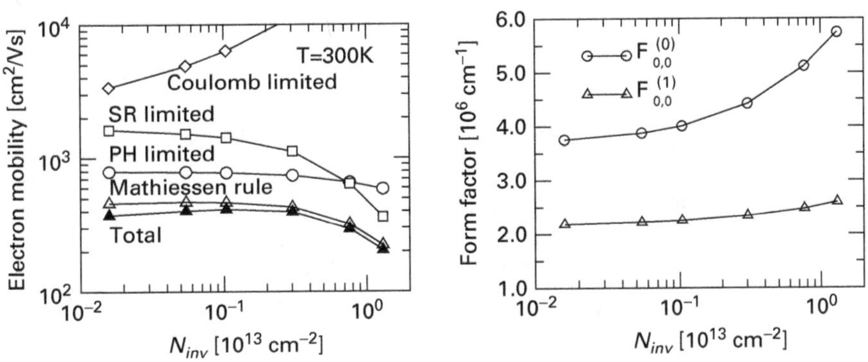

Figure 7.7 Left: phonon, surface roughness and Coulomb scattering limited mobility versus N_{inv} for the same n-MOSFET as in Fig.7.3. $N_A = 3{\cdot}10^{17}$ cm^{-3}. Also shown is the total mobility calculated with the MRT approximation and applying Mathiessen's rule to the scattering limited mobilities. Right: form factors for phonon scattering of the lowest unprimed ($F_{0,0}^{(0)}$) and the lowest primed ($F_{0,0}^{(1)}$) subbands.

The much stronger reduction in the experimental μ_{eff} than in the simulated μ_{ph} for increasing F_{eff} revealed in Fig.7.3 is explained by the observation that surface roughness scattering is already relevant at moderate effective field values, as is confirmed, for the n-MOSFET case, by the simulations in the left plot of Fig.7.7.

In the region at low F_{eff}, Coulomb scattering due to ionized substrate dopants and to charged interface states further reduces the mobility. The Coulomb scattering mechanism is weakened by the screening effect of the inversion charge. Consequently, the Coulomb scattering limited mobility increases for increasing N_{inv} until it becomes much larger than μ_{ph} and thus essentially non-influential on the total mobility. These aspects are clearly visible in the simulations of Fig.7.7 (left plot). The figure also shows with open triangles the total mobility resulting from the combination of the mobilities due to the individual scattering mechanisms according to the so-called Mathiessen's rule (Eq.5.117). The difference with respect to the total mobility deriving from the correct composition of the momentum relaxation times and subband occupation factors (curve with filled triangles) underlines once more the risks that may be incurred whenever Matthiessen's rule is applied without critical awareness of all its limitations.

7.1.3 Low field mobility in SOI devices

The excellent control of short channel effects achievable by means of ultra-thin body SOI transistor architectures makes them good candidates for ultimate nanometre scale CMOS technologies. Consequently, study of the transport properties of thin semiconductor films has attracted considerable attention in recent years [10, 37–40].

Thick SOI films (above ≈ 20 nm) exhibit essentially the same inversion layer mobility as bulk transistors, as is apparent from the close proximity of the curves with open and filled circles in the left plot of Fig.7.8. For T_{Si} below approximately 10 nm instead, experiments have repeatedly and unambiguously shown a reduction of the channel mobility and an appreciable increase of the threshold voltage V_T [38, 42–45] as

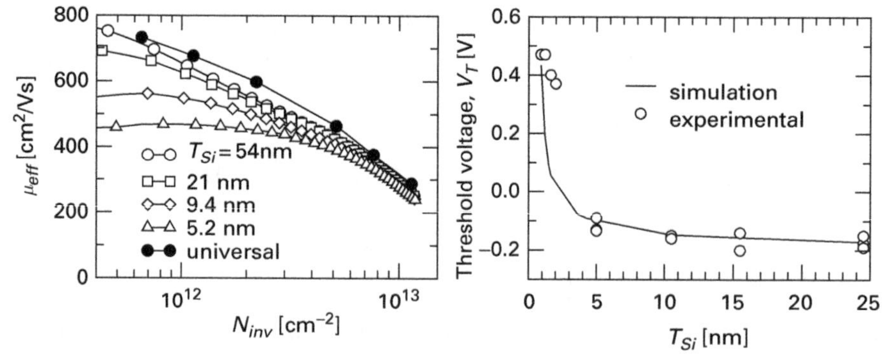

Figure 7.8 Left: experimental mobility versus inversion charge density in ultra-thin body single gate SOI MOSFETs. Data from [41]. Right: threshold voltage shift as a function of T_{Si} in ultra-thin body SOI MOSFETs. *Reprinted with permission from [42]. Copyright 2009 by the Institute of Electrical and Electronics Engineers.*

visible in Fig.7.8. These phenomena have been ascribed to a shift of the subband energy levels, change in the occupation of the subbands and modulation of the scattering rates induced by the quantum mechanical confinement of the carriers in the thin silicon layer.

The multi-subband Monte Carlo transport model for inversion layers and the MRT approximation are thus theoretical frameworks naturally suited to support the analysis and understanding of these effects. In particular, interpretation of the effective mobility curves in unstrained ultra-thin SOI MOSFETs can be carried out along the same lines followed for bulk devices. However, we should also take into account the impact of the silicon film thickness T_{Si}.

In this respect, Fig.7.9 shows experimental and simulated effective mobility as a function of the silicon film thickness for n-type and p-type SOI MOSFETs operated in single gate (SG) and double gate (DG) mode. The key feature common to these plots is the non-negligible mobility reduction for decreasing T_{Si}. A local maximum of the electron mobility is observed for T_{Si} in the 3-5 nm range.

Insight into the T_{Si} dependence of the electron μ_{eff} in n-MOSFETs can be gained thanks to the multi-subband inversion layer model [41, 47, 48]. To this purpose, Fig.7.10 shows the phonon (left graph) and surface roughness (right graph) scattering limited mobility of the n-MOS devices in the left graph of Fig.7.9, while Fig.7.11 shows the corresponding phonon scattering form factors (Eq.4.268) and the relative occupation of the unprimed and primed subbands.

We see that for large T_{Si} the primed subbands with smaller mobility contain the largest fraction of the total carrier population. The phonon limited μ_{eff} is reduced with decreasing T_{Si} below approximately 10 nm essentially because the corresponding form factors increase due to carrier confinement (left plot in Fig.7.11). At the same time, however, due to their small quantization mass, the primed subbands are pushed to high energy and are rapidly depopulated. For T_{Si} below approximately 8 nm, electrons

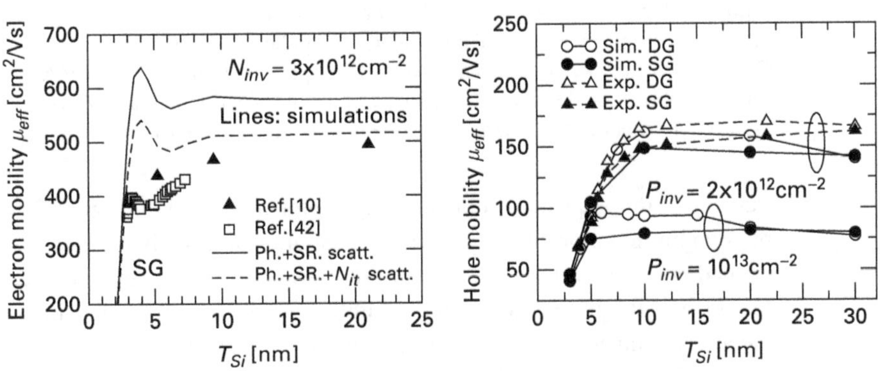

Figure 7.9 Simulated inversion layer mobility as a function of the silicon film thickness for SOI n-MOSFETs operated in single gate mode (left) and SOI p-MOSFETs operated in either single gate or double gate mode (right). n-MOS data from [10] at $N_{inv} = 3 \times 10^{12}$ cm^{-2} and from [44] at $F_{eff} = 0.3$ MV/cm as well as p-MOS data from [46] are also shown. Interface state density $N_{it} = 2.5 \times 10^{11}$ cm^{-2} for the dashed curve in the left plot. *Reprinted with permission from [10]. Copyright 2007 by the Institute of Electrical and Electronics Engineers.*

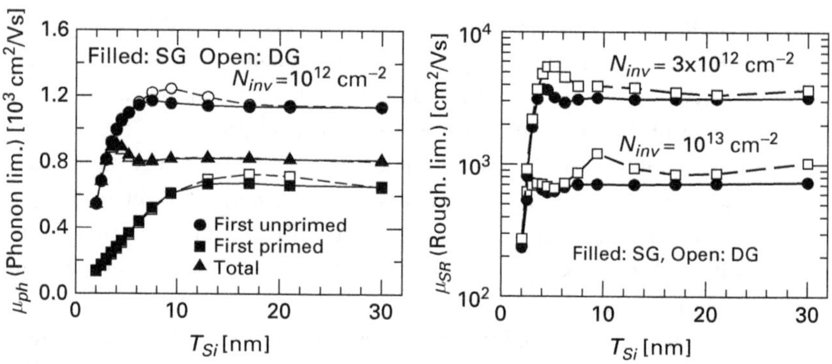

Figure 7.10 Phonon limited mobility (μ_{ph}, left) and surface roughness limited mobility (μ_{SR}, right) versus T_{Si} for the lowest unprimed and the lowest primed subbands in SOI n-MOSFETs operated in single gate (SG) and double (DG) gate mode. $N_{inv,DG} = 2N_{inv,SG}$. Calculations are performed with the EMA quantization model. *Reprinted with permission from [41]. Copyright 2003 by the Institute of Electrical and Electronics Engineers.*

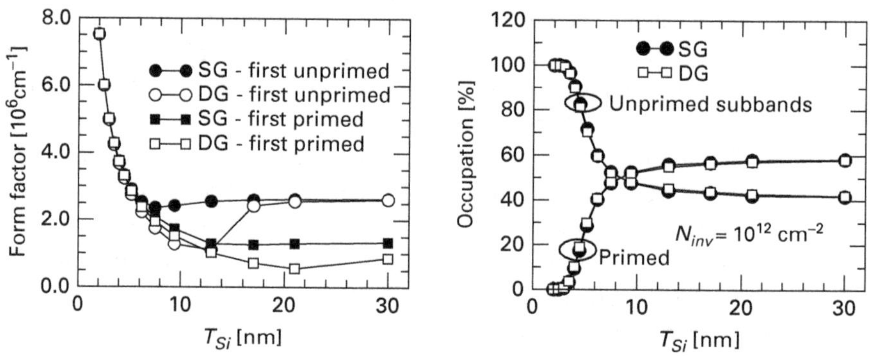

Figure 7.11 Left: intra-subband form factors for acoustic and optical phonon scattering of the lowest unprimed and the lowest primed subband (Eq.4.268) versus T_{Si} and for the same n-MOS devices as in the left graph of Fig.7.9. Right: relative population of the unprimed and primed subbands. Calculations are performed with the MRT approximation and the EMA quantization model. Filled symbols: single gate (SG) mode. Open symbols: double gate (DG) mode. *Reprinted with permission from [41]. Copyright 2003 by the Institute of Electrical and Electronics Engineers.*

transfer to the unprimed subbands (right plot in Fig.7.11). Figure 7.10 shows that the phonon limited mobility of electrons in the unprimed subbands is larger than that of the primed ones, essentially because of the lower transport mass. The increase of the form factors and the repopulation of the subbands are thus two opposing effects that make the average inversion layer mobility quite independent of T_{Si} down to about 6 nm. The surface roughness limited mobility μ_{SR} is also quite independent of T_{Si} until the 6 nm limit and therefore does not alter the physical interpretation of the μ_{eff} versus T_{Si} curves outlined so far.

For T_{Si} below approximately 6 nm the fraction of the electron population in the unprimed subbands increases quite rapidly and the total mobility slightly increases as well. Below approximately 4 nm the surface roughness limited mobility μ_{SR} decreases abruptly and the phonon form factors also rapidly increase, thus causing a sudden decay of the overall inversion layer mobility. This explains the hump and the subsequent drop observed in the μ_{eff}–T_{Si} mobility curves for $T_{Si} \leq 4$ nm. Indeed, a qualitatively similar (but quantitatively smaller) effect is observed in the experiments in the left plot of Fig.7.9 [44]. Although rather small to be of practical relevance for applications, this effect demonstrates the ability of the multi-subband model to correctly embody the physics of electron transport in a two-dimensional carrier gas.

The effective mobility limited by Coulomb scattering due to charged interface states μ_{it} is also a rapidly decreasing function of T_{Si} below approximately 10 nm [41]. This is essentially due to the exponential decay of the scattering potential from the interface towards the center of the silicon film (see Section 4.3.1 and Eq.4.134) and to the more effective confinement of the carriers at the interface for small T_{Si} values. However, unless the charge density in the interface traps is very large (an unrealistic situation for state of the art, industrial quality technologies), μ_{it} is much larger than μ_{SR}. Moreover, as discussed in Section 4.3.3, the Coulomb scattering potential is screened by the inversion charge; consequently the impact of Coulomb scattering due to interface states decreases at large N_{inv}. As a result, Coulomb scattering is expected to play a limited role in explaining the reduction of the effective mobility with T_{Si}.

Regarding the T_{Si} dependence of the inversion layer mobility of holes shown in Fig.7.9, we observe that the mobility stays constant or at most slightly increases down to $T_{Si} \approx 5$–10 nm and then drops to very small values when thicknesses in the few nanometers range are reached. The qualitative trend of the curves is similar to that for electrons. Figure 7.12 shows the form factors and the relative occupation of the lowest subbands of the p-MOS transistors, and the same qualitative features as for the n-MOSFETs of Fig.7.11 are also visible. In particular, a remarkable redistribution of the carriers among the subbands is seen for SOI films below 10 nm thickness.

The multi-subband transport model qualitatively explains also the slightly larger mobility exhibited by SOI MOSFETs with silicon thickness in the 10 nm range when

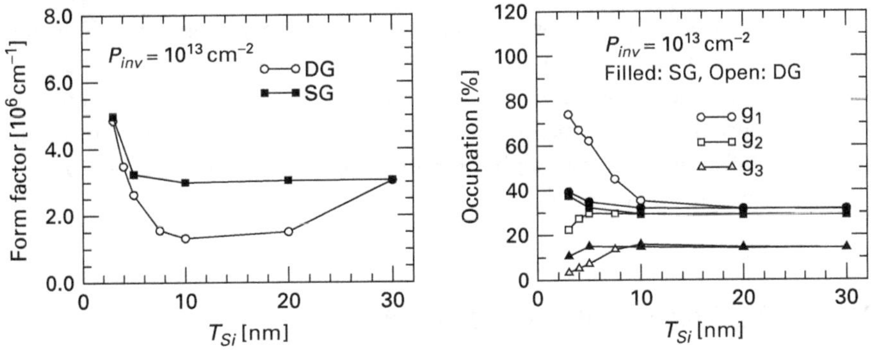

Figure 7.12 Left: form factors for phonon scattering at $\mathbf{k} = \mathbf{0}$ for the lowest subbands of the same p-MOS devices as in the right graph of Fig.7.9 as a function of T_{Si}. Right: relative occupation of the three lowest subbands (not including spin degeneracy). Calculations are carried out with the $\mathbf{k}\cdot\mathbf{p}$ quantization model. Filled symbols: single gate (SG) mode. Open symbols: double gate (DG) mode.

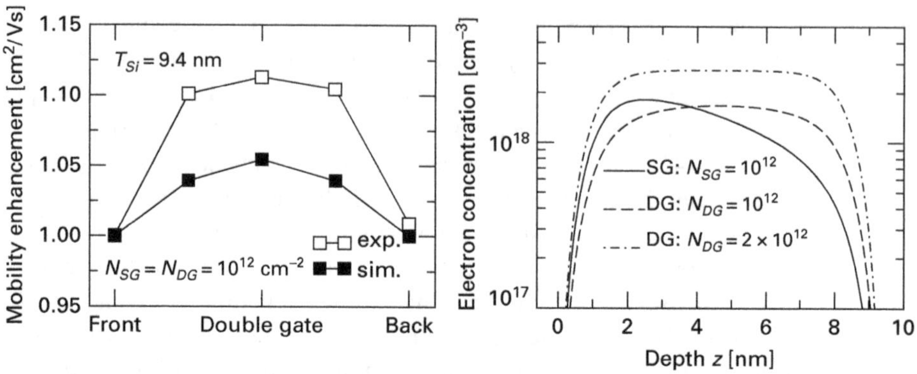

Figure 7.13 Left: experimental and simulated inversion layer mobility enhancement for ultra-thin body SOI n-MOSFETs operated in single gate at the front channel interface, double gate and single gate at the back channel interface mode. Right: simulated electron concentration profiles corresponding to the single gate and double gate modes. $T_{Si} = 9.4$ nm, EOT $= 4.5$ nm. *Reprinted with permission from [38]. Copyright 2003 by the Institute of Electrical and Electronics Engineers.*

they are operated in symmetric double gate mode as opposed to the single gate condition. Representative data of this effect are shown in the left panel of Fig.7.13 for n-MOSFETs; the effect is also visible in Fig.7.9 for p-MOS devices. To understand this experimental evidence we first note that the front channel and back channel mobility of the devices in Fig.7.13 are essentially the same (not shown, [49]) consistently with the good quality of the back interface that can be achieved in state of the art SOI wafers fabricated with the Unibond-Smartcut process [50].

In the symmetric double gate regime, however, the mobility is slightly higher than in single gate mode and the multi-subband transport simulations qualitatively reproduce this effect. This behavior is explained by the fact that as T_{Si} is reduced, the potential at the bottom of the quantum well formed by the semiconductor layer becomes flatter, so that in double gate mode the wave-functions are less confined at the interface and tend to invade the whole thickness of the silicon film, as shown in the right graph of Fig.7.13. Consequently, the form factors for phonon scattering are smaller than in single gate mode (as also shown in Figs.7.11 and 7.12) and the mobility increases. The condition where the charge occupies the whole silicon film thickness is often referred to as *volume inversion* and, as mentioned above, implies a slight increase of the mobility per gate. Unfortunately the mobility improvement over the single gate mode of operation is too small to be of practical relevance.

7.2 Far from equilibrium transport

Far from equilibrium transport conditions can be achieved in either uniform or non-uniform lateral electric field profiles parallel to the transport direction. In the following we briefly discuss the uniform case and then focus our attention on the non-uniform transport case encountered in short channel MOSFETs.

7.2.1 High field transport in uniform samples

Uniform transport at high fields can be studied with the same methodology as is used to calculate the low field mobility with the MSMC model; that is, the simulation of an inversion layer with: 1) exactly the same potential energy well in the z direction independently of x; 2) a uniform lateral driving field $F_x = d\varepsilon_i(x)/dx$ independent of x and of the subband i, as in Example 5.1; 3) the looping boundary conditions explained in Section 6.1.3 which mimic an infinitely long device. If these assumptions are fulfilled, then the subbands charge density and the **k**-space occupation function ($N_i(x)$, $f_i(x, \mathbf{k})$) are independent of x, which makes uniform transport well suited to unambiguously relate the N_i and the $f_i(\mathbf{k})$ to the driving field F_x. Such relations can represent a useful reference point for analysis of transport in more complex situations where F_x is not constant along the channel. Moreover, the entire drift velocity-field curve can be traced by sweeping the electric field value and the saturation velocity v_{sat} can thus be easily determined from the flat part of the curve. Preliminary studies based on multi-subband Monte Carlo simulations (see Fig.7.14) suggest that the saturation velocity in the electron inversion layer of ultra-thin body SOI MOSFETs depends on the inversion charge density and degrades substantially for reduced film thickness [51].

Achieving experimentally a uniform and high lateral field in inversion layers is practically impossible in a conventional MOSFET. Indeed, if a large drain-source voltage is applied to a short transistor, then the constant gate potential and the rapidly changing channel potential unavoidably result in a reduction of the vertical field while moving from the source to the drain, hence in a reduction of N_{inv} and in non-uniform physical conditions along the channel. The difficulty can be circumvented in part by resorting to the so called *resistive gate* MOSFET [52, 53]. In such a device, an attempt is made to fabricate a gate electrode with separate contacts at the source and drain edges and with a non-negligible poly-silicon sheet resistance. A current is forced in the gate poly-silicon along the channel direction, thus causing a lateral voltage drop inside the gate that should mirror as precisely as possible the one in the channel. In such a test structure

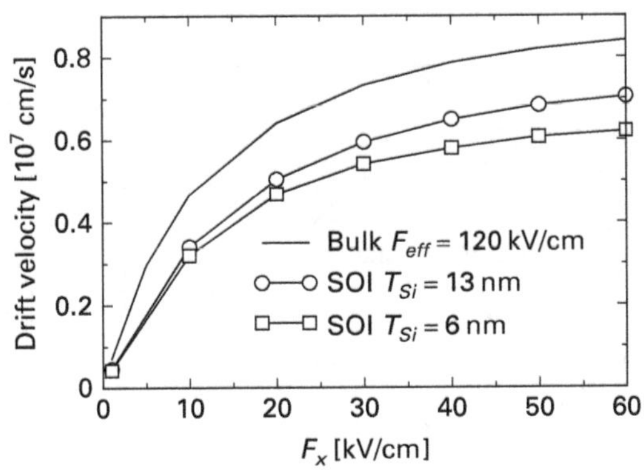

Figure 7.14 Simulated average velocity versus driving field for uniform transport conditions in a bulk n-MOSFET with $N_A = 2 \cdot 10^{12}$ cm^{-3} biased at $F_{eff} = 120$ kV/cm and for SOI n-MOSFETs at $F_{eff} = 500$ kV/cm. Data from [51].

the vertical field is expected to be fairly independent of x. Resistive gate devices have been used in the past to characterize inversion layer velocity versus field curves in bulk MOSFETs [52, 53]. A saturation velocity v_{sat} lower than the experimental value for bulk silicon has been measured.

7.2.2 High field transport in bulk and SOI devices

The lateral electric field parallel to the transport direction of short MOSFETs biased at high V_{DS} increases rapidly from source to drain on a distance scale comparable to the carrier mean free path. Consequently, strongly non-uniform transport conditions exist in most of the channel region. Differently from the uniform transport case discussed in Section 7.2.1, now the charge density and the drift velocity change with position and have a complex dependence on the device structure and bias resulting from the coupling between the electrostatics and the transport. To illustrate this aspect, we describe below a collection of simulation results on bulk and SOI MOSFETs.

Figure 7.15 shows the lateral profile $\varepsilon_i(x)$ of the lowest subbands of a bulk and an ultra-thin body SOI device. The position of the *virtual source* (x_{VS}, Section 5.6) is marked with circles. Due to the combined effect of the low doping, the fairly constant depletion charge along the channel, the small quantization mass, and relatively thin silicon film, the subbands of the SOI transistor run almost parallel to one another so that the corresponding quasi-field $\mathrm{d}\varepsilon_i/\mathrm{d}x$ (Eq.5.42) is almost the same for all the subbands (right plot). In the bulk transistor instead (left plot), an appreciable difference exists between the curves and each subband has its own virtual source coordinate $x_{VS,i}$ and quasi-field profile $\mathrm{d}\varepsilon_i/\mathrm{d}x$. This is because the depletion charge changes appreciably along the channel and so does the vertical potential energy profile determining the system of subbands.

Figure 7.16 shows the relative occupation of the lowest subbands of double gate SOI n- and p-MOSFETs. In proximity to the source the three lowest subbands account for no less than 80% of the carrier population, suggesting that a relatively small number

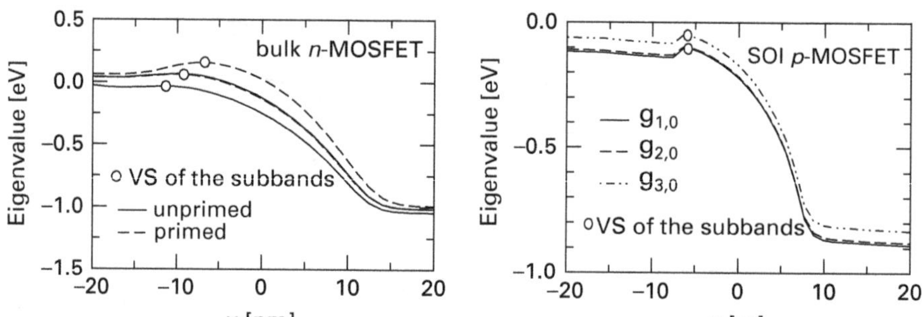

Figure 7.15 Subband profile $\varepsilon_i(x)$ along the transport direction in a bulk n-MOSFET (left, $N_A = 5 \cdot 10^{17}\mathrm{cm}^{-3}$, $L_G = 35$ nm, EOT $= 2.27$ nm, $V_{GS} = V_{DS} = 1.1$ V) and in a double gate SOI p-MOSFET (right, undoped channel, $L_G = 16$ nm, EOT $= 0.975$ nm, $T_{Si} = 9$ nm, $V_{SG} = V_{SD} = 1$ V). Circles mark the position of the virtual source in each subband.

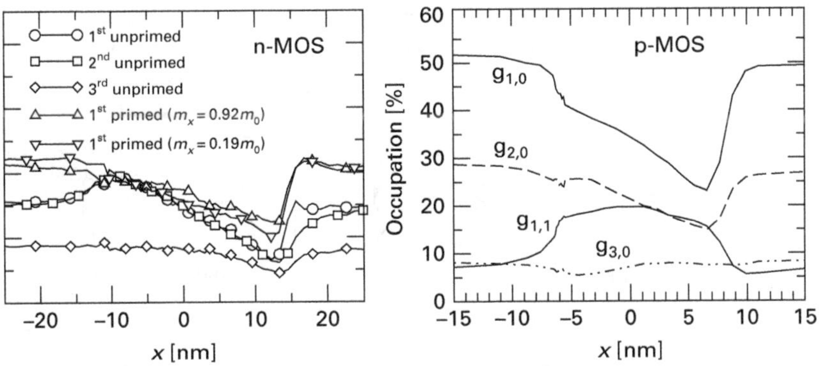

Figure 7.16 Relative occupation of the lowest subbands along the transport direction in n- (left) and p- (right) double gate SOI MOSFETs. $L = 22\,\text{nm}$, $T_{Si} = 10\,\text{nm}$, EOT $= 1.1\,\text{nm}$ for the n-MOSFET. The p-MOS device is the same as in Fig.7.15. $x_{VS} = -10.95\,\text{nm}$ for the n-MOSFET and $x_{VS} = -5.9\,\text{nm}$ for the p-MOSFET. $|V_{GS}| = |V_{DS}| = 1\,\text{V}$.

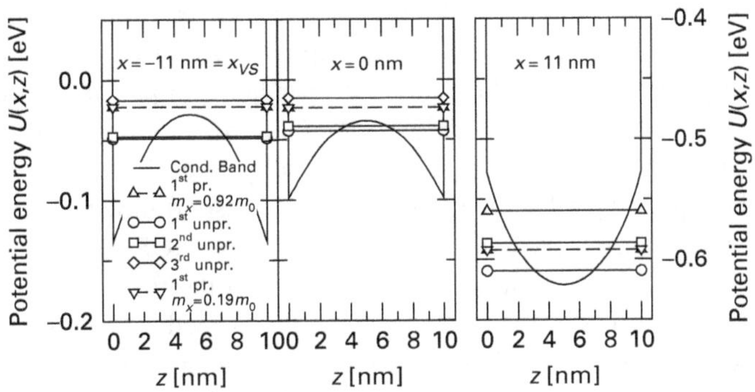

Figure 7.17 Potential energy profiles in the quantization direction and lowest eigenvalues at the beginning, the middle, and the end of the channel of the same n-MOSFET as in Fig.7.16. Note the degenerate eigenvalues of the first and second unprimed subbands and of the lowest primed subbands featuring $m_x = 0.92m_0$ and $m_x = 0.19m_0$. The energy axis on the right refers to the rightmost panel only.

of subbands is necessary in the calculations. The two lowest unprimed subbands of the n-MOS device are degenerate throughout most of the channel because the potential energy well leads to the formation of two decoupled inversion layers at the top and bottom interface, as shown in Fig.7.17. Consistently, they feature practically the same occupation probability (left plot in Fig.7.16). The lowest primed subbands are degenerate as well since they feature the same quantization mass. Consequently, they also have essentially the same occupation. However, as will be seen below, the different transport mass results in a different drift velocity. As we move beyond the virtual source the occupation of the lowest bands decreases and higher energy subbands host a larger fraction of the total carrier population. In the p-MOSFET case at the drain end of the channel some of the high energy subbands (e.g., $g_{1,1}$) become even more occupied than the low energy ones (e.g., $g_{2,0}$ and $g_{3,0}$). In the semi-classical model this situation is generated

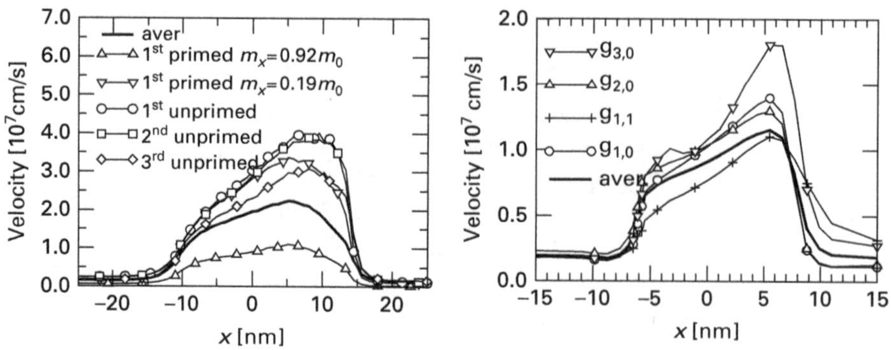

Figure 7.18 Average velocity in the x direction v_i for the lowest subbands of the same devices and bias point as in Fig.7.16.

by the transfer of carriers to higher subbands; in fact, the field provides energy to the carriers, which in turn enhances their ability to scatter into higher subbands.

Figure 7.18 shows the average velocity of the lowest subbands for the same devices as in Fig.7.16. The velocity steps up quite sharply at the virtual source, especially in the p-MOS device which has a steeper junction profile than the n-MOS. The velocity reaches values well above the saturation velocity observed in uniform transport conditions. As expected, the average velocity of each subband v_i is tightly related to the corresponding effective mass m_x in the transport direction; bands with smaller m_x feature a higher v_i. We also note that in the n-MOSFET case the first primed subband with $m_x = 0.19m_0$ has lower velocity than the first unprimed subband which has the same transport mass. This is because the primed subbands have higher density of state masses than the unprimed and therefore higher scattering rates (see Section 3.6.1 and Sections 4.6.3 to 4.6.6). In the p-MOSFET the velocity of the lowest subbands of each group ($g_{i,0}$) is higher at the drain end of the channel than for some of the higher energy subbands (e.g. the second lowest subband $g_{1,1}$ in the right plot of Fig.7.18). However, due to the large relative population of the $g_{1,1}$ subband (reported in the right plot of Fig.7.16) the average velocity is rather smaller than for the $g_{i,0}$ subbands.

7.3 Drive current

7.3.1 Ballistic and quasi-ballistic transport

It is instructive at this stage to examine the microscopic quantities identified by the ballistic and quasi-ballistic transport models of Sections 5.6 and 5.7 as the most relevant to interpret the on current I_{ON} of modern nanoscale MOSFETs.

To this purpose, we use as a vehicle device a double gate SOI MOSFET which offers the advantage that the virtual source coordinate is essentially the same for all the subbands. Figure 7.19 shows in the left plot the charge density of the carriers with positive (N^+) and negative (N^-) group velocity v_{gx}, as well as the total inversion charge, N_{inv}

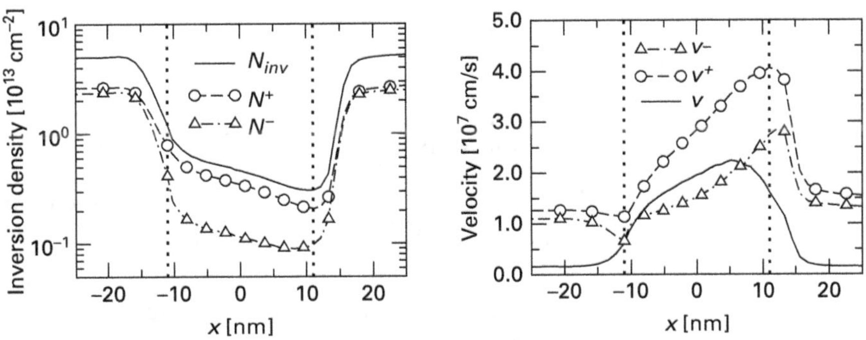

Figure 7.19 Left: inversion charge density ($N_{inv}(x)$) and charge density of the carriers with positive and negative group velocity ($N^+(x)$, $N^-(x)$). Right: average x-component of the drift velocity (v) and average velocity of the carriers with positive (v^+) and with negative (v^-) group velocity. Same single gate SOI n-MOSFET and same bias point as in Fig.7.16 and 7.18.

(Eq.5.167). Simulations account for the scattering in the channel so that the transport regime is at the best quasi-ballistic.

We see that the virtual source falls in a region of rapidly changing inversion density. Therefore, even a small error in the identification of the virtual source abscissa produces substantial uncertainty in the extraction of the carrier density $N_{inv}(x_{VS})$ required by the top-of-the-barrier models [54]. We also note that for this device N^+ and N^- are quite similar in the source and drain regions because of the almost isotropic carrier distribution therein, whereas they are quite different at the virtual source and in the channel region.

The right plot in Fig.7.19 shows the absolute value of the average velocity (v) and of the velocities v^+ and v^- of carriers having a positive or a negative group velocity, respectively. The drift velocity $v(x)$ is low in the source and drain regions because the electric field is low and the carrier concentration high. In non-degenerate conditions the values of v^+ and v^- should correspond to the weighted average thermal velocity of the primed ($v_{th} \simeq 1.2 \cdot 10^7$ cm/s) and unprimed ($v_{th} \simeq 0.55 \cdot 10^7$ cm/s) subbands (Eq.5.169). Higher values are observed in the figure, which indicate a moderate velocity increase due to carrier degeneration as expressed by Eqs.5.176 and 5.177. The drift velocity at the virtual source $v(x_{VS})$ is lower than v_{th}, which is the high voltage ballistic limit in non-degenerate conditions. We also see that v^+ grows very rapidly along the channel and exceeds v^- and the value v_{sat}. As a result, velocity overshoot is observed at the drain end of the channel ($v \approx 2 \cdot 10^7$cm/s), as already noted with regard to Fig.7.18.

In order to illustrate the importance of scattering in nano-MOSFETs, Fig.7.20 shows the absolute value of the static current per unit width due to particles with $v_{gx}>0$ (I^+) and to particles with $v_{gx}<0$ (I^-) for simulations with and without scattering in the channel region. As expected, the on-current with scattering $I_{ON} = [I^+_{ON}(x) - I^-_{ON}(x)]$ and the ballistic current without scattering $I_{BL} = [I^+_{BL}(x) - I^-_{BL}(x)]$ are independent of the position, as can be appreciated from the difference between the curves in Fig.7.20. Moreover, the $I^+(x)$ and $I^-(x)$ are quite similar in both simulations with and without scattering and take the largest value inside the source and drain, due to the large

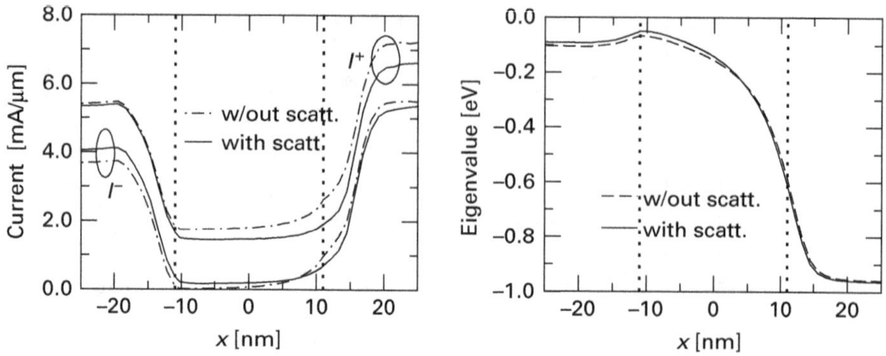

Figure 7.20 Left: MSMC simulations of the absolute value of the current per unit width due to particles with $v_{gx}>0$ (I^+) and to particles with $v_{gx}<0$ (I^-) along the channel for simulations with and without scattering. Right: simulated profile of the lowest subband $\varepsilon_0(x)$ with and without scattering. Device structure and bias as in Fig.7.19.

number of quasi-thermal electrons with energy lower than the potential energy barrier at the virtual source. However, I_{ON} is rather smaller than I_{BL}, indicating that scattering in the channel has a large impact on the on-current even in relatively short channel devices. As expected, at the large V_{DS} considered here we have $I^-_{BL}(x_{VS}) \simeq 0$ at the virtual source, since the only backward directed carriers are those injected by the drain reservoir: $I^-_{ON}(x_{VS})$ is instead appreciable in the dissipative transport regime where scattering is active.

Figure 7.20 also shows the potential energy profile of the lowest subband $\varepsilon_0(x)$ for the same device. The position of the virtual source does not change appreciably with and without scattering. However, the energy barrier at the virtual source is raised by the scattering, thus reducing the N^+. This is due to the electrostatic feedback of the back-scattered carriers that increase the electron concentration beyond the virtual source. Since the carrier density $N_{inv} = N^+ + N^-$ at the virtual source is essentially determined by the electrostatics along the vertical (z) direction and it is almost the same with and without scattering, the density of the carriers flowing toward the drain N^+ decreases with scattering because N^- becomes non-zero (Fig.7.19). For all these reasons, I^+_{ON} is rather smaller than I^+_{BL} and scattering events in the channel, even if very few, cannot be neglected.

Moreover, the subband energy profile is distorted in such a way that the KT-layer length increases as well. For realistic predictions of the I_{ON} based on Eq.5.181 it is thus not possible to correct the I_{BL} calculated with Eq.5.168 with a backscattering term $(1-r)/(1+r)$ where KT-layer length is extracted from ballistic simulations.

It is instructive at this stage to examine the distribution along x of the scattering events contributing to the backward directed carrier flux $\mathcal{F}^-$ (hence to the I^-_{ON}, see Eq.5.179), shown in Fig.7.21 for double gate transistors featuring different channel length and T_{Si}. The number of back-scattering events is evaluated in the MSMC simulations by counting the carriers that cross the virtual source with $v_{gx}<0$ and were back-scattered by a collision between x and $[x+\Delta x]$; such a number of carriers is then divided by Δx. The contribution of scattering to the I^-_{ON} decays rapidly beyond the virtual source. This

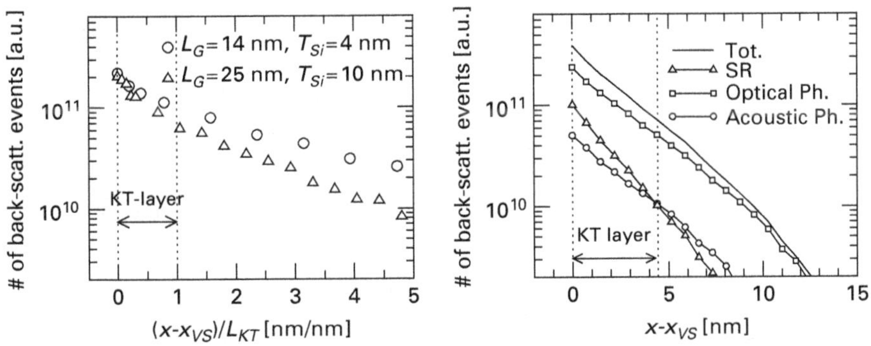

Figure 7.21 Number of scattering events contributing to the reverse carrier flux $\mathcal{F}^-$ at the virtual source (x_{VS}) as a function of the distance from the virtual source where the scattering occurred. Left: double gate SOI n-MOS devices with EOT = 0.7 nm and different L_G and T_{Si} values. Right: double gate SOI p-MOS device with $L_G = 22$ nm, EOT = 1.1 nm, $T_{Si} = 10$ nm. $|V_{GS}| = |V_{DS}| = 1$ V.

fact is in qualitative agreement with the indications of the Lundstrom model, which indeed emphasizes the role of the scattering events in a small region near the virtual source (see Eqs.5.181 and 5.187, [55, 56]).

A hint about the relative importance of the different scattering mechanisms is given for a p-MOS device in the right plot of Fig.7.21. Optical phonon and surface roughness scattering are the largest contributors to the I_{ON}^- in this device with a relatively thick silicon film, but surface roughness is expected to become the dominant back-scattering mechanism in ultra-thin body MOSFETs. Moreover, the relative importance of acoustic and optical phonon scattering is quantitatively different for n-MOSFETs than for p-MOSFETs because of the quite different values of the deformation potentials (see Table 7.1).

It is worth mentioning that in general the total number of scattering events at a given position (not shown) largely exceeds the number of those responsible for the backscattered flux which are shown in Fig.7.21. In fact, a large number of scattering events do not contribute to I_{ON}^- either because the velocity after scattering is positive (i.e., the carrier moves toward the drain) or because the velocity is negative but the carrier does not have enough energy to overcome the source/channel barrier [57, 58]. Therefore, on the one hand low backscattering is not necessarily synonymous with ballistic channel transport; on the other hand, the KT-layer length is only an approximate estimate of the portion of the channel that determines the on-current. In this regard also we point out that the scattering mechanisms considered throughout this chapter do not include long range Coulomb interactions, whose impact on the transport in ultra-short transistors is still debated and could be substantial [59, 60].

As an example of the details provided by the MSMC model for the analysis of carrier transport in nanoscale transistors, we show in Fig.7.22 the contours of the occupation function of the lowest subband of n-MOS (left) and p-MOS (right) devices in different physical conditions. The graphs (a) and (b) refer to the virtual source position and to the ballistic transport regime. We observe that in the n-MOSFET case the occupation

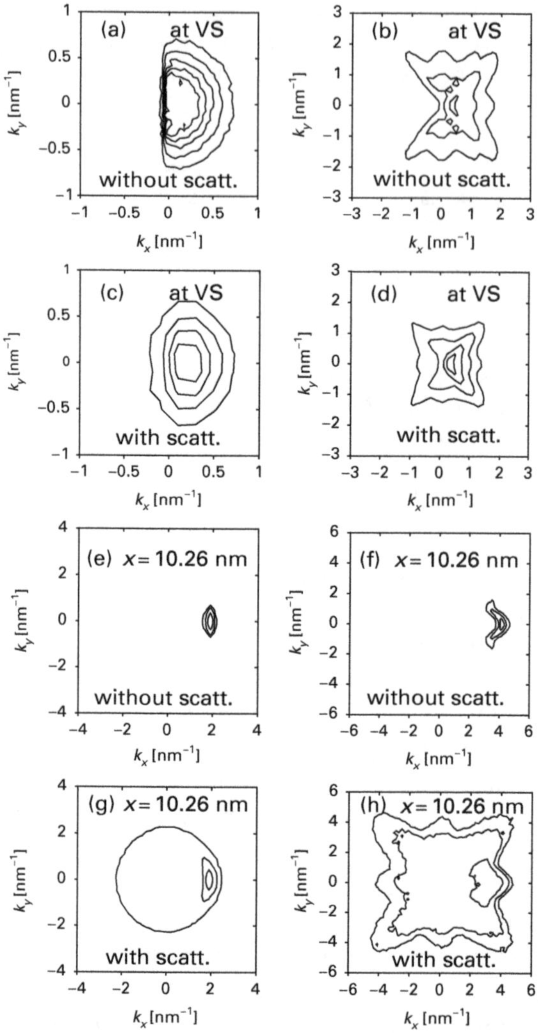

Figure 7.22 In-plane **k**-space contours of the occupation function $f_0(x, \mathbf{k})$ of the lowest subband of a double gate SOI n-MOSFET (left) and p-MOSFET (right). $L_G = 22$ nm. The p-MOS device is the same as in Figs.7.16 and 7.18. The n-MOS device has an exactly complementary structure and it is the same as in Figs.7.19 and 7.20. Note that for both n-MOS and p-MOS the distribution evolves in the positive **k** direction, consistently with the definition of $\mathbf{k}_h$ in Eq.5.22. (a) contours at $f_0 = 0.03, 0.1, 0.2, 0.3, 0.4$; (b) contours at $f_0 = 0.01, 0.1, 0.4$; (c) contours at $f_0 = 0.03, 0.1, 0.2, 0.3$; (d) contours at $f_0 = 0.03, 0.1, 0.2, 0.3$; (e) contours at $f_0 = 0.01, 0.03, 0.1$; (f) contours at $f_0 = 0.005, 0.05, 0.14$; (g) contours at $f_0 = 0.0001, 0.005, 0.03$; (h) contours at $f_0 = 0.0001, 0.001, 0.005$.

function is non-zero only for **k** states with positive k_x. In the p-MOSFET case instead, **k** states with negative k_x are also occupied because, due to the warped nature of the hole bands, a positive group velocity $v_{gx} = \hbar^{-1}\partial E(\mathbf{k})/\partial k_x$ is found also for negative k_x values (see, e.g., Fig.2.11).

The graphs (c) and (d) in the same figure illustrate the contours of the occupation function at the virtual source of the same MOSFETs as in the plots (a) and (b) in a quasi-ballistic transport regime, where phonon and surface roughness scattering are active. Although the average number of scattering events in the channel of such short MOSFETs is limited, these events are enough to redistribute the carriers in the **k**-space and to populate electron and hole states with negative v_{gx}. The **k**-space distribution, however, is far from being quasi-isotropic as assumed by several transport models, such as those based on the first momenta of the BTE and briefly described in Section 5.5. A marked anisotropy is observed, which underlines the importance of a full **k**-space treatment of carrier dynamics. The anisotropic shape of the carrier distributions is even more evident in the plots (e)–(h) of Fig.7.22, which capture the $f_i(\mathbf{r}, \mathbf{k})$ at the end of the channel in the ballistic ((e) and (f)) and quasi-ballistic ((g) and (h)) cases. It is interesting to note the similarity of the ballistic transport electron equi-occupation contours in the graphs (a) and (e) with those predicted by the ballistic model and sketched in Fig.5.11. Such a similarity is expected because the dispersion relation in the transport plane at the bottom of the unprimed valleys is indeed circular as assumed in the analytical calculations in Section 5.6.

It is also useful to examine quantitatively the particle group velocity distributions. To this purpose, Fig.7.23 shows those of the electrons at different points along the channel of a short n-MOSFET ($L_G = 14\,\text{nm}$). In the ballistic transport regime (solid lines) at the virtual source the carriers populate only the hemi-distribution with positive v_{gx}. This is because in the absence of scattering the carriers with positive v_{gx} that enter the channel are never backscattered toward the source. The centroid of the distribution shifts toward higher v_{gx} values as we consider a position close to the drain. Consistently, the peak of the ballistic distribution is located at much higher v_{gx} in the right than in the left plot of Fig.7.23.

In the dissipative transport case instead (dashed lines), carrier motion is affected by the scattering events and the negative velocity part of the distribution is populated,

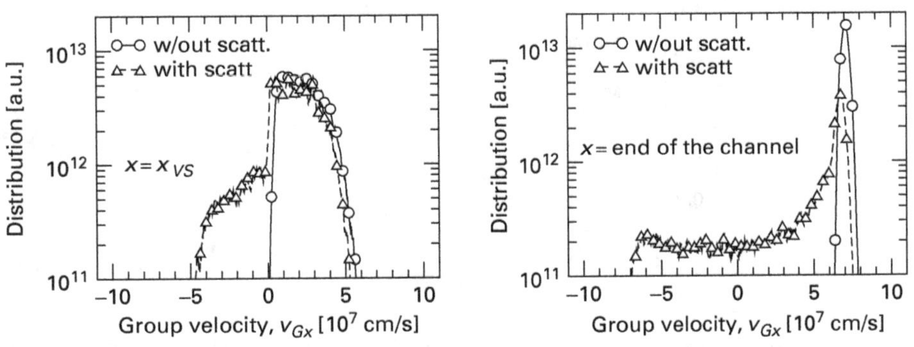

Figure 7.23 Electron distribution as a function of the electron velocity in the transport direction at the virtual source (left) and at the end of the channel immediately before entering the drain region (right) with and without scattering in the channel. n-MOSFET with $L_G = 14\,\text{nm}$, $T_{Si} = 4\,\text{nm}$, EOT $= 0.7\,\text{nm}$, $V_{GS} = V_{DS} = 1\,\text{V}$.

although much less than the positive one. A ballistic peak is still seen in the distribution at the drain end of the channel, but it is much less pronounced than under ballistic transport conditions.

7.3.2 Voltage dependence and gate length scaling

In order to develop a physical insight into low and high field transport in inversion layers, the previous sections have illustrated in detail the constituent elements of transport (scattering limited mobilities, occupation functions, average velocity, etc.) and their dependence on the position, the inversion charge, the effective field, and the silicon film thickness of SOI transistors. It is now useful to investigate also how the terminal voltages and the channel length affect the key quantities relevant for the I_{ON}, since this is one of the most important targets of nano-MOSFET device optimization. Without any ambition for completeness, in this section we exemplify how valuable insight can be achieved into the critical scaling trends for bulk and SOI transistors via MSMC transport simulations. The quasi-ballistic transport model of Section 5.7 is once again used as a guideline to select the ingredients relevant for the I_{ON} (that is N_{inv}, v^+, and r). The ability of the MSMC model to support the analysis of the performance improvements that can be expected from the technology boosters (strain, crystal orientation, channel material) is separately addressed in the following chapters.

Before starting to describe the simulation results, a short preliminary discussion is in order with respect to the maturity and dependability of the MSMC approach for the simulation of the MOSFET on-current with respect to other established transport models, either based on the BTE for the free electron gas solved using the Monte Carlo techniques described in Sections 5.1 and 6.1, or based on the momenta of the BTE discussed in Section 5.5.

In this regard, Fig.7.24 compares the trans-characteristics of a 32 nm fully depleted unstrained SOI MOSFET simulated with different transport models that were previously

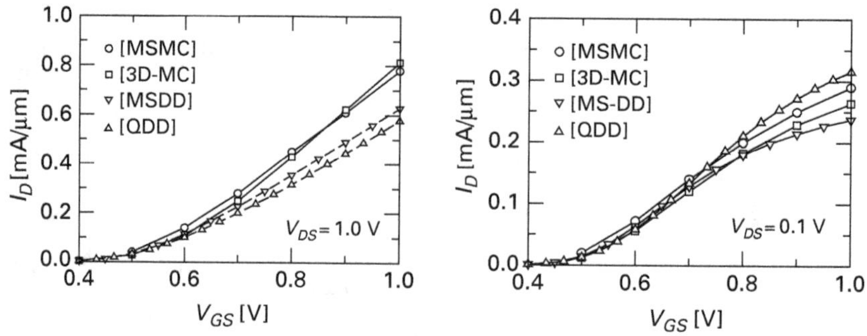

Figure 7.24 I_D-V_{GS} curve of a 32 nm SOI n-MOSFET simulated with different transport models previously calibrated to reproduce the effective mobility of [25]. $V_{DS} = 1.0$ V (left) and $V_{DS} = 0.1$ V (right). Data from [61]. The transport models are the MSMC of [62, 63] (○), the free electron gas Monte Carlo of [64, 65] (□), the quantum Drift–Diffusion of [66–68] (△) and the multi-subband Drift–Diffusion of [69, 70] (▽).

calibrated to reproduce (within approximately 15% error, [61]) the universal mobility curve of [25]. Simulations with the Monte Carlo method include phonon, surface roughness, and ionized impurity scattering. Calculations performed with drift–diffusion based transport models are also shown. Although not at all conclusive, these results represent a meaningful snapshot of the status at the time of writing of research-oriented simulation codes in this field.

The figure shows that the model predictions are fairly well aligned with one another for $V_{DS} = 1$ V. The MSMC model consistently predicts a larger drain current than the quantum Drift–Diffusion [66] or the multi-subband Drift–Diffusion [69, 70] models. This is an expected result, since the Drift–Diffusion model limits the maximum velocity to v_{sat}, whereas an average velocity higher than v_{sat} is reached in the far from equilibrium transport regime typical of nanometer scale MOSFETs, as illustrated in Figs.7.18 and 7.19. Drift–Diffusion models can also achieve good agreement with simulations based on the exact solution of the BTE provided the drift velocity is allowed to take values larger than the bulk silicon saturation velocity, e.g. by tuning the value of the corresponding model parameter [71, 72]. Such a calibration of the v_{sat}, however, is empirical and specific to each device and channel length, so that the calibrated model has limited predictive ability.

In the linear region of operation the curves spread out, in spite of the calibration against the same inversion layer mobility data, mainly because the parasitic source and drain series resistances play a significant role in determining the current. The series resistance is strongly affected by the ionized impurity scattering models, which are quite different between the simulators considered here, and that can be critical when comparing devices biased in the linear region [73].

The simulations in Fig.7.24 suggest that in unstrained silicon devices fabricated on thick silicon films, conventional Monte Carlo transport solvers based on the assumption of a three-dimensional free carrier gas yield results similar to the MSMC transport models, provided that the inversion layer mobility has been carefully calibrated against the same experimental data and on the same device structure.

Concerning the bias dependence of the ingredients of the quasi-ballistic model, Fig.7.25 shows the average velocity of the carriers injected into the channel (Eq.5.176) at the virtual source of short n- and p-MOSFETs versus the gate bias. As expected from the results in Section 5.6.4 (in particular, Fig.5.12 and Eq.5.176), we see that the injection velocity increases for increasing $|V_{GS}|$ because the carrier gas at the virtual source becomes progressively more degenerate as a consequence of the increased N_{inv}. The velocity v^+ at the VS of the p-MOSFET is smaller than that of the n-MOSFET and less sensitive to the gate bias, essentially because the carrier degeneracy is smaller in hole inversion layers (see Figs.7.5 and 7.6) due to the larger density of states (Section 3.6 and Fig.3.13).

Figure 7.26 shows the on-current I_{ON}, the average velocity of the electrons injected from the source into the channel v^+, the average velocity v and the backscattering coefficient r at the virtual source for a set of scaled double gate SOI n-MOSFETs featuring an approximately constant DIBL. The charge density at the virtual source $N_{inv}(x_{VS})$ is essentially the same for all the devices. We see that a reduction of T_{Si} increases v^+,

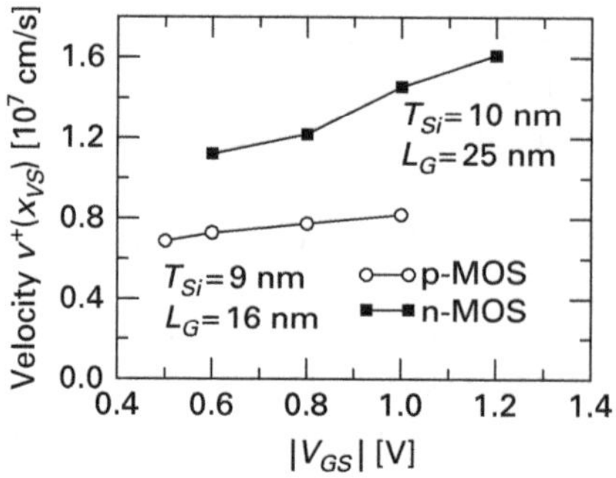

Figure 7.25 Average velocity at the virtual source for double gate SOI n- and p-MOSFETs. The p-MOS device is the same as in Fig.7.15. *Reprinted with permission from [74]. Copyright 2009 by the Institute of Electrical and Electronics Engineers.*

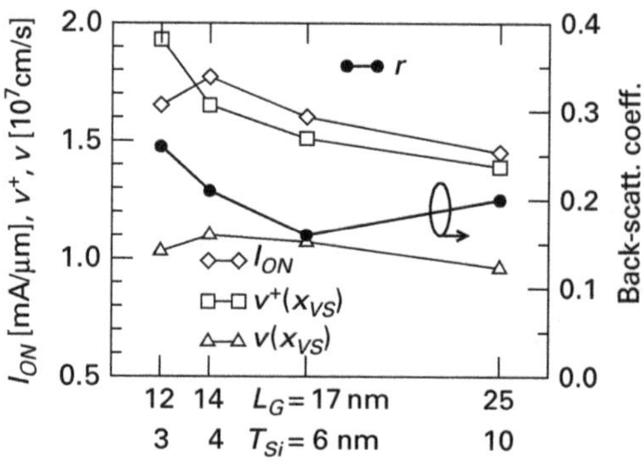

Figure 7.26 On-current (I_{ON}, diamonds), average velocity of the injected electrons ($v^+(x_{VS})$, □), average velocity ($v(x_{VS})$, △) and back-scattering coefficient (r, •) at the virtual source of scaled double gate SOI n-MOSFETs calculated with the MSMC simulator of [75]. $V_{GS} = V_{DS} = 1V$. The DIBL is approximately 100 mV/V for all the devices. $N_{inv}(x_{VS})$ ranges from $1.0 \cdot 10^{13}$ cm^{-2} to $1.1 \cdot 10^{13}$ cm^{-2}. EOT = 0.7nm. The n-MOS devices are the same as in Fig.7.21.

since the electron gas becomes more degenerate because the subband splitting reduces the density of states at small energies. However, the quantum confinement induced by the reduced film thickness increases surface roughness scattering, so that r increases when reducing T_{Si} even if L_G is scaled. As a result, in this family of devices the maximum I_{ON} is obtained for $T_{Si} = 4$ nm and $L_G = 14$ nm rather than for the slightly shorter device with $T_{Si} = 3$ nm.

The importance of carrier degeneracy is also evident in the data of Table 7.3. Here we compare the I_{ON}, N_{inv}, velocity and backscattering coefficient in simulations with and without the Pauli exclusion principle. Degeneracy is accounted for by rejecting scattering events toward occupied final states as explained in Section 6.1.2. We observe that the average velocity at the VS is much higher in the degenerate case than in the non-degenerate. However, the larger velocity observed in the degenerate case does not translate into an on-current appreciably higher than in the non-degenerate case because the back-scattering coefficient is also larger when the Pauli exclusion principle is accounted for [76].

Table 7.3 Drain current I_{ON}, inversion charge density N_{inv}, average carrier velocity v, average velocity v^+ of electrons with $v_{gx} > 0$ and reflection coefficient r at the VS of a DG SOI MOSFET with and without the Pauli exclusion principle. $T_{Si} = 4$nm, $L_G = 14$nm. $V_{GS} = V_{DS} = 1$V.

	With degen.	W/out degen.	
I_{ON}	1.77	1.75	mA/μm
N_{inv}	1.0	1.1	10^{13} cm^{-2}
v^+	1.65	1.3	10^7 cm/s
v	1.10	1.0	10^7 cm/s
r	0.21	0.11	

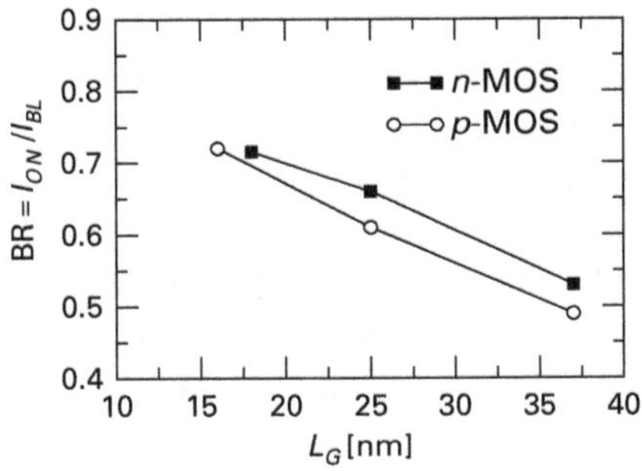

Figure 7.27 Ballistic ratio I_{ON}/I_{BL} versus gate length for scaled double gate MOSFETs. $|V_{GS}| = |V_{DS}| = 1$ V. The DIBL is approximately 100 mV/V for all the devices. *Reprinted with permission from [74]. Copyright 2009 by the Institute of Electrical and Electronics Engineers.*

Figure 7.27 compares the scaling trends of unstrained n-MOS and p-MOS double gate SOI transistors in terms of the so-called ballistic ratio, BR $= I_{ON}/I_{BL}$. The ballistic ratio provides a metric of how closely a transistor is operating to the drain current ballistic limit I_{BL}. In fact, according to the model in Section 5.7 the ballistic ratio is BR $= (1 - r)/(1 + r)$ and tends to 1 for r approaching zero. The MSMC simulations with phonon and surface roughness scattering predict a similar ballistic ratio for both n- and p-MOSFETs. The values approximately correspond to reflection coefficients between $r = 0.33$ and $r = 0.18$ for the long and short device, respectively. It should be noticed, however, that the I_{ON} of the n-MOS is higher than that of the p-MOS because the ballistic current I_{BL} is also higher due to imbalance between electron and hole velocity in unstrained silicon. This imbalance can be appreciably reduced in the strained silicon technologies discussed in Chapter 9.

7.4 Summary

This chapter has illustrated experimental and simulation results on low and high field transport in state of the art bulk and SOI unstrained CMOS technologies. A critical re-examination of the concept of effective field (Sections 7.1.1 and 7.1.2) has pointed out the empirical nature of this quantity. The physically-based justification for the widespread use of the effective field is thus rather weak and the practical usefulness of F_{eff} for the purpose of understanding the mobility of advanced devices with engineered channel materials is at risk. A transparent interpretation of the measurements is possible if mobility is plotted as a function of N_{inv}, provided that any ambiguity in the number of parallel channels considered in the normalization of the carrier density is removed. In this case, however, the universal behavior is lost.

We have shown that MSMC and MRT mobility calculations based on a two-dimensional carrier gas description are in remarkably good agreement and provide detailed insight into the physics of low field transport in inversion layers. The main features of the measured mobility in unstrained silicon bulk and ultra-thin body SOI MOSFETs are nicely explained by the multi-subband model.

We have then described the essential physics of carrier transport in ultra-short channel transistors that the multi-subband transport model allows us to infer from simulations with a basic set of scattering mechanisms (phonon, surface roughness, and ionized impurity scattering). In particular, we have seen that, due to the self-consistent loop between the transport and the electrostatics, scattering in the channel is expected to play a non-negligible role in determination of the on-current down to very short gate length even if the average number of scattering events decreases appreciably. As a result, the on-current may be rather smaller than the ballistic current calculated in the absence of scattering.

According to these simulations, the operation of short MOSFETs appears to be consistent with the elements of the model described in Section 5.7. In particular, it is confirmed that, as long as we limit the scattering terms to those considered in this chapter, a short region in proximity to the virtual source controls the drain current in nano-scale devices. For a given device and temperature the extension of this region is related to, but not coincident with, the KT-layer length.

We should remember, however, that additional scattering mechanisms such as long range Coulomb interactions may be relevant in short MOSFETs, and these have not been included in the simulations presented in this chapter, which could modify somewhat the physical picture of the device operation, at least in specific transport regimes.

References

[1] C. Canali, C. Jacoboni, F. Nava, G. Ottaviani, and A. Alberigi Quaranta, "Electron drift velocity in silicon," *Phys. Rev. B*, vol. 12, pp. 2265–2284, 1975.

[2] G. Ottaviani, L. Reggiani, C. Canali, F. Nava, and A. Alberigi Quaranta, "Hole drift velocity in silicon," *Phys. Rev. B*, vol. 12, pp. 3318–3329, 1975.

[3] C. Jacoboni and L. Reggiani, "The Monte Carlo method for the solution of charge transport in semiconductors with applications to covalent materials," *Rev. Mod. Phys.*, vol. 55, pp. 645–705, 1983.

[4] J.S. Blakemore, "Semiconducting and other major properties of Gallium Arsenide," *Journal of Applied Physics*, vol. 53, no. (10), pp. R123–R181, 1982.

[5] Y. Taur and T. Ning, *Fundamentals of Modern VLSI Devices.* Edinburgh: Cambridge University Press, 1998.

[6] G. Masetti, G. Severi, and M. Solmi, "Modeling of carrier mobility against carrier concentration in Arsenic-, Phosphorus- and Boron-doped silicon," *IEEE Trans. on Electron Devices*, vol. 30, pp. 764–769, 1983.

[7] D. Schroder, *Semiconductor Material and Device Characterization.* New York: Wiley, 1990.

[8] C.G. Sodini, T.W. Ekstedt, and J.L. Moll, "Charge accumulation and mobility in thin dielectric MOS transistors," *Solid State Electronics*, vol. 25, pp. 833–841, 1982.

[9] F. Lime, C. Guiducci, R. Clerc, *et al.*, "Characterization of effective mobility by split *C(V)* technique in n-MOSFETs with ultra-thin gate oxides," in *Proc. Int. Conf. on Ultimate Integration on Silicon (ULIS)*, pp. 23–26, 2002.

[10] D. Esseni, M. Mastrapasqua, G.K. Celler, *et al.*, "Low field electron and hole mobility of SOI transistors fabricated on ultra-thin silicon films for deep sub-micron technology application," *IEEE Trans. on Electron Devices*, vol. 48, no. 12, pp. 2842–2850, 2001.

[11] M. Rudan, S. Reggiani, E. Gnani, *et al.*, "Theory and experimental validation of a new analytical model for the position-dependent Hall voltage in devices with arbitrary aspect ratio," *IEEE Trans. on Electron Devices*, vol. 53, no. 2, pp. 314–322, 2006.

[12] H.S. Momose, M. Ono, T. Yoshitomi, *et al.*, "Tunneling gate oxide approach to ultra high current drive in small geometry MOSFETs," in *IEEE IEDM Technical Digest*, pp. 593–596, 1994.

[13] H.S. Momose, M. Ono, T. Yoshitomi, *et al.*, "1.5 nm direct-tunneling gate oxide Si MOSFETs," *IEEE Trans. on Electron Devices*, vol. 43, pp. 1233–1241, August 1996.

[14] L. Lucci, D. Esseni, J. Loo, *et al.*, "Quantitative assessment of mobility degradation by remote Coulomb scattering in ultra-thin oxide MOSFETs: Measurements and simulations ," in *IEEE IEDM Technical Digest*, pp. 463–466, December 2003.

[15] P. Palestri, D. Esseni, L. Selmi, G. Guegan, and E. Sangiorgi, "A methodology to extract the channel current of permeable gate oxide MOSFETs ," *IEEE Trans. on Electron Devices*, vol. 50, no. 5, pp. 1314–1321, 2003.

[16] Y.M. Meziani, J. Lusakowski, W. Knap, *et al.*, "Magnetoresistance characterization of nanometer Si metal-semiconductor transistors," *Journal of Applied Physics*, vol. 96, pp. 5761–5765, 2004.

[17] W. Chaisantikulwat, M. Mouis, G. Ghibaudo, *et al.*, "Magnetoresistance technique for mobility extraction in short channel FDSOI transistors," in *Proc. European Solid State Device Res. Conf.*, pp. 569–572, 2005.

[18] G. Ghibaudo, "Transport in the inversion layer of a MOS transistor: use of Kubo–Greenwood formalism," *J. of Physics C: Solid State Physics*, vol. 19, pp. 767–780, 1986.

[19] Y.M. Meziani, J. Lusakowski, F. Teppe, *et al.*, "Magnetoresistance mobility measurements in sub-0.1 μm silicon MOSFETs," in *Proc. European Solid State Device Res. Conf.*, pp. 157–160, 2004.

[20] L. Donetti, F. Gamiz, and S. Cristoloveanu, "A theoretical interpretation of magnetoresistance mobility in silicon inversion layers," *Journal of Applied Physics*, vol. 102, p. 013708, 2007.

[21] N. Rodriguez, L. Donetti, F. Gamiz, and S. Cristoloveanu, "Geometric magnetoresistance and mobility behavior in single-gate and double-gate SOI devices," in *Proc. IEEE SOI Conference*, pp. 59–60, 2007.

[22] M. Zilli, D. Esseni, P. Palestri, and L. Selmi, "On the apparent mobility in nanometric n-MOSFETs," *IEEE Electron Device Lett.*, vol. 28, no. 11, pp. 1036–1039, 2007.

[23] A.G. Sabnis and J.T. Clemens, "Characterization of the electron mobility in the inverted ⟨100⟩ Si surface," in *IEEE IEDM Technical Digest*, pp. 18–21, 1979.

[24] J. Watt and J. Plummer, "Universal mobility-field curves for electrons and holes in MOS inversion layers," in *IEEE Symposium on VLSI Technology - Technical Digest*, p. 81, 1987.

[25] S. Takagi, A. Toriumi, M. Iwase, and H. Tango, "On the universality of inversion layer mobility in Si MOSFET's: Part I-Effects of substrate impurity concentration," *IEEE Trans. on Electron Devices*, vol. 41, pp. 2357–2362, Dec 1994.

[26] A. Emrani, G. Ghibaudo, and F. Balestra, "On the universal electric field dependence of the electron and hole effective mobility in MOS inversion layers.," *Solid State Electronics*, vol. 37, no. 1, pp. 111–113, 1994.

[27] H.M. Nayfeh, C.W. Leitz, A.J. Pitera, *et al.*, "Influence of high channel doping on the inversion layer electron mobility in strained silicon n-MOSFETs," *IEEE Electron Device Lett.*, vol. 24, no. 4, pp. 248–250, 2003.

[28] K. Bennamane, M. De Michielis, G. Ghibaudo, and D. Esseni, "Extraction of η parameter characterising μ_{eff} against E_{eff} curves in strained Si nMOS devices," *Electronics Letters*, vol. 44, no. 20, pp. 1219–1220, 2008.

[29] T. Rudenko, V. Kilchytska, N. Collaert, *et al.*, "Carrier mobility in undoped triple-gate FinFET structures and limitations of its description in terms of top and sidewall channel mobilities," *IEEE Trans. on Electron Devices*, vol. 55, no. 12, pp. 3532–3541, 2008.

[30] M.V. Fischetti and S.E. Laux, "Monte Carlo study of electron transport in silicon inversion layers," *Phys. Rev. B*, vol. 48, pp. 2244–2274, Jul 1993.

[31] M.V. Fischetti and S.E. Laux, "Monte Carlo simulation of electron transport in Si: The first 20 years," in *Proc. European Solid State Device Res. Conf.*, pp. 813–820, 1996.

[32] C. Jungemann, A. Edmunds, and W.L. Engl, "Simulation of linear and nonlinear electron transport in homogeneous silicon inversion layers," *Solid State Electronics*, vol. 36, no. 11, pp. 1529–1540, 1993.

[33] S.M. Goodnick, D.K. Ferry, C.W. Wilmsen, *et al.*, "Surface roughness at the Si(100)-SiO_2 interface," *Phys. Rev. B*, vol. 32, pp. 8171–8186, 1985.

[34] A. Pirovano, A. Lacaita, A.L. Ghidini, and G. Tallarida, "On the correlation between surface roughness and inversion mobility in Si-MOSFETs," *IEEE Electron Device Lett.*, vol. 21, no. 1, pp. 34–36, 2000.

[35] Y. Zhao, H. Matsumoto, T. Sato, *et al.*, "Comprehensive understanding of surface roughness limited mobility in unstrained- and strained-Si MOSFETs by novel characterization scheme of Si/SiO_2 interface roughness," in *IEEE Symposium on VLSI Technology - Technical Digest*, pp. 18–19, 2009.

[36] T. Ando, A. Fowler, and F. Stern, "Electronic properties of two-dimensional systems," *Review of Modern Physics*, vol. 54, pp. 437–673, 1982.

[37] B. Doris, M. Ieong, T. Kanarsky, *et al.*, "Extreme scaling with ultra-thin Si channnel MOSFETs," in *IEEE IEDM Technical Digest*, p. 267, 2002.

[38] D. Esseni, M. Mastrapasqua, G.K. Celler, *et al.*, "An experimental study of mobility enhancement in ultrathin SOI transistors operated in double-gate mode," *IEEE Trans. on Electron Devices*, vol. 50, no. 3, pp. 802–808, 2003.

[39] T. Ernst, S. Cristoloveanu, G. Ghibaudo, *et al.*, "Ultimately thin double gate SOI MOSFETs," *Microelectronic Engineering*, vol. 50, no. 3, pp. 830–838, 2003.

[40] K. Shimizu and T. Hiramoto, "Mobility enhancement in uniaxially strained (110) oriented ultra-thin body single- and double-gate MOSFETs with SOI thickness of less than 4 nm," in *IEEE IEDM Technical Digest*, pp. 715–718, 2007.

[41] D. Esseni, A. Abramo, L. Selmi, and E. Sangiorgi, "Physically based modeling of low field electron mobility in ultra-thin single and double-gate SOI n-MOSFETs.," *IEEE Trans. on Electron Devices*, vol. 50, no. 12, pp. 2445–2455, 2003.

[42] M. Schmidt, M.C. Lemme, H.D.B. Gottlob, *et al.*, "Mobility extraction of UTB n-MOSFETs down to 0.9 nm SOI thickness," in *Proc. Int. Conf. on Ultimate Integration on Silicon (ULIS)*, pp. 27–30, 2009.

[43] K. Uchida, J. Koga, R. Ohba, and T.N.S. Takagi, "Experimental evidences of quantum-mechanical effects on low-field mobility, gate-channel capacitance and threshold voltage of ultrathin body SOI MOSFETs," in *IEEE IEDM Technical Digest*, pp. 633–636, 2001.

[44] K. Uchida, H. Watanabe, A. Kinoshita, *et al.*, "Experimental study on carrier transport mechanisms in ultrathin-body SOI n- and p-MOSFETs with SOI thickness less than 5nm," in *IEEE IEDM Technical Digest*, pp. 47–50, 2002.

[45] K. Uchida and S. Takagi, "Carrier scattering induced by thickness fluctuation of silicon-on-insulator film in ultrathin-body metal-oxide-semiconductor field-effect transistors," *Applied Physics Letters*, vol. 82, no. 17, pp. 2916–2918, 2003.

[46] S. Kobayashi, M. Saitoh, and K. Uchida, "Hole mobility enhancement by double-gate mode in ultrathin-body silicon-on-insulator p-type metal-oxide-semiconductor field-effect transistors," *Journal of Applied Physics*, vol. 106, p. 024511, 2009.

[47] P.J. Price, "Two-dimensional transport in semiconductor layers. I. Phonon scattering," *Annals of Physics*, vol. 133, pp. 217–233, 1981.

[48] B.K. Ridley, "The electron–phonon interaction in quasi-two-dimensional semiconductor quantum-well," *J. of Physics C: Solid State Physics*, vol. 15, pp. 5899–5917, 1982.

[49] D. Esseni, M. Mastrapasqua, C. Fiegna, *et al.*, "An experimental study of low field electron mobility in double-gate, ultra-thin SOI MOSFETs," in *IEEE IEDM Technical Digest*, pp. 445–448, 2001.

[50] M. Bruel, "Silicon on insulator material technology," *Electronics Letters*, vol. 31, p. 1201, 1995.

[51] L. Lucci, P. Palestri, D. Esseni, and L. Selmi, "Modeling the uniform transport in thin film SOI MOSFETs with a Monte Carlo simulator for the 2D electron gas," *Solid State Electronics*, vol. 49, no. 9, pp. 1529–1535, 2005.

[52] J.A. Cooper and D.F. Nelson, "High-field drift velocity of electrons at the Si–SiO_2 interface as determined by a time-of-flight technique," *Journal of Applied Physics*, vol. 54, pp. 1445–1456, 1983.

[53] A. Modelli and S. Manzini, "High-field drift velocity of electrons in silicon inversion layers," *Solid State Electronics*, vol. 21, pp. 99–104, 1988.

[54] M. Lundstrom and Z. Ren, "Essential physics of carrier transport in nanoscale MOSFETs," *IEEE Trans. on Electron Devices*, vol. 49, no. 1, pp. 133–141, 2002.

[55] M. Lundstrom, "Elementary scattering theory of the Si MOSFET," *IEEE Electron Device Lett.*, vol. 18, pp. 361–363, July 1997.

[56] C. Jeong, D.A. Antoniadis, and M.S. Lundstrom, "On backscattering and mobility in nanoscale silicon MOSFETs," *IEEE Trans. on Electron Devices*, vol. 56, no. 11, p. 2762, 2009.

[57] P. Palestri, D. Esseni, S. Eminente, *et al.*, "Understanding quasi-ballistic transport in nano-MOSFETs. Part I: Scattering in the channel and in the drain," *IEEE Trans. on Electron Devices*, vol. 52, no. 12, pp. 2727–2735, 2005.

[58] J. Saint Martin, A. Bournel, and P. Dollfus, "On the ballistic transport in nanometer-scaled DG MOSFETs," *IEEE Trans. on Electron Devices*, vol. 51, no. 7, pp. 1148–1155, 2004.

[59] K. Nakanishi, T. Uechi, and N. Sano, "Self-consistent Monte Carlo device simulations under nano-scale device structures: Role of Coulomb interaction, degeneracy, and boundary condition," in *IEEE IEDM Technical Digest*, pp. 79–82, 2009.

[60] M. Fischetti, S. Jin, T.-W. Tang, *et al.*, "Scaling MOSFETs to 10 nm: Coulomb effects, source starvation, and virtual source model," *Journal of Computational Electronics*, vol. 8, no. 2, pp. 60–77, 2009.

[61] P. Palestri, C. Alexander, A. Asenov, *et al.*, "A comparison of advanced transport models for the computation of the drain current in nanoscale nMOSFETs," *Solid State Electronics*, vol. 53, pp. 1293–1302, 2009.

[62] L. Lucci, P. Palestri, D. Esseni, L. Bergagnini, and L. Selmi, "Multi-subband Monte Carlo study of transport, quantization and electron gas degeneration in ultra-thin SOI n-MOSFETs," *IEEE Trans. on Electron Devices*, vol. 54, no. 5, pp. 1156–1164, 2007.

[63] D. Esseni, "On the modeling of surface roughness limited mobility in SOI MOSFETs and its correlation to the transistor effective field ," *IEEE Trans. on Electron Devices*, vol. 51, no. 3, pp. 394–401, 2004.

[64] D. Querlioz, J. Saint-Martin, K. Huet, *et al.*, "On the ability of particle Monte Carlo technique to include quantum effects in nano-MOSFET simulation," *IEEE Trans. on Electron Devices*, vol. 54, no. 9, pp. 2232–2242, 2007.

[65] V. Aubry-Fortuna, A. Bournel, P. Dollfus, and S. Galdin-Retailleau, "Ultra-short n-MOSFETs with strained Si: device performance and effect of ballistic transport using Monte Carlo simulation," *Semiconductor Science Technology*, vol. 21, pp. 422–428, 2006.

[66] G. Baccarani, E. Gnani, A. Gnudi, S. Reggiani, and M. Rudan, "Theoretical foundations of the quantum drift–diffusion and density-gradient models," *Solid State Electronics*, vol. 52, no. 4, pp. 526–532, 2008.

[67] S. Reggiani, L. Silvestri, A. Cacciatori, *et al.*, "Physically-based unified compact model for low-field carrier mobility in MOSFETs with different gate stacks and biaxial/uniaxial stress conditions," in *IEEE IEDM Technical Digest*, 2007.

[68] L. Silvestri, S. Reggiani, E. Gnani, A. Gnudi, and G. Baccarani, "Unified model for low-field electron mobility in bulk and SOI-MOSFETs with different substrate orientations and its application to quantum drift–diffusion simulation," in *Proc. Int. Conf. on Ultimate Integration on Silicon (ULIS)*, pp. 129–132, 2008.

[69] G. Curatola, G. Doornbos, J. Loo, Y.V. Ponomarev, and G. Iannaccone, "Detailed modeling of sub-100-nm MOSFETs based on Schrödinger DD per subband and experiments and evaluation of the performance gap to ballistic transport," *IEEE Trans. on Electron Devices*, vol. 52, pp. 1851–1858, 2005.

[70] G. Fiori and G. Iannaccone, "Three-dimensional simulation of one-dimensional transport in silicon nanowire transistors," *IEEE Trans. on Nanotechnology*, vol. 6, pp. 524–529, 2007.

[71] J.D. Bude, "MOSFET modeling into the ballistic regime," in *Proc.SISPAD*, pp. 23–26, 2000.

[72] S. Eminente, N. Barin, P. Palestri, C. Fiegna, and E. Sangiorgi, "Small-signal analysis of decananometer bulk and SOI MOSFETs for analog/mixed-signal and RF applications using the time-dependent Monte Carlo approach," *IEEE Trans. on Electron Devices*, vol. 54, pp. 2283–2292, Sept. 2007.

[73] C. Fiegna, M. Braccioli, S.C. Brugger, *et al.*, "Comparison of Monte Carlo transport models for nanometer-size MOSFETs," in *Proc.SISPAD*, pp. 57–60, 2007.

[74] M. De Michielis, D. Esseni, P. Palestri, and L. Selmi, "Semi-classical modelling of quasi-ballistic hole transport in nanoscale *p*-MOSFETs based on a multi-subband Monte Carlo approach," *IEEE Trans. on Electron Devices*, vol. 56, no. 9, pp. 2081–2091, 2009.

[75] L. Lucci, P. Palestri, D. Esseni, and L. Selmi, "Multi-subband Monte Carlo modeling of nano-MOSFETs with strong vertical quantization and electron gas degeneration," in *IEEE IEDM Technical Digest*, pp. 631–634, 2005.

[76] P. Palestri, R. Clerc, D. Esseni, L. Lucci, and L. Selmi, "Multi-subband-Monte Carlo investigation of the mean free path and of the kT layer in degenerated quasi-ballistic nanoMOSFETs," in *IEEE IEDM Technical Digest*, pp. 945–948, 2006.

8 MOS transistors with arbitrary crystal orientation

In the previous chapters we considered devices featuring a silicon channel and a (001) transport plane. These assumptions simplify calculation of the electron subband structure with the EMA approach, since one of the principal axes of the constant energy ellipsoids of the Δ valleys is aligned to the quantization direction. Furthermore, we have considered the [100] transport direction, which is again aligned to one of the principal axes of the constant energy ellipses in the transport plane.

In this chapter we show that semi-classical transport modeling in the frame of multi-subband theory can be supplemented to describe arbitrary orientations by extending quite naturally most of the theoretical concepts explained in the previous chapters. In particular we generalize to arbitrary materials and crystal orientations the EMA (for electron inversion layers) and the $\mathbf{k}\cdot\mathbf{p}$ model (for hole inversion layers) discussed in Chapter 3.

We demonstrate the application of the theory by comparing results for silicon MOSFETs with non-conventional crystal orientations, namely (110) and (111), to the ones for the conventional (001) orientation.

The general theory is also used in Chapter 10 to analyze some alternative channel materials such as germanium and gallium arsenide.

8.1 Electron inversion layers

We appreciated in Section 3.2.1 the usefulness of the effective mass approximation (EMA) for analysis of the electron inversion layer, particularly when the dispersion relationship is assumed to be parabolic and the constant energy surfaces of the bulk crystal are ellipsoids. Here we extend the EMA to arbitrary crystal orientations, considering the Δ as well as the Λ valleys.

8.1.1 Definitions

Let us consider the sketch of the MOSFET in Fig.8.1.a. The coordinate system (x, y, z) is denoted in the following discussion as the Device Coordinate System (DCS), where z is the quantization direction (the direction normal to the silicon–oxide interface) and the drain current flows in the x direction. In this chapter we use lower-case letters to indicate the components of the wave-vector in the transport plane as well as in the 3D

K space. In particular, we use $\mathbf{r} = (x, y)$ and $\mathbf{k} = (k_x, k_y)$ to indicate the position and the wave-vector in the transport plane.

Let us now consider again the Schrödinger equation in the equivalent Hamiltonian approximation (Eq.3.10), which we rewrite as

$$E_\nu\left(-\mathrm{i}\frac{\partial}{\partial x}, -\mathrm{i}\frac{\partial}{\partial y} - \mathrm{i}\frac{\partial}{\partial z}\right)\Phi(\mathbf{R}) + U(z)\,\Phi_{\nu,n}(\mathbf{R}) = E_{\nu,n}\,\Phi_{\nu,n}(\mathbf{R}), \tag{8.1}$$

where $E_\nu(k_x, k_y, k_z)$ is the energy relation around the considered conduction band minimum, expressed in the DCS coordinate system and, differently from Eq.3.10, contains the energy reference $E_{\nu 0}$ of the valley.[1] The conduction band minima of most semiconductors are well represented by ellipsoidal constant energy surfaces (see Section 2.3.1), which are easily described in the so-called ellipsoid coordinate system, ECS (see Fig.8.1.c), with the very general expression

$$E_\nu(k_{t1}, k_{t2}, k_l) = E_{\nu 0} + \frac{\hbar^2}{2}\left(\frac{k_{t1}^2}{m_{t1}} + \frac{k_{t2}^2}{m_{t2}} + \frac{k_l^2}{m_l}\right), \tag{8.2}$$

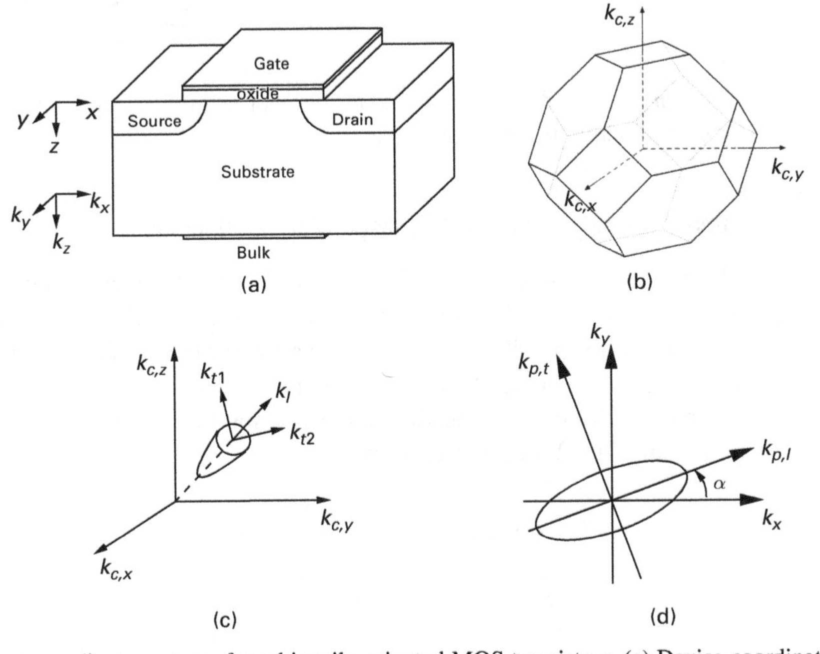

Figure 8.1 Relevant coordinate systems for arbitrarily oriented MOS transistors. (a) Device coordinate system, DCS (k_x,k_y,k_z); (b) Crystal coordinate system, CCS ($k_{c,x}$,$k_{c,y}$,$k_{c,z}$); (c) Ellipsoid coordinate system, ECS (k_{t1},k_{t2},k_l); (d) In-plane ellipse coordinate system, EpCS ($k_{p,t}$,$k_{p,l}$), in the transport plane. *Reprinted with permission from [1]. Copyright 2009 by Springer.*

[1] We thus call it E_ν instead of $E_{cb}^{(\nu)}$ as in Eq.3.10 to avoid confusion.

where $E_{\nu 0}$ is the energy of the bottom of the valley, the wave-vectors k_{t1}, k_{t2}, and k_l are referred to the minimum of the valley $\mathbf{K}_\nu$ and m_{t1}, m_{t2}, and m_l are the three effective masses in the directions k_{t1}, k_{t2}, and k_l.

In unstrained cubic semiconductors, symmetry considerations give $m_{t1} = m_{t2} = m_t$. If the DCS and the ECS are aligned to each other, Eq.8.2 reduces to Eq.2.60, with m_x, m_y, and m_z equal to either m_l or m_t depending on the direction of the principal axis of the ellipsoid.

Moreover, if the axes of the ECS coincide with those of the DCS (as in the previous chapters), we can solve Eq.8.1 by writing

$$\Phi_{\nu,n}(\mathbf{R}) = \xi_{\nu,n}(z)\,\frac{\mathrm{e}^{\mathrm{i}(k_x x + k_y y)}}{\sqrt{A}}, \tag{8.3}$$

where $\xi_{\nu,n}(z)$ is the solution of Eq.3.16:

$$-\frac{\hbar^2}{2m_z}\frac{\partial^2}{\partial z^2}\xi_{\nu,n}(z) + U(z)\xi_{\nu,n}(z) = \varepsilon_{\nu,n}\,\xi_{\nu,n}(z), \tag{8.4}$$

and the total energy is:

$$E_{\nu,n} = E_{\nu 0} + \varepsilon_{\nu,n} + \frac{\hbar^2}{2}\left(\frac{k_x^2}{m_x} + \frac{k_y^2}{m_y}\right), \tag{8.5}$$

where m_x, m_y, and m_z are equal to m_l, m_{t1}, or m_{t2} depending on which axis of the ECS is oriented along the x, y, or z direction.

8.1.2 Subband energy and in-plane dispersion relationship

The general case where the ECS and the DCS do not coincide may appear significantly more difficult to handle. However, following the approach proposed in [2] and summarized below it is possible to find an expression for the envelope wave-function $\Phi_{\nu,n}(\mathbf{R})$ similar to Eq.8.3 but applicable to arbitrary orientations.

First of all we write the dispersion relationship around the νth conduction band minimum (Eq.8.2) in a matrix form as

$$E_\nu(k_{t1}, k_{t2}, k_l) = E_{\nu 0} + \frac{\hbar^2}{2}(k_{t1}, k_{t2}, k_l)\cdot\mathbf{W}_{ECS}\cdot(k_{t1}, k_{t2}, k_l)^T, \tag{8.6}$$

where (k_{t1}, k_{t2}, k_l) is a row vector and

$$\mathbf{W}_{ECS} = \begin{bmatrix} 1/m_{t1} & 0 & 0 \\ 0 & 1/m_{t2} & 0 \\ 0 & 0 & 1/m_l \end{bmatrix}. \tag{8.7}$$

The matrix $\mathbf{W}_{ECS}$ depends on the considered conduction band minima, that is on the index ν. However, we have dropped the indication of the minima to simplify the notation.

The wave-vector in the ECS can be related to the one in the DCS by the linear relationship

$$(k_{t1}, k_{t2}, k_l)^T = \mathbf{R}_{D\to E} \cdot (k_x, k_y, k_z)^T, \tag{8.8}$$

where the columns of the matrix $\mathbf{R}_{D\to E}$ are the components in the ECS of the basis vectors of the DCS. We can thus write $E_\nu(\mathbf{K})$ as a function of the wave-vector in the DCS as

$$E_\nu(k_x, k_y, k_z) = E_{\nu 0} + \frac{\hbar^2}{2}(k_x, k_y, k_z) \cdot \mathbf{W}_{DCS} \cdot (k_x, k_y, k_z)^T, \tag{8.9}$$

where

$$\mathbf{W}_{DCS} = \begin{bmatrix} w_{11} & w_{12} & w_{13} \\ w_{21} & w_{22} & w_{23} \\ w_{31} & w_{32} & w_{33} \end{bmatrix} = \mathbf{R}^T_{D\to E} \cdot \mathbf{W}_{ECS} \cdot \mathbf{R}_{D\to E}. \tag{8.10}$$

The *ansatz* for the unknown wave-function is [2]

$$\Phi_{\nu,n}(\mathbf{R}) = \frac{\xi_{\nu,n}(z)}{\sqrt{A}} \exp\left[\mathrm{i}(k_x x + k_y y)\right] \exp\left[-\mathrm{i}\frac{(w_{13}k_x + w_{23}k_y)z}{w_{33}}\right], \tag{8.11}$$

which is the desired envelope wave-function, that is the solution of Eq.8.1 for a generic orientation of the constant energy ellipsoid. The function $\xi(z)$ is obtained by solving Eq.8.4 using $m_z = 1/w_{33}$ as quantization mass:

$$-\frac{\hbar^2 w_{33}}{2}\frac{\partial^2}{\partial z^2}\xi_{\nu,n}(z) + U(z)\xi(z) = \varepsilon_{\nu,n}\,\xi_{\nu,n}(z). \tag{8.12}$$

The subband energies $\varepsilon_{\nu,n}$ given by the solution of Eq.8.12 enter the expression of the total energy [2]

$$E_{\nu,n}(k_x, k_y) = E_{\nu 0} + \varepsilon_{\nu,n} + \frac{\hbar^2}{2}\left[\left(w_{11} - \frac{w_{13}^2}{w_{33}}\right){k_x}^2 + 2\left(w_{12} - \frac{w_{13}w_{23}}{w_{33}}\right)k_x k_y + \left(w_{22} - \frac{w_{23}^2}{w_{33}}\right){k_y}^2\right]. \tag{8.13}$$

Therefore, the calculation of the subband energy and of the in-plane energy dispersion for arbitrary orientations in the EMA framework is accomplished by solving a single **k**-independent Schrödinger-like equation (Eq.8.12) with a suitable definition of the quantization effective mass $m_z = 1/w_{33}$.

It is worth pointing out that Eq.8.12 is **k**-independent only with boundary conditions in the form $\Phi(z = z_i) = 0$, where z_i is the interface position [2]. In fact, these boundary conditions imply $\xi(z = z_i) = 0$. Different boundary conditions may result in **k**-dependent wave-functions $\xi(z)$ and eigenvalues $\varepsilon_{\nu,n}$, thus requiring the solution of Eq.8.12 for all **k** values in the transport plane. In the following we restrict our analysis to cases where only one Schrödinger equation for each valley ν is necessary.

The choice of the energy reference for the potential energy in Eq.8.12 and its relation with the $E_{\nu 0}$ value adopted throughout the book deserve some attention. As discussed in Section 3.7, we can define $U(z) = -[e\phi(z) + \chi(z)]$. Following this choice, the potential energy profile used in Eq.8.12 is the same for all valleys. The electron affinity χ in the semiconductor is determined by the position of the lowest conduction band minimum with respect to the vacuum level. This definition of $U(z)$ unequivocally sets the $E_{\nu 0}$ for the lowest conduction band. The higher conduction band minima produce valleys in the inversion layer with higher $E_{\nu 0}$ values. According to Eq.8.13, the $\varepsilon_{\nu,n}$ obtained from Eq.8.12 for a given valley must be added to the corresponding $E_{\nu 0}$.

8.1.3 Carrier dynamics

One of the most prominent advantages of the generalized EMA approach proposed in the previous section is that, when applied to a multi-subband treatment of carrier transport in the inversion layer (Section 5.2), the total energy can be separated into a **k**-independent term given by the solution of the Schrödinger equation (i.e. the term $E_{\nu 0} + \varepsilon_{\nu,n}$) and an in-plane energy component that does not depend on the position. This partitioning of the total energy significantly simplifies the treatment of carrier transport. Consistent with Section 5.2 (see Eqs.5.36 and 5.37 and the related discussion), we observe that the driving force experienced by the carriers is $\mathbb{F} = -(\mathrm{d}\varepsilon_{\nu,n}/\mathrm{d}x)$, where $\varepsilon_{\nu,n}$ is the subband energy relative to $E_{\nu 0}$ and in the most general case depends on x because the potential energy $U(z)$ entering the Schrödinger equation has a different profile in each section x.

To determine the response of the electrons to the driving force, it is convenient to rewrite the expression of the total energy (Eq.8.13) in a compact form:

$$E_{\nu,n}(k_x, k_y) = E_{\nu 0} + \varepsilon_{\nu,n} + \frac{\hbar^2}{2}\left[(k_x, k_y)\begin{bmatrix} A & C \\ C & B \end{bmatrix}(k_x, k_y)^T\right], \tag{8.14}$$

where

$$A = w_{11} - \frac{w_{13}^2}{w_{33}}, \quad B = w_{22} - \frac{w_{23}^2}{w_{33}}, \quad C = w_{12} - \frac{w_{13}w_{23}}{w_{33}}. \tag{8.15}$$

We see that the energy in Eq.8.14 is an elliptic function of $\mathbf{k} = (k_x, k_y)$, but the principal axes of the ellipse are not aligned with the k_x and k_y directions of the DCS, unless the second term in the square brackets of Eq.8.13 is null. From Eq.8.14 it is easy to determine the principal axes of the ellipse (which define the in-plane ellipse coordinate system, EpCS, represented in Fig.8.1.d), the angle α between the longitudinal axis of the ellipse and the k_x axis of the DCS (Fig.8.1.d), and the effective masses $m_{p,t}$ and $m_{p,l}$ in the EpCS, which are henceforth denoted effective masses in the transport plane or simply transport masses. In fact we can write

$$(k_x, k_y)\begin{bmatrix} A & C \\ C & B \end{bmatrix}(k_x, k_y)^T = (k_{p,l}, k_{p,t})\begin{bmatrix} 1/m_{p,l} & 0 \\ 0 & 1/m_{p,t} \end{bmatrix}(k_{p,l}, k_{p,t})^T. \tag{8.16}$$

Diagonalization of the $(A, C; C, B)$ matrix gives

$$m_{p,l} = \frac{2}{A + B - \sqrt{(A-B)^2 + 4C^2}}, \tag{8.17a}$$

$$m_{p,t} = \frac{2}{A + B + \sqrt{(A-B)^2 + 4C^2}}, \tag{8.17b}$$

$$\alpha = \arctan\left(\frac{C}{m_{p,l}^{-1} - B}\right). \tag{8.17c}$$

When evaluating carrier dynamics (e.g. the free-flights in the Monte Carlo procedure, see Chapter 6) we may choose to work in the EpCS, because it makes it easier to compute the group velocity:

$$v_{g,l} = \frac{\hbar k_{p,l}}{m_{p,l}}, \tag{8.18a}$$

$$v_{g,t} = \frac{\hbar k_{p,t}}{m_{p,t}}. \tag{8.18b}$$

On the other hand, the driving force $-(\mathrm{d}\varepsilon_{\nu,n}/\mathrm{d}x)\hat{\mathbf{x}}$ is aligned with the x direction of the DCS in Fig.8.1.a. Hence, in order to evaluate the carrier dynamics in the EpCS, we have to transform the driving force to the EpCS and write

$$\hbar\frac{\mathrm{d}k_{p,l}}{\mathrm{d}t} = -\frac{\mathrm{d}\varepsilon_{\nu,n}}{\mathrm{d}x}\cos\alpha, \tag{8.19a}$$

$$\hbar\frac{\mathrm{d}k_{p,t}}{\mathrm{d}t} = \frac{\mathrm{d}\varepsilon_{\nu,n}}{\mathrm{d}x}\sin\alpha. \tag{8.19b}$$

Finally, the velocity along the x direction of the DCS is simply

$$\frac{\mathrm{d}x}{\mathrm{d}t} = v_{g,l}\cos\alpha - v_{g,t}\sin\alpha. \tag{8.20}$$

As discussed in Section 5.2, the semi-classical model assumes that carriers do not change subband during free-flight. However, for wave-functions in the form of Eq.8.11, it has been demonstrated in [3] that, if w_{13} or w_{23} are not null, carriers can change subbands even during free-flight. This effect is difficult to include in the BTE and, to our best knowledge, it is usually neglected.

8.1.4 Change of the coordinates system

The link between the effective masses in the EpCS, the quantization mass m_z and the masses m_l, m_{t1}, and m_{t2} in the ECS is provided by the rotation matrix $\mathbf{R}_{D\to E}$. To identify this matrix it is useful to split the transition from DCS to ECS into a transition from DCS to the Crystal-Coordinate-System (CCS, Fig.8.1.b), and then from CCS to ECS. In other words we can write

$$\mathbf{R}_{D\to E} = \mathbf{R}_{C\to E}\cdot\mathbf{R}_{D\to C}. \tag{8.21}$$

It is quite straightforward to build the rotation matrix from the DCS to the CCS and from the CCS to the ECS if we recall that the columns of the transform matrix from the

Table 8.1 Rotation matrix from the DCS to the CCS for the most common crystal orientations. For a given orientation (namely a given quantization direction *z* in the DCS) the *x* direction of the DCS can be aligned with different crystal directions. The matrices given in the table correspond to the *x* direction indicated in the third column (see also the sketch in Fig.8.3).

Orientation	$\mathbf{R}_{D\to C}$	x-direction
(001)	$\begin{bmatrix} 1 & 0 & 0 \\ 0 & 1 & 0 \\ 0 & 0 & 1 \end{bmatrix}$	[100]
(111)	$\begin{bmatrix} -2/\sqrt{6} & 0 & 1/\sqrt{3} \\ 1/\sqrt{6} & -1/\sqrt{2} & 1/\sqrt{3} \\ 1/\sqrt{6} & 1/\sqrt{2} & 1/\sqrt{3} \end{bmatrix}$	$[\bar{2}11]$
(110)	$\begin{bmatrix} 0 & 1/\sqrt{2} & 1/\sqrt{2} \\ 0 & -1/\sqrt{2} & 1/\sqrt{2} \\ 1 & 0 & 0 \end{bmatrix}$	[001]

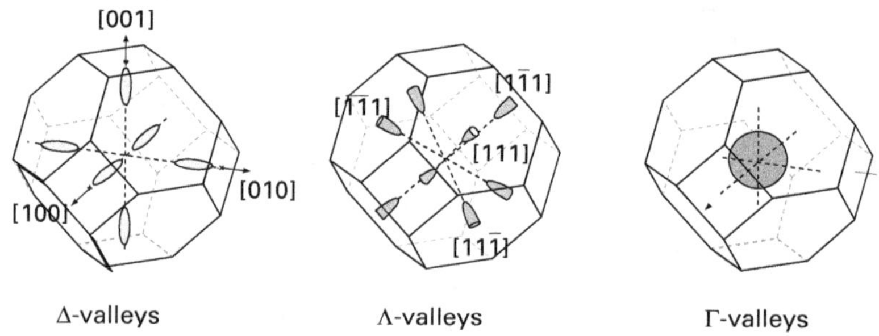

Figure 8.2 The Δ, Λ, and Γ valleys and their position in the first Brillouin zone of face-centered-cubic semiconductors.

coordinate system A to the coordinate system B are the components of the basis vectors of A in the system B.

Following this simple recipe, we can derive the transformation from the CCS to the ECS. The minima of the conduction band of most common semiconductors are of type Δ, Λ or Γ, as sketched in Fig.8.2. The case of the minimum in Γ is trivial since the constant energy surface is a sphere, and thus the quantization and the in-plane masses m_z, $m_{p,l}$, and $m_{p,t}$ are all equal to the effective mass of the minimum, regardless of the crystal orientation. This means that the rotation matrix is the identity matrix. In the other cases, each minimum has its own rotation matrix. For the Δ valleys we have six minima but only three matrices are needed for symmetry reasons. Four matrices instead are needed for the Λ valleys.

The rotation matrices from the DCS to the CCS are given in Table 8.1 for the most common crystal orientations, whereas the rotation matrices from the CCS to the ECS are given in Table 8.2 for the Δ and Λ valleys.

Table 8.2 Rotation matrix from the CCS to the ECS for the Δ and Λ valleys.

Minimum	$\mathbf{R}_{C\to E}$	Minimum	$\mathbf{R}_{C\to E}$
$\Delta^{[100]}$	$\begin{bmatrix} 0 & 1 & 0 \\ 0 & 0 & 1 \\ 1 & 0 & 0 \end{bmatrix}$	$\Lambda^{[111]}$	$\begin{bmatrix} -1/\sqrt{6} & -1/\sqrt{6} & 2/\sqrt{6} \\ -1/\sqrt{2} & 1/\sqrt{2} & 0 \\ 1/\sqrt{3} & 1/\sqrt{3} & 1/\sqrt{3} \end{bmatrix}$
$\Delta^{[010]}$	$\begin{bmatrix} 0 & 0 & 1 \\ 1 & 0 & 0 \\ 0 & 1 & 0 \end{bmatrix}$	$\Lambda^{[\bar{1}11]}$	$\begin{bmatrix} -1/\sqrt{6} & 1/\sqrt{6} & -2/\sqrt{6} \\ 1/\sqrt{2} & 1/\sqrt{2} & 0 \\ -1/\sqrt{3} & 1/\sqrt{3} & 1/\sqrt{3} \end{bmatrix}$
$\Delta^{[001]}$	$\begin{bmatrix} 1 & 0 & 0 \\ 0 & 1 & 0 \\ 0 & 0 & 1 \end{bmatrix}$	$\Lambda^{[1\bar{1}1]}$	$\begin{bmatrix} 1/\sqrt{6} & -1/\sqrt{6} & -2/\sqrt{6} \\ 1/\sqrt{2} & 1/\sqrt{2} & 0 \\ 1/\sqrt{3} & -1/\sqrt{3} & 1/\sqrt{3} \end{bmatrix}$
		$\Lambda^{[11\bar{1}]}$	$\begin{bmatrix} -1/\sqrt{6} & -1/\sqrt{6} & -2/\sqrt{6} \\ -1/\sqrt{2} & 1/\sqrt{2} & 0 \\ 1/\sqrt{3} & 1/\sqrt{3} & -1/\sqrt{3} \end{bmatrix}$

Table 8.3 Effective masses and multiplicity (μ_v) for the Δ and Λ valleys considering different orientations of the transport plane. m_l and $m_{t1} = m_{t2} = m_t$ are the effective masses of the bulk crystal ellipsoids (see Eq.8.2) .

Orientation	$m_{p,t}$	$m_{p,l}$	m_z	μ_v
		Δ valleys		
(001)	m_t	m_t	m_l	2
	m_t	m_l	m_t	4
(110)	m_t	$(m_t + m_l)/2$	$(2m_t m_l)/(m_t + m_l)$	4
	m_t	m_l	m_t	2
(111)	m_t	$(m_t + 2m_l)/3$	$(3m_t m_l)/(m_t + 2m_l)$	6
		Λ valleys		
(001)	m_t	$(m_t + 2m_l)/3$	$(3m_t m_l)/(m_t + 2m_l)$	4
(110)	m_t	$(2m_t + m_l)/3$	$(3m_t m_l)/(2m_t + m_l)$	2
	m_t	m_l	m_t	2
(111)	m_t	m_t	m_l	1
	m_t	$(m_t + 8m_l)/9$	$(9m_t m_l)/(m_t + 8m_l)$	3

Using these matrices we obtain the expression for the effective quantization mass m_z and the effective transport masses $m_{p,t}$ and $m_{p,l}$ reported in Table 8.3 for the most common crystal orientations. The table also gives the multiplicity of the different valleys, here defined as the number of valleys featuring the same values for m_z, $m_{p,t}$, and $m_{p,l}$. In equilibrium conditions these valleys are equivalent, whereas in the presence of

Table 8.4 Numerical value (in units of m_0) of the effective masses for the Δ valleys of silicon for different orientations of the transport plane.

Orientation	$m_{p,t}$	$m_{p,l}$	m_z	μ_v
(001)	0.19	0.19	0.92	2
	0.19	0.92	0.19	4
(110)	0.19	0.555	0.315	4
	0.19	0.92	0.19	2
(111)	0.19	0.74	0.26	6

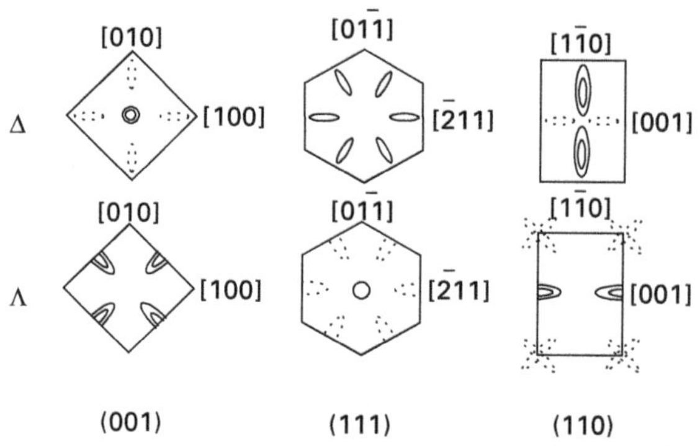

Figure 8.3 The constant energy curves corresponding to Δ and Λ valleys in inversion layers with different orientations. The figure also indicates the first Brillouin zone (1BZ) of the 2D gas. The shape of the 1BZ for the (001) case is discussed in Section 3.4. Solid line: unprimed. Dashed line: primed.

an applied electric field they are not, because they have a different value of the angle α between the channel direction in the DCS and the principal axis of the EpCS (see Fig.8.1.d). This means that valleys which are equivalent at equilibrium can feature different occupations when the system is driven out of equilibrium, so that they must be handled separately when solving the Boltzmann equation. However, provided that we use the EpCS as the coordinate system, the total scattering rate out of a given state is the same in all these valleys because they share the same m_z and thus the wave-functions are the same. This means that the matrix elements for the transition rates are also the same, see Eq.4.4. Furthermore, sharing the same $m_{p,t}$ and $m_{p,l}$, the integration of the transition rates over the final states gives the same results, since integration is performed on the same equi-energy curves.

The effective masses in Table 8.3 describe the properties of inversion layers with different orientations. Numerical values for the Δ valleys of silicon are given in Table 8.4. The corresponding constant energy curves for Δ and Λ valleys are sketched in Fig.8.3.

The case of Δ valleys (i.e. materials such as silicon) and (001) orientation of the transport plane has already been described in the Chapters 3 and 7. Figure 8.3 confirms that in this case we have 2-fold valleys (solid lines) with quantization mass $m_z = m_l$, featuring circular constant energy curves with effective mass m_t in the transport plane, and 4-fold valleys (dotted lines) with $m_z = m_t$ and elliptical constant energy curves.

Based on the discussion above, the 4-fold valleys should be grouped in pairs and treated separately when dealing with the transport.

Considering Δ valleys and a (111) orientation of the transport plane, all the valleys of the 2D gas feature the same quantization mass and constant energy curves with the same $m_{p,t}$ and $m_{p,l}$. This results in the multiplicity of six in Table 8.3; the six valleys give three different groups for transport simulations.

In (110) inversion layers, instead, the Δ valleys split in two families: 4-fold valleys with constant energy ellipses oriented along the $[1\bar{1}0]$ direction (solid lines in Fig.8.3) and 2-fold valleys oriented along the [001] direction (dotted lines in Fig.8.3). Due to their large m_z, the 4-fold valleys are the lowest subbands in the (110) inversion layer.

Considering Λ valleys (which are the dominant ones in germanium, Section 10.6), in the case of (001) inversion layers all valleys feature the same quantization and transport masses (but different α, see the lower left plot in Fig.8.3). Note that the multiplicity is 4 and not 8, because the 8 minima are exactly at the border of the first Brillouin zone, so that each of them contributes to the total density of states by 1/2 (see Section 2.3.1). Also in (110) and (111) orientations the multiplicities in Table 8.3 sum to 4, but we see two families of valleys. In the (111) inversion layer the lowest subbands have a multiplicity of 1 and are circular (solid line in Fig.8.3), while the other subbands are elliptical with multiplicity 3. Finally, in (110) inversion layers we have two families of subbands each with multiplicity 2.

8.1.5 Scattering rates

By comparing Eq.8.11 to Eq.4.5, we see that calculation of the scattering rates for electron inversion layers with arbitrary crystal orientations requires use of **k** dependent wave-functions,

$$\xi^{\mathrm{arb}}_{\nu,n\mathbf{k}}(z) = \xi_{\nu,n}(z)\ \exp\left[\frac{-\mathrm{i}\,(w_{13}k_x + w_{23}k_y)z}{w_{33}}\right], \tag{8.22}$$

in the expressions for the scattering matrix elements derived in Sections 4.3 to 4.5. Hence, the main difference with respect to the matrix elements derived in Chapter 4 is that, due to the **k** dependence of $\xi^{\mathrm{arb}}_{\nu,n\mathbf{k}}(z)$, the electron matrix elements for arbitrary crystal orientations may depend on the initial and final wave-vectors **k** and **k**′, not only on the exchanged wave-vector $\mathbf{q} = \mathbf{k}' - \mathbf{k}$. Since for the Δ valleys in (001) inversion layers we have $w_{13} = w_{23} = 0$, in such a case $\xi^{\mathrm{arb}}_{\nu,n\mathbf{k}}(z)$ simplifies to the **k** independent $\xi_{\nu,n}(z)$ used throughout Chapter 4.

The **k** dependence of the matrix elements complicates the picture significantly and increases the computational complexity such that, to the best of our knowledge, all authors have neglected the exponential term in Eq.8.22 and used a **k**-independent $\xi_{\nu,n}(z)$ to compute the scattering rates. We adopt this simplification also in the simulations reported in the following sections.

Apart from this aspect, evaluation of the scattering rates by integration of the matrix elements over the possible final states proceeds exactly as in Chapter 4, by working in the EpCS and replacing m_x and m_y with $m_{p,l}$ and $m_{p,t}$, respectively.

The treatment of screening should also be consistent with the **k** dependent wave-function in Eq.8.22, so that one should use Eqs.4.83 and 4.84 for the dielectric function. However, as in the computation of the matrix elements, to our best knowledge all authors have neglected the exponential term in Eq.8.22 and used only $\xi_{\nu,n}(z)$ in calculation of the screening effect.

8.2 Hole inversion layers

The 6×6 **k·p** Hamiltonian introduced in Section 2.2.2 and applied to hole inversion layers in Section 3.3.1 has been written in the CCS. So, when different coordinate systems are considered, we more explicitly indicate that the components of the **K** vector in the 3×3 matrix $\mathbf{H}_{\mathbf{k\cdot p}}^{3\times3}$ (henceforth denoted by lower case k) are the ones in the CCS:

$$\mathbf{H}_{\mathbf{k\cdot p}}^{3\times3} = \begin{bmatrix} Lk_{c,x}^2 + M(k_{c,y}^2 + k_{c,z}^2) & Nk_{c,x}k_{c,y} & Nk_{c,x}k_{c,z} \\ Nk_{c,x}k_{c,y} & Lk_{c,y}^2 + M(k_{c,x}^2 + k_{c,z}^2) & Nk_{c,y}k_{c,z} \\ Nk_{c,x}k_{c,z} & Nk_{c,y}k_{c,z} & Lk_{c,z}^2 + M(k_{c,x}^2 + k_{c,y}^2) \end{bmatrix}, \tag{8.23}$$

where $\mathbf{k}_c$ is the wave-vector in the CCS defined in Fig.8.1.b.

When evaluating the subband structure of the hole inversion layer, the substitution $k_z \to -\mathrm{i}\frac{\partial}{\partial z}$ holds in the DCS, so that the transformation

$$(k_{c,x}, k_{c,y}, k_{c,z})^T = \mathbf{R}_{D\to C} \cdot (k_x, k_y, k_z)^T \tag{8.24}$$

is necessary before introducing the substitution $k_z \to -\mathrm{i}\frac{\partial}{\partial z}$.

For example, for (110) inversion layers, the elements of $\mathbf{H}_{\mathbf{k\cdot p}}^{3\times3}$ obtained by substituting Eq.8.24 into Eq.8.23 using the corresponding rotation matrix given in Table 8.1 are

$$H_{\mathbf{k\cdot p},(1,1)}^{3\times3} = Mk_x^2 + \frac{L+M}{2}(k_z^2 + k_y^2) + (L-M)k_yk_z, \tag{8.25a}$$

$$H_{\mathbf{k\cdot p},(2,2)}^{3\times3} = Mk_x^2 + \frac{L+M}{2}(k_z^2 + k_y^2) + (M-L)k_yk_z, \tag{8.25b}$$

$$H_{\mathbf{k\cdot p},(3,3)}^{3\times3} = Lk_x^2 + M(k_y^2 + k_z^2), \tag{8.25c}$$

$$H_{\mathbf{k\cdot p},(1,2)}^{3\times3} = H_{\mathbf{k\cdot p},(2,1)}^{3\times3} = \frac{N}{2}(k_z^2 - k_y^2), \tag{8.25d}$$

$$H_{\mathbf{k\cdot p},(1,3)}^{3\times3} = H_{\mathbf{k\cdot p},(3,1)}^{3\times3} = \frac{N}{\sqrt{2}}k_x(k_y + k_z), \tag{8.25e}$$

$$H_{\mathbf{k\cdot p},(2,3)}^{3\times3} = H_{\mathbf{k\cdot p},(3,2)}^{3\times3} = \frac{N}{\sqrt{2}}k_x(k_z - k_y). \tag{8.25f}$$

The subbands in the 2D hole inversion layer are then found solving the differential eigenvalue problem in Eq.3.34, that can be converted to an algebraic eigenvalue problem as explained in Section 3.3.2. We thus obtain a numerical formulation identical to

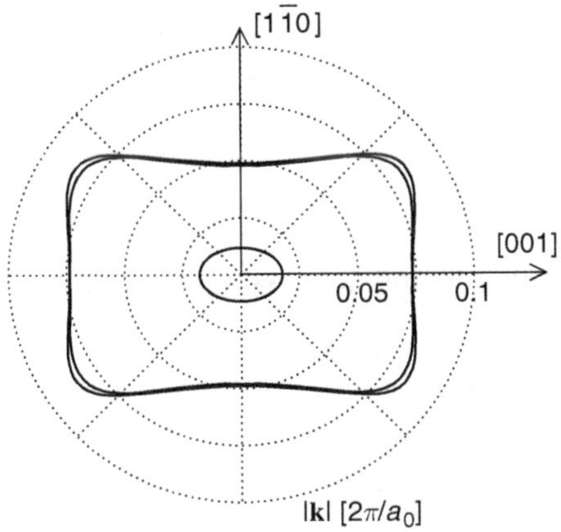

Figure 8.4 Constant energy lines for holes (100meV above the subband minimum, in an electron-like energy convention) for a (110) silicon inversion layer. The quantization well is triangular with an electric field of 1 MV/cm. We have two pairs of curves (although the two close to the point $\mathbf{k} = \mathbf{0}$ are essentially indistinguishable) since the spin-orbit coupling removes the spin degeneracy.

Eq.3.36, where the matrices $\mathbf{D}_-$, $\mathbf{D}_l$ and $\mathbf{D}_+$ have to be modified according to the $H^{3\times3}_{\mathbf{k}\cdot\mathbf{p},(i,j)}$ of Eq.8.25. An example of the results is given by the constant energy lines for a (110) silicon inversion layer in Fig.8.4.

The methodology explained above requires analytically finding the expression for $H^{3\times3}_{\mathbf{k}\cdot\mathbf{p},(i,j)}$ and then $\mathbf{D}_-$, $\mathbf{D}_l$, and $\mathbf{D}_+$ for each crystal orientation. A procedure to automate this step has been proposed in [4].

Apart from the calculation of the band structure, in the $\mathbf{k}\cdot\mathbf{p}$ approach the crystal orientation does not change the computation of the scattering rates and of the screening and it does not alter the description of the carrier dynamics with respect to that presented in the previous chapters.

When, instead, hole inversion layers are described with the analytical model of Section 3.3.3, the periodicity of the equi-energy curves as a function of the polar angle in the $\mathbf{k}$ plane changes with the orientation. For instance, it is clear from Fig.8.4 that for the (110) transport plane the angular periodicity of the equi-energy contour is π instead of $\pi/2$ as it was for the (001) plane (compare Fig.8.4 with Fig.3.5). Consequently, a change of the crystal orientation requires re-calibrating the model against the $\mathbf{k}\cdot\mathbf{p}$ results.

More details about, and results from, use of the analytical model described in Section 3.3.3 for (110) hole inversion layers are given in [5]. The need to recalibrate the model makes it inefficient for systematically studying the influence of the crystal orientation on transport in p-MOSFETs.

8.3 Simulation results

In this section we illustrate the transport properties of silicon inversion layers with crystal orientation other than (001). Electron inversion layers have been simulated with the multi-subband Monte Carlo model described in [6] (Section 6.3.3). The low field mobility of hole inversion layers has been computed with the MRT method (Section 5.4) based on the subbands and wave-functions obtained with the **k·p** method. The on-current in nanoscale p-MOSFETs, instead, has been simulated with the MSMC program described in [7].

8.3.1 Mobility in electron and hole inversion layers

The mobilities of electron inversion layers with (001), (110), and (111) crystal orientation are compared in Fig.8.5. The effective field F_{eff} used as x-axis has been computed following Eq.7.6 and by using $\eta = 1/2$ for (001) inversion layers (see Section 7.1.2), but $\eta = 1/3$ for (110) and (111) inversion layers [8].

In Fig.8.5 we see that the simulations (symbols) nicely reproduce the experimental data. A single set of parameters for phonon and surface roughness scattering has been used (see Table 7.1 in Chapter 7). The (001) orientation provides the largest mobility. This can be easily understood by using Table 8.4 and Fig.8.3: in the (001) case the dominant subbands (that is the most populated, because of the largest quantization mass) also feature a lower conduction effective mass ($m_t = 0.19m_0$) with respect to the (110) and (111) cases. Also the density of states effective mass $m_d = \sqrt{m_{p,t}m_{p,t}}$ (see Section 3.5.2) is the lowest, resulting in lower scattering rates.

Figure 8.6 demonstrates that the mobility in (001) and (111) inversion layers does not depend on the transport direction, due to the high symmetry of the constant energy curves (see Fig.8.3). The mobility in (110) electron inversion layers, instead, strongly depends on the direction of the current flow in the transport plane (circles in Fig.8.6). This is consistent with the fact that the dominant subbands feature elliptical constant energy contours elongated in the $[1\bar{1}0]$ direction (see Fig.8.3). Consistently, the lowest mobility is found for the $[1\bar{1}0]$ transport direction, whereas the highest is found for the [100].

As an important applications-related remark, we note that the sidewall interfaces of FinFET devices drawn with a Manhattan layout in {100} wafers (see Fig.8.7.a) lie on

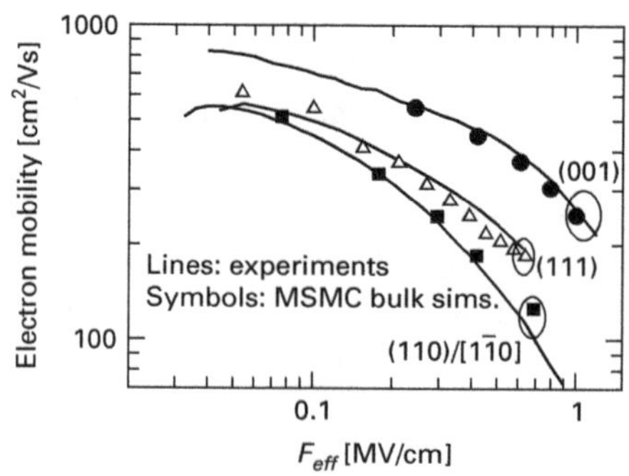

Figure 8.5 Electron effective mobility versus effective field for unstrained bulk silicon with different crystal orientations. The transport is aligned with the $[1\bar{1}0]$ direction in all cases. Lines: experiments from [8, 9] (low doping case). Symbols: MSMC simulations. *Reprinted with permission from [1]. Copyright 2009 by Springer.*

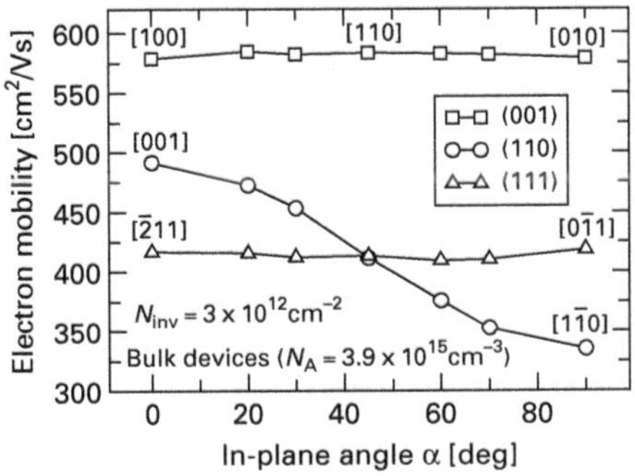

Figure 8.6 Simulated electron effective mobility versus transport direction for the (001), (110), and (111) transport planes. The in-plane angle of the transport direction is referred to the x-direction defined in the third column of Table 8.1.

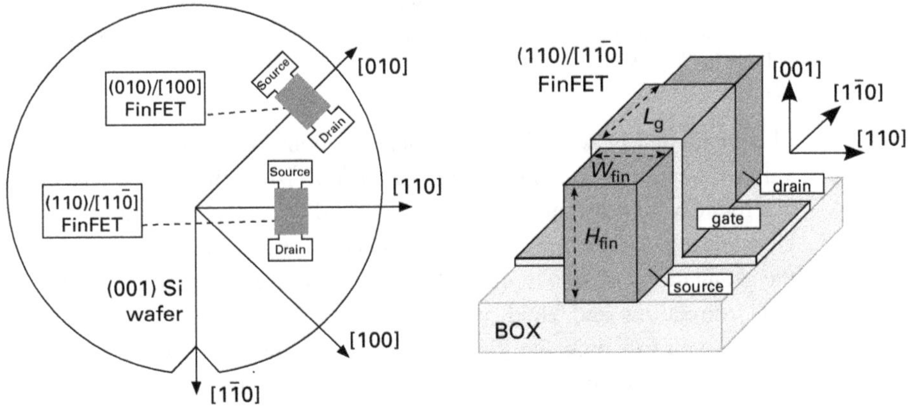

Figure 8.7 a) Possible orientations for FinFET devices drawn on a (001) silicon wafer. b) Sketch of the device and indication of the transport and quantization directions for a SOI FinFET oriented along the $[1\bar{1}0]$ direction. *Reprinted with permission from [10]. Copyright 2010 by the Institute of Electrical and Electronics Engineers.*

$\{110\}$ planes with $[1\bar{1}0]$ transport direction (see Fig.8.7.b). As a result, the corresponding mobility is smaller than in conventional (001) inversion layers, mainly because the 4-fold valleys in (110) inversion planes feature an effective mass in the $[1\bar{1}0]$ direction (the x direction in the DCS) which is significantly higher than the one for (001) inversion layers ($m_x = (m_l + m_t)/2 = 0.555m_0$ versus $m_t = 0.19m_0$ for (001) in Tables 8.3, 8.4). However, experimental mobility data for FinFETs are usually larger than the data for (110) planar MOSFETs [11], mainly due to unintentional strain in the fin, as we see in the next chapter. If FinFETs are drawn with a 45 degree orientation, instead, the sidewall interfaces correspond to (100) inversion layers.

Concerning the simulation of (110) inversion layers, it should be mentioned that the treatment of arbitrary orientations with a parabolic dispersion relationship can become inaccurate when applied to the 2-fold valleys of the (110) inversion layer. This is due to the non-parabolic and anisotropic nature of the Δ valleys in bulk silicon [12]

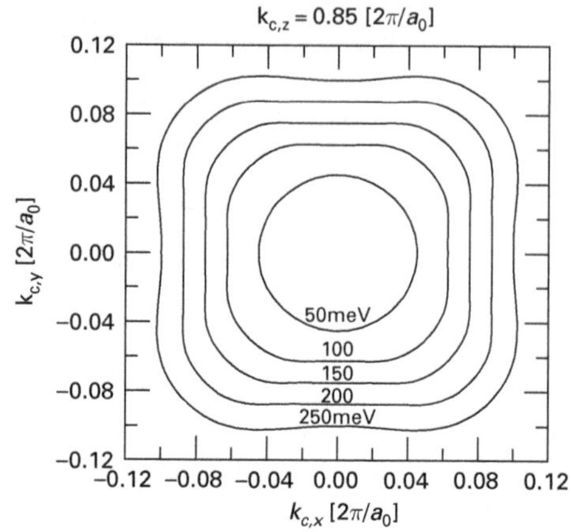

Figure 8.8 Constant energy lines for unstrained bulk silicon. Since we set $k_{c,z} = 0.85 \times 2\pi/a_0$, the point $k_{c,x} = k_{c,y} = 0$ corresponds to the minimum of one of the Δ valleys that generate the 2-fold valleys in (110) silicon inversion layers.

illustrated in Fig.2.10. In fact, when the constant energy surfaces of the silicon Δ valleys are cut with a plane normal to the $\langle 100 \rangle$ directions, constant energy curves are found which are not circles, but instead appear elongated in the $\langle 110 \rangle$ directions. Considering for example the valleys along the [001] direction, Fig.8.8 shows that the elongations of the supposedly circular equi-energy curves occur in the [011] and $[0\bar{1}1]$ directions and yield an energy dependent effective mass larger than the value $0.19m_0$ valid in both the [100] and [010] directions. Comparison between parabolic EMA and LCBB calculations suggests that an empirical adjustment of the quantization mass to $m_z = 0.23m_0$ provides a simple way to account for the effect described above [13].

The mobility in hole inversion layers is shown in Fig.8.9. Simulation parameters are not the same as in Table 7.1 because in Fig.8.9 surface roughness scattering is unscreened. The (110) crystal orientation has the largest mobility because the transport effective mass, defined as the inverse curvature of the valence band energy relation (Eq.2.64), is smallest.

Figure 8.9 illustrates the effect of the transport direction on the hole effective mobility for the (110) transport plane. Transport in the $[0\bar{1}1]$ direction has higher mobility than in the [001] direction, consistent with the anisotropy of the constant energy contours in Fig.8.4.

8.3.2 Drain current in n- and p-MOSFETs

We have seen how the changes of the energy relation associated with different crystal orientations with respect to the channel affect the carrier mobility as measured in long channel devices at low drain to source voltage V_{DS}. To illustrate the impact of the channel orientation on the drain current of nanoscale MOSFETs, we consider two complementary but otherwise identical n-MOS and p-MOS double gate SOI devices with

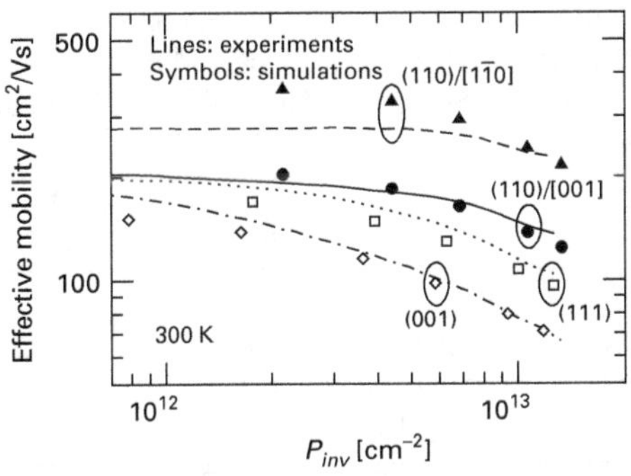

Figure 8.9 Hole mobility versus inversion charge density in silicon inversion layers with (001), (110), and (111) inversion planes. Lines: experiments from [9, 14]. Symbols: simulations. T = 300K. Simulation parameters are: $D_{ac} = 10.2$ eV, $D_{op} = 15\times10^8$ eV/cm $\hbar\omega_{op} = 62$ meV, $\Delta_{SR} = 0.24$ nm, $\Lambda_{SR} = 4$ nm. *Reprinted with permission from [1]. Copyright 2009 by Springer.*

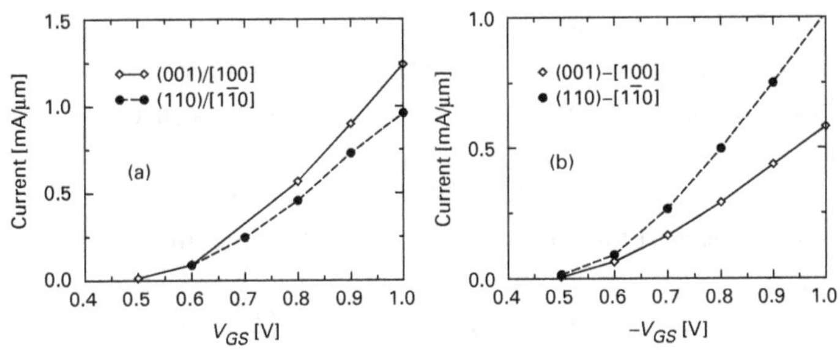

Figure 8.10 Multi-subband Monte Carlo simulations of the drain current for a double gate SOI n-MOSFET (plot a) and p-MOSFET (plot b) with $L_G = 22$ nm and different crystal orientations. $V_{DS} = 1V$.

$L_G = 22$ nm and $T_{Si} = 8$ nm. The metal gate work-function is $\Phi_M = 4.8$ eV for the n-MOS and $\Phi_M = 4.49$ eV for the p-MOS for the conventional (001) orientation, and it has been adjusted when changing the orientation in order to have the same threshold voltage for all the orientations. The gate stack EOT is 1.1 nm.

Results of multi-subband Monte Carlo simulations are shown in Fig.8.10 for the n-MOS and p-MOS devices. Concerning the n-MOS device, the quantization mass for the 2-fold valleys in the (110) inversion layer has been set to $0.23m_0$, as discussed in the previous section.

We clearly see that moving away from the conventional (001)/[100] orientation degrades the current drive of the n-MOS but enhances that of the p-MOS. This is consistent with the behavior of the low field mobility seen in the previous section.

To provide a deeper insight into the origin of these changes, we give in Table 8.5 some relevant quantities evaluated at the virtual source of the devices (see Section 5.7) for $V_{GS} = V_{DS} = 1V$.

We first note that the inversion charge density is almost the same in all cases, as expected due to the work-function adjustment. Focusing on the n-MOS, we see that

Table 8.5 On-current, inversion charge, velocity of the injected carriers and back-scattering coefficient at the virtual source of the *n*-MOS and *p*-MOS devices of Fig.8.10 for $V_{GS} = V_{DS} = 1V$. Section 5.7 defines the meaning of the symbols.

		I_{ON} [$mA/\mu m$]	N_{inv} [10^{13}cm^{-2}]	v^+ [10^7] cm/s	r
n-MOS	(001)/[100]	1.24	1.3	1.15	0.22
n-MOS	(110)/[1$\bar{1}$0]	0.96	1.25	0.92	0.25
p-MOS	(001)/[100]	0.58	0.91	0.77	0.28
p-MOS	(110)/[1$\bar{1}$0]	1.01	0.94	1.28	0.26

in the (110) case the injection velocity is slightly lower and the back-scattering coefficient is slightly higher than in the (001) case. However, the differences are not as large as one would expect by considering that the transport mass for the 2-fold valleys in the (001) case is $0.19m_0$, whereas that of the 4-fold in the (110) case is $0.555m_0$ (see Table 8.4). The main reason is that not only the lowest unprimed subbands are populated; in the (001) case 50% of the charge lies in the 2-fold valleys and 50% in the 4-fold, whereas in the (110) case 67% is in the 4-fold valleys and 33% in the 2-fold.

Focusing on the *p*-MOSFETs, we see that the main effect of the (110) orientation with respect to the (001) is to enhance the injection velocity, which is consistent with the significant reduction of the transport effective mass (in other words the inverse of the curvature of the energy relation) in the [1$\bar{1}$0] direction in (110) hole inversion layers compared to (001) inversion layers.

8.4 Summary

We have seen in this chapter that the main models, concepts and approaches discussed in the previous chapters for silicon channel transistors with conventional (001) orientation can be extended to arbitrary crystal orientations.

In particular, we have shown how to generalize the treatment for electron inversion layers based on the effective mass approximation (Section 8.1). Besides addressing the effect of the crystal orientation on the subband energies, we have explained how to handle the calculation of the free-flights and of phonon scattering in *n*-type inversion layers with different valleys arbitrarily orientated with respect to the transport direction. This allowed us to extend the multi-subband Monte Carlo approach to arbitrary materials and crystal orientations. Relevant case studies have been illustrated for (110)-silicon inversion layers, also representative of transport in FinFETs (see Section 8.3).

We have also seen (Section 8.2) that the **k**·**p** model for hole inversion layers can be naturally extended from (001) silicon inversion layers to different cases of practical relevance, such as the (110)-silicon inversion layers discussed in Section 8.3.

References

[1] D. Esseni, F. Conzatti, M. De Michielis, *et al.*, "Semi-classical transport modelling of CMOS transistors with arbitrary crystal orientations and strain engineering," *Journal of Computational Electronics*, vol. 8, pp. 209–224, 2009.

[2] F. Stern and W.E. Howard, "Properties of semiconductor surface inversion layers in the electric quantum limit ," *Phys. Rev.*, vol. 163, no. 3, pp. 816–835, 1967.

[3] D. Esseni and P. Palestri, "Theory of the motion at the band crossing points in bulk semiconductor crystals and in inversion layers," *Journal of Applied Physics*, vol. 105, no. 5, pp. 053702–1–053702–11, 2009.

[4] A.T. Pham, C. Jungemann, and B. Meinerzhagen, "Modeling of hole inversion layer mobility in unstrained and uniaxially strained Si on arbitrarily oriented substrates," *Proc. European Solid State Device Res. Conf.*, pp. 390–393, September 2007.

[5] M. De Michielis, D. Esseni, Y.L. Tsang, *et al.*, "A semianalytical description of the hole band structure in inversion layers for the physically based modeling of pMOS transistors," *IEEE Trans. on Electron Devices*, vol. 54, no. 9, pp. 2164–2173, 2007.

[6] L. Lucci, P. Palestri, D. Esseni, L. Bergagnini, and L. Selmi, "Multi-subband Monte Carlo study of transport, quantization and electron gas degeneration in ultra-thin SOI n-MOSFETs," *IEEE Trans. on Electron Devices*, vol. 54, no. 5, pp. 1156–1164, 2007.

[7] M. De Michielis, D. Esseni, P. Palestri, and L. Selmi, "Semiclassical modeling of quasi-ballistic hole transport in nanoscale pMOSFETs based on a multi-subband Monte Carlo approach," *IEEE Trans. on Electron Devices*, vol. 56, no. 9, pp. 2081–2091, 2009.

[8] S. Takagi, A. Toriumi, M. Iwase, and H. Tango, "On the universality of inversion-layer mobilty in Si MOSFETs. Part II- Effect of surface orientations," *IEEE Trans. on Electron Devices*, vol. 41, no. 12, pp. 2363–2368, 1994.

[9] S. Takagi, A. Toriumi, M. Iwase, and H. Tango, "On the universality of inversion-layer mobilty in Si MOSFETs. Part I- Effect of substrate impurity concentration," *IEEE Trans. on Electron Devices*, vol. 41, no. 12, pp. 2357–62, 1994.

[10] N. Serra and D. Esseni, "Mobility enhancement in strained n-FinFETs: Basic insight and stress engineering," *IEEE Trans. on Electron Devices*, vol. 57, no. 2, pp. 482–490, 2010.

[11] N. Serra, F. Conzatti, D. Esseni, *et al.*, "Experimental and physics based modeling assessment of strain induced mobility enhancement in FinFETs," in *IEEE IEDM Technical Digest*, pp. 71–74, 2009.

[12] K. Uchida, A. Kinoshita, and M. Saitoh, "Carrier transport in (110) nMOSFETs: subband structures, non-parabolicity, mobility characteristics, and uniaxial stress engineering," in *IEEE IEDM Technical Digest*, pp. 1019–1022, 2006.

[13] N. Serra and D. Esseni, "Basic insight about the strain engineering of n-type FinFETs," in *Proc. Int. Conf. on Ultimate Integration on Silicon (ULIS)*, (Aachen), pp. 113–116, March 2009.

[14] H. Irie, K. Kita, K. Kyuno, and A. Toriwni, "In-plane mobility anisotropy and universality under uni-axial strains in n- and p-MOS inversion layers on (100), (110), and (111) Si," in *IEEE IEDM Technical Digest*, pp. 225–228, 2004.

9 MOS transistors with strained silicon channel

Starting from the 90 nm technology node, several semiconductor companies have introduced strain as an important booster for the performance of MOS transistors; among them we can mention IBM [1], Intel [2], Texas Instruments [3], and Freescale [4]. This consideration explains the decision to devote an entire chapter of the book to transport in strained MOS devices.

Strain affects the characteristics of MOS transistors in several respects. In fact, besides its impact on carrier transport in the device channel, strain induces shifts of the band edges affecting the threshold voltage of the transistors [5], the leakage of the source and drain junctions [6], the energy barrier to the gate dielectric and consequently the gate leakage current [7], and also the transistor reliability [8]. The present chapter, however, is essentially focused on the methodologies and the models necessary to account for the strain effects on transport in MOS transistors, more precisely on the low field mobility and the drain current I_{DS}.

The chapter is organized as follows. After a concise introduction to the fabrication techniques used for strain engineering in Section 9.1, all the relevant definitions related to stress and strain in cubic crystals are described in Section 9.2. Correct evaluation of the strain tensor in the crystal coordinate system is the first step necessary to model the effects of strain on the band structure of n-type and p-type MOS transistors, which are described respectively in Section 9.3 and 9.4. Section 9.5 is devoted to the results of mobility simulations for strained MOS transistors for both uniaxial and biaxial strain conditions; a systematic comparison between the simulations and the experimental results is also discussed. Section 9.6 finally illustrates simulation results for the drain current in nanoscale strained transistors.

9.1 Fabrication techniques for strain engineering

The impact of strain on silicon and germanium resistivity, hence on mobility, has been experimentally observed since the early 1950s [9, 10] and very comprehensive theoretical studies have been devoted to this topic [11]. The massive technological exploitation of strain, however, started only after the beginning of the 1990s, when a clear experimental demonstration of electron mobility enhancements in n-MOSFETs realized on biaxially strained silicon films was reported [12–15].

Since then, several strain technologies have been devised to produce and engineer the strain in a CMOS manufacturing process flow, and nowadays strain is probably the most cost effective CMOS technology booster developed in recent years. The strain technologies are a broad topic that we will not cover systematically in this book. In the rest of this section, however, we briefly describe the basic differences between global strain techniques, which produce the same strain condition throughout the entire wafer, and local strain techniques, which, instead, can induce different strain configurations in selected regions of the wafer.

9.1.1 Global strain techniques

In the pioneering work concerning strain induced mobility enhancements in MOS transistors, the devices were realized in strained silicon layers epitaxially grown on a relaxed SiGe virtual substrate [13–15]. As sketched in Fig.9.1 and explained in more detail in Section 9.2.5, the lattice mismatch between SiGe and silicon yields a biaxial tensile strain in the epitaxial silicon layer; by changing the mole fraction x of the relaxed $Si_{1-x}Ge_x$ virtual substrate the strain magnitude in the silicon layer can be varied.

The $Si_{1-x}Ge_x$ virtual substrate technique is a global strain technique because the same strain configuration is produced throughout the wafer. Global strain techniques have also been demonstrated for SOI wafers, where the silicon is strained either by using a $Si_{1-x}Ge_x$ layer on the insulator [16–18] or directly on the insulator [19, 20].

The main limitation related to the global strain techniques is that they provide the same strain configuration for all devices. Unfortunately, n-MOS and p-MOS transistors are differently affected by the strain and, for instance, a compressive biaxial strain improves hole mobility whereas it degrades electron mobility. Such a reduced flexibility in the strain conditions impedes strain engineering and optimization, which require different strain patterns for n-MOS and p-MOS devices. In this latter respect local strain techniques are thus much preferred.

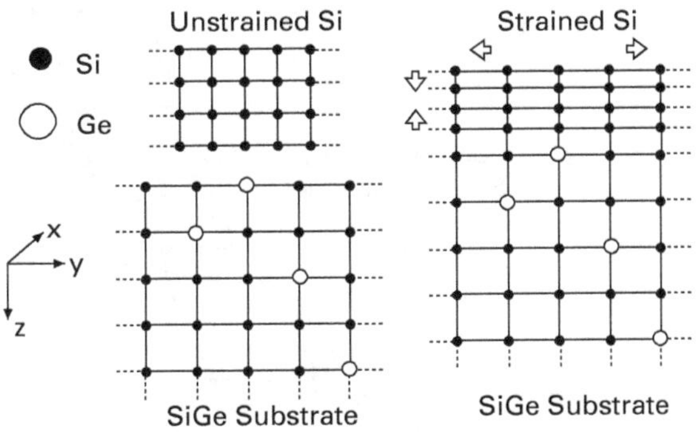

Figure 9.1 A biaxially strained silicon film grown on a SiGe virtual substrate. The difference in the lattice constants produces the strain in the silicon film.

9.1.2 Local strain techniques

Several CMOS fabrication steps produce strain in the channel of MOS transistors. Since the late 1990s it has been observed that a significant stress in the channel is produced by shallow trench isolation (STI) [21, 22], the silicidation of the source and drain regions [23], and nitride contact etch stop liners [24, 25]. Starting from these observations, much work has been devoted to developing well-controlled process steps able to induce strain magnitudes comparable to those produced by the global strain techniques, but with the ability to selectively change the strain conditions at the device level. Among the various local strain techniques, we will describe below the basic ideas behind the Contact Etch Stop Liners (CESL) technique and the uniaxial stress induced by source and drain stressors.

Contact Etch Stop Liners (CESL) technique

In this fabrication option a strongly stressed liner is deposited on some regions of the wafer after silicidation, and, depending on the material and the thickness of the liner, a significant stress can be transferred to the underlying transistors [24, 26].

In order to selectively induce the best stress patterns in n-MOS and p-MOS transistors, two types of stress liner can be used. In a dual stress liner process, for instance, a compressive silicon nitride layer can be deposited on p-type MOSFETs, while a tensile silicon nitride layer can be used for the n-MOSFETs [26, 27]. By using the CESL technique, stress components in the transistor channel up to 2 GPa have been reported, resulting in strain levels comparable to those produced by the epitaxial growth of silicon on a $Si_{1-x}Ge_x$ virtual substrate for the largest Ge contents.

SiGe or SiC source–drain stressors

An alternative and widely used local strain technique relies on the lattice mismatch between the silicon in the channel region of the MOSFET and the epitaxially grown $Si_{1-x}Ge_x$ or $Si_{1-x}C_x$ in the source–drain regions. As schematically illustrated in

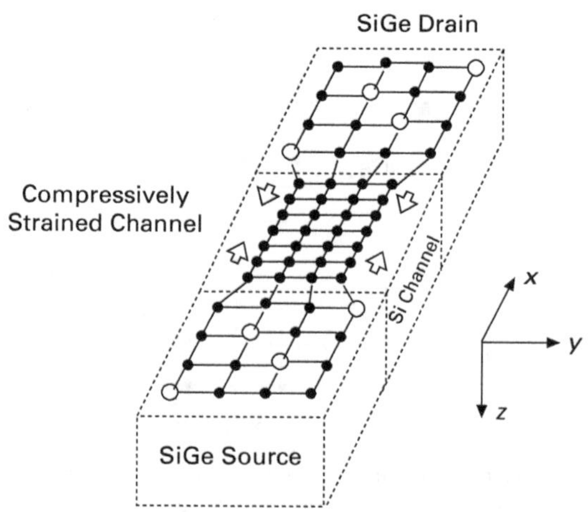

Figure 9.2 Stress produced by SiGe source–drain stressors in the channel of a MOS transistor.

Fig.9.2, the source–drain stressors induce a uniaxial stress in the transistor channel. In particular, $Si_{1-x}Ge_x$ source–drain regions are used to obtain a compressive uniaxial stress in p-type MOSFETs [1, 2, 4, 28–30], whereas the $Si_{1-x}C_x$ can produce a tensile stress in n-type transistors [31–33].

Local strain techniques turned out to be more cost effective and more versatile for the design of devices compared to global strain techniques, so that the introduction of strain engineering in mass production of integrated circuits has taken place by exploiting such techniques.

9.2 Elastic deformation of a cubic crystal

In this section we illustrate the basic concepts and introduce the notation for stress and strain in crystalline solids by limiting the treatment to the case of a small deformation, which results in a linear relation between stress and strain.

9.2.1 Stress: definitions and notation

According to elasticity theory, the deformation of a crystal is produced by internal forces between the portions of the body, which in turn may be produced by external forces exerted on the body under study. The internal forces are described by the stress (force per unit area) and Fig.9.3 visualizes the stress components acting on the faces of an infinitesimal cube of the solid. In particular, the stress acting on each face of the cube can be decomposed into two shear components (belonging to the plane of the face) and one axial component perpendicular to the face. As an example, T_{xx} is the axial stress component exerted on the face of the infinitesimal cube that is normal to the x direction, whereas T_{xy} and T_{xz} are the shear components at the same face respectively in the y and z direction.

As illustrated in Fig.9.3, the sign of the axial stress components is defined as positive when directed towards the outside of the infinitesimal cube, hence a tensile and a compressive stress in the x direction correspond respectively to a positive or a negative T_{xx} component.

The fact that the infinitesimal cube is at rest requires that the stress components acting on two opposite faces of the cube are equal and that, furthermore, the total torque applied to the cube is zero, namely

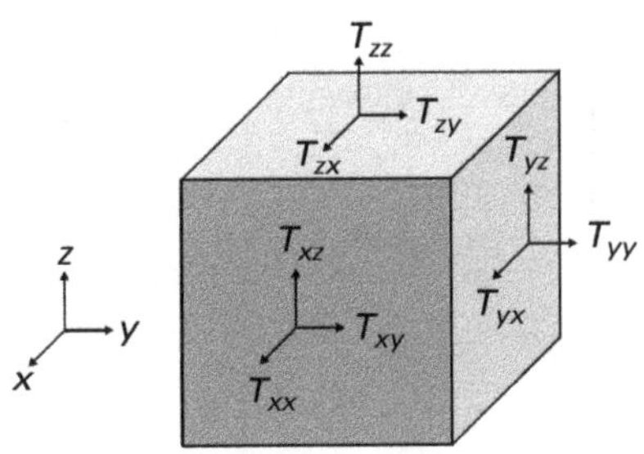

Figure 9.3 Stress components acting on an infinitesimal cubic element.

$$T_{xy} = T_{yx}, \quad T_{yz} = T_{zy}, \quad T_{zx} = T_{xz},$$

which reduces the number of independent stress components to only six. Hence the stress is completely described either by the 3×3 symmetric matrix

$$\mathbf{T}_{3\times3} = \begin{pmatrix} T_{xx} & T_{xy} & T_{zx} \\ T_{xy} & T_{yy} & T_{yz} \\ T_{zx} & T_{yz} & T_{zz} \end{pmatrix}, \tag{9.1}$$

or by the six component column vector

$$\mathbf{T}_6 = \begin{pmatrix} T_{xx} \\ T_{yy} \\ T_{zz} \\ T_{yz} \\ T_{zx} \\ T_{xy} \end{pmatrix}. \tag{9.2}$$

9.2.2 Strain: definitions and notation

The stresses in a crystal solid yield a deformation of the body and change the relative position of the unit cells. Let us define a coordinate system (with the origin taken, for example, at an atom of the unstrained solid), and then a sample point whose position in the unstrained solid is given by the column vector $\mathbf{R} = (x, y, z)^{\mathrm{T}}$. In the strained body both the origin of the coordinate system and the sample point may have changed their positions. Thus, in the strained body, the position $\mathbf{R}'$ of the sample point with respect to the origin may be written as

$$\mathbf{R}' = \mathbf{R} + \mathbf{u}(\mathbf{R}) = \mathbf{R} + [u_x(\mathbf{R}), u_y(\mathbf{R}), u_z(\mathbf{R})]^{\mathrm{T}}, \tag{9.3}$$

where $\mathbf{u}(\mathbf{R}) = [u_x(\mathbf{R}), u_y(\mathbf{R}), u_z(\mathbf{R})]^{\mathrm{T}}$ is the column vector expressing the relative displacement between the sample point and the origin produced by the strain; by definition we have $\mathbf{u}(\mathbf{0}) = \mathbf{0}$.

For a weak strain and small x, y, and z values (i.e. for a sample point close to the origin in the unstrained solid), the $\mathbf{u}(x, y, z)$ may be expressed by a first order Taylor series expansion as

$$\begin{aligned}
\begin{pmatrix} u_x \\ u_y \\ u_z \end{pmatrix} &\simeq \begin{pmatrix} \frac{\partial u_x}{\partial x} & \frac{\partial u_x}{\partial y} & \frac{\partial u_x}{\partial z} \\ \frac{\partial u_y}{\partial x} & \frac{\partial u_y}{\partial y} & \frac{\partial u_y}{\partial z} \\ \frac{\partial u_z}{\partial x} & \frac{\partial u_z}{\partial y} & \frac{\partial u_z}{\partial z} \end{pmatrix} \begin{pmatrix} x \\ y \\ z \end{pmatrix} \\
&= \begin{bmatrix} \frac{\partial u_x}{\partial x} & \frac{1}{2}(\frac{\partial u_x}{\partial y} + \frac{\partial u_y}{\partial x}) & \frac{1}{2}(\frac{\partial u_x}{\partial z} + \frac{\partial u_z}{\partial x}) \\ \frac{1}{2}(\frac{\partial u_x}{\partial y} + \frac{\partial u_y}{\partial x}) & \frac{\partial u_y}{\partial y} & \frac{1}{2}(\frac{\partial u_y}{\partial z} + \frac{\partial u_z}{\partial y}) \\ \frac{1}{2}(\frac{\partial u_x}{\partial z} + \frac{\partial u_z}{\partial x}) & \frac{1}{2}(\frac{\partial u_y}{\partial z} + \frac{\partial u_z}{\partial y}) & \frac{\partial u_z}{\partial z} \end{bmatrix} \begin{pmatrix} x \\ y \\ z \end{pmatrix} \\
&+ \begin{bmatrix} 0 & \frac{1}{2}(\frac{\partial u_x}{\partial y} - \frac{\partial u_y}{\partial x}) & \frac{1}{2}(\frac{\partial u_x}{\partial z} - \frac{\partial u_z}{\partial x}) \\ \frac{1}{2}(\frac{\partial u_y}{\partial x} - \frac{\partial u_x}{\partial y}) & 0 & \frac{1}{2}(\frac{\partial u_y}{\partial z} - \frac{\partial u_z}{\partial y}) \\ \frac{1}{2}(\frac{\partial u_z}{\partial x} - \frac{\partial u_x}{\partial z}) & \frac{1}{2}(\frac{\partial u_z}{\partial y} - \frac{\partial u_y}{\partial z}) & 0 \end{bmatrix} \begin{pmatrix} x \\ y \\ z \end{pmatrix}.
\end{aligned} \tag{9.4}$$

In Eq.9.4 we have written the Jacobian matrix of $\mathbf{u}(x, y, z)$ in terms of its symmetric and anti-symmetric part (respectively the first and the second matrix in squared brackets).

It is well known from classical kinematics that the last term in Eq.9.4 is non-null only if the body undergoes a rigid rotation [34, 35]. If this is not the case, the anti-symmetric matrix in Eq.9.4 is null, so that the relative displacement $\mathbf{u}(x, y, z)$ between the sample point and the origin produced by the body deformation can be written as

$$\begin{pmatrix} u_x \\ u_y \\ u_z \end{pmatrix} = \begin{pmatrix} \varepsilon_{xx} & \varepsilon_{xy} & \varepsilon_{xz} \\ \varepsilon_{yx} & \varepsilon_{yy} & \varepsilon_{yz} \\ \varepsilon_{zx} & \varepsilon_{zy} & \varepsilon_{zz} \end{pmatrix} \begin{pmatrix} x \\ y \\ z \end{pmatrix}, \tag{9.5}$$

where we have introduced the strain matrix

$$\boldsymbol{\varepsilon}_{3\times 3} = \begin{pmatrix} \varepsilon_{xx} & \varepsilon_{xy} & \varepsilon_{xz} \\ \varepsilon_{yx} & \varepsilon_{yy} & \varepsilon_{yz} \\ \varepsilon_{zx} & \varepsilon_{zy} & \varepsilon_{zz} \end{pmatrix}, \tag{9.6}$$

whose components are defined as

$$\varepsilon_{xx} = \frac{\partial u_x}{\partial x}, \qquad \varepsilon_{yy} = \frac{\partial u_y}{\partial y}, \qquad \varepsilon_{zz} = \frac{\partial u_z}{\partial z}, \tag{9.7a}$$

$$\varepsilon_{xy} = \frac{1}{2}\left(\frac{\partial u_x}{\partial y} + \frac{\partial u_y}{\partial x}\right) = \varepsilon_{yx}, \tag{9.7b}$$

$$\varepsilon_{yz} = \frac{1}{2}\left(\frac{\partial u_y}{\partial z} + \frac{\partial u_z}{\partial y}\right) = \varepsilon_{zy}, \tag{9.7c}$$

$$\varepsilon_{zx} = \frac{1}{2}\left(\frac{\partial u_z}{\partial x} + \frac{\partial u_x}{\partial z}\right) = \varepsilon_{xz}. \tag{9.7d}$$

Since the matrix $\boldsymbol{\varepsilon}_{3\times 3}$ is symmetric, it has only six independent components and it can be equivalently represented with the reduced or Voigt notation as the column vector

$$\boldsymbol{\varepsilon}_6 = \begin{pmatrix} \varepsilon_{xx} \\ \varepsilon_{yy} \\ \varepsilon_{zz} \\ \varepsilon_{yz} \\ \varepsilon_{zx} \\ \varepsilon_{xy} \end{pmatrix}. \tag{9.8}$$

The matrix $\boldsymbol{\varepsilon}_{3\times 3}$ and the vector $\boldsymbol{\varepsilon}_6$ consist of the same six scalar quantities ε_{xx}, ε_{yy}, ε_{zz}, ε_{xy}, ε_{yz}, ε_{zx}, hence they are completely equivalent. As illustrated in the following sections both the matrix and the vector notation are useful when dealing with the strain to stress relation and with the transformation of stress and strain between different coordinate systems.

Before we move on to discuss the relation between stress and strain, some considerations about the definition of the strain components are necessary. As discussed above, $\mathbf{u}(\mathbf{R})$ is the relative displacement between the origin and the sample point (located at $\mathbf{R} = (x, y, z)$ in the unstrained crystal). If both the strain and the x, y, z values are small, then we can express $\mathbf{u}(\mathbf{R})$ according to Eq.9.4. It should be noted, however, that for a non-uniform strain configuration the spatial derivatives of $u_x(\mathbf{R})$, $u_y(\mathbf{R})$, and $u_z(\mathbf{R})$ that define the strain components through Eq.9.7 may very well depend on the point taken as the origin. In other words, for a non-uniform strain, the strain components are local quantities, whose values are valid only in a spatial region close to the origin.

For a uniform and small strain configuration, however, the strain components are the same everywhere in the solid and Eq.9.4 is valid for any $\mathbf{R}$ value. This implies that the relative displacement of two points produced by a uniform strain depends only on the difference $\mathbf{R}$ in the positions of the two points in the unstrained crystal, but not on the absolute position of the points.

9.2.3 Strain and stress relation: the elastic constants

For sufficiently small deformations solids exhibit an elastic behavior, and Hooke's law states that the strain is proportional to the stress. If the strain and the stress are represented as 3×3 matrixes, then we need a four order tensor to relate $\boldsymbol{\varepsilon}_{3\times3}$ to $\mathbf{T}_{3\times3}$. If, instead, we use the vectorial notation $\boldsymbol{\varepsilon}_6$ and $\mathbf{T}_6$ as in Eqs.9.8 and 9.2, then we need a 6×6 matrix to express the proportionality between $\boldsymbol{\varepsilon}_6$ and $\mathbf{T}_6$.

A rank six matrix has 36 elements, but the number of independent components of the matrix relating the strain to the stress can be drastically reduced by invoking symmetry arguments [36]. In order to do that, however, we must express the strain and the stress in the crystal coordinate system, CCS, already defined in Section 8.1 and Fig.8.1. To make this latter point explicit, we henceforth denote by $\boldsymbol{\varepsilon}_{c,3\times3}$, $\boldsymbol{\varepsilon}_{c,6}$ and by $\mathbf{T}_{c,3\times3}$, $\mathbf{T}_{c,6}$ the strain and the stress in the CCS: $\varepsilon_{c,xx}$, $\varepsilon_{c,yy}$, $\varepsilon_{c,zz}$, $\varepsilon_{c,xy}$, $\varepsilon_{c,yz}$, $\varepsilon_{c,zx}$ and $T_{c,xx}$, $T_{c,yy}$, $T_{c,zz}$, $T_{c,xy}$, $T_{c,yz}$, $T_{c,zx}$ are the corresponding scalar quantities. The symbols $\boldsymbol{\varepsilon}_{3\times3}$, $\boldsymbol{\varepsilon}_6$, $\mathbf{T}_{3\times3}$, and $\mathbf{T}_6$ are instead used for strain and stress in a generic coordinate system. They will be useful in the next section, where we discuss the transformation of strain and stress between different coordinate systems.

For cubic semiconductors such as Si, Ge, or GaAs, the elastic compliance matrix $\mathbf{S}$ has only the three independent components S_{11}, S_{12}, and S_{44} and the $\boldsymbol{\varepsilon}_{c,6}$ to $\mathbf{T}_{c,6}$ relation reads [35, 36]

$$\begin{pmatrix} \varepsilon_{c,xx} \\ \varepsilon_{c,yy} \\ \varepsilon_{c,zz} \\ \varepsilon_{c,yz} \\ \varepsilon_{c,zx} \\ \varepsilon_{c,xy} \end{pmatrix} = \begin{pmatrix} S_{11} & S_{12} & S_{12} & 0 & 0 & 0 \\ S_{12} & S_{11} & S_{12} & 0 & 0 & 0 \\ S_{12} & S_{12} & S_{11} & 0 & 0 & 0 \\ 0 & 0 & 0 & S_{44}/2 & 0 & 0 \\ 0 & 0 & 0 & 0 & S_{44}/2 & 0 \\ 0 & 0 & 0 & 0 & 0 & S_{44}/2 \end{pmatrix} \cdot \begin{pmatrix} T_{c,xx} \\ T_{c,yy} \\ T_{c,zz} \\ T_{c,yz} \\ T_{c,zx} \\ T_{c,xy} \end{pmatrix} . \tag{9.9}$$

Equation 9.9 can be readily inverted to give

$$\begin{pmatrix} T_{c,xx} \\ T_{c,yy} \\ T_{c,zz} \\ T_{c,yz} \\ T_{c,zx} \\ T_{c,xy} \end{pmatrix} = \begin{pmatrix} C_{11} & C_{12} & C_{12} & 0 & 0 & 0 \\ C_{12} & C_{11} & C_{12} & 0 & 0 & 0 \\ C_{12} & C_{12} & C_{11} & 0 & 0 & 0 \\ 0 & 0 & 0 & 2C_{44} & 0 & 0 \\ 0 & 0 & 0 & 0 & 2C_{44} & 0 \\ 0 & 0 & 0 & 0 & 0 & 2C_{44} \end{pmatrix} \cdot \begin{pmatrix} \varepsilon_{c,xx} \\ \varepsilon_{c,yy} \\ \varepsilon_{c,zz} \\ \varepsilon_{c,yz} \\ \varepsilon_{c,zx} \\ \varepsilon_{c,xy} \end{pmatrix} , \tag{9.10}$$

Table 9.1 Elastic stiffness and elastic compliance constants for silicon and germanium [36–40].

	Silicon	Germanium
C_{11}[Pa]	$1.66 \cdot 10^{11}$	$1.26 \cdot 10^{11}$
C_{12}[Pa]	$6.4 \cdot 10^{10}$	$4.4 \cdot 10^{10}$
C_{44}[Pa]	$7.96 \cdot 10^{10}$	$6.77 \cdot 10^{10}$
S_{11}[1/Pa]	$8.31 \cdot 10^{-11}$	$9.69 \cdot 10^{-11}$
S_{12}[1/Pa]	$-2.56 \cdot 10^{-11}$	$-2.51 \cdot 10^{-11}$
S_{44}[1/Pa]	$1.256 \cdot 10^{-10}$	$1.477 \cdot 10^{-10}$

where C_{11}, C_{12}, and C_{44} are the three independent components of the elastic stiffness matrix **C** [35, 36]. The elastic compliance constants S_{ij} are related to the elastic stiffness constants C_{ij} by

$$S_{11} = \frac{C_{11} + C_{12}}{C_{11}^2 + C_{11}\,C_{12} - 2C_{12}^2}, \quad S_{12} = \frac{-C_{12}}{C_{11}^2 + C_{11}\,C_{12} - 2C_{12}^2}, \quad S_{44} = \frac{1}{C_{44}}. \tag{9.11}$$

As discussed in Section 9.2.4, the matrixes **S** and **C** take the simple form given in Eqs.9.9 and 9.10 only in the CCS, as implied by the symbols $\varepsilon_{c,ij}$ and $T_{c,ij}$ used to denote the strain and stress components. The numerical values for the elastic compliance and stiffness constants are given in Table 9.1 for silicon and germanium. The strain to stress relation in a coordinate system different from the CCS is discussed in the following section.

A final remark concerns the factors 1/2 and 2 entering respectively Eq.9.9 and 9.10. Such factors are necessary because the S_{44} and C_{44} constants are typically defined and experimentally determined by using the *engineering strain* rather than the strain components defined in Eq.9.7. The engineering strain components e_{ij} are defined as [35]

$$\begin{pmatrix} e_{xx} & e_{xy} & e_{xz} \\ e_{xy} & e_{yy} & e_{yz} \\ e_{xz} & e_{yz} & e_{zz} \end{pmatrix} = \begin{pmatrix} \varepsilon_{xx} & 2\varepsilon_{xy} & 2\varepsilon_{zx} \\ 2\varepsilon_{xy} & \varepsilon_{yy} & 2\varepsilon_{yz} \\ 2\varepsilon_{zx} & 2\varepsilon_{yz} & \varepsilon_{zz} \end{pmatrix}, \tag{9.12}$$

hence the shear components e_{ij} (for $i,j = x,y,z$ and $i \neq j$) of the engineering strain are twice as large as the corresponding strain components ε_{ij}.

By reconsidering Eqs.9.9 and 9.10, it is easy to see that the factors 1/2 and 2 would not have been necessary if we had used the engineering strain e_{ij} rather than the strain ε_{ij}, which is the convention used, for instance, in [36]. This is the convention also used for the experimental determination of S_{44} or C_{44}, which are extracted directly from the ratio between a given shear component e_{ij} ($i \neq j$) of the strain engineering and the corresponding shear component T_{ij} of the stress [35]. However, it is the strain components ε_{ij} that enter the band structure calculations (see Sections 9.3 and 9.4). In this sense the engineering strain is not necessary for the further developments of this chapter, hence, in order to keep the notation as unambiguous as possible, we decided not to use the engineering strain in the book (except for the present discussion).

To summarize, the factors 1/2 and 2 in Eq.9.9 and 9.10 have been introduced to keep the ε_{ij} as the only symbols used for the strain and, at the same time, to employ definitions and numerical values for the elastic constants consistent with the literature [36, 38–40].

9.2.4 Change of coordinate systems for strain and stress

As illustrated by the examples developed in the following sections, in the cases of practical interest the strain in the devices is produced by the fabrication process, hence the strain and stress components are typically set in the device coordinate system DCS (see Section 8.1 and Fig.8.1). However the strain components in the crystal coordinate system, CCS, are those that enter the calculations of the band structure for the strained devices, so that it is very important to clarify the transformation of the stress and strain between different coordinate systems.

To this purpose, let us consider the DCS and CCS illustrated in Fig.9.4, where $\mathbf{R} = (x, y, z)^T$ denotes a generic point in real space. The coordinates $\mathbf{R}_c = (x_c, y_c, z_c)^T$ of the same point in the CCS are given by

$$\begin{pmatrix} x_c \\ y_c \\ z_c \end{pmatrix} = \underbrace{\begin{pmatrix} l_1 & m_1 & n_1 \\ l_2 & m_2 & n_2 \\ l_3 & m_3 & n_3 \end{pmatrix}}_{\mathbf{R}_{D\to C}} \begin{pmatrix} x \\ y \\ z \end{pmatrix}, \tag{9.13}$$

where $\mathbf{R}_{D\to C}$ is the rotation matrix from the DCS to the CCS already introduced in Section 8.1, whose columns are the components in the CCS of the three unit vectors of the DCS.

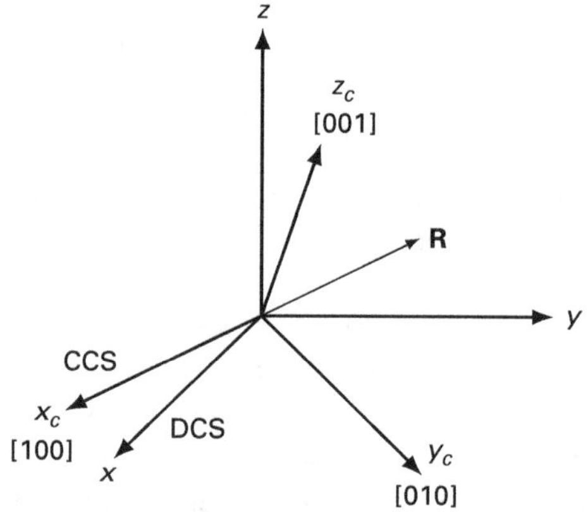

Figure 9.4 Device coordinate system, (x, y, z) DCS, and crystal coordinate system (x_c, y_c, z_c) CCS. **R** is a generic point in the three dimensional space.

Equation 9.13 can be concisely written as

$$\mathbf{R}_c = \mathbf{R}_{D\to C} \cdot \mathbf{R}, \tag{9.14}$$

and it provides the transformation of the vector $\mathbf{R}$ from the DCS to the CCS.

As can be readily verified by using Eq.9.14, any 3×3 matrix $\mathbf{M}$ representative of a linear relation in the DCS is transformed to the matrix $\mathbf{M}_c$ in the CCS according to

$$\mathbf{M}_c = \mathbf{R}_{D\to C} \cdot \mathbf{M} \cdot \mathbf{R}_{D\to C}^{-1}, \tag{9.15}$$

where the inverse matrix $\mathbf{R}_{D\to C}^{-1}$ is the same as the transpose $\mathbf{R}_{D\to C}^{T}$ because $\mathbf{R}_{D\to C}$ is a unitary matrix.

Equation 9.5 has shown that $\boldsymbol{\varepsilon}_{3\times3}$ is the matrix that governs the linear relation between the position $\mathbf{R}$ of a point in the unstrained body and its strain induced displacement, hence the transformation of $\boldsymbol{\varepsilon}_{3\times3}$ from the DCS to the CCS is given by

$$\boldsymbol{\varepsilon}_{c,3\times3} = \mathbf{R}_{D\to C} \cdot \boldsymbol{\varepsilon}_{3\times3} \cdot \mathbf{R}_{D\to C}^{-1}. \tag{9.16}$$

The stress matrix is similarly transformed as

$$\mathbf{T}_{c,3\times3} = \mathbf{R}_{D\to C} \cdot \mathbf{T}_{3\times3} \cdot \mathbf{R}_{D\to C}^{-1}. \tag{9.17}$$

Equations 9.16, 9.17 together with the strain to stress relation in the CCS given by Eqs.9.9, 9.10 allow us to handle all the cases of practical interest. In particular, they allow us to find the relation between the stress and the strain in the DCS, namely the matrixes of the elastic compliance and stiffness constants in the DCS. In order to make use of Eqs.9.9, 9.10, however, we need to write the strain and the stress in the CCS as the six component vectors $\boldsymbol{\varepsilon}_{c,6}$ and $\mathbf{T}_{c,6}$. To this purpose it is convenient to introduce the 6×6 matrix $\mathbf{R}_{6,D\to C}$ that directly transforms $\boldsymbol{\varepsilon}_6$ and $\mathbf{T}_6$ in the DCS to $\boldsymbol{\varepsilon}_{c,6}$ and $\mathbf{T}_{c,6}$ in the CCS.

For the strain this can be obtained easily by using Eq.9.16 to express $\varepsilon_{c,xx}$, $\varepsilon_{c,yy}$, $\varepsilon_{c,zz}$, $\varepsilon_{c,xy}$, $\varepsilon_{c,yz}$, and $\varepsilon_{c,zx}$ as linear combinations of ε_{xx}, ε_{yy}, ε_{zz}, ε_{xy}, ε_{yz}, and ε_{zx}. A direct calculation allows us to write the $\boldsymbol{\varepsilon}_{c,6}$ to $\boldsymbol{\varepsilon}_6$ transformation in the concise form

$$\boldsymbol{\varepsilon}_{c,6} = \mathbf{R}_{6,D\to C} \cdot \boldsymbol{\varepsilon}_6, \tag{9.18}$$

where $\mathbf{R}_{6,D\to C}$ is expressed in terms of the elements of the $\mathbf{R}_{D\to C}$ defined in Eq.9.13 as [35]

$$\mathbf{R}_{6,D\to C} = \begin{pmatrix} l_1^2 & m_1^2 & n_1^2 & 2m_1n_1 & 2n_1l_1 & 2l_1m_1 \\ l_2^2 & m_2^2 & n_2^2 & 2m_2n_2 & 2n_2l_2 & 2l_2m_2 \\ l_3^2 & m_3^2 & n_3^2 & 2m_3n_3 & 2n_3l_3 & 2l_3m_3 \\ l_2l_3 & m_2m_3 & n_2n_3 & m_2n_3+m_3n_2 & n_2l_3+n_3l_2 & m_2l_3+m_3l_2 \\ l_3l_1 & m_3m_1 & n_3n_1 & m_3n_1+m_1n_3 & n_3l_1+n_1l_3 & m_3l_1+m_1l_3 \\ l_1l_2 & m_1m_2 & n_1n_2 & m_1n_2+m_2n_1 & n_1l_2+n_2l_1 & m_1l_2+m_2l_1 \end{pmatrix}. \tag{9.19}$$

For the stress we similarly have

$$\mathbf{T}_{c,6} = \mathbf{R}_{6,D\to C} \cdot \mathbf{T}_6. \tag{9.20}$$

Equations 9.18 and 9.20 are completely equivalent to Eqs.9.16 and 9.17, respectively. However Eqs.9.18 and 9.20 are useful because, if we substitute them in Eq.9.9, then we obtain

$$\boldsymbol{\varepsilon}_6 = (\mathbf{R}_{6,D\to C}^{-1} \cdot \mathbf{S} \cdot \mathbf{R}_{6,D\to C}) \cdot \mathbf{T}_6. \tag{9.21}$$

Equation 9.21 identifies the elastic compliance matrix in the DCS as

$$\mathbf{R}_{6,D\to C}^{-1} \cdot \mathbf{S} \cdot \mathbf{R}_{6,D\to C}. \tag{9.22}$$

Equations 9.18 and 9.20 can be similarly substituted in Eq.9.10 to obtain

$$\mathbf{T}_6 = (\mathbf{R}_{6,D\to C}^{-1} \cdot \mathbf{C} \cdot \mathbf{R}_{6,D\to C}) \cdot \boldsymbol{\varepsilon}_6, \tag{9.23}$$

which defines the elastic stiffness matrix in the DCS as

$$\mathbf{R}_{6,D\to C}^{-1} \cdot \mathbf{C} \cdot \mathbf{R}_{6,D\to C}. \tag{9.24}$$

Equations 9.22 and 9.24 reiterate the fact that the elastic compliance and stiffness matrixes take the simple form given in Eqs.9.9 and 9.10 only in the CCS.

9.2.5 Biaxial strain

As discussed in Section 9.1.1, a uniform biaxial strain can be produced in a semiconductor layer by epitaxially growing it on a relaxed virtual substrate with a different lattice constant (see Fig.9.1). The resulting strain components depend on the mismatch between the lattice constant of the semiconductor to be strained and the lattice constant of the virtual substrate, that is respectively silicon and SiGe in Fig.9.1. Furthermore, the strain also depends on the crystal orientation of the substrate.

By assuming that the lattice constant $a_{||}$ in the growth plane remains the same throughout the structure, an expression for $a_{||}$ can be obtained by minimizing the elastic energy in the hetero-structure. For the silicon layer grown on the $Si_{1-x}Ge_x$ substrate sketched in Fig.9.1, for instance, the $a_{||}$ can be expressed as [41]

$$a_{||} = \frac{a_{Si}\, G_{Si}\, T_{Si} + a_{SiGe}\, G_{SiGe}\, T_{SiGe}}{G_{Si}\, T_{Si} + G_{SiGe}\, T_{SiGe}}, \tag{9.25}$$

where T_{Si} and T_{SiGe} are the thickness of respectively the silicon and the SiGe layer and G_m is defined as

$$G_m = 2(C_{11,m} + 2\, C_{12,m}) \left[1 - \frac{1}{2\sigma_m}\right], \tag{9.26}$$

where m denotes the material (i.e. Si or SiGe). The constant σ denotes the Poisson ratio defined in Eq.9.32; expressions for the Poisson ratio for the (100), (011), and (111) interfaces are given respectively in Eqs.9.36, 9.39, and 9.42.

Equation 9.25 clarifies that, for two layers with comparable thickness T_{Si} and T_{SiGe}, the $a_{||}$ coincides with neither a_{Si} nor a_{SiGe}, consequently both the silicon and the SiGe layers are strained when the hetero-structure is formed.

In practice, however, the silicon layer has a thickness T_{Si} very small compared to T_{SiGe}. This is because the T_{Si} must be kept below a critical thickness, above which misfit dislocations are produced in the strained silicon layer to alleviate the stress [42, 43]. The critical thickness decreases with increase of the Ge content x in the $Si_{1-x}Ge_x$ virtual substrate and it is about 30 nm for $x = 10\%$ and about 4 nm for $x = 50\%$ [43]. For $T_{Si} \ll T_{SiGe}$ the Eq.9.25 simplifies to $a_{||} \simeq a_{SiGe}$ and the strain in the virtual substrate tends to vanish.

Once the lattice constant in the growth plane $a_{||}$ is known, the strain components in the silicon film can be determined as explained in [38]. Let us consider the DCS (x,y,z), where the z axis is normal to the virtual substrate interface; $\boldsymbol{\varepsilon}_6$ denotes the strain vector in the DCS. The so-called parallel strain component $\varepsilon_{||}$ in the growth plane is

$$\varepsilon_{||} = \frac{a_{||} - a_{Si}}{a_{Si}} \simeq \frac{a_{SiGe} - a_{Si}}{a_{Si}}, \tag{9.27}$$

and it varies with the Ge content x in the $Si_{1-x}Ge_x$ virtual substrate through the lattice constant a_{SiGe}. The dependence of the $Si_{1-x}Ge_x$ lattice constant a_{SiGe} on x has been experimentally investigated in [44], and it can be approximately expressed as [39]

$$a_{SiGe}(x) = a_{Si}(1-x) + a_{Ge}\,x - b\,x(1-x), \tag{9.28}$$

where $a_{Si} = 0.543$ nm and $a_{Ge} = 0.565$ nm are the silicon and the germanium lattice constants, respectively, and $b = 1.88 \cdot 10^{-3}$ nm.

The strain components in the xy plane are [38]

$$\varepsilon_{xx} = \varepsilon_{yy} = \varepsilon_{||}, \qquad \varepsilon_{xy} = 0. \tag{9.29}$$

The ε_{zz}, ε_{yz}, and ε_{zx} remain to be determined. Equations 9.27 to 9.29 indicate that the silicon films grown on a $Si_{1-x}Ge_x$ virtual substrate have tensile strain components $\varepsilon_{xx} = \varepsilon_{yy}$ in the DCS, because a_{SiGe} is larger than a_{Si}.

The in-plane stress exerted by the virtual substrate is the only stress applied to the semiconductor film, hence we have

$$T_{zz} = 0, \qquad T_{yz} = 0, \qquad T_{zx} = 0, \tag{9.30}$$

while T_{xx}, T_{yy}, and T_{xy} remain to be determined.

By using Eq.9.23 to express T_{zz}, T_{yz}, and T_{zx} as a function of the strain components, Eq.9.30 provides three linear equations that allow us to determine the ε_{zz}, ε_{yz}, and ε_{zx} components for an arbitrary crystal orientation of the virtual substrate, that is for an arbitrary DCS. When all the strain components are known, then Eq.9.23 provides the stress components T_{xx}, T_{yy}, and T_{xy}. Equations.9.18 and 9.20 finally allow us to obtain the strain and stress components in the CCS.

The calculation procedure is simplified in the case of {001}, {110}, and {111} virtual substrates. In fact the high rotational symmetry of the $\langle 001 \rangle$, $\langle 110 \rangle$, and $\langle 111 \rangle$ directions yields [38]

$$\varepsilon_{yz} = \varepsilon_{zx} = 0, \tag{9.31}$$

hence ε_{zz} remains the only strain component to be determined. Any of the equations Eq.9.30 can be equivalently used to determine ε_{zz}. For these substrate orientations the Poisson ratio σ is defined as

$$\sigma = -\frac{\varepsilon_{||}}{\varepsilon_{zz}}, \tag{9.32}$$

and the strain components in the DCS can be thus summarized as

$$\varepsilon_{xx} = \varepsilon_{yy} = \varepsilon_{||}, \qquad \varepsilon_{zz} = -\frac{\varepsilon_{||}}{\sigma}, \qquad \varepsilon_{ij} = 0 \;\; i \neq j. \tag{9.33}$$

We finally reiterate that Eq.9.18 must be used to determine the strain components in the CCS needed to determine the strain induced band structure modifications.

Example 9.1: (001) virtual substrate.

The determination of ε_{zz} is particularly simple for a (001) virtual substrate, for which the device and the crystal coordinate system can be taken as coincident, so that $\boldsymbol{\varepsilon}_6 = \boldsymbol{\varepsilon}_{c,6}$ and $\mathbf{T}_6 = \mathbf{T}_{c,6}$. In this case the condition $T_{zz} = 0$ set by Eq.9.30 can be imposed by using the third of the Eqs.9.10 as

$$T_{c,zz} = C_{12}\,\varepsilon_{||} + C_{12}\,\varepsilon_{||} + C_{11}\,\varepsilon_{c,zz} = 0, \tag{9.34}$$

which gives

$$\varepsilon_{c,zz} = -\frac{2\,C_{12}}{C_{11}}\,\varepsilon_{||}. \tag{9.35}$$

Recalling Eq.9.32, the Poisson ratio for the (001) substrates is

$$\sigma_{(001)} = \frac{C_{11}}{2\,C_{12}}. \tag{9.36}$$

The overall strain vector for the (001) virtual substrates is thus

$$\varepsilon_{c,xx} = \varepsilon_{c,yy} = \varepsilon_{||}, \qquad \varepsilon_{c,zz} = -\frac{2\,C_{12}}{C_{11}}\,\varepsilon_{||}, \qquad \varepsilon_{c,xy} = \varepsilon_{c,yz} = \varepsilon_{c,zx} = 0. \tag{9.37}$$

As can be seen, the biaxial strain produced by a (001) virtual substrate results in null shear strain components in the epitaxial film.

Example 9.2: (110) virtual substrate.

For a (110) substrate a possible rotation matrix $\mathbf{R}_{D\to C}$ from the DCS to the CCS has been already introduced in Table 8.1 and reads

$$\mathbf{R}_{D\to C} = \begin{pmatrix} 0 & \frac{1}{\sqrt{2}} & \frac{1}{\sqrt{2}} \\ 0 & -\frac{1}{\sqrt{2}} & \frac{1}{\sqrt{2}} \\ 1 & 0 & 0 \end{pmatrix}, \tag{9.38}$$

which places the x and y axis of the DCS along the [001] and $[1\bar{1}0]$ directions of the CCS, respectively.

By using Eq.9.23 to impose $T_{zz} = 0$ we can determine the Poisson ratio for the (110) substrate:

$$\sigma_{(110)} = \frac{C_{11} + C_{12} + 2C_{44}}{C_{11} + 3C_{12} - 2C_{44}}, \tag{9.39}$$

and the strain components in the DCS are given by Eq.9.33.

By using Eq.9.18 we can finally obtain the strain components in the CCS as

$$\begin{aligned} \varepsilon_{c,xx} &= \varepsilon_{c,yy} = \frac{2C_{44} - C_{12}}{C_{11} + C_{12} + 2C_{44}}\varepsilon_{||}, & \varepsilon_{c,zz} &= \varepsilon_{||}, \\ \varepsilon_{c,xy} &= -\frac{C_{11} + 2C_{12}}{C_{11} + C_{12} + 2C_{44}}\frac{\varepsilon_{||}}{2}, & \varepsilon_{c,yz} &= \varepsilon_{c,zx} = 0. \end{aligned} \tag{9.40}$$

Example 9.3: (111) virtual substrate.

Recalling Section 8.1 and Table 8.1, the rotation matrix $\mathbf{R}_{D\to C}$ for a (111) substrate can be taken as

$$\mathbf{R}_{D\to C} = \begin{pmatrix} -\frac{2}{\sqrt{6}} & 0 & \frac{1}{\sqrt{3}} \\ \frac{1}{\sqrt{6}} & -\frac{1}{\sqrt{2}} & \frac{1}{\sqrt{3}} \\ \frac{1}{\sqrt{6}} & \frac{1}{\sqrt{2}} & \frac{1}{\sqrt{3}} \end{pmatrix}, \tag{9.41}$$

so that the x and y axes of the DCS lie along the $[\bar{2}11]$ and $[0\bar{1}1]$ directions of the CCS, respectively. The condition $T_{zz} = 0$ imposed through Eq.9.23 provides the Poisson ratio for the (111) substrate:

$$\sigma_{(111)} = \frac{C_{11} + 2C_{12} + 4C_{44}}{2C_{11} + 4C_{12} - 4C_{44}}. \tag{9.42}$$

The strain components in the DCS are given by Eq.9.33 and the strain components in the CCS are finally obtained from Eq.9.18 as

$$\begin{aligned} \varepsilon_{c,xx} &= \varepsilon_{c,yy} = \varepsilon_{c,zz} = \frac{4C_{44}}{C_{11} + 2C_{12} + 4C_{44}}\varepsilon_{||}, \\ \varepsilon_{c,xy} &= \varepsilon_{c,yz} = \varepsilon_{c,zx} = -\frac{C_{11} + 2C_{12}}{C_{11} + 2C_{12} + 4C_{44}}\frac{\varepsilon_{||}}{2}. \end{aligned} \tag{9.43}$$

As can be seen, for both the $\langle 110\rangle$ and the $\langle 111\rangle$ substrates we have non-null shear strain components in the CCS. The numerical values for the Poisson ratios are given in Table 9.2 for different substrate orientations and semiconductors.

9.2.6 Uniaxial strain

As discussed in Section 9.1, several fabrication techniques have been proposed to produce a uniaxial strain in both planar MOSFETs and FinFETs. Such techniques

Table 9.2 Poisson ratios for silicon and germanium and for different substrate orientations calculated by using the elastic stiffness constants given in Table 9.1

	$\sigma_{(001)}$	$\sigma_{(110)}$	$\sigma_{(111)}$
Silicon	1.297	1.958	2.271
Germanium	1.4318	2.491	3.084

frequently result in a dominant stress component in the source–drain direction (see Fig.9.2).

If we suppose that the stress components in the DCS are known, then the stress components in the CCS can be obtained equivalently with either Eq.9.17 or Eq.9.20. The strain in the CCS can finally be calculated using Eq.9.9. The procedure can be summarized by writing

$$\boldsymbol{\varepsilon}_{c,6} = \mathbf{S} \cdot \mathbf{R}_{6,D\to C} \cdot \mathbf{T}_6. \tag{9.44}$$

Equation 9.44 allows us to understand how the stress components in the DCS translate into strain components in the CCS. Some relevant examples for both planar MOSFETs and FinFETs are illustrated below.

Example 9.4: Uniaxial stress in (001) / [110] planar MOSFETs.

For planar MOSFETs in a (001) wafer the transport direction x is typically along the [110] crystallographic direction, hence the rotation matrix $\mathbf{R}_{D\to C}$ from the DCS to the CCS is

$$\mathbf{R}_{D\to C} = \begin{pmatrix} \frac{1}{\sqrt{2}} & -\frac{1}{\sqrt{2}} & 0 \\ \frac{1}{\sqrt{2}} & \frac{1}{\sqrt{2}} & 0 \\ 0 & 0 & 1 \end{pmatrix}. \tag{9.45}$$

Let us now suppose that the components of the stress in the DCS are given by:

$$T_{xx} = T_L, \qquad T_{yy} = T_W, \qquad T_{zz} = T_\perp, \qquad T_{ij} = 0, \quad \text{for } \mathrm{i} \neq \mathrm{j} \tag{9.46}$$

where the symbols T_L and T_W denote the uniaxial stress in the device length and width, respectively, while $T_\perp$ is the stress in the direction normal to the silicon–oxide interface.

Equation 9.17 or Eq.9.20 give the stress components in the CCS as

$$\begin{aligned} T_{c,xx} &= T_{c,yy} = \frac{1}{2}(T_L + T_W), \quad T_{c,zz} = T_\perp, \\ T_{c,xy} &= \frac{1}{2}(T_L - T_W), \qquad T_{c,yz} = T_{c,zx} = 0, \end{aligned} \tag{9.47}$$

and Eq.9.9 allows us to express the strain components in the CCS:

$$\varepsilon_{c,xx} = \varepsilon_{c,yy} = \frac{S_{11} + S_{12}}{2}(T_L + T_W) + S_{12}\, T_\perp,$$

$$\varepsilon_{c,zz} = S_{12}\,(T_L + T_W) + S_{11}\,T_\perp,$$
$$\varepsilon_{c,xy} = \frac{S_{44}}{4}(T_L - T_W), \quad \varepsilon_{c,yz} = \varepsilon_{c,zx} = 0. \tag{9.48}$$

Since the device length and width are along the $\langle 110 \rangle$ directions, a shear strain component $\varepsilon_{c,xy}$ exists in the CCS unless T_L and T_W are the same.

Example 9.5: Uniaxial stress in (110)/[1$\bar{1}$0] FinFETs.

As a further example we consider the FinFET structure illustrated in Fig.9.5, where the lateral interfaces are of (110)/[1$\bar{1}$0] type, so that the rotation matrix is

$$\mathbf{R}_{D\to C} = \begin{pmatrix} \frac{1}{\sqrt{2}} & 0 & \frac{1}{\sqrt{2}} \\ -\frac{1}{\sqrt{2}} & 0 & \frac{1}{\sqrt{2}} \\ 0 & -1 & 0 \end{pmatrix}. \tag{9.49}$$

If we denote with T_{fL}, T_{fW}, and T_{fH} the possible stress components respectively in the fin length, fin width, and fin height direction, then the axial stress components in the DCS can be expressed as

$$T_{xx} = T_{fL}, \qquad T_{yy} = T_{fH}, \qquad T_{zz} = T_{fW}, \qquad T_{ij} = 0.\ i \neq j \tag{9.50}$$

The stress in the CCS is then

$$T_{c,xx} = T_{c,yy} = \frac{1}{2}\,(T_{fL} + T_{fW}), \quad T_{c,zz} = T_{fH},$$
$$T_{c,xy} = \frac{1}{2}\,(T_{fW} - T_{fL}), \qquad T_{c,yz} = T_{c,zx} = 0, \tag{9.51}$$

and the strain is

$$\varepsilon_{c,xx} = \varepsilon_{c,yy} = \frac{S_{11} + S_{12}}{2}\,(T_{fL} + T_{fW}) + S_{12}\,T_{fH},$$
$$\varepsilon_{c,zz} = S_{12}\,(T_{fL} + T_{fW}) + S_{11}\,T_{fH},$$
$$\varepsilon_{c,xy} = \frac{S_{44}}{4}(T_{fW} - T_{fL}), \qquad \varepsilon_{c,yz} = \varepsilon_{c,zx} = 0. \tag{9.52}$$

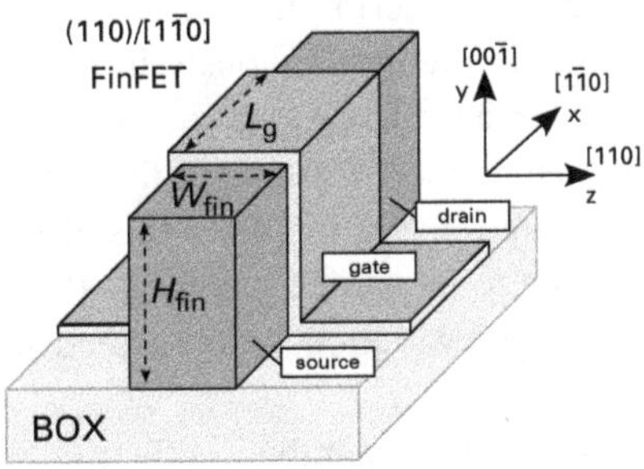

Figure 9.5 Structure of a FinFET with [110] lateral interfaces. L_G, W_{fin}, and H_{fin} denote the gate length, fin width and fin height, respectively.

The strain coordinate system

So far we have assumed that the stress simply consists of axial components T_{xx}, T_{yy}, T_{zz} in the DCS, in other words we have supposed that the stress components are aligned with the main device directions (namely the channel length and width or the fin height for FinFETs). This is not the most general case. However, the procedure described in Section 9.2.4 to determine the strain $\boldsymbol{\varepsilon}_{c,6}$ in the CCS starting from the stress in the DCS is completely general and it can be used even if the shear stress components ($T_{ij}, i \neq j$) in the DCS are non-null.

In some circumstances it may happen that the shear stress components are non-null in the DCS, but a different coordinate system exists where the stress has only axial components. We may define such a coordinate system as the stress coordinate system, SCS, because the stress is aligned with the axes of the SCS. The determination of the $\boldsymbol{\varepsilon}_{c,6}$ starting from the stress components in the SCS is straightforward on the basis of the procedure described in the previous sections. In fact, the results of Section 9.2.4 can be readily used by introducing the rotation matrix $\mathbf{R}_{S\to C}$ going from the SCS to the CCS. In particular, if $\mathbf{R}_{S\to D}$ is the rotation matrix going from SCS to the DCS, then the $\mathbf{R}_{S\to C}$ is given by

$$\mathbf{R}_{S\to C} = \mathbf{R}_{S\to D} \cdot \mathbf{R}_{D\to C}, \tag{9.53}$$

where $\mathbf{R}_{D\to C}$ is the rotation matrix from the DCS to the CCS defined in Section 9.2.4. Knowing the matrix $\mathbf{R}_{S\to C}$, the corresponding matrix $\mathbf{R}_{6,S\to C}$ is then given by Eq.9.19 and all the results of Section 9.2.4 are of immediate applicability.

9.3 Band structure in strained *n*-MOS transistors

The strain affects the band structure of bulk semiconductors and inversion layers in many respects. From a theoretical viewpoint the presence of the strain reduces the symmetry of the crystal, so that the irreducible wedge in the first Brillouin zone becomes larger with respect to the unstrained case (see Section 2.1.3 for the definition of the irreducible wedge), and, furthermore, the degeneracy of the bands at some of the symmetry points in the first Brillouin zone of the unstrained crystal may be removed [11].

A simple model to describe the effect of the strain on the energy relation is the deformation potential theory previously introduced in Section 4.6.1. The deformation potentials have been used to express the energy shifts at the band edges, however, the strain can also change the curvature of the bands close to the edges, hence the effective masses. Although in Section 4.6.1 this effect has been neglected in the scattering Hamiltonian related to the lattice vibrations (hence in the phonon scattering rates), the possible strain induced variations of the effective masses may be important in transport modeling.

In the framework of the parabolic effective mass approximation (EMA), the band structure for a valley ν of an electron inversion layer is obtained by using a limited number of parameters which describe the bands in the bulk crystal close to the minimum ν of the conduction band. Such parameters are the $E_{\nu 0}$ value for the valley, the effective masses of the constant energy ellipsoid and the information retained by the matrix $\mathbf{R}_{D\to E}$ of the transformation from the device to the ellipsoid coordinate system (see Sections 3.2.1 and 8.1). This can be summarized by recalling Eq.3.15 that, for a (100) inversion layer, expresses the energy in the valley ν versus the wave-vector $\mathbf{k} = (k_x, k_y)$ as

$$E_{\nu,n}(\mathbf{k}) = E_{\nu 0} + \varepsilon_{\nu,n} + \frac{\hbar^2}{2}\left(\frac{k_x^2}{m_x} + \frac{k_y^2}{m_y}\right). \tag{9.54}$$

Equation 9.54 is generalized to an arbitrary crystal orientation by Eq.8.13.

Thus, in the EMA picture the strain effects can be naturally accounted for, starting from the changes in the properties of the bulk crystal conduction band. One merit of this approach is that the changes in band edges and in the effective masses are useful to develop a basic insight into the way the strain affects transport, which in turn is very important for device design and optimization.

Before we move on, a few important considerations about the strain induced changes of the energy relation are in order. According to Eq.9.54, the $E_{\nu,n}(\mathbf{k})$ is given by the $E_{\nu 0}$ of the valley, plus the subband minima $\varepsilon_{\nu,n}$ due to the quantum confinement (obtained from the Schrödinger-like equation), plus the kinetic or parallel energy related to $\mathbf{k}$. In unstrained silicon the six Δ valleys are degenerate, that is the six $E_{\nu 0}$ are identical. The strain can remove such a degeneracy and change the $E_{\nu 0}$ of some valleys with respect to the corresponding value in the unstrained silicon. In this chapter we take as the reference energy the $E_{\nu 0}$ of the six Δ minima in the unstrained silicon, thus we set $E_{\nu 0} = 0$ for the unstrained silicon (see Fig.9.6). Thanks to this choice for the energy reference, the $E_{\nu 0}$ values in strained silicon coincide with the strain induced variations of the valley minima with respect to the unstrained case (see again Fig.9.6), and we can thus avoid cumbersome notations such as $\Delta E_{\nu 0}$ to denote the strain induced energy shifts. It remains understood, however, that the $E_{\nu 0}$ expressions for strained silicon discussed throughout the rest of the chapter are the strain induced energy shifts with respect to the unstrained silicon.

We finally notice that the shift of the valley minima is not the only way the strain affects the energy relation, in fact the strain may change the effective masses of the bulk silicon conduction band (see, for instance, Eqs.9.59 to 9.61), which in turn affect the subband minima $\varepsilon_{\nu,n}$ through the Schrödinger-like equation as well as the $\mathbf{k}$ dependent kinetic energy.

9.3.1 Strain effects in the bulk silicon conduction band

In this sub-section we follow quite closely the approach introduced in [40, 45] and then employed in [46–48]. The effects of the strain in the conduction band of a bulk crystal can be numerically calculated by using the pseudo-potential method [39, 40, 45, 49], and Fig.9.6 shows the changes produced by the shear strain component $\varepsilon_{c,xy}$ in the Δ_z minimum along the [001] direction of the silicon conduction band. As clearly stated

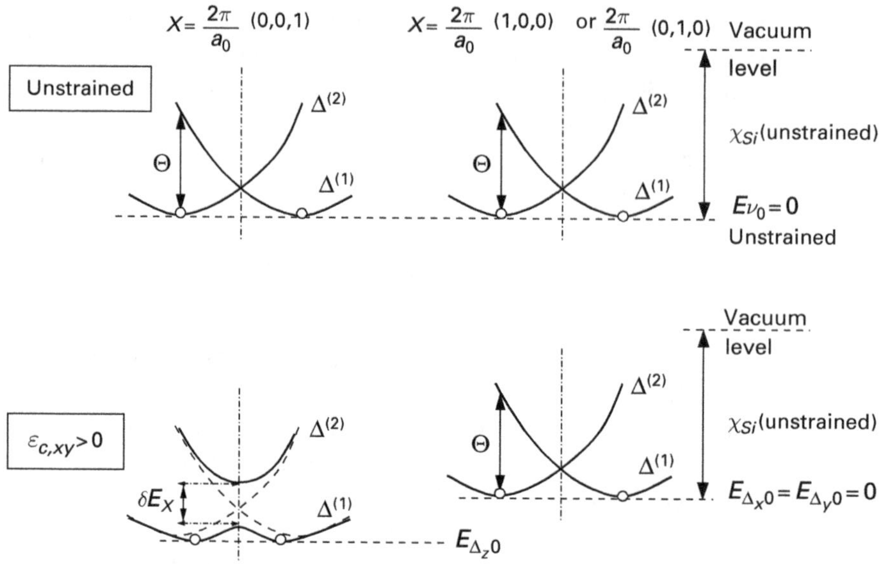

Figure 9.6 The behavior of the two lowest bulk silicon conduction bands $\Delta^{(1)}$ and $\Delta^{(2)}$ close to the X symmetry points (for the quantitative plots obtained with the pseudo-potential method see [40]). Θ is the separation between the $\Delta^{(1)}$ and $\Delta^{(2)}$ bands in the unstrained silicon at the conduction band edge in $\mathbf{K} = \frac{2\pi}{a_0}(0, 0, 0.85)$ [40]. $E_{\nu 0}$ is the energy minimum of the six degenerate Δ valleys and χ_{Si} is the electron affinity in the unstrained silicon (E_0 being the void energy); as discussed in the text, in this chapter $E_{\nu 0}$ is taken as the reference energy, hence it is set to zero. The shear strain component $\varepsilon_{c,xy}$ affects the conduction band minima Δ_z along the [001] direction by producing an energy split δE_X between the $\Delta^{(1)}$ and $\Delta^{(2)}$ bands at the X point and by changing the position, the energy, and the masses of the minimum of the $\Delta^{(1)}$ band. $E_{\Delta_z 0}$ is the band minimum along the [001] direction for $\varepsilon_{c,xy} \neq 0$. The energy minima $E_{\Delta_x 0}$, $E_{\Delta_y 0}$ along the [100] and [010] directions are not affected by $\varepsilon_{c,xy}$.

also in the figure caption, the symbols $\Delta^{(1)}$ and $\Delta^{(2)}$ in Fig.9.6 indicate the two lowest conduction bands of bulk silicon and they should not be confused with the symbols Δ_2 (or $\Delta_{0.92}$) and Δ_4 (or $\Delta_{0.19}$) denoting the unprimed and primed valley in a (001) silicon inversion layer (see Section 3.2.2 and Table 3.1).

As can be seen, the effect of $\varepsilon_{c,xy}$ is by no means limited to a rigid shift of the band edge. In fact the strain removes the degeneracy of the two lowest conduction bands $\Delta^{(1)}$ and $\Delta^{(2)}$ at the X points $\frac{2\pi}{a_0}(0, 0, \pm 1)$ and changes both the position and the energy value of the $\Delta^{(1)}$ minimum with respect to the unstrained silicon. The strain also affects the curvature of the $\Delta^{(1)}$ band close to the minimum, hence the effective masses. The changes in the effective masses produced by the shear strain components have been clearly observed in experiments [50–52].

An effective approach to describing the strain effects on the minima of the silicon conduction band is based on a two-band degenerate $\mathbf{k}\cdot\mathbf{p}$ theory around the X symmetry point [11, 50]. The interest in this approach stems from the fact that, for the two-by-two matrix of the $\mathbf{k}\cdot\mathbf{p}$ Hamiltonian, it is possible to calculate analytically the energy relation close to the X point for the two lowest conduction bands Δ_1 and Δ_2 as a function of the strain components [45, 53]. From the energy relation it is then quite straightforward to calculate analytical expressions for the position of the minimum of Δ_1, the energy

Table 9.3 Parameters related to band structure calculations in strained silicon MOSFETs. The deformation potentials l, m, and n allow us to include the strain effects in the **k·p** model for holes (see Eq.9.70); numerical values from [39]. Ξ_u and Ξ_d are the uniaxial and the dilation deformation potentials at the Δ minima of the conduction band (see also Table 4.2), and Θ is the separation between the Δ_1 and the Δ_2 bands at the same minima (see Fig.9.6). Finally η and κ are parameters entering Eqs.9.55 to 9.63; numerical values from [40].

Parameter	Value	Parameter	Value
l [eV]	−2.30	Ξ_{dX} [eV]	1.1
m [eV]	4.30	Θ [eV]	0.53
n [eV]	−9.18	η	−0.809
Ξ_{uX} [eV]	9.29	κ	0.0189

value at the minimum, and the corresponding effective masses. The separation between the Δ_1 and Δ_2 bands at the X point can also be derived. The analytical results thus obtained are in good quantitative agreement with numerical calculations based on the pseudo-potential method [45, 53].

Let us first discuss the results for the Δ_z minima along the [001] direction and sketched in the left part of Fig.9.6. In this case the shear strain component $\varepsilon_{c,xy}$ removes the band degeneracy at the X symmetry points $\mathbf{K}_X = \frac{2\pi}{a_0}(0, 0, \pm 1)$ and, furthermore, yields a deformation of the energy relation close to the Δ_z minima. In particular, the position of the Δ_z minima (located at $\mathbf{K}_X = \frac{2\pi}{a_0}(0, 0, \pm 0.85)$ in the unstrained silicon) in the strained crystal becomes [40]

$$k_{z,min} = \begin{cases} \pm(2\pi/a_0)\,[1 - 0.15\sqrt{1 - (\varepsilon_{c,xy}/\kappa)^2}], & |\varepsilon_{c,xy}| < \kappa \\ \\ \pm(2\pi/a_0), & |\varepsilon_{c,xy}| > \kappa \end{cases} \tag{9.55}$$

while the separation δE_X of the Δ_1 and Δ_2 bands at the X point is

$$\delta E_X = E_{\Delta 2}(\mathbf{K}_X) - E_{\Delta 1}(\mathbf{K}_X) = \frac{\Theta\,|\varepsilon_{c,xy}|}{\kappa}, \tag{9.56}$$

where κ is a parameter of the model and Θ is the separation between the Δ_1 and the Δ_2 band at the minima of the unstrained silicon conduction band. The numerical values of κ and Θ are given in Table 9.3.

Equation 9.55 shows that for $|\varepsilon_{c,xy}|$ larger than κ the position of the minimum remains fixed at the X point $\frac{2\pi}{a_0}(0, 0, 1)$, whereas the δE_X continues to increase (see Eq. 9.56). For most stress magnitudes of practical interest, however, the inequality $|\varepsilon_{c,xy}| < \kappa$ is fulfilled because $|\varepsilon_{c,xy}| = \kappa = 1.89\%$ corresponds to a quite large shear strain component.

Also, the energy value and the effective masses at the Δ_z minima can be calculated analytically. The band edge is given by [40, 45, 53]

$$E_{\Delta_z 0} = \Xi_{dX}\left(\varepsilon_{c,xx} + \varepsilon_{c,yy} + \varepsilon_{c,zz}\right) + \Xi_{uX}\,\varepsilon_{c,zz} + E^{\text{shear}}_{\Delta_z 0}, \tag{9.57}$$

where $E^{\text{shear}}_{\Delta_z 0}$ is related to the shear strain component $\varepsilon_{c,xy}$:

$$E^{\text{shear}}_{\Delta_z 0} = -\frac{\Theta}{4\kappa^2}\,\varepsilon^2_{c,xy}, \tag{9.58}$$

and Ξ_{dX} and Ξ_{uX} denote respectively the dilation and the shear deformation potentials for the Δ minima of the conduction band already defined in Section 4.6.1 and Table 4.2 (see Table 9.3 for the numerical values). As discussed above, the quoted $E_{\Delta_z 0}$ values are always referred to the minima of the six times degenerate Δ valleys of the unstrained silicon.

Regarding the effective masses, the strain component $\varepsilon_{c,xy}$ does not affect the m_l and m_t values of the Δ_x and Δ_y valleys (see Fig.9.6), but it *does* modify the masses of the Δ_z valleys. The longitudinal mass $m_{l,[001]}$ of the Δ_z valleys (which is the quantization mass in a (001) inversion layer), can be expressed as [40]

$$m_{l,[001]} = \begin{cases} m_l \left(1 - \varepsilon_{c,xy}^2/\kappa^2\right)^{-1}, & |\varepsilon_{c,xy}| < \kappa \\ m_l \left(1 - \kappa/|\varepsilon_{c,xy}|\right)^{-1}, & |\varepsilon_{c,xy}| > \kappa. \end{cases} \tag{9.59}$$

Besides affecting the longitudinal effective mass, the shear strain $\varepsilon_{c,xy}$ removes some of the crystal symmetries that result in revolutionary ellipsoids for the energy relation in the unstrained silicon. In the presence of $\varepsilon_{c,xy}$ the two principal directions [110] and $[\bar{1}10]$ of the energy ellipsoids have different masses, thus the ellipsoids become scalene and are described by three effective masses $m_{l,[001]}$, $m_{t,[110]}$, and $m_{t,[\bar{1}10]}$. The two transverse masses are given by [45]

$$m_{t,[110]} = \begin{cases} m_t \left(1 - \eta\varepsilon_{c,xy}/\kappa\right)^{-1}, & |\varepsilon_{c,xy}| < \kappa \\ m_t \left(1 - \mathrm{sgn}(\varepsilon_{c,xy})\eta\right)^{-1}, & |\varepsilon_{c,xy}| > \kappa \end{cases} \tag{9.60}$$

$$m_{t,[\bar{1}10]} = \begin{cases} m_t \left(1 + \eta\varepsilon_{c,xy}/\kappa\right)^{-1}, & |\varepsilon_{c,xy}| < \kappa \\ m_t \left(1 + \mathrm{sgn}(\varepsilon_{c,xy})\eta\right)^{-1}, & |\varepsilon_{c,xy}| > \kappa \end{cases} \tag{9.61}$$

where η is the unitless parameter reported in Table 9.3 and $\mathrm{sgn}(x)$ is the sign function.

Equations 9.55 to 9.61 give explicit results for the effect of the axial strain components $\varepsilon_{c,ii}$ and of the shear component $\varepsilon_{c,xy}$ on the energy relation of the Δ_z valleys. The $\varepsilon_{c,yz}$ and $\varepsilon_{c,zx}$ strain components do not affect the Δ_z valleys along the [001] direction.

The Δ_z valleys have been used as an example to illustrate the effects of the shear strain component $\varepsilon_{c,xy}$. The results can be directly applied also to the effects of $\varepsilon_{c,yz}$ on the Δ_x valleys and of $\varepsilon_{c,zx}$ on the Δ_y valleys. Consequently the $E_{\nu 0}$ energies for the Δ_x and Δ_y valleys can be written as

$$\begin{aligned} E_{\Delta_x 0} &= \Xi_{dX}\left(\varepsilon_{c,xx} + \varepsilon_{c,yy} + \varepsilon_{c,zz}\right) + \Xi_{uX}\,\varepsilon_{c,xx} + E^{\mathrm{shear}}_{\Delta_x,0}, \\ E_{\Delta_y 0} &= \Xi_{dX}\left(\varepsilon_{c,xx} + \varepsilon_{c,yy} + \varepsilon_{c,zz}\right) + \Xi_{uX}\,\varepsilon_{c,yy} + E^{\mathrm{shear}}_{\Delta_y,0}, \end{aligned} \tag{9.62}$$

where $E^{\mathrm{shear}}_{\Delta_x,0}$ and $E^{\mathrm{shear}}_{\Delta_y,0}$ are

$$E^{\text{shear}}_{\Delta_x 0} = -\frac{\Theta}{4\kappa^2}\varepsilon^2_{c,yz}, \qquad E^{\text{shear}}_{\Delta_y 0} = -\frac{\Theta}{4\kappa^2}\varepsilon^2_{c,zx}. \tag{9.63}$$

Equations 9.59 to 9.61 are similarly extended to describe the effective masses of the Δ_x and Δ_y valleys by substituting $\varepsilon_{c,xy}$ respectively with $\varepsilon_{c,yz}$ and $\varepsilon_{c,zx}$ and by choosing the appropriate directions for the principal axes of the energy ellipsoids.

9.3.2 Biaxial and uniaxial strain in *n*-MOS transistors

By using the equations discussed in Section 9.3.1 and the results of Section 9.2, which allow us to express the strain in the CCS for different stress and strain configurations, the $E_{\nu 0}$ and the effective masses for the different valleys ν in the strained silicon can be calculated for arbitrary stress and strain configurations. As discussed above, according to Eq.9.54 or Eq.8.13, both the $E_{\nu 0}$ and the effective masses affect the energy relation in the inversion layer.

In the following we discuss a few cases of practical relevance, however, the results of Section 9.3.1 and Section 9.2 are of general applicability.

Example 9.6: Biaxial strain for a $\langle 001 \rangle$ substrate.

In order to use the results of Section 9.3.1 to determine the appropriate $E_{\nu 0}$ to be used in Eqs.3.15 or 8.14, it is first of all important to remember the relation between the valleys in the inversion layer and the corresponding minima of the bulk crystal conduction band. For a (001) silicon inversion layer such a correspondence is given by

$$\Delta_{0.92} \longleftrightarrow \Delta_z, \qquad \Delta_{0.19x} \longleftrightarrow \Delta_x, \qquad \Delta_{0.19y} \longleftrightarrow \Delta_y. \tag{9.64}$$

For a biaxial strain, Eqs.9.33, 9.57, and 9.62 allow us to write

$$E_{\Delta 0.92} = \varepsilon_{||}\left[\Xi_{dX}\left(2 - \frac{1}{\sigma_{(001)}}\right) - \frac{\Xi_{uX}}{\sigma_{(001)}}\right],$$

$$E_{\Delta 0.19x} = E_{\Delta 0.19y} = E_{\Delta 0.19} = \varepsilon_{||}\left[\Xi_{dX}\left(2 - \frac{1}{\sigma_{(001)}}\right) + \Xi_{uX}\right], \tag{9.65}$$

where the Poisson ratio $\sigma_{(001)}$ is given by Eq.9.36 (see Table 9.2 for the numerical value).

For an epitaxial silicon layer grown on a $Si_{1-x}Ge_x$ virtual substrate, the parallel strain component $\varepsilon_{||}$ can be expressed as a function of the Ge mole fraction x by using Eqs.9.27 and 9.28. The $\varepsilon_{||}$ is positive for any x value and, by substituting in Eq.9.65 the values of the parameters given in Table 9.3, we can see that $E_{\Delta 0.92}$ is negative and $E_{\Delta 0.19}$ is positive. Figure 9.7 shows $E_{\Delta 0.92}$ and $E_{\Delta 0.19}$ versus $\varepsilon_{||}$ and the Ge mole fraction. Figure 9.8 shows the corresponding population of the $\Delta_{0.92}$ and $\Delta_{0.19}$ valleys for different inversion densities. The strain favors the population of the $\Delta_{0.92}$ valleys and it is thus additive with respect to the quantization effects, so that the impact on the subband population is weaker at large than at small inversion densities N_{inv}.

We finally notice that the effective masses are not affected by the biaxial strain in (001) substrates because all the shear components are null.

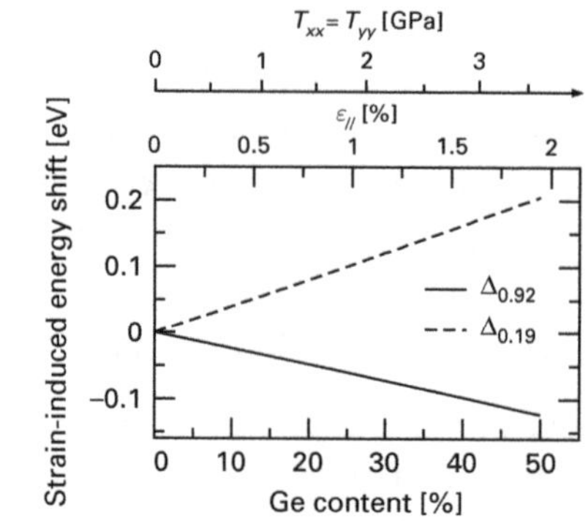

Figure 9.7 Strain induced shifts of the valley minima $E_{\Delta 0.92}$ and $E_{\Delta 0.19}$ versus the Ge mole fraction in a $Si_{1-x}Ge_x$ virtual substrate. The top x axes also show the corresponding parallel strain $\varepsilon_{||}$ as well as the parallel stress components $T_{xx} = T_{yy}$.(001) electron inversion layer.

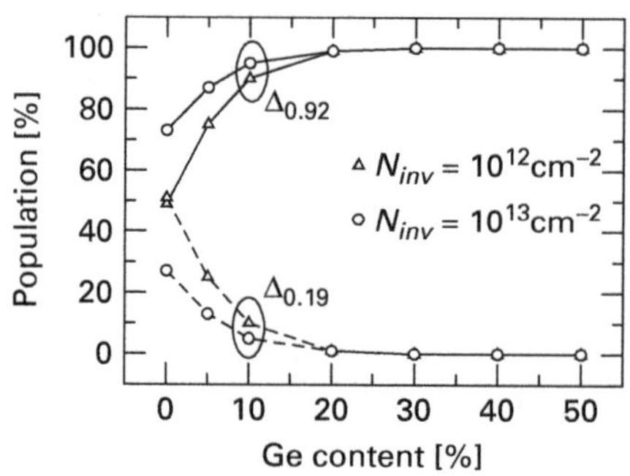

Figure 9.8 Occupation of the lowest $\Delta_{0.92}$ and lowest $\Delta_{0.19}$ subband versus the Ge mole fraction or the parallel strain $\varepsilon_{||}$ for either a small or a large inversion density N_{inv}. (001) electron inversion inversion layer in an n-MOSFET with a light channel doping $N_A = 2{\cdot}10^{16}\,\mathrm{cm}^{-3}$.

Example 9.7: Uniaxial stress in (001)/[110] MOSFETs.

Let us now consider a (001)/[110] MOSFET with a stress T_L in the source to drain direction. Section 9.2.6 has already discussed all the strain components and, in particular, showed that a non-null shear strain $\varepsilon_{c,xy}$ exists (see Eq.9.48). The corresponding $E_{\nu 0}$ energies for the $\Delta_{0.92}$ and $\Delta_{0.19}$ valleys are

$$E_{\Delta 0.92} = T_L\left[(S_{11} + 2S_{12})\,\Xi_{dX} + S_{12}\,\Xi_{uX}\right] - \frac{\Theta}{4\kappa^2}\left(\frac{S_{44}T_L}{4}\right)^2,$$

$$E_{\Delta 0.19x} = E_{\Delta 0.19y} = T_L\left[(S_{11} + 2S_{12})\,\Xi_{dX} + \left(\frac{S_{11} + S_{12}}{2}\right)\Xi_{uX}\right]. \quad (9.66)$$

For a tensile stress in the channel direction (i.e. T_L>0) the energy shift with respect to the unstrained silicon is again negative for the $\Delta_{0.92}$ and positive for the $\Delta_{0.19}$. Figure 9.9 shows $\Delta_{0.92}$ and $\Delta_{0.19}$ population versus T_L for a small and a large inversion density N_{inv}. Also in this case, as in the previous example, the strain effect on the subband population is additive with respect to the quantization.

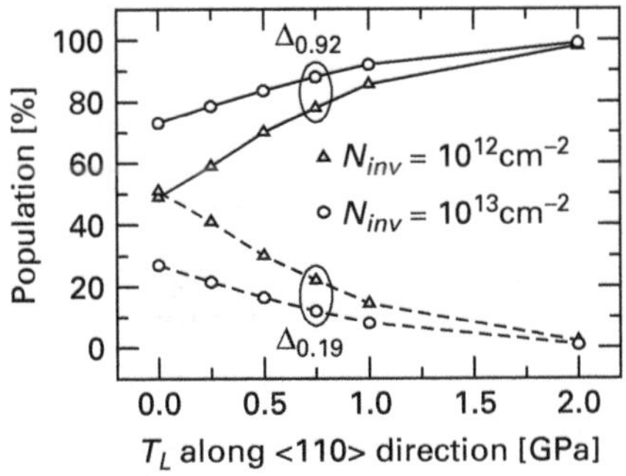

Figure 9.9 Occupation of the lowest $\Delta_{0.92}$ and lowest $\Delta_{0.19}$ subband versus the stress in the channel length direction. (001) / [110] electron inversion inversion layer.

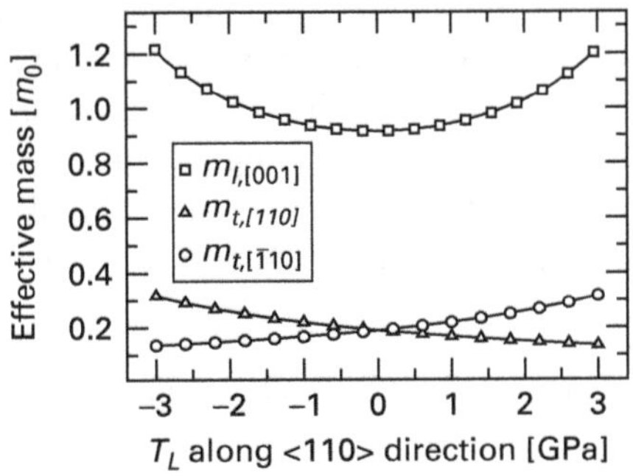

Figure 9.10 Effective masses for the $\Delta_{0.92}$ valleys versus the stress in the channel length direction. The non-null shear strain component $\varepsilon_{c,xy}$ yields a modulation of the effective masses. (001) / [110] electron inversion layer.

As illustrated in Fig.9.10, the strain component $\varepsilon_{c,xy}$ changes the effective masses of the $\Delta_{0.92}$ valleys. As can be seen, a tensile T_L reduces the effective mass $m_{t,[110]}$ in the transport direction, thus improving the transport properties in the inversion layer. The increase of the $m_{l,[001]}$ of the $\Delta_{0.92}$ valleys further contributes to enhancing the population of the valleys, in fact the $m_{l,[001]}$ is the quantization mass of the $\Delta_{0.92}$ valleys for a (001) / [110] MOSFETs.

The effect of the stress components T_W in the device width direction and $T_\perp$ in the direction normal to the silicon–oxide interface can be readily derived based on the strain components reported in Section 9.2.6.

Example 9.8: Uniaxial stress in (110)/[1$\bar{1}$0] MOSFETs and FinFETs.

The (110)/[1$\bar{1}$0] inversion layers are relevant both for planar MOSFETs and for FinFETs. The correspondence between the valleys in the inversion layer and the minima of the bulk silicon conduction band are

$$\Delta_{0.315} \longleftrightarrow \Delta_x\,,\,\Delta_y, \qquad\qquad \Delta_{0.19} \longleftrightarrow \Delta_z. \tag{9.67}$$

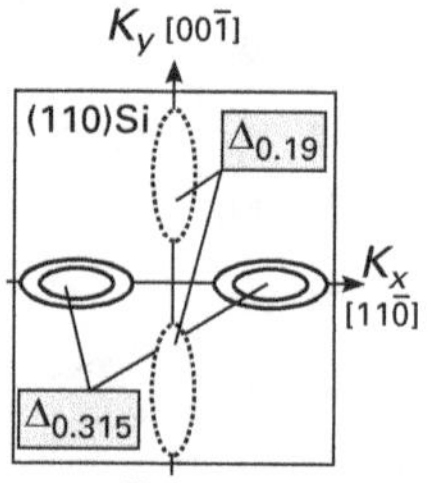

Figure 9.11 The two valleys in a (110)/1$\bar{1}$0] electron inversion layer and their position in the **k** plane. The $\Delta_{0.315}$ are the four-times degenerate lowest valleys (with the largest quantization mass $0.315m_0$), while the $\Delta_{0.19}$ are two-times degenerate and have a quantization mass $0.19m_0$.

The $\Delta_{0.315}$ are the four times degenerate and the $\Delta_{0.19}$ are the two times degenerate valleys identified in Tables 8.3 and 8.4, and sketched in Fig.9.11. We henceforth consider the effect of the stress T_L in the [1$\bar{1}$0] channel direction and of the T_{dW} stress in the device width direction, which is aligned with the [001] crystal direction. In a FinFET the T_{dW} coincides with the stress component in the fin height direction (see Fig.9.5). By substituting Eq.9.52 in Eqs.9.57 and 9.62 we obtain

$$\begin{aligned} E_{\Delta 0.19} &= T_L\left[(S_{11}+2S_{12})\,\Xi_{dX}+S_{12}\,\Xi_{uX}\right]-\frac{\Theta}{4\kappa^2}\left(\frac{S_{44}T_L}{4}\right)^2 \\ &\quad + T_{dW}\left[(S_{11}+2S_{12})\,\Xi_{dX}+S_{11}\,\Xi_{uX}\right], \\ E_{\Delta 0.315} &= T_L\left[(S_{11}+2S_{12})\,\Xi_{dX}+(S_{11}+S_{12})\,\frac{\Xi_{uX}}{2}\right] \\ &\quad + T_{dW}\left[(S_{11}+2S_{12})\Xi_{dX}+S_{12}\,\Xi_{uX}\right]. \end{aligned} \tag{9.68}$$

If we insert in Eq.9.68 the values of the parameters in Tables 9.1 and 9.3, we can see that the tensile T_L and the compressive T_{dW} tend to lower the $\Delta_{0.19}$ with respect to the $\Delta_{0.315}$ valleys. Interestingly, in this case the strain effect on the subband population is opposite to the quantization and the strain tends to re-populate the subbands that are relatively de-populated by the quantization. Figure 9.12 shows the $E_{\nu 0}$ for the $\Delta_{0.19}$ and $\Delta_{0.315}$ valleys versus T_L and T_{dW}, while Fig.9.13 shows the changes of the subband population produced by T_{dW}.

The shear strain produced by the T_L also affects the masses of the $\Delta_{0.19}$ subbands, and Fig.9.14 illustrates the corresponding modulation of the transport mass $m_{\Delta z,[1\bar{1}0]}$ and quantization mass $m_{\Delta z,[110]}$. Note that in Fig.9.14 the $m_{\Delta z,[110]}$ is $0.23m_0$ in the unstrained case, rather than $0.19m_0$ as suggested by the m_t value in Table 2.4. This value for $m_{\Delta z,[110]}$ is used to account semi-empirically for the impact on the quantization of the $\Delta_{0.19}$ subbands of the non-parabolicity in the [110] direction of bulk silicon Δ valleys (see Section 8.3.1 and Fig.8.8).

The examples above clarify that the technological exploitation of the strain can ultimately be considered an effective means for subband engineering in inversion layers.

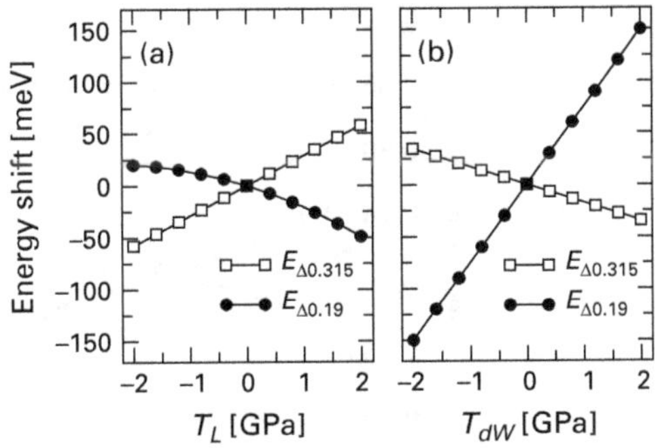

Figure 9.12 Stress induced energy shifts $E_{\Delta 0.315}$ and $E_{\Delta 0.19}$ (with respect to unstrained silicon) for a (110) silicon inversion layer versus: (a) T_L stress; (b) T_{dW} stress. The T_{dW} corresponds to the stress in the device width direction for a planar (110)/[1$\bar{1}$0] MOSFET and in the fin height direction for the (110)/[1$\bar{1}$0] FinFET in Fig.9.5.

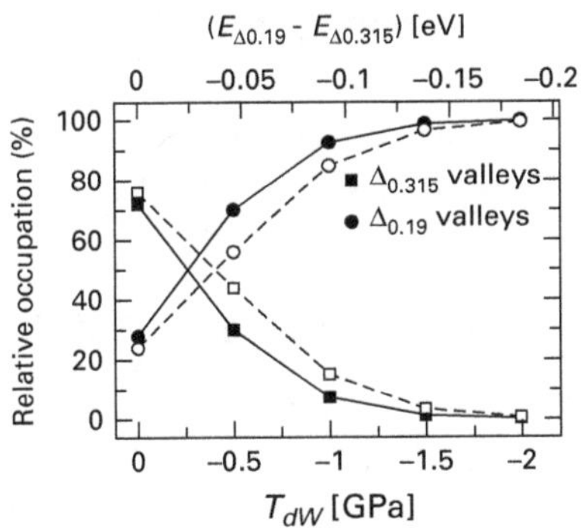

Figure 9.13 Simulated change in the relative population of the $\Delta_{0.19}$ and $\Delta_{0.315}$ valleys as a function of compressive T_{dW} stress (bottom x–axis) and for two inversion densities N_{inv} (per gate). Filled symbols: inversion density $N_{inv} = 3.2\times10^{12}\,\mathrm{cm}^{-2}$; open symbols: $N_{inv} = 1.3\times10^{13}\mathrm{cm}^{-2}$. On the top x-axis the stress induced energy split $(E_{\Delta 0.19} - E_{\Delta_{\Delta 0.315}})$ is also shown. The T_{dW} corresponds to the stress in the device width direction for a planar (110)/[1$\bar{1}$0] MOSFET and in the fin height direction for the (110)/[1$\bar{1}$0] FinFET in Fig.9.5. Results obtained with a self-consistent Schrödinger–Poisson solver [54]. (110)/[1$\bar{1}$0] SOI FET or FinFET with $T_{\mathrm{Si}} = 16\,\mathrm{nm}$.

We conclude Section 9.3.2 by reiterating that the treatment proposed for the effects of strain in the electron inversion layers has implicitly assumed that the strain and the quantization effects can be separately accounted for in a two step procedure. The fact that the overall energy dispersion in a strained inversion layer can be accurately described by using such an approach is not obvious. In this respect, the Linear Combination of Bulk Bands (LCBB) method already discussed in Section 3.4 has been used jointly with a non-local pseudo-potential solver to calculate the band structure in electron and hole inversion layers by accounting at the same time for both the strain and the quantization effects [55]. The overall procedure is computationally very demanding, but the results can be used to test the accuracy of the simpler models, such as the EMA or the **k**·**p** approach; in this latter respect, the **k**·**p** results were found to agree quite well with the LCBB calculations [55].

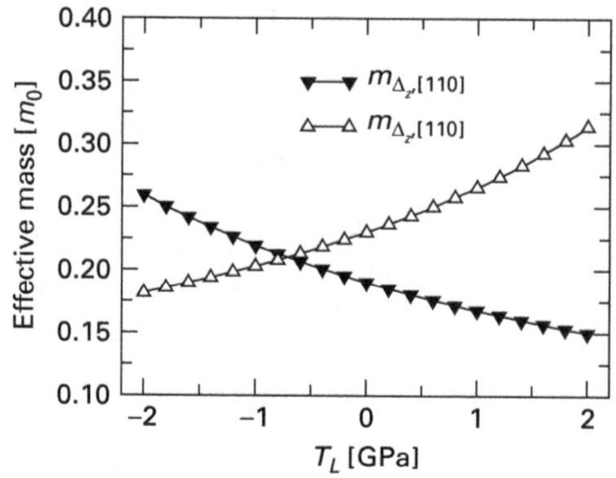

Figure 9.14 Effective masses of the $\Delta_{0.19}$ valleys versus the T_L stress component. $m_{\Delta z,[110]}$ and $m_{\Delta z,[1\bar{1}0]}$ denote respectively the quantization and the transport effective mass. The reason for $m_{\Delta z,[110]}$ being $0.23m_0$ in the unstrained case, rather than $0.19m_0$, is discussed in the text.

9.4 Band structure in strained *p*-MOS transistors

The effects of strain in *p*-MOS transistors can be described by using the **k·p** model for holes already discussed in Sections 2.2.2 and 3.3.

9.4.1 The k·p model for holes in the presence of strain

The **k·p** model can be naturally extended to account for strain both for the bulk silicon valence band (close to the Γ point), and for the hole inversion layers in *p*-MOS transistors [56–62]. Three deformation potentials are necessary to account for the strain at the Γ point, and these are typically denoted by l, m, and n [11]. More precisely, the strain can then be described by means of the algebraic operator $\mathbf{H}_\varepsilon$, which is simply given by the 6×6 matrix [56]

$$\mathbf{H}_\varepsilon = \begin{pmatrix} \mathbf{H'}_\varepsilon & \mathbf{0} \\ \mathbf{0} & \mathbf{H'}_\varepsilon \end{pmatrix}, \tag{9.69}$$

where $\mathbf{H'}_\varepsilon$ is the 3×3 matrix

$$\begin{pmatrix} l\,\varepsilon_{c,xx} + m\,(\varepsilon_{c,yy} + \varepsilon_{c,zz}) & n\,\varepsilon_{c,xy} & n\,\varepsilon_{c,zx} \\ n\,\varepsilon_{c,xy} & l\,\varepsilon_{c,yy} + m(\varepsilon_{c,xx} + \varepsilon_{c,zz}) & n\,\varepsilon_{c,yz} \\ n\,\varepsilon_{c,zx} & n\,\varepsilon_{c,yz} & l\,\varepsilon_{c,zz} + m(\varepsilon_{c,xx} + \varepsilon_{c,yy}) \end{pmatrix}. \tag{9.70}$$

The parameters l, m, n have been calibrated by comparison with the results of more rigorous pseudo-potential calculations, and the values for silicon are given in Table 9.3 [39].

For a bulk crystal the strain can be introduced by modifying Eq.2.57 according to

$$\left[\mathbf{H}_{\mathbf{k}\cdot\mathbf{p}} + \mathbf{H}_{so} + \mathbf{H}_\varepsilon\right] \mathbf{C}_\mathbf{K} = E\,\mathbf{C}_\mathbf{K}, \tag{9.71}$$

where $\mathbf{C_K}$ is a six component eigenvector. In a hole inversion layer, instead, we must change Eq.3.34 to

$$\left[\hat{\mathbf{H}}_{\mathbf{k}\cdot\mathbf{p}}\left(\mathbf{k}, -\mathrm{i}\frac{\partial}{\partial z}\right) + \mathbf{H}_{so} + \mathbf{H}_{\varepsilon} + \mathbf{I}\,U(z)\right] \xi_{n\mathbf{k}}(z) = E_n(\mathbf{k})\, \xi_{n\mathbf{k}}(z), \tag{9.72}$$

where $\mathbf{k}$ is the two component wave-vector in the transport plane and $\xi_{n\mathbf{k}}(z)$ is the six component wave-function.

The band structure for the 2D hole gas is then obtained by numerically solving the set of differential eigenvalue equations in Eq.9.72 as discussed in detail in Section 3.3.2. The $\mathbf{k}\cdot\mathbf{p}$ model turns out to be quite accurate when compared to the more rigorous LCBB calculations [55]. However, differently from the case of the electron inversion layers described with the EMA model, it is not possible to obtain simple analytical expressions for the energy shifts with respect to the unstrained silicon or for the effective transport masses of the two-dimensional subbands. Consequently it is not easy to gain an insight into the impact of strain on the band structure other than by inspecting the hole equi-energy lines (see Appendix C), as is done below for some cases of practical interest.

9.4.2 Biaxial and uniaxial strain in *p*-MOS transistors

In the case of the biaxial strain produced by a (001) virtual substrate, the hole band structure can be calculated for different Ge mole fractions by using in Eq.9.72 the strain components derived in Eq.9.37. Figure 9.15 shows the equi-energy curves for either an unstrained or a strained (001) hole inversion layer. The strain produces an appreciable deformation of the hole band structure but the in-plane symmetries are not changed by the biaxial strain configuration.

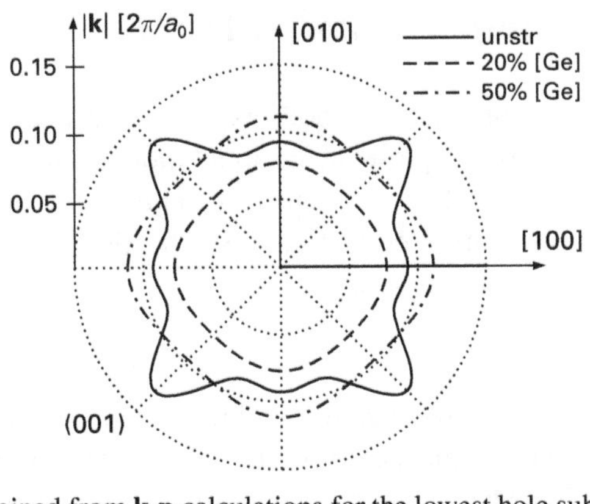

Figure 9.15 Equi-energy curves obtained from $\mathbf{k}\cdot\mathbf{p}$ calculations for the lowest hole subband at an energy of 25 meV above the subband minimum. Solid line: unstrained. Dashed line: 20% Ge; dot-dashed line 50% Ge. (001) hole inversion layer in a triangular well with a confining field $F_c = 0.7\,\mathrm{MV/cm}$.

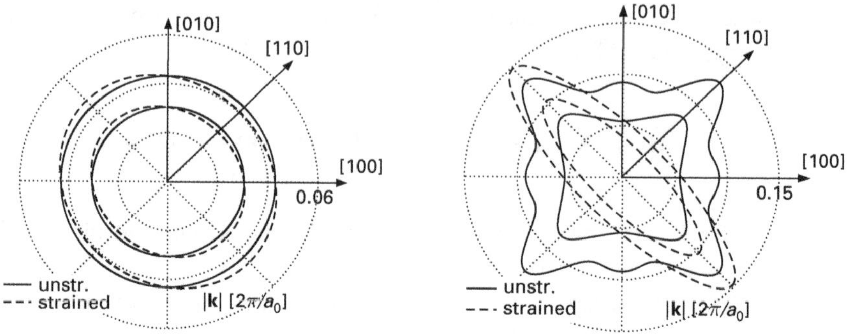

Figure 9.16 Left: equi-energy curves for the lowest unprimed subband at an energy 25 meV and 50 meV above the subband minimum. Solid line: unstrained. Dashed line: 1 GPa tensile stress along the [110] direction. The shear strain component ε_{xy} makes the subband somewhat elliptic; (001) electron inversion layer. Right: equi-energy curves for the lowest subband at an energy of 25 meV and 50 meV above the subband minimum obtained from **k·p** calculations; (001) hole inversion layer. Solid line: unstrained. Dashed line: [110] 1 GPa compressive stress. The right plot corresponds to a triangular well with a confining electric field $F_c = 0.7\,MV/cm$. *Reprinted with permission from [48]. Copyright 2009 by Springer.*

The deformation of the hole band structure is even more remarkable in the case of uniaxial strain. In this respect Fig.9.16 shows a comparison of the stress induced band structure deformation for either electrons or holes subject to a uniaxial stress equal in magnitude in the [110] direction (1 GPa tensile for electrons and 1 GPa compressive for holes). In this case the strain components are given by Eq.9.48 and then the electron effective masses can be obtained with the strain dependent expressions discussed in Section 9.3.1, while the hole equi-energy curves for holes are calculated by using Eq.9.72 directly. As can be seen, the compressive stress produces a remarkable reduction of the hole effective mass in the [110] direction. Such a deformation of the energy relation is comparatively much more pronounced than for electrons.

9.5 Simulation results for low field mobility

A vast literature exists for both experimental analysis and modeling of mobility in strained devices. We present here simulation results for a few relevant cases and discuss the comparison with experiments. The simulations are obtained by using the momentum relaxation time approximation (see Section 5.4) or the multi-subband Monte Carlo approach (see Section 6.3).

Biaxial strain

Figure 9.17 shows the electron mobility enhancement as a function of the germanium concentration or, equivalently, of the parallel strain $\varepsilon_{||}$ for biaxially strained (001)/[110] n-MOSFETs. Such a strain configuration results in a splitting between the Δ_z and the Δ_x, Δ_y valleys but does not change the effective masses, as discussed in Section 9.3. The MSMC simulations are compared to several sets of experimental data at a

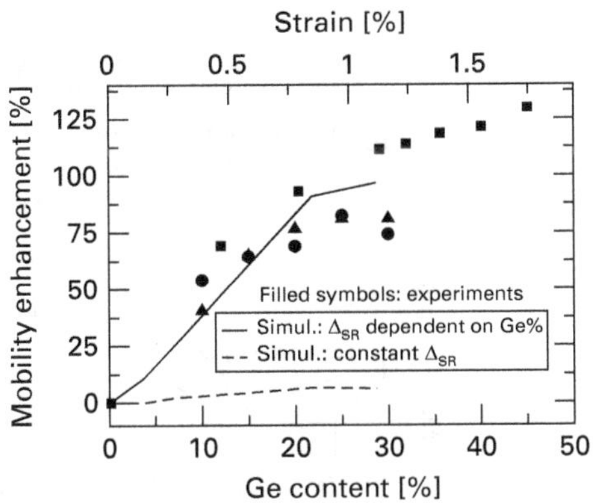

Figure 9.17 Electron mobility enhancement as a function of germanium content for biaxially strained *n*-MOSFETs. Effective field $F_{eff} = 0.7$ MV/cm. Lines are multi-subband Monte Carlo (MSMC) simulations; symbols are experiments: circles [69]; squares [20]; triangles [70]. The solid line indicates simulations with a Δ_{SR} dependent on the Ge%: $\Delta_{SR} = 0.62, 0.55, 0.5, 0.38, 0.26, 0.24$ nm for respectively 0%, 3.7%, 7.2%, 14.5%, 21.8%, 28.7% Ge content. The dashed line shows simulations with a constant $\Delta_{SR} = 0.62$ nm. (001) silicon. *Reprinted with permission from [48]. Copyright 2009 by Springer.*

moderately high effective field $F_{eff} = 0.7$ MV/cm. As can be seen, the simulated mobility enhancement is far too small compared to the experiments if the same parameters for surface roughness scattering are used at all strain levels.

In order to obtain a good agreement with the experiments it is necessary to introduce an ad hoc reduction of the r.m.s. value Δ_{SR} of the surface roughness with increasing strain [63–65]. A reduction of the Δ_{SR} in strained devices has been experimentally reported using AFM measurements [66] and theoretically predicted using *ab-initio* calculations [67, 68]. However, the possible reduction of surface roughness in strained devices is still debated and is further discussed in this section.

Figure 9.18 shows simulations and experiments for hole mobility enhancements in biaxially strained *p*-MOSFETs. As can be seen, the simulations obtained with the MRT approach and with a self-consistent $\mathbf{k}\cdot\mathbf{p}$ solver can reproduce reasonably well the experimental behavior without any change in the surface roughness parameters with the strain level. Both in some of the experiments and in the simulations we see a non-monotonic dependence of mobility on biaxial strain.

We finally notice that Figs.9.17 and 9.18 clearly show that, as anticipated in Section 9.1.1, the biaxial strain obtained by means of a SiGe virtual substrate is more effective for electron than for hole mobility enhancement, at least for Ge contents up to about 30%.

Uniaxial strain

Many combinations of surface orientation and uniaxial strain direction are possible and from here on we focus our analysis on the practically relevant case of (001) / [110] MOS transistors with a stress along the [110] transport direction. Under this stress configuration, the $\boldsymbol{\varepsilon}_{c,3\times3}$ tensor contains non-null axial strain components ($\varepsilon_{c,xx}$, $\varepsilon_{c,yy}$, $\varepsilon_{c,zz}$)

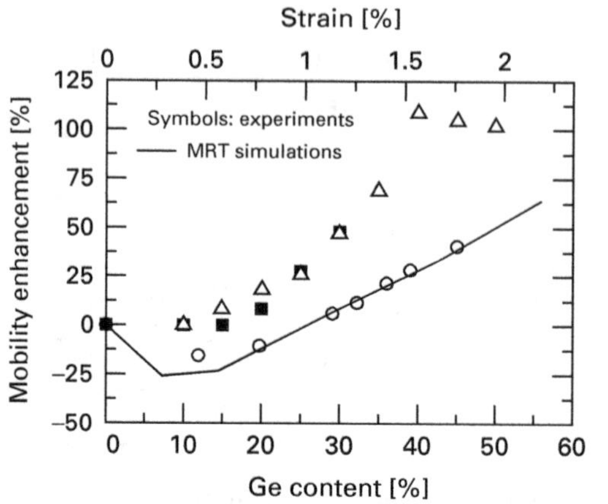

Figure 9.18 Hole mobility enhancement as a function of germanium content for biaxially strained *p*-MOSFETs. Effective field $F_{eff} = 0.7\,\text{MV/cm}$. Solid line: simulations obtained with the MRT method with a **k**·**p** energy model. Symbols are experiments: open circles [20]; filled squares [70]; open triangles [71]. (001) silicon. *Reprinted with permission from [48]. Copyright 2009 by Springer.*

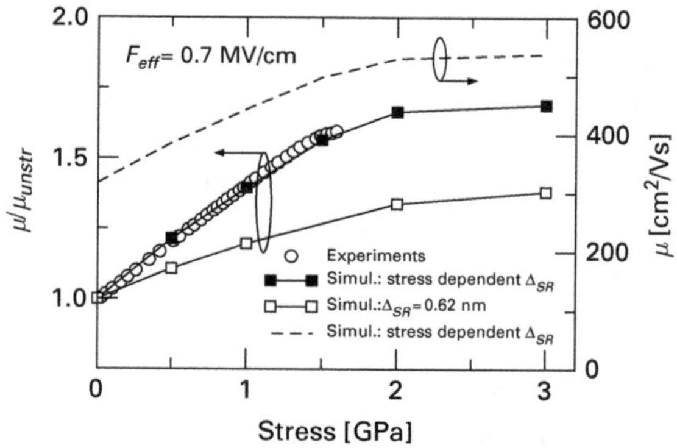

Figure 9.19 Simulated (MSMC) and experimental (open circles, [72]) electron mobility enhancement versus uniaxial tensile stress along the [110] direction for (001)/[110] *n*-MOSFETs. The dashed line shows simulated mobility values. For $\Delta_{SR} = 0.62\,\text{nm}$ the simulated mobility enhancement is smaller than in the experiments. The simulations with a stress dependent Δ_{SR} feature $\Delta_{SR} =$ 0.62, 0.572, 0.523, 0.48 nm for respectively 0 GPa, 0.5 GPa, 1 GPa, 1.5 GPa of tensile stress. $\Delta_{SR} = 0.48\,\text{nm}$ for stress larger than 1.5 GPa. *Reprinted with permission from [48]. Copyright 2009 by Springer.*

as well as a shear strain component $\varepsilon_{c,xy}$, which is responsible for a modulation of the electron effective masses.

Figure 9.19 shows simulated electron mobility enhancements compared with experiments for tensile uniaxial stress. As in the case of biaxial strain, in order to obtain a good agreement with experiment it is necessary to assume that the r.m.s. value Δ_{SR} of the surface roughness is reduced by the strain. However, the reduction of transport mass produced by the shear strain component yields an appreciable mobility improvement even if the Δ_{SR} value is kept the same as in the unstrained case.

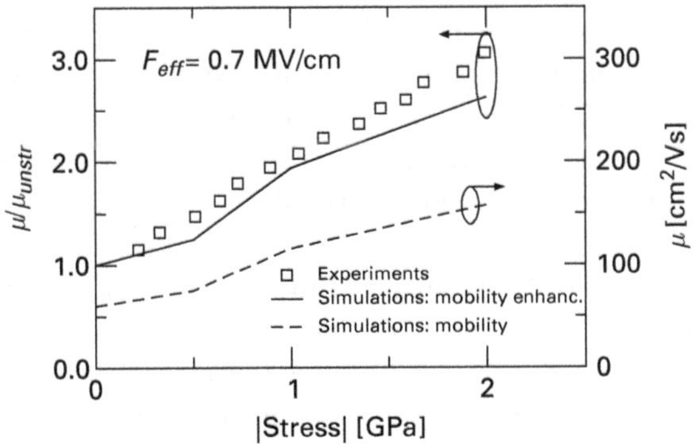

Figure 9.20 Hole mobility values and mobility enhancement versus uniaxial compressive stress along the [110] direction for (001)/[110] p-MOSFETs. Symbols: experiments [72–74]; Lines: simulations with the MSMC approach. Solid line: mobility enhancement; dashed line: absolute mobility values. In all the simulations the r.m.s. value Δ_{SR} of the surface roughness is 0.56 nm. *Reprinted with permission from [48]. Copyright 2009 by Springer.*

Figure 9.20 shows the corresponding results for hole mobility in the presence of compressive uniaxial stress in the [110] direction. The mobility simulations were obtained with the MSMC model for p-MOSFETs presented in [75], where the semi-analytical hole energy model was calibrated for uniaxially strained devices as explained in [76]. A good agreement with experiments is obtained up to a stress value of about 2 GPa and without changing the scattering parameters with the strain [76].

Figures 9.19 and 9.20 show that the stress induced mobility enhancement is considerably larger in p-MOS than in n-MOS for a uniaxial stress equal in magnitude in the channel direction. This is consistent with the large impact on the hole compared to electron band structure illustrated in Fig.9.16.

We finally note that the response of mobility to stress is linear for small stress magnitudes (up to a few hundreds MPa), hence from Figs.9.19 and 9.20 we can also infer that the electron mobility is reduced for a compressive stress (negative x-axis in Fig.9.19) and the hole mobility is reduced for a tensile stress (negative x-axis in Fig.9.20). Such behavior underlines that, as already mentioned in Section 9.1, the stress and strain configurations that improve electron and hole mobility are different, which explains why local strain techniques, able to produce a different stress in either n-MOS or p-MOS transistors, are those actually employed in mass production of CMOS integrated circuits.

Strain induced modifications of the surface roughness spectrum

The results discussed so far have shown that, for strain induced enhancement of electron mobility, simulations can be reconciled with experiments only if the r.m.s. value Δ_{SR} of the surface roughness scattering is reduced with the strain. This fact was observed some years ago [63], and deserves a few further comments. As discussed above, a reduction of Δ_{SR} in biaxially strained samples has been obtained from *ab-initio* calculations [67, 68] and observed in AFM measurements [66]. However, quantitatively speaking, the measured Δ_{SR} reductions are smaller than the ones necessary to reproduce the experiments in the simulations. Furthermore, a comparative analysis of electron and hole mobility in biaxially strained MOSFETs showed that, even at very low temperatures (4.2 K), electron mobility is enhanced by biaxial strain but hole mobility is not [77]. These observations contradict the hypothesis that strain yields only a simple reduction of Δ_{SR}. However, given the different values of the thermal wave-vectors for electrons and holes, the authors of [77] speculate that strain may change even, or maybe mostly, the correlation length of the roughness spectrum, in such a way that the electron mobility is enhanced but the hole mobility is not. This could be the case because the electrons are sensitive to a part of the roughness spectrum at smaller wave-vector values than the holes are [77] (see also Section 3.6.3 and Figs.3.15 and 3.16 for the average $k = |\mathbf{k}|$ values in electron and hole inversion layers). As pointed out in [65], strain induced electron mobility enhancement is a topic still far from being conclusively understood, and the theoretical framework described in this chapter provides a sound physical basis for understanding future developments in this field.

9.6 Simulation results for drain current in MOSFETs

The simulation results discussed so far have been focused on low field mobility, but one of the merits of the semi-classical transport approach is its ability to use the same physical models to describe both low field uniform transport and the far from equilibrium transport in nano-scale MOSFETs. As a case study for nanoscale transistors, we investigate the on-current behavior of (001)/[110] n- and p-MOSFETs with different channel lengths and in the presence of a uniaxial stress in the [110] channel direction. The stress is tensile and compressive for the n- and p-MOSFETs, respectively.

The simulated devices are double-gate SOI MOSFETs designed according to the 2007 ITRS roadmap to obtain approximately the same drain-induced barrier lowering (DIBL) of 100 mV/V. The gate work-functions were tailored to have the same off-current I_{OFF} of 100 nA/μm for the unstrained as well as for all the strained devices. The on-current, I_{ON}, simulations discussed below include the effect of the series resistances R_{SD}, which were accounted for as external lumped elements. The I_{ON} is defined as I_{DS} at $V_{GS} = V_{DS} = 1V$, where V_{GS} and V_{DS} are the extrinsic biases. The I_{DS} is always reported per unit width and per gate. The parameters of the devices are listed in Table 9.4.

Figure 9.21 shows the I_{ON} values and the ratio $[I_{ON,p}/I_{ON,n}]$ versus the stress magnitude for the $L_G = 35$ nm transistor. As can be seen, the relative I_{ON} enhancement with stress is larger for p-MOSFETs, so that the p/n I_{ON} ratio increases with stress. This is further emphasized by Fig.9.22, illustrating that, for well behaved devices, the p/n I_{ON} ratio is only weakly affected by the scaling, whereas it significantly increases upon the application of uniaxial stress.

The main reason for this behavior is the larger deformation produced by the uniaxial stress in the hole compared to the electron energy dispersion (see Fig.9.16). In fact, changes in the hole band structure not only affect the mobility but also improve significantly the average velocity v^+ at the virtual source of nano-scale p-MOSFETs. This is

Table 9.4 Parameters of the double-gate silicon-on-insulator (DG-SOI) (001) / [110] MOSFETs simulated in Section 9.6. T_{Si} and T_{ox} are the silicon and physical oxide thickness, respectively. The channel doping concentration is 10^{15} cm^{-3}. All the devices have approximately the same $I_{OFF} = 100$ nA/μm defined as the I_{DS} at $V_{GS} = 0V$ and $V_{GS} = V_{DD}$. The values for the series resistance R_{SD} have been taken from the ITRS roadmap [78]. The supply voltage is $V_{DD} = 1$V.

L_G [nm]	35	25	18
T_{Si} [nm]	17	11	8
ε_{ox}	7	7	7
T_{ox}/EOT [nm]	2.1/1.2	1.8/1.0	1.5/0.8
Gate work-function [eV]			
n-MOS/p-MOS	4.52/4.77	4.50/4.78	4.51/4.77
DIBL [mV/V]			
n-MOS/p-MOS	99/100	97/101	95/101
R_{SD} [$\Omega \cdot \mu$m]	180	180	180

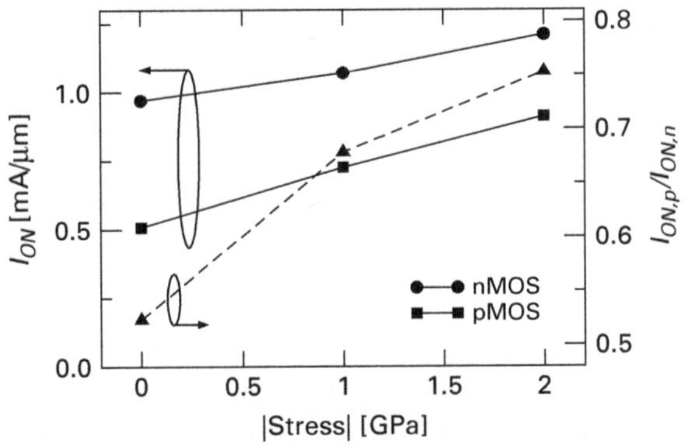

Figure 9.21 Simulated on current I_{ON} ($V_{GS} = V_{DS} = 1$V) and p/n I_{ON} ratios versus the uniaxial stress magnitude for $L_G = 35$ nm FETs (device parameters reported in Table 9.4). The stress significantly increases the p/n I_{ON} ratio. Tensile and compressive stress for n– and p-MOSFETs, respectively. *Reprinted with permission from [48]. Copyright 2009 by Springer.*

clearly illustrated by Fig.9.23, which also shows that the stress induced modulation of v^+ is small in n-MOS transistors compared to p-MOS devices.

9.7 Summary

This chapter has presented methodologies and models able to describe strain effects on the energy relation of electron and hole inversion layers and then on transport in n-MOS and p-MOS transistors. The procedure can be summarized by saying that the first step is determination of the strain components in the crystal coordinate system (CCS), starting from the strain or stress conditions typically set in the device coordinate system (DCS), which can be different with respect to the CCS. The necessary transformations have

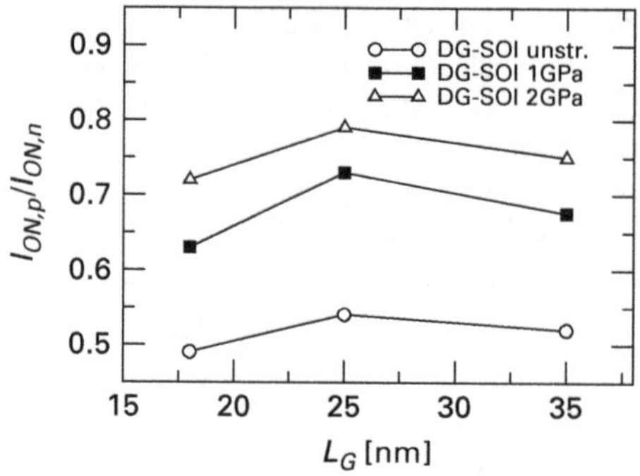

Figure 9.22 Simulated on current p/n ratios versus channel length L_G for unstrained and strained transistors with the device parameters given in Table 9.4. The stress increases the p/n I_{ON} ratios for all L_G values. Tensile and compressive stress for n– and p-MOSFETs, respectively. *Reprinted with permission from [48]. Copyright 2009 by Springer.*

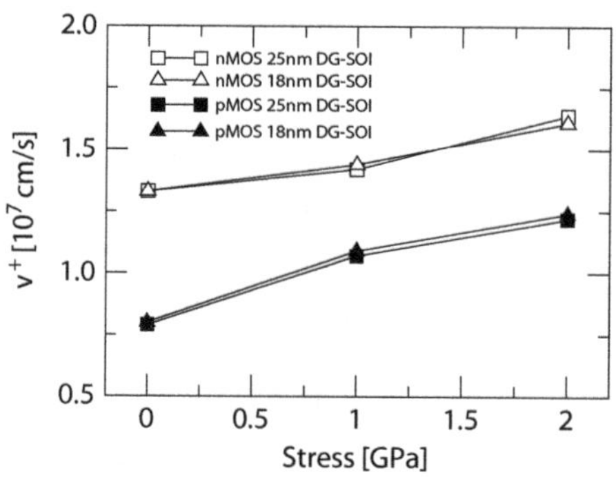

Figure 9.23 Injection velocity v^+ at the virtual source versus the uniaxial stress magnitude for n- and p-MOSFETs with different L_G (device parameters in Table 9.4). $V_{GS} = V_{DS} = 1V$.

been described in Section 9.2.4 in very general terms; they have been used in several examples of practical relevance in Sections 9.2.5 and 9.2.6.

With known strain components in the CCS, Sections 9.3 and 9.4 explain how to model the effect of the strain on the electron and hole band structure in inversion layers. This allows us to naturally describe the effects of the strain on the transport properties of n-MOS and p-MOS transistors by using the methodologies described in detail in the previous chapters. Sample simulation results were then illustrated both for low field mobility in Section 9.5 (reporting also a systematic comparison to experimental data), and for the drive currents of nanoscale MOSFETs in Section 9.6. Hence, also in the case of strained transistors, we can see that the semi-classical transport framework allows us to use the same physical models to describe both low field uniform transport in long test structures and strongly non-local transport in nanoscale transistors.

We conclude by remarking that, at the time of writing, strain engineering is a rapidly evolving field where, on the one hand, the physical mechanisms by which strain affects transport have not yet been conclusively revealed, and on the other hand exploitation of strain in CMOS technologies continues and promises further performance improvements in the near future.

References

[1] Q. Ouyang, M. Yang, J. Holt, *et al.*, "Investigation of CMOS devices with embedded SiGe source/drain on hybrid orientation substrates," in *IEEE Symposium on VLSI Technology - Technical Digest*, pp. 28–29, 2005.

[2] S.E. Thompson, M. Armstrong, C. Auth, M. Alavi, and M. Buehler, "A 90-nm logic technology featuring strained-silicon," *IEEE Trans. on Electron Devices*, vol. 53, no. 11, pp. 1790–1797, 2004.

[3] P.R. Chidambaram, C. Bowen, S. Chakravarthi, C. Machala, and R. Wise, "Fundamentals of silicon material properties for successful exploitation of strain engineering in modern CMOS manufacturing," *IEEE Trans. on Electron Devices*, vol. 53, no. 5, pp. 944–964, 2006.

[4] D. Zhang, B.Y. Nguyen, T. White, *et al.*, "Embedded SiGe S/D PMOS on thin body SOI substrate with drive current enhancement," in *IEEE Symposium on VLSI Technology - Technical Digest*, pp. 26–27, 2005.

[5] W. Zhang and J.G. Fossum, "On the threshold voltage of strained-Si-$Si_{1-x}Ge_x$ MOSFETs," *IEEE Trans. on Electron Devices*, vol. 52, no. 2, pp. 263–268, 2005.

[6] T.-J. Wang, C.-H. Ko, S.-J. Chang, S.T.-M. Kuan, and W.-C. Lee, "The effects of mechanical uniaxial stress on junction leakage in nanoscale CMOSFETs," *IEEE Trans. on Electron Devices*, vol. 55, no. 2, pp. 572–577, 2008.

[7] W. Zhao, A. Seabaugh, B. Winstead, D.Jovanovic, V. Adams, "Impact of uniaxial strain on the gate leakage currents of PD-SOI MOSFETs and ring oscillators with ultra-thin gate dielectric," in *Device Research Conference Digest*, pp. 199–200, 2005.

[8] A. Shickova, P. Verheyen, G. Eneman, *et al.*, "Reliability of strained-Si devices with post-oxide-deposition strain introduction," *IEEE Trans. on Electron Devices*, vol. 55, no. 12, pp. 3432–3441, 2008.

[9] H.H. Hall, J. Bardeen, and G.L. Pearson, "The effects of pressure and temperature on the resistance of p–n junctions in germanium," *Phys. Rev.*, vol. 84, no. 1, pp. 129–132, 1951.

[10] C.S. Smith, "Piezoresistance effect in germanium and silicon," *Phys. Rev.*, vol. 94, no. 1, pp. 42–49, 1954.

[11] G.L. Bir and G.E. Pikus, *Symmetry and Strain Induced Effects in Semiconductors.* New York: John Wiley and Sons, 1974.

[12] E.A. Fitgerald, Y.H. Xie, M.L. Green, *et al.*, "Totally relaxed Ge_xSi_{1-x} layers with low threading dislocation densities grown on Si substrates," *Journal of Applied Physics*, vol. 59, no. 7, pp. 811–813, 1991.

[13] J. Welser, J.L. Hoyt, and J.F. Gibbons, "NMOS and PMOS transistors fabricated in strained silicon/relaxed silicon–germanium structures," in *IEEE IEDM Technical Digest*, pp. 1000–1002, 1992.

[14] J. Welser, J.L. Hoyt, S. Takagi, and F. Gibbons, "Strain dependence of the performance enhancement in Strained-Si n-MOSFETs," in *IEEE IEDM Technical Digest*, pp. 373–376, 1994.

[15] J. Welser, J.L. Hoyt, and J.F. Gibbons, "Electron mobility enhancement in strained-Si n-type metal-oxide-semiconductor field-effect transistors," *IEEE Electron Device Lett.*, vol. 15, no. 3, pp. 100–102, Mar 1994.

[16] Z.-Y. Cheng, M.T. Currie, C.W. Leitz, *et al.*, "Electron mobility enhancement in strained-Si n-MOSFETs fabricated on SiGe-on-insulator (SGOI) substrates," *IEEE Electron Device Lett.*, vol. 22, pp. 321–323, 2001.

[17] B. Ghyselen, J.M. Hartmann, T. Ernst, *et al.*, "Engineering strained silicon on insulator wafers with the Smart CutTM technology," *Solid State Electronics*, vol. 48, pp. 1285–1296, 2004.

[18] F. Andrieu, T. Ernst, O. Faynot, *et al.*, "Performance and physics of sub-50 nm strained Si on $Si_{1-x}Ge_x$ on insulator (SGOI) nMOSFETs," *Solid State Electronics*, vol. 50, pp. 566–572, 2006.

[19] L.-J. Huang, J. Chu, S. Goma, *et al.*, "Carrier mobility enhancement in strained Si-on-Insulator fabricated by wafer bonding," in *IEEE Symposium on VLSI Technology - Technical Digest*, pp. 57–58, 2001.

[20] K. Rim, K. Chan, L. Shi, *et al.*, "Fabrication and mobility characteristics of ultra-thin strained Si directly on insulator (SSDOI) MOSFETs," in *IEEE IEDM Technical Digest*, pp. 49–53, 2003.

[21] G. Scott, J. Lutze, M. Rubin, F. Nouri, and M. Manley, "NMOS drive current reduction caused by transistor layout and trench isolation induced stress," in *IEEE IEDM Technical Digest*, p. 827-830, 1999.

[22] T. Matsumoto, S. Maeda, H. Dang, *et al.*, "Novel SOI wafer engineering using low stress and high mobility CMOSFET with ⟨100⟩ channel for embedded RF/analog applications," in *IEEE IEDM Technical Digest*, pp. 663–666, 2002.

[23] A. Steegen, M. Stucchi, A. Lauwers, and K. Maex, "Silicide induced pattern density and orientation dependent transconductance in MOS transistors," in *IEEE IEDM Technical Digest*, pp. 497–500, 1999.

[24] S. Ito, H. Namba, K. Yamaguchi, *et al.*, "Mechanical stress effect of etch-stop nitride and its impact on deep submicron transistor design," in *IEEE IEDM Technical Digest*, pp. 247–251, 2000.

[25] A. Shimizu, K. Hachimine, H.O.N. Ohki, *et al.*, "Local mechanical-stress control (LMC): A new technique for CMOS-performance enhancement," in *IEEE IEDM Technical Digest*, p. 433–436, 2001.

[26] H.S. Yang, R. Malik, S. Narasimha, *et al.*, "Dual stress liner for high performance sub-45nm gate length SOI CMOS manufacturing," in *IEEE IEDM Technical Digest*, p. 1075–1077, 2004.

[27] C.D. Sheraw, M. Yang, D.M. Fried, *et al.*, "Dual stress liner enhancement in hybrid orientation technology," in *IEEE Symposium on VLSI Technology - Technical Digest*, pp. 12–13, 2005.

[28] S. Eneman, P. Verheyen, R. Rooyackers, *et al.*, "Layout impact on the performance of a locally strained p-MOSFET," in *IEEE Symposium on VLSI Technology - Technical Digest*, pp. 22–23, 2005.

[29] M. Horstmann, A. Wei, T. Kammler, *et al.*, "Integration and optimization of embedded-SiGe, compressive and tensile stressed liner films, and stress memorization in advanced SOI CMOS technologies," in *IEEE IEDM Technical Digest*, pp. 233–236, 2005.

[30] P. Bai, C. Auth, S. Balakrishnan, *et al.*, "A 65 nm logic technology featuring 35 nm gate lengths, enhanced channel strain, 8 Cu interconnect layers, low-k ILD and $0.57 \mu m^2$ SRAM cell," in *IEEE IEDM Technical Digest*, pp. 657—660, 2004.

[31] K.-W. Ang, K.J. Chui, V. Bliznetsov, *et al.*, "Enhanced performance in 50 nm n-MOSFETs with silicon-carbon source/drain regions," in *IEEE IEDM Technical Digest*, pp. 1069–1071, 2004.

[32] K.-W. Ang, K.-J. Chui, V. Bliznetsov, *et al.*, "Lattice strain analysis of transistor structures with silicon–germanium and silicon–carbon source/drain stressors.," *Applied Physics Letters*, vol. 86, no. 9, p. 093102, 2005.

[33] K.J. Chui, K.W. Ang, N. Balasubramanian, M.G.S. Samudra, and Y.C. Yeo, "n-MOSFET with silicon–carbon source/drain for enhancement of carrier transport," *IEEE Trans. on Electron Devices*, vol. 54, p. 249–256, 2007.

[34] B.A. Auld, *Acoustic Fields and Waves in Solids*. Malabar, Florida: Krieger Publishing Company, 1990.

[35] M. Bao, *Micro Mechanical Transducers*, vol. 8 of *Handbook of Sensors and Actuators*. The Netherlands: Elsevier Science, 2000.

[36] C. Kittel, *Introduction to Solid State Physics*. New York: John Wiley and Sons, 1996.

[37] M. Levinshtein, S. Rumyantsev, and M. Shur, *Handbook series on Semiconductor Parameters*, vol. 1.2. London World Scientific, 1999.

[38] J.M. Hinckley and J. Singh, "Influence of substrate composition and crystallographic orientation on the band structure of pseudomorphic Si-Ge alloy films," *Phys. Rev. B*, vol. 42, no. 6, p. 35–46, 1990.

[39] M.V. Fischetti and S.E. Laux, "Band structure, deformation potentials, and carrier mobility in strained Si, Ge, and SiGe alloys," *Journal of Applied Physics*, vol. 80, p. 2234, 1996.

[40] E. Ungersboeck, S. Dhar, G. Karlowatz, *et al.*, "The effect of general strain on the band structure and electron mobility of silicon," *IEEE Trans. on Electron Devices*, pp. 2183–2190, Sept. 2007.

[41] C.G. Van de Walle and R.M. Martin, "Theoretical calculations of heterojunction discontinuities in the Si / Ge system," *Phys. Rev. B*, vol. 34, no. 8, pp. 5621–5634, 1986.

[42] D.C. Houghton, "Strain relaxation in $Si_{1-x}Ge_x$ heterostructures," *Journal of Applied Physics*, vol. 70, no. 1, pp. 2136–2151, 1991.

[43] F. Schäffler, "High-mobility Si and Ge structures," *Semiconductor Science Technology*, vol. 12, pp. 1515–1549, 1997.

[44] R.A. Logan, J.M. Rowell, and F.A. Trumbore, "Phonon spectra of Ge-Si alloys," *Phys. Rev.*, vol. 136, no. 6A, p. A1751, 1964.

[45] S.E. Ungersboeck, *Advanced Modeling of Strained CMOS Technology*. PhD thesis, Technischen Universität Wien, Wien, Austria, 2007.

[46] N. Serra and D. Esseni, "Mobility enhancement in strained n–FinFETs: basic insight and stress engineering," *IEEE Trans. on Electron Devices*, vol. 57, no. 2, pp. 482–490, 2010.

[47] N. Serra, F. Conzatti, D. Esseni, *et al.*, "Experimental and physics-based modeling assessment of strain induced mobility enhancement in FinFETs," in *IEEE IEDM Technical Digest*, pp. 71–74, 2009.

[48] D. Esseni, F. Conzatti, M. De Michielis *et al.*, "Semi-classical transport modelling of CMOS transistors with arbitrary crystal orientations and strain engineering," *Journal of Computational Electronics*, vol. 8, pp. 209–224, 2009.

[49] M.M. Rieger and P. Vogl, "Electronic-band parameters in strained $Si_{1-x}Ge_x$ alloys on $Si_{1-y}Ge_y$ substrates," *Phys. Rev. B*, vol. 48, no. 19, pp. 14276–14287, 1993.

[50] J.C. Hensel, H. Hasegawa, and M. Nakayama, "Cyclotron resonance in uniaxially stressed silicon. II. Nature of the covalent bonds," *Phys. Rev.*, vol. 138, no. 1A, p. A225, 1965.

[51] L.D. Laude, F.H. Pollak, and M. Cardona, "Effects of uniaxial stress on the indirect exciton spectrum of silicon," *Phys. Rev. B*, vol. 3, no. 8, p. 2623, 1971.

[52] K. Uchida, A. Kinoshita, and M. Saitoh, "Carrier transport in (110) nMOSFETs: subband structures, non-parabolicity, mobility characteristics, and uniaxial stress engineering," in *IEEE IEDM Technical Digest*, pp. 1019–1022, 2006.

[53] V. Sverdlov, E. Ungersboeck, H. Kosina, and S. Selberherr, "Effects of shear strain on the conduction band in silicon: An efficient two-band **k**·**p** theory," in *Proc. European Solid State Device Res. Conf.*, pp. 386–389, 2007.

[54] A. Dalla Serra, A. Abramo, P. Palestri, L. Selmi, and F. Widdershoven, "Closed- and open-boundary models for gate-current calculation in n-MOSFETs," *IEEE Trans. on Electron Devices*, vol. 48, no. 8, pp. 1811–1815, 2001.

[55] V. Sverdlov, D. Esseni, O. Baumgartner, *et al.*, "The linear combination of bulk bands-method for electron and hole subband calculations in strained silicon films and surface layers," in *International Workshop on Computational Electronics*, pp. 49–52, 2009.

[56] T. Manku and A. Nathan, "Valence energy-band structure for strained group-IV semiconductors," *Journal of Applied Physics*, vol. 73, no. 3, pp. 1205–1213, 1993.

[57] R. Oberhuber, G. Zandler, and P. Vogl, "Subband structure and mobility of two-dimensional holes in strained Si/SiGe MOSFET's," *Phys. Rev. B*, vol. 58, no. 15, pp. 9941–9948, 1998.

[58] M.V. Fischetti, Z. Ren, P.M. Solomon, M. Yang, and K. Rim, "Six-band **k**·**p** calculation of the hole mobility in silicon inversion layers: Dependence on surface orientation, strain, and silicon thickness," *Journal of Applied Physics*, vol. 94, no. 2, pp. 1079–1095, 2003.

[59] A.T. Pham, C. Jungemann, and B. Meinerzhagen, "Modeling of hole inversion layer mobility in unstrained and uniaxially strained Si on arbitrarily oriented substrates," in *Proc. European Solid State Device Res. Conf.*, pp. 390–393, 2007.

[60] D. Rideau, M. Feraille, L. Ciampolini, *et al.*, "Strained Si, Ge, and $Si_{1-x}Ge_x$ alloys modeled with a first-principles-optimized full-zone **k**·**p** method," *Phys. Rev. B*, vol. 74, no. 19, pp. 195208–195228, 2006.

[61] D. Rideau, M. Feraille, M. Michaillat, *et al.*, "On the validity of the effective mass approximation and the Luttinger **k**·**p** model in fully depleted SOI MOSFETs," *Solid State Electronics*, vol. 53, pp. 452–460, 2008.

[62] K. Huet, M. Feraille, D. Rideau, *et al.*, "Experimental and theoretical analysis of transport in uniaxially strained pMOSFETs," in *Proc. European Solid State Device Res. Conf.*, pp. 234–237, 2008.

[63] M.V. Fischetti, F. Gamiz, and W. Hansch, "On the enhanced electron mobility in strained-silicon inversion layers," *Journal of Applied Physics*, vol. 92, no. 12, pp. 7320–7324, 2002.

[64] D. Ponton, L. Lucci, P. Palestri, D. Esseni, and L. Selmi, "Assessment of the impact of biaxial strain on the drain current of decanometric n-MOSFET," in *Proc. European Solid State Device Res. Conf.*, pp. 166–169, 2006.

[65] G. Comparone, P. Palestri, D. Esseni, L. Lucci, and L. Selmi, "A better understanding of the requirements for predictive modeling of strain engineering in nMOS transistors," *Jour. of Computational and Theoretical Nanoscience*, vol. 5, no. 6, pp. 1106–1114, 2008.

[66] O. Bonno, S. Barraud, D. Mariolle, and F. Andrieu, "Measurement of conduction band deformation potential constants using gate direct tunneling current in n-type metal oxide semiconductor field effect transistors under mechanical stress," *Journal of Applied Physics*, vol. 103, pp. 063715.1–063715.9, 2008.

[67] M.H. Evans, M. Caussanel, R.D. Schrimpf, and S.T. Pantelides, "First-principles modeling of double-gate UTSOI MOSFETs," in *IEEE IEDM Technical Digest*, pp. 611–614, 2005.

[68] G. Hadjisavvas, L. Tsetseris, and S. Pantelides, "The origin of electron mobility enhancement in strained MOSFETs," *IEEE Electron Device Lett.*, vol. 28, no. 11, pp. 1018–1020, 2007.

[69] S.H. Olsen, A.G. O'Neill, P. Dobrosz, *et al.*, "Study of strain relaxation in Si / SiGe metal-oxide-semiconductor field-effect transistors," *Journal of Applied Physics*, vol. 97, p. 114504, 2005.

[70] M. Currie, C. Leitz, T. Langdo, G. Taraschi, and E. Fitzgerald, "Carrier mobilities and process stability of strained Si n- and p-MOSFETs on SiGe virtual substrates," *Journal of Vacuum Science and Technology B*, vol. 19, pp. 2268–2279, 2001.

[71] C.W. Leitz, M.T. Currie, M.L. Lee, *et al.*, "Hole mobility enhancements and alloy scattering-limited mobility in tensile strained Si/SiGe surface channel metal-oxide-semiconductor field-effect transistors," *Journal of Applied Physics*, vol. 92, pp. 3745–3751, 2002.

[72] S.E. Thompson, S. Suthram, Y. Sun, *et al.*, "Future of strained Si/semiconductors in nanoscale MOSFETs," in *IEEE IEDM Technical Digest*, pp. 681–684, 2006.

[73] L. Smith, V. Moroz, G. Eneman, *et al.*, "Exploring the limits of stress-enhanced hole mobility," *IEEE Electron Device Lett.*, vol. 26, no. 9, pp. 652–654, 2005.

[74] L. Washington, F. Nouri, S. Thirupapuliyur, *et al.*, "pMOSFET with 200% mobility enhancement induced by multiple stressors," *IEEE Electron Device Lett.*, vol. 27, no. 6, pp. 511–513, 2006.

[75] M. De Michielis, D. Esseni, P. Palestri, and L. Selmi, "Semi-classical modelling of quasi-ballistic hole transport in nanoscale p-MOSFETs based on a multi-subband Monte Carlo approach," *IEEE Trans. on Electron Devices*, vol. 56, no. 9, pp. 2081–2091, 2009.

[76] F. Conzatti, M. De Michielis, D. Esseni, and P. Palestri, "Drain current improvements in uniaxially strained p-MOSFETs: A multi-subband Monte Carlo study," *Solid State Electronics*, vol. 53, pp. 706–711, 2009.

[77] Y. Zhao, M. Takenaka, and S. Takagi, "Comprehensive understanding of surface roughness and Coulomb scattering mobility in biaxially-strained Si MOSFETs," in *IEEE IEDM Technical Digest*, pp. 577–580, 2008.

[78] *International Technology Roadmap for Semiconductors 2007.* Public home page *http://www.itrs.net.*

10 MOS transistors with alternative materials

In the previous chapters we have considered arbitrarily oriented and strained silicon devices with SiO_2 gate dielectric. In this chapter we discuss carrier transport in MOS transistors with new materials either in the gate stack or in the device channel.

We first discuss the scattering mechanisms that may be relevant for devices employing high-κ dielectrics in the gate stack, namely remote optical phonons and Coulomb scattering with fixed charges in the gate stack.

Then alternative channel materials such as germanium and gallium arsenide are analyzed using the generalization of the EMA and $\mathbf{k}\cdot\mathbf{p}$ approaches described in Chapter 8. Gallium-arsenide is also taken as a case study for polar optical phonon scattering, which was not described in Chapter 4.

10.1 Alternative gate materials

As discussed in Chapter 1, aggressive scaling of the SiO_2 dielectric has led to a substantial increase of the gate leakage current and static power dissipation. To counteract this harmful trend for the performance and reliability of CMOS devices, alternative dielectric materials, the so called *high-κ* materials with dielectric constant higher than that of SiO_2, have been extensively investigated. For optimum flat-band voltage control and improved performance, high-κ dielectrics are often integrated with metal gate materials.

Unfortunately, transistors with high-κ/metal gate stacks often exhibit a lower mobility with respect to the universal curves for thick SiO_2 dielectrics. This is illustrated in Fig.10.1, which collects experimental mobility curves for devices featuring HfO_2 and HfSiON based gate stacks. The mobility reduction strongly depends on the thickness of the interfacial SiO_2 layer, which is typically interposed between the high-κ layer and the device channel to improve the interface quality (see Fig. 10.2).

While the mobility values reported in the literature for high-κ MOSFETs slowly but steadily improve with progress of the technology, the origin of the observed mobility degradation with respect to the mobility of SiO_2 MOSFETs is still debated. In particular, two additional scattering mechanisms have been indicated as possibly responsible for the mobility reduction, namely remote phonons originating in the high-κ and propagating to the substrate, and remote Coulomb scattering with fixed charges in the gate stack. Recently, it has also been suggested that the mobility degradation might be due to

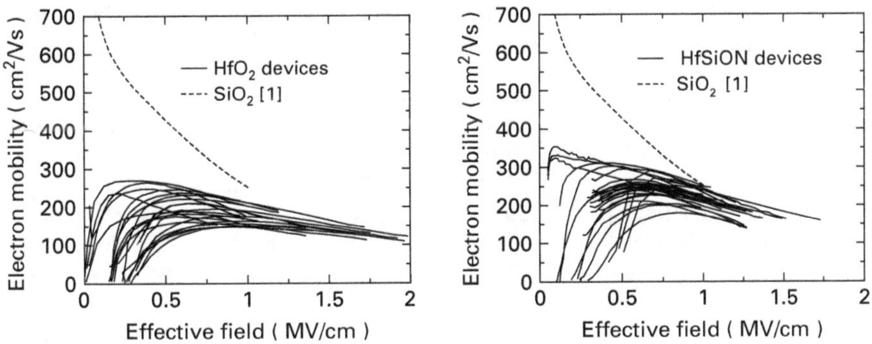

Figure 10.1 Comparison between the universal curves for thick SiO_2 gate dielectric [1] and various experimental data for HfO_2 [2–11] and HfSiON [12–20] devices with metal gate. There is a significant mobility degradation due to the introduction of the high-κ material in the gate stack.

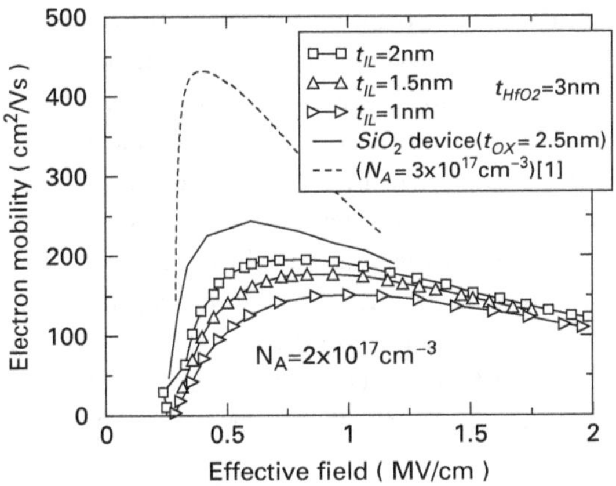

Figure 10.2 Experimental data [4] at 300K for bulk MOSFETs with doping 2×10^{17} cm^{-3}. The measurements refer to devices with 3 nm of HfO_2 and with various SiO_2 thicknesses from 0.8 nm to 2.5 nm. All devices have TiN metal gate. The reference SiO_2 device from [4] and the mobility data from [1] are also shown.

scattering with defects induced by nitrogen diffusing to the channel/dielectric interface during the fabrication process [21].

Below we illustrate the theory and models for remote phonon and remote Coulomb scattering and show several simulation results obtained with the multi-subband Monte Carlo approach to analyze the physical mechanism responsible for mobility reduction in high-κ MOSFETs.

10.2 Remote phonon scattering due to high-κ dielectrics

The term *remote phonons* indicates surface phonons originating from the polar phonon modes of the high-κ dielectrics. Indeed, the molecules of these insulators are strongly polarized, consistent with the large static electric permittivity. The thermal vibration of

Table 10.1 Parameters for the polar phonons in some high-κ materials. Data from [23].

Quantity	SiO_2	Al_2O_3	AlN	ZrO_2	HfO_2	$ZrSiO_4$
$\epsilon_{(0)}$	3.90	12.53	9.14	24.0	22.00	11.75
$\epsilon_{(int)}$	3.05	7.27	7.35	7.75	6.58	9.73
$\epsilon_{(\infty)}$	2.50	3.20	4.80	4.00	5.03	4.20
$\hbar\omega_{TO1}/e$ [meV]	55.60	48.18	81.40	16.67	12.40	38.62
$\hbar\omega_{TO2}/e$ [meV]	138.10	71.41	88.55	57.70	48.35	116.00

polar molecules generates non-stationary electric fields which penetrate into the silicon channel and rapidly decay within a short distance from the silicon/dielectric interface. For this reason, they are also called *surface optical phonons*. The label *optical* is applied because of the quite high phonon frequency, weakly dependent on the wave-vector. Moreover, they are also called *soft phonons* because the bond between the atom of the metal and the oxygen is *soft*, i.e. it allows the molecule to vibrate strongly.

The study of remote phonons is based on analysis of a transverse magnetic mode which propagates parallel to the channel/dielectric interface, thanks to the dielectric constant of the high-κ insulator becoming negative for certain frequencies. A negative dielectric constant implies that the medium is not absorbing energy, but it is re-emitting energy. This situation cannot arise in the case of stationary electric field. On the other hand, when the external field has an appropriate frequency, the molecules of a polar crystal can vibrate displacing themselves in phase opposition with the external field. Thus the material is producing an electric field that increases the external field. The dielectric constant of a polar insulator is given by [22, 23]

$$\epsilon(\omega) = \epsilon_{(\infty)} + \frac{\epsilon_{(0)} - \epsilon_{(int)}}{1 - \left(\frac{\omega}{\omega_{TO1}}\right)^2} + \frac{\epsilon_{(int)} - \epsilon_{(\infty)}}{1 - \left(\frac{\omega}{\omega_{TO2}}\right)^2}, \tag{10.1}$$

where ω_{TO1} and ω_{TO2} are the frequencies of the two polar phonons in the material, $\epsilon_{(0)}$ is the static dielectric constant, $\epsilon_{(\infty)}$ the high-frequency value, $\epsilon_{(int)}$ the dielectric permittivity at an intermediate frequency between the phonon frequencies ω_{TO1} and ω_{TO2}. The values of these parameters for some dielectrics of interest for nanoelectronics are given in Table 10.1. The frequency dependent dielectric constant for HfO_2 is plotted as an example in Fig.10.3, showing that there are frequency ranges where $\epsilon(\omega)$ is negative.

In Table 10.1 we observe that, compared to SiO_2, the high-κ materials usually feature a much larger $\epsilon_{(0)}$ but a similar $\epsilon_{(\infty)}$. We see that the strength of the scattering rate depends on the difference between $\epsilon_{(0)}$ and $\epsilon_{(\infty)}$; hence remote phonon scattering is stronger in devices with high-κ materials than in conventional SiO_2 stacks.

To understand Eq.10.1 we have to remember that an insulator is composed of many dipoles, which can rotate around their centroid. Thus, when an electric field is applied, the dipoles tend to be displaced parallel to it. If the field is stationary, all dipoles are affected and displaced against the field, so the material exhibits a large permittivity. Indeed, when all dipoles are displaced, they generate an electric field that reduces the effects of the external field. Otherwise, when the field oscillates at very high frequency, the

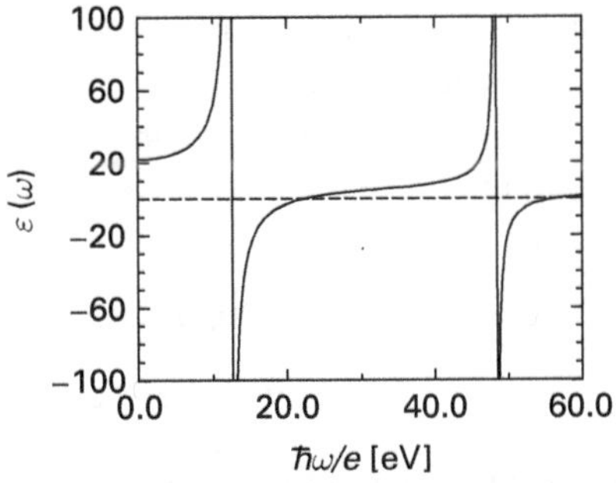

Figure 10.3 Dielectric constant vs. frequency for HfO_2. Parameters from Table 10.1

dipoles are not fast enough to fully displace and the overall electric field they generate is negligible. Thus, the dielectric constant is smaller than in the stationary case. Moreover, there is a range of intermediate frequencies where the dipoles displace in phase opposition with the oscillations of the electric field, so that the dielectric permittivity becomes negative, because the energy goes from the dipoles to the field, contrary to the previous cases. The field experienced by the material is thus the external one *plus* the dipole field. The dynamic of the dipoles with applied electric field is tightly related to the phonon oscillation modes of the material. In particular Eq.10.1 accounts for two phonon modes.

Models for remote phonon scattering have been proposed by many authors [23–29]. Some of these references also consider the coupling between remote phonons and plasmons, i.e. the fluctuations of the electric field related to plasma oscillations in the inversion layer and in the gate material, that we do not discuss in this book.

For the sake of a clear derivation of the matrix elements and of the scattering rates for remote phonons, we consider first a bulk semiconductor with an infinitely thick dielectric on top and only one phonon mode. Hence we write:

$$\epsilon(\omega) = \epsilon_{(\infty)} + \frac{\epsilon_{(0)} - \epsilon_{(\infty)}}{1 - \left(\frac{\omega}{\omega_{TO}}\right)^2}. \tag{10.2}$$

Then, we extend the calculation to two phonon modes (Eq.10.1) and finally to a stack composed of semiconductor / interfacial layer / high-κ layer /metal-gate.

10.2.1 Field propagation in the stack

Let us consider a structure which is uniform in the (x, y) plane and composed of different layers in the z direction. We also assume a piece-wise constant permittivity along z.

Maxwell's equations read [30, 31]:

$$\nabla \times \mathbf{F}(\mathbf{R}, t) = \mu(\mathbf{R})\frac{\partial \mathbf{H}(\mathbf{R}, t)}{\partial t}, \tag{10.3a}$$

$$\nabla \times \mathbf{H}(\mathbf{R}, t) = -\epsilon(\mathbf{R})\frac{\partial \mathbf{F}(\mathbf{R}, \mathbf{t})}{\partial t}, \tag{10.3b}$$

where $\nabla\times$ is the curl operator, μ the magnetic permeability of the material and ϵ its permittivity (i.e. the dielectric constant).

Let us assume x as the propagation direction, and a uniform structure along y. Since we are interested in fields that are not vanishing in the x direction, we write the electric and magnetic fields as:

$$\mathbf{F}(x,z,t) = \widetilde{\mathbf{F}}(z)\mathrm{e}^{\mathrm{i}qx}\mathrm{e}^{-\mathrm{i}\omega t} + (c.c.), \tag{10.4a}$$

$$\mathbf{H}(x,z,t) = \widetilde{\mathbf{H}}(z)\mathrm{e}^{\mathrm{i}qx}\mathrm{e}^{-\mathrm{i}\omega t} + (c.c.), \tag{10.4b}$$

where the propagation constant q is real and $\widetilde{\mathbf{F}}(z)$ and $\widetilde{\mathbf{H}}(z)$ are three dimensional vectors (each component is a complex number) which depend only on the z coordinate. For the time being we drop the complex conjugate terms in Eq.10.4. We consider them again when discussing the energy associated with the remote phonons. Substituting Eq.10.4 into the Maxwell equations (Eq.10.3) yields:

$$-\frac{\partial \widetilde{F}_y}{\partial z} = -\mathrm{i}\omega\mu\widetilde{H}_x, \tag{10.5a}$$

$$\frac{\partial \widetilde{F}_x}{\partial z} - \mathrm{i}q\widetilde{F}_z = -\mathrm{i}\omega\mu\widetilde{H}_y, \tag{10.5b}$$

$$\mathrm{i}q\widetilde{F}_y = -\mathrm{i}\omega\mu\widetilde{H}_z, \tag{10.5c}$$

$$-\frac{\partial \widetilde{H}_y}{\partial z} = \mathrm{i}\omega\epsilon\widetilde{F}_x, \tag{10.5d}$$

$$\frac{\partial \widetilde{H}_x}{\partial z} - \mathrm{i}q\widetilde{H}_z = \mathrm{i}\omega\epsilon\widetilde{F}_y, \tag{10.5e}$$

$$\mathrm{i}q\widetilde{H}_y = \mathrm{i}\omega\epsilon\widetilde{F}_z. \tag{10.5f}$$

The system of equations is composed of two decoupled sets of relations. Indeed, $\widetilde{H}_y$, $\widetilde{F}_x$, $\widetilde{F}_z$ are not coupled to $\widetilde{F}_y$, $\widetilde{H}_z$, and $\widetilde{H}_x$. These two sets are usually called modes. With respect to the direction of propagation x, we can define these modes as transverse electric (TE) mode, and transverse magnetic (TM) mode:

$$\{\mathbf{H},\mathbf{F}\} = \{\mathbf{H},\mathbf{F}\}_{TE} + \{\mathbf{H},\mathbf{F}\}_{TM}. \tag{10.6}$$

It can be demonstrated that in this particular structure, propagating TE modes $\{\mathbf{H},\mathbf{F}\}_{TE}$ are null, because their existence would require a different magnetic permeability in the different layers of the stack, which should furthermore be negative at the frequency of the TE modes in at least one layer.

For the TM mode $\widetilde{H}_x$ is null and thus, from Eqs.10.5, also $\widetilde{F}_y$ and $\widetilde{H}_z$ are null. By using Eqs.10.5f, 10.5d and 10.5b we find:

$$\frac{\partial^2 \widetilde{F}_z}{\partial z^2} - \left(q^2 - \omega^2\mu\epsilon\right)\widetilde{F}_z = 0, \tag{10.7}$$

where $1/\sqrt{\mu\epsilon} = v_l$ is the light velocity. Since the material properties (ϵ and μ) are constant in each region, the solutions of Eq.10.7 are in the form:

$$\widetilde{F}_z = \alpha\mathrm{e}^{+k_p z} + \beta\mathrm{e}^{-k_p z}, \tag{10.8}$$

where

$$k_p^2 = q^2 - \omega^2 \mu \epsilon \tag{10.9}$$

changes layer by layer. Only real values of k_p are meaningful, otherwise we would have modes propagating along z also and not only along x [32].

In the absence of conduction currents at the interface between different materials, we must impose the conservation of $\widetilde{F}_x$ and $\widetilde{H}_y$ at each interface.

10.2.2 Device structure with an infinite dielectric

We now consider for simplicity a structure consisting of two semi-infinite regions (see Fig.10.4), for example a silicon substrate with an infinite dielectric on top.

Phonon dispersion relationship

Equation 10.8 simplifies to:

$$\widetilde{F}_z(A) = A\,\mathrm{e}^{-k_A z}, \qquad \text{(silicon, } z > 0\text{)} \tag{10.10a}$$

$$\widetilde{F}_z(B) = B\,\mathrm{e}^{+k_B z}. \qquad \text{(high-}\kappa\text{, } z < 0\text{)} \tag{10.10b}$$

In each region we have, from Eqs.10.5d, 10.5f:

$$\widetilde{F}_x = -\frac{\mathrm{i}}{q}\frac{\partial \widetilde{F}_z}{\partial z}, \tag{10.11a}$$

$$\widetilde{H}_y = \frac{\omega\epsilon}{q}\widetilde{F}_z. \tag{10.11b}$$

By using Eqs.10.10, 10.11, the continuity of $\widetilde{F}_x$ and of $\widetilde{H}_y$ at the interface ($z = 0$) implies, respectively:

$$-k_A A = k_B B, \tag{10.12}$$

and

$$\epsilon_A A = \epsilon_B B, \tag{10.13}$$

where ϵ_A and ϵ_B are the dielectric constant of region A and B, respectively.

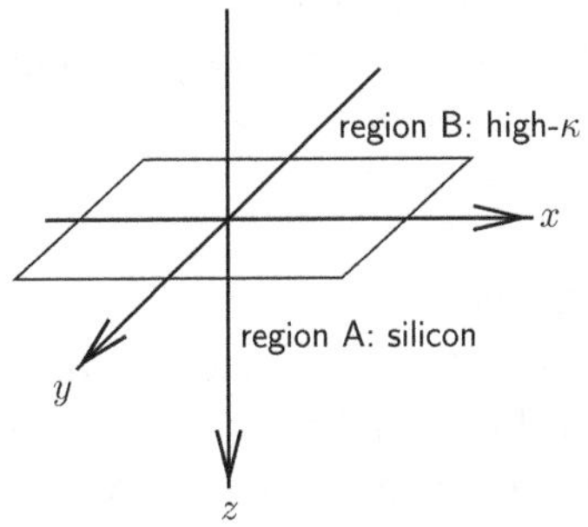

Figure 10.4 A template structure consisting of semi-infinite regions.

The dispersion relationship of the TM mode is then obtained by solving:

$$\frac{\epsilon_A(\omega)}{k_A(\omega, q)} = -\frac{\epsilon_B(\omega)}{k_B(\omega, q)}, \tag{10.14}$$

where k_A, k_B depend on ω and q according to Eq.10.9 and ϵ_A and ϵ_B may depend on ω as in Eq.10.2. Hence Eq.10.14 provides the ω to q relation describing the energy dispersion $\hbar\omega(q)$ of the surface phonon modes; unfortunately in general $\hbar\omega(q)$ cannot be expressed analytically.

Low frequency analysis

Calculation of the relation $\hbar\omega(q)$ is significantly simplified for the low frequency modes, where:

$$\omega^2 \ll \frac{q^2}{\mu\epsilon} = q^2 v_l^2. \tag{10.15}$$

Equation 10.9 gives $k_A \simeq k_B \simeq q$, hence Eq.10.14 simply gives:

$$\epsilon_A(\omega) = -\epsilon_B(\omega). \tag{10.16}$$

If we now assume that region A is silicon with $\epsilon_A = \epsilon_{Si}$ and region B is an insulator with the dielectric constant given by Eq.10.2, we get:

$$\epsilon_{(\infty)} + \frac{\epsilon_{(0)} - \epsilon_{(\infty)}}{1 - \left(\frac{\omega}{\omega_{TO}}\right)^2} = -\epsilon_{Si}, \tag{10.17}$$

where it is clear that surface phonon modes are possible when the permittivity of the insulator is negative (see Fig.10.3). By solving Eq.10.17 for ω we finally obtain the dispersion relationship for the soft-phonon mode (SO):

$$\omega_{SO} = \omega_{TO}\sqrt{\frac{\epsilon_{Si} + \epsilon_{(0)}}{\epsilon_{Si} + \epsilon_{(\infty)}}}. \tag{10.18}$$

Similarly to the case of optical phonons (Section 4.5), in the low frequency approximation ω_{SO} does not depend on q.

If instead of the static dielectric constant of silicon ϵ_{Si}, we use in Eq.10.16 the frequency dependent dielectric constant that includes the screening effect of the inversion layer (Section 4.2 and Eq.4.91), the procedure outlined above provides the energy relation for the remote phonon modes coupled to the plasmons of the inversion layer [23, 33, 34].

Use of a scalar potential

Besides the analysis of the template structure of Fig.10.4, the low frequency approximation can be used to simplify the general problem. In fact, at low ω, we can express the electric field through the scalar electrostatic potential ϕ [30, 31]:

$$\mathbf{F}(\mathbf{R}) = -\nabla\phi(\mathbf{R}). \tag{10.19}$$

By considering modes propagating along x, similarly to Eq.10.4 we can write:

$$\phi(x,z,t) = \widetilde{\phi}(z)\mathrm{e}^{\mathrm{i}qx}\mathrm{e}^{-\mathrm{i}\omega t}. \tag{10.20}$$

The complex function $\widetilde{\phi}(z)$ is related to $\widetilde{F}_z(z)$ by:

$$\widetilde{F}_z = -\frac{\mathrm{d}\widetilde{\phi}}{\mathrm{d}z}. \tag{10.21}$$

From Eq.10.7 at low frequency (i.e. $k_p \simeq q$) we obtain, inside each region:

$$\widetilde{\phi}(z) = A_1\mathrm{e}^{+qz} + A_2\mathrm{e}^{-qz}. \tag{10.22}$$

The $\widetilde{F}_x$ and $\widetilde{H}_y$ continuity conditions at the interfaces imply the continuity of $\widetilde{\phi}$ and of $\epsilon(\mathrm{d}\widetilde{\phi}/\mathrm{d}z)$, respectively (see Eqs.10.11a, 10.11b).

As a simple example, we can consider again the structure in Fig.10.4 and write

$$\widetilde{\phi}(A) = A\mathrm{e}^{-qz}, \tag{10.23a}$$

$$\widetilde{\phi}(B) = B\mathrm{e}^{+qz}. \tag{10.23b}$$

The continuity conditions (potential and $\epsilon\mathrm{d}\widetilde{\phi}/\mathrm{d}z$) at $z = 0$ give $A = B$ and $\epsilon_A A = -\epsilon_B B$, thus $\epsilon_A(\omega) = -\epsilon_B(\omega)$, that is consistent with Eq.10.16. This means that we can use Eq.10.22 in each region, impose continuity of $\widetilde{\phi}$ and of $\epsilon(\mathrm{d}\widetilde{\phi}/\mathrm{d}z)$, and obtain results that, in the low frequency limit, are the same as those derived considering the complete electromagnetic problem.

Amplitude of the potential

To compute the amplitude of the phonon modes, i.e. the value of the coefficients A and B in Eq.10.23, we proceed as in Section 4.5.2, by using the equivalence between the classical energy and the quantum mechanical energy of the mode. A rigorous evaluation of A and B requires a full quantum mechanical treatment of the field oscillations, which goes beyond the scope of this book. We here, instead, follow the simplified methodology proposed in [23].

Let us start by determining the classical energy of the mode. In all the directions where the electric field is not null (the x and z directions in our case) the energy of the system oscillates between the kinetic and the potential form. In fact the vibrations in the polar material move the atoms from their rest positions. When an atom is at the maximum distance from its rest position, it has no kinetic energy, but its potential energy (in the electrostatic form) is maximum. When the atom passes through its rest position, instead, it has no potential energy but only kinetic energy, which is at its maximum since the atom velocity is maximum.

We then start computing the electrostatic energy per unit area in each region. In region A we have:

$$W_{Ae} = \frac{1}{2}\int_0^{+\infty} \epsilon_A|\mathbf{F}|^2\mathrm{d}z, \tag{10.24}$$

where, from Eq.10.4

$$\mathbf{F} = \widetilde{\mathbf{F}}\mathrm{e}^{\mathrm{i}(qx-\omega t)} + (c.c.). \tag{10.25}$$

Thus, we duly take account of the complex conjugate term which we have neglected so far (see also Section 4.5.2).

By substituting the potential profile (Eq.10.23) into Eqs.10.21 and 10.11a, we have:

$$\widetilde{F}_x = -\mathrm{i}qA\mathrm{e}^{-qz}, \tag{10.26a}$$

$$\widetilde{F}_z = qA\mathrm{e}^{-qz}. \tag{10.26b}$$

With an appropriate choice of the time origin we can take A real without loss of generality. Therefore we derive

$$F_x = 2qA\mathrm{e}^{-qz}\sin(qx - \omega t), \tag{10.27a}$$

$$F_z = 2qA\mathrm{e}^{-qz}\cos(qx - \omega t). \tag{10.27b}$$

The squared magnitude of the electric field is then:

$$|\mathbf{F}|^2 = 4q^2A^2\mathrm{e}^{-2qz}, \tag{10.28}$$

which does not depend on time. However, if separately taken, the components F_x and F_z depend on time. When F_x is null, the kinetic energy associated with the oscillations in the x direction is maximum, but the potential energy associated with the electric field in the x direction is null. The maximum value of the kinetic energy associated with the oscillations in the x direction is equal to the maximum potential energy associated with the electric field in the x direction. The same applies for the z direction. At each time, the total energy W_A is the sum of the kinetic energies associated with the oscillations in the x and z directions and of the potential energies associated with F_x and F_z, and it is constant and equal to the maximum over time of the potential energy associate with F_x (i.e. Eq.10.27a) plus the one associated with F_z. So we can write

$$\begin{aligned} W_A &= \frac{1}{2}\int_0^{+\infty} \epsilon_A \left(|F_x^{(max)}|^2 + |F_z^{(max)}|^2\right)\mathrm{d}z \\ &= \epsilon_A \int_0^{+\infty} 4q^2A^2\mathrm{e}^{-2qz}\mathrm{d}z = 2\epsilon_A qA^2. \end{aligned} \tag{10.29}$$

Similarly, for the high-κ dielectric (region B) we find:

$$W_B = 2\epsilon_B qB^2. \tag{10.30}$$

Hence we see that, since $A = B$ and $\epsilon_A = -\epsilon_B$, the total energy of the system is null. This is not surprising, since otherwise the phonons would generate or absorb energy. To proceed further, we consider the total energy W_0 when the dipoles are completely polarized ($\epsilon_B = \epsilon_{(0)}$), and the total energy W_∞ when the dipoles are not responding to the perturbations ($\epsilon_B = \epsilon_{(\infty)}$). In the first case the amplitude of the mode is $A_0 = B_0$, while in the second case it is $A_\infty = B_\infty$.

We have:

$$W_0 = (\epsilon_{Si} + \epsilon_{(0)})2A_0^2 q, \tag{10.31a}$$

$$W_\infty = (\epsilon_{Si} + \epsilon_{(\infty)})2A_\infty^2 q. \tag{10.31b}$$

We now impose the condition, in both cases, that the classical energy is equal to the quantum mechanical one [35] (that is $(n_{SO} + 1/2)\hbar\omega_{SO}$), that has to be divided by a normalizing area denoted A_{norm}, since W_0 and W_∞ are energies per unit area. We thus obtain:

$$A_0^2 = \frac{\hbar\omega_{SO}\left(n_{SO} + \frac{1}{2}\right)}{2q(\epsilon_{Si} + \epsilon_{(0)})} \frac{1}{A_{\text{norm}}}, \tag{10.32a}$$

$$A_\infty^2 = \frac{\hbar\omega_{SO}\left(n_{SO} + \frac{1}{2}\right)}{2q(\epsilon_{Si} + \epsilon_{(\infty)})} \frac{1}{A_{\text{norm}}}. \tag{10.32b}$$

We finally compute the amplitude at the interface as [23]:

$$A = B = \sqrt{A_\infty^2 - A_0^2} = \sqrt{\frac{\hbar\omega_{SO}\left(n_{SO} + \frac{1}{2}\right)}{2qA_{\text{norm}}}\left(\frac{1}{\epsilon_{Si} + \epsilon_{(\infty)}} - \frac{1}{\epsilon_{Si} + \epsilon_{(0)}}\right)}. \tag{10.33}$$

Of course, as in Section 4.5.2, when computing the scattering rate the term $(n_{SO} + 1/2)$ should be replaced either by n_{SO} or $(n_{SO} + 1)$ for absorption and emission, respectively.

Extension to the two phonon case

So far we have considered only one phonon mode in the dielectric, i.e. we used Eq.10.2 instead of Eq.10.1 for the ω dependence of the dielectric constant. In many cases accounting only for the lowest phonon can be enough, however, a more accurate modeling of the remote phonons can be carried out considering surface modes originating from both the phonon modes in the dielectric.

The case with an infinitely thick dielectric on top of an infinite substrate can be easily extended to the two phonon case, since $\epsilon_A = -\epsilon_B$ still holds, so that the equation we need to solve is:

$$\epsilon_{(\infty)} + \frac{\epsilon_{(0)} - \epsilon_{(int)}}{1 - \left(\frac{\omega}{\omega_{TO1}}\right)^2} + \frac{\epsilon_{(int)} - \epsilon_{(\infty)}}{1 - \left(\frac{\omega}{\omega_{TO2}}\right)^2} = -\epsilon_{Si}, \tag{10.34}$$

which can be easily re-arranged as a second order equation in ω^2:

$$a\omega^4 + b\omega^2 + c = 0, \tag{10.35}$$

where:

$$a = \epsilon_{Si} + \epsilon_{(\infty)}, \tag{10.36a}$$

$$b = -a\left(\omega_{TO1}^2 + \omega_{TO2}^2\right) - \omega_{TO2}^2\left(\epsilon_{(int)} - \epsilon_{(\infty)}\right) - \omega_{TO1}^2\left(\epsilon_{(0)} - \epsilon_{(int)}\right), \tag{10.36b}$$

$$c = \left(a + \epsilon_{(0)} - \epsilon_{(\infty)}\right)\omega_{TO1}^2\omega_{TO2}^2. \tag{10.36c}$$

The data in Table 10.1 demonstrate that in many practical cases $\omega_{TO2} \gg \omega_{TO1}$, so that we can approximate:

$$b \simeq -\left(\epsilon_{(int)} + \epsilon_{Si}\right)\omega_{TO2}^2. \tag{10.37}$$

It is then straightforward to show that the frequencies of the two modes are:

$$\omega_{SO1} = \omega_{TO1}\sqrt{\frac{\epsilon_{(0)} + \epsilon_{Si}}{\epsilon_{(int)} + \epsilon_{Si}}}, \tag{10.38a}$$

$$\omega_{SO2} = \omega_{TO2}\sqrt{\frac{\epsilon_{(int)} + \epsilon_{Si}}{\epsilon_{(\infty)} + \epsilon_{Si}}}, \tag{10.38b}$$

and are thus independent of q.

To calculate the amplitude of the modes we can follow the same approach used for the single phonon case [23], that is, for each phonon we compute the difference between the squared amplitude of the potential wave when the phonon is completely on (i.e the dipole is responding) and the squared amplitude when the phonon is off (not responding). Lengthy but straightforward calculations lead us to derive two expressions for the potential amplitude in the silicon channel (one for each mode) identical to Eq.10.33, except that the term $[1/(\epsilon_{Si} + \epsilon_{(\infty)}) - 1/(\epsilon_{Si} + \epsilon_{(0)})]$ inside the square root is replaced by:

$$\frac{1}{\hat{\epsilon}_{SO1}} = \frac{1}{\epsilon_{Si} + \epsilon_{(int)}} - \frac{1}{\epsilon_{Si} + \epsilon_{(0)}}, \tag{10.39a}$$

$$\frac{1}{\hat{\epsilon}_{SO2}} = \frac{1}{\epsilon_{Si} + \epsilon_{(\infty)}} - \frac{1}{\epsilon_{Si} + \epsilon_{(int)}}, \tag{10.39b}$$

for the two different phonons.

Compared to the single phonon mode case seen previously, we now have two phonons, which can be treated as independent scattering mechanisms.

10.2.3 Device structure with ITL/high-κ/metal-gate stack

We now consider the more realistic gate stack of Fig.10.5 including a high-κ dielectric (HK) separated from the silicon channel by an interfacial layer (ITL) and finally covered by a metal gate, described from now on as an ideal conductor.

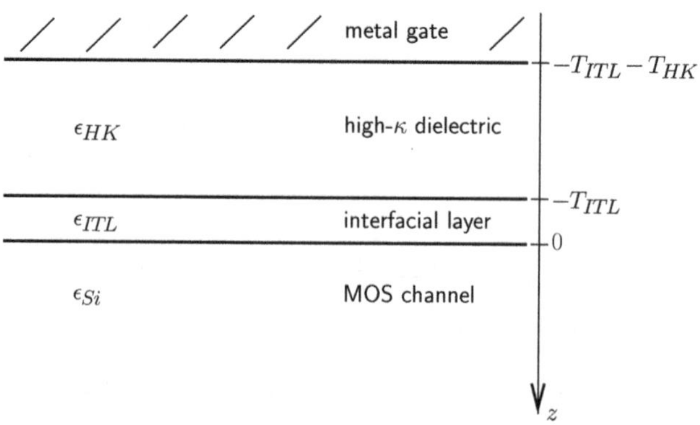

Figure 10.5 A gate stack featuring a high-κ material, an interfacial layer and a metal gate. A bulk silicon substrate is assumed.

Below we adopt the low frequency assumption as in the case of the infinite dielectric model, i.e. we assume $\omega \ll q v_l$, where v_l is the light velocity.

Phonon dispersion relationship

From Eq.10.22, we can write:

$$\widetilde{\phi}_{HK} = B_1 e^{qz} + B_2 e^{-qz}, \tag{10.40a}$$

$$\widetilde{\phi}_{ITL} = B_3 e^{qz} + B_4 e^{-qz}, \tag{10.40b}$$

$$\widetilde{\phi}_{Si} = B_5 e^{-qz}. \tag{10.40c}$$

The continuity of the parallel component of the electric and magnetic fields imposes conservation of $\widetilde{\phi}$ and $\epsilon(d\widetilde{\phi}/dz)$. Furthermore, since the metal is an ideal conductor we can impose $\widetilde{\phi}(-T_{ITL} - T_{HK}) = 0$. We then obtain:

$$B_1 e^{-q(T_{ITL}+T_{HK})} + B_2 e^{q(T_{ITL}+T_{HK})} = 0, \tag{10.41a}$$

$$B_1 e^{-qT_{ITL}} + B_2 e^{qT_{ITL}} = B_3 e^{-qT_{ITL}} + B_4 e^{qT_{ITL}}, \tag{10.41b}$$

$$B_3 + B_4 = B_5, \tag{10.41c}$$

$$\epsilon_{HK}(B_1 e^{-qT_{ITL}} - B_2 e^{qT_{ITL}}) = \epsilon_{ITL}(B_3 e^{-qT_{ITL}} - B_4 e^{qT_{ITL}}), \tag{10.41d}$$

$$\epsilon_{ITL}(B_3 - B_4) = -\epsilon_{Si} B_5, \tag{10.41e}$$

which is a linear system in the five unknown coefficients B_1, B_2, B_3, B_4, B_5. Since the system is homogeneous, in order to have non-null solutions we must impose that the determinant of the corresponding matrix is zero. After some algebraic manipulations we obtain:

$$\epsilon_{HK} = \epsilon_{ITL}\left(\frac{1 - e^{-2qT_{HK}}}{1 + e^{-2qT_{HK}}}\right)\left(1 - \frac{\epsilon_{ITL} + \epsilon_{Si}}{\epsilon_{ITL} - \epsilon_{Si}} e^{2qT_{ITL}}\right) \times \left(1 + \frac{\epsilon_{ITL} + \epsilon_{Si}}{\epsilon_{ITL} - \epsilon_{Si}} e^{2qT_{ITL}}\right)^{-1}. \tag{10.42}$$

In principle we should replace both ϵ_{HK} and ϵ_{ITL} with the corresponding frequency dependent expressions (Eq.10.1). However, we see in Table 10.1 that if the interfacial layer is SiO_2, ϵ_{HK} changes in a range of frequency where ϵ_{ITL} is essentially equal to the low frequency limit of ϵ_{SiO_2}, which we denote $\epsilon_{SiO_2,0}$. Furthermore, for the high-κ material, $\omega_{TO2} \gg \omega_{TO1}$ so that the lowest frequency mode of the stack is expected to be somehow related to ω_{TO1} (which we denote $\omega_{TO1,HK}$), which is responsible for the difference between $\epsilon_{HK,0}$ and $\epsilon_{HK,int}$. We can thus use Eq.10.42 to determine the lowest mode of the structure in Fig.10.5 by setting:

$$\epsilon_{ITL} \simeq \epsilon_{SiO_2,0}, \tag{10.43}$$

$$\epsilon_{HK} \simeq \epsilon_{HK,int} + \frac{\epsilon_{HK,0} - \epsilon_{HK,int}}{1 - \left(\frac{\omega}{\omega_{TO1,HK}}\right)^2}. \tag{10.44}$$

Substitution into Eq.10.42 yields:

$$\omega_{SO} = \omega_{TO1,HK}\left[\epsilon_{SiO_2,0}\left(\frac{1-\mathrm{e}^{-2qT_{HK}}}{1+\mathrm{e}^{-2qT_{HK}}}\right)\left(\frac{1-\frac{\epsilon_{SiO_2,0}+\epsilon_{Si}}{\epsilon_{SiO_2,0}-\epsilon_{Si}}\mathrm{e}^{2qT_{ITL}}}{1+\frac{\epsilon_{ITL}+\epsilon_{Si}}{\epsilon_{SiO_2,0}-\epsilon_{Si}}\mathrm{e}^{2qT_{ITL}}}\right)-\epsilon_{HK,0}\right]^{1/2}$$
$$\times\left[\epsilon_{SiO_2,0}\left(\frac{1-\mathrm{e}^{-2qT_{HK}}}{1+\mathrm{e}^{-2qT_{HK}}}\right)\left(\frac{1-\frac{\epsilon_{SiO_2,0}+\epsilon_{Si}}{\epsilon_{SiO_2,0}-\epsilon_{Si}}\mathrm{e}^{2qT_{ITL}}}{1+\frac{\epsilon_{ITL}+\epsilon_{Si}}{\epsilon_{SiO_2,0}-\epsilon_{Si}}\mathrm{e}^{2qT_{ITL}}}\right)-\epsilon_{HK,int}\right]^{-1/2}. \tag{10.45}$$

We see that, differently from the previous situations, ω_{SO} depends on q.

Amplitude of the potential

To determine the amplitude of the perturbation potential, we need to compute the term B_5 in Eq.10.40c. In fact, the perturbation potential in the silicon substrate is given by:

$$\phi_{Si} = B_5\mathrm{e}^{-qz}\mathrm{e}^{\mathrm{i}(\mathbf{q}\cdot\mathbf{r}-\omega t)} + (c.c.). \tag{10.46}$$

As described in Section 10.2.2, we first compute the electrostatic energy and then multiply it by 2 to include the kinetic energy of the atoms. The calculation for the silicon substrate is the same as in the infinite dielectric case:

$$W_{Si} = 2q\epsilon_{Si}B_5^2. \tag{10.47}$$

The potential in the ITL and HK layers, instead, includes both the $\mathcal{B}^-\mathrm{e}^{-qz}$ and $\mathcal{B}^+\mathrm{e}^{+qz}$ terms (where with $\mathcal{B}^+$ and $\mathcal{B}^-$ we denote respectively the terms B_1 and B_2 in the HK and B_3 and B_4 in the ITL) . It is easy to derive:

$$|F_x|^2 = 4q^2\left(|\mathcal{B}^+|^2\mathrm{e}^{+2qz}+|\mathcal{B}^-|^2\mathrm{e}^{-2qz}+2|\mathcal{B}^+||\mathcal{B}^-|\right)\cos^2(qx-\omega t), \tag{10.48a}$$
$$|F_z|^2 = 4q^2\left(|\mathcal{B}^+|^2\mathrm{e}^{+2qz}+|\mathcal{B}^-|^2\mathrm{e}^{-2qz}-2|\mathcal{B}^+||\mathcal{B}^-|\right)\sin^2(qx-\omega t), \tag{10.48b}$$

so that the total energy in a region extending from z_i to z_f and with dielectric constant ϵ_z is:

$$W = \frac{1}{2}\epsilon_z\int_{z_i}^{z_f}\left[|F_x^{(max)}|^2+|F_z^{(max)}|^2\right]\mathrm{d}z$$
$$= 4\epsilon_z q^2\int_{z_i}^{z_f}\left(|\mathcal{B}^+|^2e^{+2qz}+|\mathcal{B}^-|^2e^{-2qz}\right)\mathrm{d}z, \tag{10.49}$$

where we have used the maximum value of the time-varying electric field for the reasons discussed when deriving Eq.10.29. Equation 10.49 applied to the ITL and HK regions gives:

$$W_{ITL} = 2q\epsilon_{ITL}\left[|B_3|^2\left(1-\mathrm{e}^{-2qT_{ITL}}\right)+|B_4|^2\left(\mathrm{e}^{2qT_{ITL}}-1\right)\right], \tag{10.50}$$

and

$$W_{HK} = 2q\epsilon_{HK}\left[|B_1|^2\left(e^{-2qT_{ITL}} - e^{-2q(T_{ITL}+T_{HK})}\right) + |B_2|^2\left(e^{2q(T_{ITL}+T_{HK})} - e^{2qT_{ITL}}\right)\right]. \tag{10.51}$$

From Eq.10.41 we can express all the amplitudes as a function of B_5:

$$B_1 = \frac{(\epsilon_{ITL} - \epsilon_{Si}) + (\epsilon_{ITL} + \epsilon_{Si})e^{2qT_{ITL}}}{2\epsilon_{ITL}(1 - e^{-2qT_{HK}})} B_5, \tag{10.52a}$$

$$B_2 = -\frac{(\epsilon_{ITL} + \epsilon_{Si}) + (\epsilon_{ITL} - \epsilon_{Si})e^{-2qT_{IL}}}{2\epsilon_{ITL}(e^{2qT_{HK}} - 1)} B_5, \tag{10.52b}$$

$$B_3 = \frac{\epsilon_{ITL} - \epsilon_{Si}}{2\epsilon_{ITL}} B_5, \tag{10.52c}$$

$$B_4 = \frac{\epsilon_{ITL} + \epsilon_{Si}}{2\epsilon_{ITL}} B_5. \tag{10.52d}$$

This allows us to write the total energy of the mode as a function of $|B_5|^2$:

$$W_{SO} = W_{HK} + W_{ITL} + W_{Si} = 2q\epsilon_{eff}|B_5|^2, \tag{10.53}$$

which is equivalent to Eq.10.29 and where ϵ_{eff} is an effective dielectric constant depending on the frequency dependent dielectric constant of the high-κ material:

$$\epsilon_{eff}(\omega) = \epsilon_{HK}(\omega)\left[\left(\frac{\epsilon_{ITL} - \epsilon_{Si}}{2\epsilon_{ITL}}\right)^2 e^{-2qT_{ITL}} + \left(\frac{\epsilon_{ITL} + \epsilon_{Si}}{2\epsilon_{ITL}}\right)^2 e^{2qT_{ITL}} + 2\frac{\epsilon_{ITL}^2 - \epsilon_{Si}^2}{(2\epsilon_{ITL})^2}\right]\frac{1 + e^{-2qT_{HK}}}{1 - e^{-2qT_{HK}}} + \frac{(\epsilon_{ITL} + \epsilon_{Si})^2}{4\epsilon_{ITL}} e^{2qT_{ITL}} - \frac{(\epsilon_{ITL} - \epsilon_{Si})^2}{4\epsilon_{ITL}} e^{-2qT_{ITL}} + \epsilon_{Si}. \tag{10.54}$$

If we substitute ϵ_{HK} from Eq.10.42, we find $W_{SO} = 0$, as in the case of the infinite dielectric structure.

As in Section 10.2.2, the next step is to compute W_{SO} when the phonon is fully responding ($\epsilon_{HK} = \epsilon_{HK,0}$) and when it is not responding at all ($\epsilon_{HK} = \epsilon_{HK,int}$) to the field. In both cases we equate the energy thus derived to the corresponding quantum mechanical expression $\hbar\omega_{SO}(n_{SO} + 1/2)/A_{\text{norm}}$ (where A_{norm} is the normalization area), thus finding the amplitudes $B_5^{(0)}$ and $B_5^{(\infty)}$. We finally obtain:

$$B_5 = \sqrt{(B_5^{(0)})^2 - (B_5^{(\infty)})^2} = \sqrt{\frac{\hbar\omega_{SO}\left(n_{SO} + \frac{1}{2}\right)}{2qA_{\text{norm}}}\left(\frac{1}{\epsilon_{eff}(\epsilon_{HK,int})} - \frac{1}{\epsilon_{eff}(\epsilon_{HK,0})}\right)}, \tag{10.55}$$

where in the evaluation of $\epsilon_{eff}(\epsilon_{HK})$ we substitute ϵ_{HK} with $\epsilon_{HK,0}$ or $\epsilon_{HK,int}$ when considering the phonon either on or off. Furthermore ϵ_{ITL} is usually taken as $\epsilon_{SiO_2,0}$, although the very small thickness and partly unknown composition of the interfacial layer may result in different values.

The expression of the amplitude of the potential in the silicon channel is thus identical to the case with the infinite dielectric (Eq.10.33) except that the term $[1/(\epsilon_{Si} + \epsilon_{(\infty)}) - 1/(\epsilon_{Si} + \epsilon_{(0)})]$ is replaced by:

$$\frac{1}{\hat{\epsilon}_{SO}} = \frac{1}{\epsilon_{eff}(\epsilon_{HK,int})} - \frac{1}{\epsilon_{eff}(\epsilon_{HK,0})}. \tag{10.56}$$

Note that now $\hat{\epsilon}_{SO}$ depends on q.

10.2.4 Calculation of the scattering rates

Electron inversion layers

Let us start for simplicity by considering the structure with infinite dielectric shown in Fig.10.4 and a single phonon. From Eq.10.33 we are able to express the perturbation potential in the silicon channel associated with the soft-optical phonon. So far we have assumed x as the propagation direction. If we now assume a generic propagation vector $\mathbf{q}$ of amplitude q, similarly to Eq.4.230 for lattice vibrations we can write:

$$\phi(\mathbf{r}, z; \mathbf{q}) = \sqrt{\frac{\hbar\omega_{SO}}{2qA_{\text{norm}}}\frac{1}{\hat{\epsilon}}} \times \mathrm{e}^{-qz}\left(a_{SO}\mathrm{e}^{\mathrm{i}(\mathbf{q}\cdot\mathbf{r}-\omega_{SO}t)} + a_{SO}^{\dagger}\mathrm{e}^{-\mathrm{i}(\mathbf{q}\cdot\mathbf{r}-\omega_{SO}t)}\right), \tag{10.57}$$

where

$$\frac{1}{\hat{\epsilon}} = \frac{1}{\epsilon_{Si} + \epsilon_{(\infty)}} - \frac{1}{\epsilon_{Si} + \epsilon_{(0)}}, \tag{10.58}$$

and, as discussed above in this section and in Sections 4.5.2, 4.6.2:

$$|a_{SO}|^2 = n_{SO}, \tag{10.59a}$$

$$|a_{SO}^{\dagger}|^2 = n_{SO} + 1, \tag{10.59b}$$

With respect to the lattice vibrations studied in Section 4.5, we do not need to invoke deformation potentials, since we already know the potential associated with the phonon modes. We just need to insert $e\phi(\mathbf{r}, z; \mathbf{q})$ into Eq.4.43. It is understood that the term proportional to $\exp[\mathrm{i}(\mathbf{q}\cdot\mathbf{r} - \omega_{SO}t)]$ in Eq.10.57 must be used to compute the phonon absorption matrix element ($M_{n,n'}^{(ab)}(\mathbf{k}, \mathbf{k}')$ in Eq.4.43), and the term proportional to $\exp[-\mathrm{i}(\mathbf{q}\cdot\mathbf{r} - \omega_{SO}t)]$ to compute the corresponding emission matrix element ($M_{n,n'}^{(em)}(\mathbf{k}, \mathbf{k}')$ in Eq.4.43).

It is evident that, due to the strong reduction of the perturbation potential for large $\mathbf{q}$, remote phonon scattering produces only intra-valley transitions, so that n and n' identify subbands belonging to the same valley (Section 4.1.4). If we write the wave-functions of the initial and final state as (Eq.3.14):

$$\psi_n(\mathbf{k}) = \frac{\mathrm{e}^{\mathrm{i}\mathbf{r}\cdot\mathbf{k}}}{\sqrt{A_{\text{norm}}}}\xi_n(z), \tag{10.60a}$$

$$\psi_{n'}(\mathbf{k}') = \frac{e^{i\mathbf{r}\cdot\mathbf{k}'}}{\sqrt{A_{\rm norm}}}\xi_{n'}(z), \qquad (10.60b)$$

the matrix element for phonon emission is given by:

$$M_{n,n'}^{(em)}(\mathbf{k},\mathbf{k}') = \sqrt{\frac{\hbar\omega_{SO}}{2qA_{\rm norm}}\frac{1}{\hat{\epsilon}}}a_{SO}^{\dagger}\int \xi_n\,\xi_{n'}\,e^{-qz}dz\,\delta_{\mathbf{k}',(\mathbf{k}-\mathbf{q})}. \qquad (10.61)$$

A similar expression holds for phonon absorption, provided that we replace $a_{SO}^{\dagger}$ with a_{SO} and $\delta_{\mathbf{k}',(\mathbf{k}-\mathbf{q})}$ with $\delta_{\mathbf{k}',(\mathbf{k}+\mathbf{q})}$.

We can now compute the total scattering rate out of state (n,**k**) by using Eq.4.43, and then integrating over the final states (n′,**k**′). We get:

$$S_n(\mathbf{k}) = \frac{e^2\omega_{SO}}{4\pi}\frac{1}{\hat{\epsilon}}\left(n_{SO}+\frac{1}{2}\mp\frac{1}{2}\right)\sum_{n'}\int\frac{|I_{n,n'}|^2}{q(\mathbf{k},\mathbf{k}')}\delta(E_{n'}(\mathbf{k}')-E_n(\mathbf{k})\mp\hbar\omega_{SO})d\mathbf{k}', \qquad (10.62)$$

where the upper and lower sign correspond respectively to phonon absorption and emission, and

$$I_{n,n'} = \int_0^{\infty}\xi_n(z)\xi_{n'}(z)e^{-qz}dz, \qquad (10.63)$$

$$q = \sqrt{k^2+(k')^2-2kk'\cos\theta}, \qquad (10.64)$$

where θ is the angle between $\mathbf{k}$ and $\mathbf{k}'$.

We now recall Eq.3.15 and write the electron energy relation as:

$$\begin{aligned} E_{n'}(\mathbf{k}') &= \varepsilon_{n'} + \frac{\hbar^2(k_x')^2}{2m_l} + \frac{\hbar^2(k_y')^2}{2m_t} \\ &= \varepsilon_{n'} + \frac{\hbar^2(k')^2}{2}\left(\frac{\cos^2\beta}{m_l}+\frac{\sin^2\beta}{m_t}\right), \end{aligned} \qquad (10.65)$$

where β is the angle between $\mathbf{k}'$ and $\hat{k}_x$.

Substitution in Eq.10.62 yields:

$$\begin{aligned} S_n(\mathbf{k}) &= \frac{e^2\omega_{SO}}{4\pi\hbar^2\hat{\epsilon}}\left(n_{SO}+\frac{1}{2}\mp\frac{1}{2}\right)\sum_{n'}H_v(E_n(\mathbf{k})-\varepsilon_{n'}\pm\hbar\omega_{SO}) \\ &\times\int_0^{2\pi}\frac{|I_{n,n'}(q(\mathbf{k},k',\beta))|^2}{q(\mathbf{k},k',\beta)}m_{xy}^2(\beta)d\beta, \end{aligned} \qquad (10.66)$$

where (Eq.3.70):

$$m_{xy}(\beta) = \left(\frac{\cos^2\beta}{m_l}+\frac{\sin^2\beta}{m_t}\right)^{-1}, \qquad (10.67)$$

and $q(\mathbf{k},k',\beta)$ is given by Eq.10.64 with $\theta = \beta - \theta_{\mathbf{k}}$, as illustrated in Fig.10.6. It is evident from Eq.10.66 that only subbands n' with a minimum $\varepsilon_{n'}$ lower than $[E_n(\mathbf{k})\mp\hbar\omega_{SO}]$ contribute to the total scattering rate.

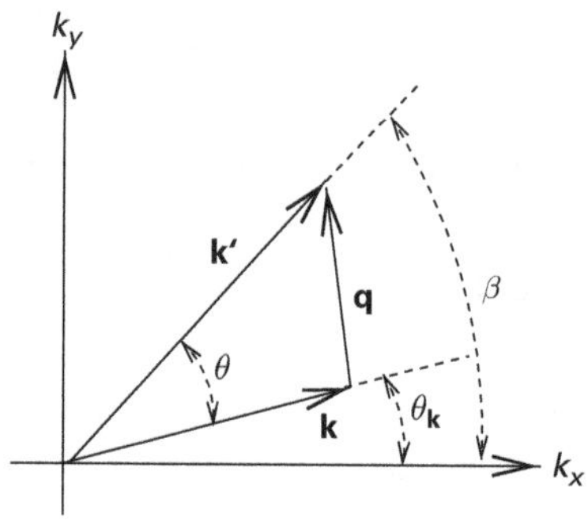

Figure 10.6 Sketch showing the wave-vectors involved in a scattering event with a remote phonon.

Unfortunately it is not possible to further simplify Eq.10.66 as was done for the acoustic and optical phonon scattering in Section 4.5, since the remote phonon scattering mechanism is anisotropic.

Equation 10.66 can be extended to non-parabolic subbands by multiplying the terms inside the summation over n' by $[1+\alpha(E_n(\mathbf{k})-\varepsilon_{n'}\pm\hbar\omega_{SO})]$, where α is the non-parabolicity factor defined in Eq.2.63.

We have derived the expression for the matrix element and for the total scattering rate considering the template structure with an infinite dielectric on top and a single phonon. However, we have seen that the more realistic gate stack structure analyzed in Section 10.2.3 leads to expressions for the perturbation potential in the channel essentially identical to the one in the structure with infinite dielectric (Eq.10.57). The only difference is that Eq.10.56 must be used to express the appropriate $\hat{\epsilon}$. The same applies for the infinite dielectric case with two phonons, where $\hat{\epsilon}$ is given by Eqs.10.39a and 10.39b.

For the realistic gate stack structure of Section 10.2.3 and in more general cases, ω_{SO}, n_{SO}, as well as $\hat{\epsilon}$ depend on q and thus they should be moved inside the integral over β in Eq.10.66. It is worth recalling that for the reasons discussed in Section 4.7, the effect of screening on remote optical phonons is very weak, and it is usually not taken into account.

Hole inversion layers

In this case, the perturbation potential is still given by Eq.10.57 and can be applied to the various gate stack structures by using the corresponding expression for $\hat{\epsilon}$, as discussed above. The calculation of the matrix elements, instead, depends on the model describing the 2D hole gas.

In the $\mathbf{k}\cdot\mathbf{p}$ approach, we end up with a $\mathbf{k}$-dependent overlap integral. Similarly to other scattering mechanisms, we can reduce the computational burden significantly by considering the wave-functions $\xi_{n\mathbf{k}}$ and $\xi_{n'\mathbf{k}'}$ of the initial and final state only for $\mathbf{k}=\mathbf{k}'=\mathbf{0}$ and approximate:

$$\psi_n(\mathbf{k}) \simeq \frac{e^{i\mathbf{r}\cdot\mathbf{k}}}{\sqrt{A_{\rm norm}}}\xi_{n\mathbf{0}}(z), \tag{10.68a}$$

$$\psi_{n'}(\mathbf{k}') \simeq \frac{e^{i\mathbf{r}\cdot\mathbf{k}'}}{\sqrt{A_{\rm norm}}}\xi_{n'\mathbf{0}}(z). \tag{10.68b}$$

We thus obtain the same expression for the matrix element (Eq.10.61) and for the overlap integral (Eq.10.63) as for the electron case. When computing the total scattering rate, the matrix elements must be integrated over the numerical energy dispersion provided by the quantized **k**·**p** procedure.

If, instead, we use the analytical model for hole inversion layers described in Section 3.3.3, the subband minima and the wave-functions are computed exactly as in the electron case, so that the expressions for the matrix element and for the overlap integral obtained for the electron case can also be directly used in the hole inversion layer.

Regarding total scattering rate, the calculation procedure is similar to the one for an electron inversion layer but the elliptical constant energy contour line must be substituted with the in-plane energy dispersion described in Section 3.3.3. We thus obtain:

$$S_n(\mathbf{k}) = \frac{\mathrm{e}^2\omega_{SO}}{4\pi\hat{\epsilon}}\left(n_{SO}+\frac{1}{2}\pm\frac{1}{2}\right)\sum_{n'} H_v(E_n(\mathbf{k})-\varepsilon_{n'}\pm\hbar\omega_{SO})$$
$$\times\int_0^{2\pi}\frac{|I_{n,n'}(q(\mathbf{k},k',\beta))|^2}{q(\mathbf{k},k',\beta)}\left[k(E_p,\beta)\frac{\partial k}{\partial E_p}(E_p,\beta)\right]\mathrm{d}\beta, \quad (10.69)$$

with $E_p=(E_n(\mathbf{k})-\varepsilon_{n'}\pm\hbar\omega_{SO})$, where, $\partial k/\partial E_p$ is given by Eq.3.84, $q(\mathbf{k},k',\beta)$ is given by Eq.10.64, $I_{n,n'}$ is given by Eq.10.63, and $\theta=\beta-\theta_{\mathbf{k}}$, as illustrated in Fig.10.6.

10.3 Scattering due to remote Coulomb centers

Significant densities of fixed charges are believed to be trapped in high-κ dielectrics because of the fabrication process and the operating conditions [36].

The treatment of Coulomb scattering provided in Section 4.3 is quite general and applicable also to calculation of the scattering rate due to Coulomb centers located in the high-κ dielectric as well as in the ITL and in the substrate. The expression for the scattering rate is exactly the same as derived in Section 4.3 (Eqs. 4.134, 4.135, 4.140). The remote location of the charges, however, requires a new derivation of the normalized point charge potential $\phi_{pc}(q,z,z_0)$ to account for the presence of a multi-layer gate stack.

10.3.1 Scattering matrix elements

With reference to the structure in Fig.10.5, we observe that Eq.4.116 can be solved in each layer providing that:

$$\phi_{pc}(q,z,z_0)_{HK} = \frac{e}{2q\epsilon_{HK}}\mathrm{e}^{-q|z-z_0|}+A_1\mathrm{e}^{qz}+A_2\mathrm{e}^{-qz}, \quad (10.70a)$$

$$\phi_{pc}(q,z,z_0)_{ITL} = \frac{e}{2q\epsilon_{ITL}}\mathrm{e}^{-q|z-z_0|}+A_3\mathrm{e}^{qz}+A_4\mathrm{e}^{-qz}, \quad (10.70b)$$

$$\phi_{pc}(q,z,z_0)_{Si} = \frac{e}{2q\epsilon_{Si}}\mathrm{e}^{-q|z-z_0|}+A_5\mathrm{e}^{-qz}. \quad (10.70c)$$

The presence of the term proportional to $e^{-q|z-z_0|}$ in all the regions is practically convenient and its correctness has been discussed in detail in Section 4.3.1.

By assuming an ideal metal gate and thus setting a null potential at the high-κ/metal gate interface (namely at $z = -(T_{ITL} + T_{HK})$, see Fig.10.5) and, furthermore, imposing the continuity of the potential and of the displacement field at each interface, we obtain:

$$A_1 e^{-q(T_{ITL}+T_{HK})} + A_2 e^{q(T_{ITL}+T_{HK})} + \frac{e}{2q\epsilon_{HK}} e^{-q(T_{ITL}+T_{HK}+z_0)} = 0, \tag{10.71a}$$

$$A_1 e^{-qT_{ITL}} + A_2 e^{qT_{ITL}} + \frac{e}{2q\epsilon_{HK}} e^{-q|T_{ITL}+z_0|} = A_3 e^{-qT_{ITL}} + A_4 e^{qT_{ITL}} + \frac{e}{2q\epsilon_{ITL}} e^{-q|T_{ITL}+z_0|}, \tag{10.71b}$$

$$A_3 + A_4 + \frac{e}{2q\epsilon_{ITL}} e^{-q|z_0|} = A_5 + \frac{e}{2q\epsilon_{Si}} e^{-q|z_0|}, \tag{10.71c}$$

$$\epsilon_{HK}\left[qA_1 e^{-qT_{ITL}} - qA_2 e^{qT_{ITL}} + sgn(T_{ITL}+z_0)\frac{e}{2\epsilon_{HK}} e^{-q|T_{ITL}+z_0|}\right] = \epsilon_{ITL}\left[qA_3 e^{-qT_{ITL}} - qA_4 e^{qT_{ITL}} + sgn(T_{ITL}+z_0)\frac{e}{2\epsilon_{ITL}} e^{-q|T_{ITL}+z_0|}\right], \tag{10.71d}$$

$$\epsilon_{ITL}\left[qA_3 - qA_4 + sgn(z_0)\frac{e}{2\epsilon_{ITL}} e^{-q|z_0|}\right] = \epsilon_{Si}\left[-qA_5 + sgn(z_0)\frac{e}{2\epsilon_{Si}} e^{-q|z_0|}\right], \tag{10.71e}$$

where $sgn(z)$ denotes the sign function.

The scattering potential in the semiconductor $\phi_{pc}(q, z, z_0)_{Si}$ is obtained by determining A_5 from the system of equations 10.71. Then the matrix element for the point charge is given by (see also Eq.4.135):

$$M^{(0)}_{n,n'}(\mathbf{k}, \mathbf{k}', z_0) = \int_z \xi^{\dagger}_{n'\mathbf{k}'}(z)\, \xi_{n\mathbf{k}}(z) \left(\frac{e}{2q\epsilon_{Si}} e^{-q|z-z_0|} + A_5 e^{-qz}\right) dz, \tag{10.72}$$

where we have assumed that the wave-functions can be written as in Eq.8.3.

It should be noted that the general treatment proposed here can also describe charges at the interface between two regions, as already discussed in Section 4.3.1 for the case of an infinite dielectric.

We can thus compute the squared matrix element due to the Coulomb centers in the gate stack following the same steps outlined in Section 4.3.2 and by invoking the random phase approximation. The squared matrix elements are integrated over z_0 leading to:

$$|M_{n,n'}(\mathbf{k}, \mathbf{k}')|^2 \simeq \frac{1}{A_{\rm norm}}\left[\int_{-T_{HK}-T_{ITL}}^{-T_{ITL}} |M^{(0)}_{n,n'}(\mathbf{k}, \mathbf{k}', z_0)|^2 N_{HK}(z_0) dz_0 + \int_{-T_{ITL}}^{0} |M^{(0)}_{n,n'}(\mathbf{k}, \mathbf{k}', z_0)|^2 N_{ITL}(z_0) dz_0\right.$$

$$+ \int_0^{z_{max}} |M_{n,n'}^{(0)}(\mathbf{k}, \mathbf{k}', z_0)|^2 N_{Si}(z_0) dz_0$$
$$+ |M_{n,n'}^{(0)}(\mathbf{k}, \mathbf{k}', -T_{ITL})|^2 N_{HK/ITL}$$
$$+ |M_{n,n'}^{(0)}(\mathbf{k}, \mathbf{k}', 0)|^2 N_{Si/ITL}\Big], \qquad (10.73)$$

where N_{HK}, N_{ITL}, and N_{Si} are the charge concentrations (per unit volume) in the high-κ layer, in the ITL, and in the silicon substrate respectively, whereas $N_{HK/ITL}$ and $N_{Si/ITL}$ are the charge densities (per unit area) of Coulomb centers at the high-κ/ITL and ITL/silicon interface, respectively.

Note that, according to this approach, in a device with the gate stack structure of Fig.10.5, the scattering produced by ionized impurities in the substrate (N_{Si} in Eq.10.73) must be computed also using the expression for $\phi_{pc}(q, z, z_0)$ given by Eq.10.70. In other words, the effect of the gate stack should be considered for the charges located in the bulk semiconductor also.

10.3.2 Effect of the screening

We have seen in Section 4.2 that computation of the dielectric function requires the expression for the normalized potential $\phi_{pc}(q, z, z_0)$ (see e.g. Eqs.4.73, 4.74, and 4.75). In Chapter 4 we give the expression for $\phi_{pc}(q, z, z_0)$ for bulk devices with infinite dielectric on top and for SOI devices. When dealing with stacks including a high-κ dielectric, we can still use the same equations for the dielectric function discussed in Section 4.2, but we have to replace the expression for $\phi_{pc}(q, z, z_0)$ with Eq.10.70c where A_5 must be determined by solving Eq.10.71.

We finally note that the boundary condition Eq.10.70a essentially defines the treatment of the screening produced by the metal gate proposed in this section. In fact, as already noted, Eq.10.70a imposes a null value for $\phi_{pc}(q, z, z_0)$ at the gate interface, corresponding to an ideal metal gate behavior.

10.4 Simulation results for MOSFETs with high-κ dielectrics

In this section we illustrate the results of mobility and drain current calculations for silicon inversion layers with high-κ dielectrics in the gate stack. Simulations have been run with the multi-subband Monte Carlo model described in [37]. The remote polar phonons (SOph) and remote Coulomb (RemQ) scattering models described in Section 10.2 and 10.3, respectively, have been implemented as described in [38].

We begin by analyzing the impact of soft phonons on mobility. We consider an idealized inversion layer in an undoped channel with an infinite dielectric on top. The simulated mobility curves for various high-κ dielectric materials as well as SiO_2 are shown in Fig.10.7. We see that soft phonon scattering has a very limited effect on overall mobility in the SiO_2 case, thus confirming that simulations for SiO_2 transistors typically do not account for this scattering mechanism. On the other hand, soft phonons

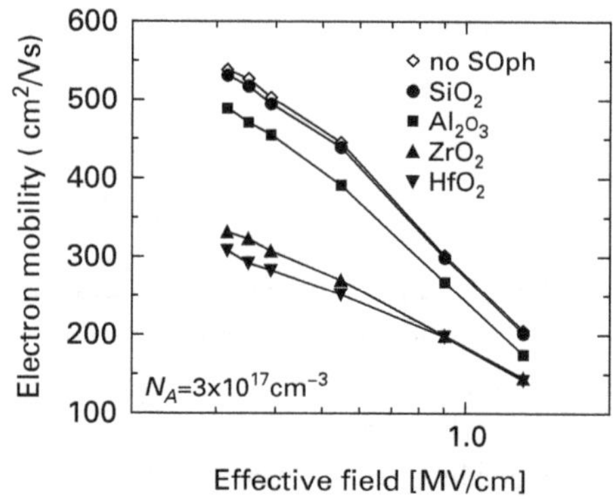

Figure 10.7 Mobility vs. effective field in n-type inversion layers obtained with MSMC simulations. Acoustic, intervalley, and remote phonons as well as surface roughness have been considered. Simulations assume an infinitely thick high-κ dielectric directly grown on top of the silicon substrate (no ITL).

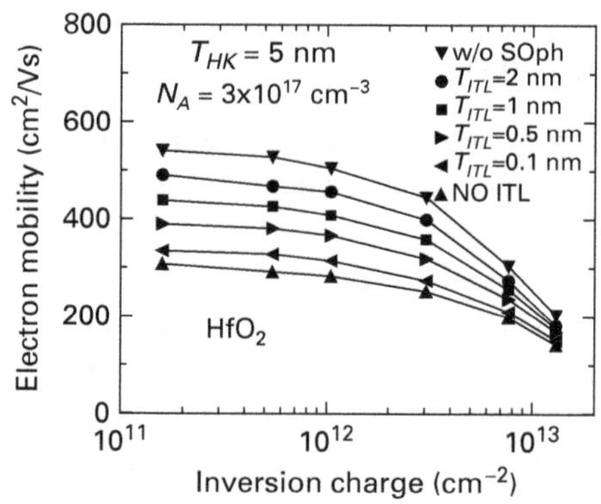

Figure 10.8 Simulated mobility vs. effective field. Scattering mechanisms include acoustic, intervalley, and remote phonons as well as surface roughness. Gate stack with ITL/high-κ(HfO_2)/MG. $T_{HK} = 5\,nm$.

originating in ZrO_2 and HfO_2 strongly degrade mobility, at least in this simple idealized structure without any interfacial layer.

To demonstrate the effect of the interfacial layer, we show in Fig.10.8 the simulated mobility for a structure featuring a HfO_2 layer on top of a SiO_2 interfacial layer (ITL). We see that for realistic interfacial layer thickness T_{ITL} (around 1nm), the mobility degradation is much less than that predicted by the simulations without ITL.

It is worth noting that modeling approaches based on a 3D electron gas description predict a much stronger influence of soft phonons on mobility [39, 40]. This is because, by neglecting the quantum mechanical set back of the charge from the semiconductor/dielectric interface enforced by the shape of the envelope wave-functions, the 3D electron gas is much more sensitive to remote scattering mechanisms.

The results of Fig.10.8 refer to an oversimplified picture, since Coulomb scattering due to interface charge and substrate doping has been neglected. A more realistic case is shown in Fig.10.9 [38]. Here the concentration of interface states has been adjusted to fit

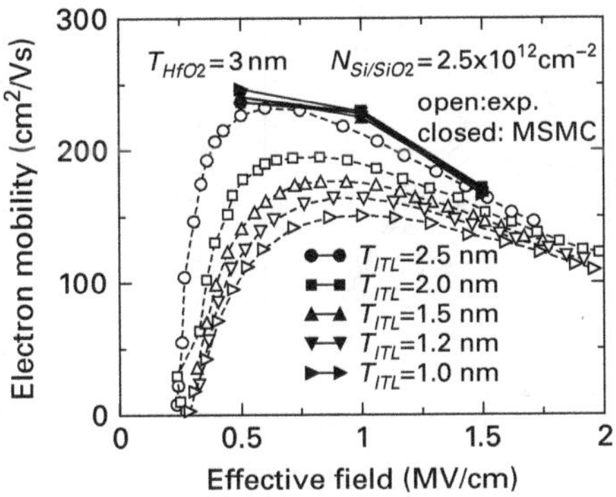

Figure 10.9 Simulated and experimental [4] mobility curves for gate stacks featuring different ITL thickness and a 3nm HfO_2 layer. Scattering mechanisms include acoustic, intervalley, and remote phonons, surface roughness and Coulomb scattering with substrate doping and interface states.

the experimental data [4] for the SiO_2 control devices. Then the soft phonon scattering mechanism has been switched on, and the mobility has been compared with the experimental data for ITL/HfO_2 stacks with different ITL thickness [4]. The mobility is much lower than in Fig.10.8 even for large T_{ITL} values because of the degradation induced by scattering with interface states. In this situation the additional contribution due to soft phonons is essentially negligible, as demonstrated by the fact that the simulated mobility is insensitive to T_{ITL}.

Soft phonon scattering has been proposed as the main mechanism responsible for the mobility reduction observed in gate stacks featuring high-κ materials. However, it is clear from the previous figures that its influence on the mobility of realistic n-MOS transistors is small. Furthermore, experimental mobility data in [4] show a temperature dependence which is not consistent with the one expected for soft phonon limited mobility, but instead with the Coulomb scattering limited mobility.

The effect of remote charges in the gate stack on electron mobility can be analyzed using the model described in Section 10.3. The results of this analysis are shown in Fig.10.10, where we consider the presence of charge at the ITL/high-κ interface and study the effect of T_{ITL} and T_{HK} on channel mobility. As expected, the most evident effect is a reduction of the mobility for small T_{ITL} (plot a), since the remote Coulomb centers get closer to the channel. Decreasing T_{HK}, instead, enhances mobility, due to a more effective screening provided by the metal gate. Significant mobility reduction is observed for a charge density at the ITL/high-κ interface in the order of 10^{13} cm^{-2}. The impact on mobility of the remote charges is much larger at small inversion density N_{inv}, because at large N_{inv} the screening produced by the inversion layer drastically reduces the Coulomb scattering.

The results in Fig.10.10 were obtained in template structures without interface charge. We now consider the devices measured in [4]. Figure 10.11 shows the simulated mobility when remote Coulomb scattering is activated; the corresponding results *without* remote Coulomb scattering were discussed in Fig.10.9. The concentration of charge at the ITL/HfO_2 interface has been adjusted in order to get the best fit with the

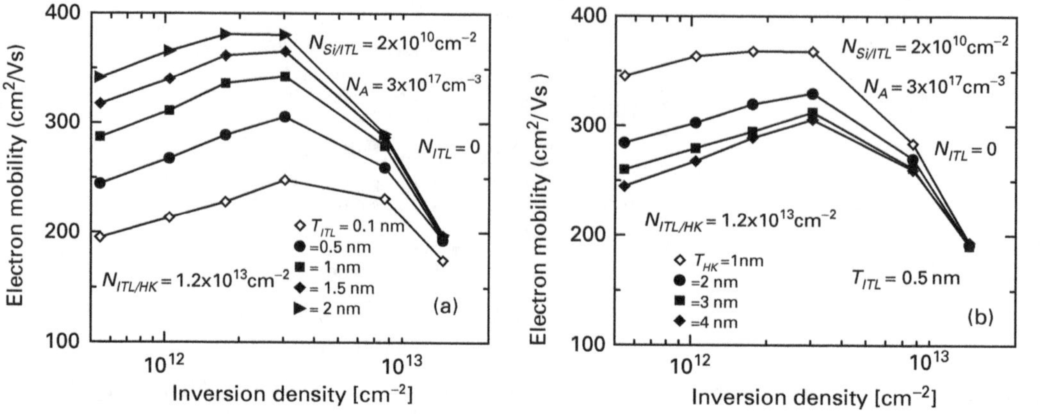

Figure 10.10 Effect of remote Coulomb scattering in a template MOS structure featuring a SiO_2 ITL, an HfO_2 layer and an ideal metal gate. Charge at the ITL/HfO_2 interface has been assumed. Other scattering mechanisms included are acoustic intra-valley and inter-valley phonons as well as surface roughness. Plot a): effect of T_{ITL}. Plot b): effect of T_{HK}. $N_{HK/ITL}$ and $N_{Si/ITL}$ are defined in Eq.10.73 and related text.

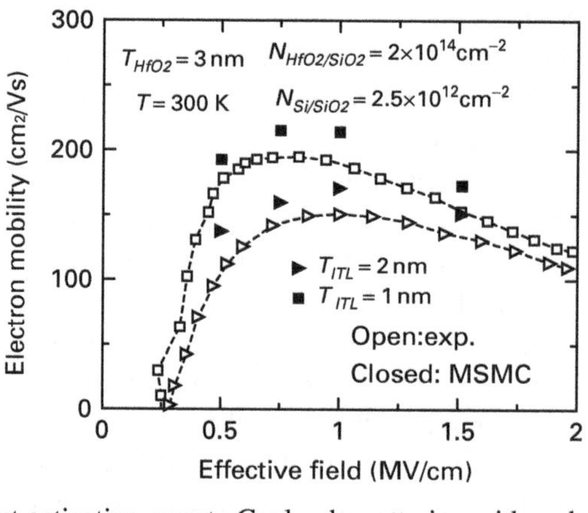

Figure 10.11 Same as in Fig.10.9, but activating remote Coulomb scattering with a charge of 2×10^{14} cm^{-2} at the ITL/HfO_2 interface.

experimental data for the high-κ devices, whereas the concentration at the Si / ITL interface has been adjusted to reproduce the SiO_2 control devices. We see that a very large charge concentration at the ITL/HfO_2 interface is required to reproduce the experiments. Assuming that all the fixed charges have the same sign, this concentration would produce a threshold voltage shift of many Volts, that has *not* been observed experimentally.

This discrepancy with the experiments might be overcome if the charge were globally neutral, so that it contributes to the scattering but it does not modify the device electrostatics. However, if, for instance, we consider dipoles oriented normal to the channel, the dipole concentration and dipole momentum needed to fit the experimental mobility would be so large that a significant threshold voltage shift would appear again, as in the single charge case [41].

Another possible explanation for mobility reduction in high-κ transistors, according to [21], is that the mobility is degraded by the presence of process-induced defect centers at the Si / ITL interface. While more work is certainly needed to assess the origin of the mobility reduction in high-κ MOSFETs, these examples demonstrate the usefulness of a detailed physically based treatment of the scattering for the modeling of the device physics. It should be emphasized, indeed, that remote Coulomb and soft-optical phonon scattering models do not have free parameters except for the density of remote charges.

In order to exemplify the effect of the remote Coulomb charge on the drive current of nanoscale devices, we show results for a 32 nm single-gate SOI device whose geometry and doping profiles have been tuned according to a realistic fabrication process [42]. A huge charge density of $2 \times 10^{14}\,\text{cm}^{-2}$ at the ITL/high-κ interface was required to reproduce the experimental low-field mobility data of long channel devices. More details on the device structure and on the simulation of the low-field mobility are provided in [43].

Figure 10.12 illustrates the effect of the charge on the drive current I_{ON}. We see that neglecting remote Coulomb scattering leads to an overestimate of the drain current with respect to the experimental values. Accounting for remote Coulomb scattering, instead, provides a better agreement with the data, especially at low V_{DS} and high V_{GS} (left plot), consistent with the fact that the charge concentration value has been calibrated on low-field mobility data. On the other hand, the effect on the sub-threshold region (right plot) is very significant, since in this region screening is very weak.

To conclude this section we underline that the results above which refer to electron inversion layers cannot be simply generalized to holes. In fact, the mobility reduction

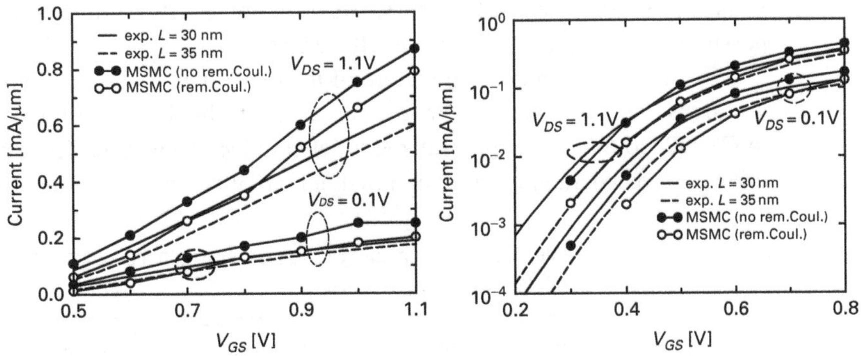

Figure 10.12 Comparison between the simulated and experimental trans-characteristics of a single gate 32 nm SOI n-MOSFET. Multi-subband Monte Carlo with remote Coulomb scattering (rem.Coul.) assumes a density of $2 \times 10^{14}\,\text{cm}^{-2}$ at the ITL/high-κ interface [43]. Note that the charges introduced as a source of remote Coulomb scattering were not treated in a self-consistent way, i.e. they were not included in the computation of device electrostatics, so we do not observe the threshold voltage shift (4.6 V) associated with them.

associated with the introduction of high-κ dielectric in p-MOSFETs is significantly smaller than in the n-MOSFET case [4]. Simulation results for hole inversion layers may be found in [41].

10.5 Alternative channel materials

In Section 8.1 we saw how to describe electron inversion layers with arbitrary crystal orientations using the EMA and considering conduction band minima of the Δ, Λ, and Γ type. In Section 8.3 we have applied the generalized EMA to silicon inversion layers, where the Δ valleys are usually the ones accounted for in the transport simulations, since they are the most populated. In the case of alternative channel materials more than one family of conduction band minima may contribute to transport. Moreover the transport modeling must account for scattering mechanisms (e.g. polar optical phonons) which have not been described in Chapter 4.

In the following we discuss transport modeling in these materials, focusing our analysis on electron transport. The treatment of holes according to the $\mathbf{k}\cdot\mathbf{p}$ formalism (Sections 2.2.2 and 8.2) is very general and can be naturally extended to different materials by adjusting the parameters L, M, and N in the $\mathbf{k}\cdot\mathbf{p}$ Hamiltonian (Eq.8.23). The $\mathbf{k}\cdot\mathbf{p}$ parameters for different materials are reported in Table 2.2.

Among the possible channel material alternatives to silicon, we consider here germanium and gallium-arsenide because they feature bulk electron and hole mobility significantly higher than silicon (Table 10.2). It has to be emphasized, however, that at present fabrication of both Ge and III-V MOSFETs still suffers from the lack of a high quality dielectric stack. Moreover, processing germanium n-type MOSFETs is problematic due to difficult activation of n-type doping in the source and drain regions [44], so that, although many techniques are currently explored to solve this issue [45, 46], Ge is most promising for p-MOSFETs. GaAs and other III-V semiconductors are instead possible silicon replacements for n-MOSFETs.

Providing a comprehensive picture of the experimental data available at the time of writing is premature because, in most cases, the technology is still immature compared to CMOS, so that the strong interest in alternative channel materials has been mainly driven by considerations based on low-field mobility data or on modeling studies.

Table 10.2 Carrier mobility in bulk intrinsic semiconductors at room temperature.

Semiconductor	Electron mobility [cm^2/Vs]	Hole mobility [cm^2/Vs]
Silicon	1417	471
Germanium	3900	1900
Gallium arsenide	8800	400

The effect of replacing the silicon channel with alternative materials has been analyzed by using Monte Carlo simulations in [47, 48]; a free electron gas was considered. In more recent years there has been a renewed interest in assessing the advantages in terms of on-current related to replacement of silicon as channel material. However, most of the simulation studies are presently limited to purely ballistic transport [49–57] and focused on assessing the advantages of the new materials for end-of-the-Roadmap CMOS technologies (i.e. gate lengths below 10nm). We therefore start with a description of ballistic transport models and their ingredients, and then move to multi-subband Monte Carlo based models.

10.5.1 Ballistic transport modeling of alternative channel devices

We provide here a brief overview of the modeling approaches and main results concerning optimization of the on-current in MOSFETs with alternative channel materials under the ballistic limit (that we introduced in Section 5.6).

The best wafer and channel orientation for germanium ballistic n-MOSFETs has been investigated in [49, 51] using fully numerical models based on quantum transport and including a multi-subband description of the electron gas in the inversion layer.

Top-of-the-barrier ballistic transport models have also been used to assess the possible advantages of III-V materials in n-MOSFETs using analytical [52] and full-band [53] descriptions of the electron energy dispersion in the inversion layer. It has been found that quantum confinement strongly modifies the band structure of such materials and tends to populate valleys at higher energy with higher masses compared to the lowest valley dominating transport in the bulk crystal.

Simple analytical models for the on-current that were presented in [50, 54, 56, 57] provide guidelines to discuss the main trade-offs associated with the use of alternative channel materials.

According to [54], the contribution of a single subband i in valley ν to the on-current in the ballistic limit is given by:

$$I_{ON,\nu,i} = \frac{e}{\sqrt{2}\hbar^2}\left(\frac{K_B T}{\pi}\right)^{\frac{3}{2}} \sqrt{m_{\mathrm{Ion}}(\alpha_\nu)}\,\mathcal{F}_{1/2}(\eta_{\nu,i}). \qquad (10.74)$$

This expression extends to an arbitrary crystal orientation the treatment we provided in Section 5.6. Here $\mathcal{F}_{1/2}(\eta_{\nu,i})$ is the Fermi integral of order $1/2$ defined in Eq.A.27, and $\eta_{\nu,i} = (E_{FS} - E_{\nu,i})/K_B T$, with $E_{\nu,i}$ denoting the subband energy and E_{FS} the Fermi level in the source. The drain current effective mass is given by:

$$m_{\mathrm{Ion}}(\alpha_\nu) = m_{p,t}\cos^2(\alpha_\nu) + m_{p,l}\sin^2(\alpha_\nu), \qquad (10.75)$$

where α_ν is the angle formed between the k_x direction of the DCS and the longitudinal axis of the constant energy ellipse in the inversion layer (see Fig.8.1.d). The total on-current is obtained by summing over all valleys and subbands the contributions given by Eq.10.74.

Since in elliptical subbands $m_{p,l}$ is by definition larger than $m_{p,t}$, Eqs.10.74 and 10.75 tell us that, for a given channel material, transport plane, and bias, the current depends on the in-plane orientation and it is maximum for $\alpha = \pi/2$. Hence the current is maximum when the transport is aligned with the direction corresponding to the lighter mass $m_{p,t}$. However, it is not straightforward to assess which material and crystal orientation maximizes the on-current by using Eq.10.74.

Let us consider the comparison between the I_{ON} of different materials at a given inversion charge, which is given by:

$$N_{inv} = \sum_{\nu,i} \frac{\mu_\nu \sqrt{m_{p,l} m_{p,t}}}{\pi \hbar^2} \left(\frac{K_B T}{2} \right) \ln(1 + e^{\eta_{\nu,i}}), \qquad (10.76)$$

where μ_ν is the multiplicity of the valley ν.

It is easy to see that for a circular subband in the quantum limit (that is when only the lowest subband is occupied, so that we can drop the indexes i and ν) the larger current at given N_{inv} is obtained for $m_{p,l} = m_{p,t}$ as small as possible. In fact, if we compute I_{ON}/N_{inv} we are left with a term $\sqrt{m_{\text{Ion}}}/\sqrt{m_{p,l} m_{p,t}}$ which becomes $m_{p,l}$ for circular subbands. Furthermore the ratio between $\mathcal{F}_{\frac{1}{2}}(\eta)$ and $\ln(1 + e^{\eta})$ increases for increasing η. This means that, at given inversion charge, a strongly degenerate inversion channel provides a higher ballistic current than a non-degenerate one. For given N_{inv}, in turn, the electron gas degeneracy is enhanced (i.e. η is larger) when the DoS effective mass $\sqrt{m_{p,l} m_{p,t}}$ is smaller.

However, when compared to silicon at the same N_{inv}, many alternative materials do not show real advantages in terms of ballistic current when strong quantum confinement and strong carrier degeneracy are considered, because bias- and size-induced quantization tend to populate satellite valleys with much larger effective masses than those of the conduction band minimum which defines the transport properties in the bulk material [56].

It is furthermore evident that a comparison at the same N_{inv} does not mean the same applied bias. In fact, considering again for simplicity the quantum limit, for smaller DoS effective mass ($\sqrt{m_{p,l} m_{p,t}}$) the same N_{inv} is obtained with higher η, i.e. higher subband energy with respect to the source Fermi level. This means that a higher gate voltage is needed to achieve the same N_{inv}.

In order to compare the on-current at a given bias, it is useful to consider the inversion capacitance. We assume again working within the quantum limit, so that we can write:

$$C_{inv} = \frac{\mathrm{d}(eN_{inv})}{\mathrm{d}V_{SO}} = e^2 \frac{\mathrm{d}E_0}{\mathrm{d}(-eV_{SO})} \left(-\frac{\mathrm{d}N_{inv}}{\mathrm{d}E_0} \right), \qquad (10.77)$$

where E_0 denotes the lowest energy eigenvalue and the term $\mathrm{d}E_0/\mathrm{d}(-eV_{SO})$ describes how closely E_0 follows the changes of the electrostatic potential V_{SO} at the semiconductor-oxide interface at the virtual source, see Fig. 10.13. The term $\mathrm{d}E_0/\mathrm{d}(-eV_{SO})$ is in general smaller than 1, but we assume $\mathrm{d}E_0/\mathrm{d}(-eV_{SO}) = 1.0$, which provides an overestimate of the value of C_{inv}. The term $(\mathrm{d}N_{inv}/\mathrm{d}E_0)$ can be readily

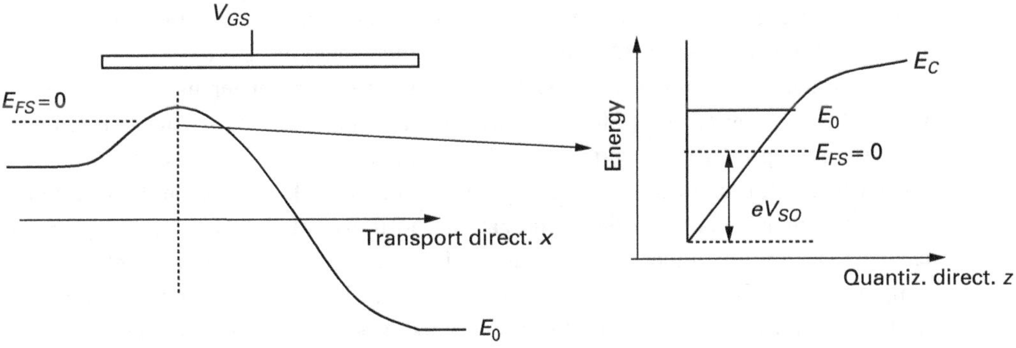

Figure 10.13 The lowest subband profile along the transport direction of a ballistic MOSFET and of the quantum well at the virtual source.

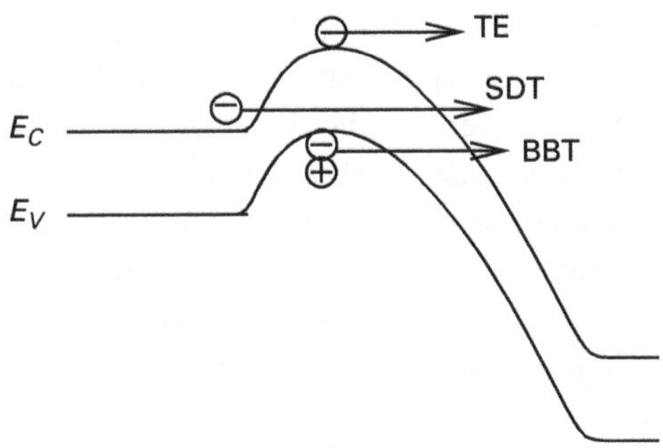

Figure 10.14 The main mechanisms determining the drain current of a MOSFET in the off state.

calculated from Eq.10.76. In particular, for a strongly degenerate electron gas, C_{inv} is indicated as *quantum capacitance* [58–60] and given by:

$$C_{inv} \simeq C_{QM} = \frac{e^2}{2}\frac{\mu_v\sqrt{m_{p,l}m_{p,t}}}{\pi\hbar^2}. \tag{10.78}$$

The inversion charge is then obtained by considering the series of the oxide capacitance C_{ox} and C_{QM}, integrated from the threshold voltage V_T up to the supply voltage V_{DD}:

$$eN_{inv} = \frac{C_{ox}C_{QM}}{C_{ox}+C_{QM}}(V_{DD}-V_T). \tag{10.79}$$

Since C_{QM} is proportional to the effective masses in the transport plane, materials with low effective masses feature lower N_{inv} at given supply voltage with respect to materials with higher masses. Consequently the drain current does not fully benefit from the reduction of the effective mass. Quantitatively speaking, due to the effect of the quantum capacitance, in the ballistic limit the advantages of unstrained III-V materials over silicon and germanium are expected to be quite limited [48, 54].

Similar conclusions are reached when comparing the on-current for a given off-current. The main reasons for that are discussed below. As illustrated in Fig.10.14, the main contributions to the off-current in an SOI device where junction leakage is negligible are given by thermionic emission above the S/D energy barrier (TE), source-to-drain-tunneling (SDT), and band-to-band-tunneling (BBT).

Thermionic emission can be modeled with a top-of-the-barrier model such as the one used for the on-current. Short channel effects controlling the barrier height in the subthreshold regime should be included for a realistic evaluation of this term.

Concerning source-to-drain-tunneling, full quantum models based on the solution of the Schrödinger equation in the quantization and transport directions are in principle needed to assess its importance in nanoscale FETs [61–63]. Simpler approaches to compute the SDT current have been proposed in [57, 64]. The general indication of these models is that, due to the relation between conduction and valence band masses and the energy dependent effective mass for tunneling, SDT is enhanced in low effective mass materials.

Band-to-band tunneling generation of electron–hole pairs is enhanced in materials with low energy gap and low effective mass [65–67], as is the case to many of the materials proposed to replace silicon in the FET channel.

A reliable assessment of the I_{ON}/I_{OFF} ratio in devices with alternative channel materials should consider all these effects. The complexity of the band-to-band tunneling process makes it difficult to draw definitive conclusions. The results in [68] suggest that the maximum on-current for given off-current is obtained in materials with energy gap and effective mass close to those of silicon.

From the analyses reported in the literature it thus appears that under ballistic transport there are limited margins of I_{ON} improvement for most of the alternative channel materials with respect to silicon. However, the situation appears different when channel length in the deca-nanometric range is assumed, and scattering is taken into account. In particular, the studies in [69, 70] indicate that, despite the limitations posed by the quantum capacitance and the strong influence of the access regions on the on-current, alternative materials can outperform silicon. The picture becomes more complex if strain is considered among the technology options for both silicon and alternative materials.

To exemplify the applicability of the multi-subband Monte Carlo technique to simulation of FETs with alternative channel materials, we present in Sections 10.6 and 10.7 the main modeling ingredients necessary to simulate the on-current of Ge and GaAs MOSFETs in the presence of scattering.

10.5.2 Energy reference in alternative channel materials

As a final remark before considering specific examples of channel materials, we consider the choice of the energy reference $E_{\nu 0}$ when accounting for more than one type of valley. We assume that the constant energy surface for a conduction band minimum can be written as in Eq.8.2, where what really matters is not the absolute value of $E_{\nu 0}$, but rather the difference between the $E_{\nu 0}$ values of the different valleys. All the $E_{\nu 0}$ can be shifted by the same amount without affecting the physical picture. However, the shift must be consistent with the energy reference used for the confining potential $U(z)$ in the Schrödinger equation (Eq.8.4) and, in the case of equilibrium simulations, for the Fermi level E_F. Similarly to the treatment in Section 3.7, one can refer $U(z)$ to the vacuum level, writing $U(z) = -[e\phi(z) + \chi]$. In this case, the $E_{\nu0}$ of the lowest valley must be set to zero, since the electron affinity is the distance between the vacuum level and the lowest minimum of the bulk conduction band. The $E_{\nu0}$ values of the other valleys are then referred to the lowest. Note that $E_{\nu0}$ does not enter the Schrödinger equation 8.4, but it enters the calculation of the subband energy with respect to the vacuum level (Eq.8.13), hence the expression of all the scattering rates. Therefore $E_{\nu0}$ must be added to the subband energies $\varepsilon_{\nu,i}$ obtained by solving the Schrödinger equation with the confinement potential $U(z)$ defined above, which is the same for all the valleys. This is the choice we adopt in the rest of this chapter. The definition of proper boundary conditions for the electrostatic potential and the identification of the Fermi level in the regions at equilibrium, instead, follows from the discussion in Section 3.7 for (001) silicon, which applies also to different channel materials.

10.6 Germanium MOSFETs

10.6.1 Conduction band and phonon parameters

The conduction band parameters for bulk germanium are given in Table 10.3. These parameters can be used together with the data in Table 8.3 to obtain the quantization and transport masses in n-type Ge inversion layers.

In Table 10.3 we see that the lowest valleys are the Λ ones, which are strongly anisotropic (i.e. m_l is much larger than m_t) and also somewhat non-parabolic. The Λ valleys are located at the border of the first Brillouin zone along the $\langle 111 \rangle$ directions (see Fig.8.2). Since only half of the states of an equi-energy ellipsoid lie inside the Brillouin zone, the multiplicity of the Λ valleys is set to 4. The Δ and Γ valleys are quite close in energy to the Λ valleys, so that they significantly contribute to electron transport in inversion layers and cannot be neglected.

Concerning phonon scattering, the expressions for the scattering rates are the same derived in Section 4.5. In particular, Eq.4.274 holds for acoustic intra-valley transitions and Eq.4.286 for inter-valley transitions.

The phonon scattering parameters for electron transport are listed in Table 10.4. With respect to the silicon case, which has only a single type of valley, we need to distinguish transitions between valleys belonging to the same family and valleys of different families. We thus have intra-valley elastic acoustic phonons with deformation potentials specific for each valley.

As can be seen in Table 10.4, inter-valley transitions between valleys of the same type should be separated into f- and g-type, as described in Section 4.6.5. However, if the initial valley is a Λ family, the g-type process is more correctly referred to as intra-valley, because the opposite valley coincides with the valley itself. This is consistent with the selection rules discussed in Section 4.6.7 that allow for optical phonon intra-valley transitions for the Λ valleys. Note that for Λ valleys the multiplicity of the final valley (i.e. the valley after scattering) is 3 for f-type and 1 for intra-valley processes (see Table 10.5).

Table 10.3 Parameters of the conduction band minima in Ge (from [71]).

Valley	m_l/m_0	m_t/m_0	$\alpha[eV^{-1}]$	$E_{\nu 0}[eV]$	Multiplicity
Λ	1.588	0.081	0.3	0	4
Δ	1.353	0.29	0	0.18	6
Γ	0.037	0.037	0	0.14	1

Table 10.4 Parameters for phonon scattering of electrons in the conduction band of germanium. Data from [71]. It is worth noting that this set does not contain f-type transitions between Δ valleys.

Phonon type	Valleys	Def.pot.	Energy [meV]
Acoustic/intra-valley	Λ	11 eV	0
Acoustic/intra-valley	Δ	9 eV	0
Acoustic/intra-valley	Γ	5 eV	0
Optical intra-valley	Λ	$5.5 \times 10^{10}\, eV/m$	37
Inter-valley (f-type)	Λ	$3 \times 10^{10}\, eV/m$	27.6
Inter-valley (f-type)	Λ	$2 \times 10^{9}\, eV/m$	10
Inter-valley (g-type)	Δ	$7.89 \times 10^{9}\, eV/m$	8.6
Inter-valley (g-type)	Δ	$9.46 \times 10^{10}\, eV/m$	37
Inter-valley	$\Delta \rightarrow \Lambda$	$4.06 \times 10^{10}\, eV/m$	27.6
Inter-valley	$\Gamma \rightarrow \Lambda$	$2 \times 10^{10}\, eV/m$	27.6
Inter-valley	$\Delta \rightarrow \Gamma$	$1 \times 10^{11}\, eV/m$	27.6

	density	$\rho = 5320\, kg/m^3$
	sound velocity	$v_s = 5.4 \times 10^3$ m/s

Table 10.5 Number of phonon processes for inter-valley phonons. $\mu_{w,\nu}$ is the parameter to be used in Eq.4.286.

Initial valley (ν)	Final valley (w)	Type	$\mu_{w,\nu}$
Λ	Λ	g or intra-valley	1
Λ	Λ	f	3
Δ	Δ	g	2
Δ	Δ	f	4
Λ or Δ	Γ		1
Λ or Γ	Δ		6
Δ or Γ	Λ		4

On the other hand, transitions between valleys belonging to different families are not separated into f- and g-type processes. In fact this distinction originates from the fact that the distance between a valley of a given family (Δ or Λ) and the opposite one differs from the distance to the other valleys of the same family. However, if we consider transitions, for instance, between a ν valley of the Δ family and a ν' valley of the Λ family, it is easy to see that all Λ valleys result in the same distance in $\mathbf{K}$ space $|\mathbf{K}_\nu - \mathbf{K}_{\nu'} + \mathbf{G}_{\nu\nu'}|$, when an appropriate reciprocal lattice vector $\mathbf{G}_{\nu\nu'}$ is used to assure that $(\mathbf{K}_\nu - \mathbf{K}_{\nu'} + \mathbf{G}_{\nu\nu'})$ belongs to the first BZ (see the discussion about *umklapp* processes in Section 4.6.5). The same applies to transitions between Λ and Γ and between Δ and Γ, so that the multiplicity of the final valley (i.e. $\mu_{w,\nu}$ in Eq.4.286) in the transitions between valleys belonging to different families is 6, 4, and 1 for final valleys of the Δ, Λ, and Γ type, respectively (see Table 10.5).

We note that for each transition in Table 10.4, the dual transition in the opposite direction exists. For example, the table lists the transitions between Γ and Λ, but that between Λ and Γ exists as well and is governed by the same parameters.

10.6.2 Electrons: velocity and low field mobility

This section describes simulation results for pure Ge inversion layers calculated with the multi-subband Monte Carlo model and the conduction band and scattering parameters of Section 10.6.1 (Tables10.3 and 10.4).

Figure 10.15 shows the velocity versus electric field curve for bulk Ge. The filled circles refer to simulations using a free-electron gas model, whereas the open circles refer to the simulation of a thick quantum well with (001) crystal orientation. The figure also shows experimental data from [71], where we observe that, despite the higher low-field mobility of germanium with respect to silicon, the saturation velocity is significantly lower than the value of approximately 10^7 cm/s measured in silicon. The simulated quantum well is large enough that its density of states is indistinguishable from the one of the free electron gas. However, we see that the velocity obtained with the

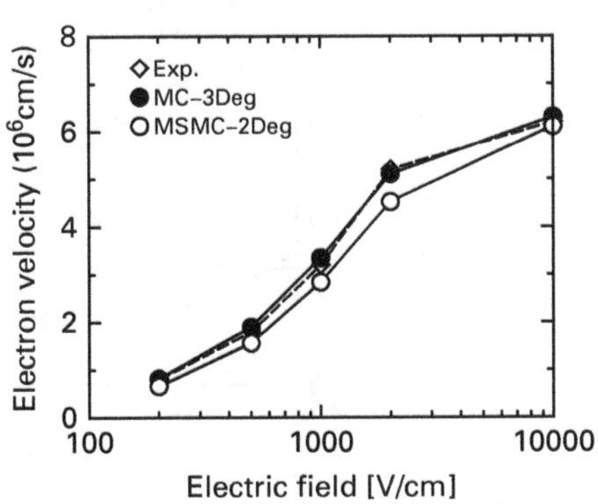

Figure 10.15 Simulated velocity/field curves using the valley and phonon parameters of Tables10.3 and 10.4. Experimental data from [71].

multi-subband (2Deg) model slightly differs from that of the free-electron gas. As discussed in Section 8.1.3, this is due to one of the basic assumptions of semi-classical modeling, namely the fact that during free-flights the electrons cannot change subbands (Section 5.2). As demonstrated in [72], this assumption breaks down in cases such as the Λ valleys in (001) germanium inversion layers, where the component w_{13} of $\mathbf{W}_{DCS}$ is non-null. However, this limitation of the semi-classical model becomes less and less relevant when the subbands get more separated in energy, as in realistic inversion layers.

Mobility results for realistic n-type germanium inversion layers are shown in Fig.10.16. The simulated data (filled circles) were obtained by including phonon and surface roughness scattering, starting from self-consistent solutions of the device electrostatics. Regarding the phonon parameters, the coupling constants for acoustic phonons in Table 10.4 have been multiplied by a factor 13/9, to mimic what is usually done in silicon inversion layers (see Section 7.1.2), where the coupling constant used in the inversion layer is $D_{AC} = 13eV$, whereas the value for bulk silicon is 9eV. Surface roughness has been accounted for by using the model of Section 4.4, where the mass m_z appearing in the expression for the matrix element (Eq.4.188) has been computed following Table 8.3. The correlation length and the r.m.s. value of the surface roughness spectrum are the same used for the Si / SiO_2 interface, which is an optimistic assumption for Ge inversion layers. Consistently, we see that the simulated mobility is significantly higher than the experimental results obtained by many groups but significantly lower than the mobility in bulk Ge (3900 $cm^2/(Vs)$). Furthermore, it is also evident that the measured mobility is well below the universal curve for silicon inversion layers. Possible explanations for this discrepancy are that: 1) most of the fabricated devices feature an interfacial silicon layer between the Ge substrate and the dielectric, thus possible transport in the Si layer cannot be ruled out; 2) the high-κ gate stack may degrade the low-field mobility due to additional scattering mechanisms, as discussed in Section 10.1. This means that the simulations in Fig.10.16 are likely to be an overoptimistic target for Ge inversion layers. It is

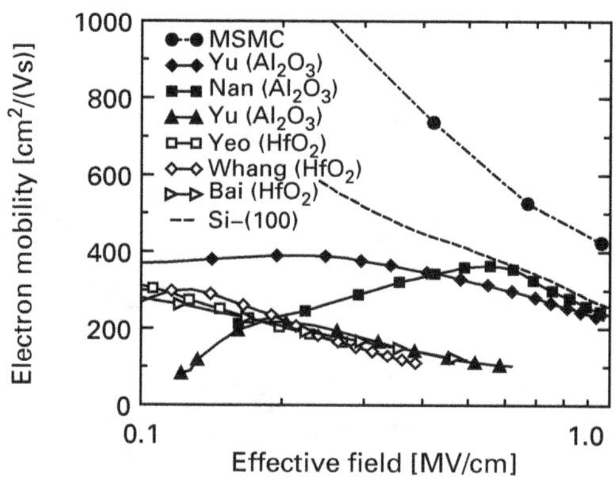

Figure 10.16 Simulated low-field mobility vs. effective field in Ge n-type inversion layers. The experimental data are from [75–80]. The gate dielectric is indicated inside the brackets. Note that the electron mobility in bulk Ge is 3900 $cm^2/(Vs)$.

worth mentioning that mobility data for germanium n-MOSFETs featuring GeO_2 gate dielectric are much closer to (and even slightly exceed) the universal curves for Si [45, 73].

More details and simulation results for n-type Ge inversion layers can be found in [74].

10.6.3 Holes: band structure and low field mobility

The energy relation of the top of the valence band of germanium can be approximated with the analytical model of Eq.2.64 with $A = -13$, $B = 8.9$ and $C = 10.3$[81].

Concerning hole inversion layers, we can proceed according to the quantized **k·p** model of Section 3.3.1; the **k·p** parameters for Ge are given in Table 2.2. Figure 10.17 shows the equi-energy lines in **k**-space for holes in a triangular well. As in the case of silicon (Figs.3.5 and 8.4), the dispersion relationship is strongly anisotropic. A direct comparison between the energy dispersion of the lowest subbands in germanium and silicon is illustrated in Fig.10.18: it is clear that, for a given $|\mathbf{k}|$, germanium has a larger

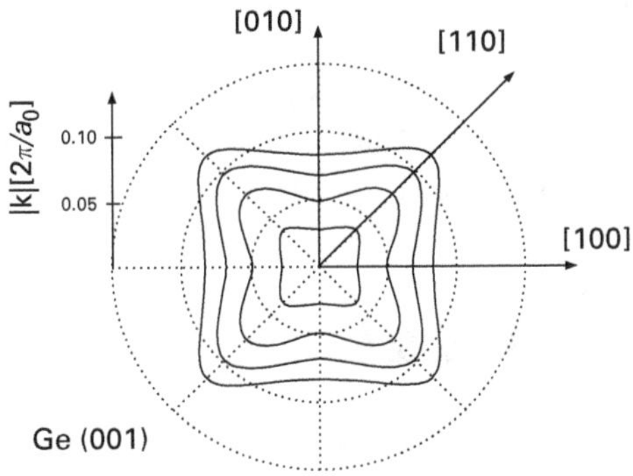

Figure 10.17 Equi-energy lines (25, 50, 75, 100 meV) for a Ge p-type inversion layer obtained with the quantized **k·p** approach. Triangular well with $F_z = 0.7$ MV/cm.

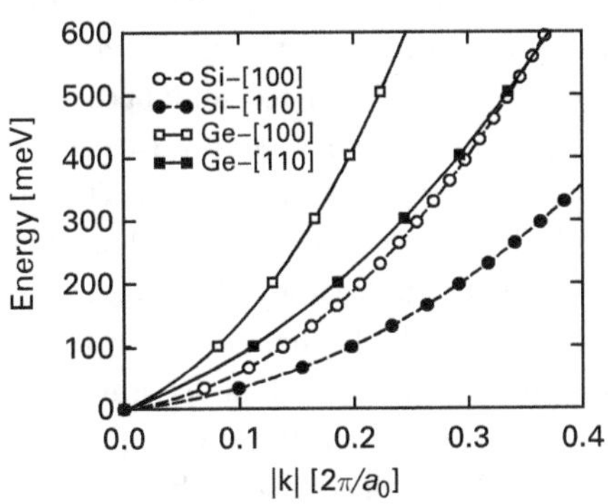

Figure 10.18 Comparison between the $E(\mathbf{k})$ relationships obtained with the quantized **k·p** approach for the lowest subbands in the valence bands of Si and Ge (001) inversion layers, considering the [100] and [110] directions.

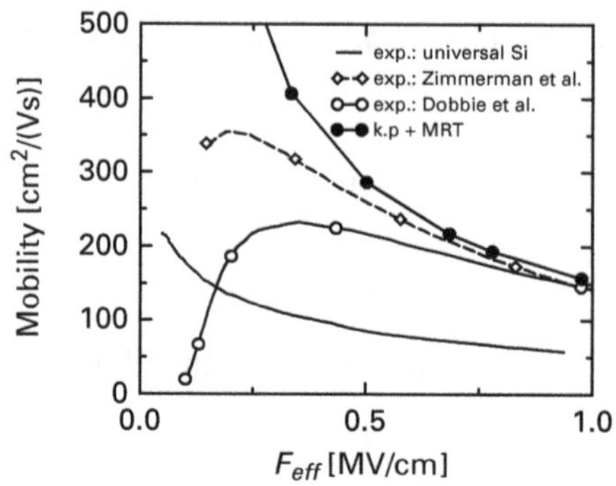

Figure 10.19 Simulated hole mobility in a germanium MOSFET and comparison with experimental data for Si [1] and Ge [82, 83]. Simulation parameters are: $D_{AC} = 11\,\text{eV}$, $D_{OP} = 6 \times 10^{10}\,\text{eV/m}$, $E_{OP} = 38\,\text{meV}$, $\Lambda_{SR} = 4\,\text{nm}$, $\Delta_{SR} = 0.368\,\text{nm}$, which are consistent with [84] except for the value of Λ_{SR} (3 nm vs. 4 nm). $T = 300\,\text{K}$.

energy than silicon for all in-plane orientations. This results in a higher hole velocity and effective mobility.

Figure 10.19 shows the simulated hole mobility in a Ge inversion layer using the momentum relaxation time approximation (see Section 5.4.3) and a description of the confined hole gas based on the **k·p** approach. We see that simulations are in good agreement with experimental data for Ge inversion layers, and significantly higher than the universal curve for silicon p-MOSFETs.

10.7 Gallium arsenide MOSFETs

10.7.1 Conduction band parameters

The parameters for the conduction band of GaAs are listed in Table 10.6. The Γ valley is the lowest and most populated, but Λ and Δ valleys are quite close in energy and cannot be neglected when modeling electron transport in inversion layers.

The effective mass of the Γ valley is very small, and this explains the very large bulk mobility (8800 $\text{cm}^2/(\text{Vs})$ for undoped GaAs at 300 K). Since the quantization mass ($m_z = m_l = m_t$) of the Γ valley is also very low, a significant subband splitting is expected in GaAs quantum wells. The effective masses for Λ and Δ valleys are significantly larger. The parabolic EMA quantization model predicts that for thin quantum wells the Λ and Δ valleys become more populated than the Γ valley because they have lower subband minima. However, it must be noted that the non-parabolicity of the Γ valley is fairly large and an accurate calculation of the energy levels requires us to treat quantization beyond the simple parabolic EMA. The problem is common to many III-V materials and has been addressed in [53, 55]. Tight-binding calculation of template quantum wells can be used to determine an effective quantization mass for the EMA model. Such an effective mass obviously depends on the thickness of the quantum well.

Table 10.6 Parameters of the conduction band minima in GaAs (from [47]).

Valley	m_t/m_0	m_l/m_0	$\alpha[eV^{-1}]$	$E_{\nu 0}[eV]$
Γ	0.063	0.063	1.16	0
Λ	0.127	1.538	0.4	0.323
Δ	0.229	1.987	0.55	0.447

10.7.2 Phonon scattering

The dominant phonon assisted transitions in GaAs, and in other III-V materials as well, are due to the polar optical phonons related to the polar nature of the bonding between the Ga and As atoms. Polar optical phonons in a 3D electron gas are discussed in [85].

In order to derive the matrix elements for a quasi-2D electron gas we start from the perturbation potential $U_{ph}^{POP}(\mathbf{R}, t)$ associated with the phonon mode as expressed in [85]. Consistently with the notation of Section 4.5, we write

$$U_{ph}^{POP}(\mathbf{R}, t) = \frac{e\sqrt{\hbar\omega_{ph}}}{\mathrm{i}\sqrt{2}\sqrt{\Omega}Q}\sqrt{\frac{1}{\epsilon_{(\infty)}} - \frac{1}{\epsilon_{(0)}}}\left[a\mathrm{e}^{\mathrm{i}(\mathbf{Q}\cdot\mathbf{R}-\omega_{ph}t)} + a^{\dagger}\mathrm{e}^{-\mathrm{i}(\mathbf{Q}\cdot\mathbf{R}-\omega_{ph}t)}\right], \tag{10.80}$$

where ω_{ph} is the phonon energy, $\epsilon_{(\infty)}$ the high-frequency dielectric constant and $\epsilon_{(0)}$ the static dielectric constant; Ω is a normalizing volume and Q is the magnitude of the phonon wave-vector, that is $Q = |\mathbf{Q}|$. As explained in Section 4.5, in Eq.10.80 we can assume $|a| = |a^{\dagger}| = \sqrt{n_{ph} + 1/2}$, so that $U_{ph}^{POP}(\mathbf{R}, t)$ is a real valued scattering potential. However, when computing the scattering rates (Eq.4.43) we must more appropriately set $|a| = \sqrt{n_{ph}}$ for phonon absorption and $|a^{\dagger}| = \sqrt{n_{ph} + 1}$ for phonon emission.

In GaAs quantum wells and GaAs MOSFETs, the wave-function is expressed by Eq.8.3. Combining Eq.8.3 and Eq.10.80, the matrix element is then given by:

$$M_{n,n'}(\mathbf{q}, q_z) = \int_{\mathbf{r}} d\mathbf{r} \int_z dz \frac{e\sqrt{\hbar\omega_{ph}}}{\mathrm{i}\sqrt{2}\sqrt{\Omega}\sqrt{q^2 + q_z^2}}\sqrt{\frac{1}{\epsilon_{(\infty)}} - \frac{1}{\epsilon_{(0)}}}$$
$$\times \sqrt{n_{ph} + \frac{1}{2} \pm \frac{1}{2}}\,\mathrm{e}^{\mp\mathrm{i}q_z z}\mathrm{e}^{\mp\mathrm{i}\mathbf{q}\cdot\mathbf{r}}\frac{\mathrm{e}^{-\mathrm{i}\mathbf{k}'\cdot\mathbf{r}}}{\sqrt{A}}\xi_{n'}^{\dagger}\frac{\mathrm{e}^{\mathrm{i}\mathbf{k}\cdot\mathbf{r}}}{\sqrt{A}}\xi_n, \tag{10.81}$$

where q_z is the component in the quantization direction of the phonon wave-vector $\mathbf{Q}$ and the upper and the lower sign are for emission and absorption, respectively. The polar optical phonon transitions are assumed here to be intra-valley, since inter-valley processes have a very low rate due to the large $\mathbf{Q}$ involved (see Section 4.1.4). The integration over $\mathbf{r}$ in Eq.10.81 yields a Kronecker delta $\delta_{\mathbf{k}',(\mathbf{k}\mp\mathbf{q})}$, so that the matrix element reads

$$M_{n,n'}(\mathbf{q},q_z) = \frac{e\sqrt{\hbar\omega_{ph}}}{\mathrm{i}\sqrt{2}\sqrt{\Omega}}\sqrt{\frac{1}{\epsilon(\infty)} - \frac{1}{\epsilon(0)}}\sqrt{n_{ph} + \frac{1}{2} \pm \frac{1}{2}}$$
$$\times\, \delta_{\mathbf{k}',(\mathbf{k}\mp\mathbf{q})} \int_z \xi^{\dagger}_{n'}(z)\xi_n(z)\frac{\mathrm{e}^{\mp \mathrm{i}q_z z}}{\sqrt{q^2+q_z^2}}\mathrm{d}z. \tag{10.82}$$

The squared matrix element can thus be written as

$$|M_{n,n'}(\mathbf{q})|^2 = \frac{L}{2\pi}\int_{q_z} |M_{n,n'}(\mathbf{q},q_z)|^2 \mathrm{d}q_z$$
$$= \frac{e^2\hbar\omega_{ph}}{4\pi A}\left(\frac{1}{\epsilon(\infty)} - \frac{1}{\epsilon(0)}\right)\left(n_{ph} + \frac{1}{2} \pm \frac{1}{2}\right)$$
$$\times \int_{q_z}\left|\int_z \xi^{\dagger}_{n'}(z)\xi_n(z)\frac{\mathrm{e}^{\mp \mathrm{i}q_z z}}{\sqrt{q^2+q_z^2}}\mathrm{d}z\right|^2 \mathrm{d}q_z, \tag{10.83}$$

where we have used the random phase approximation to write the squared magnitude of the sum of $M_{n,n'}(\mathbf{q},q_z)$ over q_z as the sum of the $|M_{n,n'}(\mathbf{q},q_z)|^2$ (see Section 4.6.2).

The multiple integral in the last line of Eq.10.83 can be identified as a sort of form factor $I_{n,n'}$ for the polar optical phonon scattering, and it can be simplified by recalling the mathematical properties

$$\left|\int_z f(z)\mathrm{d}z\right|^2 = \int_z f(z)\mathrm{d}z \int_{z'} f^{\dagger}(z')\mathrm{d}z' = \int_z\int_{z'} f(z)f^{\dagger}(z')\mathrm{d}z\,\mathrm{d}z', \tag{10.84}$$

where $f(z)$ is a generic function of the abscissa z. In fact by using Eq.10.84 for the integral over z in Eq.10.83, we can define the form factor $I_{n,n'}$ as

$$I_{n,n'} = \int_{q_z}\mathrm{d}q_z\int_z \mathrm{d}z\int_{z'}\mathrm{d}z'\xi^{\dagger}_{n'}(z)\xi_n(z)\xi^{\dagger}_{n'}(z')\xi_n(z')\frac{\mathrm{e}^{\mathrm{i}q_z(z-z')}}{q^2+q_z^2}. \tag{10.85}$$

The mathematical identity

$$\int_{-\infty}^{+\infty}\frac{1}{\pi}\frac{\beta}{\alpha^2+\beta^2}\,\mathrm{e}^{\mathrm{i}2\pi\alpha\gamma}\mathrm{d}\alpha = \mathrm{e}^{-2\pi\beta|\gamma|}, \tag{10.86}$$

and the substitutions $\alpha = q_z/(2\pi)$, $\beta = q/(2\pi)$ and $\gamma = (z-z')$ allow us to rewrite the integral over q_z in Eq.10.85 as

$$\int_{-\infty}^{+\infty}\frac{\mathrm{e}^{\mp \mathrm{i}q_z(z-z')}}{q^2+q_z^2}\mathrm{d}q_z = \frac{\pi}{q}\mathrm{e}^{-q|z-z'|}. \tag{10.87}$$

The squared matrix element is thus finally given by:

$$|M_{n,n'}(q)|^2 = \frac{e^2\hbar\omega_{ph}}{4Aq}\left(\frac{1}{\epsilon(\infty)} - \frac{1}{\epsilon(0)}\right)\left(n_{ph} + \frac{1}{2} \pm \frac{1}{2}\right)$$
$$\times \int_z \mathrm{d}z\int_{z'}\mathrm{d}z'\xi^{\dagger}_{n'}(z)\,\xi_n(z)\,\xi^{\dagger}_{n'}(z')\,\xi_n(z')\,\mathrm{e}^{-q|z-z'|}, \tag{10.88}$$

where the upper and the lower sign are for emission and absorption, respectively.

Table 10.7 Phonon scattering parameters in GaAs from [87].

Phonon type	Valleys	Def.pot.	Energy [meV]
Polar optical	Γ, Λ, Δ	n.a.	36.2
Acoustic intra-valley	Γ, Λ, Δ	7 eV	n.a.
Inter-valley	$\Lambda \to \Lambda$	5×10^{10} eV/m	29.9
Inter-valley	$\Delta \to \Delta$	1×10^{11} eV/m	29.9
Inter-valley	$\Gamma \to \Lambda$	1.8×10^{10} eV/m	29.9
Inter-valley	$\Gamma \to \Delta$	1×10^{11} eV/m	29.9
Inter-valley	$\Lambda \to \Delta$	1×10^{10} eV/m	29.9

low-frequency dielectric constant	$\epsilon_{(0)}$=12.9ϵ_0
high-frequency dielectric constant	$\epsilon_{(\infty)}$=10.9ϵ_0
density	ρ=5370 kg/m^3
sound velocity	v_s=5.2$\times$ 10^{3}m/s

The computational effort needed to calculate Eq.10.88 is significantly larger than that required for the scattering rates for acoustic and inter-valley phonons discussed in Section 4.5, essentially because the form factor $I_{n,n'}$ depends on q. Furthermore, it is clear from Eq.10.88 that the matrix element depends on $\mathbf{q}$, so that the polar optical phonon scattering is both inelastic and anisotropic. Simple expressions considering template wave-functions can be found in [86].

Due to the dynamic de-screening discussed in Section 4.7 the effect of screening on the transitions assisted by polar-optical phonons is very weak and it is usually neglected [85, 86].

Acoustic intra-valley and inter-valley phonon scattering in GaAs can be treated as in silicon and germanium, namely by using Eq.4.274 for acoustic intra-valley transitions, Eq.4.286 for inter-valley transitions and by considering all the relevant valleys.

The phonon scattering parameters for GaAs are summarized in Table 10.7. We see that polar optical phonons have to be considered as intra-valley transitions inside each family of valleys. The same phonon energy (36.2meV) is used in all cases.

10.7.3 Simulation results

Figure 10.20 shows the velocity versus field curves in a uniform GaAs slab considering a free-electron gas model (solid line) and a quasi-2D model applied to a thick quantum well (dashed line). The figure also shows experimental data from various authors.

Differently from Si and Ge, the velocity–field curve of GaAs is not monotonic. In fact, at low field almost all electrons are in the Γ valley and velocity increases with field, as expected. Above a critical electric field (approximately 4kV/cm), electrons start populating the Δ and Λ valleys that have much larger effective masses and scattering

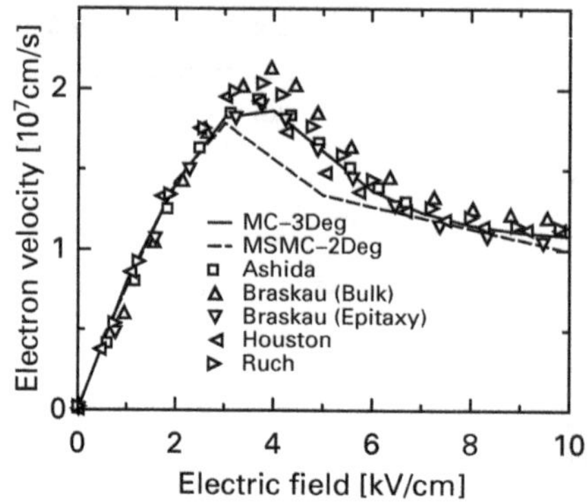

Figure 10.20 Simulated velocity vs. field curves considering a Monte Carlo approach for the free-electron gas (MC-3Deg) and a Multi-Subband Monte Carlo transport model applied to a thick quantum well (MSMC-2Deg). Experimental data are from [88–91]

rates with respect to the Γ valley. Hence the average electron velocity decreases, and the resulting value of the saturation velocity is very close to that of silicon.

The comparison between the free and the 2D electron gas reveals some discrepancies, mainly due to: 1) the problems associated with simulation of Λ valleys using the multi-subband formalism already discussed in the Ge case (see Section 10.6.2), that is related to the fact that in the semi-classical model electrons cannot change subband during free-flight; 2) the parabolic EMA model used for the Γ valley, which does not provide the same DoS as a free-electron gas with non-parabolic corrections when applied to a thick quantum well.

10.8 Summary

In this chapter we have seen that the main models, concepts and approaches discussed in the previous chapters for silicon channel transistors with SiO_2 gate dielectric can be extended to more advanced device structures, featuring channel materials other than silicon and gate stacks including high-κ dielectrics and metal gates.

We have analyzed the modeling issues associated with use of high-κ materials in the gate stack. Additional scattering mechanisms, namely remote optical phonons (Section 10.2) and remote Coulomb scattering (Section 10.3), are responsible for the mobility reduction associated with high-κ dielectrics, although the impact of remote optical phonons is very limited. On the other hand, remote Coulomb scattering can account for the experimentally observed mobility reduction, but a very large density of Coulomb centers in the stack has to be assumed in the model.

Regarding use of alternative channel materials, we have used the generalized EMA model of electron inversion layers with arbitrary crystal orientations described in Section 8.1 to analyze relevant case studies such as germanium inversion layers (Section 10.6.2) and gallium arsenide quantum wells (Section 10.7). Also, the **k·p** model for hole inversion layers described in Section 8.2 has been used for Ge

p-MOSFETs (Section 10.6.3), which are promising alternatives to conventional (001)-silicon transistors.

A comprehensive analysis of real devices requires adequate models for the off current, including contributions such as band-to-band tunneling and source-to-drain-tunneling, which are difficult to handle in the semi-classical approximation. However, we believe that the multi-subband Monte Carlo is a powerful approach to assessing the possible improvements related to use of non-conventional channel materials, and much more accurate than the fully ballistic approach adopted in many studies.

References

[1] S. Takagi, A. Toriumi, M. Iwase, and H. Tango, "On the universality of inversion-layer mobilty in Si MOSFETs. Part I- Effect of substrate impurity concentration," *IEEE Trans. on Electron Devices*, vol. 41, no. 12, pp. 2357–62, 1994.

[2] B. Guillaumot, X. Garros, F. Lime, *et al.*, "75 nm damascene metal gate and high-k integration for advanced CMOS devices," in *IEEE IEDM Technical Digest*, pp. 355–358, 2002.

[3] R. Chau, J. Brask, S. Datta, *et al.*, "Application of high-κ gate dielectrics and metal gate electrodes to enable silicon and non-silicon logic nanotechnology," *Microelectronic Engineering*, vol. 80, pp. 1–6, 2005. 14th biennial Conference on Insulating Films on Semiconductors.

[4] M. Casse, L. Thevenod, B. Guillaumot, *et al.*, "Carrier transport in HfO_2 / metal gate MOSFETs: physical insight into critical parameters," *IEEE Trans. on Electron Devices*, vol. 53, no. 4, pp. 759–768, 2006.

[5] F. Andrieu, O. Faynot, X. Garros, *et al.*, "Comparative scalability of PVD and CVD TiN on HfO2 as a metal gate stack for FDSOI cMOSFETs down to 25nm gate length and width," in *IEEE IEDM Technical Digest*, pp. 641–644, 2006.

[6] S.B. Samavedam, L.B. La, J. Smith, *et al.*, "Dual-metal gate cmos with HfO_2/ gate dielectric," in *IEEE IEDM Technical Digest*, pp. 433–436, 2002.

[7] O. Weber, F. Andrieu, M. Casse, *et al.*, "Experimental determination of mobility scattering mechanisms in Si / HfO_2 / TiN and SiGe:C/HfO_2/TiN surface channel n- and p-MOSFETs," in *IEEE IEDM Technical Digest*, pp. 867–870, 2004.

[8] S. Datta, G. Dewey, M. Doczy, *et al.*, "High mobility Si / SiGe strained channel MOS transistors with HfO_2 / TiN gate stack," in *IEEE IEDM Technical Digest*, pp. 653–656, 2003.

[9] V. Narayanan, V. K. Paruchuri, E. Cartier, *et al.*, "Recent advances and current challenges in the search for high mobility band-edge high-k/metal gate stacks," *Microelectron. Eng.*, vol. 84, no. 9-10, pp. 1853–1856, 2007.

[10] O. Weber, Y. Bogumilowicz, T. Ernst, *et al.*, "Strained Si and Ge MOSFETs with high-k/ metal gate stack for high mobility dual channel CMOS," in *IEEE IEDM Technical Digest*, pp. 137–140, 2005.

[11] Y. H. Kim, C. Cabral Jr., E.P. Gusev, *et al.*, "Systematic study of work function engineering and scavenging effect using NiSi alloy FUSI metal gates with advanced gate stacks," in *IEEE IEDM Technical Digest*, pp. 657–660, 2005.

[12] S. Inumiya, Y. Akasaka, T. Matsuki, *et al.*, "A thermally-stable sub-0.9nm EOT TaSix / HfSiON gate stack with high electron mobility, suitable for gate-first fabrication of hp45 lop devices," in *IEEE IEDM Technical Digest*, pp. 23–26, 2005.

[13] P.D. Kirsch, M.A. Quevedo-Lopez, S.A. Krishnan, *et al.*, "Band edge n-MOSFETs with high-k/metal gate stacks scaled to EOT = 0.9 nm with excellent carrier mobility and high temperature stability," in *IEEE IEDM Technical Digest*, pp. 629–632, 2006.

[14] M.A. Quevedo-Lopez, S.A. Krishnan, D. Kirsch, *et al.*, "High performance gate first HfSiON dielectric satisfying 45nm node requirements," in *IEEE IEDM Technical Digest*, pp. 437–440, 2005.

[15] M. Sato, Y. Nakasaki, K. Watanabe, *et al.*, "Impact of very low Hf concentration (Hf = 6%) cap layer on performance and reliability improvement of HfSiON -CMOSFET with EOT scalable to 1nm," in *IEEE IEDM Technical Digest*, pp. 243–246, 2006.

[16] T. Hoffmann, A. Veloso, A. Lauwers, *et al.*, "Ni-based FUSI gates: CMOS integration for 45 nm node and beyond," in *IEEE IEDM Technical Digest*, pp. 269–272, 2006.

[17] K. Terashima, K. Manabe, K. Takahashi, *et al.*, "Practical Vth control methods for Ni-FUSI/HfSiON MOSFETs on SOI substrates," in *IEEE IEDM Technical Digest*, pp. 273–276, 2006.

[18] T. Aoyama, T. Maeda, K. Torii, *et al.*, "Proposal of new HfSiON CMOS fabrication process (HAMDAMA) for low standby power device," in *IEEE IEDM Technical Digest*, pp. 95–98, 2004.

[19] H.R. Harris, H. Alshareef, H.C. Wen, *et al.*, "Simplified manufacturable band edge metal gate solution for NMOS without a capping layer," in *IEEE IEDM Technical Digest*, pp. 633–636, 2006.

[20] H.-J. Cho, H.L. Lee, S.G. Park, *et al.*, "The effects of TaN thickness and strained substrate on the performance and PBTI characteristics of poly-Si/TaN/HfSiON MOSFETs," in *IEEE IEDM Technical Digest*, pp. 503–506, 2004.

[21] X. Garros, M. Casse, G. Reimbold, *et al.*, "Performance and reliability of advanced high-K/metal gate stacks," *Microelectronic Engineering*, vol. 86, pp. 1609–1614, 2009.

[22] D. K. Ferry and S. M. Goodnick, *Transport in Nanostructures*. Cambridge: Cambridge University Press, 1997.

[23] M. Fischetti, D. Neumayer, and E. Cartier, "Effective electron mobility in Si inversion layers in metal-oxide-semiconductor systems with a high-k insulator: The role of remote phonon scattering," *Journal of Applied Physics*, vol. 90, no. 9, pp. 4587–4608, 2001.

[24] K. Hess and P. Vogl, "Remote polar phonon scattering in silicon inversion layers," *Solid State Communications*, vol. 30, no. 12, pp. 797–799, 1979.

[25] B.T. Moore and D.K. Ferry, "Remote polar phonon scattering in Si inversion layers," *Journal of Applied Physics*, vol. 51, pp. 2603–2605, May 1980.

[26] M.V. Fischetti and S.E. Laux, "Monte Carlo study of electron transport in silicon inversion layers," *Phys. Rev. B*, vol. 48, pp. 2244–2274, Jul 1993.

[27] Z. Ren, M.V. Fischetti, E.P. Gusev, E.A. Cartier, and M. Chudzik, "Inversion channel mobility in high-κ high performance MOSFETs," in *IEEE IEDM Technical Digest*, pp. 793–796, 2003.

[28] R. Kotlyar, M.D. Giles, P. Matagne, *et al.*, "Inversion mobility and gate leakage in high-k/metal gate MOSFETs," in *IEEE IEDM Technical Digest*, pp. 391–394, 2004.

[29] R. Shah and M.M. De Souza, "Impact of a nonideal metal gate on surface optical phonon-limited mobility in high-κ gated MOSFETs," *IEEE Trans. on Electron Devices*, vol. 54, no. 11, pp. 2991–2997, Nov. 2007.

[30] S. Ramo, T. Whinnery, and T. van Duzer, *Fields and Waves in Communication Electronics*. New York: John Wiley and Sons, 1994.

[31] J.D.Jackson, *Classical Electrodynamics*. New York: Wiley, 1975.

[32] S.Q. Wang and G.D. Mahan, "Electron scattering from surface excitations," *Phys. Rev. B*, vol. 6, pp. 4517–4524, 1972.

[33] M.V. Fischetti and S.E. Laux, "Long-range Coulomb interactions in small Si devices. Part I: Performance and reliability," *Journal of Applied Physics*, vol. 89, no. 2, pp. 1205–1231, 2001.

[34] M.V. Fischetti, "Long-range Coulomb interactions in small Si devices. Part II: Effective electron mobility in thin-oxide structures," *Journal of Applied Physics*, vol. 89, no. 2, p. 1232, 2001.

[35] S. Datta, *Quantum Phenomena*. New York: Addison Wesley, 1989.

[36] M. Houssa, L. Pantisano, L.-A. Ragnarsson, *et al.*, "Electrical properties of high-κ gate dielectrics: Challenges, current issues, and possible solutions," *Materials Science and Engineering, R: Reports*, vol. 51, no. 4-6, pp. 37–85, 2006.

[37] L. Lucci, P. Palestri, D. Esseni, L. Bergagnini, and L. Selmi, "Multi-subband Monte Carlo study of transport, quantization and electron gas degeneration in ultra-thin SOI n-MOSFETs," *IEEE Trans. on Electron Devices*, vol. 54, no. 5, pp. 1156–1164, 2007.

[38] P. Toniutti, P. Palestri, D. Esseni, and L. Selmi, "Revised analysis of the mobility and I_{ON} degradation in high-k gate stacks: Surface optical phonons vs. remote Coulomb scattering," in *Proc. European Solid State Device Res. Conf.*, pp. 246–249, 2008.

[39] G. Ferrari, J. Watling, S. Roy, *et al.*, "Monte Carlo study of mobility in Si devices with HfO2-based oxides," *Materials Science in Semiconductor Processing*, vol. 9, no. 6, pp. 995–999, 2006. E-MRS 2006 Spring Meeting – Symposium L: Characterization of high-k dielectric materials.

[40] G. Ferrari, J. Watling, S. Roy, J. Barker, and A. Asenov, "Beyond SiO_2 technology: Simulation of the impact of high-κ dielectrics on mobility," *Journal of Non-Crystalline Solids*, vol. 353, pp. 630–634, 2007.

[41] P. Toniutti, M. D. Michielis, P. Palestri, *et al.*, "Understanding the mobility reduction in MOSFETs featuring high-κ dielectrics," in *Proc. Int. Conf. on Ultimate Integration on Silicon (ULIS)*, pp. 65–68, 2010.

[42] C. Fenouillet-Beranger, S. Denorme, P. Perreau, *et al.*, "FDSOI devices with thin box and ground plane integration for 32nm node and below," in *Proc. European Solid State Device Res. Conf.*, pp. 206–209, 2008.

[43] P. Palestri, C. Alexander, A. Asenov, *et al.*, "A comparison of advanced transport models for the computation of the drain current in nanoscale nMOSFETs," *Solid State Electronics*, vol. 53, pp. 1293–1302, 2009.

[44] M. Caymax, G. Eneman, F. Bellenger, *et al.*, "Germanium for advanced CMOS anno 2009: a SWOT analysis," in *IEEE IEDM Technical Digest*, pp. 461–464, 2009.

[45] K. Morii, T. Iwasaki, R. Nakane, M. Takenada, and S. Takagi, "High performance GeO_2/Ge nMOSFETs with source/drain junctions formed by gas phase doping," in *IEEE IEDM Technical Digest*, pp. 681–684, 2009.

[46] H.-Y. Yu, M. Kobayashi, W.S. Jung, *et al.*, "High performance n-MOSFETs with novel source/drain on selectively grown Ge on Si for monolithic integration," in *IEEE IEDM Technical Digest*, pp. 685–688, 2009.

[47] M.V. Fischetti, "Monte Carlo simulation of transport in technologically significant semiconductors of the diamond and zinc-blende structures - Part I: homogeneous transport," *IEEE Trans. on Electron Devices*, vol. ED-38, p. 634, 1991.

[48] M.V. Fischetti and S.E. Laux, "Monte Carlo simulation of transport in technologically significant semiconductors of the diamond and zinc-blende structures - Part II: submicrometre MOSFET's," *IEEE Trans. on Electron Devices*, vol. ED-38, p. 650, 1991.

[49] T. Low, Y.T. Hou, M.F. Li, *et al.*, "Investigation of performance limits of germanium double-gated MOSFETs," in *IEEE IEDM Technical Digest*, p. 691, 2003.

[50] S. Takagi, "Re-examination of subband structure engineering in ultra-short channel MOSFETs under ballistic carrier transport," in *VLSI Symposium*, p. 115, 2003.

[51] S.E. Laux, "Simulation study of Ge n-channel 7.5nm DGFETs of arbitrary crystallographic alignment," in *IEEE IEDM Technical Digest*, p. 135, 2004.

[52] A. Pethe, T. Krishnamohan, D. Kim, *et al.*, "Investigation of the performance limits of III-V double-gate n-MOSFETs," in *IEEE IEDM Technical Digest*, pp. 619–622, 2005.

[53] A. Rahman, G. Klimeck, and M. Lundstrom, "Novel channel materials for ballistic nanoscale MOSFETs: bandstructure effects," in *IEEE IEDM Technical Digest*, pp. 615–618, 2005.

[54] M. De Michielis, D. Esseni, and F. Driussi, "Analytical models for the insight into the use of alternative channel materials in ballistic nano-MOSFETs," *IEEE Trans. on Electron Devices*, vol. 54, no. 1, pp. 115–123, 2006.

[55] K.D. Cantley, Y. Liu, H.S. Pal, *et al.*, "Performance analysis of III-V materials in a double-gate nano-MOSFET ," in *IEEE IEDM Technical Digest*, pp. 113–116, 2007.

[56] Q. Rafhay, R. Clerc, M. Ferrier, G. Pananakakis, and G. Ghibaudo, "Impact of channel orientation on ballistic current of nDGFETs with alternative channel materials," *Solid State Electronics*, vol. 52, pp. 540–547, 2008.

[57] Q. Rafhay, R. Clerc, G. Ghibaudo, and G. Pananakakis, "Impact of source-to-drain tunneling on the scalability of arbitrarily oriented alternative channel material nMOSFETs," *Solid State Electronics*, vol. 52, pp. 1474–1481, 2008.

[58] A. Rahman, J. Guo, S. Datta, and M.S. Lundstrom, "Theory of ballistic nanotransistors," *IEEE Trans. on Electron Devices*, vol. 50, no. 9, pp. 1853–1863, 2003.

[59] S. Luryi, "Quantum capacitance devices," *Applied Physics Letters*, vol. 52, no. 6, pp. 501–503, 1988.

[60] P. M. Solomon and S.E. Laux, "The ballistic FET: Design, capacitance and speed limit," in *IEEE IEDM Technical Digest*, pp. 95–98, 2001.

[61] J. Wang and M.S. Lundstrom, "Does source to drain tunneling limit the ultimate scaling of MOSFETs ?," in *IEEE IEDM Technical Digest*, pp. 707–710, 2002.

[62] Z. Ren, R. Venugopal, S. Goasguen, S. Datta, and M.S. Lundstrom, "nanomos 2.5: A two-dimensional simulator for quantum transport in double-gate MOSFETs," *IEEE Trans. on Electron Devices*, vol. 50, no. 9, pp. 1914–1925, 2003.

[63] D. Munteanu and J.L.Autran, "Two-dimensional modeling of quantum ballistic transport in ultimate double-gate SOI devices," *Solid State Electronics*, vol. 47, pp. 1219–1225, 2003.

[64] J. L. Autran, D. Munteanu, O. Tintori, E. Decarre, and A. M. Ionescu, "An analytical subthreshold current model for ballistic quantum-wire double-gate MOS transistors," *Molecular Simulation*, vol. 31, 2005.

[65] L. V. Keldish, "Behavior of non-metallic crystals in strong electric fields," *Soviet Phys. - JETP*, vol. 6, no. 4, pp. 763–770, 1958.

[66] E.O.Kane, "Zener tunneling in semiconductors," *J.Phys.Chem.Solids*, vol. 12, p. 181, 1959.

[67] E.O. Kane, "Theory of tunneling," *Journal of Applied Physics*, vol. 32, no. 1, p. 83, 1961.

[68] Q. Rafhay, R. Clerc, G. Pananakakis, and G. Ghibaudo, "Source-to-drain Vs. band-to-band tunneling in ultra-scaled nDGFETs," in *Proc. SSDM*, 2008.

[69] M.V. Fischetti, T.P. O'Regan, S. Narayanan, *et al.*, "Theoretical study of some physical aspects of electronic transport in nMOSFETs at the 10-nm gate-length," *IEEE Trans. on Electron Devices*, vol. 54, no. 9, pp. 2116–2136, 2007.

[70] S.E. Laux, "A simulation study of the switching times of 22- and 17-nm gate-length SOI nFETs on high mobility substrate and Si," *IEEE Trans. on Electron Devices*, vol. 54, no. 9, pp. 2304–2320, 2007.

[71] C. Jacoboni, F. Nava, C. Canali, and G. Ottaviani, "Electron drift velocity and diffusivity in germanium," *Phys. Rev. B*, vol. 24, p. 1014, 1981.

[72] D. Esseni and P. Palestri, "Theory of the motion at the band crossing points in bulk semiconductor crystals and in inversion layers," *Journal of Applied Physics*, vol. 105, no. 5, pp. 053702-1–053702-11, 2009.

[73] C. H. Lee, T. Nishimura, N. Saido, *et al.*, "Record-high electron mobility in Ge n-MOSFETs exceeding Si universality," in *IEEE IEDM Technical Digest*, pp. 457–460, 2009.

[74] Q. Rafhay, P. Palestri, D. Esseni, R. Clerc, and L. Selmi, "Mobility and backscattering in germanium n-type inversion layers," in *Proc. SSDM*, pp. 46–47, 2007.

[75] D. S. Yu, K.C. Chiang, C.F. Cheng, *et al.*, "Fully silicided NiSi:Hf $LaAlO_3$/ SG GOI n-MOSFETs with high electron mobility," *IEEE Electron Device Lett.*, vol. 25, no. 8, pp. 559–561, 2004.

[76] N. Wu, Q. Zhang, D.S.H. Chan, N. Balasubramanian, and C. Zhu, "Gate-first germanium nMOSFET with CVD HfO_2 gate dielectric and silicon surface passivation," *IEEE Electron Device Lett.*, vol. 27, no. 6, pp. 479–481, 2006.

[77] D. S. Yu, A. Chin, C.C. Liao, *et al.*, "Three dimensional metal gate high-κ GOI CMOSFETs on 1-poly-6-metal 0.18-μm Si devices," *IEEE Electron Device Lett.*, vol. 26, no. 2, pp. 118–120, 2005.

[78] C. C. Yeo, B.J. Cho, F. Gao, *et al.*, "Electron mobility enhancement using ultrathin pure Ge on Si substrate," *IEEE Electron Device Lett.*, vol. 26, no. 10, pp. 761–763, 2006.

[79] S. J. Whang, S.J. Lee, F. Gao, *et al.*, "Germanium p- & n-MOSFETs fabricated with novel surface passivation (plasma-PH_3 and thin AIN) and TaN/HfO_2 gate stack," in *IEEE IEDM Technical Digest*, pp. 307–310, 2004.

[80] W. P. Bai, N. Lu, A. Ritenour, *et al.*, "Ge n-MOSFETs on lightly doped substrates with high-κ dielectric and tan gate," *IEEE Electron Device Lett.*, vol. 27, no. 3, pp. 175–178, 2006.

[81] G. Dresselhaus, A.F. Kip, and C. Kittel, "Cyclotron resonance of electrons and holes in silicon and germanium crystals," *Phys. Rev.*, vol. 98, p. 368, 1955.

[82] P. Zimmerman, G. Nicholas, B. De Jaeger, *et al.*, "High performance Ge pMOS devices using a Si-compatible process flow," in *IEEE IEDM Technical Digest*, pp. 655–658, 2006.

[83] A. Dobbie, B. De Jaeger, M. Meuris, *et al.*, "Channel backscattering characteristic of high performance germanium pMOSFETs," in *Proc. Int. Conf. on Ultimate Integration on Silicon (ULIS)*, pp. 7–10, 2008.

[84] A.T. Pham, C. Jungemann, and B. Meinerzhagen, "Physics-based modeling of hole inversion-layer mobility in strained-SiGe-on-insulator," *IEEE Trans. on Electron Devices*, vol. 54, no. 9, pp. 2174–2182, 2007.

[85] M. Lundstrom, *Fundamentals of Carrier Transport.* New York: Addison Wesley, 1990.

[86] P.J. Price, "Polar-optical-mode scattering for an ideal quantum-well heterostructure," *Phys. Rev. B*, vol. 30, pp. 2234–2235, 1984.

[87] J. Pozela and A. Reklaitis, "Electron transport properties in GaAs at high electric fields," *Solid State Electronics*, vol. 23, pp. 927–933, 1980.

[88] P.A. Houston and A.G.R. Evans, "Electron drift velocity in n-GaAs at high electric fields," *Solid State Electronics*, vol. 20, no. 3, pp. 197–204, 1977.

[89] K. Ashida, M. Inoue, J. Shirafuji, and Y. Inuishi, "Energy relaxation effect of hot electrons in GaAs," *Journal of the Physical Society of Japan*, vol. 37, no. 2, pp. 408–414, 1974.

[90] N. Braslau and P.S. Hauge, "Microwave measurement of the velocity-field characteristic of GaAs," *IEEE Trans. on Electron Devices*, vol. 17, pp. 616–622, Aug 1970.

[91] J.G. Ruch and G.S. Kino, "Transport properties of GaAs," *Phys. Rev.*, vol. 174, pp. 921–931, Oct 1968.

A Mathematical definitions and properties

A.1 Fourier transform

Let us consider a possibly complex function $f(x)$ of the real variable x and suppose that $f(x)$ is integrable over the axis x, so that the integral

$$\int_{-\infty}^{\infty} |f(x)|\,\mathrm{d}x \tag{A.1}$$

takes a finite value. We define the Fourier transform of $f(x)$ as

$$F_T(k) = \mathcal{F}\{f(x)\} = \frac{1}{2\pi}\int_{-\infty}^{\infty} f(x)\,\mathrm{e}^{\mathrm{i}kx}\,\mathrm{d}x, \tag{A.2}$$

and the reverse transform is

$$f(x) = \mathcal{F}^{-1}\{F_T(k)\} = \int_{-\infty}^{\infty} F_T(k)\,\mathrm{e}^{-\mathrm{i}kx}\,\mathrm{d}k. \tag{A.3}$$

As an example, an easy evaluation of the integral in Eq.A.2 provides

$$\mathcal{F}\{\delta(px-t)\} = \frac{1}{2\pi|p|}\,\mathrm{e}^{\frac{itk}{p}}. \quad \text{for } p \neq 0 \tag{A.4}$$

Several Fourier pairs can be found in many mathematical handbooks [1].
A few useful properties follow directly from the above definitions:

$$\mathcal{F}\{f^{\dagger}(x)\} = F_T^{\dagger}(-k), \tag{A.5}$$

$$\mathcal{F}\{f(x-a)\} = \mathrm{e}^{\mathrm{i}ka}\,F_T(k), \tag{A.6}$$

$$\mathcal{F}\{f(x)\,\mathrm{e}^{\mathrm{i}k_1 x}\} = F_T(k+k_1), \tag{A.7}$$

$$\mathcal{F}\left\{\frac{\partial f(x)}{\partial x^n}\right\} = (-\mathrm{i}k)^n\,F_T(k), \tag{A.8}$$

$$\mathcal{F}\{f(x)\,g(x)\} = (F_T * G_T)(k) = \int_{-\infty}^{\infty} F_T(k')\,G_T(k-k')\,\mathrm{d}k', \tag{A.9}$$

where $(F_T * G_T)(k)$ indicates the convolution between $F_T(k)$ and $G_T(k)$.

From the previous properties we can also derive

$$\mathcal{F}\{f(x)\, g^{\dagger}(x)\} = \int_{-\infty}^{\infty} F_T(k')\, G_T^{\dagger}(k'-k)\, \mathrm{d}k', \tag{A.10}$$

$$\int_{-\infty}^{\infty} f(x)\, g^{\dagger}(x)\, \mathrm{d}x = 2\pi \int_{-\infty}^{\infty} F_T(k)\, G_T^{\dagger}(k)\, \mathrm{d}k, \tag{A.11}$$

$$\int_{-\infty}^{\infty} |f(x)|^2\, \mathrm{d}x = 2\pi \int_{-\infty}^{\infty} |F_T(k)|^2\, \mathrm{d}k. \tag{A.12}$$

Similar definitions and properties apply to a function of a vectorial variable. In particular, for a function $f(\mathbf{r})$ of the two component vector $\mathbf{r}$ the Fourier transform is

$$\mathcal{F}\{f(\mathbf{r})\} = F_{2T}(\mathbf{k}) = \frac{1}{(2\pi)^2} \int_{\mathbf{r}} f(\mathbf{r})\, \mathrm{e}^{\mathrm{i}\,\mathbf{k}\cdot\mathbf{r}}\, \mathrm{d}\mathbf{r}, \tag{A.13}$$

and the reverse transform is

$$f(\mathbf{r}) = \mathcal{F}^{-1}\{F_{2T}(\mathbf{k})\} = \int_{\mathbf{k}} F_{2T}(\mathbf{k})\, \mathrm{e}^{-\mathrm{i}\,\mathbf{k}\cdot\mathbf{r}}\, \mathrm{d}\mathbf{k}. \tag{A.14}$$

The three-dimensional Fourier transform is defined as

$$\mathcal{F}\{f(\mathbf{R})\} = F_{3T}(\mathbf{K}) = \frac{1}{(2\pi)^3} \int_{\mathbf{R}} f(\mathbf{R})\, \mathrm{e}^{\mathrm{i}\,\mathbf{K}\cdot\mathbf{R}}\, \mathrm{d}\mathbf{R}, \tag{A.15}$$

and the reverse transform is

$$f(\mathbf{R}) = \mathcal{F}^{-1}\{F_{3T}(\mathbf{K})\} = \int_{\mathbf{K}} F_{3T}(\mathbf{K})\, \mathrm{e}^{-\mathrm{i}\,\mathbf{K}\cdot\mathbf{R}}\, \mathrm{d}\mathbf{K}. \tag{A.16}$$

Furthermore, for a function $f(\mathbf{R})$ of a three component vector $\mathbf{R} = (\mathbf{r}, \mathrm{z})$, it is also possible to develop a hybrid representation $f(\mathbf{R})$ consisting of the Fourier transform with respect to only some of the $\mathbf{R}$ components. The Fourier transform thus obtained retains the dependence on the remaining spatial coordinates; in this sense it is thus a hybrid representation of $f(\mathbf{R})$.

In particular, we can define the hybrid two dimensional Fourier transform of $f(\mathbf{R})$ with respect to $\mathbf{r}$ as

$$F_{2T}(\mathbf{k}, z) = \frac{1}{(2\pi)^2} \int_{\mathbf{r}} f(\mathbf{r}, z)\, \mathrm{e}^{\mathrm{i}\,\mathbf{k}\cdot\mathbf{r}}\, \mathrm{d}\mathbf{r}, \tag{A.17}$$

and the reverse transform is given by

$$f(\mathbf{r}, z) = \int_{\mathbf{k}} F_{2T}(\mathbf{k}, z)\, \mathrm{e}^{-\mathrm{i}\,\mathbf{k}\cdot\mathbf{r}}\, \mathrm{d}\mathbf{k}. \tag{A.18}$$

By definition the three dimensional Fourier transform $F_{3T}(\mathbf{K})$ of $f(\mathbf{r}, z)$ is the Fourier transform with respect to z of the hybrid two dimensional Fourier transform $F_{2T}(\mathbf{k}, z)$, that is we have

$$F_{3T}(\mathbf{K}) = \frac{1}{2\pi} \int_{z} F_{2T}(\mathbf{k}, z)\, \mathrm{e}^{\mathrm{i}k_z z}\, \mathrm{d}z. \tag{A.19}$$

A.2 Fourier series

Let us consider a function $f(x)$ which is periodic in the interval $[-L_p/2, L_p/2]$. We define its expansion in Fourier exponential series as

$$f(x) = \sum_k C_k \, \mathrm{e}^{-\mathrm{i}kx}, \tag{A.20}$$

where the parameter k takes the discrete values

$$k = n\frac{2\pi}{L_p}. \qquad n = 0, \pm 1, \pm 2, \pm 3 \; \cdots \tag{A.21}$$

The coefficients of the series expansion are given by

$$C_k = \frac{1}{L_p}\int_{-\frac{L_p}{2}}^{\frac{L_p}{2}} f(x)\,\mathrm{e}^{\mathrm{i}kx}\,\mathrm{d}x, \tag{A.22}$$

so that we have $C_{-k}=C_k^{\dagger}$ if $f(x)$ takes real values.

The Fourier series shows that a function periodic in the interval $[-L_p/2, L_p/2]$ has non-null spectral components only for the k values defined in Eq.A.21, which are multiples of $2\pi/L_p$ (or null).

The Fourier series for a function of a vectorial variable is a natural extension of the one dimensional case. For a function $f(\mathbf{r})$ periodic over an area $A_p = (L_x L_y)$ we have

$$f(\mathbf{r}) = \sum_{\mathbf{k}} C_{\mathbf{k}}\,\mathrm{e}^{-\mathrm{i}\mathbf{k}\cdot\mathbf{r}} \qquad \text{with} \qquad C_{\mathbf{k}} = \frac{1}{A_p}\int_{A_p} f(\mathbf{r})\,\mathrm{e}^{\mathrm{i}\mathbf{k}\cdot\mathbf{r}}\,\mathrm{d}\mathbf{r}, \tag{A.23}$$

and the vector $\mathbf{k} = (K_x, K_y)$ takes the discrete values

$$k_x = n\frac{2\pi}{L_x}, \qquad k_y = m\frac{2\pi}{L_y}. \qquad n, m = 0, \pm 1, \pm 2, \pm 3 \; \cdots \tag{A.24}$$

For a function $f(\mathbf{R})$ periodic over a volume $\Omega_p = (L_x L_y L_z)$ we finally have

$$f(\mathbf{R}) = \sum_{\mathbf{K}} C_{\mathbf{K}}\,\mathrm{e}^{-\mathrm{i}\mathbf{K}\cdot\mathbf{R}} \qquad \text{with} \qquad C_{\mathbf{K}} = \frac{1}{\Omega_p}\int_{\Omega_p} f(\mathbf{R})\,\mathrm{e}^{\mathrm{i}\mathbf{K}\cdot\mathbf{R}}\,\mathrm{d}\mathbf{R}, \tag{A.25}$$

and the vector $\mathbf{K} = (K_x, K_y, K_z)$ takes the discrete values

$$K_x = n\frac{2\pi}{L_x}, \quad K_y = m\frac{2\pi}{L_y}, \quad K_z = p\frac{2\pi}{L_z}. \qquad n, m, p = 0, \pm 1, \pm 2, \pm 3 \; \cdots \tag{A.26}$$

A.3 Fermi integrals

The definition of Fermi integrals used in this book is the one given in [2]. The Fermi integral $\mathcal{F}_j(\eta)$ of order j of the real variable η is defined as

$$\mathcal{F}_j(\eta) = \frac{1}{\Gamma(1+j)}\int_0^{\infty}\frac{x^j}{1+\mathrm{e}^{(x-\eta)}}\mathrm{d}x, \tag{A.27}$$

where the gamma function $\Gamma(y)$ of the real variable y is in turn defined as

$$\Gamma(y) = \int_0^{\infty} t^{(y-1)}\, \mathrm{e}^{-t} \mathrm{d}t. \tag{A.28}$$

For calculation of the Fermi integrals of most practical interest for semiconductor devices it may be useful to know the value of the gamma function $\Gamma(y)$ in the following cases:

$$\Gamma\left(\frac{1}{2}\right) = \sqrt{\pi}, \qquad \Gamma(1) = \Gamma(2) = 1, \qquad \Gamma\left(\frac{3}{2}\right) = \frac{\sqrt{\pi}}{2}. \tag{A.29}$$

The definition of Fermi integrals given in Eq.A.27 has some convenient properties. In particular if $\mathrm{e}^{(x-\eta)} \gg 1$ (i.e. $\eta \ll 0$), then we have

$$\mathcal{F}_j(\eta) \simeq \mathrm{e}^{-\eta} \quad \forall j, \tag{A.30}$$

that is, for negative and large magnitude η values the Fermi integrals tend to $\mathrm{e}^{-\eta}$ irrespective of the order j. Furthermore we also have

$$\frac{\mathrm{d}}{\mathrm{d}\eta}\mathcal{F}_j(\eta) = \mathcal{F}_{j-1}(\eta), \tag{A.31}$$

hence the derivation links the Fermi integrals of different order.

It can be easily verified that an analytical expression exists for the Fermi integral of order $j = 0$, in fact we have

$$\mathcal{F}_0(\eta) = \ln(1 + \mathrm{e}^{\eta}). \tag{A.32}$$

References

[1] Alan Jeffrey, *Handbook of Mathematical Formulas and Integrals.* San Diego: Academic Press, 2000.

[2] J.S. Blakemore, "Approximations for Fermi-Dirac integrals, especially the function $F_{1/2}$ used to describe electron density in a semiconductor," *Solid State Electronics*, vol. 25, pp. 1067–1076, 1982.

B Integrals and transformations over a finite area A

In Section 3.1 we introduced the basic concepts concerning a quasi-2D carrier gas and, in particular, we discussed the allowed values of the wave-vector $\mathbf{k}$ in relation to the finite normalization area A of the physical system in the transport plane. More precisely, by imposing periodic boundary conditions for the wave-function at the boundaries of the area A, the $\mathbf{k}$ values were expressed by Eq.3.3 as

$$k_x = n\,\frac{2\pi}{L}, \qquad k_y = m\,\frac{2\pi}{L}, \qquad n, m = 0, \pm 1, \pm 2, \pm 3\ \cdots \tag{B.1}$$

with $L = \sqrt{A}$.

The assumption of periodic boundary conditions for the wave-function makes the entire problem periodic over the area A. Such a periodicity is clearly artificial, but as long as A is large enough and we are not interested in the properties at the edges of the area A, the periodic boundary conditions are adequate for analysis of the system. The use of a finite normalization area A, however, has a few implications worth mentioning.

As an example, Eq.4.6 implicitly stated

$$U_{2T}(\mathbf{k}, z) = \frac{1}{(2\pi)^2}\int_A U_{sc}(\mathbf{R})\, e^{i\,\mathbf{k}\cdot\mathbf{r}}\, d\mathbf{r}, \tag{B.2}$$

where $U_{2T}(\mathbf{k}, z)$ is the Fourier transform of the scattering potential $U_{sc}(\mathbf{R})$ with respect to $\mathbf{r} = (x, y)$ defined in Eq.A.17. However, one may argue that the Fourier transform defined in Eq.A.17 requires an integral over the entire $\mathbf{r}$ plane, rather then over a finite area A.

In fact, strictly speaking, the integrals over the area A in Eq.B.2 correspond to the coefficients of the Fourier series defined in Eq.A.23. This is the consequence of the periodic boundary conditions which make every quantity in the system periodic, including the scattering potential $U_{sc}(\mathbf{R})$. As discussed above, such a periodicity is artificial. However, if we consider the potential of a Coulomb center and suppose that A is very large compared to the region where the scattering potential is appreciable, then artificial repetition of the potential outside the area A has no effect on the spectral components of the potential practically relevant for calculation of the scattering rates.

In order to be more explicit about the relation between Fourier series and Fourier transform, let us now consider a function $g(\mathbf{r})$ that *is not periodic* over the area A, but is vanishing for large enough $\mathbf{r}$ values. Under these circumstances, if the area A is large enough, then an integral over A is an arbitrarily good approximation of the integral over the entire $\mathbf{r}$ plane. In other words, for a large enough area A, we have

$$\int_A g(\mathbf{r})\, \mathrm{e}^{\mathrm{i}\,\mathbf{k}\cdot\mathbf{r}}\, \mathrm{d}\mathbf{r} \simeq \int_{\mathbf{r}} g(\mathbf{r})\, \mathrm{e}^{\mathrm{i}\,\mathbf{k}\cdot\mathbf{r}}\, \mathrm{d}\mathbf{r}. \tag{B.3}$$

If we now recall Eqs.A.13 and A.23, we see that the Fourier series coefficients $C_{\mathbf{k}}$ of $g(\mathbf{r})$ are simply proportional to the values of the Fourier transform $G_{2T}(\mathbf{k})$ of $g(\mathbf{r})$ at the $\mathbf{k}$ values given by Eq.B.1. More precisely we have

$$C_{\mathbf{k}} \simeq \frac{(2\pi)^2}{A}\, G_{2T}(\mathbf{k}). \tag{B.4}$$

We also notice that, by virtue of Eq.B.4, the expression for $g(\mathbf{r})$ given by its series expansion in Eq.A.23 is perfectly consistent with the inverse Fourier transform in Eq.A.14. In fact, by inserting Eq.B.4 in Eq.A.23, we have

$$g(\mathbf{r}) = \sum_{\mathbf{k}} C_{\mathbf{k}}\, \mathrm{e}^{-\mathrm{i}\,\mathbf{k}\cdot\mathbf{r}} \simeq \frac{(2\pi)^2}{A} \sum_{\mathbf{k}} G_{2T}(\mathbf{k})\, \mathrm{e}^{-\mathrm{i}\,\mathbf{k}\cdot\mathbf{r}} \simeq \int_{\mathbf{k}} G_{2T}(\mathbf{k})\, \mathrm{e}^{-\mathrm{i}\,\mathbf{k}\cdot\mathbf{r}} \mathrm{d}\mathbf{k}, \tag{B.5}$$

for the $\mathbf{k}$ values given by Eq.B.1.

It should be noted, however, that the identity between the Fourier series and the Fourier transform of the *non-periodic* function $g(\mathbf{r})$ holds only for $\mathbf{r}$ values inside the area A. In fact the Fourier series is by definition periodic for $\mathbf{r}$ values outside A, whereas the Fourier transform is not. However, if A is large enough to include the entire region where $g(\mathbf{r})$ takes non-negligible values, then the two representations of $g(\mathbf{r})$ are practically equivalent.

A correct understanding of the role played by the finite normalization area A is also important for evaluation of the integral

$$\frac{1}{A} \int_A \mathrm{e}^{\mathrm{i}\,\mathbf{k}\cdot\mathbf{r}}\, \mathrm{d}\mathbf{r} \tag{B.6}$$

for the $\mathbf{k}$ values defined in Eq.B.1. Such an integral frequently enters calculations of the physical properties of a quasi-2D carrier gas (see Eqs.4.72 and 4.244, for example). Calculation of the integral is straightforward and, for $L_x = L_y = L = \sqrt{A}$ and $k_x,\, k_y \neq 0$, we obtain

$$\begin{aligned} \frac{1}{A} \int_A \mathrm{e}^{\mathrm{i}\,\mathbf{k}\cdot\mathbf{r}}\, \mathrm{d}\mathbf{r} &= \frac{1}{A} \int_{-0.5L}^{0.5L} \mathrm{e}^{\mathrm{i}\,k_x x}\, \mathrm{d}x \int_{-0.5L}^{0.5L} \mathrm{e}^{\mathrm{i}\,k_y y}\, \mathrm{d}y \\ &= \frac{4}{A}\, \frac{\sin(0.5\, k_x L)}{k_x}\, \frac{\sin(0.5\, k_y L)}{k_y}. \end{aligned} \tag{B.7}$$

Hence the integral in Eq.B.6 evaluates to zero for any $\mathbf{k}$ value defined in Eq.B.1, whereas it evaluates to 1.0 for $\mathbf{k} = \mathbf{0}$. We can concisely express these results by writing

$$\frac{1}{A} \int_A \mathrm{e}^{\mathrm{i}\,\mathbf{k}\cdot\mathbf{r}}\, \mathrm{d}\mathbf{r} = \delta_{\mathbf{k},\mathbf{0}}, \tag{B.8}$$

where the Kronecker symbol $\delta_{\mathbf{k},\mathbf{0}}$ is one if $\mathbf{k}$ is null and zero otherwise. It should be reiterated that Eq.B.8 holds only for the set of discrete $\mathbf{k}$ values defined in Eq.B.1.

C Calculation of the equi-energy lines with the k·p model

The eigenvalue problem derived from the **k·p** model for either the bulk semiconductors or the inversion layers is typically formulated in such a way that the energy is calculated for given values of the wave-vector **K** or **k**. In many circumstances, however, it is very informative to calculate and inspect the equi-energy lines, namely those surfaces or curves (in the 3D and 2D case, respectively) which correspond to a given energy. This can be accomplished conveniently by rearranging the **k·p** model in order to obtain an eigenvalue problem for the magnitude of the wave-vector in a given direction, rather than for the energy.

The case for a bulk semiconductor is easier and is discussed first.

C.1 Three dimensional hole gas

By recalling Sections 2.2.2 and 9.4.1, the **k·p** eigenvalue problem for a three dimensional hole gas can be written as

$$\left[\mathbf{H}_{\mathbf{k}\cdot\mathbf{p}} + \mathbf{H}_0\right] \mathbf{C}_{\mathbf{K}} = E\mathbf{C}_{\mathbf{K}}, \tag{C.1}$$

where E is the energy and $\mathbf{H}_0$ indicates the **K** independent matrix

$$\mathbf{H}_0 = \mathbf{H}_{so} + \mathbf{H}_{\varepsilon} \tag{C.2}$$

accounting for the spin-orbit interaction and the strain; $\mathbf{C}_{\mathbf{K}}$ is the six component eigenvector. The 6×6 matrix $\mathbf{H}_{\mathbf{k}\cdot\mathbf{p}}$ is defined according to Eqs.2.58 and 2.55.

The important observation is now that, if we write **K** in spherical coordinates

$$\begin{cases} K_x = K\cos(\phi)\cos(\theta) \\ K_y = K\cos(\phi)\sin(\theta), \\ K_z = K\sin(\phi) \end{cases} \tag{C.3}$$

where ϕ and θ are respectively the polar and azimuthal angle, then Eq.2.54 shows that the **k·p** Hamiltonian matrix can be rewritten as

$$\mathbf{H}_{\mathbf{k}\cdot\mathbf{p}} = K^2\mathbf{H}_2, \tag{C.4}$$

where $\mathbf{H}_2$ is the six by six matrix

$$\mathbf{H}_2 = \begin{pmatrix} \mathbf{H}' & \mathbf{0} \\ \mathbf{0} & \mathbf{H}' \end{pmatrix}. \tag{C.5}$$

Direct substitution of Eq.C.3 into Eqs.2.55 and 2.58 shows that the elements of the $\mathbf{H}'$ matrix are

$$\begin{aligned}
H'_{1,1} &= L\cos^2(\phi)\cos^2(\theta) + M[\cos^2(\phi)\sin^2(\theta) + \sin^2(\phi)],\\
H'_{2,2} &= L\cos^2(\phi)\sin^2(\theta) + M[\cos^2(\phi)\cos^2(\theta) + \sin^2(\phi)],\\
H'_{3,3} &= L\sin^2(\phi) + M\cos^2(\phi),\\
H'_{1,2} &= H'_{2,1} = N\cos^2(\phi)\cos(\theta)\sin(\theta),\\
H'_{1,3} &= H'_{3,1} = N\cos(\phi)\cos(\theta)\sin(\phi),\\
H'_{2,3} &= H'_{3,2} = N\cos(\phi)\sin(\theta)\sin(\phi),
\end{aligned} \tag{C.6}$$

hence $\mathbf{H}_2$ depends on the angles ϕ and θ that identify the direction of $\mathbf{K}$ but it is independent of the magnitude K of $\mathbf{K}$.

By substituting Eq.C.4 into Eq.C.1 we obtain

$$\left[K^2\,\mathbf{H}_2 + \mathbf{H}_0\right]\mathbf{C}_\mathbf{K} = E\,\mathbf{C}_\mathbf{K}. \tag{C.7}$$

We can now collect at the r.h.s. of the equation all the K dependent terms and write

$$[\mathbf{H}_0 - E\,\mathbf{I}]\,\mathbf{C}_\mathbf{K} = K^2\,\mathbf{H}_2\,\mathbf{C}_\mathbf{K}, \tag{C.8}$$

where $\mathbf{I}$ is the 6×6 identity matrix. Equation C.8 can now be easily cast in the form of an eigenvalue problem for the magnitude K of the wave-vector $\mathbf{K}$. In fact, by left multiplying both sides of the equation by the inverse of the matrix $\mathbf{H}_2$, we finally obtain

$$\mathbf{M}(E,\phi,\theta)\,\mathbf{C}_\mathbf{K} = K^2\,\mathbf{C}_\mathbf{K}, \tag{C.9}$$

where $\mathbf{M}(E,\phi,\theta)$ is

$$\mathbf{M}(E,\phi,\theta) = \mathbf{H}_2^{-1}\cdot[\mathbf{H}_0 - E\,\mathbf{I}]. \tag{C.10}$$

Only the positive eigenvalues of Eq.C.9 are meaningful values for K^2.

As implied by the notation, the matrix $\mathbf{M}(E,\phi,\theta)$ depends on the energy E and on the angles ϕ and θ of the wave-vector $\mathbf{K}$. Hence, for any given energy E, one can solve Eq.C.10 for the magnitude K by varying ϕ and θ. By doing so the equi-energy surfaces in the three-dimensional $\mathbf{K}$ space can be calculated.

C.2 Two dimensional hole gas

The solution by means of the finite differences method of the **k·p** problem for a hole inversion layer has been discussed in detail in Sec.3.3.2. In particular the equations from Eq.3.36 to 3.40 summarize the matrixes entering the eigenvalue problem, where $\mathbf{k} = (k_x, k_y)$ is the two component wave-vector. If we now express $\mathbf{k}$ in polar coordinates

$$\begin{cases} k_x = k\cos(\theta) \\ k_y = k\sin(\theta) \end{cases}, \tag{C.11}$$

then Eqs.3.36 to 3.40 show that the problem can be written in the form

$$(k^2\,\mathbf{H}_2 + k\,\mathbf{H}_1 + \mathbf{H}_0)\,\boldsymbol{\xi}_\mathbf{k} = E\,\boldsymbol{\xi}_\mathbf{k}, \tag{C.12}$$

where $\boldsymbol{\xi}_\mathbf{k}$ is the $6N_z$ vector representing the envelope wave-function (N_z being the number of discretization points). The expression for $\mathbf{H}_2$, $\mathbf{H}_1$, and $\mathbf{H}_0$ can be obtained by substituting Eq.C.11 into Eqs.3.36 to 3.40; the matrixes depend on the angle θ but are independent of the magnitude k of $\mathbf{k}$.

Equation C.12 cannot be directly cast in the form of an eigenvalue problem for k because, differently from Eq.C.9, we have one term depending on k and one term depending on k^2. In order to proceed further we re-write Eq.C.12 in the equivalent form

$$(\mathbf{H}_0 - E\,\mathbf{I})\,\boldsymbol{\xi}_\mathbf{k} = -k\,(k\,\mathbf{H}_2 + \mathbf{H}_1)\,\boldsymbol{\xi}_\mathbf{k}. \tag{C.13}$$

We now introduce the auxiliary unknown vector

$$\boldsymbol{\psi}_\mathbf{k} = (k\,\mathbf{H}_2 + \mathbf{H}_1)\,\boldsymbol{\xi}_\mathbf{k}, \tag{C.14}$$

so that Eq.C.13 becomes

$$\begin{cases} (k\,\mathbf{H}_2 + \mathbf{H}_1)\boldsymbol{\xi}_\mathbf{k} - \mathbf{I}\,\boldsymbol{\psi}_\mathbf{k} = \mathbf{0} \\ (\mathbf{H}_0 - E\mathbf{I})\boldsymbol{\xi}_\mathbf{k} + k\,\mathbf{I}\,\boldsymbol{\psi}_\mathbf{k} = \mathbf{0} \end{cases}, \tag{C.15}$$

where $\mathbf{I}$ is the $6N_z{\times}6N_z$ identity matrix. As can be seen, Eq.C.14 allowed us to cast the eigenvalue problem in a form where only k enters the equations explicitly, thus Eq.C.15 can be rearranged as an eigenvalue problem for k.

To this purpose we bring to the r.h.s. of the equation all the terms including k by writing

$$\begin{pmatrix} \mathbf{H}_1 & -\mathbf{I} \\ \mathbf{H}_0 - E\,\mathbf{I} & \mathbf{0} \end{pmatrix} \begin{pmatrix} \boldsymbol{\xi}_\mathbf{k} \\ \boldsymbol{\psi}_\mathbf{k} \end{pmatrix} = -k \begin{pmatrix} \mathbf{H}_2 & \mathbf{0} \\ \mathbf{0} & \mathbf{I} \end{pmatrix} \begin{pmatrix} \boldsymbol{\xi}_\mathbf{k} \\ \boldsymbol{\psi}_\mathbf{k} \end{pmatrix}. \tag{C.16}$$

If $\mathbf{H}_2$ is invertible, then Eq.C.16 can be finally rewritten as

$$\begin{pmatrix} -\mathbf{H}_2^{-1}\cdot\mathbf{H}_1 & \mathbf{H}_2^{-1} \\ E\,\mathbf{I} - \mathbf{H}_0 & \mathbf{0} \end{pmatrix} \begin{pmatrix} \boldsymbol{\xi}_\mathbf{k} \\ \boldsymbol{\psi}_\mathbf{k} \end{pmatrix} = k \begin{pmatrix} \boldsymbol{\xi}_\mathbf{k} \\ \boldsymbol{\psi}_\mathbf{k} \end{pmatrix}, \tag{C.17}$$

which provides the eigenvalue problem for the magnitude k of the wave-vector that we wish to derive. More precisely, for any given energy E, one can vary the angle θ that enters the matrixes $\mathbf{H}_1$ and $\mathbf{H}_2$ and then use Eq.C.17 to obtain the magnitude k of the wave-vector. By doing so the equi-energy lines for the 2D hole gas can be obtained.

It should be noted that Eq.C.17 is an eigenvalue problem with twice as many equations as the direct problem Eq.3.36 that determines the energy. Furthermore, the matrix of the system is not Hermitian, so that Eq.C.17 may provide complex eigenvalues. The complex as well as the real but negative eigenvalues of Eq.C.17 are not acceptable values for k, hence appropriate checks on the eigenvalues must be implemented when Eq.C.17 is used to calculate the equi-energy lines for a 2D hole gas.

D Matrix elements beyond the envelope function approximation

The main purpose of this appendix is to derive for the integral I_Ω defined in Eq.4.13 the expression given in Eq.4.14. Since I_Ω can be recognized as the matrix element of a scattering potential between two Bloch states, we start by evaluating the matrix element

$$\langle n\mathbf{k}k_z|U_{sc}(\mathbf{r},z)|n'\mathbf{k}'k_z'\rangle = \int_\Omega u^\dagger_{n\mathbf{k}k_z} u_{n'\mathbf{k}'k_z'} U_{sc}(\mathbf{r},z) \mathrm{e}^{\mathrm{i}(\mathbf{k}'-\mathbf{k})\cdot\mathbf{r}} \mathrm{e}^{\mathrm{i}(k_z'-k_z)z}\, \mathrm{d}\mathbf{r}\, \mathrm{d}z, \tag{D.1}$$

where $u_{n\mathbf{k}k_z}(\mathbf{r},z)$ is the periodic part of the Bloch wave-function $|n\mathbf{k}k_z\rangle$ and Ω is the normalization volume.

Given its periodicity over the crystal unit cell, the $u_{n\mathbf{k}k_z}(\mathbf{r},z)$ can be expressed by means of a Fourier series defined in Eq.A.25. In particular we can write

$$u_{n'\mathbf{k}'k_z'}(\mathbf{r},z) = \sum_{(\mathbf{g},g_z)} B_{n'\mathbf{k}'k_z'}(\mathbf{g},g_z)\, \mathrm{e}^{-\mathrm{i}\mathbf{g}\cdot\mathbf{r}}\, \mathrm{e}^{-\mathrm{i}g_z z}, \tag{D.2}$$

where $\mathbf{G} = (\mathbf{g}, g_z)$ is a reciprocal lattice vector and $B_{n\mathbf{k}k_z}$ are the coefficients

$$B_{n'\mathbf{k}'k_z'}(\mathbf{g},g_z) = \frac{1}{\Omega_{cell}} \int_{\Omega_{cell}} u_{n'\mathbf{k}'k_z'}(\mathbf{r},z)\, \mathrm{e}^{\mathrm{i}\mathbf{g}\cdot\mathbf{r}}\, \mathrm{e}^{\mathrm{i}g_z z}\, \mathrm{d}\mathbf{r}\, \mathrm{d}z, \tag{D.3}$$

with Ω_{cell} being the volume of the crystal unit cell defined in Sec.2.1, which should not be confused with the normalization volume Ω of the crystal.

Equation D.2 can be inserted into Eq.D.1 to obtain

$$\begin{aligned}\langle n\mathbf{k}k_z|U_{sc}(\mathbf{r},z)|n'\mathbf{k}'k_z'\rangle &= \sum_{(\mathbf{g}_2,g_{2z})} \sum_{(\mathbf{g}_1,g_{1z})} B^\dagger_{n\mathbf{k}k_z}(\mathbf{g}_1,g_{1z}) B_{n'\mathbf{k}'k_z'}(\mathbf{g}_2,g_{2z}) \\ &\times \int_\Omega U_{sc}(\mathbf{r},z) \mathrm{e}^{\mathrm{i}(\mathbf{k}'-\mathbf{k}-\mathbf{g}_2+\mathbf{g}_1)\cdot\mathbf{r}} \mathrm{e}^{\mathrm{i}(k_z'-k_z-g_{2z}+g_{1z})z}\, \mathrm{d}\mathbf{r}\, \mathrm{d}z.\end{aligned} \tag{D.4}$$

By defining $(\mathbf{g}_3, g_{3z}) = (\mathbf{g}_1 - \mathbf{g}_2, g_{1z} - g_{2z})$, Eq.D.4 can be rewritten as

$$\begin{aligned}\langle n\mathbf{k}k_z|U_{sc}(\mathbf{r},z)|n'\mathbf{k}'k_z'\rangle &= \sum_{(\mathbf{g}_2,g_{2z})} \sum_{(\mathbf{g}_3,g_{3z})} B^\dagger_{n\mathbf{k}k_z}(\mathbf{g}_2+\mathbf{g}_3, g_{2z}+g_{3z}) B_{n'\mathbf{k}'k_z'}(\mathbf{g}_2,g_{2z}) \\ &\times (2\pi)^3\, U_{3T}(\mathbf{k}'-\mathbf{k}+\mathbf{g}_3, k_z'-k_z+g_{3z}),\end{aligned} \tag{D.5}$$

where $U_{3T}(\mathbf{q},q_z)$ is the three dimensional Fourier transform of $U_{sc}(\mathbf{r},z)$ defined in A.15.

Since the U_{3T} in Eq.D.5 does not depend on ($\mathbf{g_2}$,g_{2z}), we can finally write

$$\langle n\mathbf{k}k_z|U_{sc}(\mathbf{r},z)|n'\mathbf{k}'k_z'\rangle$$
$$= \frac{(2\pi)^3}{\Omega} \sum_{(\mathbf{g_3},g_{3z})} S^{(n,n')}_{\mathbf{k}k_z\mathbf{k}'k_z'}(\mathbf{g_3},g_{3z})U_{3T}(\mathbf{k}'-\mathbf{k}+\mathbf{g_3},k_z'-k_z+g_{3z}), \quad \text{(D.6)}$$

where we have introduced

$$S^{(n,n')}_{\mathbf{k}k_z\mathbf{k}'k_z'}(\mathbf{g},g_z) = \Omega \sum_{(\mathbf{g}',g_z')} B^{\dagger}_{n\mathbf{k}k_z}(\mathbf{g}'+\mathbf{g},g_z'+g_z)B_{n'\mathbf{k}'k_z'}(\mathbf{g}',g_z'). \quad \text{(D.7)}$$

In order to proceed with the derivation of Eq.4.14 we must now verify that the $S^{(n,n')}_{\mathbf{k}k_z\mathbf{k}'k_z'}(\mathbf{g},g_z)$ defined in Eq.D.7 is the overlap integral between the periodic parts of the Bloch functions indicated in Eq.4.15.

In this part of the derivations we do not need to indicate $\mathbf{K}$ as ($\mathbf{k}$,k_z) and $\mathbf{G}$ as ($\mathbf{g}$,g_z), hence we simplify the notation by using $\mathbf{K}$ and $\mathbf{G}$ for the three component vectors. To proceed further we recall Eq.2.33 which, for any $\mathbf{K}$ and for any reciprocal lattice vector $\mathbf{G}$, allows us to write

$$u_{n(\mathbf{K}+\mathbf{G})}(\mathbf{R}) = \mathrm{e}^{-\mathrm{i}\mathbf{G}\cdot\mathbf{R}}\, u_{n\mathbf{K}}(\mathbf{R}),$$

which leads to

$$u_{n(\mathbf{K}+\mathbf{G})}(\mathbf{R}) = \mathrm{e}^{-\mathrm{i}\mathbf{G}\cdot\mathbf{R}} \sum_{\mathbf{G_1}} B_{n\mathbf{K}}(\mathbf{G_1})\mathrm{e}^{-\mathrm{i}\mathbf{G_1}\cdot\mathbf{R}}$$
$$= \sum_{\mathbf{G_1}} B_{n\mathbf{K}}(\mathbf{G_1})\mathrm{e}^{-\mathrm{i}(\mathbf{G_1}+\mathbf{G})\cdot\mathbf{R}}, \quad \text{(D.8)}$$

where we have used the expansion in Fourier series of $u_{n\mathbf{K}}(\mathbf{R})$ already introduced in Eq.D.2.

Since $\mathbf{G_2} = (\mathbf{G_1}+\mathbf{G})$ is just another reciprocal lattice vector, then we can rewrite Eq.D.8 as

$$u_{n(\mathbf{K}+\mathbf{G})}(\mathbf{R}) = \sum_{\mathbf{G_2}} B_{n\mathbf{K}}(\mathbf{G_2}-\mathbf{G})\mathrm{e}^{-\mathrm{i}\mathbf{G_2}\cdot\mathbf{R}}. \quad \text{(D.9)}$$

Thus the overlap integral between $u_{n(\mathbf{K}+\mathbf{G})}$ and the generic $u_{n'\mathbf{K}'}$ is given by

$$\langle u_{n(\mathbf{K}+\mathbf{G})}(\mathbf{R})|u_{n'\mathbf{K}'}(\mathbf{R})\rangle = \sum_{\mathbf{G_1},\mathbf{G_2}} B^{\dagger}_{n\mathbf{K}}(\mathbf{G_2}-\mathbf{G})B_{n'\mathbf{K}'}(\mathbf{G_1}) \int_{\Omega} \mathrm{e}^{\mathrm{i}(\mathbf{G_2}-\mathbf{G_1})\cdot\mathbf{R}}\mathrm{d}\mathbf{R}$$
$$= \Omega \sum_{\mathbf{G_1}} B^{\dagger}_{n\mathbf{K}}(\mathbf{G_1}-\mathbf{G})B_{n'\mathbf{K}'}(\mathbf{G_1}), \quad \text{(D.10)}$$

where the last equality has been obtained by noting that the integral over Ω gives

$$\int_{\Omega} \mathrm{e}^{\mathrm{i}(\mathbf{G_1}-\mathbf{G_2})\cdot\mathbf{R}}\mathrm{d}\mathbf{R} = \Omega\, \delta_{\mathbf{G_1},\mathbf{G_2}}, \quad \text{(D.11)}$$

because $\mathbf{G_1}$ and $\mathbf{G_2}$ are reciprocal lattice vectors. Equation D.11 can be readily verified by noting that, according to Eq.2.1, the points at the boundaries of the normalization volume Ω (as well as all the points of the direct lattice), can be expressed as

$$\mathbf{R_n} = \sum_{i=1}^{3} n_i \mathbf{a}_i, \tag{D.12}$$

where the n_i are integer numbers. Thus, by recalling Eq.2.9, we see that all the points $\mathbf{R_n}$ at the boundaries of Ω satisfy the equation

$$\exp(\mathrm{i}\mathbf{G_m} \cdot \mathbf{R_n}) = 1. \tag{D.13}$$

By virtue of Eq.D.13 a direct evaluation of the integral in Eq.D.11 gives zero for $\mathbf{G_1} \neq \mathbf{G_2}$. The integral, instead, evaluates Ω for $\mathbf{G_1} = \mathbf{G_2}$.

By comparing Eq.D.7 with Eq.D.10 we obtain

$$S^{(n,n')}_{\mathbf{KK'}}(\mathbf{G}) = \langle u_{n(\mathbf{K}-\mathbf{G})}(\mathbf{R}) | u_{n'\mathbf{K'}}(\mathbf{R}) \rangle, \tag{D.14}$$

which finally demonstrates Eq.4.15.

In order to complete the derivation of Eq.4.14, we now note that the integral

$$I_\Omega = \int_\Omega \left(u_w(\mathbf{R})\, \mathrm{e}^{\mathrm{i}\,\mathbf{K}_w \cdot \mathbf{R}} \right)^\dagger U_{sc}(\mathbf{R})\, \mathrm{e}^{\mathrm{i}(\mathbf{k}-\mathbf{k'})\cdot\mathbf{r}}\, \mathrm{e}^{\mathrm{i}(k'_z - k_z)z}\, u_\nu(\mathbf{R})\, \mathrm{e}^{\mathrm{i}\,\mathbf{K}_\nu \cdot \mathbf{R}}\, \mathrm{d}\mathbf{R} \tag{D.15}$$

can be written as

$$I_\Omega = \langle n_{c1}\mathbf{K}_w | U_{sc}(\mathbf{r}, z)\, \mathrm{e}^{\mathrm{i}(\mathbf{k}-\mathbf{k'})\cdot\mathbf{r}}\, \mathrm{e}^{\mathrm{i}(k'_z - k_z)z} | n_{c1}\mathbf{K}_\nu \rangle, \tag{D.16}$$

where $|n_{c1}\mathbf{K}_\nu\rangle$ and $|n_{c1}\mathbf{K}_w\rangle$ are the Bloch states at the minima ν and w of the lowest conduction band.

By recalling Eq.A.6 (applied to a three-dimensional Fourier transform) as well as Eq.D.6, we can finally write I_Ω as

$$I_\Omega = \frac{(2\pi)^3}{\Omega} \sum_{\mathbf{G}=(\mathbf{g},g_z)} S_{w,\nu}(\mathbf{g}, g_z) U_{3T}(-\Delta\mathbf{k}_{w\nu} - \mathbf{q} + \mathbf{g}, -\Delta k_{w\nu,z} + k'_z - k_z + g_z), \tag{D.17}$$

where $S_{w,\nu}(\mathbf{g}, g_z)$ is the overlap integral

$$S_{w,\nu}(\mathbf{g}, g_z) = \langle u_{n_{c1}(\mathbf{K}_w - \mathbf{G})}(\mathbf{R}) | u_{n_{c1}\mathbf{K}_\nu}(\mathbf{R}) \rangle, \tag{D.18}$$

and, consistently with Eq.4.16, we have introduced $\Delta\mathbf{k}_{w\nu}$ and $\Delta k_{w\nu,z}$ as

$$(\Delta\mathbf{k}_{w\nu}, \Delta k_{w\nu,z}) = \mathbf{K}_w - \mathbf{K}_\nu.$$

Equation D.17 coincides with Eq.4.14.

E Charge density produced by a perturbation potential

This appendix derives an expression for the charge density $\rho_{ind}(\mathbf{r}, z)$ produced in an inversion layer by a stationary perturbation potential $U_p(\mathbf{r}, z)$. To this purpose, let us denote the *unperturbed* envelope wave-function for the subband n in the valley ν as

$$\Phi_{\nu,n,\mathbf{k}}(\mathbf{R}) = \xi_{\nu,n,\mathbf{k}}(z)\,\frac{e^{i\,\mathbf{k}\cdot\mathbf{r}}}{\sqrt{A}}, \tag{E.1}$$

where $\xi_{\nu,n,\mathbf{k}}(z)$ may depend on the wave-vector $\mathbf{k}$. The index ν can be dropped if there is no need to consider different valleys in the system. For a $\mathbf{k}\cdot\mathbf{p}$ Hamiltonian the wave-function in Eq.E.1 is a six-component vector. In this appendix we do not use the boldface symbol for the wave-function, but *all the derivations are valid also for a vectorial wave-function.*

Using static perturbation theory, the *perturbed* wave-function for the state (ν,n,$\mathbf{k}$) can be written as

$$\Phi^{(p)}_{\nu,n,\mathbf{k}}(\mathbf{R}) = \Phi_{\nu,n,\mathbf{k}}(\mathbf{R}) + \sum_{n',\mathbf{q}} b_{\nu,n,n'}(\mathbf{k},\mathbf{k}+\mathbf{q})\;\Phi_{\nu,n',(\mathbf{k}+\mathbf{q})}(\mathbf{R}), \tag{E.2}$$

where the coefficients $b_{\nu,n,n'}(\mathbf{k},\mathbf{k}+\mathbf{q})$ are given by

$$b_{\nu,n,n'}(\mathbf{k},\mathbf{k}+\mathbf{q}) = \frac{M_{\nu,n,n'}(\mathbf{k},\mathbf{k}+\mathbf{q})}{E_{\nu,n'}(\mathbf{k}+\mathbf{q}) - E_{\nu,n}(\mathbf{k})}. \tag{E.3}$$

The matrix element $M_{\nu,n,n'}(\mathbf{k},\mathbf{k}+\mathbf{q})$ is defined according to Eq.4.6 as

$$M_{\nu,n,n'}(\mathbf{k},\mathbf{k}+\mathbf{q}) = \frac{(2\pi)^2}{A}\int_z \xi^{\dagger}_{n'\mathbf{k}'}(z)\cdot\xi_{n\mathbf{k}}(z)\,U_{2T}(-\mathbf{q}, z)\,dz, \tag{E.4}$$

where $U_{2T}(\mathbf{q}, z)$ is the Fourier transform of $U_p(\mathbf{r}, z)$ with respect to $\mathbf{r} = (x, y)$ defined in Eq.A.17. For a $\mathbf{k}\cdot\mathbf{p}$ quantization model the dot sign in Eq.E.4 denotes the scalar product defined in Eq.4.40.

Equation E.3 is valid for non-degenerate states, namely for $E_{\nu,n'}(\mathbf{k}+\mathbf{q}) \neq E_{\nu,n}(\mathbf{k})$. We neglect here the complications related to use of perturbation theory for degenerate states that, strictly speaking, should be used to determine the coefficients $b_{\nu,n,n'}(\mathbf{k},\mathbf{k}+\mathbf{q})$ for $E_{\nu,n'}(\mathbf{k}+\mathbf{q}) = E_{\nu,n}(\mathbf{k})$.

As can be seen, the matrix element $M_{\nu,n,n'}(\mathbf{k},\mathbf{k}+\mathbf{q})$ depends on the $(-\mathbf{q})$ spectral component of $U_p(\mathbf{r}, z)$. Since the perturbation potential is real, then $U_{2T}(-\mathbf{q}, z) = U^{\dagger}_{2T}(\mathbf{q}, z)$ and the matrix elements satisfy the identity

$$M^{\dagger}_{\nu,n,n'}(\mathbf{k}, \mathbf{k}+\mathbf{q}) = M_{\nu,n',n}(\mathbf{k}+\mathbf{q}, \mathbf{k}). \tag{E.5}$$

The matrix elements defined in Eq.E.4 are intra-valley matrix elements, as is implied by the notation; in fact, as discussed in Section 4.1.4, the inter-valley electron transitions can be practically neglected for the stationary perturbation potentials representative of either Coulomb or surface roughness scattering.

The perturbation of the wave-functions yields a variation of the charge density with respect to the unperturbed case that is given by

$$\rho_{ind}(\mathbf{r}, z) = -e \sum_{\nu,n,\mathbf{k}} f_{\nu,n}(\mathbf{k}) \left\{ |\Phi^{(p)}_{\nu,n,\mathbf{k}}(\mathbf{R})|^2 - |\Phi_{\nu,n,\mathbf{k}}(\mathbf{R})|^2 \right\}, \tag{E.6}$$

where $f_{\nu,n}(\mathbf{k})$ is the occupation function of the subband (ν,n) such that

$$\sum_{\mathbf{k}} f_{\nu,n}(\mathbf{k}) = N_{\nu,n}, \tag{E.7}$$

with $N_{\nu,n}$ being the inversion density in the subband (ν,n).

Recalling Eqs.E.1 and E.2 we see that, for any state (ν,n,$\mathbf{k}$), the term in the bracket of Eq.E.6 can be written

$$|\Phi^{(p)}_{\nu,n,\mathbf{k}}(\mathbf{R})|^2 - |\Phi_{\nu,n,\mathbf{k}}(\mathbf{R})|^2 \approx \frac{1}{A} \sum_{\mathbf{q}} \left\{ \sum_{n'} b_{\nu,n,n'}(\mathbf{k}, \mathbf{k}+\mathbf{q})\, \xi^{\dagger}_{\nu,n,\mathbf{k}}(z) \right.$$
$$\left. \cdot\, \xi_{\nu,n',(\mathbf{k}+\mathbf{q})}(z)\, e^{i\mathbf{q}\cdot\mathbf{r}} + b^{\dagger}_{\nu,n,n'}(\mathbf{k}, \mathbf{k}+\mathbf{q})\, \xi_{\nu,n,\mathbf{k}}(z) \cdot \xi^{\dagger}_{\nu,n',(\mathbf{k}+\mathbf{q})}(z) e^{-i\mathbf{q}\cdot\mathbf{r}} \right\}, \tag{E.8}$$

where the second order terms with respect to $b_{\nu,n,n'}$(**k**,**k**+**q**) have been neglected because we assume a small perturbation.

If we now take the contribution of only one generic **q** value in Eq.E.8 and substitute it in Eq.E.6 we obtain

$$\rho_{-\mathbf{q}}(\mathbf{r}, z) = \frac{-e}{A} \sum_{\nu,n,\mathbf{k},n'} f_{\nu,n}(\mathbf{k}) \left[b_{\nu,n,n'}(\mathbf{k}, \mathbf{k}+\mathbf{q})\, \xi^{\dagger}_{\nu,n,\mathbf{k}}(z) \cdot \xi_{\nu,n',(\mathbf{k}+\mathbf{q})}(z)\, e^{i\mathbf{q}\cdot\mathbf{r}} \right.$$
$$\left. + b^{\dagger}_{\nu,n,n'}(\mathbf{k}, \mathbf{k}+\mathbf{q})\, \xi_{\nu,n,\mathbf{k}}(z) \cdot \xi^{\dagger}_{\nu,n',(\mathbf{k}+\mathbf{q})}(z) e^{-i\mathbf{q}\cdot\mathbf{r}} \right], \tag{E.9}$$

where $\rho_{-\mathbf{q}}(\mathbf{r}, z)$ denotes the charge density produced by the spectral component $U_{2T}(-\mathbf{q}, z)$ of the perturbation potential, to which $b_{\nu,n,n'}$(**k**,**k**+**q**) is proportional (see Eqs.E.3 and E.4). By substituting Eq.E.3 in Eq.E.9 we obtain

$$\rho_{-\mathbf{q}}(\mathbf{r}, z) = \frac{-e}{A} \sum_{\nu,n,\mathbf{k},n'} f_{\nu,n}(\mathbf{k}) \left[\frac{\xi^{\dagger}_{\nu,n,\mathbf{k}}(z) \cdot \xi_{\nu,n',(\mathbf{k}+\mathbf{q})}(z) M_{\nu,n,n'}(\mathbf{k}, \mathbf{k}+\mathbf{q})}{E_{\nu,n'}(\mathbf{k}+\mathbf{q}) - E_{\nu,n}(\mathbf{k})} e^{i\mathbf{q}\cdot\mathbf{r}} \right.$$
$$\left. + \frac{\xi_{\nu,n,\mathbf{k}}(z) \cdot \xi^{\dagger}_{\nu,n',(\mathbf{k}+\mathbf{q})}(z) M^{\dagger}_{\nu,n,n'}(\mathbf{k}, \mathbf{k}+\mathbf{q})}{E_{\nu,n'}(\mathbf{k}+\mathbf{q}) - E_{\nu,n}(\mathbf{k})} e^{-i\mathbf{q}\cdot\mathbf{r}} \right], \tag{E.10}$$

where the second term in the bracket of Eq.E.10 is the complex conjugate of the first term.

Equation E.10 shows that the spectral component $U_{2T}(-\mathbf{q}, z)$ of the perturbation potential produces both a $\mathbf{q}$ and a $(-\mathbf{q})$ component of the induced charge density $\rho_{ind}(\mathbf{r}, z)$. We can thus understand that the spectral component $U_{2T}(\mathbf{q}, z)$ must similarly yield a $\mathbf{q}$ and a $(-\mathbf{q})$ component of $\rho_{ind}(\mathbf{r}, z)$. More precisely, the charge density produced by the $U_{2T}(\mathbf{q}, z)$ is given by

$$\rho_{\mathbf{q}}(\mathbf{r}, z) = \frac{-e}{A} \sum_{\nu,n,\mathbf{k},n'} f_{\nu,n}(\mathbf{k}) \left[\frac{\xi^{\dagger}_{\nu,n,\mathbf{k}}(z) \cdot \xi_{\nu,n',(\mathbf{k}-\mathbf{q})}(z) M_{\nu,n,n'}(\mathbf{k}, \mathbf{k}-\mathbf{q})}{E_{\nu,n'}(\mathbf{k}-\mathbf{q}) - E_{\nu,n}(\mathbf{k})} \mathrm{e}^{-\mathrm{i}\mathbf{q}\cdot\mathbf{r}} \right.$$
$$\left. + \frac{\xi_{\nu,n,\mathbf{k}}(z) \cdot \xi^{\dagger}_{\nu,n',(\mathbf{k}-\mathbf{q})}(z) M^{\dagger}_{\nu,n,n'}(\mathbf{k}, \mathbf{k}-\mathbf{q})}{E_{\nu,n'}(\mathbf{k}-\mathbf{q}) - E_{\nu,n}(\mathbf{k})} \mathrm{e}^{\mathrm{i}\mathbf{q}\cdot\mathbf{r}} \right]. \tag{E.11}$$

Hence the charge density produced by $U_{2T}(\mathbf{q}, z)$ and $U_{2T}(-\mathbf{q}, z) = U^{\dagger}_{2T}(\mathbf{q}, z)$ is obtained by summing Eqs.E.10 and E.11.

We start by considering the first term in Eq.E.10 and the second term in Eq.E.11, whose sum produces the $\mathbf{q}$ component of the charge density. Before summing such two terms, we rewrite the second term in Eq.E.11 by exchanging n with n' in the sum over the subbands and by using $\mathbf{k}' = (\mathbf{k} - \mathbf{q})$ for the sum over the wave-vectors. By doing so the second term in the square bracket of Eq.E.11 can be written as

$$\sum_{\nu,n',\mathbf{k}',n} f_{\nu,n'}(\mathbf{k}'+\mathbf{q}) \left[\frac{\xi_{\nu,n',(\mathbf{k}'+\mathbf{q})}(z) \cdot \xi^{\dagger}_{\nu,n,\mathbf{k}'}(z) M^{\dagger}_{\nu,n',n}(\mathbf{k}'+\mathbf{q}, \mathbf{k}')}{E_{\nu,n}(\mathbf{k}') - E_{\nu,n'}(\mathbf{k}'+\mathbf{q})} \mathrm{e}^{\mathrm{i}\mathbf{q}\cdot\mathbf{r}} \right]. \tag{E.12}$$

If we now use Eq.E.5 and rename $\mathbf{k}'$ as $\mathbf{k}$ in Eq.E.12, then we can write the sum of the first term in Eq.E.10 and the second term in Eq.E.11 as

$$\sum_{\nu,n,\mathbf{k},n'} \xi_{\nu,n',(\mathbf{k}'+\mathbf{q})}(z) \cdot \xi^{\dagger}_{\nu,n,\mathbf{k}}(z) M_{\nu,n,n'}(\mathbf{k}, \mathbf{k}+\mathbf{q}) \left[\frac{f_{\nu,n}(\mathbf{k}) - f_{\nu,n'}(\mathbf{k}+\mathbf{q})}{E_{\nu,n'}(\mathbf{k}+\mathbf{q}) - E_{\nu,n}(\mathbf{k})} \right] \mathrm{e}^{\mathrm{i}\mathbf{q}\cdot\mathbf{r}}, \tag{E.13}$$

which clearly represents the $\mathbf{q}$ spectral component of the charge density.

The sum of the second term in Eq.E.10 and the first term in Eq.E.11 produces the complex conjugate of Eq.E.13, which is the $(-\mathbf{q})$ spectral component of the charge density. The overall charge density $\rho_{\pm\mathbf{q}}(\mathbf{r}, z)$ produced by the $U_{2T}(\mathbf{q}, z)$ and $U_{2T}(-\mathbf{q}, z)$ spectral components of the perturbation potential is thus given by

$$\rho_{\pm\mathbf{q}}(\mathbf{r}, z) = \frac{e}{A} \sum_{\nu,n,\mathbf{k},n'} \xi^{\dagger}_{\nu,n,\mathbf{k}}(z) \cdot \xi_{\nu,n',(\mathbf{k}+\mathbf{q})}(z)\, M_{\nu,n,n'}(\mathbf{k}, \mathbf{k}+\mathbf{q})$$
$$\times \left[\frac{f_{\nu,n'}(\mathbf{k}+\mathbf{q}) - f_{\nu,n}(\mathbf{k})}{E_{\nu,n'}(\mathbf{k}+\mathbf{q}) - E_{\nu,n}(\mathbf{k})} \right] \mathrm{e}^{\mathrm{i}\mathbf{q}\cdot\mathbf{r}} + (c.c.), \tag{E.14}$$

where a minus sign has been absorbed in the numerator of the squared bracket, where $f_{\nu,n'}(\mathbf{k}+\mathbf{q})$ and $f_{\nu,n}(\mathbf{k})$ have been swapped with respect to Eq.E.13.

If the wave-functions do not depend on the wave-vector $\mathbf{k}$, then the matrix elements do not depend on $\mathbf{k}$ and we can indicate them as $M_{\nu,n,n'}(\mathbf{q})$. Furthermore, in Eq.E.14

only the occupation functions and the energies depend on $\mathbf{k}$ and thus we can rewrite $\rho_{\pm\mathbf{q}}(\mathbf{r}, z)$ as

$$\rho_{\pm\mathbf{q}}(\mathbf{r}, z) = e \sum_{\nu,n,n'} \xi^{\dagger}_{\nu,n}(z) \cdot \xi_{\nu,n'}(z)\, \Pi_{\nu,n,n'}(\mathbf{q})\, M_{\nu,n,n'}(\mathbf{q})\, \mathrm{e}^{\mathrm{i}\mathbf{q}\cdot\mathbf{r}} + (c.c.), \tag{E.15}$$

where we have introduced the polarization factor

$$\Pi_{\nu,n,n'}(\mathbf{q}) = \frac{1}{A} \sum_{\mathbf{k}} \frac{f_{\nu,n'}(\mathbf{k}+\mathbf{q}) - f_{\nu,n}(\mathbf{k})}{E_{\nu,n'}(\mathbf{k}+\mathbf{q}) - E_{\nu,n}(\mathbf{k})}. \tag{E.16}$$

We conclude by reiterating that, for a $\mathbf{k}\cdot\mathbf{p}$ quantization model, the dot sign in Eqs.E.14 and E.15 denotes the scalar product of wave-functions defined in Eq.4.40.

Index

For EU product safety concerns, contact us at Calle de José Abascal, 56–1°,
28003 Madrid, Spain or eugpsr@cambridge.org.

www.ingramcontent.com/pod-product-compliance
Lightning Source LLC
LaVergne TN
LVHW062203200926
848484LV00006B/1307